THE PHYSICS OF LIQUID AND SOLID HELIUM

PART I

ROBERT E. KRIEGER PUBLISHING COMPANY INC.
MALABAR
FLORIDA 32950

INTERSCIENCE MONOGRAPHS AND TEXTS IN PHYSICS AND ASTRONOMY

Found by **R. E. MARSHAK**

Volume I:	E. R. Cohen, K. M. Crowe, and J. W. M. DuMond, **The Fundamental Constants of Physics**
Volume II:	G. J. Dienes and G. H. Vineyard, **Radiation Effects in Solids**
Volume III:	N. N. Bogoliubov and D. V. Shirkov, **Introduction to the Theory of Quantized Fields**
Volume IV:	J. B. Marion and J. L. Fowler, *Editors*, **Fast Neutron Physics** *In two parts* **Part 1: Techniques Part II: Experiments and Theory**
Volume V:	D. M. Ritson, *Editor*, **Techniques of High Energy Physics**
Volume VI:	R. N. Thomas and R. G. Athay, **Physics of the Solar Chromosphere**
Volume VII:	Lawrence H. Aller, **The Abundance of the Elements**
Volume VIII:	E. N. Parker, **Interplanetary Dynamical Processes**
Volume IX:	Conrad L. Longmire, **Elementary Plasma Physics**
Volume X:	R. Brout and P. Carruthers, **Lectures on the Many-electron Problem**
Volume XI:	A. I. Akhiezer and V. B. Berestetskii, **Quantum Electrodynamics**
Volume XII:	John L. Lumley and Hans A. Panofsky, **The Structure of Atmospheric Turbulence**
Volume XIII:	Robert D. Heidenreich, **Fundamentals of Transmission Electron Microscopy**
Volume XIV:	G. Rickayzen, **Theory of Superconductivity**
Volume XV:	Raymond J. Seeger and G. Temple, *Editors*, **Research Frontiers in Fluid Dynamics**
Volume XVI:	C. S. Wu and S. A. Moszkowski, **Beta Decay**
Volume XVII:	Klaus G. Steffen, **High Energy Beam Optics**
Volume XVIII:	V. M. Agranovich and V. L. Ginsburg, **Spacial Dispersion in Crystal Optics and the Theory of Excitons**
Volume XIX:	A. B. Migdal, **Theory of Finite Fermi Systems and Applications to Atomic Nuclei**
Volume XX:	B. L. Moiseiwitsch, **Variational Principles**
Volume XXI:	L. S. Shklovsky, **Supernovae**
Volume XXII:	S. Hayakawa, **Cosmic Ray Physics**
Volume XXIII:	J. H. Piddington, **Cosmic Electrodynamics**
Volume XXIV:	R. E. Marshak, Riazuddin, and C. P. Ryan, **Theory of Weak Interactions in Particle Physics**
Volume XXV:	Norman Austern, **Direct Nuclear Reaction Theories**
Volume XXVI:	Gérard G. Emch, **Algebraic Methods in Statistical Mechanics and Quantum Field Theory**
Volume XXVII:	William Jones and Norman H. March, **Theoretical Solid State Physics** *In two volumes* **Volume 1: Perfect Lattices in Equilibrium** **Volume 2: Non-equilibrium and Disorder**
Volume XXVIII:	L. Allen and J. H. Eberly, **Optical Resonance and Two-Level Atoms**
Volume XXIX:	K. H. Bennemann and J. B. Ketterson, Editors, **The Physics of Liquid and Solid Helium, Part I.**

THE PHYSICS OF LIQUID AND SOLID HELIUM

Part I

edited by

K. H. BENNEMANN

Free University of Berlin

J. B. KETTERSON

Northwestern University
Argonne National Laboratory

A WILEY-INTERSCIENCE PUBLICATION

John Wiley & Sons New York · London · Sydney · Toronto

Library of Congress Cataloging in Publication Data:
Main entry under title:

The Physics of liquid and solid helium.

(Interscience monographs and texts in physics and astronomy; v. 29)
"A Wiley-Interscience publication."
Includes bibliographical references and indexes.
1. Liquid helium. 2. Solid helium. I. Bennemann, K. H. II. Ketterson, J. B.
QC145.45.H4P49 546'.751 75–20235
ISBN 0–471–06600–1

Printed in the United States of America

10 9 8 7 6 5 4 3 2 1

Preface

The study of the properties of liquid and solid helium continues to be a fascinating area of condensed matter research. Theorists have long been drawn to study the condensed isotopes of helium (and their mixtures) because these liquids and solids are model many-body systems, i.e., they are fertile proving grounds for various quantum many-body formalisms. As Landau has emphasized, these systems are amenable to theoretical attack because, when studied at relatively low temperatures, they are only weakly excited from their ground state; a description in terms of weakly interacting elementary excitations is then appropriate. The elementary excitation concept has had a profound influence on many branches of physics.

Liquid ^{3}He is the only neutral Fermi liquid available for laboratory study (this, of course, ignores the very indirect information available for the neutron Fermi liquid from astronomical observations on neutron stars). The isotropic property of this liquid introduces simplifications not possible for its cousin, the metallic electron liquids. Liquid ^{4}He is the only Bose liquid of finite rest mass in nature; its property of superfluidity is a most dramatic example of a macroscopic quantum effect. The lattice dynamics of solid helium are of interest because of the extreme importance of anharmonic effects resulting from the large zero-point motion. The very-low-temperature properties of solid ^{3}He are dominated by exchange effects. Liquid or solid solutions of ^{3}He and ^{4}He allow the study of systems with mixed Fermi and Bose characteristics.

Helium is also of considerable practical importance; the rapidly growing industry surrounding the various applications of superconductivity relies on liquid helium as a coolant or refrigerant. The ^{3}He–^{4}He dilution refrigerator, already an extremely important research tool, is sure to find application in a future industry based on very-low-temperature technology.

A number of excellent books on the properties of helium have appeared over the years (e.g., those by Atkins, Keesom, Keller, and Wilks). However, a number of very important advances, theoretical and experimental, have been made in recent years. The most dramatic has been the discovery by Osheroff, Richardson, and Lee of two superfluid phases of liquid ^{3}He at temperatures near 2×10^{-3} K and the rapid growth of experimental and theoretical activity following this event. In considering the form that a new book on condensed helium might take, the editors quickly concluded that a multiple author format, with experts selected from the various subfields, would be the optimum way to present the information. The range of phenomena that have been explored and the diversity of theoretical machinery that has been applied make it virtually impossible for a single author to cover the subject adequately. It also developed that the material was far more extensive than could be covered in a single volume. In addition to the subjects listed in the contents of this volume, the following chapters are planned for a second volume: (1) Phenomenological Theory of Liquid ^{3}He, (2) Phenomenological Theory of ^{3}He–^{4}He Solutions, (3) Theory of Superfluid ^{3}He, (4) Experimental Properties of Superfluid ^{3}He, (5) Helium Monolayers, (6) Helium in Restricted Geometries, and (7) Neutron Scattering in Helium. It is inevitable, when trying to coordinate the treatment of such a wide scope of material, that some subjects will receive double coverage while others are overlooked. Except in cases for which the addition of a few sentences would suffice, we have not attempted to correct this defect.

The editors have encouraged the authors to treat their subject matter in a way that will make the material of use to graduate students as well as research workers. It is our feeling that the chapters in this book (and those in the second volume to follow) clearly show that the study of condensed helium is a very active and exciting area of research.

We thank Dr. O. C. Simpson, former director of the Solid State Science Division of Argonne National Laboratory, for his help and encouragement in the preparation of this work. The excellent secretarial help of Carol Poile and Midge Thompson of the Solid State Science Division is also acknowledged. Finally, one of us (J. B. K.) thanks his colleagues, B. M. Abraham, Y. Eckstein, and P. R. Roach, for discussions and suggestions.

K. H. Bennemann
J. B. Ketterson

February 1975

Contents

THE PHYSICS OF LIQUID AND SOLID HELIUM

PART I

CHAPTER 1

Phenomenological Theory of Superfluid ^{4}He*†

I.M. Khalatnikov
The Academy of Sciences of the USSR
The Landau Institute for Theoretical Physics
Moscow, USSR

*Portions of this chapter reproduced from *Introduction to the Theory of Superfluidity*, written by I. M. Khalatnikov, with the permission of publishers, W. A. Benjamin Inc., Advanced Book Program, Reading, Massachusetts, U. S. A.

†Portions of this chapter were reproduced with the permission of the American Institute of Physics from I. M. Khalatnikov and D. M. Chernikova, Soviet Physics JETP **22**, 1336 (1966) and **23**, 274 (1966); copywrite American Institute of Physics.

1.1. QUANTUM LIQUID AND SUPERFLUIDITY

1.1.1. Excitations in a Superfluid Liquid

Liquid helium is unique in that it does not freeze down to the absolute zero of temperature (at the normal pressure). At $T = 2.18$ K, it undergoes a second-order phase transition. At temperatures below the phase transition point, liquid helium has a number of unusual properties, of which the most important is superfluidity. Superfluidity, the ability of a liquid to flow through narrow capillaries without friction, was discovered by P. Kapitza [1].

These unusual properties of liquid helium were explained by L. Landau on the basis of a quantum liquid concept [2]. It is easy to convince oneself that at temperatures of the order of 1 or 2 K, the de Broglie wavelength of helium atoms is comparable to the interatomic distance. In a system with such properties one should not consider the motion and states of separate atoms, but quantum states of the whole system of interacting atoms. Then it becomes evident why liquid helium is a nonfreezing liquid. A relatively weak interaction of helium atoms results in its transition to a quantum state before crystallization occurs. Helium has two stable isotopes, ^{4}He and ^{3}He. The liquid that exhibits superfluidity at relatively high temperatures is the one formed from the atoms of ^{4}He. Since ^{4}He atoms obey the Bose statistics such a quantum liquid is sometimes called a Bose liquid. ^{3}He atoms also form a quantum liquid which, however, exhibits superfluidity at much lower temperatures ($\sim 10^{-3}$ K). A liquid made up of Fermi particles is called a Fermi liquid. The high temperature superfluidity property is exhibited by a quantum Bose liquid; the superfluidity of ^{3}He will not be discussed in this chapter.

The properties of a quantum superfluid liquid can be well understood when the energy spectrum of such a system is known. At absolute zero the quantum liquid is in the ground state; at temperatures different from zero the liquid is in an excited state.

According to quantum mechanics any system of particles with arbitrary interactions can, in its weakly excited states, be looked upon as a set of distinct elementary excitations. Each elementary excitation behaves like a quasiparticle, capable of motion throughout the body. It has a definite energy and momentum. The function that characterizes the dependence of the energy on the momentum is called the energy spectrum of the body.

Let us denote by ε the energy of an elementary excitation in liquid helium as a function of its momentum p. The form of the energy spectrum for small values of the momentum p is easily determined. Small momenta correspond to long wavelength excitations which, in a liquid, are obviously just longitudinal sound waves. The corresponding elementary excitations are, therefore, sound quanta or phonons, whose energy is a linear function of the momentum

$$\varepsilon = cp, \tag{1.1.1}$$

where c is the velocity of sound. As the momentum increases the curve $\varepsilon(p)$ departs from a straight line. Its subsequent behavior, however, cannot be obtained from general considerations. To explain the experimental values obtained for the thermodynamic functions of liquid helium, L. Landau proposed the energy spectrum shown in Figure 1.1. It turned out that the

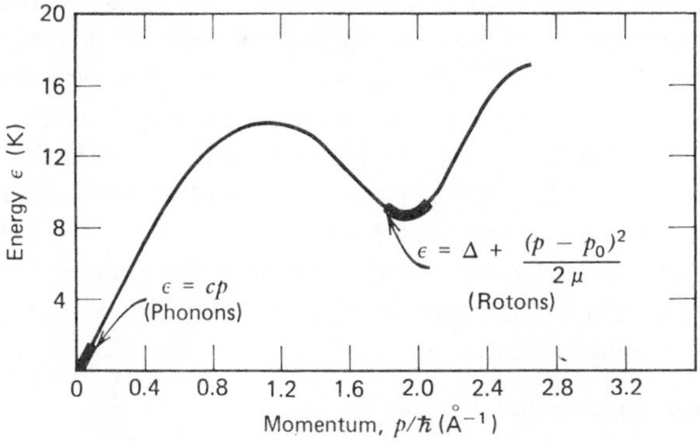

FIGURE 1.1. The dispersion curve for excitations in liquid He(II) showing the phonon and roton regions.

phonons alone were not sufficient to explain the temperature dependence and the absolute values of such thermodynamic quantities as the specific heat. It is easy to see that elementary excitations with energies close to the minimum of the curve in Figure 1.1 give a contribution to all thermodynamic quantities that competes with the contribution of the phonons. The corresponding excitations were called rotons and their energy can be represented near the minimum by the form

$$\varepsilon = \Delta + \frac{(p - p_0)^2}{2\mu}. \tag{1.1.2}$$

Here p_0 is the value of the momentum at which the function ε has a minimum equal to Δ. The exact values of the parameters characterizing the energy spectrum of liquid helium were found by neutron-scattering experiments. Monochromatic neutrons emit or absorb elementary excitations in helium. By measuring the energies of neutrons scattered at given angles, one can determine the whole spectrum of elementary excitations. In this manner the following values of the spectrum parameters were obtained [3]

$$\frac{\Delta}{k} = 8.6\ K, \qquad \frac{p_0}{\hbar} = 1.9\ A^{-1}, \qquad \mu = 0.15\, m_{\mathrm{He}}.$$

The quantity μ, which has the dimensions of mass, is usually called the effective mass of the roton.

The concept of elementary excitations can be used if relatively few of these are present so that their interaction energy is small compared to their own energy. In this case, the gas of elementary excitations can be looked upon as an ideal gas. Since upon excitation of the liquid phonons and rotons can appear one at a time, it is obvious that they should have integer spin and therefore obey Bose statistics. Thus at equilibrium the phonon and roton gases are described by the equilibrium functions of Bose statistics. The roton energies contain the large quantity Δ, and therefore the Bose distribution can, for rotons, be replaced by a Boltzmann distribution. In this manner, the model of an ideal gas of excitations is appropriate at temperatures that are not too near the λ point. Near the λ point, there are many excitations, and their interactions begin to be important; the lifetime, resulting from the collisions between elementary excitations, then becomes small and the indeterminacy in the energy becomes comparable with the energy itself. Therefore the concept of elementary excitations is not applicable near the λ point. However, in practice, already at temperatures of the order of 1.7–1.8 K, one can consider the phonon and roton gases as ideal.

1.1.2. Superfluid and Normal Motion

Let us now show how the property of superfluidity follows from the notion of elementary excitations introduced above. We first assume that the helium is at zero temperature, that is, that it occupies the ground state of energy. Let the liquid now flow through a capillary with velocity v. If this flow were accompanied by friction, then a part of the kinetic energy would be dissipated and would be transformed into thermal energy. If the helium heats up, this means that it makes transitions to excited states. But we know that a quantum liquid cannot receive energy in a continuous fashion. In order for such a liquid to go to the lowest excited state, an elementary excitation must be created. Let the energy of such an excitation be $\varepsilon(p)$ and the corresponding momentum be $\mathbf{p}$ in the frame moving with the liquid. Then in the fixed frame of reference the energy of the system has changed by the following amount:

$$\varepsilon(p) + \mathbf{p}\cdot\mathbf{v}. \tag{1.1.3}$$

The transition to an excited state will be energetically favorable if the condition

$$\varepsilon(p) + \mathbf{p}\cdot\mathbf{v} < 0 \tag{1.1.4}$$

is fulfilled. Obviously, the most favorable situation is one in which the momentum of the created excitation is directed opposite to the velocity. In that case, it follows from (1.1.4) that

$$v > \frac{\varepsilon(p)}{p}. \tag{1.1.5}$$

Obviously, an excitation can appear in liquid helium if condition (1.1.5) is fulfilled at least at that point in the spectrum where the ratio $\varepsilon(p)/p$ has its minimum value. We thus obtain the necessary condition for the creation of an excitation:

$$v > \min \frac{\varepsilon(p)}{p}. \tag{1.1.6}$$

Superfluidity will occur if

$$v_{\text{cr}} = \min \frac{\varepsilon(p)}{p} \neq 0. \tag{1.1.7}$$

In that case, for values of the velocity of motion of the helium which are less than v_{cr}, the creation of an excitation will be energetically unfavorable and the liquid will flow without dissipation of energy, i.e., without friction. Frictionless flow will not slow down; it will be superfluid. On the other hand, at velocities greater than v_{cr}, the motion of the helium will cause the creation of excitations and consequently dissipation of energy. As can be seen from Figure 1.1 the spectrum of elementary excitations of helium (II) satisfies condition (1.1.7). The minimum of the ratio ε/p can be found by the extremal condition,

$$\frac{d}{dp}\frac{\varepsilon}{p} = \frac{1}{p}\frac{d\varepsilon}{dp} - \frac{\varepsilon}{p^2} = 0,$$

to occur at the point where

$$\frac{\varepsilon}{p} = \frac{d\varepsilon}{dp}, \tag{1.1.8}$$

i.e., at the point where a straight line going through the origin of coordinates is tangent to the curve $\varepsilon(p)$. For the spectrum of liquid helium, this point lies near the minimum of the curve $\varepsilon(p)$ and is numerically equal to about 60 m/sec. This value is several orders of magnitude larger than the actually observed critical velocity in He (II). This appears to be explainable by the fact that, as we shall see below, phonons and rotons are not the only excitations that can appear in superfluid helium; there can exist excitations of another type—so-called quantum vortices. Although these excitations have too little statistical weight to play a role in the thermodynamics, they are easily created and consequently are important in determining hydrodynamic properties such as critical velocities*

We may note that in deriving criterion (1.1.6) we did not require that there be no gaslike excitations already present in the liquid. The presence of such excitations would not make the derivation invalid, and therefore criterion (1.1.6) is also true at finite temperatures. However, the presence of excitations already in the liquid at finite temperatures introduces special features into the flow of He (II) through capillaries. The excitations present in the liquid will be reflected at the walls and will transfer part of their momentum to the walls. Because of this, that part of the liquid that is carried along by the motion of the excitations will behave like a normal viscous liquid and will be slowed down by friction with the walls. Therefore at $T = 0$, the whole liquid flows through the capillary without friction, but for $T \neq 0$, only part of the liquid does. In this manner we obtain the remarkable situation that in a superfluid two independent motions can take place simultaneously. One part, which is carried along by the excitations, behaves like a normal liquid; the remainder—the superfluid—does not experience friction and as long as its velocity is less than some critical value, it moves independently of the normal part. There are thus two simultaneous motions possible in He (II)—superfluid motion with velocity v_s and normal motion with velocity v_n. To each one of these motions there corresponds a different effective mass. The sum of the normal and superfluid masses is the total mass of the liquid. The two motions and independent of one another (at least in the region where v_s and v_n are smaller than certain critical values at which momentum transfer between the two parts becomes possible). The momentum per unit volume $\mathbf{j}$ of He (II) is thus divided into two parts:

$$\mathbf{j} = \rho_s \mathbf{v}_s + \rho_n \mathbf{v}_n. \tag{1.1.9}$$

The coefficient ρ_n is called the normal density and ρ_s the superfluid density. Their sum is the total density of the liquid

$$\rho = \rho_n + \rho_s. \tag{1.1.10}$$

The ratio ρ_n/ρ is unity at the λ point; as the temperature decreases, it decreases, and it is zero at $T = 0$.

*The critical velocity connected with the creation of quantum vorticies depends on the diameter of the capillary according to the relation

$$v_{\text{cr}}\, d \sim \frac{\hbar}{m}.$$

This is why one must use narrow capillaries in order to observe superfluidity. In thick pipes the value of v_{cr} is too small for superfluidity to be observable.

Let us suppose that in helium there are two simultaneous translational motions with velocities $\mathbf{v}_n$ and $\mathbf{v}_s$. The energy of an elementary excitation in the rest frame, $E(p)$, can be expressed in terms of its energy $\varepsilon(p)$ in the frame in which the superfluid is at rest by the relation

$$E(p) = \varepsilon(p) + \mathbf{p}\cdot\mathbf{v}_s. \tag{1.1.11}$$

The normal motion of the liquid is associated with the translational motion of the gas of excitations, occurring with the velocity $\mathbf{v}_n$. The distribution function n of the gas of elementary excitations depends on the energy of relative motion

$$E' = E - \mathbf{p}\cdot\mathbf{v}_n = \varepsilon(p) + \mathbf{p}\cdot\mathbf{v}_s - \mathbf{p}\cdot\mathbf{v}_n. \tag{1.1.12}$$

In moving He (II) the energy distribution of phonons is, therefore, determined by the Planck function,

$$n = \left[\exp\left(\frac{\varepsilon + \mathbf{p}\cdot\mathbf{v}_s - \mathbf{p}\cdot\mathbf{v}_n}{kT}\right) - 1\right]^{-1}, \tag{1.1.13}$$

where the function ε is given in (1.1.1)

The energy distrubution of rotons is determined by the Boltzmann function

$$n = \exp\left(-\frac{\varepsilon + \mathbf{p}\cdot\mathbf{v}_s - \mathbf{p}\cdot\mathbf{v}_n}{kT}\right), \tag{1.1.14}$$

where ε is given by (1.1.2).

1.1.3. The Thermodynamic Functions of Helium (II)

At temperatures not too near the λ point the phonon and roton densities are small and, as was already mentioned, one may consider the gases of excitations as ideal gases. In that case all thermodynamic functions consist of two parts, one due to phonons and the other due to rotons.

In order to calculate the thermodynamic functions, we use the distrubtion for phonons and rotons (1.1.13) and (1.1.14). As a first approximation, we may neglect the dependence of the distribution functions on the relative velocity $\mathbf{v}_n - \mathbf{v}_s$. This dependence only begins to be important for a range of values of $\mathbf{v}_n - \mathbf{v}_s$ at which, under usual conditions, the destruction of superfluidity begins to set in. Only when sound waves of large amplitude are propagating through helium does the relative velocity $\mathbf{v}_n - \mathbf{v}_s$ attain large values. In treating that problem it is essential to take into account the quadratic terms in $\mathbf{v}_n - \mathbf{v}_s$ in the thermodynamic functions.

Let us first calculate the thermodynamic functions for He (II) at rest. These can be obtained by using the general formulas of Bose statistics. As we stated previously, the phonons obey Bose statistics, and the distribution function of the rotons is independent of statistics because of the presence of the large constant term $\Delta \gg kT$ in the roton energy.

The free energy of a Bose gas is given by

$$F = -kT \int \ln(1+n)d\tau_{\mathbf{p}}, \tag{1.1.15}$$

where $d\tau_{\mathbf{p}} = p^2\, dpdo/(2\pi\hbar)^3$ is the volume element in p space and do the element of solid angle. After integrating (1.1.15) by parts, we obtain

$$F = -\frac{1}{3}\int np\frac{\partial \varepsilon}{\partial p}\, d\tau_{\mathbf{p}}, \tag{1.1.16}$$

which allows one to calculate the free energy of the excitation gas. The entropy of the excitations is found by differentiating the free energy with respect to temperature

$$S = -\frac{\partial F}{\partial T} = -\frac{1}{3kT^2}\int n'\varepsilon p\frac{\partial \varepsilon}{\partial p}\, d\tau_{\mathbf{p}}, \tag{1.1.17}$$

where n' is the derivative of the distribution function with respect to its arguments.

(*a*) *The Free Energy, Entropy, and Specific Heat of the Phonon Gas.* If we carry out the integral in (1.1.16) using the distribution function (1.1.13), we find the free energy of the phonon gas ($\varepsilon = cp$).

$$F_{\text{ph}} = -\frac{1}{3}\int (e^{\varepsilon/kT}-1)^{-1}\, p\frac{\partial \varepsilon}{\partial p}\, d\tau_{\mathbf{p}} = -\frac{1}{3}E_{\text{ph}}. \tag{1.1.18}$$

The energy per unit volume of the phonons is equal to

$$E_{\text{ph}} = \frac{4\pi^5}{15}\left(\frac{kT}{2\pi\hbar c}\right)^3 kT \approx \frac{\pi^4}{36}kTN_{\text{ph}}, \tag{1.1.19}$$

where $N_{\text{ph}} = 2.4\cdot 4\pi(kT/2\pi\hbar c)^3$ is the number of phonons per unit volume in He (II). The entropy per unit volume can be found either by the general formula (1.1.17) or directly by differentiating the relation obtained for the free energy. We thus find

$$S_{\text{ph}} = -\frac{\partial F_{\text{ph}}}{\partial T} = \frac{16\pi^5}{45}k\left(\frac{kT}{2\pi\hbar c}\right)^3. \tag{1.1.20}$$

We may, furthermore, calculate the specific heat of the phonon gas

$$C_{\text{ph}} = T\frac{\partial S_{\text{ph}}}{\partial T} = \frac{16\pi^5}{15}k\left(\frac{kT}{2\pi\hbar}\right)^3. \tag{1.1.21}$$

(*b*) *The Free Energy, Entropy, and Specific Heat of the Roton Gas.* The free energy per unit volume for the roton gas can be found by the general formula (1.1.16) using the distribution function (1.1.14). In carrying out the integration, it is necessary to take into account the fact that the roton momenta are in magnitude close to p_0. In this manner we find that the following obtains:

$$F_{\mathrm{r}} = -kTN_{\mathrm{r}}, \tag{1.1.22}$$

where N_{r} is the number of rotons per unit volume of helium, i.e.,

$$N_{\mathrm{r}} = \int n\, d\tau_{\mathrm{p}} = \frac{2p_0^2(\mu kT)^{1/2}e^{-\Delta/T}}{(2\pi)^{3/2}\hbar^3} \tag{1.1.22'}$$

By differentiating relation (1.1.22′), we find the entropy of the roton gas

$$S_{\mathrm{r}} = -\frac{\partial F_{\mathrm{r}}}{\partial T} = kN_{\mathrm{r}}\left(\frac{\Delta}{T} + \frac{3}{2}\right). \tag{1.1.23}$$

Let us, furthermore, calculate the specific heat of the roton gas

$$C_{\mathrm{r}} = T\frac{\partial S_{\mathrm{r}}}{\partial T} = kN_{\mathrm{r}}\left(\frac{\Delta^2}{T^2} + \frac{\Delta}{T} + \frac{3}{4}\right). \tag{1.1.24}$$

By summing the results found in (1.1.20), (1.1.21), (1.1.23), and (1.1.24), we find expressions for the entropy and specific heat per unit volume of He (II)

$$S = S_{\mathrm{r}} + S_{\mathrm{ph}} = kN_{\mathrm{r}}\left(\frac{\Delta}{T} + \frac{3}{2}\right) + \frac{16\pi^5 k}{45}\left(\frac{kT}{2\pi\hbar c}\right)^3, \tag{1.1.25}$$

$$C = C_{\mathrm{r}} + C_{\mathrm{ph}} = kN_{\mathrm{r}}\left(\frac{\Delta^2}{T^2} + \frac{\Delta}{T} + \frac{3}{4}\right) + \frac{16\pi^5 k}{15}\left(\frac{kT}{2\pi\hbar c}\right)^3. \tag{1.1.26}$$

1.1.4. The Normal Density

The momentum per unit volume of He (II) in the frame of reference moving with the superfluid part can by (1.1.9) be written as

$$\mathbf{p} = \mathbf{j} - \rho\mathbf{v}_s = \rho_n(\mathbf{v}_n - \mathbf{v}_s). \tag{1.1.27}$$

On the other hand, this momentum can be represented in the form of an integral

$$\int \mathbf{p}\, n(\varepsilon + \mathbf{p}\cdot\mathbf{v}_s - \mathbf{p}\cdot\mathbf{v}_n)\, d\tau_{\mathrm{p}} \tag{1.1.28}$$

taken over all elementary excitations. Comparing (1.1.27) and (1.1.28), we obtain the relation defining the normal density

$$\rho_n(\mathbf{v}_n - \mathbf{v}_s) = \int \mathbf{p}\, n(\varepsilon + \mathbf{p}\cdot\mathbf{v}_s - \mathbf{p}\cdot\mathbf{v}_n)\, d\tau_{\mathrm{p}}. \tag{1.1.29}$$

In the case of small values of the difference $\mathbf{v}_n - \mathbf{v}_s$, one can expand the distribution function n in powers of this difference. The zeroth order term on the right-hand side of (1.1.29) is zero. The first-order term in the expansion gives

$$\rho_n = -\frac{1}{3kT}\int p^2 n'\, d\tau_{\mathrm{p}}. \tag{1.1.30}$$

Formula (1.1.30) holds for small values of the difference $\mathbf{v}_n - \mathbf{v}_s$.

Let us first calculate the phonon part of the normal density. Inserting the distribution function (1.1.13) into (1.1.29), we obtain

$$\rho_{n\text{ph}}(\mathbf{v}_n - \mathbf{v}_s) = \int \mathbf{p}\left[\exp\left(\frac{\varepsilon + \mathbf{p}\cdot\mathbf{v}_s - \mathbf{p}\cdot\mathbf{v}_n}{kT}\right) - 1\right]^{-1} d\tau_{\mathbf{p}}. \tag{1.1.31}$$

Performing the simple integration and dividing both sides by the factor $\mathbf{v}_n - \mathbf{v}_s$, we find

$$\rho_{n\text{ph}} = \frac{4}{3}\frac{E_{\text{ph}}}{c^2}\left(1 - \frac{w^2}{c^2}\right)^{-3} \tag{1.1.32}$$

where $w = \mathbf{v}_n - \mathbf{v}_s$.

For rotons we may perform an analogous calculation. We get

$$\rho_{n\text{r}}(\mathbf{v}_n - \mathbf{v}_s) = \int \mathbf{p}\exp\left(-\frac{\varepsilon + \mathbf{p}\cdot\mathbf{v}_s - \mathbf{p}\cdot\mathbf{v}_n}{kT}\right) d\tau_{\mathbf{p}},$$

from which we easily find

$$\rho_{n\text{r}} = N_{\text{r}}\frac{kT}{w^2}\left(\cosh\frac{p_0 w}{kT} - \frac{kT}{p_0 w}\sinh\frac{p_0 w}{kT}\right). \tag{1.1.33}$$

The normal density of helium ρ_n is equal to the sum of $\rho_{n\text{ph}}$ and $\rho_{n\text{r}}$. According to (1.1.29) and (1.1.33) we have

$$\rho_n = N_{\text{r}}\frac{kT}{w^2}\left(\cosh\frac{p_0 w}{kT} - \frac{kT}{p_0 w}\sinh\frac{p_0 w}{kT}\right) + \frac{4}{3}\frac{E_{\text{ph}}}{c^2}\left(1 - \frac{w^2}{c^2}\right)^{-3}. \tag{1.1.34}$$

For small values of the velocity of relative motion of the normal and superfluid components, we can neglect the dependence of $\rho_{n\text{ph}}$ and $\rho_{n\text{r}}$ on w. In that case, formulas (1.1.32) and (1.1.33) give

$$\rho_{n\text{ph}} = \frac{4}{3}\frac{E_{\text{ph}}}{c^2}, \tag{1.1.35}$$

$$\rho_{n\text{r}} = \frac{p_0^2}{3kT}N_{\text{r}}, \tag{1.1.36}$$

$$\rho_n = \frac{4}{3}\frac{E_{\text{ph}}}{c^2} + \frac{p_0^2}{3kT}N_{\text{r}}. \tag{1.1.37}$$

In the temperature range of 0.8 to 2 K, it is the rotons that contribute most to the thermodynamic functions. However, the relative contribution of rotons to these functions decreases rapidly with decreasing temperature. This is because the roton contributions decrease exponentially with temperature whereas the phonon contributions go as T^3.

In practice, one need not in most problems take into account the dependence of thermodynamic quantities on the relative velocities $\mathbf{w} = \mathbf{v}_n - \mathbf{v}_s$. This is because the ratios w/c and wp_0/kT are very small in the region of velocities where the phonomenon of superfluidity is observable. It is only in treating the problem of the propagation of large amplitude sound waves that one may not neglect this dependence. Finally let us remark that since $c > kT/p_0$, the dependence on w is more pronounced for rotons than for phonons.

1.1.5. The Connection of the Energy Spectrum of Excitations with the Structure Factor of a Liquid

In recent experiments on neutron scattering, Cowly and Woods [3] have observed the presence of two branches in the spectrum of liquid helium ^4He below the λ point. The experimental spectrum of excitations is shown in Figure 1.2 by solid lines. The lower branch corresponds to the elementary excitations of the Landau spectrum (phonons and rotons). The upper branch is due to the bound states (two-excitation), which are formed from the Landau elementary excitations.

Let us find the relation between the liquid form factor and the spectrum of elementary excitations (taking account of the multiparticle excitations),

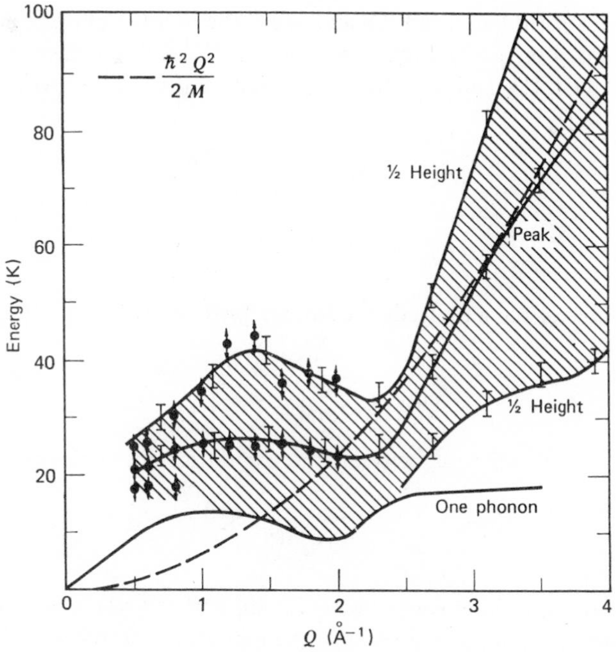

FIGURE 1.2. The energy wave-vector dependence of the scattering at 1.1 K. Shown are the one-phonon dispersion curve, the upper and lower energies corresponding to half the peak intensity, and the mean energy, of the multiphonon peak.

where the density fluctuation ρ_q is given as a sum of contributions from a real single excitation and a real multiple excitation (two-phonon)

$$\rho_q = A_q^+ a_q^+ + A_{-q} a_{-q} + B_q^+ b_q^+ + B_{-q} b_{-q}. \tag{1.1.38}$$

Here a_q^+ is the creation operator of a single excitation (one-phonon) with momentum q and energy $\omega_1(q)$, b_q^+ is the creation operator of two-phonon excitations with momentum q and energy $\omega_2(q)$, and A_q and B_q are the corresponding amplitudes.*

The dynamical form factor $S(q, \omega)$ is equal to [4]

$$S(q, \omega) = \frac{\sum_n e^{-\beta\omega_n} \langle n | \rho_q \delta(\omega - H) \rho_q^+ | n \rangle}{\sum_n e^{-\beta\omega_n}}. \tag{1.1.39}$$

The sum rules for the finite temperature can be given as follows:

$$\int_{-\infty}^{\infty} d\omega S(q, \omega) \text{cth} \frac{\beta\omega}{2} = NS_q = |A_q|^2 \text{cth} \frac{\beta\omega_1}{2} + |B_q|^2 \text{cth} \frac{\beta\omega_2}{2}, \tag{1.1.40}$$

$$\int_{-\infty}^{+\infty} d\omega\omega S(q, \omega) = \frac{Nq^2}{2m} = \omega_1 |A_q|^2 + \omega_2 |B_q|^2. \tag{1.1.41}$$

Here S_q is the form factor, and N and m are the number and the mass of the helium atoms. For low enough temperatures cth can be replaced by unity in (1.140) and (1.1.41), and then these equations can be solved with respect to $|A_q|^2$ and $|B_q|^2$

$$|A_q|^2 = NS_q \frac{\omega_2 - q^2/2mS_q}{\omega_2 - \omega_1}, \tag{1.1.42}$$

$$|B_q|^2 = NS_q \frac{q^2/2mS_q - \omega_1}{\omega_2 - \omega_1}. \tag{1.1.43}$$

As the squares of the amplitude moduli are positive, we obtain the following inequality [5]

$$\omega_2 > \frac{q^2}{2mS_q} > \omega_1. \tag{1.1.44}$$

The Feynman [6] relation between the excitation energy ω and the form factor S_q is

$$\omega = \frac{q^2}{2mS_q}. \tag{1.1.45}$$

Formula (1.1.45) follows easily from (1.1.40) and (1.1.41) if we put $B_q = 0$, that is, when one branch is present in the excitation spectrum. In Figure 1.3

*In the perturbation theory, the operators b_q^+ can be given as a sum of $\sum a_p^+ a_{q-p}^+$, $\sum a_p^+ a_{-q+p}$, or $\sum a_{-p} a_{p-q}$, in which the states corresponding to the scattering of the excitation should be excluded.

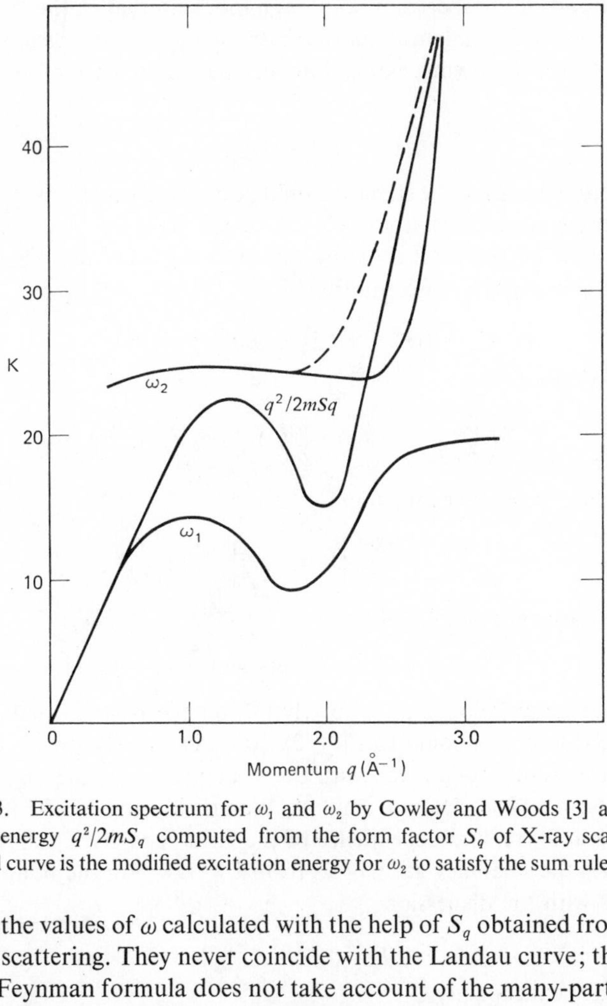

FIGURE 1.3. Excitation spectrum for ω_1 and ω_2 by Cowley and Woods [3] and the mean excitation energy $q^2/2mS_q$ computed from the form factor S_q of X-ray scattering data. The dotted curve is the modified excitation energy for ω_2 to satisfy the sum rules.

we show the values of ω calculated with the help of S_q obtained from the data of X-ray scattering. They never coincide with the Landau curve; the reason is that the Feynman formula does not take account of the many-particle excitations. Taking many particle excitations into account one gets the inequality (1.1.44) which is confirmed by experiments for all momentum values less than 2.2 A^{-1}. For larger values of the momentum, the discrepancy between (1.1.44) and the data probably can be explained by the fact that the calculations given above do not take into account the multiple excitations higher than the two-phonon contributions.

1.1.6. On the Dispersion of the Phonon Branch of the Spectrum

The initial part of the elementaıy excitation spectrum is linear (phonon) in the first approximation. For a number of effects, however, the dispersion of the phonon part of the spectrum becomes important, that is, the deviation of $\varepsilon(p)$ from a linear dependence. On the basis of the liquid isotropy, we may already write the following expression for the initial part of the phonon spectrum:

$$\varepsilon = cp(1 - \Upsilon p^2). \tag{1.1.46}$$

For the phonon branch to be stable it would be necessary that $\Upsilon > 0$. However, recent evidence suggests that some parts of the phonon spectrum are not stable. We may try to evaluate Υ by interpolating over the whole energy curve using a four-term representation

$$\varepsilon^2 = A_1 p^2 + A_2 p^4 + A_3 p^6 + A_4 p^8, \tag{1.1.47}$$

and the four conditions

$$\left(\frac{\partial \varepsilon}{\partial p}\right)_{p=0} = c, \quad \left(\frac{\partial \varepsilon}{\partial p}\right)_{p=p_0} = 0, \quad \left(\frac{\partial^2 \varepsilon}{\partial p^2}\right)_{p=p_0} = \frac{1}{\mu}, \quad \varepsilon(p_0). = \Delta. \tag{1.1.48}$$

In this manner we get the expression for Υ:

$$\Upsilon = -\frac{A_2}{2c^2} = \frac{3}{2p_0^2}\left\{1 - 2\left(\frac{\Delta}{p_0 c}\right)^2 - \frac{\Delta}{12\mu c^2}\right\}, \tag{1.1.49}$$

from which the estimation of Υ follows:

$$\Upsilon \approx 2.5 \cdot 10^{37} \ \sec^2/g^2 cm^2.$$

On the other hand, Woods and Cowely [3], on the basis of neutron data, have deduced that Υ is equal to $(0 \pm 2) \cdot 10^{36}$. As was shown by Pines and Woo [17], this result agrees well with the structure factor $S(q)$ measured in X-ray scattering. The data of Phillips, Waterfield, and Hoffer [8], on the temperature and pressure dependence of the heat-capacity C_v allows one to obtain Υ from the small deviation of C_v from a T^3 law. For the noninteracting phonon gas with the dispersion

$$\varepsilon = cp(1 - \Upsilon p^2 - \delta p^4), \tag{1.1.50}$$

we obtain the heat capacity C_v*:

$$C_v = \frac{16\pi^5 k^4}{15(2\pi\hbar c)^3}\left[T^3 + \frac{25}{7}\Upsilon\left(\frac{2\pi k}{c}\right)^2 T^5 + 7(4\Upsilon^2 + \delta)\left(\frac{2\pi k}{c}\right)^4 T^7\right]. \tag{1.1.51}$$

The experimental data [8] agree well with the formula (1.1.51) when Υ is negative

$$\Upsilon = -4.10^{37} \ cgs.$$

Thus at present we have some evidence that the dispersional constant Υ may

*Taking into account the phonon interactions should not change the T^5 term in (1.1.51).

be negative. In this case the phonon spectrum will be unstable. The phonon lifetime will be finite due to the decay of one phonon into two phonons (a three-phonon process). The lifetime can be easily calculated:

$$\frac{1}{\tau_{3ph}} = \frac{(u+1)^2}{240\pi}\frac{p^5}{\rho\hbar^3}; \quad u = \frac{\rho}{c}\frac{\partial c}{\partial \rho}. \tag{1.1.52}$$

The value $1/\tau_{3ph}$ is negligible for long-wave length phonons; however, it should define the linewidth for the scattered neutron, for example:

$$\frac{\hbar}{\tau_{3ph}} \approx 3.5\cdot 10^{-3}\,K \quad \text{for } \frac{p}{\hbar} = 0.2\,\AA^{-1},$$

$$\frac{\hbar}{\tau_{3ph}} \approx 0.85\,K \quad \text{for } \frac{p}{\hbar} = 0.6\AA^{-2}.$$

It should be noted that in the experiments [3] any change of the linewidth proportional to p^5 has not been observed. There remains another possibility when the constant γ changes signs at certain parts of the spectrum. Such a possibility was pointed out by Jackle and Kehr [9]. They analyzed the experiments of Roach, Ketterson, and Kuchnir [10], who measured the high-frequency ultrasound attenuation in superfluid helium under pressure. At very low temperatures an absorption proportional to T^4 was always observed which is expected for a three-phonon process. The corresponding absorption coefficient is equal to†

$$\alpha_1 = \frac{3\pi}{8}(u+1)^2\frac{\omega}{c}\frac{\rho_{nph}}{\rho}; \quad (\rho_{nph} \sim T^4). \tag{1.1.53}$$

When the temperature is in the vicinity of 0.5 K the attenuation increases more slowly than the T^4 dependence of equation (1.1.53) and a shoulder appears in the data. The authors [9] suggested that the phonon part of the curve $\varepsilon(p)$ for $q < q_c$ is concave ($\gamma < 0$), which is replaced by a convexity at $q > q_c$. In this case formula (1.1.53) is modified and there appears an additional factor

$$\frac{F(z)}{F(\infty)}, \tag{1.1.54}$$

where

$$F(z) = \int_0^z dx\, x^4 n(x)[1 + n(x)] \tag{1.1.55}$$

and $n(x)$ is a Bose distribution function; $z = cq_c/kT$ for $kT > cq_c$. Formula (1.1.53) with the additional factor (1.1.54) produce the following:

†Compare with [2], where the coefficient differs by 2. This difference is due to the fact that the three-phonon process was allowed owing to the widening of the spectrum and not because of the dispersion, as it was in the case of the formula (1.1.53).

$$\alpha_1' = \alpha_1 \left(\frac{cq_c}{kT}\right)^3 \sim \omega T. \tag{1.1.56}$$

The linear temperature dependence near the shoulder is apparently confirmed by experiments. The change of the location of the shoulder with pressure can be explained by [9] the dependence of q_c on pressure. Therefore, a model with $\gamma < 0$ for long-wavelength phonons explains the ultrasound attenuation; however, if it is accepted, it is not clear whether the magnitude of the deviation of heat capacity C_v from the T^3 law can also be explained.

1.1.7. On the Dispersion of the Roton Branch of the Excitation Spectrum

In the neutron-scattering experiments [11] there has been observed a damping of the roton branch of the spectrum which increases with the momentum to the right of the minimum. The linewidth of the scattered neutrons at $p = 2.2 \cdot 10^{-19}$ gm cm sec^{-1} reached the value of about 1 K (0.1 μeV). The large damping may be caused by the presence of a region of the curve $\varepsilon(p)$ where the roton velocity $v = \partial\varepsilon/\partial p$ exceeds the phonon velocity c. Obviously, such a region begins and ends by "sound" points where $v = c$. Rotons that possess energies in this interval can emit and absorb phonons. A simple calculation enables us to find the probability for phonon emission and the linewidth of the roton branch. The lifetime of the phonon with momentum q, resulting from the process, is determined in terms of the perturbation theory with the phonon–roton interaction energy in the form [see (1.1.12)]

$$H_{\text{int}} = \mathbf{p} \cdot \mathbf{v}_q, \tag{1.1.57}$$

where p is the roton momentum. Let the deviation of the phonon distribution function from equilibrium be equal to δn. Then the collision integral is easily reduced to the following form:

$$I = -\delta n \int \frac{2\pi}{\hbar} (\mathbf{p}|H|\mathbf{p} + \mathbf{q})^2 \delta(E(\mathbf{p}) + c\mathbf{q} - E(\mathbf{p} + \mathbf{q})) \tag{1.1.58}$$

$$\cdot \frac{d^3p}{(2\pi\hbar)^3} (N(\mathbf{p}) - N(\mathbf{p} + \mathbf{q})).$$

Here N is the roton distribution function, and $(\mathbf{p}|H|\mathbf{p} + \mathbf{q})$ is a transition matrix element for the absorption of the phonon by the roton, which according to [12] is equal to

$$(\mathbf{p}|H|\mathbf{p} + \mathbf{q}) = (\mathbf{p} \cdot \mathbf{q}) \sqrt{\frac{c}{2\rho q}}. \tag{1.1.59}$$

Having integrated over the angles, we obtain the characteristic time of the phonon emission from formulas (1.1.58) and (1.1.59) as follows:

$$\frac{1}{\tau} = (1 - e^{-cq/T}) \int \frac{|\mathbf{p} + \mathbf{q}|(\mathbf{p}\cdot\mathbf{q})^2}{4pq^2(\partial E/\partial p)_{p+q}} \frac{2\pi c}{\hbar\rho} \frac{4\pi p^2\, dp}{(2\pi\hbar)^3} N(E). \qquad (1.1.60)$$

The angle between the p and q vectors can be found from the energy conservation law

$$E(\mathbf{p}) + cq = E(\mathbf{p} + \mathbf{q}). \qquad (1.1.61)$$

We get the maximum value of q from the relation (1.1.61) for collinear p and q. For the case $qc \ll T$ (1.1.60) becomes

$$\frac{1}{\tau} = \frac{p_c^4 q}{4\pi\rho\hbar^4} e^{-\Delta_c/T}, \qquad (1.1.62)$$

where p_c and Δ_c are the momentum and the energy of the roton at the sound point. In formula (1.1.61), it is necessary to replace qc by T for $qc \gg T$.

Thus the decay of rotons leads not only to the widening of the roton branch, but to the widening of the phonon part of the spectrum as well.

The lifetime of phonons appears to be comparable to the phonon-scattering time by rotons, since Δ_c differs slightly from Δ in the minimum. Due to the last process, an equilibrium is established between the phonon and roton gases, and, as is well known, the latter is responsible for the absorption and dispersion of sound in superfluid helium at higher temperatures [13]. To what extent does the phonon emission by supersonic rotons influence this equilibrium? The estimations show, that in so far as the energy of the phonons emitted by the rotons is small, the contribution of this process is negligible. At low temperatures, where the factor $e^{-\Delta_c/T}$ is small, the process of phonon absorption by rotons is also not important for the ultrasound absorption in superfluid helium for the region of temperatures where the sound wavelength is less than the mean free path of excitations.

Let us consider now the width of the roton branch of the spectrum, caused by the emission of phonons. A simple determination analogous to that obtained above for the lifetime of rotons gives the following expression:

$$\frac{1}{\tau_r} = \frac{p_c^2|p - p_c|^3}{24\pi\rho\hbar^4}. \qquad (1.1.63)$$

Therefore, the damping of rotons increases proportionally to $|p - p_c|^3$ [14]. This regularity is qualitatively similar to the one obtained in [11]. Unfortunately, a quantitative agreement cannot yet be obtained because the damping of the roton branch observed in the experiments [11] exceeds considerably the effect calculated above.

According to the form of the roton spectrum [11], there must be two sound points on the curve $\varepsilon(p)$. Then, in the vicinity of the supersonic region the spectrum must have the following schematic form:

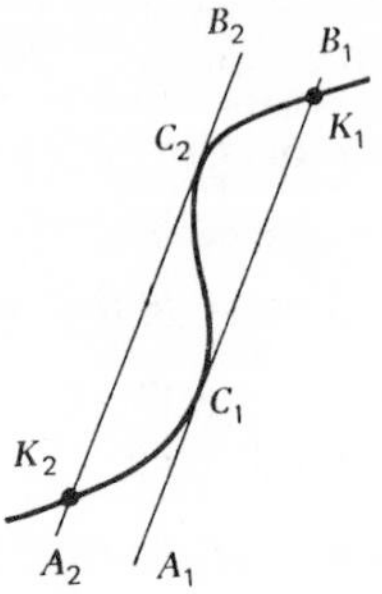

The straight lines A_2, B_2 and A_1B_1 have a slope equal to the sound velocity c. Then it can be easily seen that phonons can be emitted and absorbed by the rotons lying on the curve in the interval K_1K_2 (rather than C_1C_2, as one might have thought). This phenomenon must be taken into consideration while analyzing the experimental data.

1.1.8. Bound States Formed by Excitations

Another confirmation of the presence of the second branch in the excitation spectrum of liquid helium was obtained by Geytak et al. [15], who investigated the Raman light scattering in helium. They discovered that the energy that is necessary for the formation of two rotons is less than twice the energy of a single roton. This result can be explained by the possible existence of a bound state of two rotons. The observed binding energy is equal to 0.37 ± 0.10 K. For the formation of roton pairs to be possible, it is necessary to suppose an attraction in the roton interaction. A theoretical consideration was performed analogous to that used in superconductivity theory, where electron pairs are formed. Such a bound state may be found by analyzing the amplitude (the vertex part) for roton–roton scattering. Near $\varepsilon = 2\Delta$, the sum of singular terms of the perturbation series gives the result [14]

$$\Upsilon_2 = \frac{Q}{1 + Q \ln(\alpha/(2\Delta - \varepsilon))}; \quad \begin{matrix} \alpha(p) \to 0 \\ p \to 2p_0 \end{matrix} \tag{1.1.64}$$

where Q is a roton–roton interaction constant. If the interaction constant $Q < 0$, i.e., if the rotons are attracted, the vertex has a pole at the value

$$\varepsilon = 2\Delta - \alpha \exp \frac{1}{Q},$$

and there appears a new spectrum branch in helium with an energy less than 2Δ. The presence of an attraction between rotons influences the form of the single-excitation one-particle curve near $\varepsilon = 2\Delta$. The Green's function in this region is [14]

$$G^{-1} = \left[\omega - \varepsilon_0(p) + \beta \ln\left(\frac{\alpha}{2\Delta - \varepsilon}\right)\left(1 + Q \ln \frac{\alpha}{2\Delta - \varepsilon}\right)^{-1}\right], \tag{1.1.65}$$

where $\varepsilon_0(p)$ is the unperturbed energy of the roton ($\beta > 0$). Comparing (1.1.64) and (1.1.65), we find that the single-excitation and the two-roton branches of the spectrum never intersect, since the equations $\Upsilon_2^{-1} = 0$, and $G^{-1} = 0$, have no common solutions. Therefore there exist two possibilities for the two-roton spectrum [16]. With increasing momentum, the energy of the diroton changes from ε_2 up to 2Δ, and the branch comes to an end earlier than that of the single excitations (at that value of momentum at which the roton interaction constant changes sign). Both branches go almost horizontally up to the point $p = 2p_0$. Since $\alpha \to 0$ at $p \to 2p_0$, the two-roton branch energy at this point is equal to 2Δ and then this branch comes to an end. The single-excitation branch may continue above this value. It may be that it is this branch that transforms into the spectrum of vortex rings. If $\beta/|Q| < \Delta$ in (1.1.65), then there may exist another branch of a two-roton origin also. This branch goes above the two-roton one but below $\varepsilon = 2\Delta$, and reaches 2Δ when the momentum has the value given by the equation $\varepsilon_0(p) = 2\Delta - \beta/|Q|$.

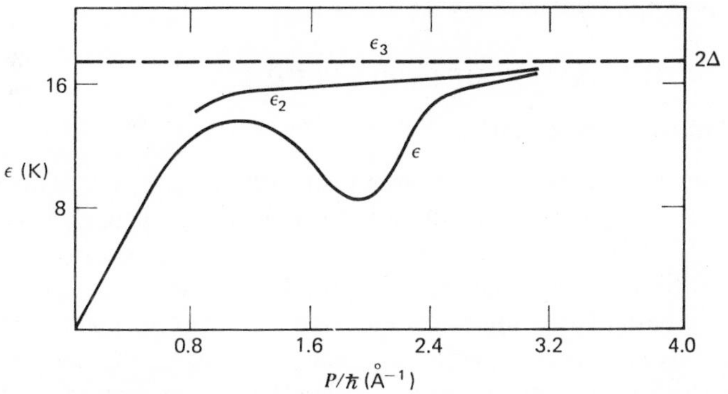

Special attention must be paid to the form of the two-roton spectrum at small values of the momentum, which are interesting in connection with the fact that two-roton excitations were found in the Raman scattering experiment [15]. The momentum of the light quantum is insignificantly small in comparison to the roton momentum, and the diroton obtained has practically a zero momentum; that is, the corresponding diroton is made of two rotons with equal and opposite momenta. In this case, in the vertex characterizing roton scattering, there arises not the logarithmic singularity near $\varepsilon = 2\Delta$, but the square-root singularity

$$\Gamma \sim \frac{1}{\sqrt{\varepsilon - 2\Delta}}. \qquad (1.1.66)$$

As a result, the density of states measured in the Raman scattering experiments in the vicinity of $\varepsilon = 2\Delta$ has the form

$$\rho_2(p = 0, \varepsilon) = \frac{p_0^2}{4\pi^2}\left(\frac{\mu}{\varepsilon - 2\Delta}\right)^{\frac{1}{2}}, \tag{1.1.67}$$

where μ is the effective mass of a roton.

The given picture occurs when the complete stability of the two-roton branch of excitations is supposed. According to [3], the energy of two-roton excitations exceeds twice the energy of the roton Δ. In the region $\varepsilon_2 > 2\Delta$, there is large damping that indicates a possibility of the decay of the two-roton excitation into two rotons.

It should be noted that the comparison of the experimental data [15] on Raman scattering with the theoretical analysis of the problem [17, 18] enables us to show that two-roton excitations are in D states, that is, they have an angular momentum $l = 2$. The last fact indicates that the roton interaction is anisotropic, i.e., it depends on the angle between the momenta of the interacting rotons. The formation of a diroton in a D state indicates that at large distance the roton interaction is attractive.

1.2. HYDRODYNAMICS OF SUPERFLUIDITY

1.2.1. Hydrodynamic Equations

Using the microscopic description of a superfluid that was outlined in the preceding sections, one can construct the complete system of hydrodynamic equations. The fundamental propositions upon which our whole analysis will be based are the following: the ordered motion of the excitations carries along with it only part of the liquid, characterized by the "normal" density ρ_n. The remaining part, the "superfluid," is characterized by the density $\rho_s = \rho - \rho_n$, and performs an independent motion. This latter motion has the important property of being irrotational. Thus, in a superfluid, there exist two simultaneous but independent motions, one normal and the other superfluid, with velocities v_n and v_s, with

$$\operatorname{curl} \mathbf{v}_s = 0. \tag{1.2.1}$$

Condition (1.2.1), which states that superfluid motion is irrotational, will not be violated until the flow velocities have reached certain critical values at which the normal and superfluid parts of the liquid begin to interact.

We may now proceed to the derivation of the complete system of hydrodynamic equations, taking as our starting point the conservation laws and the Galilean relativity principle.

The conservation laws have the same form for all physical quantities; they are differential equations stating that the time derivative of the conserved quantity is equal to the divergence of some vector.

The law of conservation of mass is the so-called continuity equation, which relates the total density ρ and the current $\mathbf{j}$ (the momentum per unit volume):

$$\dot{\rho} + \operatorname{div} \mathbf{j} = 0. \tag{1.2.2}$$

The momentum conservation law is the equation of motion

$$\dot{j}_i + \frac{\partial \Pi_{ik}}{\partial r_k} = 0, \tag{1.2.3}$$

where Π_{ik} is the momentum flux density tensor.

We shall not at present consider dissipative processes. Then the flow is reversable and the entropy S is also conserved. We may, therefore, write

$$\dot{S} + \operatorname{div} \mathbf{F} = 0, \tag{1.2.4}$$

where $\mathbf{F}$ is the entropy flux vector. Since the entropy is only connected with the excitations, it should be carried along with the normal motion. Consequently the entropy flux is equal to $S\mathbf{v}_n$. However, we shall not at this point make any reference to the microscopic picture; later on, we shall show that the relation $\mathbf{F} = S\mathbf{v}_n$ follows from the conservation laws. For the moment, therefore, $\mathbf{F}$ remains some unknown vector to be determined.

In a superfluid there can be two different motions and, consequently, there should be two hydrodynamic equations of motion. The first one is equation (1.2.3) and the second one an equation for the time derivative of $\mathbf{v}_s$. Since $\operatorname{curl} \mathbf{v}_s = 0$, we can write

$$\dot{\mathbf{v}}_s + \nabla\left(\varphi + \frac{\mathbf{v}_s^2}{2}\right) = 0, \tag{1.2.5}$$

where φ is some scalar function.

Equations (1.2.2)–(1.2.5) represent the complete system of hydrodynamic equations of a superfluid. In order to make them completely meaningful, we must still determine the form of the unknown terms Π_{ik}, F, and φ. In order to do this, we use the energy conservation law which can be written in differential form as

$$\frac{\partial E}{\partial t} + \operatorname{div} \mathbf{Q} = 0, \tag{1.2.6}$$

where E is the energy per unit volume of liquid, and $\mathbf{Q}$ the energy flux. It is necessary to choose the unknown terms in equations (1.2.2)–(1.2.5) in such a way that equation (1.2.6) be automatically satisfied. We may further use the Galilean relativity principle, which allows us to determine the dependence of all quantities on the velocity $\mathbf{v}_s$ for a given value of the difference $\mathbf{v}_n - \mathbf{v}_s$.

In what follows, it will be useful to consider a new frame of reference (K_0) in which the velocity of superfluid motion of a given element of the liquid is equal to zero. The frame K_0 moves at a velocity $\mathbf{v}_s$ with respect to the original frame K. The values of quantities that interest us in the different frames are related by the well-known formulas*

$$\mathbf{j} = \rho \mathbf{v}_s + \mathbf{j}_0, \tag{1.2.7}$$

$$\Pi_{ik} = \rho v_{si} v_{sk} + v_{si} j_{0k} + v_{sk} j_{0i} + \pi_{ik}, \tag{1.2.8}$$

$$E = \frac{\rho \mathbf{v}_s^2}{2} + \mathbf{v}_s \cdot \mathbf{j}_0 + E_0, \tag{1.2.9}$$

$$\mathbf{Q} = \left(\frac{\rho v_s^2}{2} + \mathbf{v}_s \cdot \mathbf{j}_0 + E_0 \right) \mathbf{v}_s + \frac{v_s^2}{2} \mathbf{j}_0 + \vec{\pi} \cdot \mathbf{v}_s + \mathbf{q}, \tag{1.2.10}$$

$$\mathbf{F} = S \mathbf{v}_s + \mathbf{f}. \tag{1.2.11}$$

Here $\mathbf{j}_0$ is the momentum, π_{ik} the momentum flux tensor, E_0 the energy, $\mathbf{q}$ the energy flux vector, and $\mathbf{f}$ the entropy flux in the frame K_0.

In the frame K_0 the liquid moves with the velocity $\mathbf{v}_n - \mathbf{v}_s$, and obviously all quantities ($\mathbf{j}_0$, π_{ik}, E_0, $\mathbf{q}$, $\mathbf{f}$) may only depend on this difference. The energy E_0 satisfies the thermodynamic identity

$$dE_0 = T\,dS + \mu\,d\rho + (\mathbf{v}_n - \mathbf{v}_s) \cdot d\mathbf{j}_0. \tag{1.2.12}$$

Here μ is the chemical potential and T the temperature. The third term in (1.2.12)simply expresses the fact that the velocity is the derivative of the energy with respect to the momentum and should be looked upon as a definition of the velocity $\mathbf{v}_n$. By symmetry considerations, it follows that the vector $\mathbf{j}_0$ may only be directed along $\mathbf{v}_n - \mathbf{v}_s$, so that we may write

$$\mathbf{j}_0 = \rho_n (\mathbf{v}_n - \mathbf{v}_s). \tag{1.2.13}$$

Relation (1.2.13) should be considered as the definition of the normal density ρ_n. From (1.2.7) and (1.2.13) we then find

$$\mathbf{j} = \rho_s \mathbf{v}_s + \rho_n \mathbf{v}_n, \tag{1.2.14}$$

where $\rho_s = \rho - \rho_n$.

The ensuing calculation may be summed up as follows: we differentiate the energy E with respect to time, and express all time derivatives of thermodynamic quantities, namely of $\mathbf{j}$ and $\mathbf{v}_s$, by means of equations (1.2.2)–(1.2.5). We then calculate div $\mathbf{Q}$ using (1.2.10) and insert E and div $\mathbf{Q}$ into equation (1.2.6). After a considerable amount of rearrangement we obtain

$$\begin{aligned} \operatorname{div} \mathbf{q} = &- (\vec{\pi} \cdot \nabla) \mathbf{v}_s + (\mathbf{v}_n - \mathbf{v}_s) \cdot (\nabla \cdot \vec{\pi}) + \mathbf{j}_0 \cdot ((\mathbf{v}_n - \mathbf{v}_s) \cdot \nabla) \mathbf{v}_n \\ &+ (\mathbf{j}_0 - \rho(\mathbf{v}_n - \mathbf{v}_s)) \cdot \nabla(\varphi - \mu) - \nabla T \cdot (\mathbf{f} - S(\mathbf{v}_n - \mathbf{v}_s)) \\ &+ \operatorname{div} (\mathbf{f} T + \mathbf{j}_0 \mu). \end{aligned} \tag{1.2.15}$$

Instead of π_{ik} it is convenient to introduce another tensor as follows:

*Formulas (1.2.7)–(1.2.11) follow immediately from the Galilean principle of relativity.

$$m_{ik} = \pi_{ik} + [E_0 - TS - \mu\rho - (\mathbf{v}_n - \mathbf{v}_s)\cdot\mathbf{j}_0]\,\delta_{ik}. \tag{1.2.16}$$

In the absence of energy dissipation, the quantities m_{ik} **q**, **f**, and φ are functions of the theremodynamic variables and the velocities, but do not depend on their space or time derivatives. This enables us to obtain from (1.2.15) unique expressions for the desired quantities

$$m_{ik} = j_{0i}(v_{nk} - v_{sk}), \quad \mathbf{f} = S(\mathbf{v}_n - \mathbf{v}_s), \quad \varphi = \mu$$

$$\mathbf{q} = T\mathbf{f} + \mu\mathbf{j}_0 - ((\mathbf{v}_n - \mathbf{v}_s)\cdot\mathbf{v}_s)\mathbf{j}_0 + (\mathbf{v}_n - \mathbf{v}_s)(\mathbf{v}_n\cdot\mathbf{j}_0). \tag{1.2.17}$$

Finally we may write

$$\mathbf{F} = \mathbf{f} + S\mathbf{v}_s = S\mathbf{v}_n, \tag{1.2.18}$$

$$\Pi_{ik} = \rho v_{si}v_{sk} + v_{si}j_{0k} + v_{nk}j_{0i} - [E_0 - TS - \mu\rho - (\mathbf{v}_n - \mathbf{v}_s)\cdot\mathbf{j}_0]\,\delta_{ik}, \tag{1.2.19}$$

$$\mathbf{Q} = \left(\mu + \frac{v_s^2}{2}\right)(\mathbf{j}_0 + \rho\mathbf{v}_s) + ST\mathbf{v}_n + \mathbf{v}_n(\mathbf{v}_n\cdot\mathbf{j}_0). \tag{1.2.20}$$

The expression in square brackets in (1.2.19) is the pressure; by definition it is equal to the derivative of the total energy with respect to the volume when the total mass, the total entropy, and the total momentum of relative motion are all held constant.

$$p = -\frac{\partial(E_0V)}{\partial V} = -E_0 + TS + \mu\rho + ((\mathbf{v}_n - \mathbf{v}_s)\cdot\mathbf{j}_0). \tag{1.2.21}$$

Inserting the expressions for **F**, φ, and Π_{ik} into equations (1.2.3)–(1.2.5), we obtain the complete system of hydrodynamic equations for superfluids

$$\dot{\rho} + \operatorname{div}\mathbf{j} = 0, \tag{1.2.22}$$

$$\frac{\partial\mathbf{j}}{\partial t} + \mathbf{v}_s\operatorname{div}\mathbf{j} + (\mathbf{j}\cdot\nabla)\mathbf{v}_s + \mathbf{j}_0\operatorname{div}\mathbf{v}_n + (\mathbf{v}_n\cdot\nabla)\mathbf{j}_0 + \nabla\varphi = 0, \tag{1.2.23}$$

$$\dot{S} + \operatorname{div} S\mathbf{v}_n = 0, \tag{1.2.24}$$

$$\dot{\mathbf{v}}_s + \nabla\left(\frac{v_s^2}{2} + \mu\right) = 0 \tag{1.2.25}$$

The momentum flux tensor Π_{ik} may, by (1.2.14), be put into the form

$$\Pi_{ik} = \rho_n v_{ni}v_{nk} + \rho_s v_{si}v_{sk} + p\delta_{ik}, \tag{1.2.19'}$$

where the first term is the momentum flux of the normal motion, and the second term the momentum flux of the superfluid motion. The hydrodynamic equations we have obtained in (1.2.22)–(1.2.25) are rather complicated since the quantitites μ, ρ_n, S, etc., which appear in them, are functions of the relative velocity $\mathbf{v}_n - \mathbf{v}_s$, whose form can only be determined by going to the microscopic theory.

These general hydrodynamic equations are considerably simplified in the case of small velocities. It must be noted that the property of superfluidity is destroyed when the velocities are greater than certain critical values. However, under nonstationary conditions, eg. sound propagation at high amplitude, the velocities may become considerably larger than these critical values. There exists, therefore, a region of applicability of the general equations where their nonlinear character becomes apparent. If we restrict ourselves to quadratic terms in the velocities we may neglect the dependence of ρ_n and ρ_s on the velocities. Let us choose as our independent thermodynamic variables the pressure p and the temperature T. Let us write the thermodynamic identity satisfied by the chemical potential, which by (1.2.12) and (1.2.21) can be written

$$d\mu = -\sigma\, dT + \frac{1}{\rho}\, dp - \frac{\rho_n}{\rho}(\mathbf{v}_n - \mathbf{v}_s)\cdot d(\mathbf{v}_n - \mathbf{v}_s), \quad \left(\sigma = \frac{S}{\rho}\right). \tag{1.2.26}$$

From this relation it is easy to find the dependence of the entropy σ and the density ρ on the relative velocity $\mathbf{w} = \mathbf{v}_n - \mathbf{v}_s$. From the general relations from derivatives

$$\frac{\partial \sigma}{\partial (w^2)} = \frac{1}{2}\frac{\partial}{\partial T}\frac{\rho_n}{\rho}, \quad \frac{\partial \rho}{\partial (w^2)} = \rho^2 \frac{1}{2}\frac{\partial}{\partial p}\frac{\rho_n}{\rho}, \tag{1.2.27}$$

we can find the first terms in the expansion of σ and ρ in powers of w^2:

$$\sigma(p, T, w) = \sigma(p, T) + \frac{1}{2} w^2 \frac{\partial}{\partial T}\frac{\rho_n}{\rho}, \tag{1.2.28}$$

$$\rho(p, T, w) = \rho(p, T) + \frac{1}{2}\rho^2 w^2 \frac{\partial}{\partial p}\frac{\rho_n}{\rho}. \tag{1.2.29}$$

Inserting these expressions into the general equations (1.2.22)–(1.2.25), we obtain equations that are valid up to second order in the velocities

$$\frac{\partial}{\partial t}\left(\rho + \frac{1}{2} w^2 \rho^2 \frac{\partial}{\partial p}\frac{\rho_n}{\rho}\right) + \operatorname{div}\mathbf{j} = 0, \quad \mathbf{j} = \rho_s \mathbf{v}_s + \rho_n \mathbf{v}_n, \tag{1.2.30}$$

$$\frac{\partial j_i}{\partial t} + \frac{\partial \Pi_{ik}}{\partial r_k} = 0, \quad \Pi_{ik} = \rho_s v_{si} v_{sk} + \rho_n v_{ni} v_{nk} + p\delta_{ik}, \tag{1.2.31}$$

$$\frac{\partial}{\partial t}\left[S + \frac{\rho w^2}{2}\left(\frac{\partial}{\partial T}\frac{\rho_n}{\rho} + S\frac{\partial}{\partial p}\frac{\rho_n}{\rho}\right)\right] + \operatorname{div} S\mathbf{v}_n = 0 \tag{1.2.32}$$

$$\frac{\partial \mathbf{v}_s}{\partial t} + \nabla\left[\mu + \frac{v_s^2}{2} - \frac{\rho_n w^2}{2\rho}\right] = 0 \tag{1.2.33}$$

In the formula (1.2.33) there appears the chemical potential μ of the liquid at rest. In the above approximate equations, the velocities are considered small with respect to the velocities of first and second sound. The general equations may also be considerably simplified in the case in which the velocities are small only in comparison with the velocity of first sound, but are comparable to the velocity of second sound. In this case it is possible to find a general solution in the form of a one-dimensional running wave. The considerable simplification is due to the fact that the normal density ρ_n depends only weakly on the square of the relative velocity w^2.

Let us now consider the question of the boundary conditions satisfied by the thermodynamic quantities. Obviously, the normal component of the current j should vanish at the walls, since there can be no transport of matter across this boundary. As for the velocity of normal motion $\mathbf{v}_n$, it is linked to the motion of the gas of excitations, which has all the properties of a normal viscous liquid. Consequently, the tangential component of $\mathbf{v}_n$ should vanish at the boundary of a solid body. The normal component of $\mathbf{v}_n$ (directed along the z-axis) is not equal to zero, but determines the heat flux from the liquid to the solid body, which according to (1.2.20), for $j_z = 0$ is equal to STv_{nz}. The normal component of the heat flux is continuous across the boundary between the liquid and the solid body. The temperature has a discontinuity across the boundary, which is proportional to the magnitude of the heat flux.

Let us choose the x- and y- axes to be along the surface of the solid and the z- axis perpendicular to it. Then the above boundary conditions are written as follows:

$$\rho_s v_{sz} + \rho_n v_{nz} = 0,\ v_{nx} = v_{ny} = 0, \tag{1.2.34}$$

$$STv_{nz} = -\kappa_s\left(\frac{\partial T}{\partial z}\right)_s, \quad T_l - T_s = Kq_z. \tag{1.2.35}$$

Here κ_s is the thermal conductivity of the solid.

In many cases it is possible to neglect the thermal conductivity of the solid and to set κ_s equal to zero. This gives

$$v_{nz} = 0, \quad v_{ns} = 0, \tag{1.2.34}$$

i.e., the boundary conditions for $\mathbf{v}_n$ are those of a normal liquid and for $\mathbf{v}_s$ those of an ideal liquid.

A moving superfluid in the presence of a current normal to the walls causes tangential forces acting on the surface of a solid body. This can be seen from the fact that the component Π_{xz} of the momentum flux tensor is nonzero in this case. Indeed, from the first of relations (1.2.34) we can find this component

$$\Pi_{xz} = \rho_s v_{sx} v_{sz} + \rho_n v_{nx} v_{nz} = \rho_n v_{nz}(v_{nx} - v_{sx}).$$

Expressing the component v_{nz} in terms of the normal component of the heat flux q_z, we finally obtain

$$\Pi_{xz} = \frac{\rho_n q_z}{ST}(v_{nx} - v_{sx}). \tag{1.2.37}$$

1.2.2. The Dissipative Terms in the Hydrodynamic Equations

The system of equations (1.2.22)–(1.2.25) obtained in the preceding paragraph describes the motion of a superfluid in the absence of energy dissipation. In reality, the lack of equilibrium leads to the appearance, in all the fluxes, of terms depending on the derivatives of the velocities and thermodynamic variables with respect to the coordinates. It must be noted that under nonequilibrium conditions the usual definitions of thermodynamic quantities lose their meaning and must be made more precise. If as before we denote by ρ the mass per unit volume of liquid and by $\mathbf{j}$ its momentum, then the continuity equation retains its usual form

$$\dot{\rho} + \operatorname{div} \mathbf{j} = 0. \tag{1.2.38}$$

We shall further denote by $\bar{E}_0$ the energy per unit volume in the frame of reference in which the superfluid part is at rest. The remaining thermodynamic variables are defined to have the same functional dependence on the density ρ, the energy E, and the relative velocity w as they do in thermodynamic equilibrium. Here the entropy $S(p, E, w)$ will not be the true entropy, whose integral $\int S\, dV$ must always increase with time. However, for situations close to equilibrium, when the gradients of all quantities are small, the entropy defined in this manner will be almost identical to the true entropy. Indeed, it is easy to see that there can be no terms linear in the gradients in the power series expansion of the entropy, since these are odd. The entropy attains its maximum value in the equilibrium state; consequently, its expansion in powers of the gradients begins with quadratic terms, which may be neglected in the approximation we are considering.

The law of conservation of momentum can, as previously, be written in the form

$$\dot{j}_i + \frac{\partial}{\partial r_k}(\Pi_{ik} + \tau_{ik}) = 0, \tag{1.2.39}$$

with the difference that in the momentum flux, along with the usual terms, there occurs an unknown dissipative term τ_{ik}, which must still be determined. In a similar manner we may add another term ∇h in the equation of superfluid flow (1.2.25):

$$\dot{\mathbf{v}}_s + \nabla\left(\mu + \frac{v_s^2}{2} + h\right) = 0 \tag{1.2.40}$$

(but as before curl $\mathbf{v}_s = 0$).

Obviously, the expression for the heat flux in the law of conservation of energy is also changed by some quantity $\mathbf{Q}'$

$$\frac{\partial E}{\partial t} + \operatorname{div}(\mathbf{Q} + \mathbf{Q}') = 0. \tag{1.2.41}$$

As for the entropy equation, it now does not have the form of a continuity equation, since entropy is not conserved, but increases. We will use the requirement that the entropy should increase, in order to determine the unknown dissipation coefficients.

Let us differentiate expression (1.2.9) for E with respect to time; this yields

$$\frac{\partial E}{\partial t} = \left(\frac{v_s^2}{2} + \mu\right)\dot{\rho} + \mathbf{v}_s\cdot\dot{\mathbf{j}} + \mathbf{v}_n\cdot\frac{\partial \mathbf{j}_0}{\partial \mathbf{t}} + T\dot{S}. \tag{1.2.42}$$

We may then eliminate the time derivatives by equations (1.2.38)–(1.2.40), obtaining

$$\frac{\partial E}{\partial t} - \operatorname{div}\{\mathbf{Q} + \mathbf{q} + h(\mathbf{j} - \rho\mathbf{v}_n) + (\overset{\leftrightarrow}{\tau}\cdot\mathbf{v}_n)\} = T\left(\dot{S} + \operatorname{div}\left(S\mathbf{v}_n + \frac{\mathbf{q}}{T}\right)\right)$$

$$+h\operatorname{div}(\mathbf{j} - \rho\mathbf{v}_n) + \tau_{ik}\frac{\partial v_{ni}}{\partial r_k} + \frac{1}{T}\mathbf{q}\cdot\nabla T. \tag{1.2.43}$$

Since in our nonequilibrium situation we may also have an additional term in the entropy flux, we here added to both sides of (1.2.6) a term div $\mathbf{q}$ containing the unknown quantity $\mathbf{q}$.

Comparing (1.2.43) and (1.2.41), we obtain the equation that determines the rate of increase of entropy

$$T\left(\frac{\partial S}{\partial t} + \operatorname{div}\left(S\mathbf{v}_n + \frac{\mathbf{q}}{T}\right)\right) = -h\operatorname{div}(\mathbf{j} - \rho\mathbf{v}_n) - \tau_{ik}\frac{\partial v_{ni}}{\partial r_k} \tag{1.2.44}$$

$$-\frac{1}{T}\mathbf{q}\cdot\nabla T,$$

and the expression for the additional dissipative flux of heat

$$\mathbf{Q}' = \mathbf{q} + h(\mathbf{j} - \rho\mathbf{v}_n) + \overset{\leftrightarrow}{\tau}\cdot\mathbf{v}_n. \tag{1.2.45}$$

The expression on the right-hand side of (1.2.44) is the dissipative function of the superfluid. If the spatial derivatives of the velocities and thermodynamic variables are not too large, then in the first approximation all additions to the equations (τ_{ik}, h, q) are linear functions of these derivatives. From the law of increase of entropy, it then follows that the dissipative function must be a positive definite quadratic form in these same derivatives.* From this requirement we may immediately determine the form of the unknown terms

$$\tau_{ik} = -\eta\left(\frac{\partial v_{ni}}{\partial r_k} + \frac{\partial v_{nk}}{\partial r_i} - \frac{2}{3}\delta_{ik}\frac{\partial v_{nl}}{\partial r_l}\right)$$

$$-\delta_{ik}[\xi_1\operatorname{div}(\mathbf{j} - \rho\mathbf{v}_n) + \xi_2\operatorname{div}\mathbf{v}_n], \tag{1.2.46}$$

*Here we mean that the relative velocity $\mathbf{v}_n - \mathbf{v}_s$ is small in comparison with the sound velocity. In ^{3}He-^{4}He solutions, where it is possible to have large temperature and concentration gradients, other dissipative terms containing the relative velocity $\mathbf{v}_n - \mathbf{v}_s$ may appear.

$$h = -\xi_3 \operatorname{div}(\mathbf{j} - \rho \mathbf{v}_n) - \xi_4 \operatorname{div} \mathbf{v}_n, \tag{1.2.47}$$

$$q = -\kappa \nabla T. \tag{1.2.48}$$

As usual, in the momentum flux τ_{ik} we separate the combination of derivatives of $\mathbf{v}_n$ that has zero trace (first viscosity).

According to the Onsager symmetry principle for kinetic coefficients, we have the relation

$$\xi_1 = \xi_4. \tag{1.2.49}$$

The coefficients ξ_1, ξ_2, ξ_3, ξ_4 are the coefficients of second viscosity. There are, therefore, in all three independent coefficients of second viscosity; η is the coefficient of first viscosity and is entirely connected with the normal motion, and κ is the thermal conduction coefficient. As was to be expected, there is no coefficient analogous to the first viscosity for the superfluid motion.

Let us now write in final form the hydrodynamic equations for a superfluid taking into account the dissipative terms

$$\dot{\rho} + \operatorname{div} \mathbf{j} = 0, \tag{1.2.50}$$

$$\frac{\partial j_i}{\partial t} + \frac{\partial \Pi_{ik}}{\partial r_k} = \frac{\partial}{\partial r_k}\left\{\eta\left(\frac{\partial v_{ni}}{\partial r_k} + \frac{\partial v_{nk}}{\partial r_i} - \frac{2}{3}\delta_{ik}\frac{\partial v_{nl}}{\partial r_l}\right)\right.$$
$$\left. + \delta_{ik}\xi_1 \operatorname{div}(\mathbf{j} - \rho\mathbf{v}_n) + \delta_{ik}\xi_2 \operatorname{div}\mathbf{v}_n\right\}, \tag{1.2.51}$$

$$\dot{\mathbf{v}}_s + \nabla\left(\mu + \frac{v_s^2}{2}\right) = \nabla\{\xi_3 \operatorname{div}(\mathbf{j} - \rho\mathbf{v}_n) + \xi_4 \operatorname{div}\mathbf{v}_n\}, \tag{1.2.52}$$

$$\dot{S} + \operatorname{div}\left(S\mathbf{v}_n + \frac{\mathbf{q}}{T}\right) = \frac{1}{T}R. \tag{1.2.53}$$

The dissipative function is equal to

$$R = \xi_2(\operatorname{div}\mathbf{v}_n)^2 + \xi_3(\operatorname{div}(\mathbf{j} - \rho\mathbf{v}_n))^2 + 2\xi_1 \operatorname{div}\mathbf{v}_n \operatorname{div}(\mathbf{j} - \rho\mathbf{v}_n)$$
$$+ \frac{1}{2}\eta\left(\frac{\partial v_{ni}}{\partial r_k} + \frac{\partial v_{nk}}{\partial r_i} - \frac{2}{3}\delta_{ik}\frac{\partial v_{nl}}{\partial r_l}\right)^2 + \kappa\frac{(\nabla T)^2}{T}. \tag{1.2.54}$$

In order to ensure that the function R will be positive the kinetic coefficients η, ξ_2, ξ_3, and κ must be positive and ξ_1 must satisfy the inequality

$$\xi_1^2 \leq \xi_2\xi_3. \tag{1.2.55}$$

1.2.3. The Propagation of Sound in a Superfluid

In a sound wave the velocities $\mathbf{v}_n$ and $\mathbf{v}_s$ are assummed to be small*, and the thermodynamic quantities almost equal to their equilibrium values. The propagation of sound in He (II) is described by the system of hydrodynamic equations (1.2.22)–(1.2.25), which in this case may be linearized; this yields

*Here we mean that $\mathbf{v}_n$ and $\mathbf{v}_s$ are small compared to the velocity of sound.

$$\frac{\partial \rho}{\partial t} + \operatorname{div} \mathbf{j} = 0, \tag{1.2.56}$$

$$\frac{\partial \rho\sigma}{\partial t} + \rho\sigma \operatorname{div} \mathbf{v}_n = 0, \quad (\sigma\rho = S) \tag{1.2.57}$$

$$\frac{\partial \mathbf{j}}{\partial t} + \nabla p = 0, \tag{1.2.58}$$

$$\frac{\partial \mathbf{v}_s}{\partial t} + \nabla \mu = 0. \tag{1.2.59}$$

Let us eliminate the momentum **j** from (1.2.56) and (1.2.58); we get

$$\frac{\partial^2 \rho}{\partial t^2} = \Delta p. \tag{1.2.60}$$

Furthermore, let us eliminate the velocities v_n and v_s from the three equations (1.2.57)–(1.2.59). For this, let us take the time derviative of (1.2.57) and the divergence of (1.2.58) and (1.2.59). Eliminating the terms $\partial/\partial t \operatorname{div} \mathbf{v}_n$ and $\partial/\partial t \operatorname{div} \mathbf{v}_s$ from the equations thus obtained, we find

$$\rho_s \Delta\mu - \Delta p + \frac{\rho_n}{\rho\sigma} \frac{\partial^2}{\partial t^2} (\rho\sigma) = 0. \tag{1.2.61}$$

Let us express the quantity $\partial^2\rho/\partial t^2$ occuring in the above equation by means of (1.2.60) and use the thermodynamic identity (1.2.26). We have, finally,

$$\frac{\partial^2 \sigma}{\partial t^2} = \frac{\rho_s}{\rho_n} \sigma^2 \Delta T. \tag{1.2.62}$$

Equations (1.2.61) and (1.2.62) determine the changes in the thermodynamic quantities in the presence of a sound wave.

Let us use, in these equations, the independent variables p and T, which we can represent in the form $p = p_0 + p'$, $T = T_0 + T'$. The quantities with subscript zero denote the equilibrium values and the primed quantities the deviations from equilibrium due to the sound wave. As a result, equations (1.2.61) and (1.2.62) take the form

$$\frac{\partial \rho}{\partial p} \frac{\partial^2 p'}{\partial t^2} - \Delta p' + \frac{\partial \rho}{\partial T} \frac{\partial^2 T'}{\partial t^2} = 0, \tag{1.2.63}$$

$$\frac{\partial \sigma}{\partial p} \frac{\partial^2 p'}{\partial t^2} + \frac{\partial \sigma}{\partial T} \frac{\partial^2 T'}{\partial t^2} - \frac{\sigma^2 \rho_s}{\rho_n} \Delta T' = 0. \tag{1.2.64}$$

Let us look for a solution of the system (1.2.63)–(1.2.64), representing a plane wave propagating in some direction. In such a wave the quantities p' and T' vary as $\exp[-i\omega(t - x/u)]$ (the x-axis can be chosen as the direction of propagation of the wave, ω is the frequency and u the velocity of sound). The system of equations (1.2.63)–(1.2.64) may now be written in the following form:

$$\left(\frac{\partial \rho}{\partial p} u^2 - 1\right) p' + \frac{\partial \rho}{\partial T} u^2 T' = 0, \tag{1.2.65}$$

$$\frac{\partial \sigma}{\partial p} u^2 p' + \left(\frac{\partial \sigma}{\partial T} u^2 - \frac{\sigma^2 \rho_s}{\rho_n}\right) T' = 0. \tag{1.2.66}$$

As usual, the above two equations will be compatible if the determinant of their coefficients is equal to zero. Expanding the determinant, we find

$$u^4 \frac{\partial(\sigma,\rho)}{\partial(T,p)} - u^2 \left(\frac{\partial \sigma}{\partial T} + \sigma^2 \frac{\rho_s}{\rho_n} \frac{\partial \rho}{\partial p}\right) + \frac{\rho_s}{\rho_n} \sigma^2 = 0, \tag{1.2.67}$$

which after some simple transformations yields

$$u^4 - u^2 \left[\left(\frac{\partial p}{\partial \rho}\right)_\sigma + \frac{\rho_s}{\rho_n} \sigma^2 \left(\frac{\partial T}{\partial \sigma}\right)_\rho\right] + \frac{\rho_s}{\rho_n} \sigma^2 \left(\frac{\partial T}{\partial \sigma}\right)_\rho \left(\frac{\partial p}{\partial \rho}\right)_T = 0. \tag{1.2.68}$$

Eq. (1.2.68) determines two possible velocities of sound in He (II). The coefficient of thermal expansion $(\partial \rho/\partial T)_p$ is, in practice, very small for all substances, For He (II) it is even anomalously small. Therefore, according to the well-known thermodynamic relations, one can consider the two specific heats C_p and C_V to be practically equal in He (II). However, in that case the derivatives $(\partial p/\partial \rho)_T$ and $(\partial p/\partial \rho)_\sigma$ which are related by the equation $(\partial p/\partial \rho)_\sigma = (C_p/C_V)\,(\partial p/\sigma \rho)_T$ can also be considered equal to a high degree of accuracy. This considerably simplifies equation (1.2.68), whose roots become equal to

$$u_1 = c = \sqrt{\left(\frac{\partial p}{\partial \rho}\right)_\sigma}, \tag{1.2.69}$$

$$u_2 = \sqrt{\frac{\sigma^2 \rho_s}{\rho_n (\partial \sigma/\partial T)}} \tag{1.2.70}$$

The first root determines the velocity of ordinary (first) sound in He (II). From (1.2.60) we see that it is with this velocity that oscillations of pressure (density) are propagated in He (II). The second root u_2 determines the velocity of so-called second sound. From (1.2.62) we see that oscillations of temperature (entropy) are propagated with this velocity. The ability to propagate undamped temperature waves is a property specific to He (II). The temperature dependence of the velocity of second sound, calculated from formula (1.2.70), is presented in Figure 1.4. At the λ point $\rho_s = 0$ and the velocity u_2 also goes to zero. At sufficiently low temperatures, when all thermodynamic quantities are determined by the phonons (below 0.5 K) the velocity u_2 tends toward the limit $c/\sqrt{3}$.

Second sound can be looked upon as a compressional wave in the gas of excitations. This follows immediately from the fact that oscillations of temperature bring about oscillations in the density of excitations. The velocity of second sound is therefore the velocity of ordinary sound in the excitation gas.

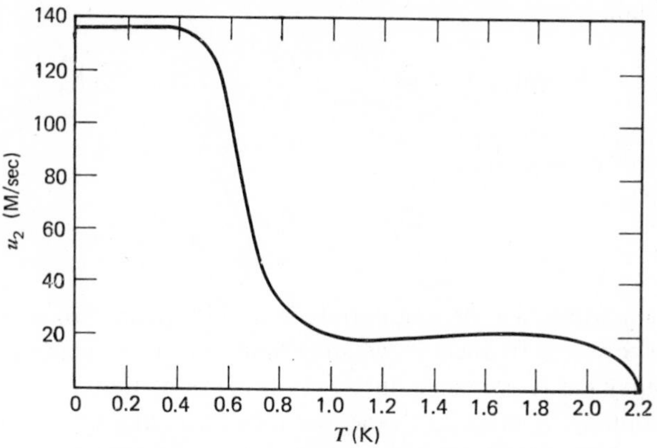

FIGURE 1.5. The velocity of second sound in He (II).

The limiting value $u_2 = c/\sqrt{3}$ follows immediately from the well-known result on the velocity of sound in a gas whose spectrum is given by $\varepsilon = cp$.*

1.2.4. The Absorption of Sound

The presence of dissipative processes in a superfluid leads to absorption or damping of sound. In order to investigate the question of sound propagation in the presence of dissipation, let us write the general equations (1.2.50)–(1.2.53) in the following linearized from:

$$\dot{\rho} + \operatorname{div} \mathbf{j} = 0, \tag{1.2.71}$$

$$\frac{\partial j_i}{\partial t} + \frac{\partial p}{\partial r_i} = \eta \frac{\partial}{\partial r_k}\left(\frac{\partial v_{ni}}{\partial r_k} + \frac{\partial v_{nk}}{\partial r_i} - \frac{2}{3}\delta_{ik}\frac{\partial v_{nl}}{\partial r_l}\right) + \frac{\partial}{\partial r_i}\{\xi_1 \operatorname{div}(\mathbf{j} - \rho \mathbf{v}_n) + \xi_2 \operatorname{div} \mathbf{v}_n\}, \tag{1.2.72}$$

$$\dot{\mathbf{v}}_s + \nabla \mu = \nabla \{\xi_3 \operatorname{div}(\mathbf{j} - \rho \mathbf{v}_n) + \xi_4 \operatorname{div} \mathbf{v}_n\}, \tag{1.2.73}$$

$$T\{(\dot{\sigma}\rho) + \sigma\rho \operatorname{div} \mathbf{v}_n\} = \kappa \Delta T. \tag{1.2.74}$$

In a sound wave the velocities $\mathbf{v}_n$ and $\mathbf{v}_s$ and the varying parts of the thermodynamic variables ρ' and σ' (which we choose as our independent variables) vary as $e^{-i\omega(t-x/u)}$ (x is the direction of propagation and ω the frequency of the sound wave). The velocity of sound will in this case be a complex quantity u, its imaginary part determining the damping. In order to simplify the calculations let us use the fact that the thermal expansion in He (II) is very small. Eliminating the variables $\mathbf{v}_n$ and $\mathbf{v}_s$ from (1.2.71)–

*Thus, for instance, in a gas of photons the velocity of sound is $c/\sqrt{3}$ where c is the velocity of light.

(1.2.74), we obtain

$$\left(u^2 - \frac{\partial p}{\partial \rho}\right)\rho' = i\omega\left\{\left(\frac{4}{3}\eta + \xi_2\right)\frac{\rho'}{\rho} + \left(\frac{4}{3}\eta + \xi_2 - \rho\xi_1\right)\frac{\sigma'}{\sigma}\right\}, \quad (1.2.75)$$

$$\left(\sigma\frac{\partial T}{\partial \sigma} - \frac{\rho_n}{\rho_s\sigma}u^2\right)\sigma' = i\omega\left\{\left(\xi_4 - \frac{1}{\rho}\xi_2 - \frac{4}{3\rho}\eta\right)\frac{\rho'}{\rho}\right.$$

$$\left. + \left(\xi_4 - \rho\xi_3 - \frac{1}{\rho}\xi_2 + \xi_1 - \frac{4}{3\rho}\eta\right)\frac{\sigma'}{\sigma} - \frac{\rho_n}{\rho_s}\frac{\kappa}{\rho\sigma T}\frac{\partial T}{\rho\sigma}\sigma'\right\}. \quad (1.2.76)$$

These two equations will be compatible if the detrminant of their coefficients vanishes. For low frequencies we may limit ourselves to terms linear in ω. But as we may easily convince ourselves, this means that we can neglect the terms containing σ' in (1.2.75) and the terms containing ρ' in (1.2.76). As a result, we obtain two independent homogeneous equations from which the equation for the velocity of sound follows

$$u^2 - \left(\frac{\partial p}{\partial \rho}\right)_\sigma = \frac{i\omega}{\rho}\left(\frac{4}{3}\eta + \xi_2\right), \quad (1.2.77)$$

$$\left(\sigma\frac{\partial T}{\partial \sigma} - \frac{\rho_n}{\rho_s\sigma}u^2\right) = \frac{i\omega}{\rho\sigma}\left\{\rho(\xi_1 + \xi_4) - \rho^2\xi_3 - \xi_2 - \frac{4}{3}\eta\right.$$

$$\left. - \frac{\rho_n}{\rho_s}\frac{\kappa}{T}\frac{\partial T}{\partial \sigma}\right\}. \quad (1.2.78)$$

The root of (1.2.77) determines the velocity of first sound, when damping is taken into account

$$u_1^2 = \left(\frac{\partial p}{\partial \rho}\right)_\sigma + \frac{i\omega}{\rho}\left(\frac{4}{3}\eta + \xi_2\right). \quad (1.2.79)$$

The root of (1.2.78) determines the velocity of second sound, also including damping

$$u_2^2 = \sigma^2\frac{\partial T}{\partial \sigma}\frac{\rho_s}{\rho_n} + \frac{i\omega}{\rho}\cdot\frac{\rho_s}{\rho_n}\left\{\xi_2 + \rho^2\xi_3 - 2\rho\xi_1 + \frac{4}{3}\eta\right.$$

$$\left. + \frac{\rho_n}{\rho_s}\frac{\kappa}{T}\frac{\partial T}{\partial \sigma}\right\}. \quad (1.2.80)$$

The velocities u_1 and u_2 are complex quantities and, therefore, the wave vectors will also be complex. The real part of the wave vector determines the change in the phase with distance and the imaginary part is the coefficient of sound absorption. For first sound it is equal to

$$\alpha_1 = \operatorname{Im}\frac{\omega}{u_1} = \frac{\omega^2}{2\rho u_1^3}\left(\frac{4}{3}\eta + \xi_2\right). \quad (1.2.81)$$

Thus the coefficient of absorption of first sound depends only on two kinetic coefficients, the first viscosity η and the second viscosity ξ_2. The other coefficients do not appear in our expression because we neglected the thermal expansion of He (II). Similarly, the thermal conductivity leads to an additional term in α_1 of form

$$\frac{\omega^2}{2u_1^3}\frac{\kappa}{C}\left(\frac{C_p}{C_v}-1\right). \tag{1.2.82}$$

We may note that the most important contribution to the absorption of the first sound comes from the coefficient ξ_2, as can be seen by calculating η and ξ_2.

The coefficient of absorption of second sound is equal to the imaginary part of the wave vector calculated with the aid of expression (1.2.80)

$$\alpha_2 = \mathrm{Im}\frac{\omega}{u_2} = \frac{\omega^2}{2\rho u_2^3}\frac{\rho_s}{\rho_n}\left\{\frac{4}{3}\eta + (\xi_2 + \rho^2\xi_3 - 2\rho\xi_1)\right. \\ \left. + \frac{\rho_n}{\rho_s}\frac{\kappa}{T}\frac{\partial T}{\partial \sigma}\right\}. \tag{1.2.83}$$

A detailed analysis shows that the major contribution comes from the thermal conductivity, whose effect is described by the last term in (1.2.83).

1.3. THE KINETIC EQUATION AND RELAXATION PHENOMENA IN SUPERFLUID HELIUM

1.3.1. The Kinetic Equation for Elementary Excitations [18]

The distribution function of the elementary excitations satisfies the kinetic equation

$$\frac{\partial n}{\partial t} + \frac{\partial n}{\partial \mathbf{r}}\cdot\frac{\partial H}{\partial \mathbf{p}} - \frac{\partial n}{\partial \mathbf{p}}\cdot\frac{\partial H}{\partial \mathbf{r}} = J(n), \tag{1.3.1}$$

where $J(n)$ is the collision integral whose concrete form is not essential for us at this time. The Hamiltonian H of a quasiparticle in the presence of superfluid motion with velocity v_s is equal to

$$H = \varepsilon(p) + \mathbf{p}\cdot\mathbf{v}_s \tag{1.3.2}$$

where $\varepsilon(p)$ is the energy in the frame of reference in which the superfluid component is at rest. Starting with the kinetic equation, one may derive the hydrodynamic equations for a superfluid. In doing this one encounters expressions for the momentum and energy fluxes, which are just those needed for calculating the kinetic coefficients. Let us multiply (1.3.1) by the momentum component p_i and integrate over all of momentum space. According to the law of conservation of energy, the integral on the right-hand side vanishes, so that we obtain

$$\frac{\partial}{\partial t}\int n p_i\, d\tau_p + \int p_i \frac{\partial n}{\partial r_k}\frac{\partial H}{\partial p_k}\, d\tau_p - \int p_i \frac{\partial n}{\partial p_k}\frac{\partial H}{\partial r_k}\, d\tau_p = 0,$$

which after some simple manipulations yields

$$\frac{\partial}{\partial t}\overline{np_i} + \frac{\partial}{\partial r_k}\overline{np_i\frac{\partial H}{\partial p_k}} + n\overline{\frac{\partial H}{\partial r_i}} = 0. \tag{1.3.3}$$

Here and in what follows a bar signifies integration over momentum space. Inserting (1.3.2) into (1.3.3), we obtain the equation of motion

$$\frac{\partial}{\partial t}\overline{np_i} + \frac{\partial}{\partial r_k}\overline{np_i\left(\frac{\partial \varepsilon}{\partial p_k} + v_{sk}\right)} + n\overline{\left(\frac{\partial \varepsilon}{\partial r_i} + \frac{\partial}{\partial r_i}\mathbf{p}\cdot\mathbf{v}_s\right)} = 0, \tag{1.3.4}$$

which determines the time rate of change of the momentum of relative motion of the superfluid and normal components.

Let us now write the equation of motion for the total momentum

$$\mathbf{j} = \overline{\mathbf{p}n} + \rho\mathbf{v}_s. \tag{1.3.5}$$

Since the total momentum is conserved its time derivative is equal to the divergence of a symmetric tensor, the momentum flux Π_{ik} [of (1.2.3)]

$$\frac{\partial}{\partial t} j_i + \frac{\partial \Pi_{ik}}{\partial r_k} = 0. \tag{1.3.6}$$

As in (1.2.8), we may write the momentum flux tensor Π_{ik} in the rest frame in terms of its value Π_{ik} in the frame of reference moving with the velocity v_s

$$\Pi_{ik} = \pi_{ik} + v_{sk}\,\overline{np_i} + v_{si}\,\overline{np_k} + \rho v_{si}v_{sk}. \tag{1.3.7}$$

Let us subtract (1.3.4) from (1.3.6) and use the continuity equation

$$\frac{\partial \rho}{\partial t} + \frac{\partial}{\partial r_k}(\overline{np_k} + \rho v_{sk}). \tag{1.3.8}$$

Finally, we obtain

$$\rho\frac{\partial v_{si}}{\partial t} + \rho v_{sk}\frac{\partial v_{si}}{\partial t} + \frac{\partial \pi_{ik}}{\partial r_k} - n\overline{\frac{\partial \varepsilon}{\partial r_i}} - \frac{\partial}{\partial r_k}\overline{np_i\frac{\partial \varepsilon}{\partial p_k}} = 0. \tag{1.3.9}$$

From the condition curl $\mathbf{v}_s = 0$, it follows that the sum of the last three terms in (1.3.9) should be the product of the density ρ with the gradient of some function. The derivative $\partial\varepsilon/\partial r_i$ may obviously be written in the form $\partial\varepsilon/\partial\rho\cdot\partial\rho/\partial r_i$. Furthermore, at absolute zero, in the absence of excitations, the tensor π_{ik} should be equal to $p_0\delta_{ik}$ (p_0 is the pressure in superfluid helium at $T = 0$). From these requirements the form of π_{ik} follows uniquely

$$\pi_{ik} = \overline{np_i\frac{\partial \varepsilon}{\partial p_k}} + \delta_{ik}\left(p_0 + n\frac{\partial \varepsilon}{\partial \rho}\rho\right). \tag{1.3.10}$$

According to the thermodynamic identity we have

$$\frac{\partial p_0}{\partial r_i} = \rho\frac{\partial \mu_0}{\partial r_i} \tag{1.3.11}$$

(μ_0 is the value of the chemical potential at $T = 0$).

If we insert (1.3.10) into (1.3.9) and take into account (1.3.11), we obtain

$$\frac{\partial \mathbf{v}_s}{\partial t} + \nabla\left\{\mu_0 + \overline{n\frac{\partial \varepsilon}{\partial \rho}} + \frac{v_s^2}{2}\right\} = 0. \tag{1.3.12}$$

In this manner we have derived the equation of superfluid motion. Comparing (1.3.12) and (1.2.25), we may see that the chemical potential μ is equal to

$$\mu = \mu_0 + \int n \frac{\partial \varepsilon}{\partial \rho}\, d\tau_p. \tag{1.3.13}$$

The second term in this formula is due to the excitations.

Let us now derive the expression for the energy flux vector $\mathbf{Q}$. To do this we calculate the time derivative of the total energy and reexpress all the time derivatives by means of the equations of motion, the equation of continuity, and the kinetic equation, in terms of derivatives with respect to the coordinates. If we then group all terms into a divergence we find an expression for $\mathbf{Q}$. The total energy consists of the kinetic energy [of (1.2.9)]

$$E_k = \rho \frac{v_s^2}{2} + \mathbf{v}_s \cdot \overline{n\mathbf{p}}, \tag{1.3.14}$$

the internal energy of the excitation gas

$$E_k = \overline{n\varepsilon}, \tag{1.3.15}$$

and the zero-point energy E_0 (at $T = 0$), determined by the identity

$$dE_0 = \mu_0\, d\rho. \tag{1.3.16}$$

The time derivation of the kinetic energy is by (1.3.4), (1.3.8), and (1.3.12) equal to

$$\frac{\partial}{\partial t} E_k = -(\rho v_{si} + \overline{np_i}) \frac{\partial}{\partial r_i}\left(\mu_0 + \overline{\frac{\partial \varepsilon}{\partial \rho} n} + \frac{v_s^2}{2}\right) - \frac{v_s^2}{2}\frac{\partial}{\partial r_i}(\rho v_{si} + \overline{np_i})$$
$$- v_{si}\frac{\partial}{\partial r_k}\overline{np_i\left(\frac{\partial \varepsilon}{\partial p_k} + v_{sk}\right)} - v_{si}\left(\overline{n\frac{\partial \varepsilon}{\partial r_i}} + \overline{np_k}\frac{\partial v_{sk}}{\partial r_i}\right). \tag{1.3.17}$$

In order to find the time derivative of the internal energy we multiply both sides of the kinetic equation by ε and integrate over all space. The integral of the right-hand side vanishes since the energy is conserved during collisions, and we obtain

$$\overline{\varepsilon \frac{\partial n}{\partial t}} + \overline{\varepsilon\frac{\partial n}{\partial r_k}\frac{\partial H}{\partial p_k}} - \overline{\varepsilon\frac{\partial n}{\partial p_k}\cdot\frac{\partial H}{\partial r_k}} = 0, \tag{1.3.18}$$

which may be transformed to

$$\frac{\partial}{\partial t}\overline{n\varepsilon} + \frac{\partial}{\partial r_k}\overline{n\varepsilon\left(\frac{\partial \varepsilon}{\partial p_k} + v_{sk}\right)} - v_{sk}\overline{n\frac{\partial \varepsilon}{\partial r_k}} - \overline{n\frac{\partial \varepsilon}{\partial p_k}p_i}\frac{\partial v_{si}}{\partial r_k}$$
$$+ \overline{n\frac{\partial \varepsilon}{\partial \rho}}\frac{\partial}{\partial r_k}(\overline{np_k} + \rho v_{sk}) = 0. \tag{1.3.19}$$

Making use of (1.3.16), (1.3.17), and (1.3.19), we find the time derivative of the total energy

$$\frac{\partial E}{\partial t} = -\frac{\partial}{\partial r_k}\left\{\overline{(np_k + \rho v_{sk})\left(\mu_0 + n\frac{\partial \varepsilon}{\partial \rho} + \frac{v_s^2}{2}\right)} + \overline{nH\frac{\partial}{\partial p_k}H}\right\}. \quad (1.3.20)$$

Thus the energy flux vector is equal to

$$\mathbf{Q} = \overline{(n\mathbf{p} + \rho\mathbf{v}_s)\left(\mu_0 + n\frac{\partial \varepsilon}{\partial \rho} + \frac{v_s^2}{2}\right)} + \overline{nH\frac{\partial}{\partial \mathbf{p}}H}. \quad (1.3.21)$$

1.3.2. Scattering of Elementary Excitations

Let us consider the scattering of elementary excitations in He (II). A change in the number of elementary excitations in an element of phase space can take place by the following scattering processes: phonons by phonons, rotons by rotons, phonons by rotons, and rotons by phonons; in addition, there can be absorption and emission of phonons and rotons from inelastic collisions with one another and between themselves. The fundamental laws which characterize these scattering processes were investigated in references [19] and [20].

The scattering of phonons takes place mainly in the collision of phonons with nearly equal directions of momenta. The matrix element of the transition, H'_{AF}, (where A and F are the initial and final states respectively) in second-order perturbation theory is given by

$$H'_{AF} = \sum_i \frac{(V_3)_{Ai}(V_3)_{iF}}{E_A - E_i} + (V_4)_{AF},$$

$$V_3 = \frac{\mathbf{v}\cdot\rho'\mathbf{v}}{2} + \frac{1}{3!}\frac{\partial}{\partial \rho}\left(\frac{c^2}{\rho}\right)\rho'^3, \quad V_4 = \frac{1}{4!}\frac{\partial^2}{\partial \rho^2}\left(\frac{c^2}{\rho}\right)\rho'^4 \quad (1.3.22)$$

where v is the velocity of the liquid and ρ' is the deviation of the density from the value for the motionless liquid. The largest contribution is made by terms whose denominators vanish if the dispersion of the energy of the phonons is neglected when the angle between the momenta of the colliding phonons is equal to zero [19]. Thus, for example, for a specific intermediate state in which there is a phonon with momentum $p + p_1$, the energy difference $E_A - E_i$ is equal to

$$\varepsilon(p) + \varepsilon(p_1) - \varepsilon(|\mathbf{p} + \mathbf{p}_1|),$$

and it vanishes for $|\mathbf{p} + \mathbf{p}_1| = p + p_1$, i.e., when the angle between the momenta of the colliding phonons p and p_1 is equal to zero. Therefore, in the expression for the energy of the phonons, it is necessary to consider also terms that are cubic in the momenta

$$\varepsilon(p) = cp(1 - \gamma p^2). \quad (1.3.23)$$

Since $\gamma p^2 \ll 1$, equation (1.3.23) reaches the maximum value at small angles between the momenta of the colliding phonons. Thus, in our case of phonon scattering, the principal role is played by collisions among nearly parallel phonons. According to the laws of conservation of momentum and energy, these collisions are not accompanied by an appreciable change in the directions of the momenta of the colliding phonons, and, consequently, they can only lead to the establishment of an energy equilibrium between the phonons moving in a given direction. The characteristic time t_{pp}, of such a scattering process, according to (1.3.22) is given in the case of the scattering of a phonon with high energy (and in the following we shall be interested only in this case) by the relation ($x = pc/kT$)

$$\frac{1}{t_{pp}} = \frac{4.15(u+1)^4}{192\pi^3\gamma c\rho^2}\left(\frac{kT}{\hbar c}\right)^7 x^2\left(1 + 14.7\frac{1}{x}\right). \tag{1.3.24}$$

In the calculation of $1/t_{pp}$ we used the following parameters:

$$\gamma = 3\cdot 10^{37}\ \mathrm{sec}^2/g^2\mathrm{cm}^2,\ u = \frac{\rho}{c}\frac{\partial c}{\partial \rho} = 2.7. \tag{1.3.25}$$

This value of γ was computed by interpolation over the entire energy curve including rotons [19]. It is obviously difficult at present to obtain any information on this value from existing neutronographic scattering data because of its smallness. However, the data of Woods and Cowley [3] point to a very insignificant dispersion of the energy of the phonons. These data show that γ must be appreciably smaller than $10^{37}\ \mathrm{sec}^2/g^2\ \mathrm{cm}^2$. We shall see below, that this assumption confirms the data of our research on the absorption of first sound at low temperatures. Consequently, the real value of the time t_{pp} is evidently much less than that determined from equation (1.3.24) with the value of γ from (1.3.25).*

In what follows we shall need the values of the derivatives of the density with respect to the parameters p_0 and Δ (p_0, Δ, and μ are parameters of the energy of the rotons $\mathscr{E} = \Delta + ((p - p_0)^2/2\mu)$; ρ is the momentum of the roton). From current data [22–24], the most probable values of the derivatives are

$$\frac{\rho}{p_0}\frac{\partial p_0}{\partial \rho} = 0.4,\quad \frac{\rho}{\Delta}\frac{\partial \Delta}{\partial \rho} = -0.57,\quad \frac{\rho^2}{\Delta}\frac{\partial^2 \Delta}{\partial \rho^2} = -5.2. \tag{1.3.26}$$

The values of the derivatives of p_0 were taken from the neutron-scattering data of Henshaw and Woods [22]. These same authors obtained $(\rho/\Delta)\,\partial\Delta/\partial\rho =$

*In the range of temperatures below 0.5 K, phonons with small energies are essentially in the region of the energy curve where the anomalous dispersion ($\gamma < 0$) takes place. In this case, instead of the four-phonon process which we are discussing the three-phonon process plays a similar role.

– 1 for the first derivative of the parameter Δ. Henshaw and Woods did not consider that Δ has a large second derivative with repsect to the density. Therefore, we shall use for the first derivative of Δ the value $(\rho/\Delta)\partial\Delta/\partial\rho = -0.57$, computed by Atkins and Edwards [23] from data on the temperature dependence of the coefficient of thermal expansion of He (II), while for the second derivative, we shall use the value $(\rho^2/\Delta)\partial\Delta/\partial\rho^2 = -5.2$ computed by us from the data of Henshaw and Woods.

As has already been pointed out above, the scattering of phonons takes place principally by means of collisions of particles with nearly equal directions of momenta. In comparison with this small angle scattering, collisions of phonons where there is a large change in direction of momenta of the colliding particles have a low probability. Nevertheless, for kinetic phenomena one must take into account this wide angle scattering of phonons by phonons since in the temperature region below 0.9 K it becomes significant along with the scattering of phonons by rotons.

We shall need below the differential effective cross section for the process under consideration. It is represented by the relation

$$d\sigma_{pp}(\mathbf{p}, \mathbf{p}_1, \mathbf{p}^*, \mathbf{p}_1^*) = \left(\frac{2\pi}{\hbar c}\right) |H'_{AF}|^2\, \delta(\varepsilon + \varepsilon_1 - \varepsilon^* - \varepsilon_1^*)\, d\tau_{p^*}, \quad (1.3.27)$$

where H'_{AF} is the matrix element of the transition between the initial state A and the final state F in second-order perturbation theory determined from equation (1.3.22), and $d\tau_{p^*}$ is the element of volume in p^* space ($d\tau_{p^*} = p^{*2}\, dp^*\, do^*/(2\pi\hbar)^3$ where do^* is the element of solid angle). The quantities without and with asterisks in (1.3.27) refer to incident and scattered phonons, respectively.

The scattering of a phonon by a roton in some sense is analogous to the scattering of a light particle by a heavy one. This analogy is associated with the fact that the phonon possesses a momentum that is much smaller than the momentum of the roton. The law of conservation of energy for the given process is written in the form

$$cp + \frac{(P - P_0)^2}{2\mu} = cp^* + \frac{(|\mathbf{P} + \mathbf{p} - \mathbf{p}^*| - P_0)^2}{2\mu}$$

(p and P are the momenta of the phonon and the roton before collsion, p^* is the momentum of the phonon after collision); hence, after several transformations, taking into account the smallness of the phonon momenta ($p, p^* \ll P_0$), we get

$$c(p - p^*) = \frac{\mathbf{P} \cdot (\mathbf{p} - \mathbf{p}^*)}{2\mu P_0^2}.$$

We now make use of the fact that the energy of the phonons is $\varepsilon = cp < 3\mu c^2$ (just such an inequality, as we shall see below, is essential in the interval of collision). This allows us to describe the energy difference of interest to us in the following way:

$$\varepsilon - \varepsilon^* = \frac{p^2}{2\mu}[\mathbf{m}\cdot(\mathbf{n} - \mathbf{n}^*)]^2, \tag{1.3.28}$$

where **m**, **n**, and **n*** are unit vectors directed along the vectors **P**, **p**, and **p***. Taking into account what has been said regarding the smallness of the quantity $\varepsilon - \varepsilon^*(\varepsilon - \varepsilon^* < \varepsilon$ and $\varepsilon^*)$, the matrix element H'_{AF} can, in agreement with [19], be written in the form

$$H'_{AF} = \frac{P_0 p}{2\mu}\left\{(\mathbf{n} + \mathbf{n}^*)\cdot\mathbf{m}(\mathbf{n}\cdot\mathbf{n}^*) + \frac{P_0}{\mu c}(\mathbf{n}\cdot\mathbf{m})^2(\mathbf{n}^*\cdot\mathbf{m})^2 + A\right\},$$

$$A = \frac{\rho^2}{P_0 c}\left[\frac{\partial^2 \Delta}{\partial\rho^2} + \frac{1}{\mu}\left(\frac{\partial P_0}{\partial\rho}\right)^2\right]. \tag{1.3.29}$$

The value of the parameter A computed from the values of the derivatives with respect to the density of P_0 and Δ, given in (1.3.26), turns out to be very small and approximately equal to -0.1.

The differential effective scattering cross section of a phonon by a roton is equal to

$$d\sigma_{pr} = \left(\frac{2\pi}{\hbar c}\right)|H'_{AF}|^2\delta(\mathscr{E} + cp - \mathscr{E}^* - cp^*)\,d\tau_{p^*} \tag{1.3.30}$$

($\mathscr{E}$ and $\mathscr{E}^*$ are the energies of the rotons before and after collision). We substitute (1.3.29) in (1.3.30) and integrate over the momentum of the scattered phonon; as a result, we have

$$d\sigma_{pr} = \left(\frac{P_0 p^2}{4\pi\hbar^2\rho c}\right)^2[(\mathbf{n} + \mathbf{n}^*)\cdot\mathbf{m}(\mathbf{n}\cdot\mathbf{n}^*) + \frac{P_0}{\mu c}(\mathbf{n}\cdot\mathbf{m})^2(\mathbf{n}^*\cdot\mathbf{m})^2 + A]\,do^*.$$

In the following, we need the cross section $d\sigma_{pr}$ averaged over all the angles of the colliding particles. It is equal to

$$d\sigma_{pr} = \left(\frac{P_0 p^2}{4\pi\hbar^2\rho c}\right)^2\left[\frac{2}{9} + \frac{1}{25}\left(\frac{P_0}{\mu c}\right)^2 + \frac{2A}{9}\frac{P_0}{\mu c} + A^2\right]do^*. \tag{1.3.31}$$

In contrast with phonon–phonon scattering and phonon–roton scattering (for which one can calculate the effective cross section exactly) the process of roton–roton scattering is considered with the help of some interaction model. The total effective cross section σ_{rr} of roton–roton scattering has been computed under the assumption that the interaction of the rotons has a δ-type character. According to [19],

$$\sigma_{rr} = \frac{P_0\mu|V_0|^2}{\left|\dfrac{\partial\mathscr{E}}{\partial\mathbf{P}} - \dfrac{\partial\mathscr{E}_1}{\partial\mathbf{P}_1}\right|\hbar^4\cos^2(\psi/2)} \tag{1.3.32}$$

Here $|\partial\mathscr{E}/\partial\mathbf{P} - \partial\mathscr{E}_1/\partial\mathbf{P}_1|$ is the relative velocity of the colliding rotons, ψ is the angle between $\mathbf{P}$ and $\mathbf{P}_1$($\mathscr{E}$, $\mathbf{P}$ and $\mathscr{E}_1$, $\mathbf{P}_1$ are the energy and momentum of the colliding rotons), V_0 is the amplitude which determines the interaction of the rotons ($V = V_0\delta\,(r_1 - r)$ where V is the energy of interaction of the rotons and r_1 and r are their radius vectors). The quantity V_0 can be computed from the experimental values of the coefficient of viscosity of He (II) [25–29]. It is seen to be of the order of $25 \cdot 10^{-38}$ erg cm^3.

The reciprocal of the mean time between two collisions of a roton, t_{rr}, is obtained from (1.3.32) by multiplication by the total roton current with subsequent averaging over all angles formed by the momenta of the colliding rotons:

$$\frac{1}{t_{rr}} = 2P_0\mu|V_0|^2\frac{N_r}{\hbar^4}, \quad N_r = \frac{2P_0^2\mu(kT)^{1/2}\,e^{-\Delta/T}}{(2\pi)^{3/2}\hbar^3} \tag{1.3.33}$$

(N_r is the number of rotons per unit volume).

Among the inelastic scattering processes of phonons and rotons, the transformation of three phonons into two (five-phonon process) has the greatest probability [19]. The five-phonon process, similar to the process of phonon–phonon scattering considered above, is not accompanied by any change in the directions of the momenta of the colliding particles. Actually, the desired matrix element of a transition from the initial state A to the final state F in second-order perturbation theory contains the sum

$$\sum_{\mathrm{I,\,II}}\frac{(V_3)_{A\mathrm{I}}(V_3)_{\mathrm{I\,II}}(V_3)_{\mathrm{II}F}}{(E_A - E_{\mathrm{I}})(E_A - E_{\mathrm{II}})},$$

some terms of which contain in their denominators factors that vanish in the absance of dispersion when the angles between the momenta of the colliding phonons are equal to zero. Thus the probability of the process $(3 \to 2)$ has a maximum for collisions of phonons at small angles. Such collisions, as pointed out above, do not lead to an appreciable change in the directions of the colliding particles; consequently, they can lead to the establishment of equilibrium between phonons moving in a given direction.

Calculation of the probability of a five-phonon process is very cumbersome. However, since the expression for this probability contains the quantity γ^2 in the denominator, and the value of γ is only known with low accuracy, one can limit oneself to obtaining a symbolic formula with an adjustable numerical coefficient [19]. With such an expression we can obtain the temperature dependence of the coefficient Γ_p, which determines the rate of change in the number of phonons N_p,

$$\dot{N}_p = -\alpha\Gamma_p$$

(α is some chemical potential; see below). According to [19], Γ_p has the form

$$\Gamma_p = \Lambda T^{11}, \tag{1.3.34}$$

where Λ is a temperature-independent coefficient.

In addition to the five-phonon process, other inelastic processes that lead to a change in the number of phonons and rotons are also possible. Let us say a few words about one of them which we have considered previously [19]. When an energetic phonon (with energy of the order of Δ) collides with a roton it may transform into a roton. The probability of this process, however, is small and certainly much smaller than the probability for scattering of a phonon by a roton.

Having established the basic laws that characterize the interaction of the elementary excitations, we can describe the form of the nonequilibrium distribution function. We begin with the phonons. Figure 1.6*a* shows the temperature dependence of the various times characterizing the phonon processes. In a comparison of the times τ_{pr} and $\tau_{3\to2}$, which describe the scatter-

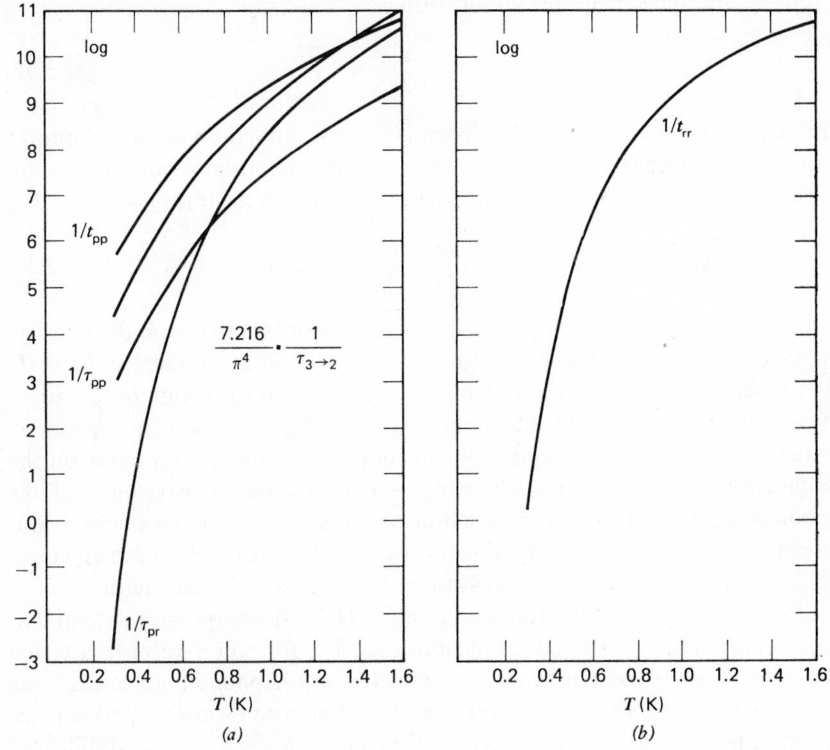

FIGURE 1.6. Temperature dependence of the times. (*a*) t_{pp}, τ_{pr}, τ_{pp}, $\tau_{3\to2}$;(*b*) t_{rr}.

ing of a phonon by a roton and the transformation of three phonons into two, it must be kept in mind that these processes enter into the kinetic equation, as we shall see below, with different weights, since the energetic phonons play more of a role in the scattering of phonons by rotons than in the five-phonon process. In this connection a graph of the quantity

$$\frac{7\cdot 216}{\pi^4}\cdot\frac{1}{\tau_{3\to 2}}$$

has been plotted in Figure 1.6*a*. This curve should be compared with the quantity $1/\tau_{pr}$.

It is seen from Figure 1.6*a* that in the temperature region below 1.5 *K*, in which we are most interested, the four-phonon and five-phonon processes guarantee a rapid establishment of equilibrium both in energies and in the number of phonons, for phonons moving in a given direction. Therefore, the phonons moving in a given direction can be characterized by a certain direction dependent temperature T_p. Thus, the distribution function of the phonons is some quasi-equilibrium function

$$n = \left[\exp\frac{\varepsilon}{kT_p} - 1\right]^{-1}, \tag{1.3.35}$$

in which the temperature T_p depends on the direction of the phonon's momentum. Departure of the distribution function from a constant-equilibrium value n_0 (for small departures) is represented in the form

$$n - n_0 = \frac{\partial n_0}{\partial \varepsilon}\left(\frac{\partial \varepsilon}{\partial \rho}\rho' + \varepsilon\nu\right). \tag{1.3.36}$$

The first term reflects the dependence of the energy spectrum on the density, the second is obtained by expanding T_p in a series in the difference $T_p - T_0$ (T_0 is the constant-equilibrium temperature); ν is some function of the direction of the momentum of the phonon, which is found by solving the kinetic equation. Generally speaking, the function n depends not on ε but on the difference $\varepsilon - \mathbf{p}\cdot(\mathbf{v}_{np} - \mathbf{v}_s)$, where $\mathbf{v}_{np} - \mathbf{v}_s$ is some relative velocity of the phonon gas, which is also a function of the direction of the phonon momentum. However, since ε is a linear function of p, the function ν in equation (1.3.36) automatically takes into account the presence of such a term.

We now proceed to the rotons. Equation (1.3.33) determines the temperature dependence of the characteristic time, t_{rr}, for roton–roton scattering (Figure 1.6*b*). A comparison of t_{rr} with the roton–phonon scattering time shows that at temperatures above 0.6 K the roton–roton scattering takes place more rapidly, which assures the establishment of a certain local equilibrium in the roton gas. Thus the rotons can be described by some quasi-equilibrium distribution function with a temperature T_r and a relative velocity $\mathbf{v}_{nr} - \mathbf{v}_s$, which is a local function

$$N = \exp\left[- \frac{\mathscr{E} - \mathbf{P}\cdot(\mathbf{V}_{nr} - \mathbf{V}_s)}{kT_r} \right]. \qquad (1.3.37)$$

Expanding this function in a series in the deviations of all the quantities from their constant equilibrium values, we get (for small deviations)

$$N - N_0 = \frac{\partial N}{\partial \mathscr{E}} \left[\frac{\partial \mathscr{E}}{\partial \rho} \rho' - \mathscr{E} \frac{T_r'}{T_0} - \mathbf{P}\cdot(\mathbf{V}_{nr} - \mathbf{V}_s) \right]. \qquad (1.3.38)$$

Here T_r' is the deviation of the temperature of the roton gas from a constant equilibrium value T_0 and N_0 is the constant equilibrium distribution function.

We note that, for generality, the square bracket in equation 1.3.38 should have contained a term that corresponds to some chemical potential α_r. However, inasmuch as the roton spectrum is such that its energy is almost constant and equal to Δ, such a term is impossible to distinguish (by its dependence on momentum and energy) from the term $\mathscr{E}T_r'/T_0 \approx \Delta T_r'/T_0$. This circumstance permits us to combine both terms from the very beginning. The physical meaning of this circumstance is also simple. Any local change in the number of rotons can be described by the local change in temperature.

1.3.3. Kinetic Equations for Sound Propagation

At low frequencies ($\omega\tau \ll 1$ where ω is the sound frequency, and τ is some characteristic time) the study of sound absorption is carried out by means of the hydrodynamic equations of the superfluid with dissipation terms [30]. The problem of the propagation of sound in a superfluid at high frequencies ($\omega\tau \gg 1$) obviously cannot be considered in the hydrodynamic approximation and can be solved only on the basis of the kinetic equation.

The free paths of the phonons and rotons, in the general case of an arbitrary value of the parameter $\omega\tau$, can be shown to be comparable with and even exceed the sound wavelength. Therefore, it is impossible to apply the hydrodynamic equations in the normal part of the fluid and one must use kinetic equations for the phonon distribution function $n(p,r,t)$ and roton distribution function $N(P, r, t)$:

$$\frac{\partial n}{\partial t} + \frac{\partial n}{\partial \mathbf{r}} \cdot \frac{\partial H_p}{\partial \mathbf{p}} - \frac{\partial n}{\partial \mathbf{p}} \cdot \frac{\partial H_p}{\partial \mathbf{r}} = J(n), \qquad (1.3.39)$$

$$\frac{\partial N}{\partial t} + \frac{\partial N}{\partial \mathbf{r}} \cdot \frac{\partial H_r}{\partial \mathbf{P}} - \frac{\partial H_r}{\partial \mathbf{r}} \cdot \frac{\partial N}{\partial \mathbf{P}} = J(N). \qquad (1.3.40)$$

Here $J(n)$ and $J(N)$ are collision integrals, which arise from the elastic and inelastic scattering of phonons and rotons. The Hamiltonians H_p and H_r are equal to

$$H_p = \varepsilon(\rho) + \mathbf{p}\cdot\mathbf{v}_s, \quad H_r = \mathscr{E}(P) + \mathbf{P}\cdot\mathbf{v}_s,$$

where $\varepsilon(p)$ and $\mathscr{E}(P)$ are energies of the phonons and rotons in a reference system moving with a velocity v_s and depend on the liquid density ρ. The quantities ρ and v_s play the role of external conditions for the excitation gas.

In order to get a complete set of equations, it is necessary to add to equations (1.3.39) and (1.3.40) the equation of mass continuity

$$\frac{\partial \rho}{\partial t} + \operatorname{div} \mathbf{j} = 0, \tag{1.3.41}$$

and also the equation of superfluid motion

$$\frac{\partial \mathbf{v}_s}{\partial t} + \nabla\left(\mu + \frac{\mathbf{v}_s^2}{2}\right) = 0, \tag{1.3.42}$$

where j is the momentum density of the liquid and $\mu = \partial E/\partial \rho$ where E is the energy density in a set of coordinates in which $v_s = 0$. The derivative of the energy with respect to density is taken for a constant distribution of the excitations, i.e., for constant $n(p)$ and $N(P)$.

To find j, we note that the momentum in a system of coordinates where the superfluid portion is at rest is equal to

$$\int \mathbf{p} n \, d\tau_p + \int \mathbf{P} N \, d\tau_P.$$

Using the well-known Galilean transformation formula, we get

$$\mathbf{j} = \rho \mathbf{v}_s + \int \mathbf{p} n \, d\tau_p + \int \mathbf{P} N \, d\tau_P. \tag{1.3.43}$$

For the energy E, we have

$$E = E_0 + \int \varepsilon n \, d\tau_p + \int \mathscr{E} N \, d\tau_P,$$

where E_0 is the energy at absolute zero.

Carrying out differentiation of E with respect to the density ρ, and using (1.3.41)–(1.3.43), we get the desired equation

$$\frac{\partial \rho}{\partial t} + \operatorname{div}\left(\rho \mathbf{v}_s + \int \mathbf{p} n \, d\tau_p + \int \mathbf{P} N \, d\tau_P\right) = 0, \tag{1.3.44}$$

$$\frac{\partial \mathbf{v}_s}{\partial t} + \nabla\left(\mu_0 + \int \frac{\partial \varepsilon}{\partial \rho} n \, d\tau_p + \int \frac{\partial \mathscr{E}}{\partial \rho} N \, d\tau_P\right) = 0 \tag{1.3.45}$$

where μ_0 is the chemical potential at $T = 0$.

In a plane sound wave, all the thermodynamic quantities are composed of constant equilibrium components and small additions that change according to the law $\exp\left[i(\mathbf{k}\cdot\mathbf{r} - \omega t)\right]$ ($\mathbf{k}$ is the wave vector). The velocitics $\mathbf{v}_s$, $\mathbf{v}_{np}$ and $\mathbf{v}_{nr}$ in the sound wave are also small quantities* which vary according to the

*One must keep in mind the smallness of the velocities $\mathbf{v}_s$, $\mathbf{v}_{np}$, and $\mathbf{v}_{nr}$ in comparison with the velocities of first and second sound.

same law. Therefore, one can set

$$n = n_0 + n', \quad N = N_0 + N', \quad \rho = \rho_0 + \rho',$$

in equations (1.3.39), (1.3.40), (1.3.44), and (1.3.45). Here n_0, N_0, ρ_0 are the constant equilibrium values and n', N', ρ', and also $\mathbf{v}_s$ are small corrections proportional to $\exp[i(\mathbf{k}\cdot\mathbf{r} - \omega t)]$. After linearization these equations become

$$\left(1 - \frac{(\mathbf{k}\cdot\mathbf{v})}{\omega}\right)n' + \frac{(\mathbf{k}\cdot\mathbf{v})}{\omega}\frac{\partial n_0}{\partial \varepsilon}\left(\frac{\partial \varepsilon}{\partial \rho}\rho' + (\mathbf{p}\cdot\mathbf{v}_s)\right) = -\frac{1}{i\omega}J(n), \quad (1.3.46)$$

$$\left(1 - \frac{(\mathbf{k}\cdot\mathbf{V})}{\omega}\right)N' + \frac{(\mathbf{k}\cdot\mathbf{V})}{\omega}\frac{\partial N_0}{\partial \mathscr{E}}\left(\frac{\partial \mathscr{E}}{\partial \rho}\rho' + (\mathbf{P}\cdot\mathbf{v}_s)\right) = -\frac{1}{i\omega}J(N), \quad (1.3.47)$$

$$-\omega\rho' + k\rho v_s + \int(\mathbf{k}\cdot\mathbf{p})n'\,d\tau_p + \int(\mathbf{k}\cdot\mathbf{P})N'\,d\tau_P = 0, \quad (1.3.48)$$

$$-\omega v_s + k\left(\frac{c^2}{\rho} + \int\frac{\partial^2\varepsilon}{\partial\rho^2}n_0\,d\tau_p + \int\frac{\partial^2\mathscr{E}}{\partial\rho^2}N_0\,d\tau_P\right)\rho'$$
$$+ k\int\frac{\partial\varepsilon}{\partial\rho}n'\,d\tau_p + k\int\frac{\partial\mathscr{E}}{\partial\rho}N'\,d\tau_P = 0, \quad (1.3.49)$$

where $\mathbf{v} = \partial\varepsilon/\partial\mathbf{p}$ and $\mathbf{V} = \partial\mathscr{E}/\partial\mathbf{P}$. We drop the zero subscript on the density and temperature* and use the fact that $d\mu_0 = c^2 d\rho$.

We introduce a spherical set of coordinates with the polar axis along the vector $\mathbf{k}$. The angle between k and p will be denoted by θ and between k and P by Θ. In this set of coordinates n' and N', in accord with (1.3.36) and (1.3.38), are equal to

$$n' = \frac{\partial n_0}{\partial\varepsilon}\left(\frac{\partial\varepsilon}{\partial\rho}\rho' + \varepsilon\nu(\cos\theta)\right), \quad (1.3.50)$$

$$N' = \frac{\partial N_0}{\partial\mathscr{E}}\left(\frac{\partial\mathscr{E}}{\partial\rho}\rho' - \mathscr{E}\frac{T'_r}{T} - Pw_r\cos\theta\right) \quad (1.3.51)$$

$(w_r = v_{nr} - v_s)$. We shall seek the unknown function $\nu(\cos\theta)$ in the form

$$\nu(\cos\theta) = \sum_{i=0}^{\infty}\nu_i P_i(\cos\theta)$$

($P_i(\cos\theta)$ are spherical harmonics).

We substitute (1.3.50) and (1.3.51) in equations (1.3.48) and (1.3.49) and integrate over all p and P space:

$$-\tilde{\omega}\rho' + \left(v_s + \frac{\rho_n}{\rho}w_r\right) - \frac{\rho_{np}}{\rho}(\nu_1 + w_r) = 0, \quad (1.3.52)$$

$$-\tilde{\omega}v_s + \frac{\rho}{c^2}\left(\frac{\partial\mu}{\partial\rho}\right)_T\rho' + \frac{T}{c^2}\left(\frac{\partial\mu}{\partial T}\right)_\rho T'_r - 3u\frac{\rho_{np}}{\rho}(\nu_0 + T'_r) = 0. \quad (1.3.53)$$

*In what follows we shall drop the zero subscript everywhere for equilibrium thermodynamic quantities.

Here ρ_{np} is the phonon part of the normal density, μ is the chemical potential, and ν_0 and ν_1 are coefficients for the zeroth and first harmonics in the expansion of $\nu(\cos\theta)$ in spherical harmonics. In equations (1.3.52) and (1.3.53), we have introduced the notation $\tilde{\omega} = \omega/kc$ and transformed to nondimensional quantities.†

The collision integral $J(n)$ on the right side of equation (1.3.46) consists of four components: $J_{pr}(n)$, due to scattering of phonons by rotons, $J_{pp}(n)$ due to phonon–phonon scattering, $J'_{pp}(n)$ due to the specific phonon–phonon scattering at small angles, and finally $J_{3\to 2}(n)$ which characterizes the change in the number of phonons which occurs by way of the five-phonon process.

The total energy of the phonons moving in a given direction, for a five-phonon process and for scattering of phonons by phonons at small angles is conserved. We integrate the left and right sides of equation (1.3.46) [with account of (1.3.50)] over all possible energies and divide by

$$\int \frac{\partial n_0}{\partial \varepsilon}\, \varepsilon^2 p^2\, dp.$$

In accord with the above, the integrals

$$\int J_{pp'}(n)\varepsilon p^2\, dp, \quad \int J_{3\to 2}\, \varepsilon p^2\, dp$$

are equal to zero, and we have

$$(\tilde{\omega} - \cos\theta)\nu(\cos\theta) + \tilde{\omega}u\rho' + \cos^2\theta\, v_s$$
$$= -\frac{\omega}{i\tilde{\omega}}\left(\frac{\int J_{pr}(n)\varepsilon p^2\, dp}{\int \partial n_0/\partial\varepsilon\, \varepsilon^2 p^2\, dp} + \frac{\int J_{pp}(n)\varepsilon p^2\, dp}{\int \partial n_0/\partial\varepsilon\, \varepsilon^2 p^2\, dp}\right). \quad (1.3.54)$$

The collision integral $J(N)$ on the right side of equation (1.3.47) consists of two parts: $J_{rp}(N)$ which is due to the scattering of rotons by phonons, and $J(_{rr}N)$ due to roton–roton scattering. We multiply the left and right sides of equation (1.3.47) first by $\mathscr{E}$, and then by P, and integrate in both cases with account of (1.3.51) over all P space.

Inasmuch as

$$\int J_{rp}(N)\mathscr{E}\, d\tau_P = -\int J_{pr}(n)\varepsilon\, d\tau_p,$$

$$\int J_{rp}(N)\mathbf{P}\, d\tau_P = -\int J_{pr}(n)\mathbf{p}\, d\tau_p,$$

†Here we shall everywhere denote by ρ' and T_r' the ratios of the deviations of the liquid density and the temperature of the roton gas, respectively, from ρ_0 and T_0 and by v_s and w_r the ratios of the velocity of the superfluid part of the liquid and the relative velocity of the roton gas to c.

in accordance with the laws of conservation of energy and momentum, and

$$\int J_{rr}(N)\mathscr{E}\, d\tau_P = 0, \quad \int J_{rr}(N)\mathbf{P}\, d\tau_P = 0,$$

we have

$$- \tilde{\omega} T\left(\frac{\partial S_r}{\partial T}\right)_\rho T'_r - \tilde{\omega}\rho\left(\frac{\partial S_r}{\partial \rho}\right)_T \rho' + S_r(w_r + v_s)$$

$$= - \frac{\tilde{\omega}}{i\omega T}\int J_{pr}(n)\varepsilon\, d\tau_p, \tag{1.3.55}$$

$$- \tilde{\omega}\rho_{nr} w_r + \frac{TS_r}{c^2} T'_r = - \frac{\tilde{\omega}}{i\omega c}\int J_{pr}(n) p\cos\theta\, d\tau_p, \tag{1.3.56}$$

where ρ_{nr} is the roton part of the normal density and S_r is the entropy density of the roton gas. The values of the integrals encountered in (1.3.52), (1.3.53), and (1.3.56) are given below:

$$\int \frac{\partial N}{\partial \mathscr{E}}\mathscr{E}^2\, d\tau_P = - T\left(\frac{\partial}{\partial T}\int \mathscr{E} N_0\, d\tau_P\right)_\rho = - T^2\left(\frac{\partial S_r}{\partial T}\right)_\rho,$$

$$\int \frac{\partial N_0}{\partial \mathscr{E}}\frac{\partial \mathscr{E}}{\partial \rho}\mathscr{E}\, d\tau_P = - T\left(\frac{\partial}{\partial T}\int N_0\frac{\partial \mathscr{E}}{\partial \rho}\, d\tau_P\right)_\rho = - T\left(\frac{\partial \mu_r}{\partial T}\right)_\rho = T\left(\frac{\partial S_r}{\partial \rho}\right)_T,$$

$$\frac{1}{3}\int \frac{\partial N_0}{\partial \mathscr{E}} PV\mathscr{E}\, d\tau_P = - TS_r, \quad \frac{1}{3}\int \frac{\partial N_0}{\partial \mathscr{E}} P^2\, d\tau_P = - \rho_{np},$$

$$\frac{1}{3}\int \frac{\partial n_0}{\partial \varepsilon} p^2\, d\tau_p = - \rho_{np},$$

$$\frac{c^2}{\rho} + \int \frac{\partial^2 \varepsilon}{\partial \rho^2} n_0\, d\tau_p + \int \frac{\partial n_0}{\partial \varepsilon}\left(\frac{\partial \varepsilon}{\partial \rho}\right)^2 d\tau_p + \int \frac{\partial^2 \mathscr{E}}{\partial \rho^2} N_0\, d\tau_P$$

$$+ \int \frac{\partial N_0}{\partial \mathscr{E}}\left(\frac{\partial \mathscr{E}}{\partial \rho}\right)^2 d\tau_P = \left[\frac{\partial}{\partial \rho}\left(\mu_0 + \int \frac{\partial \varepsilon}{\partial \rho} n_0\, d\tau_p\right.\right.$$

$$\left.\left. + \int \frac{\partial \mathscr{E}}{\partial \rho} N_0\, d\tau_P\right)\right]_T = \left(\frac{\partial \mu}{\partial \rho}\right)_T = \frac{1}{\rho}\left(\frac{\partial P}{\partial \rho}\right)_T.$$

We now calculate the collision integrals $J_{pr}(n)$ and $J_{pp}(n)$ entering into equations (1.3.54)–(1.3.56):

$$J_{pr}(n) = - \int c\, d\sigma_{pr}\{nN(n^* + 1) - n^*N^*(n + 1)\}\, d\tau_p. \tag{1.3.52}$$

Here the quantities with the asterisks refer to scattered phonons and rotons, and $d\sigma_{pr}$ is the differential effective scattering cross section of a phonon by a roton, determined by equation (1.3.47).

Substituting n, n^*, N, and N^* into (1.3.57) expressed in the form of sums of the constant equilibrium values and the additions (1.3.50) and (1.3.51), and keeping only terms that are linear in ν, ρ', T'_r, and w_r, we get

$$J_{pr}(n) = -\int c\, d\sigma_{pr} N_0 \frac{\partial n_0}{\partial \varepsilon} \{\varepsilon[\nu(\cos\theta) - \nu(\cos\theta^*)$$
$$+ w_r(\cos\theta - \cos\theta^*)] + (\varepsilon - \varepsilon^*)[\nu(\cos\theta^*) + T'_r]\}\, d\tau_p. \quad (1.3.58)$$

The quantity $\varepsilon - \varepsilon^*$ is small in comparison with the energy ε. However, the calculation of this difference in the collision integral (1.3.58) is seen to be important for establishing the energy equilibrium between the phonon and roton gases. As we shall see below, the relative slowness of this latter process leads to an appreciable dissipation and dispersion of sound and plays a fundamental role in all subsequent considerations. It is natural in this case that we shall consider the difference $\varepsilon - \varepsilon^*$ only in those terms in the collision integral (1.3.58) that correspond to the establishment of the energy equilbrium mentioned.

Calculation of the collision integral (1.3.58) with $d\sigma_{pr}$ from (1.3.30) leads to a very cumbersome expression that contains the first six harmonics of the function ν (cos θ^*). However, there is hardly any need for such a calculation, since all the harmonics involved except the first make a trivial contribution to a final result. Therefore, without great error in accuracy, one can substitute in (1.3.58) the phonon–roton scattering cross section (1.3.30) averaged over the angles of the scattering particles.* The collision integral (1.3.58), after substitution of $d\sigma_{pr}$ [which is determined by (1.3.31)] and $\varepsilon - \varepsilon^*$ from (1.3.28) and integration over all P space and over all the scattering angles of the phonon, is equal to

$$J_{pr}(n) = -\frac{1}{\tau_{pr}(p)}\frac{\partial n_0}{\partial \varepsilon}\left\{\varepsilon[\nu(\cos\theta) - \nu_0 + \cos\theta w_r] + \frac{\varepsilon^2}{3\mu c^2}(\nu_0 + T'_r)\right\},$$

$$\frac{1}{\tau_{pr}(p)} = 4\pi c N_r \left(\frac{P_0 p^2}{4\pi\hbar^2 \rho c}\right)^2\left[\frac{2}{9} + \frac{1}{25}\left(\frac{P_0}{\mu c}\right)^2 + \frac{2A}{9}\frac{P_0}{\mu c} + A^2\right]. \quad (1.3.59)$$

Taking account of (1.3.59) the integrals containing $J_{pr}(n)$ on the right sides of equations (1.3.54)–(1.3.56) are easily computed and are equal to

$$\int \frac{J_{pr}(n)\varepsilon p^2\, dp}{\int \partial n_0/\partial\varepsilon\, \varepsilon^2 p^2\, dp} = -\frac{1}{\tau_{pr}}[\nu(\cos\theta) - \nu_0 + \cos\theta w_r + \beta(\nu_0 + T'_r)],$$

*We also note that a more exact calculation would not have any meaning, for even in the derivation of the formula for the matrix element of the transition H'_{AF} in the second order perturbation theory a number of terms comparable with this effect, which can make a contribution to the higher harmonics, were neglected. Thus, for example, in the derivation of (1.3.29) terms were omitted which contain $P - P_0$. Moreover, the previously mentioned inequality $\varepsilon - \varepsilon^* < 3\mu c^2$ is satisfied with small margin, and this also limits the accuracy of the calculations.

$$\frac{1}{T}\int J_{pr}(n)\varepsilon\, d\tau_p = \frac{1}{\tau_{pr}}\beta C_p(\nu_0 + T'_r),$$

$$\frac{1}{c}\int J_{pr}(n)p\cos\theta\, d\tau_p = \frac{1}{\tau_{pr}}\rho_{np}(\nu_1 + w_p), \tag{1.3.60}$$

where $\beta = 3kT/\mu c^2$ and τ_{pr} is the characteristic time for the process of scattering of phonons by rotons:

$$\frac{1}{\tau_{pr}} = \frac{4\pi^3}{c}N_r\left[\frac{P_0(kT/c)^2}{\rho\hbar^2}\right]^2\left[\frac{2}{9} + \frac{1}{25}\left(\frac{P_0}{\mu c}\right)^2 + \frac{2A}{9}\frac{P_0}{\mu c} + A^2\right]. \tag{1.3.61}$$

Finally, we calculate $\int J_{pp}(n)\varepsilon p^2\, dp$, which turns out to be convergent without account of dispersion of the phonons:

$$\int J_{pp}(n)\varepsilon p^2\, dp = -\int c\, d\sigma_{pp}\,\{nn_1(n^* + 1)(n_1^* + 1) - n^*n_1^*(n + 1)(n_1 + 1)\}\,\varepsilon p^2\, dp\, d\tau_{p_1}. \tag{1.3.62}$$

Here the quantities with asterisks refer to scattered phonons and $d\sigma_{pp}$ is the differential effective scattering cross section of a phonon with momentum p_1 by a phonon with momentum p, equation (1.3.27).

Substituting in (1.3.62) the distribution functions of the incident and scattered phonons, expressed in the form of sums of the constant equilibrium values and the variations given by (1.3.50), and keeping only terms linear in ν and ρ', we get

$$\int J_{pp}(n)\varepsilon p^2\, dp = \int c\, d\sigma_{pp}\, n_0 n_{01}(n_0^* + 1)(n_{01}^* + 1) \times \{(\varepsilon + \varepsilon_1 - \varepsilon^* - \varepsilon_1^*)u\rho' + \varepsilon\nu(\cos\theta) + \varepsilon_1\nu(\cos\theta_1) - \varepsilon^*\nu(\cos\theta^*) - \varepsilon_1^*\nu(\cos\theta_1^*)\}\frac{\varepsilon}{kT}p^2\, dp\, d\tau_{p_1}. \tag{1.3.63}$$

As a consequence of the laws of conservation of momentum and energy, the terms with ν_0 and ν_1 in the curly brackets of (1.3.63) are cancelled, and as a result the spherical harmonics which remain begin with the second. As we shall see below, phonon–phonon scattering plays a role only at low temperatures (below 0.9 K) and in order not to complicate the caclulations, we alter the integral (1.3.63) somewhat, writing it in the form

$$\int J_{pp}(n)\varepsilon p^2\, dp = -\frac{1}{\tau_{pp}}[\nu(\cos\theta) - \nu_0 - \nu_1(\cos\theta)]\int\frac{\partial n_0}{\partial\varepsilon}\varepsilon^2 p^2\, dp, \tag{1.3.64}$$

where the time τ_{pp} is computed accurate to the second harmonic [19]

$$\frac{1}{\tau_{pp}} = \frac{9(13)!(n+1)^4}{2^{13}(2\hbar\pi)^7\rho^2 c}\left(\frac{kT}{c}\right)^9. \tag{1.3.65}$$

The appreciable error that we have introduced here does not change the picture qualitatively.

1.4. DISPERSION OF FIRST AND SECOND SOUND IN SUPERFLUID HELIUM [13]

1.4.1. The Basic Equations

As was pointed out above, the problem of the propagation of sound in a superfluid at high frequencies ($\omega\tau \gg 1$) obviously cannot be considered in the hydrodynamic approximation, and can be solved only on the basis of the kinetic equation. In this case, as a consequence of the very favorable situation with the establishment of energy equilibrium in the excitation gas, it is possible essentially to solve the problem exactly. The fact is that the effective scattering cross section of rotons by rotons is sufficiently large, and, therefore, in a roton gas one always has a local equilibrium in practice; that is, the roton gas can be described by a quasi-equilibrium distribution function with a temperature T_r and a relative velocity $v_{nr} - v_s$ which varies from point to point (v_s is the velocity of the superfluid part of the liquid).

The scattering of a phonon by a phonon is not accompanied by a change in direction of the momenta of the colliding phonons and is anomalously large for collisions of phonons at small angles. This process is more rapid than all the other processes taking place with phonons and leads to the result that there is always an energy equilibrium for phonons moving in a given direction. Moreover, as we shall see, the five-phonon process for phonons colliding at small angles also takes place more rapidly than the process of scattering of phonons by rotons, and therefore the phonon-number equilibrium is established. Thus the relatively slowest process is the scattering of phonons by rotons. The slowness of this process leads to the result that the establishment of energy equilibrium between the photon and roton gases is made more difficult, although in each of them separately it occurs. This circumstance leads to the appearance in the hydrodynamic approximation of a certain second viscosity, while in the general case, it leads to absorption and dispersion of first and second sound. In this case, the theory contains only one characteristic time τ_{pr}, which is associated with the scattering of a phonon by a roton, and which can be computed exactly. The results obtained will naturally be valid for all frequencies ($\omega\tau_{pr} \gtrsim 1$), provided they do not exceed the reciprocal of the characteristic time for establishing equilibrium in the excitation gas.

Let us consider the problem of the limits of applicability of the results obtained with respect to the sound frequency It is evident that two circumstance are significant for all the considerations that have been made: first, the presence of energy equilibrium for phonons moving in a given direction;

second, the presence of a local equilibrium in the roton gas. Therefore, in order that these two conditions not be violated, it is obviously necessary that

$$\omega t_{rr} \ll 1, \quad \omega t_{pp} \ll 1,$$

where the time t_{rr} is determined by equation (1.3.33) and the time t_{pp} by equation (1.3.24). To estimate t_{pp}, it is necessary to substitute the energy of those phonons which play an important role in the range of collisions and are significant for the phenomenon of sound dispersion (as we shall see, $\varepsilon \approx (7\text{–}8) \cdot kT_0$). Here it must be kept in mind that the real value of the time t_{pp}, as shown above, is much smaller than that determined by equation (1.3.24) with the value of γ from (1.3.25).

An appreciable dispersion takes place for first sound when $\omega t_{pr} \sim 1$, and for second sound, when $\omega \tau_{pr} \sim u_2/u_1$. Inasmuch as $u_2/u_1 \lesssim 1$, the dispersion of second sound sets in at much lower frequencies than for first sound. It is easy to see that in this and in the other case, dispersion begins when the wavelength of the corresponding sound is comparable with the mean free path of the phonon. It is seen from Figure 1.6*a* and 1.6*b* that there is a large range of frequencies for which observation of sound dispersion is possible.

It is seen from Figure 1.6*a* that at comparatively high temperatures (above 0.9 K) the phonon–phonon scattering does not have to be considered in comparison with the scattering of phonons by rotons, inasmuch as $t_{pp} \gg t_{pr}$. In the temperature range from 0.6 to 0.9 K, $t_{pp} \sim t_{pr}$ and both scattering processes are important. Finally, at very low temperatures (below 0.6 K) $t_{pp} \ll t_{pr}$ and in this case, only the effect of phonon–phonon scattering is important.

We shall first consider the most interesting temperature region of 0.9. K. In this region of temperatures the second term on the right side of equation (1.3.54) is much smaller in magnitude than the first and it can be neglected. After some transformations, and substituting the integrals (1.3.60), equations (1.3.52)–(1.3.56), can be represented in the form

$$\frac{S_r}{C_r}\frac{\rho_{np}}{\rho}(\nu_1 + w_r) - \tilde{\omega}T_r' + \frac{\rho_s}{\rho}\frac{S_r}{C_r}w_r + \frac{C_p}{C_r}\beta(\tilde{\omega} - \tilde{z}_{pr})$$

$$\times(\nu_0 + T_r') - \frac{S_p}{C_r}\tilde{\omega}(3u + 1 + \delta)\rho' = 0, \qquad (1.4.1)$$

$$-\tilde{\omega}w_r + \frac{TS_r}{\rho_{nr}c^2}T_r' + \frac{\rho_{np}}{\rho_{nr}}(\tilde{\omega} - \tilde{z}_{pr})(\nu_1 + w_r) = 0, \qquad (1.4.2)$$

$$-\tilde{\omega}\rho' + j_r - \frac{\rho_{np}}{\rho}(\nu_1 + w_r) = 0 \qquad (1.4.3)$$

$$- \tilde{\omega} j_r + \frac{1}{c^2}\left(\frac{\partial P}{\partial \rho}\right)_T \rho' + \frac{\rho_{np}}{\rho}\{(\tilde{\omega} - \tilde{z}_{pr})(\nu_1 + w_r)$$

$$- 3u(\nu_0 + T_r') - (1 + \delta) T_r'\} = 0, \tag{1.4.4}$$

$$- (\tilde{z}_{pr} - \cos\theta)\, \nu(\cos\theta) = \tilde{\omega} u \rho' + \cos^2\theta\left(j_r - \frac{\rho_n}{\rho} w_r\right)$$

$$+ (\tilde{\omega} - \tilde{z}_{pr})[\nu_0 - \cos\theta\, w_r + \beta(\nu_0 + T_r')]. \tag{1.4.5}$$

We have introduced the notation P = pressure,

$$j_r = \frac{\rho_s}{\rho} v_s + \frac{\rho_n}{\rho} v_{nr}, \quad C_r = T\left(\frac{\partial S_r}{\partial T}\right)_\rho, \quad \tilde{z}_{pr} = \tilde{\omega}\left(1 - \frac{1}{i\omega\tau_{pr}}\right),$$

$$\delta = \frac{T}{\rho}\left(\frac{\partial \rho}{\partial T}\right)_P \Big/ \frac{\rho_{np}}{\rho} =$$

$$- 3u - 1 - \frac{S_r}{S_p}\left\{1 - 2\frac{\rho}{P_0}\frac{\partial P_0}{\partial \rho} - \frac{1}{2}\frac{\rho}{\mu}\frac{\partial \mu}{\partial \rho} + \frac{\Delta}{T}\frac{\Delta/T + 1/2}{\Delta/T + 3/2}\frac{\rho}{\Delta}\frac{\partial \Delta}{\partial \rho}\right\}.$$

We multiply the left and right sides of equation (1.4.5) first by unity and then by $(\tilde{z}_{pr} - \cos\theta)^{-1}$ and average in each case over $\cos\theta$. The resulting equations

$$3\tilde{\omega} u \rho' - 3\tilde{\omega} T_r' + j_r + \frac{\rho_s}{\rho} w_r$$

$$+ 3[\tilde{\omega} - \beta(\tilde{\omega} - z_{pr})](\nu_0 + T_r') - (\nu_1 + w_r) = 0, \tag{1.4.6}$$

$$[2 + (1 - \beta)(\tilde{\omega} - \tilde{z}_{pr}) \ln \tilde{a}](\nu_0 + T_r') + \tilde{\omega} u \ln \tilde{a} \rho'$$

$$- [2 + (\tilde{\omega} - \tilde{z}_{pr}) \ln \tilde{a}] T_r' + \left[\tilde{z}_{pr} j_r - \left(\tilde{\omega} - \frac{\rho_s}{\rho}\tilde{z}_{pr}\right) w_r\right]$$

$$\times (- 2 + z_{pr} \ln \tilde{a}) = 0, \quad \tilde{a} = \frac{(\tilde{z}_{pr} + 1)}{(\tilde{z}_{pr} - 1)}, \tag{1.4.7}$$

together with (1.4.1)–(1.4.4) represent the complete set of six linear homogeneous equations. Equations (1.4.7)–(1.4.4) contain the very small quantity $\rho_{np}/\rho = 1.28 \cdot 10^{-4} T^4$. We keep terms of small order not higher than ρ_n/ρ. In this approximation, equations (1.4.3) and (1.4.4) describe the propagation of first sound, while equations (1.4.1)–(1.4.2) describe second sound. The last pair (1.4.6) and (1.4.7) gives us a dispersion equation (in zero approximation)

$$2 + (1 - \beta)(\tilde{\omega} - \tilde{z}_{pr}) \ln \tilde{a} = 0,$$

which for $T < 0.6$ K does not have undamped acoustic solutions. At temperatures below 0.6 K, equations (1.4.1) and (1.4.2) are invalid (the rotons are practically absent), while equations (1.4.6) and (1.4.7) made up from terms containing $J_{pp}(n)$ [see equation (1.3.54)], as we shall see below, describe undamped temperature waves in a phonon gas (second sound).

We note that in the derivation of equations (1.4.1)–(1.4.5), (1.4.6) and (1.4.7) we have everywhere assumed the energy of phonons to be equal to $\varepsilon = cp$. As simple calculations show, the term containing γp^2 in the expression for ε can be neglected if the following condition is satisfied:

$$\frac{1}{\omega\tau_{pr}}\left(1 + \frac{t_{pr}}{t_{pp}}\right) \gg 3\gamma\left(2\pi\frac{kT}{c}\right)^2\frac{B_3}{B_2}, \tag{1.4.8}$$

where B_2 and B_3 are Bernouilli numbers: $B_2 = \frac{1}{30}$, $B_3 = \frac{1}{42}$. For the frequency region under consideration (ωt_{rr} $\omega t_{pp} \ll 1$) the condition (1.4.8) is virtually always satisfied for temperatures above 0.6 K.

1.4.2. The Dispersion of First Sound

(*a*) *Region of Temperatures above 0.9 K.* This temperature region is characterized by the fact that the interaction of phonons with one another is negligibly small, and the principal role is played by the scattering of phonons by rotons. Equations (1.4.3) and (1.4.4), which describe the propagation of first sound to first order in ρ_n/ρ, have the form

$$-\tilde{\omega}\rho' + j_r - \frac{\rho_{np}}{\rho}\nu_1 = 0 \tag{1.4.9}$$

$$-\tilde{\omega}j_r + \frac{1}{c^2}\left(\frac{\partial P}{\partial \rho}\right)_T\rho' + \frac{\rho_n}{\rho}\left[(\tilde{\omega} - \tilde{z}_{pr})\nu_1 - 3u\nu_0\right] = 0,$$

where ν_0 and ν_1 are given by

$$\nu_0 = -\frac{\tilde{\omega}u\ln\tilde{a}\rho' + \tilde{z}_{pr}(-2 + \tilde{z}_{pr}\ln\tilde{a})j}{2 + (1 - \beta)(\tilde{\omega} - \tilde{z}_{pr})\ln\tilde{a}} \tag{1.4.10}$$

$$\nu_1 = 3\tilde{\omega}u\rho' + j_r + 3[\tilde{\omega} - \beta(\tilde{\omega} - \tilde{z}_{pr})]\nu_0, \tag{1.4.11}$$

$$\tilde{\omega} = \frac{\omega}{kc}, \quad \tilde{z}_{pr} = \tilde{\omega}\left(1 - \frac{1}{i\omega\tau_{pr}}\right), \quad \tilde{a} = \frac{\tilde{z}_{pr} + 1}{\tilde{z}_{pr} - 1}.$$

From the condition for the existence of a nontrivial solution of (1.4.9), we obtain the complex velocity of first sound:

$$\frac{\omega}{k} = u_{10} - \frac{1}{2}c\frac{\rho_{np}}{\rho}\varphi(z_{pr}), \tag{1.4.12}$$

where $u_{10} = (\partial P/\partial\rho)_T^{1/2}$ is the velocity of propagation of first sound in the region of low frequencies $\omega\tau_{pr} \ll 1$, while the function $\varphi(z_{pr})$ is equal to*

$$\varphi(z_{pr}) = z_{pr} - 3\frac{u^2\ln a + \{2uz_{pr} + z_{pr}^2[1 - \beta(1 - z_{pr})]\}(-2 + z_{pr}\ln a)}{2 + (1 - \beta)(1 - z_{pr})\ln a}$$

$$z_{pr} = 1 - \frac{1}{i\omega\tau_{pr}}, \qquad a = \frac{z_{pr} + 1}{z_{pr} - 1}. \tag{1.4.13}$$

*Since the second component on the right side of (1.4.12) is much smaller than the first, and $u_{10} \approx c$, we shall set $\tilde{\omega} = 1$ everywhere in the function $\varphi(z_{pr})$.

The real part of (1.4.12) is the velocity of first sound

$$u_1 = u_{10} - \frac{1}{2} c \frac{\rho_{np}}{\rho} \operatorname{Re} \varphi(z_{pr}). \tag{1.1.14}$$

The absorption coefficient of first sound α_1 is the imaginary part of the wave vector. According to (1.4.9), it is

$$\alpha_1 = \frac{1}{2} \frac{\omega}{c} \frac{\rho_{np}}{\rho} \operatorname{Im} \varphi(z_{pr}). \tag{1.4.15}$$

Equations (1.4.14) and (1.4.15) determine the dispersion and absorption of ordinary sound, due to the comparatively slow process of scattering of phonons by rotons. Analysis of the function $\varphi(z_{pr})$ shows that the principal dispersion of ordinary sound sets in when the parameter $\omega\tau_{pr}$ becomes of the order of unity.

Let us consider the region of low frequencies satisfying the condition

$$\omega\tau_{pr} \ll 1.$$

In this region $1/z_{pr} \ll 1$, and the expression in (1.4.13) which contains $\ln a$ can be expanded in a power series in the quantity $1/z_{pr}$:

$$\ln a = \frac{2}{z_{pr}}\left(1 + \frac{1}{3}\frac{1}{z_{pr}^2}\right),$$

$$-2 + z_{pr} \ln a = \frac{2}{z_{pr}^2}\left(\frac{1}{3} + \frac{1}{5} \cdot \frac{1}{z_{pr}^2}\right). \tag{1.4.16}$$

Substituting (1.4.16) in (1.4.13) and keeping terms linear in $\omega\tau_{pr}$, we get

$$\varphi = i\omega\tau_{pr}\left[\frac{4}{15} + \frac{(3u+1)^2}{3\beta}\right],$$

whence, in accord with (1.4.14) and (1.4.15), it follows that in the region of low frequencies

$$u_1 = u_{10}, \quad \alpha_1 = \frac{\omega^2\tau_{pr}}{c}\frac{\rho_{np}}{\rho}\left[\frac{2}{15} + \frac{(3u+1)}{6\beta}\right]. \tag{1.4.17}$$

The first term in the absorption of sound corresponds to the phonon part of the coefficient of ordinary viscosity

$$\eta_p = \frac{1}{5} c^2 \rho_{np} \tau_{pr}, \tag{1.4.18}$$

and the second term to the coefficient of second viscosity

$$\xi_2 = \frac{1}{3\beta}(3u+1)^2 c^2 \rho_{np} \tau_{pr}. \tag{1.4.19}$$

As has already been noted above, the second viscosity is due to the fact that energy balance between the phonon and roton gases is established slowly relative to the equilibirum times of the seperate gases.

Substitution in (1.4.7) of the numerical values of all the parameters shows that the second term is much greater than the first in all regions of temperatures taken into consideration. Thus at low frequencies ($\omega\tau_{pr} \ll 1$), the absorption of ordinary sound is chiefly due to the second viscosity.

In the region of high frequencies, satisfying the equation

$$\omega\tau_{pr} \gg 1,$$

z_{pr} becomes of the order of unity. Setting the quantity z_{pr} in (1.4.14) and (1.4.15) equal to unity, we get

$$u_1 = u_{10} + c\frac{\rho_{np}}{\rho}\left[\frac{3}{4}(u+1)^2 \ln(2\omega\tau_{pr}) - 3u - 2\right], \qquad (1.4.20)$$

$$\alpha_1 = \frac{3}{8}\pi(u+1)^2 \frac{\omega}{c}\frac{\rho_{np}}{\rho}. \qquad (1.4.21)$$

(*b*) *Region of Temperatures from 0.6 to 0.9 K.* In this temperature region, in addition to the scattering of phonons by rotons, phonon–phonon scattering is also important. Therefore, we must keep the collision integral $J_{pp}(n)$ in equation (1.4.54). The inclusion of $J_{pp}(n)$ affects only the form of the function φ, which now depends on the two parameters z_{pr} and $z_{pp} = 1 - 1/i\omega\tau_{pp}$

$$\varphi(z_{pr}, z_{pp}) = z_{pr} - 3\{u^2 \ln a + [2uz_{pr} + z_{pr}^2[1 - \beta(1 - z_{pr})] + 3u^2(1 - z_{pp})][-2 + (z_{pr} + z_{pp} - 1)\ln a]\}/\{2 + [1 - z_{pp} + (1 - \beta)(1 - z_{pr})]\ln a + 3(1 - z_{pp})[1 - \beta(1 - z_{pr})] \times [-2 + (z_{pr} + z_{pp} - 1)\ln a]\},$$

$$a = \frac{(z_{pr} + z_{pp})}{(z_{pr} + z_{pp} - 2)}. \qquad (1.4.22)$$

The velocity and the absorption coefficient of the first sound are determined as before by equations (1.4.14) and (1.4.15), but with the functions $\varphi(z_{pr}, z_{pp})$ from (1.4.22):

$$u_1 = u_{10} - \frac{1}{2}c\frac{\rho_{np}}{\rho}\operatorname{Re}\varphi(z_{pr}, z_{pp}), \qquad (1.4.23)$$

$$\alpha_1 = \frac{1}{2}\frac{\omega}{c}\frac{\rho_{np}}{\rho}\operatorname{Im}\varphi(z_{pr}, z_{pp}). \qquad (1.4.24)$$

In equation (1.4.9) we took into account only terms of leading order in ρ_{np}/ρ. If we include the next order of the expansion of (1.4.9), then a component appears in the expressions for (1.4.13) and (1.4.22) containing the derivative $(\partial\rho/\partial T)_p$ (that is, δ). Account of this component is important only in the region of low frequencies $\omega\tau_{pr}$, $\omega\tau_{pp} \ll 1$, and at temperatures below 0.9 K. The velocity and absorption coefficient in this case are equal to

$$u_1 = u_{10} + \frac{1}{2} c\delta^2 \frac{S_p}{C} \frac{\rho_{np}}{\rho}, \tag{1.4.25}$$

$$\alpha_1 = \frac{\omega^2\tau_{pr}}{c} \frac{\rho_{np}}{\rho} \left\{ \frac{2/15}{1 + \tau_{pr}/\tau_{pp}} + \frac{1}{6\beta}\left(3u + 1 + \delta \frac{C_p}{C}\right)^2 + \frac{1}{2}\delta^2\left(\frac{S_p}{C}\right)^2\left(1 - \frac{TS}{\rho_n c^2}\right)^2\right\}. \tag{1.4.26}$$

As can be shown, the first term in equation (1.4.26) is due to the phonon part of the first viscosity coefficient

$$\eta_p = \frac{1}{5} c^2 \rho_{np} \frac{\tau_{pr}}{1 + \tau_{pr}/\tau_{pp}}, \tag{1.4.27}$$

the second term to the coefficient of second viscosity

$$\xi_2 = \frac{1}{3\beta}\left(3u + 1 + \delta \frac{C_p}{C}\right)^2 c^2 \rho_{np}\tau_{pr}, \tag{1.4.28}$$

and the third to the phonon part of the coefficient of thermal conductivity

$$\kappa_p = c^2 S_p \tau_{pr} \left(1 - \frac{TS}{\rho_n c^2}\right)^2. \tag{1.4.29}$$

For high frequencies, $\omega\tau_{pr}$, $\omega\tau_{pp} \gg 1$, in the temperature region under consideration, the velocity of first sound is equal to

$$u_1 = u_{10} + c \frac{\rho_{np}}{\rho} \left\{ \frac{3}{4}(u + 1)^2 \ln \frac{2\omega\tau_{pr}}{1 + \tau_{pr}/\tau_{pp}} - 3u - 2\right\},$$

while the absorption coefficient is determined as before by expression (1.4.21).

1.4.3. Second Sound

As will be shown below, the dispersion feature of second sound begins at frequencies satisfying the condition

$$\omega\tau_{pr} \sim \frac{u_2}{u_1}$$

(u_2 is the velocity of second sound). In spite of the smallness of the ratio u_2/u_1, up to the present time temperature oscillations of such a frequency have not been achieved. However, according to Notarys and Pellam [32], this range of frequencies can be obtained for second sound.

(*a*) *Region of Temperatures above 0.9 K.* Equations (1.4.1) and (1.4.2), which determine the propagation of second sound, have the following form to leading order in ρ_n/ρ:

$$-\tilde{\omega}T_r' + \frac{\rho_s}{\rho}\frac{S_r}{C_r}w_r + \frac{C_p}{C_r}\beta(\tilde{\omega} - \tilde{z}_{pr})(\nu_0 + T_r') + \frac{S_r}{C_r}\frac{\rho_{np}}{\rho}(\nu_1 + w_r) = 0,$$

$$-\tilde{\omega}w_r + \frac{TS_r}{\rho_{nr}c^2}T_r' + \frac{\rho_{np}}{\rho_{nr}}(\tilde{\omega} - \tilde{z}_{pr})(\nu_1 + w_r) = 0, \qquad (1.4.30)$$

where

$$\nu_0 + T_r' = \frac{[2 + (\tilde{\omega} - \tilde{z}_{pr})\ln\tilde{a}]T_r' + \left(\tilde{\omega} - \frac{\rho_s}{\rho}\tilde{z}_{pr}\right)(-2 + \tilde{z}_{pr}\ln\tilde{a})w_r}{2 + (1-\beta)(\tilde{\omega} - \tilde{z}_{pr})\ln\tilde{a},}$$

$$\nu_1 + w_r = -3\tilde{\omega}T_r' + \left(\frac{\rho_s}{\rho}\right)w_r$$

$$+ 3[\tilde{\omega} - \beta(\tilde{\omega} - \tilde{z}_{pr})](\nu_0 + T_r'). \qquad (1.4.31)$$

It actually follows from equation (1.4.31) that the important dispersion of second sound begins when $\omega\tau_{pr}$ becomes of the order of u_2/u_1. This circumstance favors observation of the phenomenon described, since (inasmuch as $u_2/u_1 \lesssim 1$) the dispersion of the second sound begins at much lower frequencies than for first sound. The condition $\omega\tau_{pr} \sim u_2/u_1$ can be rewritten in the form $l_p \sim \lambda$, where l_p is the mean free path of the phonon and λ is the wavelength of second sound. Thus the dispersion of second sound begins (as should be expected) when the wavelength of the second sound is comparable with the mean free path of the phonons, that is, we are dealing with spatial dispersion.

The condition for the existence of a nontrivial solution of the system (1.4.30) is obtained by setting its determinant equal to zero. Expanding the determinant, we obtain an equation for the determination of the complex velocity of second sound ω/k. For frequencies of the order of $\omega\tau_{pr} \sim u_2/u_1$, this equation is as follows:

$$\left(\frac{\omega}{k}\right)^2\left\{1 - \beta\frac{C_p}{C_r}(1 - z_{pr})\frac{2 + (\tilde{\omega} - \tilde{z}_{pr})\ln\tilde{a}}{2 + (1-\beta)(\tilde{\omega} - \tilde{z}_{pr})\ln\tilde{a}}\right\} = u_{2\infty}^2, \qquad (1.4.32)$$

$$u_{2\infty} = \left[\left(1 - \frac{\rho_n}{\rho}\right)\frac{TS_r^2}{\rho_{nr}C_r}\right]^{\frac{1}{2}} \qquad (1.4.33)$$

$(\tilde{a} = \tilde{z}_{pr} + 1)/(\tilde{z}_{pr} - 1)$. The expression for $u_{2\infty}$ is indentical with the well-known expression for the velocity of second sound in which all the thermodynamic quantities for the roton gas have been substituted. In (1.4.32) we have omitted all terms the ratio of which to the last term does not exceed $\beta^{-1}(k\Delta/c\rho_0)^2 \ll 1$. (This inequality is satisfied for all temperatures above 0.3 K).

Equation (1.4.32) for the frequencies $\omega \sim \tau_{pr}^{-1}u_2/u_1$ cannot be solved in algebraic form; its solution requires numerical calculations.

In the region of high frequencies,

$$\omega\tau_{pr} \gg \frac{u_2}{u_1}$$

equation (1.4.32) is materially simplified and one easily obtains

$$\frac{\omega}{k} = u_{2\infty}\left(1 - \frac{\beta C_p/C_r}{i\omega\tau_{pr}}\right)^{-\frac{1}{2}}. \tag{1.4.34}$$

It is seen from equation (1.4.34) in the case of very high frequencies $\omega\tau_{pr} \gg \beta C_p/C_r$ that second sound propagates only via the roton gas, with a velocity equal to $u^{2\infty}$ (1.4.33), while the absorption coefficient does not depend on the frequency and is equal to

$$\alpha_2 = \frac{1}{2}\,\frac{\beta C_p/C_r}{u_{2\infty}\tau_{pr}}. \tag{1.4.35}$$

In the region of low frequencies, satisfying the condition

$$\omega\tau_{pr} \ll \frac{u_2}{u_1}$$

$1/\tilde{z}_{pr} \ll 1$, one can expand the expression in (1.4.31) containing ln $\tilde{a}$ in a power series in the quantity $1/\tilde{z}_{pr}$ as done in equation (1.4.16). As the result of lengthy calculations, the condition of consistency of the set of equations (1.4.30) takes the following form:

$$\left(\frac{\omega}{k}\right)^2 = u_{20}^2 - i\omega\tau_{pr}c^2\,\frac{\rho_s}{\rho_n}\,\frac{\rho_{np}}{\rho}\left\{\frac{4}{15} + \frac{\rho_n}{\rho_s}\,\frac{\rho c^2}{TC}\left(1 - \frac{TS}{\rho_n c^2}\right)^2 + \frac{1}{3\beta}\left(1 - 3\,\frac{S}{C}\right)^2\right\}, \tag{1.4.36}$$

$$u_{20} = \left(\frac{\rho_s}{\rho_n}\,\frac{TS^2}{\rho C}\right)^{\frac{1}{2}}. \tag{1.4.37}$$

For the frequencies considered, the imaginary terms on the right side of (1.4.36) are small in comparison with the real terms. In this case, the velocity of propagation of second sound is equal to u_{20} (1.4.37), while its absorption coefficient is determined by the expression

$$\alpha_2 = \frac{\omega^2 c^2\tau_{pr}}{u_{20}^3}\,\frac{\rho_s}{\rho_n}\,\frac{\rho_{np}}{\rho}\left\{\frac{2}{15} + \frac{1}{2}\,\frac{\rho_n}{\rho_s}\,\frac{\rho c^2}{TC}\left(1 - \frac{TS}{\rho_n c^2}\right)^2 + \frac{1}{6\beta}\left(1 - 3\frac{S}{C}\right)^2\right\}. \tag{1.4.38}$$

The first two terms in (1.4.38) are due to the appearance, respectively, of the phonon part of the first viscosity coefficient η_p (1.4.18) and the phonon part of the coefficient of thermal conductivity κ_p (1.4.29). The third term in (1.4.38) corresponds to a combination of the coefficients of second viscosity $(\xi_2 + \rho^2\xi_3 - 2\rho\xi_1)$. As can be shown the coefficients ξ_1 and ξ_3 are equal to [31]

$$\rho\xi_1 = \frac{1}{\beta}(3u + 1)\left(u + \frac{S}{C}\right)c^2\rho_{np}\tau_{pr},$$

$$\rho^2\xi_3 = \frac{3}{\beta}\left(u + \frac{S}{C}\right)^2 c^2\rho_{np}\tau_{pr}.$$

Substitution of the numerical values of all parameters in (1.4.38) shows that the second term greatly exceeds the other two in magnitude. Thus, in the region of low frequencies ($\omega\tau_{pr} \ll u_2/u_1$) the absorption of second sound is essentially determined by the phonon part of the coefficient of thermal conductivity. The temperature dependences of the velocities u_{20} (1.4.37) and $u_{2\infty}$ (1.4.33) are drawn in Figure 1.7. For a fixed temperature, as the frequency is increased, we gradually go over from the curve u_{20}, which describes the equilibrium second sound, to the curve $u_{2\infty}$, which describes the roton second sound.

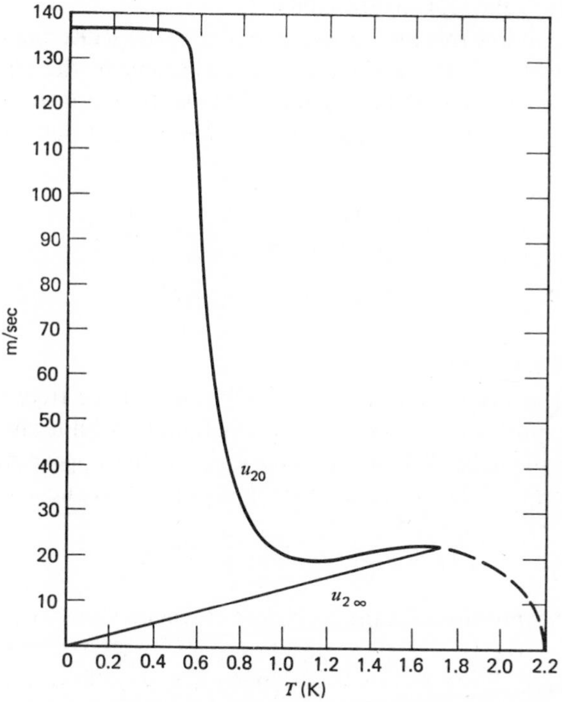

FIGURE 1.7. Temperature dependence of the velocities u_{20} and $u_{2\infty}$.

(*b*) *Region of Temperatures from 0.6 to 0.9 K.* In this temperature region, phonon–phonon scattering becomes important and in equation (1.3.54) one must take into consideration the collision integral $J_{pp}(n)$ which characterizes this scattering process. Consideration of $J_{pp}(n)$ transforms the equation (1.4.32) to the form

$$\begin{aligned}\left(\frac{\omega}{k}\right)^2\Big[1 - \beta\frac{C_p}{C_r}(1 - z_{pr})\{2 + (\tilde{\omega} - \tilde{z}_{pr})\ln\tilde{a}\\ + 3\tilde{\omega}(\tilde{\omega} - \tilde{z}_{pp})[-2 + (\tilde{z}_{pr} + \tilde{z}_{pp} - \tilde{\omega})\ln\tilde{a}\}\\ \times\{2 + [\tilde{\omega} - \tilde{z}_{pr} + (1 - \beta)(\tilde{\omega} - \tilde{z}_{pr})\ln\tilde{a}]\\ + 3(\tilde{\omega} - \tilde{z}_{pp})[\tilde{\omega} - \beta(\tilde{\omega} - \tilde{z}_{pr})][-2\\ + (\tilde{z}_{pr} + \tilde{z}_{pp} - \tilde{\omega})\ln\tilde{a}\}^{-1}\Big] = u_{2\infty}^2 \qquad (1.4.39)\end{aligned}$$

(where $\tilde{a} = (\tilde{z}_{pr} + \tilde{z}_{pp})/(\tilde{z}_{pr} + \tilde{z}_{pp} - 2)$, $\tilde{z}_{pp} = \tilde{\omega}(1 - 1/i\omega\tau_{pp})$. As we have pointed out above, in the general case this can only be solved numerically.

For high frequencies $\omega\tau_{pr}, \omega\tau_{pp} \gg 1$, equation (1.4.34) is again obtained from equation (1.4.39). For sufficiently low temperatures, the ratio $\beta C_p/C_r$ can become large and, as is seen from equation (1.4.34), the roton second sound will be strongly damped. At very high frequencies, $\omega\tau_{pr} \gg \beta C_p/C_r$; equations (1.4.33) and (1.4.35) are valid as before.

In the region of low frequencies $\omega\tau_{pr}, \omega\tau_{pp} \ll 1$, the scattering of phonons by phonons does not give a significant contribution to the absorption of second sound, inasmuch as the account of the scattering mentioned has an effect only on the form of the term in equation (1.4.38) that is smallest in magnitude:

$$\begin{aligned}\alpha_2 = \frac{\omega^2c^2\tau_{pr}}{u_{20}^3}\frac{\rho_s}{\rho_n}\frac{\rho_{np}}{\rho}\Big\{\frac{2/15}{1 + \tau_{pr}/\tau_{pp}} + \frac{1}{2}\frac{\rho_n}{\rho_s}\frac{\rho c^2}{TC}\\ \times\left(1 - \frac{TS}{\rho_n c^2}\right)^2 + \frac{1}{6\beta}\left(1 - 3\frac{S}{C}\right)^2\Big\} \qquad (1.4.40)\end{aligned}$$

corresponding to (1.4.27).

In the temperature region under consideration, for the frequencies $\omega\tau_{pr}$, $\omega\tau_{pp} \ll 1$, one must take into account in equation (1.4.30) terms containing the derivative $(\partial\rho/\partial T)_p \sim \delta)$. Its account leads to the appearance of additional components in the expression for the velocity of second sound:

$$u_2 = u_{2\infty}\left(1 - \frac{1}{2}\delta^2\frac{S_p}{C}\frac{\rho_{np}}{\rho}\right),$$

and also in the expressions for the coefficient of second viscosity (1.4.28):

$$\rho\xi_1 = \frac{1}{\beta}\left(3u + 1 + \delta\frac{C_p}{C}\right)\left(u + \frac{S}{C} + \delta\frac{S_p}{C}\right)c^2\rho_{np}\tau_{pr},$$

$$\rho^2\xi_3 = \frac{3}{\beta}\left(u + \frac{S}{C} + \delta\frac{S_p}{C}\right)^2 c^2\rho_{np}\tau_{pr}. \tag{1.4.41}$$

The fact that the derivative $(\partial\rho/\partial T)_p$ does not enter into the expression for the absorption of second sound is connected with the fact that the combination of the coefficients of second viscosity (1.4.28), (1.4.41) entering into (1.4.40) remains the same as before for $T < 0.9$ K and is equal to

$$\xi_2 + \rho^2\xi_3 - 2\rho\xi_1 = \frac{(1 - 3S/C)^2}{3\beta}.$$

1.4.4. Region of Temperatures below 0.6 K

In the region of temperatures below 0.6 K, the contribution of rotons to all the phenomena becomes unimportant and one can consider a purely phonon gas. The equations describing the propagation of sound in a phonon gas, according to (1.3.46), (1.3.48), and (1.3.49), have the form

$$\begin{aligned}
&-\tilde{\omega}\rho' + j = 0,\\
&-\tilde{\omega}j + \frac{1}{c^2}\left(\frac{\partial p}{\partial\rho}\right)_T\rho' + \frac{\rho_{np}}{\rho}(3uT' + \tilde{\omega}w) = 0,\\
&-\tilde{\omega}T' + \frac{1}{3}\frac{\rho_{sp}}{\rho}w + \tilde{\omega}\frac{3u+1}{3}\rho' = 0,\\
&-[2 + (\tilde{\omega} - \tilde{z}_{pp})\ln\tilde{a}]T' - \left(\tilde{\omega} - \frac{\rho_{sp}}{\rho}\tilde{z}_{pp}\right)\\
&\times(-2 + \tilde{z}_{pp}\ln\tilde{a})w + u\tilde{\omega}\ln\tilde{a}\rho' + \tilde{z}_{pp}(-2 + \tilde{z}_{pp}\ln\tilde{a})j = 0
\end{aligned} \tag{1.4.42}$$

($\tilde{a} = (\tilde{z}_{pp} + 1)/(\tilde{z}_{pp} - 1)$). In (1.4.42), we have introduced the notation

$$j = \left|\frac{\rho_{sp}}{\rho}\mathbf{v}_s + \frac{\rho_{np}}{\rho}\mathbf{v}_n\right|$$

$$\frac{\rho_{sp}}{\rho} = 1 - \frac{\rho_{np}}{\rho}, \quad \nu_0 = -T', \quad \nu_1 = -w$$

(T' and $|v_n - v_s|$ are the ratios of the departure of the temperature and the relative velocity, determined for the equilibrium state of He (II), to T_0 and c, respectively).

The relations $\nu_0 = -T'$, $\nu_1 = -w$ follow from the requirements that

$$\int\varepsilon(n - n^0)\,d\tau_p = 0, \quad \int\mathbf{p}(n - n^0)\,d\tau_p = 0,$$

that is, that the nonequilibrium distribution function of the phonons n must lead to the same value for the total energy and the total momentum as the equilibrium function

$$n^0 = \left\{\exp\left[\frac{\varepsilon - \mathbf{p}\cdot(\mathbf{v}_n - \mathbf{v}_s)}{kT}\right] - 1\right\}^{-1}. \tag{1.4.43}$$

The condition for the consistency of the set (1.4.42) is the vanishing of its determinant. By expanding the determinant, we get an equation which, with account of only the terms linear in ρ_{np}/ρ, splits into two equations* ($a = (z_{pp} + 1)/(z_{pp} - 1)$)

$$\left(\frac{\omega}{k}\right)^2 - u_{10}^2 + c^2 \frac{\rho_{np}}{\rho} \varphi(z_{pp}) = 0, \tag{1.4.44}$$

$$\varphi(z_{pp}) = 1 - 3\,\frac{u^2 \ln a + \{2u + 1 + 3u^2(1 - z_{pp})\}(-2 + z_{pp} \ln a)}{2 + (1 - z_{pp}) \ln a + 3(1 - z_{pp})(-2 + z_{pp} \ln a)},$$

$$2 + (\tilde{\omega} - \tilde{z}_{pp}) \ln \tilde{a} + 3\tilde{\omega}(\tilde{\omega} - \tilde{z}_{pp})(-2 + \tilde{z}_{pp} \ln \tilde{a}) \tag{1.4.45}$$

$$- 3\,\frac{\rho_{np}}{\rho}\,\frac{u^2 + 2u\tilde{\omega}^2 + \tilde{\omega}^2}{\tilde{\omega}^2 - 1}(-2 + \tilde{z}_{pp} \ln \tilde{a}) = 0$$

Equation (1.4.44) determines the complex velocity of ordinary sound, the velocity and the coefficient of absorption of which are equal to

$$u_1 = u_{10} - \frac{1}{2}\, c\, \frac{\rho_{np}}{\rho} \operatorname{Re} \varphi(z_{pp}), \tag{1.4.46}$$

$$\alpha_1 = \frac{1}{2}\,\frac{\omega}{c}\,\frac{\rho_{np}}{\rho} \operatorname{Im} \varphi(z_{pp}). \tag{1.4.47}$$

It is not difficult to establish the fact that the resultant expressions for u_1 and α_1, and also (1.4.14) and (1.4.15), are the result of the more general formulas (1.4.23) and (1.4.24).

Equation (1.4.45) has an undamped acoustic solution only in the rergion of low frequencies satisfying the condition $\omega\tau_{pp} \ll 1/\sqrt{3}$. In this region of frequencies, one can expand $\ln \tilde{a}$ and $-2 + z_{pp} \ln a$ in a power series in the function $1/z_{pp}$ (see (1.4.16) as a result of which (1.4.45) takes the form

$$\left(\frac{\omega}{k}\right)^2 = \frac{c^2}{3}\left[1 - \frac{3}{2}(3u^2 + 2u + 1)\frac{\rho_{np}}{\rho} - \frac{4}{5}\, i\omega\tau_{pp}\right].$$

The root of this equation determines the complex velocity and consequently the velocity and the absorption coefficient of second sound:

$$u_2 = \frac{c}{\sqrt{3}}\left[1 - \frac{3}{4}(3u^2 + 2u + 1)\frac{\rho_{np}}{\rho}\right], \quad \alpha_2 = \frac{2\sqrt{3}\,\omega^2\tau_{pp}}{5c}$$

As was to be expected, in a purely phonon gas, the quantity u_2 at low temperatures approaches the limit $c/\sqrt{3}$, while its absorption depends only on

*Inasmuch as the last component in equation (1.4.44) is much smaller than the first and $u_{10} \approx c$, we can everywhere set $\tilde{\omega} = 1$ in the function $\varphi(z_{pp})$.

the coefficient of ordinary viscosity

$$\eta_p = \frac{1}{5} c^2 \rho_{np} \tau_{pp}. \tag{1.4.48}$$

In the region of low frequencies, $\omega\tau_{pp} \ll 1$, the velocity and absorption coefficient of first sound, in accord with (1.4.46) and (1.4.47), are equal to

$$u_1 = u_{10} + c\frac{1}{4}(3u + 1)^2 \frac{\rho_{np}}{\rho},$$

$$\alpha_1 = \frac{3}{10}(u + 1)^2 \frac{\omega^2 \tau_{pp}}{c} \frac{\rho_{np}}{\rho}.$$

In a purely phonon gas,

$$u_{10} = c\left[1 - \frac{3}{2}\left(u^2 - \frac{1}{4}\frac{\rho^2}{c}\frac{\partial^2 c}{\partial \rho^2}\right)\frac{\rho_{np}}{\rho}\right],$$

and consequently for $T = 0$, $u_1 = c$, as it must be. The absorption of ordinary sound is connected with the coefficient of ordinary viscosity (1.4.48).* The term $(u + 1)^2$ arises because we have taken into account the derivative term $(\partial\rho/\partial T)_p$ $(\delta = -3u - 1)$ in equation (1.4.42).

For high frequencies, $\omega\tau_{pp} \gg 1$,

$$u_1 = u_{10} + c\frac{\rho_{np}}{\rho}\{\tfrac{3}{4}(u + 1)^2 \ln(2\omega\tau_{pp}) - 3u - 2\}, \tag{1.4.49}$$

$$\alpha_1 = \frac{3}{8}\pi(u + 1)^2 \frac{\omega}{c}\frac{\rho_{np}}{\rho}. \tag{1.4.50}$$

As has already been pointed out previously [31], in the very high frequencies

$$\frac{1}{\omega\tau_{pp}} \ll 3\gamma\left(2\pi\frac{kT}{c}\right)^2 \frac{B_3}{B_2},$$

Terms cubic in the momentum must be taken into consideration in the expression for the energy ε entering into equations (1.3.46), (1.3.48), and (1.3.49). Account of γp^2 gives the following formulas for u_1 and α_1†:

$$u_1 = u_{10} - c\frac{\rho_{np}}{\rho}\left\{\frac{3}{4}(u + 1)^2 \ln\left[\frac{3}{2}\gamma\left(2\pi\frac{kT}{c}\right)^2\frac{B_3}{B_2}\right] + 3u + 2\right\}, \tag{1.4.51}$$

$$\alpha_1 = \frac{3}{4}(u + 1)^2 \frac{\rho_{np}}{\rho}\left[3\gamma\left(2\pi\frac{kT}{c}\right)^2\frac{B_3}{B_2}c\tau_{pp}\right]^{-1} \tag{1.4.52}$$

*As has been shown in [31] all kinetic coefficients in a phonon gas are equal to zero except η_p (1.4.48).

†In the work of Andreev and the author [33], the case of the high-frequency limit for which $\omega t_{pp} \gg 1$ was considered.

1.4.5. Sound in Helium (II) near Zero Temperature [33,34]

Let us investigate the phenomenon of sound propagation in He (II) near zero temperature.

Near the absolute zero the mean free path of the excitations increases rapidly and may be larger than the wavelength of sound. In this temperature range only the phonons play a significant role and the rotons may be neglected. Obviously, under these conditions the equations of hydrodynamics for the excitation gas are not applicable any more. The properties of such a gas may be described by the kinetic equation, in which the collision integral is negligibly small. Thus the phonon distribution function satisfies the equation

$$\frac{\partial n}{\partial t} + \frac{\partial n}{\partial \mathbf{r}} \cdot \frac{\partial H}{\partial \mathbf{p}} - \frac{\partial n}{\partial \mathbf{p}} \cdot \frac{\partial H}{\partial \mathbf{r}} = 0. \tag{1.4.53}$$

Here $H = \varepsilon(p) + \mathbf{p}\cdot\mathbf{v}_s$ and ε is the energy of the phonons which depends on the density; taking the dispersion of the phonons into account we will write ε in the form

$$\varepsilon(p) = cp(1 - \Upsilon p^2). \tag{1.4.54}$$

The quantities p and v_s, which play the role of external conditions for the phonons, are determined by the two equations [cf. (1.3.8) and (1.3.12)]

$$\frac{\partial \rho}{\partial t} + \operatorname{div}\left(\rho \mathbf{v}_s + \int \mathbf{p} n \, d\tau_p\right) = 0, \tag{1.4.55}$$

$$\frac{\partial \mathbf{v}_s}{\partial t} + \nabla\left(\mu_0 + \frac{\mathbf{v}_s^2}{2} + \int \frac{\partial \varepsilon}{\partial p} n \, d\tau_p\right) = 0, \tag{1.4.56}$$

where μ_0 is the value of the chemical potential at absolute zero, and p is the momentum of the excitation in the frame of reference in which $\mathbf{v}_s = 0$. Let us set $n = n_0 + n'$, $\rho = \rho_0 + \rho'$; n_0 and ρ_0 are equilibrium values and n', $\mathbf{v}_s$, and ρ' small quantities, varying as $\exp(-i\omega t + i\mathbf{k}\cdot\mathbf{r})$. Linearizing equations (1.4.53), (1.4.55), and (1.4.56), we obtain the system

$$(\omega - \mathbf{k}\cdot\mathbf{v})n' + \mathbf{v}\cdot\mathbf{k}\, \frac{\partial n_0}{\partial \varepsilon}\left(\frac{\partial \varepsilon}{\partial \rho}\rho' + \mathbf{p}\cdot\mathbf{v}_s\right) = 0,$$

$$\omega\rho' - \rho\mathbf{k}\cdot\mathbf{v}_s - \mathbf{k}\cdot\int \mathbf{p}n' \, d\tau_p = 0,$$

$$-\omega\mathbf{v}_s + \mathbf{k}\left(\frac{c^2}{\rho} + \int n_0 \frac{\partial^2 \varepsilon}{\partial \rho^2} d\tau\right)\rho' + \mathbf{k}\int \frac{\partial \varepsilon}{\partial \rho} n' d\tau = 0. \tag{1.4.57}$$

Here $\mathbf{v} = \partial\varepsilon/\partial\mathbf{p}$, and we have used the fact that, $d\mu = dp_0/\rho = c^2\, d\rho/\rho$.*

*In general, just as for Fermi liquids, there exists a functional dependence of the excitation energy on the density of excitations of the form $\delta\varepsilon = \int f \delta n \, d\tau$. However, in the present case this effect is small. Its contribution to our results will be of order fN_{ph}/kT. A simple estimate shows that $fN_{ph}/kT \sim (\tau kT/\hbar)^{-1/2} \ll 1$, since $\omega\tau \gg 1$ and $kT/\hbar \gg \omega$.

Let us denote by θ the angle between the vectors $\mathbf{k}$ and $\mathbf{p}$ and solve the first of equation (1.4.57) for n':

$$n' = -\frac{\partial n_0}{\partial \varepsilon} p \frac{kv\cos\theta}{\omega - kv\cos\theta}\left(\frac{\partial c}{\partial \rho}\rho' + v_s\cos\theta\right). \tag{1.4.58}$$

Inserting n' into the other two equations in (1.4.58) and integrating over θ we find

$$\left\{\frac{\omega}{k} - \frac{\partial c}{\partial \rho}\int_0^\infty \frac{p^4\,dp}{4\pi^2\hbar^3}\frac{\partial n_0}{\partial \varepsilon}\left[2\frac{\omega}{kv} - \left(\frac{\omega}{kv}\right)^2 \ln\frac{\omega + kv}{\omega - kv}\right]\right\}\rho'$$

$$-\left\{\rho + \int_0^\infty \frac{p^4\,dp}{4\pi^2\hbar^3}\frac{\partial n_0}{\partial \varepsilon}\left[2\left(\frac{\omega}{kv}\right)^2 + \frac{2}{3} - \left(\frac{\omega}{kv}\right)^3 \ln\frac{\omega + kv}{\omega - kv}\right]\right\} v_s = 0,$$

$$\left\{\frac{c^2}{\rho} + \frac{\partial^2 c}{\partial \rho^2}\int_0^\infty \frac{p^4\,dp}{2\pi^2\hbar^3} n_0 + \left(\frac{\partial c}{\partial \rho}\right)^2 \int_0^\infty \frac{p^4\,dp}{4\pi^2\hbar^3}\frac{\partial n_0}{\partial \varepsilon}\left[2\right.\right.$$

$$\left.\left. - \frac{\omega}{kv}\ln\frac{\omega + kv}{\omega - kv}\right]\right\}\rho' + \left\{-\frac{\omega}{k} + \frac{\partial c}{\partial \rho}\int_0^\infty \frac{p^4\,dp}{4\pi\hbar^3}\frac{\partial n_0}{\partial \varepsilon}\left[2\frac{\omega}{kv}\right.\right.$$

$$\left.\left. - \left(\frac{\omega}{kv}\right)^2 \ln\frac{\omega + kv}{\omega - kv}\right]\right\} v_s = 0. \tag{1.4.59}$$

In spite of the fact that $\omega \approx kc$, $v \approx c$, the logarithmic accuracy does not seem to be sufficient when equation (1.4.59) is written. As the results show, the logarithms obtained are not large.

At low temperatures the dependence $\omega(k)$ has the form $\omega = (c + \delta c)k$, where δc is a small temperature-dependent correction. From this condition for solvability of the set of homogeneous equations (1.4.59) we find

$$\delta c = (2\rho c^4)^{-1}\left\{(u+1)^2 \int_0^\infty \frac{\varepsilon^4\,d\varepsilon}{4\pi^2\hbar^3}\frac{\partial n_0}{\partial \varepsilon}\left[2 - \ln\frac{2c^2}{3\gamma\varepsilon^2}\right]\right.$$

$$\left. + \frac{2}{3}\int_0^\infty \frac{\varepsilon^4\,d\varepsilon}{4\pi^2\hbar^3}\frac{\partial n_0}{\partial \varepsilon} + w\int_0^\infty \frac{\varepsilon^3\,d\varepsilon}{2\pi^2\hbar^3} n_0\right\}, \tag{1.4.60}$$

where $u = (\rho/c)\,(\partial c/\partial \rho)$, $w = (\rho^2/c)\,(\partial^2 c/\partial \rho^2)$, and γ is the coefficient accounting for the dispersion of phonons: $\varepsilon = cp(1 - \gamma p^2)$.

Carrying out the simple integration in equation (1.4.60), we finally obtain

$$\delta c = \frac{\pi^2}{30\rho\hbar^3}\left(\frac{T^4}{c}\right)(u+1)^2\left\{\ln\frac{2c^2}{3\gamma T^2} - \frac{27}{6}\right.$$

$$\left. + 2C - 2\phi(4) + \frac{3w - 4}{6(u+1)^2}\right\}. \tag{1.4.61}$$

Here C is the Euler constant, equal to $0.577\cdots$, and $\phi(x) = \zeta'(x)/\zeta(x)$ is the log derivative of the Riemann ζ function: $\phi(4) = -(90/\pi^4)\sum_{n=2}^{\infty}(\ln n/n^4)$.

Thus, the velocity of sound at low temperatures is greater than at absolute zero by the amount δc, given in (1.4.61). It varies with the temperature as $T^4 \ln T$. It turns out that the argument of the logarithm contains a small numerical coefficient that has a significant influence on the magnitude and the behavior of the temperature dependence of the effect in the experimental temperature range. The temperature dependence of the velocity of sound (1.4.61) has been found under the following conditions: (i) the temperature is low enough that the collisions between thermal phonons should not be neglected, and (ii) the sound wave is classical, that is, the condition $\hbar\omega \ll T$ holds. These conditions are valid at temperatures below 0.4 K for sound frequencies in the tens of megacycles. The experiments have qualitatively confirmed the $T^4 \ln T$ law [35, 36], but a more detailed comparison is rendered impossible due to the difficulty of determining the magnitude of w from the independent experiments.

1.5. THE KINETIC PHENOMENA

1.5.1. The Kinetic Coefficients [37]

The distribution function of the excitations in uniform motion and in an equilibrium state is equal to

$$n_0 = \left[\exp\left(\frac{\varepsilon + \mathbf{p}\cdot\mathbf{v}_s - \mathbf{p}\cdot\mathbf{v}_n}{kT}\right) - 1\right]^{-1}. \tag{1.5.1}$$

The equilibrium function n_0 satisfies the kinetic equation with a vanishing collision integral. When the equilibrium is destroyed, the distribution function differs from its equilibrium value and is found by solving the kinetic equation. This problem, which is in general not soluble, is simplified when one considers states that differ only slightly from equilibrium. In this case the departure from equilibrium is completely determined by the first derivatives of the velocities $\mathbf{v}_n$ and $\mathbf{v}_s$ and the thermodynamic variables, with respect to the coordinates, all of which are assumed to be small quantities. In other words, the velocities $\mathbf{v}_n$ and $\mathbf{v}_s$ and the thermodynamic variables are slowly varying functions throughout the system, so that one may neglect all higher derivatives and higher powers of the first derivatives in the kinetic equation.

Let us consider a state of the system that differs little from the equilibrium state. We may then assume that in any small portion of the system we have an approximate local equilibrium, i.e., that the distribution function may be represented in the form

$$n = n_0 + n_1, \tag{1.5.2}$$

where $n_1 \ll n_0$ and n_0 is the equilibrium function (1.5.1), depending on the local values of the velocities and thermodynamic variables. When inserting the distribution function (1.5.1) into the left-hand side of the kinetic equation (1.3.1), it is sufficient to differentiate the function n_0, since the small additive contribution n_1 itself contains first derivatives and differentiating it would only bring in higher derivatives that can be neglected. The right-hand side, the collision integral, vanishes for $n = n_0$ and we may limit ourselves to terms linear in n_1. In this manner the problem has been linearized; we have obtained a linear integral equation whose left-hand side is given, and is expressible in terms of the first derivatives of the velocities and the thermodynamic variables. Let us insert the function n_0 into the left-hand side of the kinetic equation. Along with the spatial derivatives we shall consider the velocity difference $\mathbf{v}_n - \mathbf{v}_s$ small. This in no way limits the validity of our analysis since the velocity difference $\mathbf{v}_n - \mathbf{v}_s$ should in any case be small compared to the velocity of first and second sound. As is well known, superfluidity is destroyed long before this limit is reached. Let us first compute the time derivative; according to (1.3.1) we have*

$$\frac{\partial n_0}{\partial t} = n' \left\{ \frac{1}{kT} \left(\frac{\partial \varepsilon}{\partial t} - \mathbf{p} \cdot \frac{\partial \mathbf{v}_n}{\partial t} + \mathbf{p} \cdot \frac{\partial \mathbf{v}_s}{\partial t} \right) - \frac{\varepsilon}{kT^2} \frac{\partial T}{\partial t} \right\}. \tag{1.5.3}$$

Let us choose as our independent variables the density ρ and the entropy S. We may then express the time derivatives in terms of space derivatives; in the linear approximation we have, according to equations 1.2.22 and 1.2.25

$$\frac{\partial \rho}{\partial t} = -\operatorname{div} \mathbf{j}, \quad \frac{\partial S}{\partial t} = -S \operatorname{div} \mathbf{v}_n, \quad \rho_n \frac{\partial}{\partial t} (\mathbf{v}_n - \mathbf{v}_s) = -S \nabla T. \tag{1.5.4}$$

Using (1.5.4) we may transform (1.5.3) to

$$\frac{\partial n_0}{\partial t} = \frac{n'}{kT} \left\{ \operatorname{div}(\mathbf{j} - \rho \mathbf{v}_n) \left(\frac{1}{T} \frac{\partial T}{\partial \rho} \varepsilon - \frac{\partial \varepsilon}{\partial \rho} \right) + \operatorname{div} \mathbf{v}_n \right. \tag{1.5.5}$$
$$\left. \times \left[\frac{1}{T} \left(\frac{\partial T}{\partial \rho} \rho + \frac{\partial T}{\partial S} S \right) \varepsilon - \frac{\partial \varepsilon}{\partial \rho} \rho \right] + \frac{ST}{\rho_n} \mathbf{p} \cdot \frac{\nabla T}{T} \right\}.$$

Let us now calculate the Poisson bracket on the left-hand side of the kinetic equation

$$\frac{\partial n}{\partial \mathbf{r}} \cdot \frac{\partial H}{\partial \mathbf{p}} - \frac{\partial n}{\partial \mathbf{p}} \cdot \frac{\partial H}{\partial \mathbf{r}} = \frac{n'}{kT} \left\{ -\frac{\partial \varepsilon}{\partial \mathbf{p}} \cdot \nabla (\mathbf{p} \cdot \mathbf{v}_n) - \varepsilon \frac{\nabla T}{T} \cdot \frac{\partial \varepsilon}{\partial \mathbf{p}} \right\}. \tag{1.5.6}$$

Collecting all the terms in (1.5.5) and (1.5.6), we obtain the kinetic equation in the approximation that interests us

$$\frac{n'}{kT} \left\{ \left(\frac{1}{T} \frac{\partial T}{\partial \rho} \varepsilon - \frac{\partial \varepsilon}{\partial \rho} \right) \operatorname{div}(\mathbf{j} - \rho \mathbf{v}_n) + \left[\frac{1}{T} \left(\frac{\partial T}{\partial \rho} \rho + \frac{\partial T}{\partial S} S \right) \varepsilon \right. \right.$$
$$\left. \left. - \frac{\partial \varepsilon}{\partial \rho} \right] \operatorname{div} \mathbf{v}_n + \frac{\nabla T}{T} \cdot \left(\mathbf{p} \frac{ST}{\rho_n} - \varepsilon \frac{\partial \varepsilon}{\partial \mathbf{p}} \right) - \frac{\partial \varepsilon}{\partial \mathbf{p}} \cdot \nabla (\mathbf{p} \cdot \mathbf{v}_n) \right\} = J(n_1). \tag{1.5.7}$$

*We denote with a prime differentiation of a function with respect to its argument, $n' = -n(1 + n)$.

For what follows it is convenient to separate from the $\partial\varepsilon/\partial\mathbf{p}\cdot\nabla(\mathbf{p}\cdot\mathbf{v}_n)$ term in the curly brackets a term of the form div $\mathbf{v}_n$, and to symmetrize the rest; this yields

$$\frac{n'}{kT}\left\{\left(\frac{1}{T}\frac{\partial T}{\partial\rho}\varepsilon - \frac{\partial\varepsilon}{\partial\rho}\right)\operatorname{div}(\mathbf{j} - \rho\mathbf{v}_n) + \left[\frac{1}{T}\left(\frac{\partial T}{\partial\rho}\rho + \frac{\partial T}{\partial S}S\right)\varepsilon\right.\right.$$

$$\left. - \frac{\partial\varepsilon}{\partial\rho} - \frac{1}{3}\frac{\partial\varepsilon}{\partial\mathbf{p}}\cdot\mathbf{p}\right]\operatorname{div}\mathbf{v}_n + \frac{\nabla T}{T}\cdot\left(\mathbf{p}\frac{ST}{\rho_n} - \varepsilon\frac{\partial\varepsilon}{\partial\mathbf{p}}\right)$$

$$\left. - \left(\frac{\partial\varepsilon}{\partial p_i}p_k - \frac{1}{3}\delta_{ik}\frac{\partial\varepsilon}{\partial\mathbf{p}}\cdot\mathbf{p}\right)\left(\frac{\partial v_{ni}}{\partial r_k} + \frac{\partial v_{nk}}{\partial r_i} - \frac{2}{3}\delta_{ik}\operatorname{div}\mathbf{v}_n\right)\right\} = J(n_1) \quad (1.5.8)$$

An analysis of equation (1.5.8) shows that, in agreement with Section 1.2.1, terms containing div $(\mathbf{j} - \rho\mathbf{v}_n)$ and div $\mathbf{v}_n$ determine the second viscosity in the superfluid (in all, this involves three coefficients); the term with a temperature gradient ∇T determines the heat conduction, and finally the last term determines the first viscosity. We shall perform the calculation of the corresponding kinetic coefficients below, using the general equation (1.5.8).

1.5.2. The Thermal Conductivity

When a temperature gradient is present in superfluid helium, there arises in the liquid a macroscopic motion that can be described by using the equations of motion. The macroscopic motion of the normal fluid is accompanied by a transport of heat in the opposite direction. Moreover, apart from this macroscopic heat flux there is also an irreversible heat flow, which according to the analysis of section 1.2.2 is expressible in terms of the coefficient of thermal conduction κ by the formula

$$\mathbf{q} = -\kappa\nabla T. \quad (1.5.9)$$

In order to calculate the coefficient κ it is necessary to solve the kinetic equation, which in the presence of a temperature gradient may by (1.5.8) be written in the form

$$\frac{n'}{kT}\nabla T\cdot\left(\mathbf{p}\frac{ST}{\rho_n} - \varepsilon\frac{\partial\varepsilon}{\partial\mathbf{p}}\right) = J(n_1). \quad (1.5.10)$$

The phenomenon of thermal conduction we are here considering has much in common with thermal transport in a classical liquid. However, in this case there are certain specific features connected with the fact that the effect is due to excitations having an unusual energy spectrum. Indeed, for a pure phonon gas the left-hand side of equation (1.5.10) is identically zero and consequently the corresponding coefficient of thermal conduction would vanish.

We may perform our calculation by separating the current (1.5.9) into two parts, due to rotons and phonons, respectively. The coefficient of thermal conduction κ consists also of two parts, the roton κ_r and the phonon κ_{ph}:

$$\kappa = \kappa_r + \kappa_{ph}.$$

The roton part of the thermal conduction coefficient. The coefficient κ_r is determined primarily by roton–roton collision processes. For relatively high temperatures, (above 0.9 K), the interaction of phonons with rotons turns out to be negligible. For lower temperatures this interaction begins to play a role; however, as we shall see, in this region of temperatures the main contribution to the thermal conduction comes from the phonons, so that we may merely neglect this roton–phonon effect. We shall only calculate κ_r in order of magnitude in order to determine the temperature dependence of this coefficient. There is no point in attempting a more accurate calculation since the character of the roton–roton interaction is not known (sec 1.3.2). For our purposes it is possible to simplify the problem by replacing the collision integral by the quantity

$$J(n) \to -\frac{n - n_0}{t_r}, \tag{1.5.11}$$

where t_r is some characteristic time which we shall later identify with the collision time of rotons determined by formula (1.3.33). Let us insert (1.5.11) into the right-hand side of (1.5.10) and solve the kinetic equation for $n - n_0$; this yields

$$n - n_0 = -\frac{n'}{kT^2} \nabla T \cdot \left(\mathbf{p}\frac{ST}{\rho_n} - \varepsilon\frac{\partial\varepsilon}{\partial\mathbf{p}}\right) t_r \tag{1.5.12}$$

Note, however, that in this form the solution of the kinetic equation (1.5. 12) does not satisfy the condition $\int \mathbf{p}(n - n_0)\, d\tau_p = 0$. In fact, in the presence of a temperature gradient there exists a macroscopic motion of the liquid with the relative velocity $\mathbf{v}_n - \mathbf{v}_s$ such that $\int \mathbf{p}(n - n_0)\, d\tau_p + \rho_n(\mathbf{v}_n - \mathbf{v}_s) = 0$. Therefore the equilibrium distribution functions of phonons and rotons will depend not on the energy ε, but on the combination $\varepsilon - \mathbf{p}\cdot(\mathbf{v}_n - \mathbf{v}_s)$. Expanding the equilibrium functions in a power series in $\mathbf{v}_n - \mathbf{v}_s$, we obtain the contributions $-(\partial n/\partial\varepsilon)\, \mathbf{p}\cdot(\mathbf{v}_n - \mathbf{v}_s)$, where $\mathbf{v}_n - \mathbf{v}_s$ is found from the former condition. Thus, the total contribution to the phonon distribution function equals

$$\delta n = n - n_0 + \frac{\partial n}{\partial\varepsilon}\frac{1}{\rho_n}\mathbf{p}\cdot\int (n - n_0)\,\mathbf{p}'\, d\tau_{p'} \tag{1.5.13}$$

where the difference $n - n_0$ is defined by formula (1.5.12). In a similar way for the roton distribution function we have

$$\delta n_p = \frac{\partial n_p}{\partial\varepsilon}\frac{1}{\rho_n}\mathbf{p}\cdot\int (n - n_0)\,\mathbf{p}'\, d\tau_{p'}$$

This leads to the general expression for the energy flow due to rotons:

$$\mathbf{Q} = \int \delta n \, \varepsilon \frac{\partial \varepsilon}{\partial \mathbf{p}} \, d\tau_p + \int \delta n_p \, \varepsilon \frac{\partial \varepsilon}{\partial \mathbf{p}} \, d\tau_p. \tag{1.5.14}$$

Comparing (1.5.14) with (1.5.9) we get the expression for the heat conductivity coefficient

$$\kappa_r = - t_r \frac{1}{3kT^2} \int n' \left(\varepsilon \frac{\partial \varepsilon}{\partial \mathbf{p}} - \mathbf{p} \frac{ST}{\rho_n} \right)^2 d\tau_p \tag{1.5.15}$$

If we calculate the integral (1.5.15) with the roton distribution function, we obtain the following result

$$\kappa_r = \frac{\Delta^2 t_r N_r}{3\mu T}. \tag{1.5.16}$$

In calculating the integral (1.5.15), we neglected the terms of the order $\mu kT/p_0^2$.

Thus the roton part of the coefficient of thermal conduction increases slowly with decreasing temperature, as the inverse power of T. The phonon part of the coefficient of thermal conductivity, according to (1.4.29) is equal

$$\kappa_p = c^2 S_p \tau_{pr} \left(1 - \frac{ST}{\rho_n c^2} \right)^2. \tag{1.5.17}$$

Since the number of rotons N_r depends exponentially on the temperature, the time τ_{pr} (1.3.61) increases rapidly with the decreasing of temperature, so that the main temperature dependence of κ_p is $e^{\Delta/T}$. Below 1. 4 K the phonon part of the thermal conductivity κ_p is already considerably greater than the roton part κ_r.

1.5.3. The First Viscosity [39]

The problem of calculating the coefficient of first viscosity is very similar to the one we just discussed, concerning the thermal conductivity. Let us consider the properties of superfluid helium, in which the normal velocity $\mathbf{v}_n$ is a function of the coordinates.

For the calculation of the viscosity the kinetic equation (1.5.8) has the following form

$$- \frac{n'}{kT} \left(\frac{\partial \varepsilon}{\partial p_i} p_k - \frac{1}{3} \delta_{ik} \frac{\partial \varepsilon}{\partial \mathbf{p}} \cdot \mathbf{p} \right) \left(\frac{\partial v_{ni}}{\partial r_k} + \frac{\partial v_{nk}}{\partial r_i} - \frac{2}{3} \delta_{ik} \frac{\partial v_{nl}}{\partial r_l} \right) = J(n) \tag{1.5.18}$$

We must calculate the momentum flux that arises, and for this purpose we divide it into two parts, one referring to rotons and the other to phonons.

As in Section 1.5.2 we may calculate the roton part of the viscosity coefficient η_r by changing $J(n)$ into

$$- \frac{n - n_0}{t_r}. \tag{1.5.19}$$

The time t_r that occurs in this formula is in general not the same as the time t_r used in Section 1.4.4 for the calculation of the roton part of the thermal conductivity. The two times would only coincide if the roton–roton scattering amplitude had a narrow maximum for small angles. However, since we are only interested in the temperature dependence of η_r, the approximation made in (1.5.19), with t_r given by (1.3.33) will be sufficient. Let us insert (1.5.19) into the right-hand side of (1.5.18), calculate $n - n_0$, and then find the momentum flux tensor; this yields

$$\pi_{lm} = \int \frac{\partial \varepsilon}{\partial_{pl}} p_m(n - n_0)\, d\tau_p = -\frac{t_r}{15kT} \int n' \left(\frac{\partial \varepsilon}{\partial p}\right)^2 p^2 d\tau_p$$
$$\times \left(\frac{\partial v_{nl}}{\partial r_m} + \frac{\partial v_{nm}}{\partial r_l} - \frac{2}{3}\delta_{lm} \operatorname{div} \mathbf{v}_n\right). \tag{1.5.20}$$

Performing the necessary integration in (1.5.20), we find for the roton part of the viscosity

$$\eta_r = \frac{p_0^2 t_r N_r}{15\mu} = \frac{\hbar^4 p_0}{30\mu^2 |V_0|^2}. \tag{1.5.21}$$

Expression (1.5.21) contains no terms depending on the temperature, so that η_r turns out to be a constant. The phonon part of the coefficient of first viscosity according to (1.4.27) is equal

$$\eta_p = \frac{1}{5} c^2 \rho_{np} \frac{\tau_{pr}}{1 + \tau_{pr}/\tau_{pp}}. \tag{1.5.22}$$

A numerical analysis of (1.5.22) [using (1.3.61) and (1.3.65)] gives the following result: The phonon part of the viscosity increases exponentially with the decreasing of temperature as $e^{\Delta/T}$; for the temperature below 0.7 K, when the only important effect comes from phonon–phonon scattering, this temperature dependence is changed to T^{-5}. For temperatures below 1.4 K the constant contribution of the roton viscosity is negligibly small in comparison with the phonon part.

When calculating the roton–roton scattering (necessary for the roton parts of the kinetic coefficients), we have chosen a pseudopotential interaction of the form $V_0\delta\,(r_1 - r_2)$; further, the calculations were performed in the Born approximation. In this approximation the scattering matrix is equal to

$$(P_1 P_2 | T | P_1' P_2') = 2V_0\, \delta(\mathbf{P}_1 + \mathbf{P}_2 - \mathbf{P}_1' - \mathbf{P}_2').$$

From this formula the collision frequency can be shown to be

$$t_r^{-1} = 2\mu p_0 V_0^2 N_r.$$

The temperature dependence in t_r is completely defined by the factor N_r.

Finally, this results in the temperature independence of the viscosity coefficient η ($\eta_r \sim \rho_{nr} v^2 t_r \sim (N_r/T)\, T(1/N_r) = \text{const.}$

Recently, the validity of the Born approximation has been discussed for those values of the parameters V_0, μ, and p_0, which are characteristic of this problem [10,41]; indeed, the exact expression for T matrix for our pseudopotential has the following form

$$(P_1P_2\,|T|\,P_1'P_2') = \frac{2V_0\,\delta(\mathbf{P}_1 + \mathbf{P}_2 - \mathbf{P}_1' - \mathbf{P}_2')}{1 + V_0F(\mathbf{P}_1 + \mathbf{P}_2\, E(P_1) + E(P_2))}, \tag{1.5.23}$$

where

$$F(p,E) = \frac{\mu p_0^2}{2\pi p}\left[\ln\frac{1 + (1 - 4E\mu/p^2)^{1/2}}{1 - (1 - 4E\mu/p^2)^{1/2}} + i\pi\right]. \tag{1.5.24}$$

Here

$$\mathbf{p} = \mathbf{P}_1 + \mathbf{P}_2, \quad E = E(P_1) + E(P_2) - 2\Delta.$$

Using the optical theorem, according to which

$$t_r^{-1} = -2\sum \operatorname{Im}(P_1P_2\,|T|\,P_1P_2)\, N, \tag{1.5.25}$$

the lower limit for the collision frequency can be obtained

$$t_{r\min}^{-1} = \frac{32}{3\mu p_0} N_r. \tag{1.5.26}$$

This lower limit does not depend on the value V_0, and gives a value of t_r that is four times greater than necessary to explain the observed viscosity. It should be noted that the pseudopotential used here is short-range and causes s-state scattering only. The effect of a bound state can also be investigated, but the coupling energy happens to be small in comparison with kT and, therefore, the influence of this effect on the roton scattering is negligibly small. This fact indicates that the long-range attraction between rotons, which is very important in the formation of bound states, does not seem to be essential for scattering processes. Thus, the scattering is defined by the short-range behavior, and since the potential is large, the collision frequency does not depend on its magnitude, but does depend on its range l. In reference [40] numerical calculations of $t_{r\min}^{-1}$ for various short-range potentials have been performed, and it was shown that

$$t_{r\min}^{-1} = \frac{\pi\alpha l}{\mu} N_r \tag{1.5.27}$$

where

$$\alpha = (8\pi)^{1/2} \text{ for } V = V_0 \exp\left(\frac{-r^2}{l^2}\right),$$

$$\alpha = 4\pi \text{ for } V = V_0(r^2 + l^2)^{-2},$$

$$\alpha = \tfrac{10}{3} \text{ for } V(r) = V_0\, r < l,\ V(r) = 0,\ r > l.$$

Equation (1.5.27) differs from (1.5.26) by a factor that is a measure of the relative angular momentum of two rotons. It follows from the experimental value of t_r^{-1} that $l = 1$–2 Å, and this seems reasonable enough. The use of equation (1.5.27) for t_r does not change the temperature dependence of roton parts of the kinetic coefficients obtained here.

In so far as the final aim is to calculate the temperature dependence of the kinetic coefficients, it was noted in the original work that the result does not depend on the detailed form of the short range roton pseudopotential employed [39]. This fact is totally confirmed by the analysis given above.

1.6. ABSORPTION AND DISPERSION OF SOUND IN A SUPERFLUID LIQUID NEAR THE λ POINT

1.6.1. The Hydrodynamic Equations of a Superfluid Liquid near the λ Point

The problem of sound absorption near the λ point in superfluid helium was considered in [42] on the basis of the Landau theory for second-order phase transitions [43]. In this theory [43] it is assumed that the thermodynamic potential can be expanded in a power series in the "order" parameter. This theory has not been confirmed experimentally [44]; the heat capacity at the λ point has a singularity in contradiction with this assumption. However, it can be shown that the study of sound propagation near the λ point can be carried out without the use of the Landau thermodynamic theory.

For the study of sound propagation near the λ point, we use the hydrodynamic equations obtained by Pitaevskii [45]. Since the derivation of these equations and their form require some refinement, we shall begin with their derivation.

Closeness to the λ point is characterized by a small parameter—the density of the superfluid part of the liquid, ρ_s. Introduction of a certain complex function $\psi(r, t) = \eta \exp i\varphi$ is convenient; this function is defined so that*

$$\rho_s = m\,|\psi|^2,\ \mathbf{v}_s = \frac{\hbar}{m}\nabla\varphi. \tag{1.6.1}$$

For small values of v_n and v_s, we expand the energy per unit volume E in a series in v_n and $\nabla\psi$:

$$E = (\rho - m\,|\psi|^2)\,\frac{\mathbf{v}_n^2}{2} + \frac{\hbar^2}{2m}\,|\nabla\psi|^2 + E_0\,(\rho, S, |\psi|^2) \tag{1.6.2}$$

*The notation is the same as in [45].

Further, expressing ψ in terms of ρ_s and v_s, we get

$$E = \rho_n \frac{\mathbf{v}_n^2}{2} + \rho_s \frac{\mathbf{v}_s^2}{2} + \frac{\hbar^2}{8m^2} \frac{(\nabla \rho_s)^2}{\rho_s} + E_0; \quad \rho_n = \rho - m\,|\psi|^2. \quad (1.6.3)$$

We transform to a coordinate system moving with the velocity of normal motion v_n, and introduce the momentum of the relative motion of the liquid in this system:

$$\mathbf{p} = \mathbf{j} - \rho \mathbf{v}_n = \frac{i\hbar}{2} (\psi \nabla \bar{\psi} - \bar{\psi} \nabla \psi) - m\,|\psi|^2 \mathbf{v}_n. \quad (1.6.4)$$

The energy E is then, in accord with the Galilean transformation, equal to [46]

$$E = \rho \frac{v_n^2}{2} + \mathbf{p} \cdot \mathbf{v}_n + E_{\mathrm{rel}}, \quad (1.6.5)$$

where E_{rel} is the sum of the internal energy and the energy of relative motion. Comparing (1.6.5) with (1.6.2), we find

$$E_{\mathrm{rel}} = \frac{m}{2} \left| \left(- \frac{i\hbar}{m} \nabla - \mathbf{v}_n \right) \psi \right|^2 + E_0(\rho, S, |\psi|^2). \quad (1.6.6)$$

For finding the equilibrium value of ψ it is necessary to minimize the thermodynamic potential of the system Φ which, in accord with (1.6.6), is equal to

$$\Phi = \frac{m}{2} \left| \left(- \frac{i\hbar}{m} \nabla - \mathbf{v}_n \right) \psi \right|^2 + \Phi_0(T, \rho, |\psi|^2). \quad (1.6.7)$$

The minimum must be sought for fixed values of the thermodynamic variables T, p and the relative velocity $\mathbf{v}_s - \mathbf{v}_n = \mathbf{p}/\rho_s$. We introduce the Lagrangian multiplier $\mathbf{u}$ and vary the integral $\int (\mathbf{p} + \rho_s^{-1} \mathbf{u} \cdot \mathbf{p})\, dV$ successively in ψ and v_n. In this way we find the two conditions

$$\frac{m}{2} \left(- \frac{i\hbar}{m} \nabla - \mathbf{v}_n \right)^2 \psi + \frac{\partial \Phi_0}{\partial \rho_s} m\psi - \frac{i\hbar}{2} \left(\nabla \cdot \frac{\mathbf{u}}{\rho_s} \psi + \frac{\mathbf{u}}{\rho_s} \cdot \nabla \psi \right)$$

$$- \frac{1}{\rho_s} m \mathbf{v}_n \cdot \mathbf{u} \psi - \frac{\mathbf{u} \cdot \mathbf{p}}{\rho_s^2} m\psi = 0, \quad (1.6.8)$$

$$\mathbf{p} + \mathbf{u} = 0. \quad (1.6.9)$$

Eliminating the Lagrangian multiplier u from these two conditions, we obtain an equation defining the equilibrium value of Φ:

$$\left[\frac{1}{2} \left(- \frac{i\hbar}{m} \nabla - \mathbf{v}_n \right)^2 + \mu_s + \frac{i\hbar}{2m\rho_s} \operatorname{div} \mathbf{p} \right] m\psi = 0, \quad (1.6.10)$$

where

$$\mu_s = \left(\frac{\partial \Phi_0}{\partial \rho_s} \right)_{T,p} = \left(\frac{\partial E_0}{\partial \rho_s} \right)_{\rho,S}. \quad (1.6.11)$$

Equation (1.6.10) recalls the fundamental equation of the Ginzburg–Landau theory for superconductors. However, in the theory of superconductivity, the function ψ (its modulus and phase) is completely determined by the given magnetic field, and therefore the minimum of the thermodynamic potential is found for fixed values of the vector potential of the electromagnetic field A.

In the work of Pitaevskii [45] the equation defining the equilibrium value of ψ was found by variation of the total energy E in terms of ψ for fixed values of the momentum $\mathbf{j} = \mathbf{p} + \rho\mathbf{v}_n$. In the equation thus obtained, the term $(i\hbar/2m\rho_s)$ div $\mathbf{p}$ is absent in comparison with (1.6.10). The equation obtained in [45] is equivalent in two conditions: Its real part is identical with *Eq* (1.6.10) and the imaginary part gives the extraneous condition div $\mathbf{p} = 0$, for which there is no foundation. Such a result appeared as the result of an unfounded assumption on the extremum of E relative to ψ for fixed $\mathbf{j}$. The equilibrium value of ρ_s can be found from the condition of minimum energy. Here it is necessary to choose the density ρ, the entropy S, the momentum $\mathbf{j}$ and the velocity of superfluid motion $\mathbf{v}_s$ as independent variables, the values of which ought to be fixed. The fact that such a choice of variables appears natural can be established in the limiting case when the gradients of ρ_s and specific quantum effects are small. Here the equations of hydrodynamics can be obtained from the law of energy conservation. The energy in this case is expressed as a function of ρ, S, $\mathbf{j}$, and $\mathbf{v}_s$, and the laws of conservation are used for these quantities [46].

In the given problem a new independent variable ρ_s is present, the equilibrium value of which can be found from the condition of the minimum of the total energy $\int E\, dV$ relative to ρ_s for fixed values of all the remaining variables. By varying $\int E\, dV$ in ρ_s, we obtain the desired condition

$$-\frac{\hbar^2}{4m^2}\Delta\rho_s + \frac{\hbar^2}{8m^2\rho_s}(\nabla\rho_s)^2 + \rho_s\frac{(\mathbf{v}_s - \mathbf{v}_n)^2}{2} + \rho_s\mu_s = 0.$$

Expressing ρ_s and v_s in these equations as functions of ψ with the help of (1.6.1), we write the resultant condition in the form of an equation for ψ:

$$\left[\frac{1}{2}\left(-\frac{i\hbar}{m}\nabla - \mathbf{v}_n\right)^2 + \mu_s + \frac{i\hbar}{2m\rho_s}\,\mathrm{div}\;\mathbf{p}\right]m\psi = 0.$$

Thus we again obtain condition (1.6.10). The given derivation indicates that the equilibrium value of the modulus of the function ψ (the density ρ_0) corresponds to the minimum of the energy for the given phase of φ (the velocity $\mathbf{v}_s$).

In the nonstationary case, it is assumed that the state of the system is defined by ψ (the same as for the other thermodynamic variables), i.e., ψ satisfies a linear differential equation. By analogy with quantum mechanics, the equation for ψ is written in the form

$$i\hbar \frac{\partial \psi}{\partial t} = \hat{L}\psi, \tag{1.6.12}$$

where $\hat{L}$ is some linear operator. Since the value of ρ_s can relax, the operator $\hat{L}$ contains a non-Hermitian part. The Hermitian part of the operator $\hat{L}$, by analogy with the Schrödinger equation, is written in the form

$$-\frac{\hbar^2}{2m}\Delta + U, \tag{1.6.13}$$

where

$$U = \left(\frac{\partial E_0}{\partial |\psi|^2}\right)_{\rho_n, S} = (\mu + \mu_s)m. \tag{1.6.14}$$

Here μ and μ_s are defined by the thermodynamic identity for E_0:

$$dE_0 = TdS + \mu d\rho + \mu_s d\rho_s. \tag{1.6.15}$$

U represents the potential energy of the superfluid part of the liquid.

The anti-Hermitian part describes the approach of ρ_s to its equilibrium value; for small departures from equilibrium we can write

$$i\Lambda\left[\frac{1}{2}\left(-\frac{i\hbar}{m}\nabla - \mathbf{v}_n\right)^2 + \mu_s + \frac{i\hbar}{2m\rho_s}\operatorname{div}\mathbf{p}\right]m\psi, \tag{1.6.16}$$

where Λ is some dimensionless kinetic coefficient. Finally, we have the following equation for ψ :

$$i\hbar\frac{\partial\psi}{\partial t} = -\frac{\hbar^2}{2m}\Delta\psi + (\mu + \mu_s)m\psi - i\Lambda\left[\frac{1}{2}\left(-\frac{i\hbar}{m}\nabla - \mathbf{v}_n\right)^2 + \mu_s + \frac{i\hbar}{2m\rho_s}\operatorname{div}\mathbf{p}\right]m\psi. \tag{1.6.17}$$

The coefficient Λ must be real, since otherwise a transport of the superfluid part of the liquid with the normal velocity would occur. The remaining equations of hydrodynamics are written as usual in the form of conservation laws:

mass:

$$\frac{\partial \rho}{\partial t} + \operatorname{div}\mathbf{j} = 0; \tag{1.6.18}$$

momentum:

$$\frac{\partial j_i}{\partial t} + \frac{\partial \Pi_{ik}}{\partial x_k} = 0, \tag{1.6.19}$$

$$\Pi_{ik} = \frac{\hbar^2}{2m}\left(\frac{\partial\psi}{\partial x_i}\frac{\partial\bar{\psi}}{\partial x_k} - \psi\frac{\partial^2\bar{\psi}}{\partial x_i \partial x_k} + c.c.\right) + \rho_n v_{ni} v_{nk} + p\delta_{ik},$$

$$p = -E_0 + TS + \mu\rho + \mu_s\rho_s; \qquad (1.6.20)$$

and, finally, the law of entropy growth

$$\frac{\partial S}{\partial t} + \operatorname{div} S\mathbf{v}_n = \frac{R}{T}, \qquad (1.6.21)$$

in which the dissipative function is found from the energy conservation law in the form

$$R = \frac{2\Lambda}{\hbar}\left|\left[\frac{1}{2}\left(-\frac{i\hbar}{m}\nabla - \mathbf{v}_n\right)^2 + \mu_s + \frac{i\hbar}{2m\rho_s}\operatorname{div}\mathbf{p}\right]m\psi\right|^2. \qquad (1.6.22)$$

In equations (1.6.19) and (1.6.21), the dissipative terms that are due to viscosity and thermal conductivity are omitted. These terms have the usual form [46]. As will be seen from the following, they are not the important dissipative processes for sound propagation in superfluid helium near the λ point.

For the case of small gradients we replace $-i\hbar m^{-1}\nabla\psi$ everywhere by $\dot{\mathbf{v}}_s$. As a result, equations (1.6.17) and (1.6.21) take the following simple form:

$$\dot{\mathbf{v}}_s + \nabla\left(\frac{\mathbf{v}_s^2}{2} + \mu + \mu_s\right) = 0, \qquad (1.6.23)$$

$$\dot{\rho} + \operatorname{div}(\rho_s\mathbf{v}_s + \rho_n\mathbf{v}_n) = 0, \qquad (1.6.24)$$

$$\frac{\partial}{\partial t}(\rho_s v_{si} + \rho_n v_{ni}) + \frac{\partial}{\partial x_k}(\rho_n v_{ni} v_{nk} + \rho_s v_{si} v_{sk} + p\delta_{ik}) = 0, \qquad (1.6.25)$$

$$\dot{S} + \operatorname{div} S\mathbf{v}_n = \frac{2\Lambda m}{\hbar}\left[\mu_s + \frac{(\mathbf{v}_n - \mathbf{v}_s)^2}{2}\right]^2\rho_s, \qquad (1.6.26)$$

$$\dot{\rho}_s + \operatorname{div}(\rho_s\mathbf{v}_s) = -\frac{2\Lambda m}{\hbar}\left[\mu_s + \frac{(\mathbf{v}_n - \mathbf{v}_s)^2}{2}\right]\rho_s. \qquad (1.6.27)$$

We will apply these equations to the study of sound propagation. In equations (1.6.23) and (1.6.26) dissipative terms (e.g. second viscosity) containing the coefficient Λ, are absent; this differs from the result obtained in reference [45].

1.6.2. The Relaxation Mechanism near the λ Point

The relaxation of the superfluid component of ρ_s is described by the right-hand side of equation (1.6.27). Here the rate of approach to equilibrium is determined by the kinetic coefficient Λ. Let us consider the mechanism for the dissipation of sound energy near the λ point. The system is characterized by a certain correlation length ξ. For distances of the order of ξ the correlation of the phase φ is damped out. According to equation (1.6.23) this length will also be characteristic of the temperature correlations. It is, therefore, evident that second sound can be propagated only in the case for which $k\xi \ll 1$, where k is the wave vector of second sound, i.e., when the wavelength of second sound is greater than the correlation length ξ. Waves of second sound with wavelengths of the order of ξ will be completely dissipated. Thus there is some characteristic time $\tau = \xi/u_2$ (u_2 is the speed of second sound) which determines the rate of energy dissipation of second sound [47, 48]. The energy dissipation in first sound then takes place by decay into waves of second sound and is characterized by the same time.

By giving such a dissipative mechanism, we easily find the temperature dependence of the dimensionless coefficient Λ. It is evident that Λ can depend only on u_2 and ξ, while the relaxation rate should be proportional to the speed of second sound. Using the dimensional constants $\hbar$ and m that are at our disposal, we find

$$\Lambda \approx \frac{mu_2\xi}{\hbar}. \tag{1.6.28}$$

Such a determination of the value of Λ is equivalent to the introduction of the characteristic time $\tau = \xi/u_2$. Actually, it follows from (1.6.19) that the characteristic relaxation time τ is determined by the relation

$$\frac{1}{\tau} = \frac{2\Lambda m}{\hbar} \frac{\partial \mu_s}{\partial \rho_s} \rho_s. \tag{1.6.29}$$

The basic part of the potential μ_s, by definition of ξ, is of the order of $\hbar^2/m^2\xi^2$. It therefore follows from (1.6.28) and (1.6.29) that

$$\frac{1}{\tau} \cong \frac{u_2\xi m^2}{\hbar^2} \frac{\hbar^2}{m^2\xi^2} \sim \frac{u_2}{\xi}. \tag{1.6.30}$$

It has been shown in [49] that a relaxation time defined in this way has a temperature dependence which does not depend on the critical indices of scaling theory [50], i.e., independent of the form of the singularity in the thermodynamic functions near the λ point. Actually, from the expression for u_2 [46],

$$u_2 = \sqrt{\frac{\rho_s}{\rho_n} \frac{\sigma^2 T}{C}}$$

and the obvious relations for the principal part of the heat capacity C

$$C \sim \frac{T_c}{\rho} \frac{\partial^2 \Phi}{\partial T^2} \sim \frac{\rho_s}{\rho} \frac{\hbar^2}{m^2\xi^2} \frac{1}{T_c\epsilon^2}; \epsilon = \frac{T_c - T}{T_c}$$

we get

$$\frac{1}{\tau} \sim \frac{m\sigma T_c}{\hbar} \epsilon \sim \epsilon. \tag{1.6.31}$$

Thus the relaxation time τ is inversely proportional to the difference $T_c - T$. It is curious that equation (1.6.31) gives the correct order of magnitude of the time τ, in excellent agreement with experiment ($1/\tau \sim 10^{11}\epsilon$).

We note that the ordinary and thermal conductivities, which are due to collisions of the excitations, will be characterized by very small times of an entirely different order [46], and therefore their contibution to the energy dissipation of the sound will be insignificant.

1.6.3. Sound Dispersion

For the study of sound propagation, we transform the set of equations (1.6.23)–(1.6.26) to linearized form, after which we eliminate the velocities $\mathbf{v}_n$ and $\mathbf{v}_s$. As a result, we obtain two wave equations:

$$\ddot{\rho} - \Delta p = 0, \tag{1.6.32}$$

$$\ddot{\sigma} - \frac{\rho_s}{\rho}\sigma^2 \Delta T + \frac{\rho_s}{\rho}\sigma \Delta \mu_s = 0, \tag{1.6.33}$$

in which, in view of the smallness of ρ_s, we have set $\rho_n = \rho$ everywhere. Equation (1.6.27), which describes the relaxation of ρ_s, takes the following form:

$$\dot{\rho}_s + \frac{\rho}{\sigma}\dot{\sigma} = -\frac{2\Lambda m}{\hbar}\mu_s' \rho_s. \tag{1.6.34}$$

where μ_s' is the variable part of the potential μ_s, since the condition of equilibrium in the linear approximation changes to the requirement $\mu_s = 0$.

The thermodynamic identity for the potential $w = \mu + T\sigma$, according to (1.6.15), is written in the form

$$dw = T\,d\sigma + \frac{1}{\rho}\,dp - \frac{\rho_s}{\rho}\,d\mu_s.$$

We use this identity for establishing the connection between the derivatives of the thermodynamic functions. If we choose p and the entropy per unit mass σ as independent variables, then equations (1.6.32) and (1.6.33) are completely uncoupled, thanks to the extraordinary smallness of the difference $C_p - C_v$ and the ratio of the squares of the velocities u_2^2/u_1^2. We emphasize that this does not mean the neglect of the difference $C_p - C_v$ in the final formulas. Only terms of relative order $(C_p/C_v - 1)u_2^2/u_1^2$ are thrown away. With this approximation, we find an expression from equation (1.6.32) for the square of the speed of first sound

$$\frac{1}{u_1^2} = \frac{k^2}{\omega^2} = \left(\frac{\partial \rho}{\partial p}\right)_\sigma, \tag{1.6.35}$$

and an expression for the square of the speed of second sound from (1.6.33)

$$u_2^2 = \frac{\rho_s}{\rho}\sigma^2\left[\left(\frac{\partial T}{\partial \sigma}\right)_p - \frac{1}{\sigma}\left(\frac{\partial \mu_s}{\partial \sigma}\right)_p\right]. \tag{1.6.36}$$

We now make use of equation (1.6.34). From equation (1.6.35) we find the expression for the dispersion of the speed of first sound

$$\frac{1}{u_1^2} = \left(\frac{\partial \rho}{\partial p}\right)_{\sigma, \mu_s} + \left(\frac{\partial \rho}{\partial \mu_s}\right)_{\sigma, p}\frac{\partial \mu_s}{\partial p}$$

$$= \frac{1}{u_2^{10}} - \left(-\frac{1}{u_{10}^2} + \frac{1}{u_{1\infty}^2}\right)\frac{i\omega\tau}{1 - i\omega\tau}, \tag{1.6.37}$$

where u_{10} is the equilibrium speed of sound as $\omega \to 0$ which is given by

$$\frac{1}{u_{10}^2} = \left(\frac{\partial \rho}{\partial p}\right)_{\sigma, \mu_s} = \frac{C_p}{C_v}\left(\frac{\partial \rho}{\partial p}\right)_{T, \mu_s}. \tag{1.6.38}$$

$u_{1\infty}$ is the speed of sound in the limit $\omega\tau \gg 1$; in this limit the equilibrium value of ρ_s lags behind the sound wave. The difference $u_{1\infty} - u_{10}$ is found from the relation

$$\frac{1}{u_{10}^2} - \frac{1}{u_{1\infty}^2} = \left(\frac{\partial \rho_s}{\partial p}\right)_\sigma \frac{\partial \rho}{\partial \mu_s} = \rho^2 \frac{\partial \rho_s}{\partial p}\frac{\partial}{\partial p}\left(\frac{\rho_s}{\rho}\right)\left(\frac{\partial \mu_s}{\partial \rho_s}\right)_{p, \sigma}. \tag{1.6.39}$$

Here we have used the identity (1.6.35), according to which

$$\frac{\partial \rho}{\partial \mu_s} = \rho^2 \frac{\partial}{\partial p}\left(\frac{\rho_s}{\rho}\right). \tag{1.6.40}$$

The time τ is determined by the relation

$$\frac{1}{\tau} = \frac{2\Lambda m}{\hbar}\left(\frac{\partial \mu_s}{\partial \rho_s}\right)_{p, \sigma}\rho_s. \tag{1.6.41}$$

In the variables p, T, equation (1.6.39) takes the form

$$\frac{1}{u_{10}^2} - \frac{1}{u_{1\infty}^2} = \left[\frac{\partial}{\partial T}\left(\frac{\rho_s}{\rho}\right)\frac{\partial \rho}{\partial T}\middle| \frac{\partial \sigma}{\partial T} - \rho\frac{\partial}{\partial p}\left(\frac{\rho_s}{\rho}\right)\right]^2 \frac{1}{\rho}\left(\frac{\partial \mu_s}{\partial \rho_s}\right)_{p, \sigma}. \tag{1.6.42}$$

The singular part of the derivative $(\partial/\partial p)$ (ρ_s/ρ) is evidently equal to $(\partial/\partial T)$ (ρ_s/ρ) $(\partial T_c/\partial p)$, i.e., it is expressed in terms of the derivative along the curve T_λ (p). Due to this fact, the singular parts of both components in the square bracket in (1.6.42) are reduced and the bracket is of the order of ϵ^α (ρ_s/ϵ), where α is an index characterizing the singularity in the heat capacity $C \sim \epsilon^{-\alpha}$. The derivative $\partial\mu_s/\partial\rho_s$ is of order $\Phi_0/\rho_s^2 \sim \epsilon^{2-\alpha}/\rho_s^2$. Thus, the difference $u_{10} - u_{1\infty}$ is of the order of ϵ^α, i.e., it tends to zero.

Equation (1.6.37) describes the dispersion of first sound, i.e., it is valid in the region $\omega\tau \gtrsim 1$. Actually, in this region, where $\omega\tau \sim ku_1\xi/u_2 \sim 1$, we have $k\xi \ll 1$ since $u_2/u_1 \ll 1$. Thus, in the region where $\omega\tau \gtrsim 1$, the wavelength of the sound is still greater than the correlation length.

We now return to second sound. In this case, the condition $\omega\tau \sim 1$ is identical with the condition $k\xi \sim 1$; therefore, it is legitimate to consider only the case $\omega\tau \ll 1$, i.e., the case of small damping. From equations (1.6.36) and (1.6.34), we obtain in this limit

$$u_2^2 = u_{20}^2\left[1 - i\omega\tau\left(\frac{\partial\mu_s}{\partial\rho_s}\right)\left(\frac{\partial\rho_s}{\partial\sigma} + \frac{\rho}{\sigma}\right)\left(\frac{\partial}{\partial\sigma}\left(\frac{\rho_s}{\rho}\right) + \frac{1}{\sigma}\right)\right]. \tag{1.6.43}$$

It follows from scaling theory that the function ρ_s changes with temperature according to the law $1/\xi \sim \epsilon^{(2-\alpha)/3}$; therefore, $\partial\rho_s/\partial\sigma \gg \rho/\sigma$.*

We introduce the heat capacity $C_{\rho_s} = T(\partial\sigma/\partial T)_{\rho_s}$, which is characteristic of fast processes for which the density ρ_s does not follow the temperature. The connection between C_{ρ_s} and the equilibrium value of the heat capacity $C_0 = T(\partial\sigma/\partial T)_{\mu_s}$ follows from the thermodynamic identity (1.6.35):

$$\frac{C_0}{C_{\rho_s}} - 1 = \left(\frac{\partial\rho_s}{\partial\sigma}\right)\frac{\partial}{\partial\sigma}\left(\frac{\rho_s}{\rho}\right)\left(\frac{\partial\sigma}{\partial T}\right)_{\mu_s}\left(\frac{\partial\mu_s}{\partial\rho_s}\right)_\sigma \tag{1.6.44}$$

With the help of (1.6.44), we can rewrite equation (1.6.33) in the following simple form:

$$u_2^2 = u_{20}^2\left[1 - i\omega\tau\left(\frac{C_0}{C_{\rho_s}} - 1\right)\right]. \tag{1.6.45}$$

Equations (1.6.37) and (1.6.45) allow us to compute the damping coefficients of first and second sound, respectively. We have

$$\alpha_1 = \operatorname{Im}\frac{\omega}{u_1} = \frac{\omega^2\tau}{1 + \omega^2\tau^2}\frac{1}{2u_{10}}\left(1 - \frac{u_{10}^2}{u_{1\infty}^2}\right), \tag{1.6.46}$$

and

$$\alpha_2 = \operatorname{Im}\frac{\omega}{u_2} = \omega^2\tau\frac{1}{2u_{20}}\left(\frac{C_0}{C_{\rho_s}} - 1\right). \tag{1.6.47}$$

Since the difference $u_{1\infty} - u_{10} \sim \epsilon^\alpha$ and $\tau \sim 1/\epsilon$, we have $\alpha_1 \sim \epsilon^{-1+\alpha}$ which agrees well with the temperature dependence of α_1 obsreved in the experiments of Barmatz and Rudnick [51].†

Experiments on the damping of second sound below the λ point, carried

*Only in the limiting case of the thermodynamic theory of Landau is $\rho_s \sim \epsilon$ and the terms given have the same order of magnitude.

†The experiments of Buckingham and Fairbank [44] indicate that the singularity in the heat capacity is logarithmic or a power law with a very small exponent α, so that one can set $\alpha = 0$ in comparison with experiment.

out by Tyson, [47] confirm the temperature dependence of α_2 which follows from (1.6.47). The difference $C_0/C_{\rho_s} - 1$ does not depend on the temperature for small ϵ. Actually, according to (1.6.44), we have

$$\frac{C_0}{C_{\rho s}} - 1 \sim \frac{1}{C_0}\left(\frac{\partial \rho_s}{\partial T}\right)^2 \frac{\partial \mu_s}{\partial \rho_s} \sim \frac{1}{\epsilon^{-\alpha}} \frac{\rho_s^2}{\epsilon^2} \frac{\epsilon^{2-\alpha}}{\rho_s^2} \sim \text{const.}$$

In [47] the damping of second sound is characterized by the damping constant, which is determined in the following way:

$$\mathscr{D}_2 = \text{Im}\,\frac{\omega}{k^2}$$

and, in accord with (1.6.47), is equal to

$$\mathscr{D}_2 = u_{20}^2\tau\left(\frac{C_0}{C_{\rho s}} - 1\right). \tag{1.6.48}$$

The temperature dependence follows from (1.6.48):

$$\mathscr{D}_2 \sim \frac{\rho_s}{C_0}\frac{1}{\epsilon} \sim \frac{\epsilon^{(2-\alpha)/3}}{\epsilon^{1-\alpha}} \sim \epsilon^{(-1+2\alpha)/3}$$

and agrees well with the experimental data [47] and with the predictions of the theory of dynamic scaling. We note that for damping of first sound the theory of dynamic scaling is generally not applicable. The dependence $\alpha_1 \sim \epsilon^{-1+\alpha}$ cannot be obtained from considerations of scaling only.

The problem of the dispersion of first sound is entirely analogous to the problem of the dispersions of sound in the presence of slow processes of approach to the state of equilibrium in ordinary hydrodynamics. Equation (1.6.37) has the typical form that the theory of Mandel'shtam–Leontovich gives in ordinary hydrodynamics. [52]. So far as second sound is concerned, here, due to the presence of third term in equation (1.6.34) containing μ_s explicitly, the situation differs somewhat from the ordinary. It is true that this term is small in the case of superfluid helium; however, in the limiting case of the Landau theory of phase transitions, it has the same order of magnitude as the first two terms in (1.6.33).

The system of hydrodynamic equations for a superfluid liquid near the λ point contains an extraneous equation for the function ρ_s in comparison with the usual equations of two-component hydrodynamics. It might appear that this should lead to the possibility of propagation of a new type of oscillation. However, equation (1.6.27) has the form of the equation of continuity and, eliminating div $\mathbf{v}_s$ from it and equations (1.6.24) and (1.6.26), we obtain only the connection between changes of the thermodynamic quantities and ρ_s characteristic of sound waves. A new wave equation does not appear, and therefore there are no new types of oscillation.

REFERENCES

1. R. Ya. Kapitza, *Nature* **141,** 74 (1937); *JETP* **11,** 1(1941); *JEPT* **11,** 581 (1941).
2. L. D. Landau, *JETP* **11,** 592 (1941).
3. A. D. B. Woods and R. A. Cowley, *Can. J. Phys.* **49,** 177(1971); *Phys. Rev. Lett.* **24,** 646 (1970).
4. I. Nagata, T. Soda, and K. Sawada, *Progr. Theoret. Phys.* **38,** 1023 (1967); **44,** 574 (1970).
5. T. Soda, K. Sawada, and T. Nagata, *Progr. Theoret. Phys.* **44,** 860, (1970).
6. R. P. Feynman, *Phys. Rev.* **94,** 262 (1954).
7. D. Pines and C. W. Woo, *Phys. Rev. Lett.* **24,** 1044 (1970).
8. N. E. Phillips, C. G. Waterfield, and J. K. Hoffer, *Phys. Rev. Lett.* **25,** 1260 (1970).
9. J. Jäckle and K. Kehr, *Phys. Rev. Lett.* **27,** 654 (1971).
10. P. R. Roach, J. B. Ketterson, and M. Kuchnir, *Phys. Rev.* A, 2205 (1972).
11. O. W. Dietrich, E. H. Graf, C. H. Huang, and L. Passell (preprint).
12. I. M. Khalatnikov, *The Theory of Superfluidity* (W. A. Benjamin, 1965).
13. I. M. Khalatnikov and D. M. Chernikova, *JETP* **49,** 1957 (1965).
14. L. P. Pitaevsky, *JETP* **36,** 1168 (1959).
15. T. J. Greytak, R. Woerner, J. Yan, and R. Benjamin, *Phys. Rev. Lett.* **25,** 1547 (1970).
16. L. P. Pitaevsky, *JETP Lett.* **12,** 118 (1970).
17. J. Ruvalds and A. Zaawadowsky, *Phys. Rev. Lett.* **25,** 333 (1970).
18. I. M. Khalatnikov *JEPT* **23,** 8 (1952).
19. L. D. Landau and I. M. Khalatnikov, *JETP* **19,** 637 (1949).
20. I. M. Khalatnikov, *JETP* **20,** 243 (1950).
21. S. G. Eckstein, unpublished
22. D. Henshaw and A. Woods, *Phys. Rev.* **121,** 266 (1961).
23. K. R. Atkins and M. H. Edwards, *Phys. Rev.* **97,** 1429 (1955).
24. J. de Boer, in *Phonons and Phonon Interactions*, edited by T. A. Bak (New York, W. A. Benjamin, 1964), p. 395.
25. L. D. Landau and I. M. Khalatnikov, *JETP* **19,** 709 (1949).
26. A. de Troyer, A. Van Itterbeek, and A. Van Denberg, *Physica* **17,** 50 (1951).
27. A. C. Hollis-Hallet, *Proc. Roy. Soc. (London)* **210,** 404 (1952).
28. W. J. Heikkila and A. C. Hollis-Hallet, *Can. J. Phys.* **33,** 420 (1955).
29. K. N. Zinov'eva, *JETP* **31,** 31 (1956).
30. I. M. Khalatnikov, *JETP* **23,** 21 (1952).
31. I. M. Khalatnikov and D. M. Chernikova, *JETP* **49,** 1957 (1965); *JETP* **50,** 411 (1965).
32. H. A. Notarys and J. R. Pellam, in *Proceedings of the Ninth International Conference on Low Temperature Physics* (New York, Plenum Press, 1964).
33. A. F. Andreev and I. M. Khalatnikov, *JETP* **44,** 2058 (1963); (*Sov. Phys. —JETP* **17,** 1384 (1963).
34. A. F. Andreev and I. M. Khalatnikov, *J. Low Temp. Phys.* **2,** 173 (1970).
35. W. H. Whitney and C. E. Chase, *Phys. Rev. Lett.* **9,** 243 (1962), *Phys. Rev.* **158,** 200 (1967).
36. B. M. Abraham, Y. Eckstein, J. B. Ketterson, M. Kuchnir, and P. R. Roach, *Phys. Rev. A* **1,** 250 (1970); **2,** 550 (1970).
37. I. M. Khalatnikov, *JETP* **23,** 8 (1952).
38. I. M. Khalatnikov, *JETP* **23,** 21 (1952).
39. L. D. Landau and I. M. Khalatnikov, *JETP* **19,** 637, 709 (1949).
40. J. Yan and M. Stephen, *Phys. Rev. Lett.* **27,** 482 (1971).
41. F. Iwamoto, *Progr. Theoret. Phys.* **44,** 1121 (1970).

42. L. D. Landau and I. M. Khalatnikov, *Dokl. Akad. Nauk SSSR* **96,** 469 (1954).
43. L. D. Landau, *Zh. Eksp. Teor. Fiz.* **7,** 19 (1937).
44. M. I. Buckingham and W. M. Fairbank, *Progress in Low Temperature Physics* (North-Holland Publishing Co., Amsterdam, 1961).
45. L. P. Pitaevskii, *Zh. Edsp. Teor. Fiz.* **35,** 408 (1958) [English transl.: *Sov. Phys.—JETP* **8,** 282 (1959)].
46. I. M. Khalatnikov, *Introduction to the Theory of Superfluidity* (Benjamin, New York, 1965).
47. J. Tyson, *Phys. Rev. Lett.* **21,** 1235 (1968); J. Tyson and D. Douglass, Jr., *Phys. Rev. Lett.* **21,** 1308 (1968).
48. J. Swift and L. Kadanoff (preprint, 1968).
49. V. L. Pokrovskii and I. M. Khalatnikov, *ZhETF Pis. Red.* **9,** 255 (1969) [English transl.: *JETP Lett.* **9,** 149 (1969)].
50. B. I. Halperin and P. C. Hohenberg, *Phys. Rev. Lett.* **19,** 700 (1967).
51. M. Barmatz and I. Rudnick, *Phys. Rev.* **170,** 224 (1968).
52. L. I. Mandel'shtam and M. A. Leontovich, *Zh. Eksp. Teor. Fiz.* **7,** 438 (1937).

CHAPTER 2

Experiments near the Superfluid Transition in ^4He and ^3He-^4He Mixtures

Guenter Ahlers

Bell Laboratories, Murray Hill, New Jersey

At the superfluid transition many of the properties of liquid helium have singularities. The primary purpose of this chapter is to survey the existing experimental information about these singularities, and to compare the results of diverse experiments with each other by means of the presumably exact relations that are provided by thermodynamics and two-fluid hydrodynamics. Although much of the experimental information can be tested for consistency by means of these relations, the detailed behavior of any one property is, of course, not given. We shall, therefore, also compare the singularities which are encountered with the more specific predictions based on theories of critical phenomena [1]. These theories are not necessarily exact, and the experiments near the superfluid transition provide a welcome opportunity to put them to a test.

We will draw freely upon the predictions of theory when they are needed, but in general we will not discuss theoretical results pertinent to the superfluid transition for their own sake. Nor will we attempt to be complete in the coverage of all experimental work; in particular, the older results that have been summarized elsewhere [2–4] may in part be neglected. Further, the coverage of this chapter will rather arbitrarily and because of space limitations be restricted primarily to results that fall within the range of thermodynamics and linear hydrodynamics. However, the behavior of the characteristic modes in the nonhydrodynamic regime very near the transition will also be discussed to some extent. Regretfully, a number of interesting problems remain outside the scope of this summary. Perhaps foremost amongst the ommitted material are the very beautiful experiments by Reppy and his collaborators on the decay of persistent superfluid currents [5], and on intrinsic superfluid critical velocities near T_λ [6, 7]. Fortunately, much of that work has recently been reviewed [8]. Further omissions include studies pertaining to the so-called mutual friction and other dissipative processes associated with superfluid flow at large velocities, and experiments on the depression of the transition temperature by a velocity field. [9–24]

2.1. THERMODYNAMIC PROPERTIES

2.1.1. General Relations

For the proper use and evaluation of most thermodynamic measurements near lines of critical singularities, certain thermodynamic relations are essential. Before proceeding to a systematic discussion of experimental results we shall, therefore, review the thermodynamics of λ lines. Much of the subject matter in this section is sufficiently general to be applicable to systems other than the superfluid transition.

Although it is possible to perform a large number of different experimental measurements of various derivatives of the free energy, many of the thermodynamic response functions are related to each other, and very different measurements often yield, in principle, the same information about the transition. The thermodynamic relations that exist between the response functions were first discussed by Rice [25] and Pippard [26, 27], and were later developed in more detail by Buckingham and Fairbank [28]. We shall refer to these relations as Pippard–Buckingham–Fairbank (PBF) relations. They make no assumptions or predictions about the functional form of the singularities in a particular response function, but they often make it possible to compare measurements of pairs of thermodynamic derivatives and to establish whether independent experimental results are consistent with each other. Alternatively, PBF relations may be used to calculate a particular thermodynamic derivative from other more readily available experimental information. They also help to clarify questions of a more fundamental nature. For example, it has been difficult to establish from experiment whether the isothermal compressibility κ_T of ^{4}He diverges at T_λ [29–30]. This problem arises from the relatively large regular contribution to κ_T which tends to obscure the singularity in experimental measurements. Nonetheless, there need not be any uncertainty about the functional form of κ_T sufficiently near the transition if that of the isobaric thermal expansion coefficient α_P or the specific heat at constant pressure C_P is known, for it can be shown that κ_T is an asymptotically linear function of α_P or C_P.

We shall examine in detail general thermodynamic relations involving the temperature T, and the volume V with its conjugate "field" $(-P)$. The choice V and $(-P)$ was made only because this pair is most pertinent to the experimental work on pure ^{4}He. We could equally well consider a different pair of conjugate variables (provided they are not the order parameter and its conjugate field), and develop relations equivalent to those given in this section. Thus, in dealing with an antiferromagnetic transition, one might want to consider the magnetization M and the magnetic field H, and make the substitution $V \rightarrow M$ and $(-P) \rightarrow H$ in the equations that follow. In the case of mixtures of ^{3}He and ^{4}He to be discussed later, it will be convenient to consider the molar concentration X of ^{3}He, and its conjugate Φ which is a chemical potential.

The most commonly encountered response functions are the specific heats C_P and C_V at constant pressure and volume, the isothermal and adiabatic compressibilities κ_T and κ_S, the isobaric thermal expansion coefficient α_P, and the pressure coefficient β. They are defined by

$$C_P = T\left(\frac{\partial S}{\partial T}\right)_P, \tag{2.1.1}$$

$$C_V = T\left(\frac{\partial S}{\partial T}\right)_V, \tag{2.1.2}$$

$$\kappa_T = -V^{-1}\left(\frac{\partial V}{\partial P}\right)_T, \tag{2.1.3}$$

$$\kappa_S = -V^{-1}\left(\frac{\partial V}{\partial P}\right)_S, \tag{2.1.4}$$

$$\alpha_P = V^{-1}\left(\frac{\partial V}{\partial T}\right)_P, \tag{2.1.5}$$

$$\beta = \left(\frac{\partial P}{\partial T}\right)_V. \tag{2.1.6}$$

Here V and S are the volume and entropy per unit mass. In order to find relations between the singularities in different response functions, we examine a particular example [31] in detail, and consider the entropy as a function of P and T. The total differential

$$dS = \left(\frac{\partial S}{\partial P}\right)_T dP + \left(\frac{\partial S}{\partial T}\right)_P dT, \tag{2.1.7}$$

with the aid of the Maxwell relation $(\partial S/\partial P)_T = -(\partial V/\partial T)_P$ and equations (2.1.1.) and (2.1.5), yields

$$dS = -V\alpha_P\, dP + T^{-1}C_P\, dT. \tag{2.1.8}$$

From this, the temperature derivative of S along an arbitrary thermodynamic path

$$\left(\frac{\partial S}{\partial T}\right)_t = -V\left(\frac{\partial P}{\partial T}\right)_t \alpha_P + T^{-1}C_P \tag{2.1.9}$$

is obtained. After rearranging, one has

$$C_P = T\left(\frac{\partial S}{\partial T}\right)_t + VT\left(\frac{\partial P}{\partial T}\right)_t \alpha_P. \tag{2.1.10}$$

Equation (2.1.10) is the PBF relation between the specific heat at constant pressure and the isobaric expansion coefficient. Similarly, by starting with the total differential of $V(P,T)$, one can obtain

$$\alpha_P = V^{-1}\left(\frac{\partial V}{\partial T}\right)_t + \left(\frac{\partial P}{\partial T}\right)_t \kappa_T\,. \tag{2.1.11}$$

From $S(V,T)$, one has

$$\frac{C_V}{T} = \left(\frac{\partial S}{\partial T}\right)_t - \left(\frac{\partial V}{\partial T}\right)_t\left(\frac{\partial P}{\partial T}\right)_V\,. \tag{2.1.12}$$

We can combine some of these results and obtain, for instance,

$$\kappa_T = \left[VT\left(\frac{\partial P}{\partial T}\right)_t^2\right]^{-1} C_P - \frac{[(\partial S/\partial T)_t/(\partial P/\partial T)_t^2 + (\partial V/\partial T)_t/(\partial P/\partial T)_t]}{V}, \tag{2.1.13}$$

for κ_T and C_P.

It remains to choose the path of differentiation t for the coefficients in equations (2.1.10)–(2.1.13) in a manner that is particularly convenient. We want to choose t in such a way that any singularities at T_λ in the coefficients are as mild as possible because we are interested in a relation between the singularities in the functions given by equations (2.1.1)–(2.1.6). Therefore, it will be useful to choose the path of differentiation (along which t is constant) in some sense "parallel" to the λ line. The value of t then becomes a measure of the "distance" from T_λ. More specifically, if we approach the λ line along a particular path (an isobar, isochore, or isotherm, for instance), we will choose t equal to a constant in the sense that it should have the same value on neighboring paths of the same type. Figure 2.1 is a schematic representation of a typical transition line in the P–T plane with a negative temperature derivative of the transition pressure. The three most common paths of approach towards T_λ are shown explicitly. A measurement at the point (P_1, V_1,T_1) in the figure would be a "distance"

$$t_P(T_1) = T_1 - T_\lambda(P_1) \tag{2.1.14}$$

from T_λ if we follow an isobar. Along an isochore, it may be more convenient to use

$$t_V(T_1) = T_1 - T_\lambda(V_1), \tag{2.1.15}$$

which differs considerably from $t_P(T_1)$. Similarly, $t_T(P_1)$ can be defined, but will not be needed here. If desired, t_P, t_V, and t_T can be made dimensionless by dividing by T_λ or P_λ. One can see from Figure 2.1 that along an isochore

$$t_P = t_V - \left(\frac{\partial T}{\partial P}\right)_\lambda \int_0^{t_V} \left(\frac{\partial P}{\partial T}\right)_V dt, \tag{2.1.16}$$

if we regard the λ line as locally linear. It follows that t_P/t_V in general will not be a constant and may be singular at T_λ. Therefore, the coefficients in

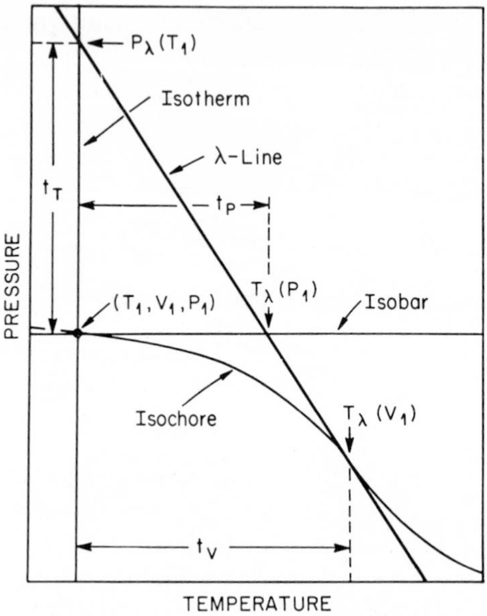

FIGURE 2.1. Schematic diagram of typical experimental paths near the λ line.

equations (2.1.10)–(2.1.13) will depend upon the detailed definition of t. However, when t vanishes, they become independent of the choice of t and equal to derivatives along the λ line. Since S, V, and P are all finite at T_λ, any definite integral of $(\partial S/\partial T)_\lambda$,$(\partial P/\partial T)_\lambda$, and $(\partial V/\partial T)_\lambda$ must be finite. This is possible only if $(\partial S/\partial T)_\lambda$, $(\partial P/\partial T)_\lambda$, and $(\partial V/\partial T)_\lambda$ are themselves finite or diverge along the λ line sufficiently mildly at most at isolated singular points. It is known from experiment that the λ line in ^{4}He does not contain any such singular points [32, 33]. We conclude that the derivatives at constant t that are pertinent to pure ^{4}He go to a finite limit at T_λ. This conclusion is not necessarily applicable to other systems, however, and we shall see that in ^{3}He–^{4}He mixtures, for instance, the pertinent derivative $(\partial \Phi/\partial T)_\lambda$ diverges at zero concentration so strongly that its integral diverges. When the derivatives are finite, the same limit is obtained regardless of whether t vanishes from above or below T_λ.

Next, let us examine the t dependence of $(\partial S/\partial T)_t$, $(\partial P/\partial T)_t$, and $(\partial V/\partial T)_t$, and compare the strength of any singularities in these coefficients with the strength of singularities in the response functions that they connect via equations (2.1.10)–(2.1.13). Consider, for instance, $(\partial S/\partial T)_t$ along an isobar. If $C_P \sim t_P^{-\alpha}$, then $S \sim \int C_P \, dt_P \sim t_P^{1-\alpha}$. Differentiating at constant t_P does not alter this behavior, and we have $(\partial S/\partial T)_{t_P} \sim t_P^{1-\alpha}$, which is a milder singularity than that of C_P. Similarly, along isobars, $(\partial V/\partial T)_{t_P}$ has the same singularity as the molar volume, and therefore is less singular than α_P. Along an isobar, $(\partial P/\partial T)_{t_P} \equiv (\partial P/\partial T)_\lambda$ and is independent of t_P. Along isochores, one finds that $(\partial S/\partial T)_{t_V} \sim S \sim \int C_V\, dt_V$, $(\partial P/\partial T)_{t_V} \sim P$, and $(\partial V/\partial T)_{t_V} \equiv (\partial V/\partial T)_\lambda$. Since $(\partial S/\partial T)_t$, $(\partial P/\partial T)_t$, and $(\partial V/\partial T)_t$ are generally finite and less singular than the response functions that they relate, they may be regarded as constant sufficiently near T_λ. Equations (2.1.10)–(2.1.13) then imply that α_P, for instance, is an asymptotically linear function of C_P. Furthermore, α_P approaches *the same* linear function of C_P above and below T_λ as t vanishes.

It follows from equations (2.1.10)–(2.1.13) that C_P, α_P, and κ_T exhibit the same leading singularities as the λ line is approached. For a divergent C_P, we therefore expect that

$$\left(\frac{\partial P}{\partial T}\right)_V = \frac{\alpha_P}{\kappa_T} \tag{2.1.17}$$

is less singular than α_P, C_P, and κ_T because the leading singularities will cancel in the ratio. According to equation (2.1.12) C_V is asymptotically linear in $(\partial P/\partial T)_V$, and thus C_V is also expected to be less singular than α_P, κ_T, or C_P. The relation

$$\kappa_S = \frac{\kappa_T}{\gamma}, \tag{2.1.18}$$

with

$$\gamma = \frac{C_P}{C_V}, \tag{2.1.19}$$

suggests a weaker singularity in κ_S than in κ_T. However, it is often difficult to establish from experiment the difference in the asymptotic behavior of C_P, α_P, and κ_T, on the one hand, and C_V, $(\partial P/\partial T)_V$, and κ_S, on the other. Thus, for instance, the ratio γ [equation (2.1.19)], which is expected to have the same singularity as C_P, is equal to 1.037 at vapor pressure and $t_P \cong 10^{-5}$ K in ^{4}He. This is only very slightly larger than the smallest value ($\gamma = 1$) permitted by thermodynamic stability arguments. From the behavior of C_P [28, 34], we know that γ becomes considerably larger than unity at T_λ, but this would be difficult to establish from direct measurements. Similarly, we expect C_V in ^{4}He to be finite at T_λ, but no measurement yields a result nearly as large as this finite limit. It would be useful to have some estimate of how closely the transition must be approached before the limiting behavior in a particular variable can be observed. Alternately, if the limiting behavior is not observable, it may be more convenient to calculate it from that of other response functions. Let us therefore, as an example, calculate the limiting behavior of C_V in more detail. We start with

$$C_P - C_V = \frac{\alpha_P^2 VT}{\kappa_T}, \tag{2.1.20}$$

which, with equations (2.1.10) and (2.1.13), may be written as

$$C_V = C_P + \left[-\left(\frac{\partial S}{\partial P}\right)_t + \frac{1}{T}\left(\frac{\partial T}{\partial P}\right)_t C_P\right]^2 T \times \left[\left(\frac{\partial S}{\partial P}\right)_t\left(\frac{\partial T}{\partial P}\right)_t + \left(\frac{\partial V}{\partial P}\right)_t - \frac{1}{T}\left(\frac{\partial T}{\partial P}\right)_t^2 C_P\right]^{-1}, \tag{2.1.21}$$

and which we shall abbreviate to

$$C_V = C_P + T\,\frac{(a + bC_P)^2}{d + eC_P}. \tag{2.1.22}$$

If $C_P > d/e$, the denominator can be expanded in C_P^{-1}. If C_P diverges at a λ line, and if we exclude possible isolated points on the λ line that are singular in the sense that d/e diverges there, then this expansion will always be valid sufficiently near T_λ. We obtain

$$\frac{C_V}{T} = \frac{2ab}{e} - \frac{b^2 d}{e^2} + \left(\frac{a^2}{e} + \frac{b^2 d^2}{e^3} - \frac{2abd}{e^2}\right) C_P^{-1} + \mathcal{O}(C_P^{-2}) + \cdots. \tag{2.1.23}$$

Let us assume, for the moment, that C_P diverges. Then the first two terms on the right yield a constant, and the other terms vanish at T_λ. Thus, $C_V(T_\lambda) = 2abT/e - b^2\, dT/e^2$, which in terms of the derivatives at constant t yields

$$C_V(T_\lambda) = T_\lambda\left[\left(\frac{\partial S}{\partial T}\right)_\lambda - \left(\frac{\partial V}{\partial T}\right)_\lambda\left(\frac{\partial P}{\partial T}\right)_\lambda\right]. \tag{2.1.24}$$

This result was obtained by Buckingham and Fairbank [28] via a slightly different route. Note that we still have made no assumptions about the nature of the singularities of any of the response functions, except that those with the strongest singularities (i.e., α_P, κ_T, and C_P) diverge. Since the derivatives along the λ line are usually finite, equation (2.1.24) implies that C_V is finite also, except possibly at isolated singular points on a λ line. If C_P is finite, equation (2.1.23) leads to an inequality with $C_V(T_\lambda)$ less than the right of equation (2.1.24). A detailed and rigorous discussion of this upper bound for C_V along a line of singularities has been given by Wheeler and Griffiths [35].

It has been recognized [28], in addition, that C_V has a sharp cusp at T_λ, and that near this cusp it varies as C_P^{-1}. This behavior of C_V also follows from the more general "renormalization" scheme introduced more recently [36, 37] by Lipa and Buckingham, and by Fisher. We already have seen this behavior of C_V in terms of C_P explicitly in equation (2.1.23), which is valid if

$$C_P > -T\left[\left(\frac{\partial S}{\partial T}\right)_\lambda + \left(\frac{\partial V}{\partial T}\right)_\lambda\left(\frac{\partial P}{\partial T}\right)_\lambda\right]. \tag{2.1.25}$$

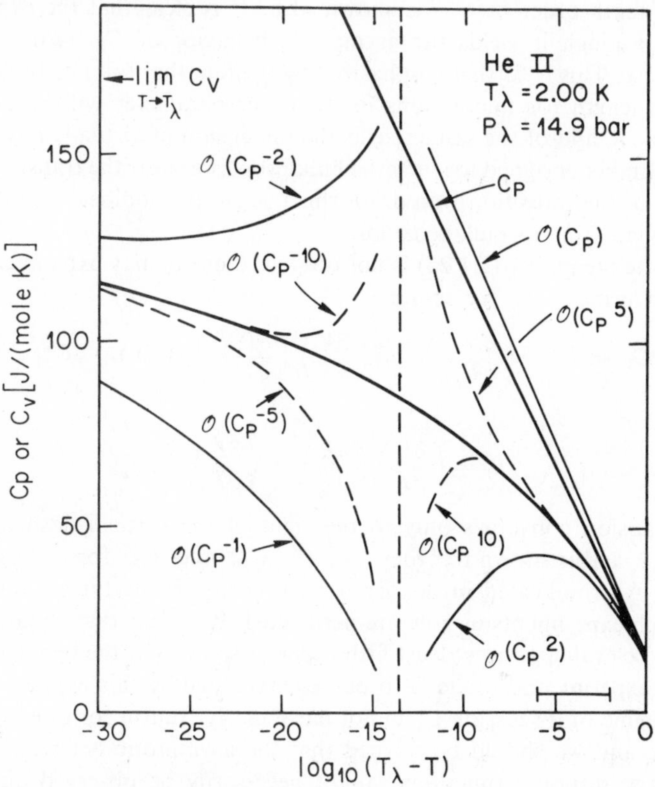

FIGURE 2.2. Expansion of C_V to various orders in C_P^{-1} and in C_P. The expansion in C_P^{-1} converges to the left of the vertical dashed line, whereas the expansion in C_P converges to the right of this line. The solid curved line which extends over the whole width of the figure is the exact C_V.

Unfortunately, in real cases, the inequality equation (2.1.25) usually is obeyed only when T is prohibitively near T_λ. This is demonstrated in Figure 2.2, where we show a logarithmically divergent C_P, which in the experimentally accessible range of $T_\lambda - T$ is a good approximation of C_P for ^{4}He below T_λ at a pressure of about 15 bar. Also shown is the exact result for C_V that was obtained from equations (2.1.20), (2.1.13), and (2.1.10), and the λ line parameters of ^{4}He at 15 bar. Note that even for $t = 10^{-30}$ K, C_V has not nearly reached its finite value at T_λ that is indicated in the upper left corner. The inequality equation (2.1.25) is satisfied only to the left of the vertical dashed line near $t = 10^{-14}$ K in Figure 2.2, and therefore the expansion equation (2.2.23) diverges to the right of this line. The lowest line in the left half of Figure 2.2 represents the expansion of C_V in C_P^{-1} to first order, and is not even nearly correct at any value of $t > 10^{-30}$ K. Equation (2.1.23) to second, fifth, and tenth order in C_P^{-1} is shown also. It is clear that the expansion (2.1.23), although it yields the asymptotic behavior of C_V, is not of any practical use. This is further emphasized by the fact that for $t \cong 10^{-14}$ K the coherence length for fluctuations in the order parameter will be approximately 10 cm, and of the same size as the dimension of virtually any experimental chamber designed to study the bulk properties near the transition. The character of the transition would, for this reason, be modified, and C_P itself would deviate from its bulk behavior.

When the inequality (2.1.25) is not obeyed, then we may expand equation (2.1.22) in terms of C_P. We obtain

$$C_V = TX^{-1}\left(\frac{\partial S}{\partial T}\right)_t^2 + T^2\left(\frac{\partial V}{\partial T}\right)_t^2\left(\frac{\partial P}{\partial T}\right)_t^2 \sum_{n=1}^{\infty} (TX)^{-n-1}C_p^n, \tag{2.1.26}$$

with

$$X = \left(\frac{\partial V}{\partial T}\right)_t\left(\frac{\partial P}{\partial T}\right)_t + \left(\frac{\partial S}{\partial T}\right)_t. \tag{2.1.27}$$

This expansion converges only to the right of the vertical dashed line in Figure 2.2, and is shown to order C_P, C_P^2, C_P^5, and C_P^{10} for our particular example. Also indicated in the lower right corner is the typical range of t over which experiments usually are performed. It is clear that equation (2.1.26) must be evaluated to at least fifth order or so in C_P if the behavior of C_V over the experimental range is to be approximated by this expansion. This entire scheme of discussing C_V is not particularly fruitful for most practical purposes, and we should be warned that the asymptotic behavior of thermodynamic response functions cannot necessarily be observed directly in real experiments.

From equation (2.1.24) and the relation, equation (2.1.12), between C_V and $(\partial P/\partial T)_V$, we have

$$\lim_{T\to T_\lambda}\left(\frac{\partial P}{\partial T}\right)_V = \left(\frac{\partial P}{\partial T}\right)_\lambda, \tag{2.1.28}$$

if C_P diverges. In addition, it is not hard to show for a divergent C_P that

$$\lim_{T\to T_\lambda}\left(\frac{\partial P}{\partial T}\right)_S = \left(\frac{\partial P}{\partial T}\right)_\lambda, \tag{2.1.29}$$

and

$$\lim_{T\to T_\lambda}\kappa_S = V^{-1}\left[\left(\frac{\partial T}{\partial P}\right)_\lambda\left(\frac{\partial S}{\partial P}\right)_\lambda - \left(\frac{\partial V}{\partial P}\right)_\lambda\right]. \tag{2.1.30}$$

The last equation is useful because the velocity of sound u_1 is related to κ_S by $u_1^2 = V/\kappa_S$, and thus for a system with a divergent C_P,

$$\lim_{T\to T_\lambda} u_1 = V\left[\left(\frac{\partial T}{\partial P}\right)_\lambda\left(\frac{\partial S}{\partial P}\right)_\lambda - \left(\frac{\partial V}{\partial P}\right)_\lambda\right]^{-\frac{1}{2}}. \tag{2.1.31}$$

By starting with equation (2.1.22), equations (2.1.28)–(2.1.31) are easily modified for the case where C_P remains finite at T_λ.

We have seen that there exist general asymptotically linear relations between various thermodynamic derivatives, and we have used these relations to calculate the finite values of several response functions, like C_V and $(\partial P/\partial T)_V$, at T_λ. In the case of C_V, we saw explicitly that experiments may never yield the asymptotic behavior, but in general we do not know how closely T_λ must be approached before our asymptotic relations are a good approximation. Let us, therefore, examine some experimental measurements that can be compared with each other via equations (2.1.10)–(2.1.13). In Figure 2.3 the measured values [38, 39] of $(\partial P/\partial T)_V$ are shown as a function

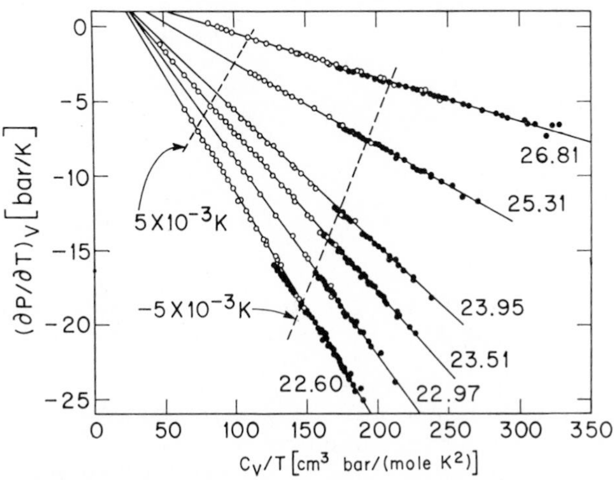

FIGURE 2.3. $(\partial P/\partial T)_V$ as a function of C_V/T at the molar volumes indicated by the numbers in the figure. The dashed lines show where $T - T_\lambda$ has reached 5×10^{-3} K [He (I)] and -5×10^{-3} K [He (II)]. Open circles, He (I); solid circles, He (II). After Reference [39].

of C_V/T along six isochores in ^{4}He near T_λ. According to equation (2.1.12), $(\partial P/\partial T)_V$ is asymptotically linear in C_V/T. Indeed, the measurements at a given volume above and below the transition fall on the same straight line, as required by equation (2.1.12). Only when $t \approx 10^{-2}$K do deviations from the asymntotic behavior become appreciable. These small deviations can be explained quantitatively [39] on the basis of the mild dependence upon t_V of $(\partial S/\partial T)_{t_V}$.

According to equation (2.1.12), the slopes of the lines in Figure 2.3 are supposed to be equal to $-(\partial V/\partial T)_\lambda^{-1}$. Figure 2.4 shows $(\partial \rho/\partial T)_\lambda = -(4.004/V^2)(\partial V/\partial T)_\lambda$, where ρ is the density in g/cm^3 and where V is per mole of ^{4}He. The solid symbols were obtained from the slopes of the lines in Figure 2.3, and the open ones are from direct measurements [32]. The agreement clearly

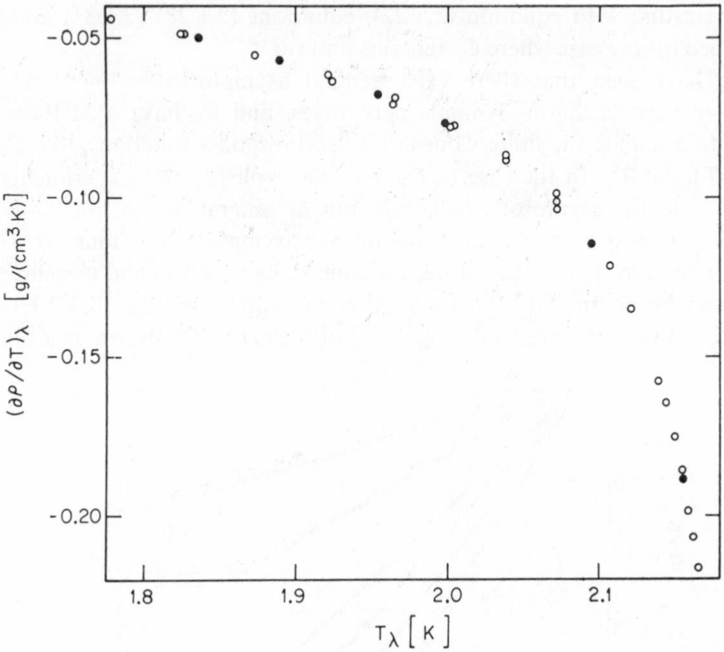

FIGURE 2.4. $(\partial\rho/\partial T)_\lambda$ as a function of T_λ. Open circles are direct measurements from Reference [32], and solid circles are derived from the slopes of the lines of Figure 2.3. After Reference [39].

is very satisfactory, and differences generally do not exceed 1%. The other coefficient $(\partial S/\partial T)_t$ that can be obtained from equation (2.1.12) is quite consistent with calorimetric entropy data [40–42], and will be discussed in Section 2.1.2(a). It can be concluded that at least some of the asymptotic relations are indeed useful at accessible values of t, but that it is difficult to know in advance, without a more detailed calculation, the maximum values of t for which the asymptotic behavior is a good approximation to real measurements.

As a further example, we consider the behavior of the isentropic sound velocity near T_λ in ^{4}He. One has

$$u_1^2 = \frac{V}{\kappa_S} = \frac{C_P V}{(\kappa_T C_V)}, \tag{2.1.32}$$

which yields, with equation (2.1.20),

$$u_1^{-2} = \frac{\kappa_T}{V} - \frac{\alpha_P^2 T}{C_P}. \tag{2.1.33}$$

Using equations (2.1.10) and (2.1.13), which give κ_T and α_P in terms of C_P, we obtain

$$u_1^{-2} = V^{-2}\left[\left(\frac{\partial T}{\partial P}\right)_t\left(\frac{\partial S}{\partial P}\right)_t - \left(\frac{\partial V}{\partial P}\right)_t - T\left(\frac{\partial S}{\partial P}\right)_t^2 C_P^{-1}\right]. \qquad (2.1.34)$$

If C_P diverges at T_λ, equation (2.1.31) for $u_1(T_\lambda)$ is obtained from equation (2.1.34). If the derivatives at constant t were independent of t, one could subtract $[u_1(T_\lambda)]^{-2}$ from equation (2.1.34) and obtain [43]

$$u_1^{-2} - [u_1(T_\lambda)]^{-2} = -\, TV^{-2}\left(\frac{\partial S}{\partial P}\right)_\lambda^2 C_P^{-1}. \qquad (2.1.35)$$

This can be further simplified when $u_1 - u_1(T_\lambda) \ll u_1$, and finally,

$$u_1 - u_1(T_\lambda) \cong \left(\frac{u_\lambda^2}{2V_\lambda^2}\right) T\left(\frac{\partial S}{\partial P}\right)_\lambda^2 C_P^{-1}, \qquad (2.1.36)$$

which is expected to be asymptotically correct. Although it is not feasible to measure $u_1 - u_1(T_\lambda)$ directly because of gravity effects and dispersion, this quantity can be readily inferred [44] from experimental measurements [31] at vapor pressure and is shown in Figure 2.5 as a function of the right side of

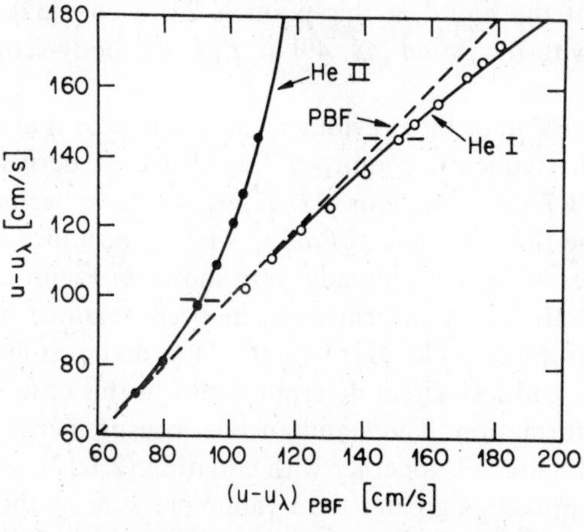

FIGURE 2.5. The difference u-u_λ in the first-sound velocity as a function of the asymptotically correct velocity difference $(u$-$u_\lambda)_{\mathrm{PBF}}$ calculated in the PBF approximation from equation (2.1.36). The dashed line has a slope of unity. The data are from Reference [31], and should approach the dashed line for sufficiently small u-u_λ. The solid lines represent calculations which include higher-order corrections to the PBF approximation and are from Reference [44]. The two horizontal marks near the data indicate where, in going from left to right, $|T_\lambda - T|$ has increased to 10^{-3} K. After Reference [44].

equation (2.1.36). The measurements are expected to approach the dashed line with slope of unity as T approaches T_λ and $u_1 - u_1(T_\lambda)$ becomes smaller. Although the measurements are not inconsistent with this expectation, we note that even for $|T - T_\lambda| = 10^{-3}$ K the deviations from the asymptotic behavior are appreciable. This seems very different from Figure 2.3, in which the asymptotic behavior of equation (2.1.12) is shown to be well satisfied when $|T - T_\lambda| \lesssim 10^{-2}$ K. The deviations in Figure 2.5 of $u_1 - u_1(T_\lambda)$ from the asymptotic behavior are explained by contributions which arise from the t dependence of the parameters involved in the derivation of equation (2.1.36). When these higher order contributions are included explicitly [44] in a calculation of u_1, then one obtains the two solid lines in Figure 2.5 that agree very well with the data. Once more this experience indicates that the known asymptotic behavior should not necessarily be expected to hold over a large experimentally accessible range of t.

2.1.2. Pure ^{4}He

(*a*) *The λ Line.* The superfluid transition occurs at 2.1720 K on the 1958 ^{4}He vapor pressure scale of temperatures [32, 45] when the pressure of the liquid is equal to the saturated vapor pressure, or 0.050 bar. At this point the molar volume is 27.38 cm^3 [32, 46]. At sufficiently high pressures, the liquid freezes, and the transition no longer exists. The λ line meets the solidification curve at 1.7633 K [47, 48] and at a pressure of 30.23 bar [47]. The molar volume of the liquid at this point is 22.18 cm^3 [32], and the phase in equilibrium with the liquid [48, 49] at T_λ is the body-centered cubic [50] solid.

From the discussion of the previous section, it is clear that thermodynamic calculations will frequently require a knowledge of derivatives along the λ line. Recent PVT measurements [32, 33, 47, 51] have been sufficiently precise to determine $(\partial V/\partial T)_\lambda$ and $(\partial P/\partial T)_\lambda$ with an accuracy of 1%. The experimental values of $(\partial \rho/\partial T)_\lambda$ already were shown in Figure 2.4. In contrast, comparatively little direct information has been obtained about $(\partial S/\partial T)_\lambda$. Calorimetric entropy data [40–42] along the λ line do not suffice to define $(\partial S/\partial T)_\lambda$ within 1%, and less direct determinations in this case yield somewhat more precise information. The simultaneous measurements of C_V and $(\partial P/\partial T)_V$ shown in Figure 2.3, together with equation (2.1.12), yielded six values of $(\partial S/\partial T)_\lambda$ for pressures greater than vapor pressure. Although one might wish for more points, it is likely that a smooth line through these data is not in error by more than 1 or 2% for $P \gtrsim 1.6$ bar.

Over the narrow pressure range near vapor pressure that is not covered by the C_v and $(\partial P/\partial T)_v$ data ($P \lesssim 1.6$ bar), $(\partial S/\partial T)_\lambda$ varies rapidly with P and changes by about 30%. Therefore additional measurements in this range should be obtained. At this time, results exist only at saturated vapor pressure. Even there, however, two separate sets of experiments are not entirely consistent with each other. A value of 10.3 J/mole K^2 was obtained from an analysis [44] of the isentropic sound-velocity results [31] in Figure 2.5 with careful consideration of higher-order corrections to equation (2.1.36). This value, and the specific heat at constant pressure, can be used to calculate the temperature at which the liquid at saturated vapor pressure has a density maximum. Then $\alpha_s = 0$, and one has [see also equation (2.1.52) below]

$$\frac{C_p}{T} = \left(\frac{\partial S}{\partial T}\right)_t - \frac{(\partial V/\partial T)_t(\partial P/\partial T)_s}{1 - (\partial P/\partial T)_s/(\partial P/\partial T)_t}.$$

Here $\alpha_s = V^{-1}(\partial V/\partial T)_s$ is the expansion coefficient at saturated vapor pressure, $(\partial P/\partial T)_s$ is the slope of the vapor pressure line, and the derivatives at constant t (see Section 2.1.1) have values very close to those at T_λ and are known. The predicted value of C_p agrees with the measured C_p [Section 2.1.2(c)] when $T - T_\lambda = 0.0104$ K. Therefore α_s is expected to vanish about 10 mK above T_λ. Contrary to this, direct measurements of α_s [see Section 2.1.2(c)] indicate that α_s vanishes when $T - T_\lambda = 0.005 \pm 0.001$ K. If alternately this latter result is used to obtain $(\partial S/\partial T)_t$, one obtains 11.9 ± 0.4 J/mole K^2, which differs from the value obtained from the sound velocity by 15%.

Obviously additional determinations of $(\partial S/\partial T)_\lambda$ at vapor pressure would be useful. A more accurate determination of the temperature difference between T_λ and the temperature T_M at which the density reaches a maximum probably would be useful for this purpose. Rather precise determinations of $T_M - T_\lambda$ have indeed been made by Peshkov and Borovikov [51a], and by Barmatz and Rudnick [51b]. They find $T_M - T_\lambda = 6.5 \pm 0.5$ mK and 6.3 ± 0.1 mK, respectively. Their measurements are based upon determinations of changes in thermal gradients associated with the onset or reversal of convective currents in the fluid at the temperature where some appropriate thermal expansion coefficient changes sign. Although their results correspond to a T_M at which the density along a particular experimental path reaches a maximum, it is difficult to decide whether, for instance, $\alpha_P = V^{-1}(\partial V/\partial T)_P$ or $\alpha_s = V^{-1}(\partial V/\partial T)_s$ vanishes here. If one assumes $\alpha_s = 0$, one obtains $(\partial S/\partial T)_t = 11.4$ J/mole K^2. If, however, $\alpha_P = 0$, one obtains $(\partial S/\partial T)_t = 10.9$ J/mole K^2. The real experiment may, of course, correspond to yet another choice; therefore, it seems difficult to use these measurements for an accurate determination of $(\partial S/\partial T)_\lambda$.†

It has been difficult in the past [32, 51] to obtain simple analytic expressions for the λ-line derivatives as a function of $T_{\lambda 0} - T_\lambda$, where $T_{\lambda 0}$ is the λ temperature at vapor pressure. For example, in order to fit results for $(\partial P/\partial T)_\lambda$ within precision of experimental data to a function of $T_{\lambda 0} - T_\lambda$, a six-parameter empirical equation was needed [32]. The reason for this is that the derivatives vary extremely rapidly with T_λ near vapor pressure and give the appearance of an approaching singularity along the λ line. Over the existence

†A recent direct and more accurate determination [C. T. Van Degrift, Ph.D. Thesis, University of California at Irvine, Irvine, Calif., 1974, (unpublished)] yields $T - T_\lambda = 7.56 \times 10^{-3}$ K for the temperature where α_s vanishes. With equations (2.1.47)–(2.1.49) below for C_p, this yields 10.9 J mole^{-1}K^{-2} for $(\partial S/\partial T)_t$. The uncertainty is probably no larger than 2%.

range of the fluid, however, the derivatives remain finite [32, 33], and a singular point is never reached before vaporization intervenes. Nonetheless, this behavior suggests that a singular function of $T_0 - T_\lambda$, with $T_0 > T_{\lambda 0}$, with a relatively small number of parameters might fit the data well. Indeed the data for $(\partial S/\partial T)_\lambda$ can be represented by [39]

$$\left(\frac{\partial S}{\partial T}\right)_\lambda = A\varepsilon_\lambda^{-x} + B, \tag{2.1.37}$$

with

$$\varepsilon_\lambda \equiv \frac{T_0 - T_\lambda}{T_0},$$

and

$$x = 1.380, \qquad T_0 = 2.260 \text{ K},$$
$$A = 0.0938 \text{ J/mole K}^2, \quad B = 1.97 \text{ J/mole K}^2. \tag{2.1.38}$$

At vapor pressure, this equation yields the value 10.3 J/mole K².† Equation (2.1.37) is easily integrated and, in conjunction with calorimetric entropy data at vapor pressure [42, 52], yields the solid line in Figure 2.6. The indi-

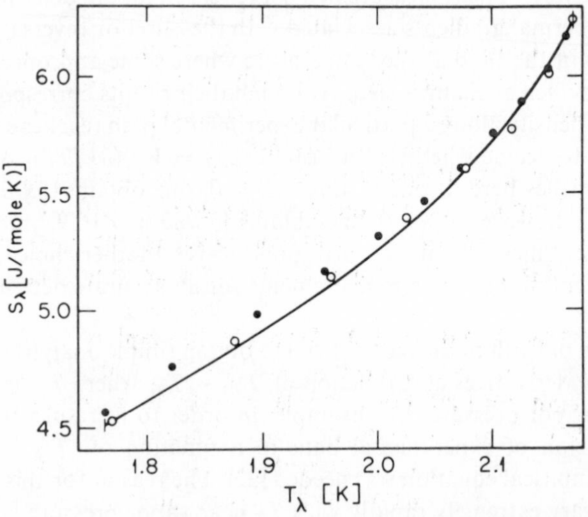

FIGURE 2.6. The entropy S_λ along the λ line as a function of T_λ. Open circles, calorimeric S_λ from Reference [41]; solid circles, calorimetric S_λ from Reference [42]; solid line, based on $(\partial S/\partial T)_\lambda$ derived from the intercepts at $(\partial P/\partial T)_V = 0$ of the lines in Figure 2.3. After Reference [39].

vidual symbols are calorimetric entropies, and agree with the line within combined errors. Similarly, the measured $(\partial\rho/\partial T)_\lambda$ [32, 39] shown in Figure

2.4 can be represented by [39]

$$\left(\frac{\partial \rho}{\partial T}\right)_\lambda = A\,(\varepsilon_\lambda^{-x} - 1), \tag{2.1.39}$$

with

$$\begin{aligned} T_0 &= 2.194 \text{ K},\\ x &= 0.383,\\ A &= -0.05020 \text{ g/cm}^3 \text{ K}. \end{aligned} \tag{2.1.40}$$

An equally simple form for $(\partial P/\partial T)_\lambda$ that fits the data within their precision has not yet been proposed. However, $(\partial \rho/\partial P)_\lambda$ [32], which may be obtained from the measured $(\partial P/\partial T)_\lambda$ [32] and equation (2.1.39), can be represented [39] with maximum deviations of 1% which occur only near freezing by

$$\left(\frac{\partial \rho}{\partial P}\right)_\lambda = A\varepsilon_\lambda^{-x} + B, \tag{2.1.41}$$

with

$$\begin{aligned} T_0 &= 2.206 \text{ K},\\ x &= 0.403,\\ A &= 3.975 \times 10^{-4} \text{ g/cm}^3 \text{ bar},\\ B &= 0.118 \times 10^{-4} \text{ g/cm}^3 \text{ bar}. \end{aligned} \tag{2.1.42}$$

Other empirical equations with a larger number of parameters are given elsewhere [32].

If C_P diverges at T_λ, then the values of C_V, $(\partial P/\partial T)_V$, κ_S, and u_1, at T_λ can be obtained easily from equations (2.1.24), (2.1.28), (2.1.30), (2.1.31), and (2.1.37)–(2.1.42). Measurements of C_V and $(\partial P/\partial T)_V$ would have to be made prohibitively near T_λ to yield a result of the predicted size. For C_V, this necessity is evident from Figure 2.2. Direct results for κ_S do not exist, but measurements of the closely related isentropic sound velocity can be compared with the calculated value. At vapor pressure, equation (2.1.31) yields $u_\lambda = 217.1$ m/sec, in excellent agreement with the measurements by Barmatz and Rudnick [31], which gave the value 217.3 m/sec. Under pressure, the available measurements [53] are not sufficiently near T_λ to warrant a comparison.

(*b*) *The Gravitational Inhomogeneity.* It is well known that two phases that are connected by a transition of higher than first order cannot coexist in a homogeneous system [27]. This is no longer true, however, when a field gradient is applied, and the system is inhomogeneous. Thus, in a tall liquid sample in the gravitational field, where the pressure is not constant, an equilibrium phase boundary can be established even if the transition between the phases

†However, see footnote on p. 95

is of higher order. On the basis of the phase diagram, one expects this boundary to rise vertically when the temperature is increased, at the rate [54]

$$\frac{dT_\lambda}{dh} = \rho g \left(\frac{\partial T}{\partial P}\right)_\lambda, \tag{2.1.43}$$

where h is the height of the boundary. At vapor pressure, one finds that $dT_\lambda/dh = (1.273 \pm 0.003) \times 10^{-6}$ K/cm, and near melting (≈ 30 bar) $dT_\lambda/dh \cong 3.3 \times 10^{-6}$ K/cm. The coexistence of He (I) and He (II) has indeed been demonstrated experimentally [54]. This was done by noting that, over a measurable temperature range, the top of a tall sample had the very high effective thermal conductivity characteristic of He (II), while the bottom had an extremely small thermal conductivity characteristic of He (I). The measured temperature range $\Delta T_\lambda = 19.3 \times 10^{-6}$ K over which the phase boundary moved vertically the distance $H = 15.6$ cm was used, in conjunction with the density ρ of the liquid [32] and the gravitational acceleration g, to calculate $(\partial P/\partial T)_\lambda = \rho g H/\Delta T_\lambda = (115.4 \pm 4.7)$ bar/K. This result is in fine agreement with (112.5 ± 1) bar/K obtained by other means [32, 33], and assures us that the He (II)–He (I) transition proceeds essentially according to classical expectations based upon the phase diagram. This had been questioned at one time on theoretical grounds [55] which, however, were later found to be not applicable [56].

We saw that the inhomogeneous sample has a transition temperature range ΔT_λ. One usually wants to obtain the behavior of the homogeneous system, and it may be necessary to apply corrections to experimental measurements

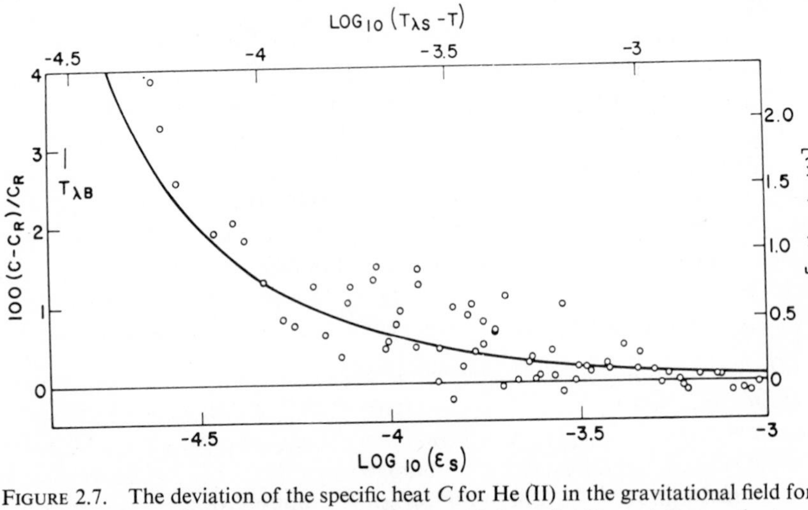

FIGURE 2.7. The deviation of the specific heat C for He (II) in the gravitational field for a sample of height 24.3 cm from the gravity-free specific heat C_R. $T_{\lambda S}$ and $T_{\lambda B}$ are the transition temperatures at the sample surface and bottom respectively.

with a temperature resolution greater than ΔT_λ. We expect that a very good approximation can be produced by assuming that the inhomogeneous system will have local properties characteristic of a homogeneous system with T_λ equal to the local T_λ. In that case we can simply average over the height, and we expect, for instance,

$$\langle C \rangle = H^{-1} \int_0^H C_R(T_\lambda)\, dh; \; T_\lambda = T_\lambda(h) \tag{2.1.44}$$

for the average heat capacity $\langle C \rangle$ of a tall sample. Here C_R is a model or reference function for the heat capacity of the homogeneous system. Deviations of experimental values [57] for $\langle C \rangle$ from C_R for a sample of height 24.3 cm are shown in Figure 2.7. For the tall sample, T_λ was chosen as $T_{\lambda S}$ at the surface of the sample. The solid line is calculated from equation (2.1.44) and agrees well with the data. This agreement gives us confidence in gravity corrections that we may apply to other measurements usually on much shorter samples. The calculated effect of gravity upon the isentropic velocity of sound at vapor pressure for a sample with $H = 4.44$ cm is shown in Figure 2.8. The minimum measurable velocity, which occurs when the transition first takes place at the bottom of the sample, in this case is about 0.6 m/sec larger than the gravity-free value at T_λ. This estimate is consistent with experimental results [31].

One of the virtually unexplored problems associated with the superfluid

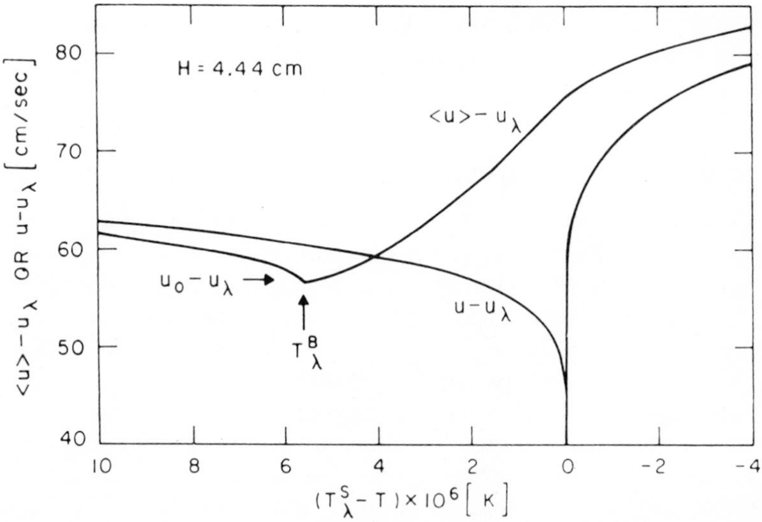

FIGURE 2.8. Comparison of the differences in first-sound velocity u-u_λ for a gravity-free sample, and of $\langle u \rangle - u_\lambda$ for a sample of height 4.44 cm in the gravitational field. T_λ^S and T_λ^B are the transition tempetratures at the surface and bottom of the sample respectively. After Reference [44].

transition is the experimental study of the nature of the He (II)–He (I) interface. We have implied so far that the inhomogeneous sample near T_λ can be understood by examining the phase diagram. This approach implies the existence of a sharp boundary in the form of a two-dimensional interface between the two phases. Actually, however, one would expect the coexistence of the two phases to be associated with a proximity effect that causes the order parameter to decay to zero gradually over a nonzero vertical characteristic length δh. This δh may be regarded as the "thickness" of the interface. The best theoretical estimate [56, 58, 58a, 58b] of δh yields 6×10^{-3} cm, and is based on a modified [59] mean field theory [60]. Although the modifications to the mean field theory were made in such a fashion as to yield agreement with some of the bulk properties near T_λ, this value of δh need not be very reliable. So far, experiments [61] have yielded only an upper limit $\delta h < 2 \times 10^{-2}$ cm. Higher resolution measurements are, of course, very difficult because the interface moves vertically by its estimated thickness if the temperature is changed by about 10^{-8} K. The use of second sound for the study of the superfluid density near the interface has been discussed theoretically [61a].

(c) Saturated Vapor Pressure. The first definitive information about the nature of the singularities in thermodynamic response functions at T_λ came from the by-now classic measurements of the heat capacity at saturated vapor pressure C_s by Fairbank, Buckingham, and Kellers [28, 34, 62–64]. Their data show that C_s still increases with decreasing $|T_\lambda - T|$ when $|T_\lambda - T| \cong 10^{-5}$ K, and thus that the transition is "sharp" at least to a few parts in 10^6 of the absolute temperature. These early results for C_s could be represented very well by a logarithmic divergence with equal amplitudes above and below T_λ, i.e.,

$$C_s = -A_{0s} \ln |T_\lambda - T| + B_{0s};\ A_{0s} = A'_{0s} \tag{2.1.45}$$

where the primed coefficient refers to $T < T_\lambda$ [He (II)], and the unprimed coefficients to $T > T_\lambda$ [He (I)]. It was not until about a decade later that almost simultaneously two independent sets of new experimental data [65–69] of sufficient precision became available to require a refinement of the early conculsions. Figure 2.9 shows C_s from a variety of sources [28, 52, 66, 70] as a function of $\log_{10} |T_\lambda - T|$ over six decades in $|T_\lambda - T|$.

The measurements by Buckingham, Fairbank, and Kellers could be represented within their precision by equation (2.1.45) for $|T_\lambda - T| \lesssim 6 \times 10^{-2}$ K on both sides of the transition. The more recent measurements [65–69] revealed, however, that deviations from equation (2.1.45), although small, are in fact systematic; if the logarithmic functional form is to be retained, then deviations from this assumed asymptotic behavior are appreciable at a much smaller value of $|T_\lambda - T|$ than the previous data had indicated. For He (II), noticeable deviations from equation (2.1.45) exist

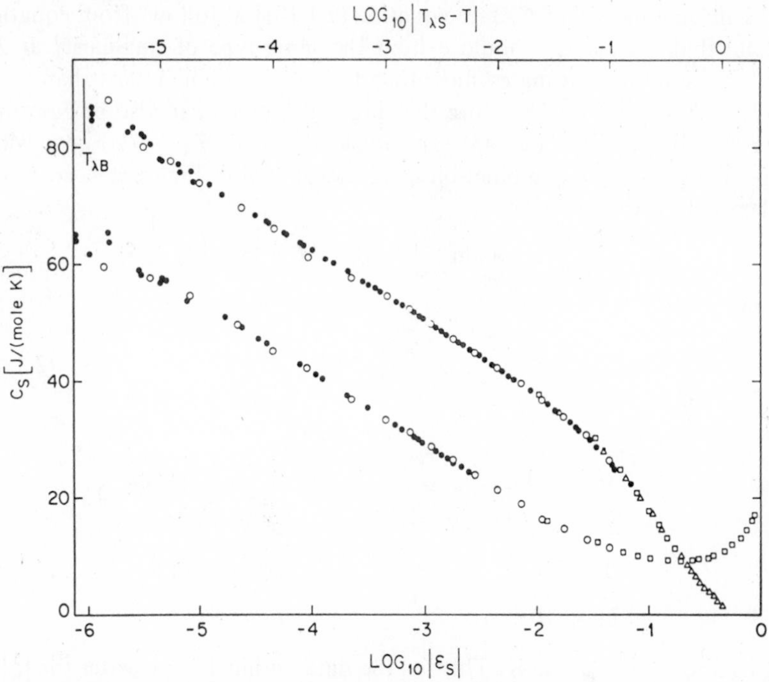

FIGURE 2.9. The heat capacity at saturated vapor pressure as a function of $\log_{10}|T_{\lambda S}-T|$. $T_{\lambda S}$ is the transition temperature at the sample surface. Open circles, References [28, 34]; open squares, Reference [52]; open triangles, Reference [70]; solid circles, Reference [66]. Upper set of data He (II); lower set, He (I). After Reference [66].

when $T_\lambda - T \gtrsim 10^{-3}$ K! After our experience in the previous section with deviations from asymptotic relations (see Figures 2.2 and 2.5), we should, perhaps, not be too surprised at this observation. However, an additional, more serious problem is associated with the recent data. These measurements are no longer consistent with equal amplitudes of the logarithmic term in equation (2.1.45) above and below the transition. If equation (2.1.45) is assumed to be valid asymptotically, both sets of data [65–69] yield

$$1.04 \leq \frac{A_{0s}}{A'_{0s}} \leq 1.06. \tag{2.1.45a}$$

Unequal amplitudes of a logarithmic divergence are contrary to theoretical predictions based on the Widom–Kadanoff scaling laws [71–75], which require $A_{0s}/A'_{0s} \equiv 1$. We shall return to the asymmetry implied by the inequality (2.1.45a) later on in this section.

It is easy to show that

$$C_P = C_s + TV\left(\frac{\partial P}{\partial T}\right)_s \alpha_P, \tag{2.1.46}$$

where $(\partial P/\partial T)_s$ is the slope of the vapor-pressure line. Since α_P is an asymptotically linear function of C_P [see equation (2.1.10)], it follows from equation (2.1.46) that C_s and C_P should exhibit the same type of singularity at T_λ. Further, numerical estimates show that C_P and C_s differ by less than 0.1% for $T_\lambda - T < 10^{-2}$ K. Therefore the singularity of C_P can also be described quite well by equation (2.1.45) for sufficiently small $T_\lambda - T$. For a wider range of $T_\lambda - T$, the combination of results shown in Figure 2.9 can be described well by the equation

$$C_s \cong C_p = -A_0 \ln|\varepsilon| + B_0 + D_0\varepsilon \ln|\varepsilon| + E_0\varepsilon, \qquad (2.1.47)$$

where

$$\varepsilon \equiv \frac{T - T_\lambda}{T_\lambda}, \qquad (2.1.48)$$

with

$$\begin{aligned} A_0 &= 5.355, & A_0' &= 5.100, \\ B_0 &= -7.77, & B_0' &= 15.52, \\ D_0 &= 14.5, & D_0' &= -14.5, \\ E_0 &= 103, & E_0' &= 69, \end{aligned} \qquad (2.1.49)$$

and C_s and C_P in J/mole K. This fits the data within 1% or better for $|\varepsilon| \lesssim 5 \times 10^{-2}$.*

The thermal expansion coefficient at saturated vapor pressure α_s has been measured by several investigators [46, 76–79]. The data usually are fitted to a function that is linear in $\ln|T_\lambda - T|$ like equation (2.1.45) for C_P. The results are not as precise as those for C_P, and a comparison of the coefficients obtained by different investigators suggests that the amplitude of $\ln|T_\lambda - T|$ is uncertain by about 10%. Most sets of data agree reasonably near $|T_\lambda - T| \cong 10^{-2}$ K, and the different amplitudes are due to relatively large uncertainties for smaller $T_\lambda - T$. In principle it should be possible to calculate α_s more reliably from C_P than it has been measured. We have

$$\alpha_S = \alpha_P - \left(\frac{\partial P}{\partial T}\right)_s \kappa_T, \qquad (2.1.50)$$

with α_P and κ_T given in terms of C_P by equations (2.1.10) and (2.1.13), re-

*An equally good fit to the data is given by the equation

$$C_p = \frac{A}{\alpha} |\varepsilon|^{-\alpha} [1 + D|\varepsilon|^x] + B + E\varepsilon$$

with $\varepsilon = (T - T_\lambda)/T_\lambda$, $\alpha = \alpha' = -0.025$, $x = x' = 0.5$, $A/A' = 1.1039$, $D/D' = 1.11$, $A' = 6.1572$ J mole^{-1}K^{-1}, $D' = -0.020$, $B = B' = 257.77$ J mole^{-1}K^{-1} and $E = E' = 95.73$ J mole^{-1}K^{-1}. This function with these parameters is more consistent with theory and other measurements near T_λ (see, for instance, the footnotes on pp. 117 and 158), and predicts a finite C_p, equal to B or B', at the phase transition.

spectively. We obtain in detail

$$\alpha_s = a_s C_P + b_s, \tag{2.1.51}$$

with

$$a_s = \frac{[1-(\partial P/\partial T)_s/(\partial P/\partial T)_t]}{[VT(\partial P/\partial T)_t]}, \tag{2.1.52}$$

$$b_s = \frac{-(\partial S/\partial P)_t[1-(\partial P/\partial T)_s/(\partial P/\partial T)_t]+(\partial V/\partial P)_t(\partial P/\partial T)_s}{V}$$

The values $(\partial P/\partial T)_s = 0.124$ bar/K, $(\partial P/\partial T)_\lambda = -112.5$ bar/K, $(\partial V/\partial T)_\lambda = 43.82$ cm^3/mole K, $(\partial S/\partial T)_\lambda = 102.4$ cm^3 bar/mole K^2, $V_\lambda = 27.38$ cm^3/mole, and $T_\lambda = 2.172$ K yield $a_{s\lambda} = -1.496 \times 10^{-3}$ moles/J and $b_{s\lambda} = 3.15 \times 10^{-2}$ K^{-1} for a_s and b_s at T_λ. For α_P, one has equivalent coefficients $a_\lambda = -1.495 \times 10^{-3}$ and $b_\lambda = 3.32 \times 10^{-2}$. The difference between α_P and α_s happens to be practically independent of $T_\lambda - T$. At vapor pressure, $(\partial P/\partial T)_t$ is virtually independent of $T_\lambda - T$, and $(\partial V/\partial P)_t$ does not contribute strongly anyway, but for $|T_\lambda - T|$ larger than about 10^{-2} K, the dependence upon $|T_\lambda - T|$ of $(\partial S/\partial T)_t$ must be taken into account for an accurate estimate. The accuracy of α_s predicted by equation (2.1.15), with C_P given by equation (2.1.47), is unfortunately limited by the large uncertainty, discussed in Section 2.1.2 (a), in $(\partial S/\partial T)_\lambda$ at vapor pressure. Very precise direct measurements of α_s in principle appear possible, and would provide a valuable confirmation of the apparent asymmetry about T_λ revealed by C_s and expressed in equation (2.1.45a). They would also provide a more accurate value of $(\partial S/\partial T)_\lambda$ when used in equation (2.1.51) together with the known C_P.†

No accurate direct measurements exist of the isothermal compressibility κ_T near T_λ at vapor pressure, but κ_T may be calculated easily and reliably from C_P, and we know from equation (2.1.13) that κ_T has the same type of singularity near T_λ as C_P. In this case, however, the constant contribution to κ_T is much larger at any experimentally accessible nonzero value of $t = T_\lambda - T$ than the singular part. Therefore it is very difficult to establish experimentally the asymptotic behavior of κ_T and even very precise measurements of κ_T would reveal little about the singular behavoir of the response functions near T_λ. In an accurate calculation of κ_T, one must be particularly careful about the t dependence of the various derivatives in equation (2.1.13) if the t dependence or the singular part of κ_T, rather than its overall magnitude, is of primary interest.

We already compared in Figure 2.5 the measured sound velocity u_1 with C_P. We found that relation (2.1.36) is obeyed only for $t \lesssim 10^{-4}$ K! Correction terms to order $t \ln t$ and t are obtainable [44] from known thermodynamic information [38, 39], and are adequate for $t \lesssim 10^{-2}$ K.

†See footnote on p. 99. Van Degrift's measurements of α_S yield $A_{0S}/A'_{0S} \cong 1.06$, consistent with equation (2.1.45a).

The surface tension σ near T_λ at SVP was measured by Atkins and Narahara [80] both above and below the transition. Although their results revealed the existence of a singularity in σ at T_λ, the singular part makes only a small contribution to the total, and its functional form cannot be determined well from the data. The measurements can be represented within the experimental scatter of perhaps $\pm$ 0.1 % of σ by

$$\sigma = \sigma_\lambda - a\varepsilon. \tag{2.1.53}$$

Equation (2.1.53) fits the measurements for $|\varepsilon| \lesssim 0.02$ with $\sigma_\lambda = 0.307$ erg/cm, $a = 0.161 \pm 0.01$ erg/cm, and $a' = 0.202 \pm 0.01$ erg/cm, but the linear dependence upon ε should not be taken too seriously. More recently, the surface tension was measured with considerably higher resolution by Gasparini et al. [80a]. These results have a precision of 10^{-2} %, but they are only for $T < T_\lambda$. They agree well with the earlier measurements, and can also be represented by equation (2.1.53). They yield practically the same coefficient a' as the earlier work [80].

It has been realized recently [81–83] that the singularity of σ in principle contains interesting information about the critical behavior near T_λ which has not readily been accessible by other measurements. It is predicted [83] on the basis of scaling arguments that $\sigma - \sigma_\lambda \sim \varepsilon^{2-\alpha-\nu}$ if $C_P \sim \varepsilon^{-\alpha}$. The exponent ν (or ν' below T_λ)describes the divergence of the coherence length $\xi = \xi_0\varepsilon^{-\nu}$ for fluctuations in the order parameter. There are no *direct* measurements of ν or ν', but measurements of the superfluid density ρ_s and scaling laws indicate that ν' is about equal to 0.67. It would be most interesting to see if ν and ν', as obtained from a surface property, agree with the value deduced from the bulk ρ_s. The measurements of σ that are available at this time tend to indicate that the exponent of $\sigma - \sigma_\lambda$ is somewhat greater than unity, but they cannot definitely distinguish between the function equation (2.1.53) and an exponent $2 - \alpha - \nu \cong 1.36$. However, Gasparini et al. [80a] have made a detailed comparison of their measurements for He [II] and those of Atkins and Narahara [80] for He (I) and He (II) with the predictions based on scaling [83], and found consistency between theory and measurement for both the exponent and the amplitude of $\sigma - \sigma_\lambda$.*

(*d*) *Isochores, $(\partial P/\partial T)_V$, and C_V.* Along isochores, measurements of $(\partial P/\partial T)_V$ [29, 38, 39, 47, 84] and C_V [38–42, 85, 86] are sufficiently complete to yield a good description of the thermodynamics of the transition along its entire existence range. We already saw in Figure 2.3 the thermodynamic consistency in terms of equation (2.1.12) of some of these measurements [38, 39, 86].

Although it is known that C_V and $(\partial P/\partial T)_V$ must be finite at T_λ, measurements by Kierstead of $(\partial P/\partial T)_V$ [84] revealed that these data could, within their fairly high prevision, be represented by the following diverging function:

*Measurements of σ with a precision of 10^{-3} %, both above and below T_λ, have been made recently by J. H. Magerlein and T. M. Sanders (to be published).

$$\left(\frac{\partial P}{\partial T}\right)_V = -a_0 \ln |t_V| + b_0. \tag{2.1.54}$$

The measurements were made near freezing, where the known finite limiting value $(\partial P/\partial T)_\lambda$ of $(\partial P/\partial T)_V$ is smallest. Nonetheless, they gave no indication of the finite limit. All other measurements are of lesser or similar precision, and can be represented equally well by equation (2.1.54). However, the recent C_V measurements [38, 39, 86] used in Figure 2.3. are an order of magnitude more precise than any data for $(\partial P/\partial T)_V$, and they indicate that C_V, and therefore $(\partial P/\partial T)_V$, do not diverge at T_λ. To be sure, C_V can be represented well [39] by the equation, similar to equation (2.1.54),

$$C_V = -A_{0V} \ln |t_V| + B_{0V}, \tag{2.1.55}$$

with the coefficients

$$\begin{aligned} A_{0V} &= 5.252 \ [1-0.01903P_\lambda + 0.0001578P_\lambda^2], \\ B_{0V} &= -3.00 \ [1+0.00595P_\lambda], \\ A'_{0V} &= 4.904 \ [1-0.03060P_\lambda + 0.0002343P_\lambda^2], \\ B'_{0V} &= 20.00 \ [1-0.00425P_\lambda] \end{aligned} \tag{2.1.56}$$

where P_λ is in bar. Here the primed coefficients refer to the low-temperature phase [He (II)]. This relation fits C_V at all densities [38, 39], and over the temperature range 10^{-4} K $\leq t \leq 10^{-2}$ K within 1 or 2%, but deviations of the data are nontheless systematic [86]. These systematic deviations are largest at the highest pressures, where the known finite $C_V(T_\lambda)$ is smallest. The deviation from equation (2.1.55) at 22.60 cm³/mole ($P_\lambda \cong 25.9$ bars, $T_\lambda \cong 1.84$ K) are shown in Figure 2.10 and demonstrate that equation (2.1.55) is only an approximation. It is, however, a very good approximation, and may be used for most thermodynamic calculations over the range 10^{-4} K $\leq t \leq 10^{-2}$ K.

In order to see more explicitly the tendency of C_V to approach a finite value, the measurements [38, 39] may be fitted to the following power law:

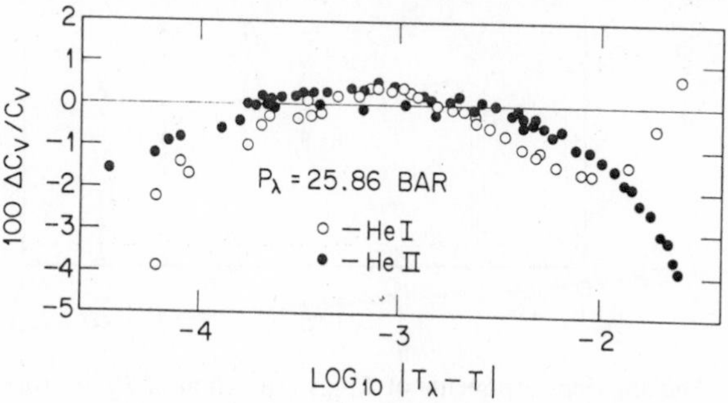

FIGURE 2.10. The deviations of experimental values for C_v from a least-squares fit to a linear function of log $|T_\lambda - T|$. After Reference [86].

$$C_V = \left(\frac{A_V}{\alpha_V}\right)\left[\left(\frac{t_V}{T_\lambda}\right)^{-\alpha_V} - 1\right] + B_V. \tag{2.1.57}$$

In the limit as α_V vanishes, equation (2.1.57) yields (2.1.55), and the subscript used on the coefficients in equation (2.1.55) indicates explicitly that the logarithmic functional form corresponds to $\alpha_V = 0$. If $\alpha_V \geq 0$, then equation (2.1.57) diverges. However, for $\alpha_V < 0$,

$$\lim_{t_V \to 0} C_V = B_V - \frac{A_V}{\alpha_V}, \tag{2.1.58}$$

so that a negative α_V corresponds to a finite C_V. It should be emphasized that equation (2.1.57) does not give the correct asymptotic dependence upon $T_\lambda - T$ of C_V near T_λ. We already saw in equation (2.1.23) that sufficiently near T_λ, C_V is linear in C_P^{-1}, but we also saw that this behavior is not observable in real experiments. Nonetheless, $\alpha_V < 0$ is indicative of a finite $C_V(T_\lambda)$, and α_V is shown in Figure 2.11 [39] as a function of P_λ. The solid symbols are for He (II), and the open ones for He (I). As is customary, the exponent for the low-temperature phase [He (II)] is also distinguished by a prime. One sees that α_V' is negative and monotonically decreasing with increasing P_λ. This reflects the known monotonic decrease of $C_V(T_\lambda)$ with increasing P_λ. Surprisingly, α_V, although negative and consistent with a finite $C_V(T_\lambda)$, is not monotonic in P_λ. We shall return to this anomaly when we discuss C_P along isobars.

Equation (2.1.55) is adequate only for $|t_V| \lesssim 10^{-2}$ K. Further away from the transition, the measurements of C_V by Lounasmaa and Kojo [40, 41] cannot be fitted by this equation, but the deviations extrapolate to zero at T_λ and, therefore indicate an overall consistency of the different sets of data. The results with $|t_V| \lesssim 0.1$ K can be represented within a few percent by

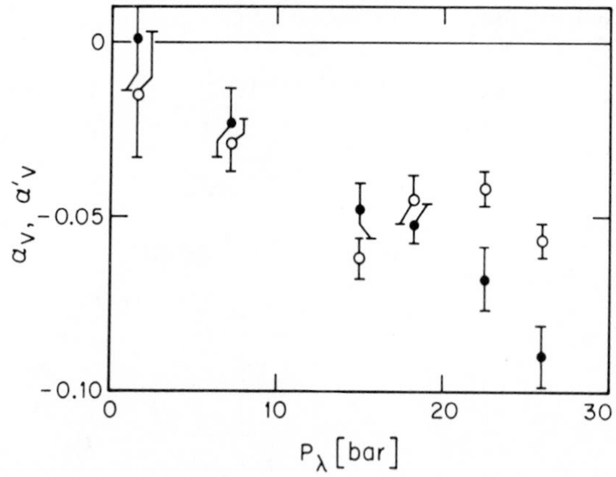

FIGURE 2.11. The apparent exponents of C_V as a function of P_λ. α_V (open circles) is for He (I), and α_V' (solid circles) for He (II). After Reference [39].

$$C_V = [-A_{0V} \ln |t_V| + B_{0V}]\,(1 + at_V), \tag{2.1.59}$$

with A_{0V} and B_{0V} given by equation (2.1.56), and $a = 1.7\ \mathrm{K}^{-1}$ for both phases.

From the dependence of C_V upon P_λ, we can calculate the dependence of the parameters in the PBF relations upon t_V at least to order $t_V \ln |t_V|$ and t_V, and we are no longer dependent upon using these relations only in the asymptotic regime. These calculations will not be reproduced here, but they have been given elsewhere in detail [39]. By considering the dependence upon t_V of the parameters, one can see explicitly the size of initial deviations from the asymptotic behavior; when these deviations are not prohibitively large, they can be reliably calculated. We saw an example of such a calculation in the case of the isentropic sound velocity [44] in Figure 2.5.

(e) κ_T, α_P, C_P, and Scaling. The three functions κ_T, α_P, and C_P are permitted to diverge along a λ line. In principle, this divergence may be studied along any experimentally convenient thermodynamic path, and the asymptotically dominant behavior can be shown to be independent of this path. Only the size of contributions of higher order in $T_\lambda - T$ than the leading term will be dependent upon the direction from which T_λ is approached. Nonetheless, it seems more natural to consider C_P and α_P along isobars, for these functions are thermodynamic derivatives at constant pressure. However, experimental considerations frequently render this choice inconvenient. Thus, there have been no direct measurements of C_P (except at SVP where the measured C_s is nearly equal to C_P), and κ_T generally has been measured along isotherms. [29, 84, 87].

The only direct results along isobars of a possibly divergent response function are those for the isobaric thermal expansion coefficient α_P [51]. These results are consistent with a logarithmic divergence of α_P similar to that suggested for the heat capacity at SVP in equation (2.1.45), i.e.,

$$\alpha_P = -A_{0\alpha} \ln |\varepsilon| + B_{0\alpha}. \tag{2.1.60}$$

This, of course, implies by virtue of equations (2.1.10) and (2.1.11) that C_P and κ_T are also of this functional form. In addition, $A_{0\alpha}/A'_{0\alpha} = A_0/A'_0$, where the primed coefficients are for He (II) and the unprimed ones for He (I), and where A_0 and A'_0 are the amplitudes of a logarithmic C_P. The measured α_P suggested [51] that $A_{0\alpha}/A'_{0\alpha} \cong 0.9 < 1$, at least for the intermediate and high-pressure range above 10 bar. As we saw earlier in connection with the results at SVP, unequal amplitudes for a logarithmically divergent C_P or α_P are contrary to scaling [71–75]; but in equation (2.1.45a) we saw, at least at SVP, that $A_0/A'_0 > 1$.*

*Recent, more precise, measurement of α_p over a wide pressure range [K. H. Mueller, F. Pobell, and G. Ahlers, *Phys. Rev. Lett.* **34**, 513 (1975); and K. H. Mueller, G. Ahlers, and F. Pobell, *Phys. Rev.*, (to be published)] have yielded $A_{0\alpha}/A'_{0\alpha} = 1.1$, consistent with C_p at vapor pressure.

It has been difficult to obtain measurements of the isothermal compressibility that are precise enough to shed much light upon the problem raised by the thermal expansion measurements, because the singular contribution to κ_T, under most circumstances, is smaller than a regular background, and thus difficult to measure with high precision. Nonetheless, available results along isotherms [29, 84, 87] definitely reveal the existence of a singularity in κ_T. The measurements nearest T_λ, and at the highest pressures where the singular part of κ_T is largest [84], are indicative of a logarithmic divergence of κ_T. These data suggest $A_{0\kappa}/A'_{0\kappa} \cong 1.16 \pm 0.1$, but we cannot be sure that they definitely rule out equal amplitudes.

The result $A_{0\kappa}/A'_{0\kappa} > 1$, which would be in conflict with the measured $A_{0\alpha}/A'_{0\alpha} < 1$, finds substantial support from independent results for non-divergent response functions. The general thermodynamic methods which lead to equations (2.1.10)–(2.1.13) can be used to show that

$$\kappa_T = -V^{-1}\left(\frac{\partial V}{\partial P}\right)_t\left[1-\left(\frac{\partial T}{\partial P}\right)_t\left(\frac{\partial P}{\partial T}\right)_V\right]^{-1}. \tag{2.1.61}$$

There are independent measurements [84] of $(\partial P/\partial T)_V$ along an isochore that meets the λ line at the same point $P_\lambda = 29.95$ bar and $T_\lambda = 1.768$ K as the isotherm for the κ_T results [84], and these are consistent with the measured κ_T and equation (2.1.61). They also yield $A_{0\kappa}/A'_{0\kappa} > 1$.

Much information is obtainable about the possibly divergent response functions along isobars from the rather plentiful and precise measurements of C_V and $(\partial P/\partial T)_V$ [38, 39, 86] that were discussed in Section 2.1.2(b). The thermodynamic relation

$$C_P = C_V - T\left(\frac{\partial P}{\partial T}\right)_V^2\left(\frac{\partial V}{\partial P}\right)_T \tag{2.1.62}$$

can be used to calculate C_P from the measured C_V and $(\partial P/\partial T)_V$, if κ_T, is known. Since κ_T is not available directly at all densities, we use equation (2.1.13) simultaneously with (2.1.62), and obtain

$$C_P = \left\{C_V - T\left(\frac{\partial P}{\partial T}\right)_V^2\left[\left(\frac{\partial S}{\partial T}\right)_t\left(\frac{\partial T}{\partial P}\right)_t^2 + \left(\frac{\partial V}{\partial T}\right)_t\left(\frac{\partial T}{\partial P}\right)_t\right]\right\}\left[1-\left(\frac{\partial P}{\partial T}\right)_V^2\left(\frac{\partial T}{\partial P}\right)_t^2\right]^{-1}. \tag{2.1.63}$$

Both C_V and $(\partial P/\partial T)_V$ are known along several isochores, and the resulting C_P will be *along the same isochores*. For these paths, $(\partial V/\partial T)_t \equiv (\partial V/\partial T)_\lambda$. The dependence upon t_V of $(\partial S/\partial T)_t$ and $(\partial P/\partial T)_t$ can be caluclated from C_V. From C_P and $(\partial P/\partial T)_V$ along isochores, one can obtain C_P along isobars. This involves primarily a caluclation of the "distance" t_P along the temperature axis, which is given by equation (2.1.16). In addition, there is a small correction for the pressure change along the isochore and the small pressure dependence at constant t_P of C_P. The resulting data for C_P and t_P are sufficiently precise to warrant a detailed analysis. [39].

Rather than establishing only qualitatively that the singularity in C_P is consistent with a logarithmic divergence, one can usefully analyze the data in terms of the more general power law

$$C_P = \frac{A}{\alpha}\left[|\varepsilon|^{-\alpha} - 1\right] + B; \; \varepsilon \equiv \frac{t_P}{T_\lambda} \tag{2.1.64}$$

which, in the limit as α vanishes, yields a logarithmic C_P. Equation (2.1.64) is meant to represent the asymptotic behavior of C_P. One might assume that contributions from omitted terms are only of order ε and negligible for $|\varepsilon| \lesssim 10^{-3}$. This assumption would be true if these terms came from contributions to the free energy that are regular functions of the temperature. The data for small $|\varepsilon|$ and equation (2.1.64) yield [38, 39] the α and α' shown in Figure 2.12. We see that, consistent with a logarithmic divergence, $\alpha' = 0$ within the experimental error at all P except SVP. For He (I), however, $\alpha = 0$ within experimental error only for small P, and for $P \gtrsim 15$ bar $\alpha > \alpha' = 0$. The pressure dependence of α is already reflected in the nonmonotonic behavior of α_V shown in Figure 2.11. The results for α, if they reflect the asymptotic behavior of C_P, are contrary to predictions based upon scaling which require $\alpha = \alpha'$.*

If *large* correction terms (i.e., contributions that are appreciable even for $\varepsilon \cong 10^{-3}$) to the asymptotic behavior of C_P are allowed in the data analysis, then the permitted statistical errors of α and α' become large enough to include zero at all P because of the high correlation between the leading exponents and the necessary additional parameters. Thus, the data can be conveniently described for $|\varepsilon| \lesssim 10^{-2}$ by

$$C_P = -A_0 \ln |\varepsilon| + B_0 + D_0\varepsilon \ln |\varepsilon| + E_0\varepsilon, \tag{2.1.65}$$

with the coefficients

$$A_0' = 5.102 - 0.05652P_\lambda + 9.643 \times 10^{-4}P_\lambda^2,$$

$$B_0' = 15.57 - 0.3601P_\lambda + 4.505 \times 10^{-3}P_\lambda^2,$$

$$D_0' = -14.5 + 6.119P_\lambda,$$

$$E_0' = 69.0 + 19.08P_\lambda,$$

*The recent thermal expansion measurements by Mueller et al. (see footnote on p. 111), when fitted to the equivalent of equation (2.1.64)., but with the equality $\alpha = \alpha'$ imposed as a constraint, yielded values of $\alpha = \alpha'$ which varied from near zero for $P = 5$ bar to 0.06 for $P = 30$ bar. This pressure dependence of α is contrary to the theoretically expected universal behavior [88–90] of the dimensionless critical point parameters, and is related to the unexpected behavior of the specific heat exponent.

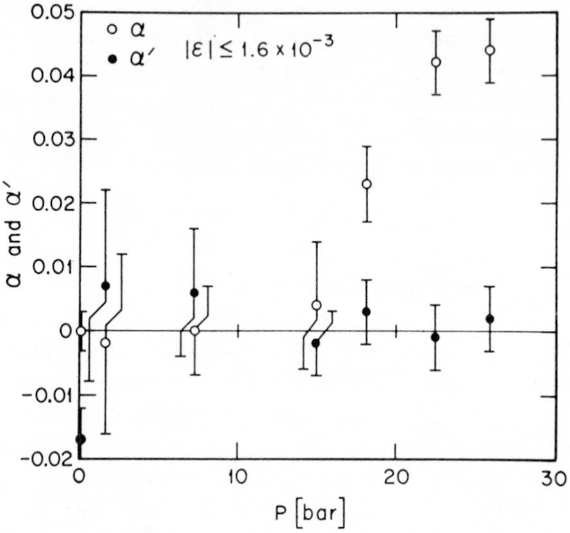

FIGURE 2.12. The exponent obtained by fitting C_P for $|\varepsilon| \leq 1.6 \times 10^{-3}$ to a pure power law. After Reference [39].

$$\begin{aligned} A_0 &= 5.357 - 0.03465P_\lambda + 8.447 \times 10^{-4}P_\lambda^2, \\ B_0 &= -7.75 - 0.362P_\lambda - 4.535 \times 10^{-4}P_\lambda^2, \\ D_0 &= 14.5 - 6.203P_\lambda, \\ E_0 &= 102 - 16.55P_\lambda, \end{aligned} \tag{2.1.66}$$

if P_λ is in bars and C_P in J/mole K. These equations are suitable for thermodynamic calculation, but the reader should be cautioned immediately that equation (2.1.65) is not necessarily representative of the asymptotic behavior of C_P. Other functions with four pressure-dependent parameters for each phase, but with $\alpha = \alpha' \neq 0$, could fit the data equally well. Equation (2.1.65) corresponds to $\alpha = \alpha' = 0$, and is consistent with theory in so far as the exponents are equal. However, according to equation (2.1.66), A_0 is greater than A_0', and for a logarithmic C_P, this is contrary to theory.

Since the difference in amplitude above and below T_λ that is revealed by the thermodynamic results appears contrary to theoretical predictions [71–73], it is useful to examine the difference between C_P above and below T_λ at the same value of $|\varepsilon|$, for the $|\varepsilon|$ dependence of this difference should be sensitive to any asymmetry about T_λ that may exist. This was done [38, 39, 65, 66] by representing the heat capacity of He (II) by a smooth function of ε, and by subtracting this function from the measured C_P of He (I). If $\alpha = \alpha' = 0$, then we expect

$$\Delta C_P = -(A_0 - A_0') \ln |\varepsilon| + (B - B'), \tag{2.1.67}$$

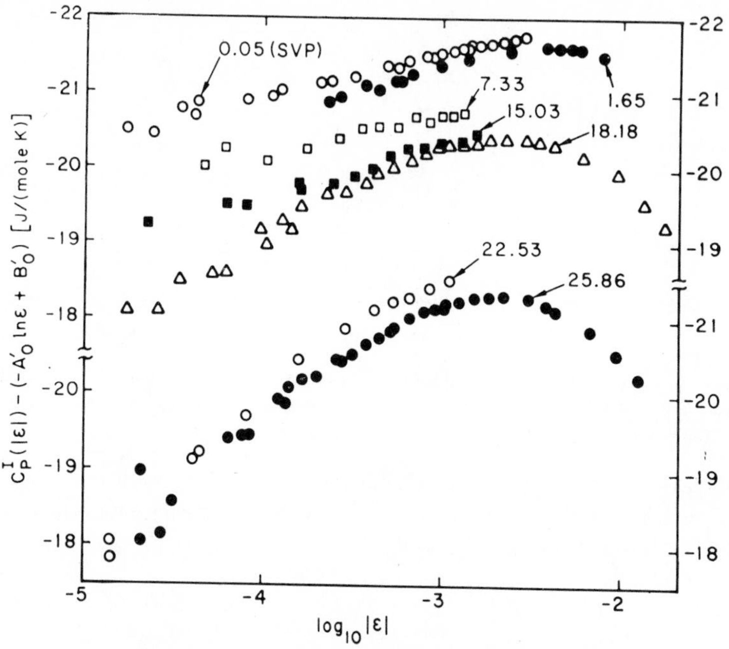

FIGURE 2.13. C_P for He (I) minus C_p for He (II) at the same $|\varepsilon|$, as a function of $\log_{10}|\varepsilon|$, on several isobars. The numbers are the pressures in bars. After Reference [39].

where $\Delta C_P \equiv C_P$ [He (I)] $-$ C_P[He (II)]. Scaling requires in this case $A_0 = A_0'$. Thus the amplitude of ΔC_P should vanish, and ΔC_P should be independent of $\ln |\varepsilon|$ and equal to $B - B'$ at all ε that are sufficiently small. Figure 2.13 shows the experimental ΔC_P [38, 39] as a function of $\ln |\varepsilon|$. We see that ΔC_P depends upon $|\varepsilon|$ at all P. This is true even near $\varepsilon \cong 2 \times 10^{-5}$, which is the smallest ε at which sufficiently precise measurements have been possible so far. For sufficiently small ε, ΔC_P appears linear in $\ln |\varepsilon|$, and this implies that equation (2.1.45) fits and that $\alpha = \alpha' = 0$ is consistent with the data. The slopes of straight lines through the data yield the A_0/A_0' shown in Figure 2.14. As we already know, this ratio is greater than unity at all P, but Figure 2.14 includes reasonable estimates of probable errors in the ratio for each pressure. In addition to $A_0/A_0' > 1$, the data also reveal that A_0/A_0' is not independent of P. A pressure dependence for this ratio is surprising and contrary to a predicted universality [88–90] of critical point parameters. A change in P is not expected to alter the basic character of the transition, and, therefore, should leave these parameters unchanged.*

It is clear that the assumption of a logarithmically divergent C_P, or a C_P

*The recent thermal expansion measurements by Mueller et al. (see footnote on p. 111) have a universal A_0/A_0'.

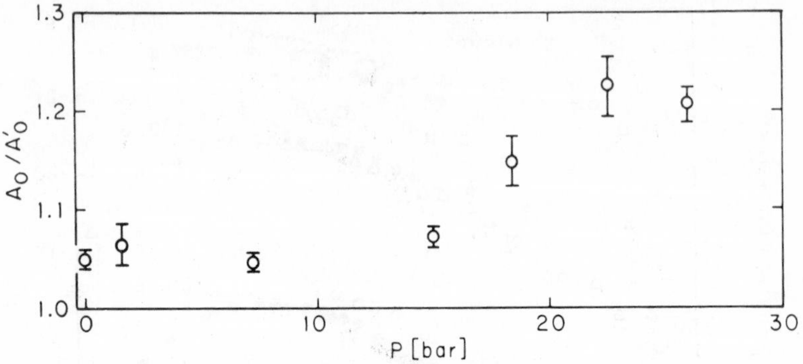

FIGURE 2.14. The ratio of the amplitude of C_P above T_λ to that below T_λ. Although A_0/A_0' was determined by assuming a logarithmic divergence, any exponent near zero will yield virtually the same ratios, provided $\alpha = \alpha'$. After Reference [39].

described by the pure power law (2.1.64), in conjunction with the measurements leads to a conflict with theory. For this reason, the data have also been analyzed in terms of the function

$$C_P = \frac{A}{\alpha}|\varepsilon|^{-\alpha}\left[1 + D|\varepsilon|^x\right] + B; \quad x > 0 \tag{2.1.68}$$

which is similar to the power law equation (2.1.64), but includes *singular* corrections to the leading singularity. These corrections may be sizable even for small $|\varepsilon|$. Equation (2.1.68), with the constraints $\alpha = \alpha'$ and $x = x'$ that are imposed by theory, together with the data yields

$$-0.04 \lesssim \alpha = \alpha' \lesssim 0.02,$$
$$0.5 \lesssim x = x' \lesssim 0.9. \tag{2.1.69}$$

The result equation (2.1.69) is valid at all pressures. It leaves sufficient latitude in α and α' to permit a leading exponent different from zero. In that case, scaling no longer requires $A/A' = 1$. However, if C_P diverges, then $\alpha = \alpha' \cong 0.02$. In that case, one still obtains $A/A' > 1$, and this implies that the ΔC_P [equation (2.1.67)] shown in Figure 2.13 will become positive for sufficiently small $|\varepsilon|$. The heat capacity above T_λ would then be larger than C_P below T_λ at the same $|\varepsilon|$. Such a behavior appears unexpected, but is not excluded by any rigorous arguments. Equation (2.1.69), of course, also permits $\alpha = \alpha' < 0$. Then again, $A/A' > 1$ is permitted by scaling. For negative α and α', an amplitude ratio greater than unity is not at all surprising and does not lead to unusual behavior of C_P.

Recently, a method has been developed [91–93] for calculating exponents and the equation of state by means of an expansion in $4 - d$, where d is the dimensionability of the system. These calculations have revealed the existence of singular corrections to the leading term for C_P, and for the correction exponent they yielded [94] $x \cong 0.5$. When this constraint for x is applied in the data analysis, then

$$-0.04 \lesssim \alpha = \alpha' \lesssim 0 \tag{2.1.70}$$

is obtained. This result, scaling, and $A/A' > 1$, permit only negative values for α and α'. At T_λ, C_P would then be finite. The extrapolated value of C_P at T_λ depends rather strongly upon the exact value of $\alpha = \alpha'$. For $\alpha = \alpha' \cong -0.02$, $C_P(T_\lambda)/R \cong 35$ and almost independent of P, and for an exponent of -0.04, C_P/R becomes only about as large as 20.*

The ratio A/A' is essentially independent of the value chosen for $\alpha = \alpha'$, at least over the range of equation (2.1.69). Thus, regardless of the precise value of the exponent, the results shown in Figure 2.14 are representative of the amplitude ratio. Although the departure from unity of A/A' is consistent with theory if C_P does not diverge logarithmically, the pressure dependence of A/A' is contrary to universality arguments [88–90] and explicit calculations [93].

For a negative leading exponent (i.e., a finite C_P) one might expect C_P to be continuous at T_λ because a discontinuity in some sense corresponds to a leading exponent equal to zero. The data and the assumption of a universal $\alpha = \alpha' < 0$ yield a C_P at T_λ that is continuous at most over part of the pressure range. The discontinuity at some pressures is an additional departure from the theoretically expected behavior.†

The coefficients of the expansion in $(4 - d)$ of exponents and of A/A' are functions only of such general properties as the number of degrees of freedom n of the order parameter [91–93]. We would thus expect $\alpha = \alpha'$ and A/A' to be smoothly varying functions of n. Therefore, we have collected in Figure 2.15 what appear to be the best available experimental estimates of $\alpha = \alpha'$ and A/A' for different types of systems with $n = 1, 2$, and 3. For $n = 2$, the superfluid transition yields $\alpha = \alpha' \cong -0.02$ and $A/A' = 1.06$ at low P. For $n = 3$, the isotropic Heisenberg antiferromagnet $RbMnF_3$ has $\alpha = \alpha' = -0.14 \pm 0.01$ and $A/A' \cong 1.4$ [95]. For $n = 1$, liquid–gas critical points [96, 97, 97a] and anisotropic antiferromagnets [97b] yield $\alpha = \alpha' \cong \frac{1}{8}$ and $A/A' \cong 0.53$. There are no other values of n for which experiments exist, but $n = \infty$ corresponds to the exactly soluble spherical model [1] that has $\alpha = -1$. The solid lines in Figure 2.15 are drawn through the ex-

*It is expected from theory that the constants B and B' in equation (2.1.68) should be equal to each other. With the constraint $B = B'$, the thermal expansion data by Mueller et al. (see footnote on p. 111) yield $\alpha = \alpha' = -0.026 \pm 0.004$, $A/A' = 1.11 \pm 0.02$, and $D/D' = 1.27 \pm 0.2$. All three of these parameters are pressure independent within experimental error, as predicted by theory [88]. It is clear now that consistency between scaling and the experimental results can be obtained only if α is negative, i.e. if C_p remains finite at T_λ. There is no obvious explanation at this time for the difference between the results obtained for A/A' from C_p and from α_p.

†See, however, the preceding footnote.

perimental data. For α, the line continues smoothly for small $1/n$ to the value -1 of the spherical model. Near $1/n = 0$, this line is drawn with the limiting slope $32/\pi^2$ which is predicted by theory in three dimensions from an expansion of α in $1/n$ [98, 99]. It is evident that the theoretical results for large n and the experimental data for $n = 1$, 2, and 3 yield a smooth dependence of α upon n. In particular, the very slightly negative α for $n = 2$, which was obtained for the superfluid transition, is quite consistent with the known theoretical and experimental results for other values of n. The dashed curve

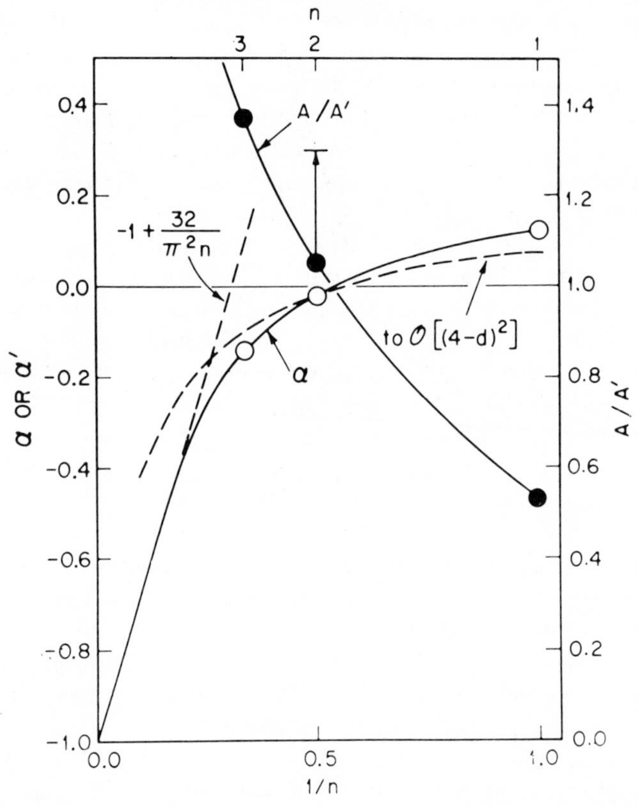

FIGURE 2.15. The exponent $\alpha = \alpha'$ and the amplitude ratio A/A' for the heat capacity as a function of the number of degrees of freedom of the order parameter. The data points are experimental results for liquid–gas critical points [96,97] ($n = 1$), for the superfluid transition at moderate pressure [39] ($n = 2$), and for a Heisenberg antiferromagnet [95] ($n = 3$). The spherical model has [1] $1/n = 0$, and yields $\alpha = -1$. The solid lines are drawn smoothly through these data, with a limiting slope for α equal to the theoretical value [98,99] of $32/\pi^2$. The dashed curve is an expansion [91–93] in $4 - d$, where d is the dimensionality (i.e., 3) of the system. The arrow ends at the A/A' for the superfluid transition under higher pressure. After Reference [39].

in Figure 2.15 corresponds to

$$\alpha = \alpha' = -\frac{n-4}{2(n+8)}(4-d) - \frac{n^3+32n^2+116n+112}{4(n+8)^3}(4-d)^2, \quad (2.1.70a)$$

which was obtained [99a] from the expansion in the dimensionality to second order in $(4-d)$ [91–93]. This estimate agrees with the experiment values.

As we saw for the exponents, the values of A/A' also fall on a smooth curve. This curve yields $A/A' = 1$ at the vaule of $1/n$ where $\alpha = \alpha' = 0$, and, therefore, the scaling prediction $A/A' = 1$ if $\alpha = \alpha' = 0$ is consistent with experiment. The larger values of A/A' obtained for the superfluid transition at the higher pressures are indicated by the arrow on the point for $1/n = 0.5$. They could also be connected with the results for other n by a smooth function, but the curve through the large A/A' for $n = 2$ and through the results for $n = 1$ and $n = 3$ would not yield $A/A' = 1$ when $\alpha = \alpha' = 0$, and would, therefore, be inconsistent with scaling.* Recently, a calculation of A/A' has also become available [99b, 99c]. Although explicitly only to first order in $4-d$, it is based on the second-order expansion of the equation of state, and can written as

$$\frac{A}{A'} = \frac{n}{4} + \left[\frac{n}{4} + \frac{(4-n)n}{8(8+n)}\ln 2\right](4-d) + \mathcal{O}\left[(4-d)^2\right] + \cdots. \quad (2.1.70b)$$

Equation (2.70b) yields $A/A' = 0.53$, 1.03, and 1.52 for $n = 1$, 2 and 3, respectively, in excellent agreement with the experimental results, shown in Figure 2.15, of 0.53, 1.06, and 1.4. The agreement with the result for helium at vapor pressure once more indicates that the high-pressure result $A/A' \cong 1.3$ is in some sense anomalous.*

Since the pressure-dependent amplitude ratio of C_P seems in conflict with theoretical predictions, one could attempt to obtain greater harmony between theory and experiment by assuming that perhaps C_P, α_P, or κ_T are not the variables to which the predictions pertain. However, the only other response function that readily comes to mind is the heat capacity at constant chemical potential C_μ, where μ is the molar chemical potential of ^{4}He; it can be shown that C_μ has the same asymptotic singularity as C_P. This can be seen by using equation (2.1.7) to obtain

$$\left(\frac{\partial S}{\partial T}\right)_\mu = \left(\frac{\partial S}{\partial T}\right)_P + \left(\frac{\partial S}{\partial P}\right)_T \left(\frac{\partial P}{\partial T}\right)_\mu, \quad (2.1.71)$$

which yields

$$C_\mu = C_P - T\left(\frac{\partial V}{\partial T}\right)_P \left(\frac{\partial P}{\partial T}\right)_\mu. \quad (2.1.72)$$

*See, however, footnote on p. 117.

Now, since $(\partial P/\partial T)_\mu = S/V$, we obtain with equation (2.1.10) for $(\partial V/\partial T)_P$, after some rearrangement,

$$C_\mu = \left[1 - \left(\frac{S}{V}\right)\left(\frac{\partial T}{\partial P}\right)_\lambda\right] C_P + \left(\frac{ST}{V}\right)\left(\frac{\partial S}{\partial P}\right)_t. \tag{2.1.73}$$

At the same $|T_\lambda - T| \lesssim 10^{-2}$ K, C_μ differs from C_P only by a few percent at all P. It has the same amplitude ratio as C_P.

We saw in this section that most of the thermodynamic results for pure ^{4}He near T_λ are consistent with theoretical predictions if C_P is finite at T_λ, and if singular correction terms to the asymptotic power law behavior exist. Even then, however, the amplitude ratio should be independent of pressure; this expectation is in disagreement with experiment. It would be extremely desirable to obtain experimental confirmation of the behavior of A/A' shown in Figure 2.14, particularly since the thermal expansion measurements [51] have yielded a different result. This confirmation could most easily be obtained from additional, possibly more precise, thermal expansion measurements, although perhaps the sound velocity under pressure would also be useful from this point of view.†

2.1.3 ^{3}He–^{4}He Mixtures

(*a*) *General Relations.* A general review of the thermodynamics of multi-component systems has been given, for instance, by Guggenheim [100]. We will define the molar concentration of ^{3}He as

$$X \equiv \frac{N_3}{(N_3 + N_4)}, \tag{2.1.74}$$

where N_3 and N_4 are the number of moles of ^{3}He and ^{4}He respectively. The molar chemical potentials μ_3 and μ_4 of ^{3}He and ^{4}He are given by

$$\mu_3 = \left(\frac{\partial \tilde{G}}{\partial N_3}\right)_{P,\,T,\,N_4} \tag{2.1.75}$$

and

$$\mu_4 = \left(\frac{\partial \tilde{G}}{\partial N_4}\right)_{P,\,T,\,N_3}, \tag{2.1.76}$$

where $\tilde{G}$ is the Gibbs free energy of the solution. The thermodynamic conjugate $(\partial G/\partial X)_{P,\,T,\,N_3+N_4}$ of X is the difference in molar chemical potential

$$\Phi = \mu_3 - \mu_4. \tag{2.1.77}$$

Here $G = \tilde{G}/(N_3 + N_4)$ is the Gibbs free energy per mole of solution.

With concentration as an additional degree of freedom, the superfluid transition is now confined to a surface. The transition in pure ^{4}He discussed in Section 2.1.2 occurs along the line on this surface that is defined by the

†See footnote on p. 117.

constraint $X = 0$. In this section, we shall consider primarily the transition along another line defined by the constraint that the pressure be constant (although occasionally we shall also consider the pressure dependence of thermodynamic parameters). The choice of constant P is dictated by the fact that almost all experimental results in mixtures have been obtained at vapor pressure. The difference between derivatives at SVP and at a constant pressure equal to SVP at T_λ is generally quite small. We shall ignore it, and use experimental results at SVP as though they were at constant P.

The response functions to be considered are $C_{\Phi P} \equiv T(\partial S/\partial T)_{\Phi P}$, $C_{XP} = T(\partial S/\partial T)_{XP}$, $(\partial X/\partial \Phi)_{TP}$, $(\partial X/\partial T)_{\Phi P}$, and $(\partial \Phi/\partial T)_{XP}$. They are equivalent to some of those defined by equations (2.1.1)–(2.1.6), which pertained to the conjugate variables $(-P)$ and V. As in Section 2.2.2, we can derive the relations

$$C_{\Phi P} = \left[T\left(\frac{\partial S}{\partial T}\right)_{tP}\right] - \left[T\left(\frac{\partial \Phi}{\partial T}\right)_{tP}\right]\left(\frac{\partial X}{\partial T}\right)_{\Phi P}, \tag{2.1.78}$$

$$\left(\frac{\partial X}{\partial T}\right)_{\Phi P} = \left[\left(\frac{\partial X}{\partial T}\right)_{tP}\right] - \left[\left(\frac{\partial \Phi}{\partial T}\right)_{tP}\right]\left(\frac{\partial X}{\partial \Phi}\right)_{TP}, \tag{2.1.79}$$

$$C_{\Phi P} = \left[T\left(\frac{\partial S}{\partial T}\right)_{tP} - T\left(\frac{\partial \Phi}{\partial T}\right)_{tP}\left(\frac{\partial X}{\partial T}\right)_{tP}\right] + \left[T\left(\frac{\partial \Phi}{\partial T}\right)_{tP}^2\right]\left(\frac{\partial X}{\partial \Phi}\right)_{TP}, \tag{2.1.80}$$

$$\left(\frac{\partial \Phi}{\partial T}\right)_{XP} = \left[\left(\frac{\partial \Phi}{\partial T}\right)_{\lambda P}\right] - \left[\left(\frac{\partial X}{\partial T}\right)_{\lambda P}\right]\left(\frac{\partial \Phi}{\partial X}\right)_{TP}, \tag{2.1.81}$$

$$C_{XP} = \left[T\left(\frac{\partial S}{\partial T}\right)_{tP}\right] + \left[T\left(\frac{\partial X}{\partial T}\right)_{tP}\right]\left(\frac{\partial \Phi}{\partial T}\right)_{XP}, \tag{2.1.82}$$

which are similar to equations (2.1.10)–(2.1.13). We should emphasize that, in this case, the pressure is held constant for all of the partial derivatives. Thus, for instance, the quantity $(\partial S/\partial T)_{tP}$ that occurs here is very different from the entropy derivative at constant t that was encountered in equation (2.1.10). That derivative for the pure system might be written more explicitly as $(\partial S/\partial T)_{tX}$, with $X = 0$.

(b) The Phase Diagram and Properties at T_λ. Addition of ^{3}He to ^{4}He will reduce the superfluid transition temperature. The T–X plane of the phase diagram at vapor pressure [101–110] is shown in Figure 2.16. At large ^{3}He concentration, the λ line terminates at [107, 110]

$$X_t = 0.675 \pm 0.001,$$
$$T_t = 0.8671 \pm 0.001 \text{ K}, \tag{2.1.83}$$

because a first-order transition intervenes. The first-order transition is a phase

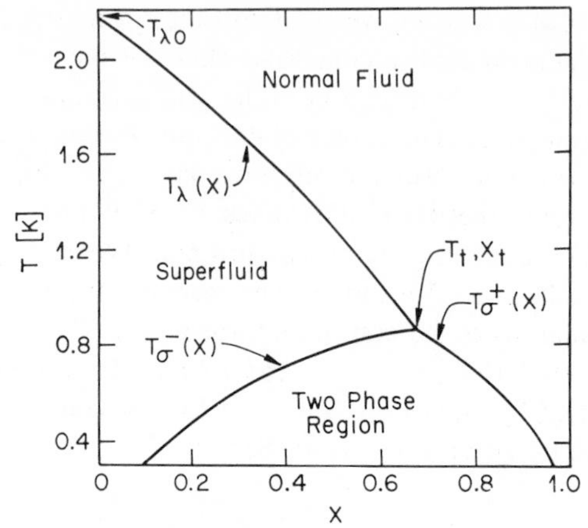

FIGURE 2.16. The phase diagram of He^3–He^4 mixtures in the T–X plane.

separation [3, 111] into a ^{4}He-rich superfluid and a ^{3}He-rich normal phase which will occur when a mixture of concentration $X \geq 0.06$ [112] is cooled below a temperature $T_\sigma(X)$. Although we shall not discuss the first-order transition for its own sake, we will be concerned about what happens in the immediate vicinity of the point (T_t, X_t) where the λ line meets the phase-separation curve. Junctions between first-order and λ-transition lines of the type encountered in ^{3}He–^{4}He mixtures have become known [113] as tricritical points, and are very interesting from the viewpoint of phase transitions.

We shall now consider the derivatives at T_λ that occur in equations (2.1.78) –(2.1.82). We believe them to be generally finite since they must be integrable [except for $(\partial\Phi/\partial T)_\lambda$ at $X = 0$, as we shall see], but we will examine the possibility that, on the λ line in the X–T plane, a singular point exists at which one or several of the derivatives may diverge. Particularly, let us consider the point with $X = 0$. We know that $(\partial X/\partial T)_{\lambda P}$ and $(\partial S/\partial T)_{\lambda P}$, in order to be integrable, must diverge less rapidly than $(T_{\lambda 0} - T_\lambda)^{-1}$, where $T_{\lambda 0}$ is T_λ at $X = 0$. Although there are no rigorous arguments, one expects on the basis of several theortical approaches [3, 101] that $(\partial X/\partial T)_{\lambda P}$ will remain finite at $X = 0$. For an ideal mixture of a degenerate Bose–Einstein gas (^{4}He) and a Fermi–Dirac gas (^{3}He), for instance [114],

$$T_\lambda(X) = T_{\lambda 0}(1 - X)^{2/3}. \tag{2.1.84a}$$

This yields

$$\left(\frac{\partial X}{\partial T}\right)_{\lambda P} = \frac{-3(1 - X)^{1/3}}{2T_{\lambda 0}}, \tag{2.1.84b}$$

which is finite and nonzero for all $X < 1$. This result is, of course, not exact for real mixtures of ^{3}He and ^{4}He; but experimental measurements [101, 102] reveal that, near $X = 0$, $(\partial X/\partial T)_{\lambda P} = -0.71$. Equation (2.1.84b) yields -0.69, and thus is remarkably accurate.

On thermodynamic grounds, one can be more specific for $(\partial \Phi/\partial T)_{\lambda P}$ than for $(\partial X/\partial T)_{\lambda P}$. We have [100]

$$\Phi = \Phi^0 + \Phi^E + RT \ln\left[\frac{X}{(1 - X)}\right], \tag{2.1.85}$$

where Φ^0 is the difference of the chemical potentials of the pure components, $RT \ln[X/(1-X)]$ is the ideal mixing contribution, and Φ^E is the excess chemical potential that contains the contributions due to deviations from ideal solution behavior. At $X = 0$, Φ diverges as $\ln X$. For $(\partial\Phi/\partial T)_{\lambda P}$, we have

$$\left(\frac{\partial \Phi}{\partial T}\right)_{\lambda P} = \left(\frac{\partial X}{\partial T}\right)_{\lambda P}\left\{\left[\frac{\partial(\Phi^0 + \Phi^E)}{\partial X}\right]_{\lambda P} + R\ln\left(\frac{X}{1 - X}\right)\left(\frac{\partial T}{\partial X}\right)_{\lambda P} + RT\,X^{-1}(1 - X)\right\}. \tag{2.1.86}$$

We do not expect any strong singularities in $(\partial\Phi^E/\partial X)_{\lambda P}$ and $(\partial\Phi^0/\partial X)_{\lambda P}$, but the ideal mixing term yields a $(\partial\Phi/\partial T)_{\lambda P}$ that diverges as X^{-1} at $X = 0$. Similar arguments can be presented to show that $(\partial S/\partial T)_{\lambda P}$ diverges at $X = 0$ as $\ln X$ because the entropy of mixing vanishes near $X = 0$ as $X \ln X$. Specifically,

$$\left(\frac{\partial S}{\partial T}\right)_{\lambda} = \left[\frac{\partial(S^0 + S^E)}{\partial T}\right]_{\lambda P} - R\left(\frac{\partial X}{\partial T}\right)_{\lambda P}\ln\left(\frac{X}{1 - X}\right). \tag{2.1.87}$$

Thermodynamic calculations like those made in Section 2.1.2. for pure ^{4}He are possible if measurements of one response function are available, and if the derivatives $(\partial S/\partial T)_{\lambda P}$, $(\partial X/\partial T)_{\lambda P}$, and $(\partial\Phi/\partial T)_{\lambda P}$ are known. Therefore, good numerical results must be obtained for these parameters. Much of the pertinent experimental information [102, 103, 115–117] at vapor pressure was surveyed recently by Gasparini [68]. Additional measurements near the tricritical concentration [104–110] have become available since then, and our best estimates based upon a combination of all those results are given in Table 2.1. Also given are the transition temperature and the entropy at T_λ. The uncertainties in these data are difficult to estimate, but approximately $(\partial X/\partial T)_{\lambda P}$, and $(\partial\Phi/\partial T)_{\lambda P}$ not too near the tricritical point, may be uncertain by perhaps 2%. Near the tricritical point, errors in $(\partial\Phi/\partial T)_{\lambda P}$ could be as large as 10%. The uncertainty for $(\partial S/\partial T)_{\lambda P}$ is roughly 1 J/mole K or 10%, whichever is larger. The entropy is probably known within about 2%.

Of the several response functions under consideration, we expect C_{XP} and $(\partial\Phi/\partial T)_{XP}$ to be finite at T_λ, except possibly at $X = 0$. The upper bound for C_{XP}

$$C_{XP}(T_\lambda) \le T_\lambda\left[\left(\frac{\partial S}{\partial T}\right)_{\lambda P} + \left(\frac{\partial X}{\partial T}\right)_{\lambda P}\left(\frac{\partial \Phi}{\partial T}\right)_{\lambda P}\right] \tag{2.1.88}$$

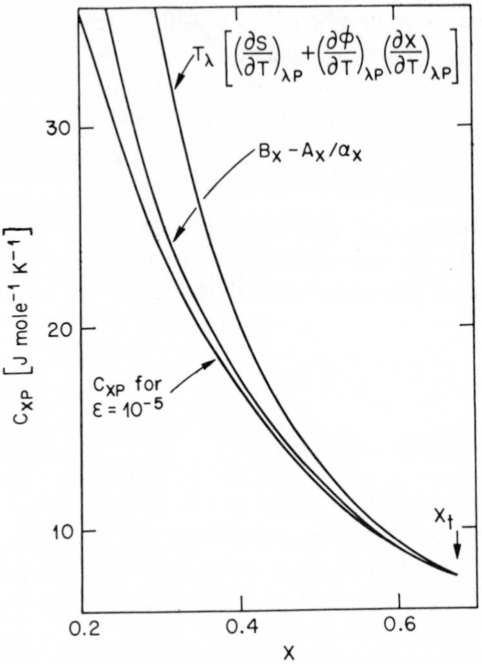

FIGURE 2.17. Three estimates of the largest value of C_{PX} as a function of X. Upper curve, thermodynamic upper bound; middle curve, power-law extrapolation of the measured C_{XP} to T_λ; lower curve, C_{XP} for $|1 - T/T_\lambda| \cong 10^{-5}$.

is analogous to equation (2.1.24) for C_V in pure He^4. It diverges at $X = 0$ [see equation (2.1.86)], but $C_{XP}\,(T_\lambda)$ does not become infinite unless $C_{\Phi P}$ diverges. For a divergent $C_{\Phi P}$, equation (2.1.88) becomes an equality. Similarly, $|(\partial\Phi/\partial T)_{XP}| \leq |(\partial\Phi/\partial T)_{\lambda P}|$ becomes an equality for $T = T_\lambda$ only if $C_{\Phi P}$ diverges. The upper bound for C_{XP} is shown in Figure 2.17 as a function of concentration. Also shown are the value $B_X - A_X/\alpha_X$ of a power-law extrapolation with $\alpha < 0$ of C_{XP} to T_λ [see equations (2.1.89) and (2.1.94) below] and an estimate of C_{XP} when $\varepsilon \cong 10^{-5}$. Except at the tricritical concentration, the experimental estimate of C_{XP} at T_λ appears lower than the upper bound, implying that $C_{\Phi P}$ is finite at T_λ; however the errors in the estimates are appreciable, and the comparison of $B_X - A_X/\alpha_X$ with equation (2.1.88) is not really conclusive. This is true particulary since the asymptotic dependence of C_{XP} upon ε is a linear function of $C_{\Phi P}^{-1}$ [see equation (2.1.23)], and not the simple power law that was used to get the estimate $C_{XP}\,(T_\lambda) \cong B_X - A_X/\alpha_X$. Nonetheless, when this estimate and the parameters in Table 2.1 are used to calculate $C_{\Phi P}\,(T_\lambda)$ from equation (2.1.96) to be given below, one obtains values between 300 and 130 J/mole K over the range $0.1 \leq X \leq 0.5$. These results are much larger than any measured C_{XP} (see

TABLE 2.1. Parameters of the λ line in the T–X plane at vapor pressure

X	T_λ [K]	S_λ [J/mole K]	$-(\partial X/\partial T)_{\lambda P}$ [K^{-1}]	$-(\partial \Phi/\partial T)_{\lambda P}$ [J/mole K]	$(\partial S/\partial T)_{\lambda P}$ [J/mole K^2]
0.00	2.172	6.24	0.714	–	–
0.05	2.100	8.22	0.692	275	–17.0
0.10	2.026	9.34	0.669	140	–12.2
0.15	1.950	10.23	0.645	93	–9.3
0.20	1.871	10.92	0.620	69.2	–7.1
0.25	1.788	11.44	0.592	53.7	–5.3
0.30	1.701	11.82	0.560	43.9	–3.7
0.35	1.609	12.07	0.525	35.7	–2.3
0.40	1.510	12.19	0.494	29.0	–0.9
0.45	1.404	12.19	0.466	24.0	0.3
0.50	1.294	12.07	0.444	19.8	1.4
0.55	1.173	11.83	0.429	16.1	2.5
0.60	1.051	11.47	0.418	12.8	3.6
0.65	0.928	10.99	0.412	10.0	4.6
0.675	0.867	10.70	0.409	8.8	5.2

Figure 2.17), and are quite consistent with the estimate of the possibly finite C_P at T_λ in pure ^{4}He that was given in Section 2.1.2(e) just below equations (2.1.70).

At $X = X_t$, the calculated upper bound agrees with the measured C_{XP} at T_t [105, 106]. This has often been interpreted to imply that $C_{\Phi P}$ diverges at T_λ near T_t, but, in fact, the uncertainties in all the parameters that are involved allow a $C_{\Phi P}$ (T_λ) as small as perhaps 40 J/mole K.

(*c*) *Response Functions Near* T_λ. One would like to obtain experimental information about the singularity at T_λ in the response functions $C_{\Phi P}$, $(\partial X/\partial \Phi)_{TP}$, or $(\partial X/\partial T)_{\Phi,P}$, which are permitted by thermodynamics to diverge along a λ line. Unfortunately, these quantities are, for the most part, not readily accessible to direct measurement. However, results for the generally finite C_{XP} can be used to obtain $C_{\Phi P}$ by means of thermodynamic calculations [117a] and $C_{\Phi P}$ can then be examined for its behavior near T_λ. High-resolution measurements of C_{XP} at six concentrations with $0 \le X \le 0.39$ have been made by Gasparini and Moldover [67–69]. In addition, another recent set of C_{XP} [105, 106] with $0.53 \le X \le 0.73$ and somewhat lesser temerature resolution covers in some detail the region near the tricritical point. The results for $X \le 0.53$, when fitted to a power law like equation (2.1.57), which we already used for C_V, yielded [68, 69] negative exponents α_X and α'_X. As we discussed for C_V, this is consistent with the finite C_{XP} (T_λ), which is required by thermodynamics. For large X, the measurements [105] for C_{XP} are shown as a function of T in Figure 2.18. At $X = 0.53$, it is still easy to recognize the cusped C_{XP} at T_λ, which for this X occurs at 1.224 K, but the largest measured heat capacities are already a factor of about 8 smaller than those in pure ^{4}He. At this concentration, a fit of the data [105] to equation (2.1.57) yields [69] $\alpha_X \cong \alpha'_X \cong -0.29$. In addition to the λ anomaly there is also a discontinuity ΔC_{XP} in C_{XP} for $X = 0.53$ that occurs at $T_\sigma = 0.81$ K where, upon heating, the system leaves the two-phase region (see Figure 2.16). As the concentration increases to-

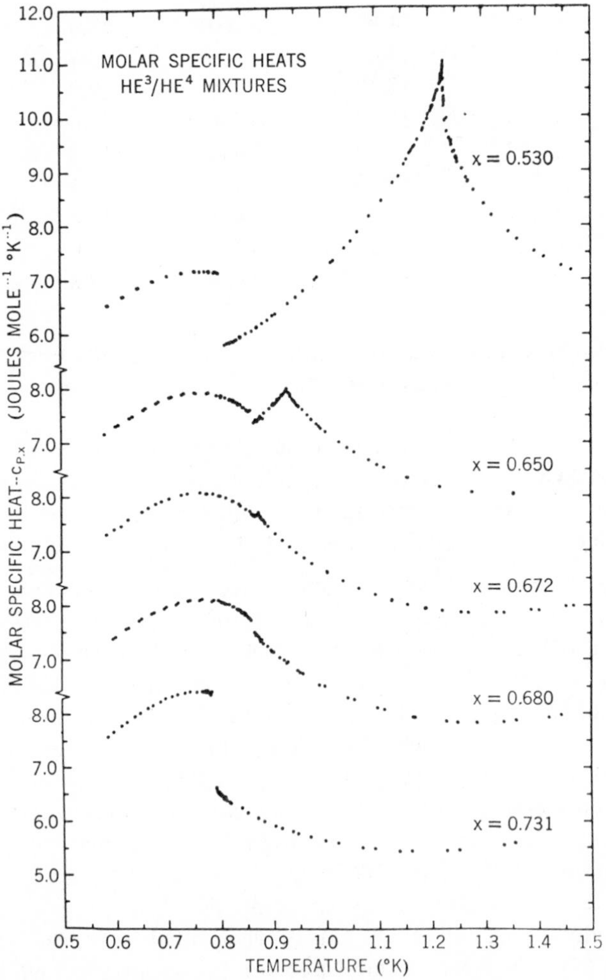

FIGURE 2.18. The measured C_{XP} as a function of T on several lines of constant X near X_t. Note the breaks in the vertical scale. After Reference [105].

wards X_t, both the λ anomaly and the discontinuity at T_σ become quite small, and probably vanish for $X = X_t$, leaving a regular heat capacity. It would be of interest to establish with greater precision the detailed behavior of the singularities in C_{XP} near X_t, for they have a bearing upon the form of the equation of state near the tricritical point. For $X > X_t$, there is, of course, no λ anomaly, but ΔC_{XP} at T_σ, again becomes appreciable.

For many computational purposes, it is useful to have a closed-form expression for C_{XP} similar to the one that we presented for C_V in equations (2.1.55) and (2.1.56). In this case, however, $C_{XP}(T_\lambda)$ is so obviously finite for large X that we can no longer retain the logarithmic functional form. Instead, the data for He (I) will be represented by the more general power law

$$C_{XP} = \left(\frac{A_X}{\alpha_X}\right)\left[\left|\frac{t}{T_\lambda(X)}\right|^{-\alpha_X} - 1\right] + B_X, \tag{2.1.89}$$

where $t \equiv T - T_\lambda(X)$ is measured along a line of constant X. We will approximate these three parameters by

$$\alpha_X = a_1 + b_1 X + c_1 X^2, \tag{2.1.90}$$

$$A_X = a_2 + b_2 X + c_2 X^2, \tag{2.1.91}$$

$$B_X = a_3 + b_3 X + c_3 X^2. \tag{2.1.92}$$

The same functions with primed coefficients will be used for He (II). The choice*

$$a_1 = a_1' = 0 \tag{2.1.93}$$

will guarantee, consistent with equation (2.1.65), a logarithmically divergent C_{XP} for $X = 0$. Equation (2.1.89) with $\alpha_X < 0$ yields

$$C_{XP}(T_\lambda) = B_X - \frac{A_X}{\alpha_X}. \tag{2.1.94}$$

For vanishing X, equations (2.1.90) and (2.1.93) assure that $C_{XP}\,(T_\lambda)$ diverges as X^{-1}, consistent with the known behavior of the upper bound equation (2.1.88) [see also equation (2.1.86)] to $C_{XP}(T_\lambda)$. We will also choose

$$a_2 = 5.355, \qquad a_2' = 5.100,$$

$$a_3 = -7.77, \text{ and } a_3' = 15.52,$$

which assures consistency with equations (2.1.65) and (2.1.66) for the pure system. For the remaining coefficients, we will assume, on the basis of the best experimental information [105, 106], that $A_X = A_X' = 0$ for $X = X_t$, and

*It is now established that $\alpha <$ for 0 pure ^{4}He (see, for instance, the footnote on p. 117). Parameters for equations (2.1.90)–(2.1.92) with $a_1 = a'_1 - 0.015$ are given in Reference 117a. They yield values for C_{XP} which differ very little from those obeained with the parameters in Table 2.2.

that $B_X = B'_X = C_{XP} = 7.60$ J/mole K at the tricritical point. This assumption results in two additional constraints and leaves four free parameters for each phase. These were least-squares adjusted so as to fit smooth data with $10^{-4} \leq t/T_\lambda \leq 10^{-2.4}$ generated from the power laws given in Reference [68] for $0.01 \leq X \leq 0.39$ and taken from Figure 9 in Reference [105] for $X = 0.53$. We collect all coefficients in Table 2.2. Equations (2.1.89)–(2.1.92) with these coefficients reproduce the input data to the least-squares fit within $\pm 1\%$.

It is not difficult, in principle, to obtain $C_{\Phi P}$ from C_{XP} and the derivatives at constant $t = T - T_\lambda(X)$. We follow a procedure very similar to that used in Section 2.1.2(e) to relate C_P to C_V in pure ^{4}He; except that for the pure system we also had $(\partial P/\partial T)_V$, equivalent to $(\partial \Phi/\partial T)_X$, from experiment. We have, analogous to the usual relations between C_P and C_V [see equation (2.1.20), for instance],

$$C_{\Phi P} = C_{XP} - T\left(\frac{\partial \Phi}{\partial T}\right)_{XP}\left(\frac{\partial X}{\partial T}\right)_{\Phi P}. \tag{2.1.95}$$

Using equations (2.1.82) and (2.1.78) we obtain $(\partial \Phi/\partial T)_{XP}$ and $(\partial X/\partial T)_{\Phi P}$ as linear functions of C_{XP} and $C_{\Phi P}$, respectively, with coefficients that are derivatives at constant t. Some manipulation results in

$$C_{\Phi P} = \left\{C_{XP}\left[T\left(\frac{\partial X}{\partial T}\right)_{tP}\left(\frac{\partial \Phi}{\partial T}\right)_{tP} - T\left(\frac{\partial S}{\partial T}\right)_{tP}\right] + \left[T\left(\frac{\partial S}{\partial T}\right)_{tP}\right]^2\right\} \times \left[T\left(\frac{\partial S}{\partial T}\right)_{tP} + T\left(\frac{\partial X}{\partial T}\right)_{tP}\left(\frac{\partial \Phi}{\partial T}\right)_{tP} - C_{XP}\right]^{-1}. \tag{2.1.96}$$

Along lines of constant concentration, $(\partial X/\partial T)_{tP} \equiv (\partial X/\partial T)_{\lambda P}$. In principle, the dependence upon t of $(\partial S/\partial T)_{tP}$ can be obtained by integrating C_{XP}/T, which is available from equation (2.1.89), and differentiating at constant t. Similarly, C_{XP} and equation (2.1.82) yield $(\partial \Phi/\partial T)_{XP}$, which may be integrated to yield the t dependence of Φ. This in turn may be differentiated at

TABLE 2.2. Coefficients of equations (2.1.90)–(2.1.92)

	He(I)	He(II)
a_1	0	0
b_1	−0.42642	−0.54603
c_1	−0.28486	−0.35733
a_2	5.355	5.100
b_2	0.043	−4.376
c_2	−11.923	−4.781
a_3	−7.773	15.52
b_3	25.992	−28.198
c_3	−4.608	24.422

constant t to yield $(\partial\Phi/\partial T)_{tP} - (\partial\Phi/\partial T)_{\lambda P}$. This procedure would be similar to that followed to obtain the necessary derivatives at constant t for the pure system. In practice, it was not employed in the calculation of $C_{\Phi P}$ from C_{XP}. Instead, Gasparini [68] used equation (2.1.96) and assumed that the three derivatives are all constant and equal to their values at T_λ, which are given in Table 2.1. The lesser precision of the derivatives at T_λ may well warrant this simplification, and on the basis of the experience with the pure system one would not expect serious errors from this source. The measured values of t along a path of constant X were changed to those appropriate to a path of constant Φ. This is similar to using equation (2.1.16) for the pure system.

For large X, $C_{\Phi P}/C_{XP}$ becomes quite large and the derivatives at T_λ are not known as accurately as in ^{4}He. Thus, the uncertainties in $C_{\Phi P}$ become more appreciable than in the pure system. Taking these possible errors into consideration, Gasparini and Moldover [68, 69] obtained the exponents α_Φ for $C_{\Phi P}$ with the error bars that are shown in Figure 2.19 from a least-squares fit to a power law like equation (2.1.89). On the basis of universality arguments [88–90], one would expect α_Φ and α'_Φ to be equal to α and α' for the pure system, and scaling [71–75] requires $\alpha_\Phi = \alpha'_\Phi$. Although apparently significant differences between α_Φ and α'_Φ seem to occur for small X, we are not inclined to take them too seriously because their existence is dependent upon the strict applicability of a pure power law over the range of $t/T_\lambda \lesssim 4 \times 10^{-3}$ used in the fit. We already saw in Section 2.1.2(e) how higher-order singular contributions can result in apparently different exponents above and below the transition when a pure power law is assumed (see Figure 2.12). The experimental values for the exponents in Figure 2.19 also are consistent with equation (2.1.69) for the pure system. They therefore tend to support the principle of universality, because the addition of the impurity did not change the exponent of $C_{\Phi P}$. The amplitudes A_Φ and A'_Φ of the singularities in $C_{\Phi P}$ have not been determined by a least-squares fit under the constraint that $\alpha_\Phi = \alpha'_\Phi$, but when the values at $t/T_\lambda = 10^{-4}$ and 10^{-3} of the power laws quoted [68] for $C_{\Phi P}$ are used to calculate the coefficients of $C_{\Phi P}$

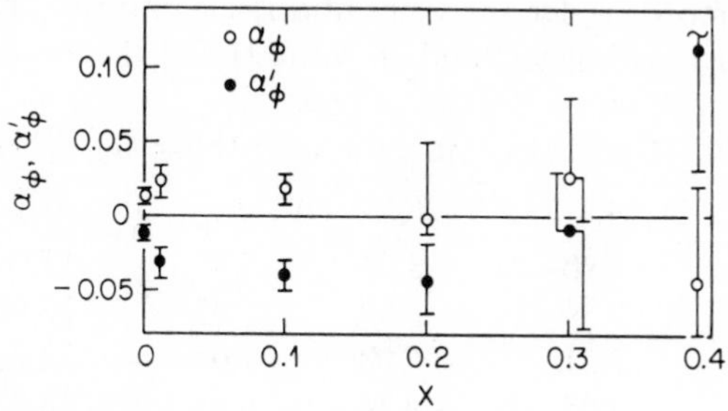

FIGURE 2.19. The exponents of $C_{\Phi P}$ above (α_Φ) and below (α'_Φ) T_λ as a function of X. The data are from Reference [68].

$= - A_{0\Phi} \ln |t/T_\lambda| + B_{0\Phi}$, one obtains the results in Table 2.3. We saw from a detailed analysis for pure ^{4}He that A/A' is not sensitive to the exact value of α, provided $\alpha = \alpha'$. Therefore, Table 2.3 implies that

$$\frac{A_\Phi}{A'_\Phi} = 1.06 \pm 0.02, \text{ for } 0 \le X \le 0.3. \tag{2.1.97}$$

This result is consistent with the amplitude ratio obtained in the pure system, and once more supports the universality hypothesis. The result, equation (2.1.97), is consistent with scaling only if $\alpha_\Phi = \alpha'_\Phi \neq 0$. Figure 2.15 would suggest that $A_\Phi/A'_\Phi > 1$ implies $\alpha_\Phi < 0$. For $X \cong 0.4$, the results on the mixtures seem to imply $A_{0\Phi}/A'_{0\Phi} < 1$. At this high concentration the difference between C_{XP} and $C_{\Phi P}$ becomes large, and it is not obvious that the permitted errors in the ratio are small enough to rule out the result equation (2.1.97). On the other hand, it is conceivable that at this concentration a change in the behavior of $C_{\Phi P}$, due to the approach towards the tricritical point, is already noticeable.

(*d*) *The Phase Diagram near the Tricritical Point in the $T-X$ Plane.* Terminations of λ lines at first-order transition lines, such as the tricritical point in ^{3}He–^{4}He mixtures, were considered already by L. D. Landau [118], and occurred naturally within the framework of his theory of phase transitions. The Landau theory assumes that the free energy can be expanded about the transition temperature as a power series in the order parameter. This assumption of analyticity leads to predictions about the phase diagram and the behavior of response functions near critical points that generally do not pertain in detail to real systems; [119] however many of the qualitative features are given correctly. Therefore Griffiths [113] has suggested that we may expect the Landau theory to serve as a valuable qualitative guide to the phase diagram near the tricritical point with generally correct overall features, but, of course, we have no a priori reason to expect the theory to be quantitatively applicable.

It is convenient, in analogy with the usual practice for critical points in liquid mixtures, to consider the concentration along the phase separation curve as an order parameter, and to write [119a] the following formula:

TABLE 2.3. The coefficients of a logarithmic $C_{\Phi P}$, [J/mole K]

X	$A_{0\Phi}$	$B_{0\Phi}$	$A'_{0\Phi}$	$B'_{0\Phi}$	$A_{0\Phi}/A'_{0\Phi}$
0.0110	5.80	−8.33	5.35	17.04	1.08
0.0997	7.39	−6.99	7.02	23.90	1.05
0.2000	7.31	−6.08	7.05	23.36	1.04
0.3012	7.05	−9.34	6.53	18.82	1.08
0.3900	4.52	−0.12	4.84	14.14	0.93

$$\left| \frac{X_t - X_\sigma}{X_t} \right| = B\left(\frac{T_t - T_\sigma}{T_t}\right)^\beta. \tag{2.1.98}$$

We may distinguish B and β for the ^{3}He-rich and ^{4}He-rich phases by using $+$ and $-$, respectively, as subscripts. Figure 2.20 shows the phase diagram near the tricritical point in the temperature–concentration plane. We have omitted most of the experimental measurements for the sake of clairty. Although there is not fully quantitative agreement among the several experiments [104–110], the major features of the phase diagram are common to all the recent results. The Landau theory predicts that the first-order transition line [i.e., the phase-separation curve $T_\sigma(X)$ in Figure 2.20] should have an angular top at T_t. In terms of equation (2.1.98), this requires

$$\beta_+ = \beta_- = 1. \tag{2.1.99}$$

Equation (2.1.99) is consistent with all recent experiments. Its validity was first discovered by Graf, Lee, and Reppy [104]. A fit of recent data [107] to equation (2.1.98) yields $\beta_- = 1.00$; the best estimate [108] of β_+ is 0.97. There are, however, significant differences in estimates of the amplitude B_-. These are probably attributable to the effect of gravity upon the experimental results that were obtained with samples of several centimeters' height. The influence of the gravitational inhomogeneity upon measurements near the tricritical point is sizable, and will be discussed in Section 2.1.3(g). If equations (2.1.98) and (2.1.99) are assumed valid, then the best, essentially gravity-free, estimates of the coefficients in equation (2.1.98) are

$$B_- = 4.19, \quad B_+ = 0.88. \tag{2.1.100}$$

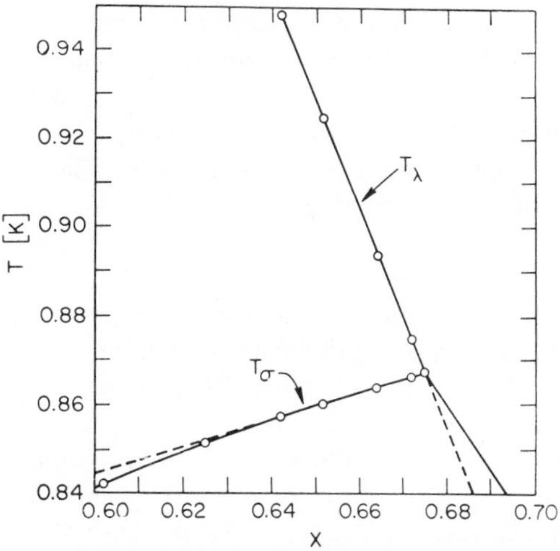

FIGURE 2.20. The phase diagram of He^3–He^4 mixtures in the T–X plane near the tricritical point. The open circles are data points from Reference [107].

These values are consistent with the three sets of measurements that employed sufficiently short samples [107–110]. It is difficult to judge at this time how reliably equations (2.1.98)–(2.1.100) represent the asymptotic behavior at T_t, X_t, because the smallest $|1 - X_\sigma/X_t|$ at which precise measurements have been made is about 5×10^{-3}, and a change in the effective β inside this region is not out of the question. However, it seems unlikely that equation (2.1.99) is grossly incorrect. For $0.64 \leq X \leq X_t$, the data points shown explicitly in Figure 2.20 [107] fall within 10^{-4} K of the straight solid lines which correspond to equations (2.1.98)–(2.1.1.100).

The Landau theory also predicts that the λ line should terminate at the top of the coexistence curve, and again is in good agreement with experiment. However, the λ line is also predicted to have the same slope as the ^{3}He-rich side of the phase-separation curve; i.e., $B_\lambda = B_t$, where B_λ is defined by

$$\frac{X_t - X_\lambda}{X_t} = B_\lambda\left(\frac{T_\lambda - T_t}{T_t}\right), \tag{2.1.101}$$

for X_λ near X_t. In reality, the data [104–110] yield [107]

$$B_\lambda = 0.52, \tag{2.1.102}$$

which differs significantly from B_+ given by equation (2.1.100).

(*e*) *The Phase Diagram Near the Tricritical Point in the T-ϕ Plane.* In order to account for the known departures from the Landau theory, a more general view of junctions between λ lines and first-order transition lines has recently been developed [113, 120, 121]. The new approach retains many of the qualitative features of the phase diagram that were predicted by Landau, but homogeneity assumptions are made for the free energy that are similar to those that underlie scaling theories for ordinary critical points [71–75]. These assumptions are more general than those of analyticity. For this purpose it is convenient to consider the phase diagram in terms of T and the intensive variable Φ, rather than T and X as in Figure 2.20. The λ line in the T–Φ plane is shown as a dashed curve in Figure 2.21. We see from equation (2.1.85) that in the limit of pure ^{4}He ($X = 0$) Φ tends towards $-\infty$. In that limit, $T_\lambda(\Phi)$ will have reached $T_{\lambda 0}$. For larger Φ, $T_\lambda(\Phi) < T_{\lambda 0}$, and at $T_\lambda(\Phi) = T_t$, the λ line terminates and meets the phase-separation curvey $T_\sigma(\Phi)$. Whereas in the T-X plane the phase-separation curve $T_\sigma(X)$ had two distinct branches, $T_\sigma(\Phi)$ is a single curve because the two phases in equilibrium have the same chemical potential.

Within the Landau theory, $T_\sigma(\Phi)$ is a smooth continuation of $T_\lambda(\Phi)$ in the T–Φ plane. This seems true also for the real system and largely follows from the experimental results [105, 106, 108–110] that $(\partial\Phi/\partial X)_{TP}$ vanishes at T_t, X_t. From equation (2.1.81), $(\partial\Phi/\partial T)_{XP} = (\partial\Phi/\partial T)_{\lambda P}$ if $(\partial\Phi/\partial X)_{TP} = 0$. But equation (2.1.81) is equally valid near T_σ; thus, $(\partial\Phi/\partial T)_{XP} = (\partial\Phi/\partial T)_{\sigma P}$

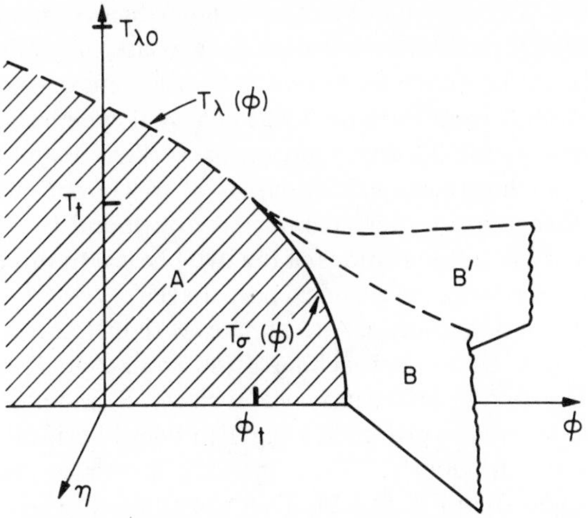

FIGURE 2.21. Schematic phase diagram of He^3–He^4 mixtures. η is the field conjugate to the superfluid order parameter, and Φ the field conjugate to the concentration order parameter. Based on Reference [113].

if $(\partial\Phi/\partial X)_{TP} = 0$. Therefore, upon approaching T_t along the λ line and along the phase-separation curve, $(\partial\Phi/\partial T)_{\lambda P}$ and $(\partial\Phi/\partial T)_{\sigma P}$ approach the same limit $(\partial\Phi/\partial T)_{XP}$ provided $(\partial\Phi/\partial X)_{TP}$ vanishes at T_t, X_t. There is as yet only little experimental information about higher derivatives, but the existence of a cusp or discontinuity at T_t and Φ_t in the second derivative of the phase-transition line is strongly suggested by the experimental data [108].

To every point in Figure 2.21 on the shaded surface A that lies below T_λ (Φ) and $T_\sigma(\Phi)$ in the T–Φ plane, there corresponds a superfluid state. The order parameter for this state may be defined such that $|\Psi|^2$ is proportional to the number of helium atoms in the zero-momentum state. We presume that $|\Psi| > 0$ everywhere on the surface A, and that it vanishes continuously at $T_\lambda(\Phi)$. At $T_\sigma(\Phi)$, the order parameter is expected to be multivalued, being nonzero for the superfluid ^{4}He-rich phase and zero for the normal-fluid ^{3}He-rich phase. This phase diagram is not unique to ^{3}He–^{4}He mixtures. Quite similar situations [113, 120] are believed to occur, for instance, in magnetic systems where an antiferromagnetic transition line meets a line of first-order transitions to a metamagnetic state. In that case, the internal magnetic field takes the place of Φ, and the magnetization corresponds to X. The order parameter equivalent to $|\Psi|$ is a sublattice magnetization.

It is convenient to define a field η that is conjugate to $|\Psi|$ in the sense that η is the derivative of the free energy with respect to $|\Psi|$. For $|\eta| > 0$, the superfluid transition will cease to exist, and there will be no temperature $T_\lambda(\Phi)$ at which the properties of the system will have singularites. The field η is not variable experimentally and is always equal to zero in the physical system.

Nonetheless, it is a useful concept. The Landau theory makes additional predictions about the qualitative features of the phase diagram for nonzero η. It predicts [113] the existence of two first-order coexistence surfaces B and B′, which extend from the line $T_\sigma(\Phi)$ at $\eta = 0$ symmetrically into the regions $\eta > 0$ and $\eta < 0$. To any point on the surfaces B and B′, there corresponds a state of the system which consists of a ^{3}He-rich and a ^{4}He-rich phase, and another nonzero order parameter χ on B and B′ is conveniently defined in terms of the concentrations of the two phases. In this case, the field conjugate to χ is, of course, the chemical potential Φ.

We already saw that the surface A terminates with increasing T in a line of critical points $T_\lambda(\Phi)$. Upon approaching $T_\lambda(\Phi)$ from $T < T_\lambda$, $|\Psi|$ vanishes continuously. The first-order coexistence surfaces B and B′, according to Landau, likewise terminate with increasing T in lines of critical points along which χ vanishes continuously. These lines are shown as the dashed top boundaries of B and B′ in Figure 2.21. They are allowed to exist for $|\eta| > 0$, because η is not a field conjugate to the pertinent order parameter χ. They meet each other and $T_\lambda(\Phi)$ at T_t and Φ_t when $\eta = 0$. Because of this junction of *three* lines of critical singularities the name *tri*critical point was suggested [113].

The phase diagram in the physically accessible $\eta = 0$ plane is shown once more in Figure 2.22. Riedel [121] has proposed that the most natural "fields" for the theory of the tricritical point are not the temperature T and the chemical potential difference Φ, but rather the "scaling fields" μ_1 and μ_2 that are shown in Figure 2.22. In terms of these, the tricritical point is at $\mu_1 = \mu_2 = 0$. The μ_2 axis is tangential to T_λ at T_t, Φ_t, and μ_1 is perpendicular to μ_2. Both μ_1 and μ_2 can be written as linear combinations of T–T_t and $\Phi - \Phi_t$.

The equation of the λ line can be written in terms of the scaling fields as

$$\mu_1 = - r_\lambda \mu_2^{1/\varphi}, \tag{2.1.103}$$

where r_λ is a positive constant and φ is the so-called "crossover exponent." If the phase-transition line is analytic at the tricritical point, then $1/\varphi$ is an integer greater than one, and the most likely value of φ is $\frac{1}{2}$. Explicit calculations by Riedel and Wegner [122] also yield this value, but give in addition small logarithmic correction [123] to equation (2.1.103) with $1/\varphi = 2$. Experimental results for $(\partial\Phi/\partial T)_{\lambda P}$ have been given in Table 2.1, and the most detailed studies of this derivative [108] indicate that it smoothly approaches a constant near T_t. This is consistent with equation (2.1.103) and $1/\varphi = 2$. No experiments so far have had the resolution to reveal the existence of the predicted [123] logarithmic corrections to equation (2.1.103).

Riedel [121] divides the phase diagram in the $\eta = 0$ plane into three regions labeled I, II, and III in Figure 2.22. In region I, the properties of the system are characteristic of being near the first-order transition. In region II, the

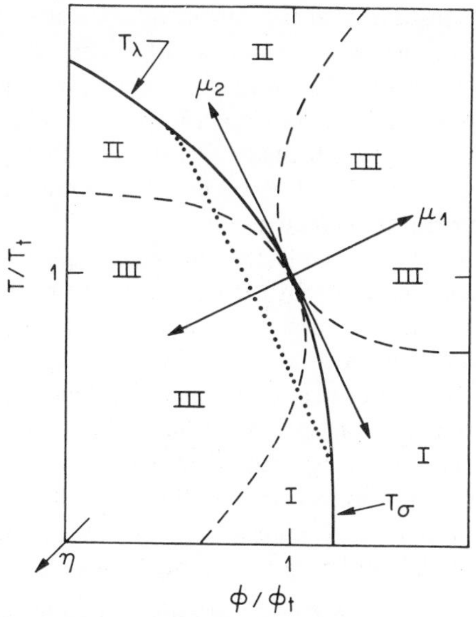

FIGURE 2.22. Schematic details of the phase diagram of He³–He⁴ mixtures in the T–Φ plane at $\eta = 0$. The fields μ_1 and μ_2 are the scaling fields proposed in Reference [121]. The dashed lines are the crossover lines of Reference [121]. The dotted line is a path of constant concentration in He (II) from the phase-separation curve to the λ line.

properties of the system are dominated by the proximity of the superfluid transition. A change in the chemical potential Φ in region II is not expected to change the character of the transition, and we believe a principle of universality [88–90] to apply. Such parameters as the specific-heat amplitude ratio A_Φ/A'_Φ or the exponents α_Φ and α'_Φ should be equal to those of the pure system and remain unaltered by a change in Φ, just as they were expected to be unaltered by a change in P for pure ^{4}He [see Section 2.1.2(e)]. To a large extent, the universal behavior in region II already is verified by the exponents in Figure 2.19, and by equation (2.1.97). In region III, the properties of the system are dominated by the tricritical behavior, and a new set of parameters (exponents, etc.) prevails, which may be different from that in region II. The regions I, II, and III are divided from each other by the dashed lines in Figure 2.22. According to Riedel [121] along these crossover lines the proportionality

$$\mu_1 \propto \mu_2^{1/\varphi} \tag{2.1.104}$$

is satisfied. Here φ is the same exponent as that in equation (2.1.103) for the λ line. Thus, the crossover lines are tangential to the phase-transition line. We note also that in the T–X plane of Figure 2.20 they are straight lines originating at T_t, X_t.

The dotted line in Figure 2.22 is a schematic representation of a typical experimental path at constant X in the superfluid. It starts at the phase-separation curve, follows a line with monotonically increasing T through the He (II) region to the left of the phase-transition curve, and meets the λ line. At the junction with T_λ, it will have the slope $(\partial\Phi/\partial T)_{\lambda P}$ if $C_{\Phi P}$ and $(\partial X/\partial\Phi)_{TP}$ diverge there [see equations (2.1.80) and (2.1.81)]. Such an experimental path will cross all three regions, and the behavior of the system should change accordingly as the temperature is increased at constant X.

Although it is convenient to divide the different regions in Figure 2.22 by lines, it should be emphasized that the transition from one region to another is not a sharp one. In fact, recent measurements of the superfluid density [107], as well as numerical calculations [123a] based upon certain models, show that the change from tricritical behavior to the behavior characteristic of the λ line is an extremely gradual one. Nonetheless, one can, of course, define a line along which this change has taken place to a specified, constant extent. In practical terms, however, it may well be difficult to make mesaurements which, for instance, are truly characteristic of the asymptotic behavior near T_λ, even when X already differs considerably from X_t.

(*f*) *Response Functions near the Tricritical Point.* In addition to the phase diagram, one would like to know the behavior of the response functions near the phase boundaries in the vicinity of T_t in order to test predictions which pertain to tricritical points. We already discussed the fact that any singular contribution to C_{XP} becomes very small at X_t [105, 106], and that the experimental data are consistent with a C_{XP} that has a continuous temperature derivative at T_t and X_t. This behavior is contained in equations (2.1.89)–(2.1.92) with the parameters in Table 2.2 for C_{XP} and has been demonstrated explicitly by Islander and Zimmerman [106]. These authors find that

$$C_{X,P}(T_t) - C_{X,P}(T) \sim (T\text{-}T_t). \tag{2.1.105}$$

Thus, $C_{X,P}$ is a regular function of T for $T > T_t$. Islander and Zimmerman note that equation (2.1.105) corresponds to $\alpha_X = -1$. In this connection it should be pointed out that equation (2.1.90) yields $\alpha_X \cong \alpha'_X \cong -0.5$, but C_{XP} [equation (2.1.89)] has a vanishing amplitude [equation (2.1.91)] for the singularity at X_t. This behavior seems different from that given by equation (2.1.105), which implies a finite amplitude, presumably at all X, and a removal of the singularity from C_{XP} for $X = X_t$ by virtue of α_X becoming equal to -1. Experiments do not at present distinguish clearly between these alternatives. However, they tend to favor an interpretation in which α_X remains larger than -1, and in which the amplitude vanishes at X_t. In any event, equations (2.1.89)–(2.1.92) with Table 2.2 are a good representation of C_{XP} very near T_λ even for X near X_t.

Since C_{XP} seems to be a regular function of T at T_t and X_t, $(\partial\Phi/\partial T)_{XP}$ is expected also to be essentially smooth because of equation (2.1.82). These response functions then by themselves do not reveal a great deal about the singular part of the equation of state of the system or about the possibly divergent response functions. Very near T_λ, $(\partial\Phi/\partial T)_{XP}$ can be calculated from C_{XP} and equation (2.1.82). Over a wider temperature range, both functions have been tabulated by Islander and Zimmerman [106] for $0.54 \leq X \leq 0.72$ and $T_\sigma \leq T \leq 1.4$ K. These authors also give similar very useful tabulations of the entropy and of $\Phi - \Phi_t$ for the same range of X and T.

The discontinuity ΔC_{XP} at T_σ that is apparent in Figure 2.18 is related to $(\partial X/\partial\Phi)_{TP}$. This derivative is a particularly important one for the tricritical point, for it is the derivative of one of the order parameters with respect to its conjugate field. It plays a role similar to that of the compressibility near a liquid–gas critical point and would be expected to be strongly divergent at T_t, X_t. It can be shown [105,124] that

$$\Delta C_{XP} = -T_\sigma\left(\frac{\partial X}{\partial T}\right)^2_{\sigma P}\left(\frac{\partial\Phi}{\partial X}\right)_{TP}. \qquad (2.1.106)$$

Of course ΔC_{XP} yields $(\partial\Phi/\partial X)_{TP}$ only on the phase-separation curve. In Figure 2.23 the results for $(\partial\Phi/\partial X)_{TP}$ along the coexistence curves, obtained [105] from equation (2.1.106) and the measured ΔC_{XP}, are shown as open circles for $X > X_t$ and as solid circles for $X < X_t$ as a function of

$$\varepsilon_+ \equiv \frac{T_t - T_\sigma^+}{T_t} \qquad (2.1.107)$$

or

$$\varepsilon_- \equiv \frac{T_t - T_\sigma^-}{T_t} \qquad (2.1.108)$$

on logarithmic scales. None of the data are for $|\varepsilon| < 10^{-2}$, and clearly we cannot as yet obtain answers to the type of detailed question that we asked about pure ^{4}He or even the mixtures at lower concentrations; however since log $[(\partial\Phi/\partial X)_{TP}]$ decreases with decreasing $|\varepsilon|$ even for $|\varepsilon| \cong 10^{-2}$, it seems extremely likely that $(\partial\Phi/\partial X)_{TP}$ vanishes at T_t, and thus that $(\partial X/\partial\Phi)_{TP}$ diverges as we expected.

Additional information about $(\partial X/\partial\Phi)_{TP}$ has become available from quite a different set of thermodynamic results. The vapor pressure of mixtures [102] has been measured in considerable detail [108,109,125] in the vicinity of (T_t,X_t). It can be shown that

$$\left(\frac{\partial\Phi}{\partial X}\right)_{TP} \cong RT(1 - X)^{-1}\left(\frac{\partial \ln\mathscr{R}}{\partial X}\right)_{TP}, \qquad (2.1.109)$$

where $\mathscr{R}$ is the ratio of the vapor pressure of the solution to that of pure ^{3}He at the same temperature. At the coexistence curve, these results for $(\partial\Phi/\partial X)_{TP}$ are consistent with those obtained from ΔC_{XP} and equation (2.1.106). They are shown in Figure 2.23 as open and solid squares. The vapor-pressure data,

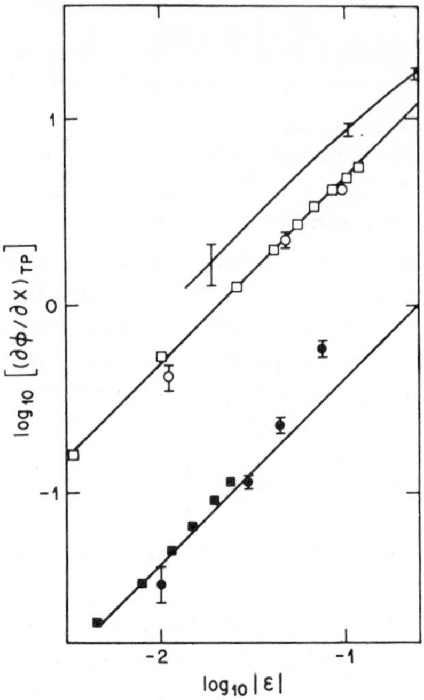

FIGURE 2.23. $(\partial\Phi/\partial X)_{TP}$ in J / mole as a function of $|\varepsilon| = |1 - T/T_t|$ on logarithmic scales. The symbols are measurements on the phase-separation curve. Solid, $X < X_t$; open, $X > X_t$. The upper line fits the data from Reference [125] for $T > T_t$ and $X = X_t$. Squares, Reference [125]; circles, Reference [105].

however, yield $(\partial\Phi/\partial X)_{TP}$ along other thermodynamic paths as well. In particular, the derivative has been evaluated for $T > T_t$ along a line of constant concentration equal to X_t. The smooth results [125] are shown in Figure 2.23 as the uppermost solid line. The estimated uncertainty in the smoothed value is indicated by the bars. The results reveal the same general behavior that was observed along the phase-separation curve.

Although the results in Figure 2.23 are not as detailed and as near T_t as might be desired, it seems justified to use them to obtain rough estimates of critical exponents. The divergence of the derivative of the order parameter with respect to its conjugate field is generally described by an exponent γ. Therefore we define exponents γ, γ'_+, and γ'_-, and amplitudes Γ, Γ'_+, and Γ'_- by

$$\left(\frac{\partial\Phi}{\partial X}\right)_{TP} = \Gamma'_+\varepsilon_+{}^{\gamma'_+} \tag{2.1.110}$$

along the ^{3}He-rich branch of the coexistence curve, by

$$\left(\frac{\partial\Phi}{\partial X}\right)_{TP} = \Gamma'_-\varepsilon_-{}^{\gamma'_-} \tag{2.1.111}$$

along the ^{4}He-rich branch of the coexistence curve, and by

$$\left(\frac{\partial \Phi}{\partial X}\right)_{TP} = \Gamma \varepsilon_t^{\gamma}; \ \varepsilon_t \equiv \frac{T - T_t}{T_t} \qquad (2.1.112)$$

along the line of constant $X = X_t$ for $T > T_t$. The results in Figure 2.23 are consistent with

$$\gamma'_+ = \gamma'_- = \gamma = 1.0 \pm 0.1. \qquad (2.1.113)$$

If these exponents are indeed equal to unity, then

$$\Gamma'_+ = -48 \text{ J/mole},$$
$$\Gamma'_- = 4.0 \text{ J/mole}$$
$$\Gamma = 91 \text{ J/mole}. \qquad (2.1.114)$$

These coefficients, and equation (2.1.113), correspond to the solid straight lines in Figure 2.23. There is, however, some danger in assuming that the asymptotic behavior of $(\partial\Phi/\partial X)_{TP}$ does indeed correspond to $0.9 \leq \gamma \leq 1.1$, because all measurements are for rather large ε, and none of them is of very high precision.

Recently, the technique of light scattering has been applied to the study of ^{3}He–^{4}He mixtures near the tricritical point [110]. One expects the total intensity of the light that is scattered by the mixture at a fixed angle to an incident laser beam to be proportional to $(\partial X/\partial \Phi)_{TP}$ (except for small corrections due to the compressibility of the fluid). Thus, a strongly divergent $(\partial X/\partial \Phi)_{TP}$ should result in critical opalescense. This has indeed been observed by Watts and Webb [110], and preliminary measurements have yielded $[\partial X/\partial \Phi]_{TP}$ along the two branches of the phase-separation curve, and for $X = X_t$ and $T > T_t$. These results are of a precision similar to that obtained with the vapor-pressure measurements, and are in reasonable agreement with those shown in Figure 2.23. They yield $\gamma \cong \gamma'_+ \cong 1$. However, they have been interpreted to imply $\gamma'_- > 1$; if this tentative conclusion is confirmed, it will be a most unexpected result.*

In addition to the measurements shown in Figure 2.23, $(\partial X/\partial \Phi)_{TP}$ has been obtained along several paths of constant X [108–110]. Some of the results for $X > X_t$ that were derived from the vapor-pressure measurements [108] are shown in Figure 2.24. Note that the temperature scale is linear, but that, because of the enormous change in $(\partial X/\partial \Phi)_{TP}$, the vertical axis is on a logarithmic scale. Along the phase-separation curve and for $X = 0.675 \cong X_t$, these data are nearly the same as some of those shown in Figure 2.23. For $X < X_t$, similar results are shown in Figure 2.25. In this case, the results near the phase-separation curve look similar in shape to those in Figure 2.24

*Recent more accurate light scattering results [P. Leiderer, D. R. Watts, and W. W, Webb, *Phys, Rev. Lett.* **33**, 483 (1974)] are consistent with $\gamma' = 1.0$.

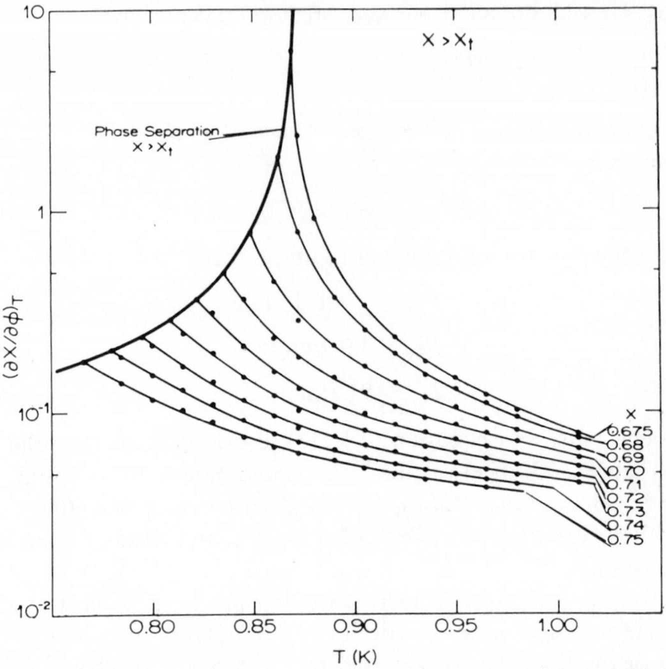

FIGURE 2.24. $(\partial X/\partial \Phi)_{TP}$ in mole/J as a function of T along several lines of constant X for $X < X_t$. Note the logarithmic vertical scale. After Reference [108].

for $X > X_t$, but for $X < X_t$ they are about an order of magnitude larger at the same T. Further away from T_σ, the experimental path crosses T_λ. This crossing manifests itself primarily in a rather sudden drop in $(\partial X/\partial \Phi)_{TP}$ by about an order of magnitude. We expect, of course, from equation (2.1.80) that $(\partial X/\partial \Phi)_{TP}$ will exhibit the same behavior near T_λ as that of $C_{\Phi P}$, but the anticipated increase with decreasing $T_\lambda - T$ is too small to be noticeable in these experiments, although it has been indicated in the figure by the dotted peaks. Smoothed values of $(\partial \Phi/\partial X)_{TP}$ have been tabulated [109].

No thorough attempt has been made so far to express the data in Figures 2.22–2.24 in terms of an equation of state in closed form. However, for the superfluid region, $(\partial \Phi/\partial X)_{TP}$ can be represented by the equation

$$\left(\frac{\partial \Phi}{\partial X}\right)_{TP} = A(T - T_0) + B(T - T_0)^2, \qquad (2.1.115)$$

with $T_0 \leq T_t$. The parameters can be fitted by the linear functions

$$\begin{aligned} T_0 &= t_0 + t_1(X_t - X); \qquad t_1 < 0 \\ A &= a_0 + a_1(X_t - X), \\ B &= b_0 + b_1(X_t - X), \end{aligned} \qquad (2.1.116)$$

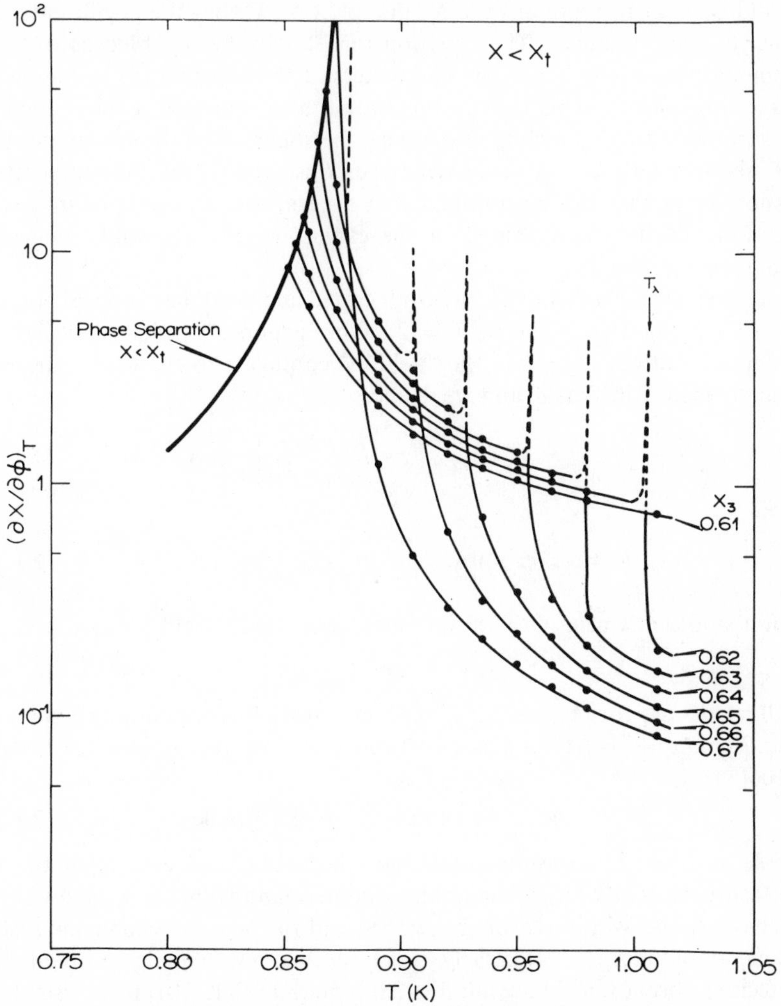

FIGURE 2.25. $(\partial X/\partial \Phi)_{TP}$ in mole/J as a function of T along several lines of constant X for $X < X_t$. Note the logarithmic vertical scale. After Reference [108].

of $X_t - X$ with the coefficients

$$\begin{aligned} t_0 &= T_t, & t_1 &= -0.418, \\ a_0 &= 7.818, & a_1 &= 50.7, \\ b_0 &= -21.6, & b_1 &= 53.3, \end{aligned} \tag{2.1.117}$$

if Φ is in J/mole. Equations (2.1.115)–(2.1.117) reproduce the experimental data between T_σ and T_λ for $0.60 \leq X \leq X_t$ within a few percent, and therefore are useful in thermodynamic calculations. A similar function probably could be used to represent the results for $X > X_t$, but the coefficients have not yet been obtained. The equation for T_0 may be regarded as that of a spinodal line, along which the parameters of the supercooled system would have singularities. This temperature would also represent a lower limit for the temperature to which the system can be supercooled. Some supercooling of mixtures with $X < X_t$ has indeed been observed [125a], but temperatures as low as T_0 have not been obtained in experiments. It would be interesting to study the thermodynamics of the experimentally accessible metastable state in more detail.

The results for $(\partial \Phi / \partial X)_{TP}$ can be integrated and yield the chemical potential $\Phi - \Phi_t$ along the isotherm $T = T_t$. The exponent which describes the singularity at constant T in the "field" Φ conjugate to the order parameter usually is identified as δ, and we define

$$\left| \Phi(X) - \Phi(X_t) \right|_{T=T_t} = D_+ \left| \frac{X_t - X}{X_t} \right|^{\delta_+}; \ X > X_t \qquad (2.1.118)$$

and

$$\left| \Phi(X) - \Phi(X_t) \right|_{T=T_t} = D_- \left| \frac{X_t - X}{X_t} \right|^{\delta_-}; \ X < X_t \qquad (2.1.119)$$

for X sufficiently near X_t. The experimental data [125] yield

$$\delta_+ \cong \delta_- \cong 2.0 \pm 0.25. \qquad (2.1.120)$$

All results are for $\left| 1 - X/X_t \right| \gtrsim 0.02$, and again it is not entirely certain that the asymptotic behavior has been obtained. For the amplitudes the data [106] yield

$$D_+ = 60 \ \text{J/mole}, \ D_- = 0.8 \ \text{J/mole}, \qquad (2.1.121)$$

if $\delta_+ = \delta_- = 2.0$. Results for Φ have been obtained also from the C_{XP} measurements [105] by Islander and Zimmermann [106]. These show similar behavior, and within reasonable errors lead to the same exponents and the same amplitudes [125b]. Clearly, it is desirable, but difficult, to obtain more precise information. The result $\delta = 2.0$ [equation (2.1.120)] is consistent with the experimental value $\gamma = 1.0$, the scaling prediction [90,121]

$$\gamma = (\delta - 1)\beta, \qquad (2.1.122)$$

and the result $\beta = 1.0$ discussed below equation (2.1.99). Smoothed values of $\Phi - \Phi_t$ have been tabulated [106, 109].

The available experimental information can be compared also with the scaling law [121]

$$\frac{1}{\delta} = 1 - \alpha_t, \qquad (2.1.123)$$

where α_t is the exponent that describes the divergence of $C_{\Phi P}$ along a path of constant chemical potential equal to Φ_t. Note that α_t may be very different from the exponent α, which describes the behavior of $C_{\Phi P}$ along a path of constant Φ near $T_\lambda(\Phi)$, and which should be equal to α for the pure system [see equation (2.1.69)]. Approximately α_t pertains in region III of Figure 2.22, whereas α pertains in region II. Values of $C_{\Phi P}$ near the tricritical point were obtained by Islander and Zimmermann, and these are consistent with $\alpha_t = \alpha_t' \cong \frac{1}{2}$, which would agree with equations (2.1.123) and (2.1.120).

The applicability of power laws is usually restricted to a narrow range of X and T, and, in addition, requires that all data be in the same of the regions indicated in Figure 2.22. However, it follows from the homogeneity assumption of scaling that even measurements in a crossover region can be represented by a unique function (which, of course, is no longer a simple power law), provided that they are converted into appropriate dimensionless reduced variables. Goellner et al. have presented the measurements that are shown in Figures (2.24) and (2.25) in a parametric form which tends to demonstrate the existence of such a unique scaling function. They show $Y \equiv (\partial\Phi/\partial X)_{TP}/|1 - X/X_t|^{\delta-1}$ as a function of $Z \equiv (T/T_t - 1)/|1 - X/X_t|^{1/\beta}$. Their results [108] are reproduced in Figure 2.26, and fall on a unique curve for all T which has different branches for $X > X_t$ and $X < X_t$. The two branches approach the same asymptote near $X = X_t$. According to the scaling formulation of Riedel [121], one would have expected this behavior

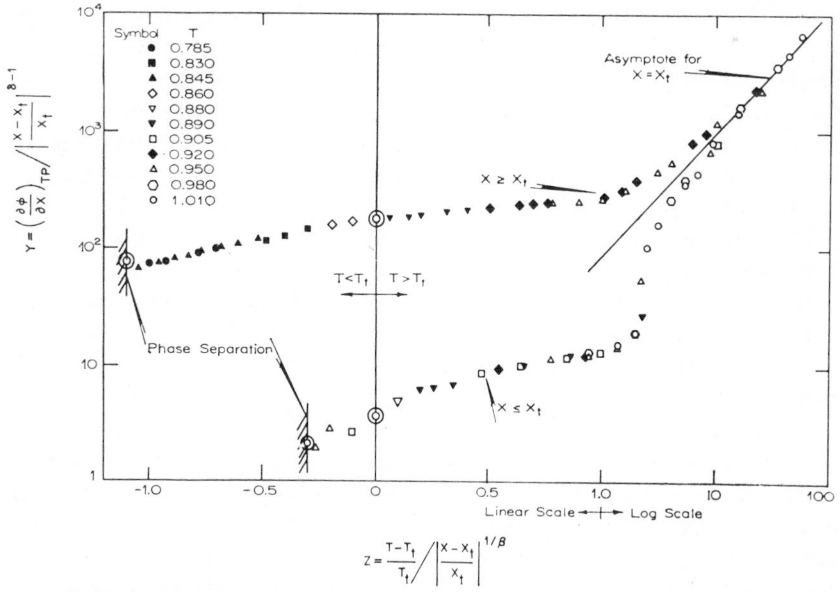

FIGURE 2.26. The scaling variable Y as a function of its argument Z for several T. After Reference [108].

for the reduced variables that correspond to the scaling fields μ_1 and μ_2 in Figure 2.22. The experimental data reveal that to a large extent μ_2 and μ_1 may be identified with $(T/T_t - 1)$ and $(\Phi/\Phi_t - 1)$, respectively, and strongly support the scaling hypothesis near the tricritical point.

(*g*) *The Effect of Gravity near the Tricritical Point*. One would expect that sufficiently near the tricritical point the inhomogeneity due to the gravitational field should affect experimental measurements on samples of nonzero height. A change in height dH in a liquid sample will be accompanied by changes $d\Phi$ and dP in the chemical potential difference [see Section 2.1.3(a)] and the pressure, which are given by

$$d\Phi = -(M_3 - M_4)g\, dH \tag{2.1.124}$$

and

$$dP = \rho g\, dH. \tag{2.1.125}$$

Here g is the gravitational acceleration. Both Φ and P change by roughly temperature-independent increments $\Delta\Phi$ and ΔP over the total sample height. However, since $(\partial X/\partial\Phi)_{TP}$ diverges at T_t [see Section 2.1.3(f)], the constant $\Delta\Phi$ contributes to ΔX an amount which increases rapidly near T_t. Although it is less obvious, a similar rapidly increasing contribution to ΔX arises from the constant ΔP [see equation (2.1.127) below].

The gravity-induced concentration profile was examined by Watts [110], who used a laser beam and an interferrometric technique to study the change in the index of refraction of the fluid near the tricritical point over a 0.4-cm height change. His measurements are consistent with expectations based on the equation of state, and he used them to correct older measurements of the phase diagram [104, 106] for gravity effects. The corrected phase diagrams are in considerably better agreement with more recent, practically gravity-free measurements [107] [see Section 2.1.3(d)].

The large inhomogeneity in X has also been observed in the form of "rounding" [107] of the second-sound velocity u_2 [see Section 2.2.2(c) below] at the junctions of the phase-separation curve and the lines of constant concentration. In the absence of the inhomogeneity, the velocity, although continuous, would have a discontinuous derivative at these points. The experimentally observed rounding is apparent in Figure 2.27, which is a detailed plot of u_2 near T_t as a function of T. At an average concentration of 0.6718, corresponding to $1 - X/X_t \cong 0.004$ and $1 - T_\sigma/T_t \cong 0.0014$, the grvaity effect is quite pronounced and the measured velocity at T_σ can be estimated to be as much as 1.5 times the gravity-free value. At $X = 0.6421$, corresponding to $1 - X_\sigma/X_t \cong 0.04$, the gravitational rounding, although observable with high-resolution data, is small and affects the velocity by at most 1 or 2%. In analogy to a criterion suggested by Hohenberg and Barmatz

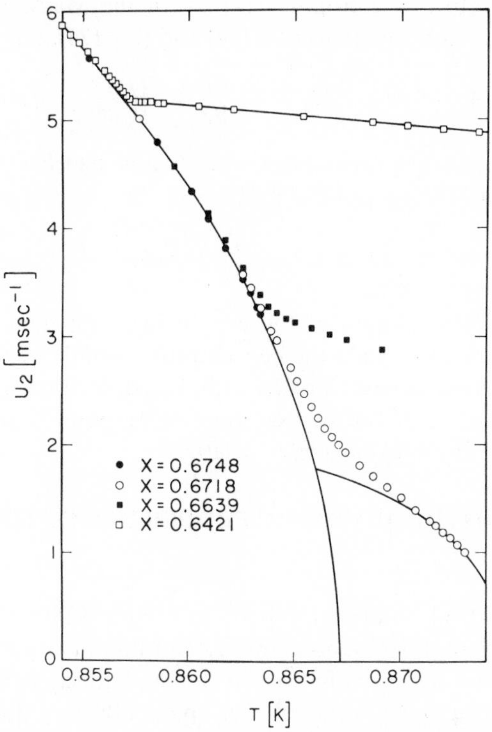

FIGURE 2.27. The velocity of second sound in ^{3}He–^{4}He mixtures as a function of T along the coexistence curve, and along several lines of constant X. The figure covers only the region very near T_t, X_t. Particularly for $X = 0.6718$, the effect of gravity is revealed by the "rounding" near the junction of the constant X data and the phase separation curve data. After Reference [107].

[125c] for liquid–gas critical point, one can estimate [107] semiquantitatively the range of $1 - X_\sigma/X_t$ over which gravity effects are "important" by equating $\Delta\Phi$ with $(X_t - X_\sigma)\,(\partial\Phi/\partial X)_{P,T}$ on the phase-separation curve. This yields $(1 - X_\sigma/X_t) \cong 0.012h^{1/2} \cong 0.006$ for the results in Figure 2.27, where $h \cong 0.3$ cm. On the basis of a somewhat different, but also only semiquantitative criterion proposed by Berestov et al. [125*d*], one obtains the remarkably similar value [107] $1 - X_\sigma/X_t = 0.015h^{1/2}$ for this range. These estimates tend to be in agreement with the experimental result. More detailed calculations of the effect of gravity similar to those for liquid–gas critical points [125c] are possible with the aid of the measured $(\partial X/\partial\Phi)_{TP}$ (see Figures 2.24 and 2.25). Let us consider the total differential of X in the form

$$dX = \left(\frac{\partial X}{\partial P}\right)_{\Phi T} dP + \left(\frac{\partial X}{\partial \Phi}\right)_{PT} d\Phi, \qquad (2.1.126)$$

where the temperature is constant. Then, with the Maxwell relation $(\partial X/\partial P)_{\Phi T} = (\partial V/\partial \Phi)_{PT}$ and equations (2.1.124) and (2.1.125), one obtains

$$dX = -\left[M_3 - M_4 - \rho\left(\frac{\partial V}{\partial X}\right)_{PT}\right]\left(\frac{\partial X}{\partial \Phi}\right)_{PT} g\, dh. \qquad (2.1.127)$$

The bracket on the right varies little with T, and one has approximately, from the known $(\partial V/\partial X)_{PT}$ [3,125e],

$$dX \cong 1.95g\left(\frac{\partial X}{\partial \Phi}\right)_{PT} dh, \qquad (2.1.128)$$

where the units of Φ are J/mole and where h is in centimeters. Equation (2.1.128) can be integrated to yield the concentration profile of the sample. An equation of state such as equation (2.1.115), based, for instance, on the data in Figures 2.24 and 2.25, can be used for this purpose. Equation (2.1.127) has been obtained also by Goellner et al. [109].

2.2. THERMO-HYDRODYNAMIC PROPERTIES OF HE (II) NEAR T_λ

2.2.1. Pure ^{4}He

The thermo-hydrodynamics of superfluid helium [126] yields predictions [127] for a number of different measurable quantities in terms of the thermodynamic parameters discussed in Section 2.1, and the superfluid density ρ_s. We shall consider here the results of four apparently very different types of experiments. One can use each of these, in conjunction with the appropriate hydrodynamic prediction and, when necessary, the known thermodynamic parameters, to derive the superfluid density. The agreement within permitted experimental error that exists between ρ_s obtained from this variety of experiments strongly supports the validity of the hydrodynamics even extremely near T_λ, where, in an experimentally accessible range of $T_\lambda - T$, ρ_s has decreased to about 0.1 % of its low-temperature value.

In addition to the bearing which the measurements have upon hydrodynamics, one may regard the ρ_s that is derived from them as a variable that is rather closely related [128] to the order parameter for the superfluid transition. We shall, therefore, discuss in some detail particularly the relation between the singularities in ρ_s and C_P at T_λ in terms of predictions based on theories of critical phenomena.

(*a*) *The Andronikashvili Experiment.* In this experiment [129], a number of parallel disks is mounted horizontally on a vertical central post, with a small gap of thickness d between each pair of them to be occupied by liquid helium. The central post is suspended by a torision fiber, and a measurement of the torsional oscillation period will yield the moment of inertia of the system. If the disk spacing d and the viscous penetration depath [130] λ

satisfied the relation

$$d \ll \lambda = \left(\frac{2\eta}{\rho_n\omega}\right)^{1/2}, \tag{2.2.1}$$

where ω is the angular frequency of the pendulum and η the normal-fluid viscosity, then in a classical fluid all the mass between the disks would contribute to the moment of inertia. However, since the superfluid motion is to be irrotational, one expects in this case that only a fraction of the fluid equal to ρ_n/ρ, where ρ_n is the normal-fluid density, will contribute. It is thus possible to determine the temperature dependence of ρ_n/ρ from the measured termperature dependence of the pendulum period. Typical periods for an experimental apparatus would be 10^1–10^2 sec, and equation (2.2.1) yields $d \ll 10^{-1}$ cm. This condition is easily satisfied without affecting the bulk character of the fluid even very near T_λ. Measurements with this technique have been made at vapor pressure, [129, 131–133] and recently also at pressures up to 25 bar [134].

It is difficult to obtain very precise information from this experiment about ρ_s near T_λ, because $\rho_s/\rho = 1 - \rho_n/\rho$, and ρ_n/ρ is nearly equal to one. In addition, it may be necessary to correct for small amounts of the superfluid fraction which are carried along with the disk pile because of geometrical imperfections [134, 135], and for a slight dependence of the measurements upon the normal-fluid viscosity, because d is not sufficiently much smaller than λ [see equation (2.2.1)] and helium outside the disk pile over a radial distance λ tends to participate in the oscillatory motion. In spite of these problems, a least-squares fit of the rather precise data by Tyson and Douglass [132, 133] to the equation

$$\frac{\rho_s}{\rho} = k\,|\varepsilon|^{\zeta}; \quad \varepsilon \equiv \frac{T - T_\lambda}{T_\lambda} \tag{2.2.2}$$

yielded quite small statistical errors for the parameters. The values

$$k = 2.40, \quad \zeta = 0.666 \pm 0.006 \tag{2.2.3}$$

were obtained from measurements with $3 \times 10^{-5} \leq \varepsilon \leq 2.3 \times 10^{-2}$ at saturated vapor pressure. For pressures greater than SVP, the measurements by Romer and Duffy [134] are for $|\varepsilon| \gtrsim 2 \times 10^{-3}$, but near T_λ the possible errors in ρ_s/ρ become large and even for $|\varepsilon| \cong 10^{-2}$ they are already perhaps 10% of ρ_s/ρ. Nonetheless, even though it is not possible to derive precise values of k and ζ from the measurements, the data show that the behavior of ρ_s/ρ under pressure is similar to that observed at vapor pressure.

(*b*) *The Angular Momentum of Persistent Superfluid Currents.* Long-lived persistent currents have been established in He (II) by several investigators [135–139], and decay constants longer than 10^{11} times that expected for

He (I) in the same geometry have been obtained [138]. In order for persistent currents of any appreciable velocity to exist, it was found necessary, however, to confine the helium in a restricted geometry with a small characteristic length *d*. This increased the *d*-dependent critical velocity ω_C^G above which dissipation will occur. This geometry-limited critical velocity is not an intrinsic property of the fluid, and has no bearing on the superfluid transition. Reppy and Depatie [139] investigated persistent angular velocity fields in He (II) confined between parallel plates, 0.3 mm apart, and mounted perpendicular to the rotation axis. They demonstrated that a peristent current with ω_C^G had an angular momentum whose temperature dependence was about the same as that of the superfluid density. This was found to be true even when $(T_\lambda - T) \cong 1.5 \times 10^{-2}$ K, where ρ_s/ρ was only about 15% of its low-temperature value. This result is expected [127, 138, 139] if, consistent with two-fluid hydrodynamics, curl $v_s = 0$, and if ω_C^G is independent of temperature as suggested by other experiments [140]. Their experimental method was very tedious because each measurement required the destruction of the current. A much more elegant approach to the problem was developed more recently independently by Reppy [141], and by Mehl and Zimmermann [138, 142]. These authors constructed superfluid gyroscopes whose angular momenta could be measured nondestructively. This was done by determining the torque required to rotate the angular-momentum vector of the gyroscope (i.e., of the He persistent current). With these new instruments, earlier observations of flow without dissipation and of reversibility of the angular momentum upon thermal cycling (provided $T < T_\lambda$) were refined [138, 141–143]. Reppy concentrated particularly on the behavior near T_λ, and demonstrated that persistent currents flowing at the temperature-independent critical velocity ω_C^G are stable and have an angular momentum with reversible temperature dependence even when $T_\lambda - T \cong 10^{-5}$ K. The temperature dependence of the angular momentum was the same within permitted experimental error as that measured by Dash and Taylor [131] for ρ_s/ρ by the Andronikashvili technique. Measurements with even greater precision were made later by Clow and Reppy [143] at velocities considerably less than ω_C^G, which were created by colling the gyroscope from $T > T_\lambda$ while rotating at the desired $\omega < \omega_C^G$.

The experiments discussed so far have demonstrated that the angular momentum *L*, and therefore the velocity field of a peristent superfluid current, is a reversible function of the temperature, and the comparison of *L* with the results of the Andronikashvili experiment [132, 133] suggests strongly that the velocity field is indeed indepedent of temperature. If it is assumed, constistent with two-fluid hydrodynamics, that any $\omega < \omega_C^G$ is rigorously independent of *T*, then the refined angular-momentum measurements of Clow and Reppy [143] yield rather precise values of ρ_s/ρ within a normalization constant that can be determined at low *T*, where ρ_s/ρ approaches one.

The data, when fitted to equation (2.2.2), yield

$$k = 2.41, \tag{2.2.4}$$

$$\zeta = 0.67 \pm 0.03, \tag{2.2.5}$$

for $3 \times 10^{-5} \leq \varepsilon \leq 5 \times 10^{-2}$. The authors do not state explicitly how they estimated the value and uncertainty of ζ. They do not give error estimates of k, although it seems that equation (2.2.2) with k from equation (2.2.4) should yield ρ_s/ρ to perhaps $\pm$ 1%. This result is in very fine agreement with that obtained by Tyson [132, 133], using the Andronikashvili technique, but the estimated errors in ζ are slightly larger than Tyson's. The good agreement between the two results supports the assumption that ω was independent of T in the angular-momentum experiment, and that the superfluid remains at rest in the Andronikashvili experiment.

The persistent-current angular-momentum technique has not yet been used for the measurement of ρ_s under pressure, but there appears to be no reason why this should not be possible [144]. Some improvement in the precision also appears within reach [144].

(*c*) *The Velocities of First and Second Sound.* The two previous methods of determining ρ_s had in common a time-independent superfluid velocity field. We shall now consider the propagating modes that are predicted by two-fluid hydrodynamics. Measurements of ρ_s based on studying these modes may be regarded to be in a slightly different category from the Andronikashvili and persistent-current techniques because oscillatory velocity fields are involved, but there seems to be no particular hydrodynamic reason to regard one experiment as more fundamental than the others. From the viewpoint of critical phenomena, measurements of first- and second-sound velocities, in principle, have the advantage that they can be performed on the bulk system, whereas the previous techniques both required that the samples have some small characteristic length d. In practice, d could be made large enough, however, to render size effects negligible compared to current experimental errors from other sources. Sound-velocity measurements have the practical advantage that they can usually be made with very high precision.

The hydrodynamic equations of superfluid helium [2, 3, 126, 127] are greatly simplified [127] for cases in which dissipation processes are absent, and in which the equations may be linearized because the velocities are sufficiently small. For the sake of convenience, it is also often assumed that the difference between C_P and C_V may be neglected. The above assumptions are applicable to a wide range of circumstances that are encountered in experimental studies of superfluid helium, but they may well not pertain in experiments near the superfluid transition. At T_λ, the superfluid density vanishes, and any experiment that probes the system by inducing a temperature-independent nonzero mass or heat flux will involve large superfluid velocities sufficiently near T_λ, which diverge at the transition. In addition, the largest velocity at which dissipative processes seem absent appears to vanish at T_λ [6]. Nonetheless, the simplifications of the hydrodynamic equations that exist in the absence of dissipation and higher-than-quadratic terms in the velocities warrant retaining the results for this case, but in any comparison of experimental results with predictions of the linear two-fluid hydrodynamics extra care is needed to assure that measurements correspond in some appropriate sense to a zero-amplitude and zero-frequency limit.

The third simplifying condition $C_P \cong C_V$ that was mentioned above is known not to apply sufficiently near T_λ [see Section 2.1.2(e)] because

$$\gamma \equiv \frac{C_P}{C_V} \tag{2.2.6}$$

is singular, but it is not difficult to obtain the velocities of the propagating modes for the case of $\gamma > 1$. For bulk samples of He (II), the velocities are given by the roots of the equation [3, 127]

$$u^4 - (\gamma c_1^2 + c_2^2)u^2 + c_1^2 c_2^2 = 0, \tag{2.2.7}$$

where

$$c_1^2 \equiv \left(\frac{\partial P}{\partial \rho}\right)_T, \tag{2.2.8a}$$

and

$$c_2^2 \equiv \left(\frac{S^2 T}{C_V}\right)\left(\frac{\rho_s}{\rho_n}\right). \tag{2.2.8b}$$

When $\gamma = 1$, c_1^2 and c_2^2 are the roots of equation (2.2.7), and c_1 and c_2 are the desired velocities of first and second sound, respectively. In general, however,

$$u^2 = \frac{1}{2}(\gamma c_1^2 + c_2^2) \pm \frac{1}{2}(\gamma c_1^2 - c_2^2)\left\{1 + 4(\gamma - 1)\frac{c_1^2 c_2^2}{(\gamma c_1^2 - c_2^2)^2}\right\}^{1/2} \tag{2.2.9}$$

For $c_1^2 \gg c_2^2$, the square root can be expanded in a power series, and the second-sound velocity u_2 is given by

$$u_2^2 = u_{20}^2\left[1 - (\gamma - 1)\frac{u_{20}^2}{u_{10}^2} + \cdots\right], \tag{2.2.10}$$

where

$$u_{20}^2 \equiv \frac{c_2^2}{\gamma}, \tag{2.2.11a}$$

and

$$u_{10}^2 \equiv c_1^2 \gamma. \tag{2.2.11b}$$

Even for large γ, the term with u_{20}^2/u_{10}^2 is small at all T, and vanishes at T_λ.

The approximation

$$u_2^2 \cong u_{20}^2 = \left(\frac{S^2T}{C_P}\right)\left(\frac{\rho_s}{\rho_n}\right) \tag{2.2.12}$$

is valid within 0.01 % of u_2^2 for T anywhere near T_λ. Similarly, one obtains to order u_{20}^2/u_{10}^2 for the velocity of first sound

$$u_1^2 = u_{10}^2\left[1 + (\gamma - 1)\frac{u_{20}^2}{u_{10}^2} + \cdots\right], \tag{2.2.13}$$

which is always well approximated by

$$u_1^2 \cong u_{10}^2 = \left(\frac{\partial P}{\partial \rho}\right)_S. \tag{2.2.14}$$

Equation (2.2.14) is, of course, the same relation that is expected to hold above T_λ for the velocity of ordinary sound.

We note that u_1 is nearly independent of ρ_s. It, therefore, cannot be used to study this thermo-hydrodynamic parameter. Nonetheless, we expect the measured u_1 to agree with the prediction equation (2.2.14), and to reflect the singularities at T_λ of the pertinent thermodynamic parameters. Before proceeding to the measurements of u_2, we shall, therefore, first discuss the experiments pertaining to u_1.

Measurements of the velocity of first sound at saturated vapor pressure sufficiently near T_λ to be of interest here were made first by Chase [145] at a frequency of 10^6 Hz. For $(T_\lambda - T) > 3 \times 10^{-4}$ K, these measurements are virtually equal to the velocity calculated [44] from equation (2.2.14) and the known thermodynamic properties. For T nearer T_λ, however, the measured velocity is larger [44, 145, 146] than that given by equation (2.2.14), and this is attributable to the nonzero frequency used in the measurement. This dispersion is, of course, of interest for its own sake, and will be discussed in Section 2.3.1(d). For the purpose of comparing equation (2.2.14) with thermodynamic measurements, however, lower frequencies must be employed near T_λ. This was done by Rudnick and Shaprio [147] and, with even greater temperature resolution, by Barmatz and Rudnick [31]. Although initially some problems arose [147–150] in the detailed interpretation of the measurements, it is now clear [44] that at a frequency of 2.2×10^4 Hz the data [31] agree extremely well with equation (2.2.14), and that dispersion effects for $T_\lambda - T \gtrsim 2 \times 10^{-5}$ K are less than one part in 10^5 of the velocity [146]. The measured velocity at 2.2×10^4 Hz as a function of $T - T_\lambda$ is hown in Figure 2.28. Earlier, in Figure 2.5, these data were compared with calculations based on (2.2.14).

Except for the early work by Atkins and Stasior [53], which does not extend very close to T_λ, no low-frequency first-sound-velocity measurements have been made under pressure near T_λ.

The first measurements of the second-sound velocity sufficiently near T_λ to be of interest here were made by Pearce, Lipa, and Buckingham [151] at

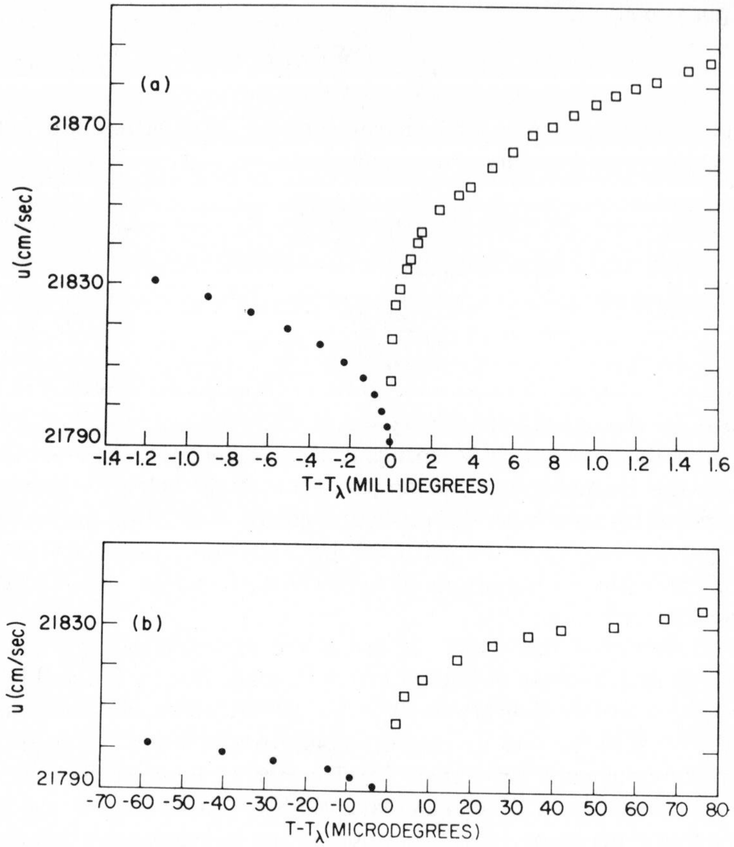

FIGURE 2.28. The velocity of first sound in He^4 at vapor pressure as a function of $T - T_\lambda$. After Reference [31].

vapor pressure. These data are for $10^{-4} \le \varepsilon \le 0.2$, and for $\varepsilon \le 4 \times 10^{-2}$ they could be represented within their scatter by

$$u_2^2 = a\varepsilon^b, \tag{2.2.15}$$

with

$$a = 2190 \pm 88 (\text{m/sec})^2, \quad b = 0.772 \pm 0.005. \tag{2.2.16}$$

The authors pointed out, however, that equation (2.2.15) is unlikely to be the correct asymptotic functional form for u_2 because of the dependence upon C_P implied by Equation (2.2.12). Almost simultaneously, measurements for $4 \times 10^{-5} \le \varepsilon \le 0.17$, also at SVP, were made by Tyson and Douglass [152], who demonstrated that their results were consistent with equation (2.2.12), the known C_P [Section 2.1.2 (c)], and the ρ_s/ρ determined by the

Andronikashvili technique [132, 133] within combined permitted errors. When they fitted their data to equation (2.2.15), they obtained $b = 0.773$, in agreement with Equation (2.2.16).

Measurements of u_2 for even smaller ε, but still only at SVP, were made by Johnson and Crooks [153, 154], and by Williams et al. [155]. Some of these results [153] initially appeared to indicate an unexpected [156] departure from equation (2.2.12) for very small ε. More extensive additional measurements [154, 155], however, revealed complete agreement between equation (2.2.12), the ρ_s/ρ [132, 133] discussed in Section 2.2.1 (a), and the thermodynamic parameters within combined allowed errors even for ε as small as 3×10^{-6}.

Very recently, new measurements for $2 \times 10^{-5} \leq \varepsilon \leq 3 \times 10^{-2}$, with errors in the velocity estimated to be only about 0.1 % and with high resolution in ε, were made by Greywall and Ahlers [157, 158] at pressures up to the melting pressure. Similar results, but with somewhat less resolution in ε, were obtained simultaneously by Terui and Ikushima [159]. At SVP, the data of Reference [157] with $\varepsilon < 10^{-2}$, when fitted to equation (2.2.15), yielded [158]

$$a = 2142 \text{ (m/sec)}^2, \quad b = 0.7740 \pm 0.001, \tag{2.2.17}$$

in good agreement with equation (2.2.16) and most other previous results. Over the entire existence range of the superfluid for $2 \times 10^{-5} \leq \varepsilon \leq 10^{-2}$, the velocity can be represented [158] within about 0.5 % by the empirical equation

$$u_2 = h(P)\, \varepsilon^{1/3} \left[1 + b(P)\, \varepsilon^{1/6} \right]; \quad \varepsilon \equiv 1 - \frac{T}{T_\lambda(P)} \tag{2.2.18}$$

with

$$h = h_0 + h_1 P, \tag{2.2.19a}$$

and

$$b = b_0 + b_1 P + b_2 P^2. \tag{2.2.19b}$$

The parameters are given by

$$\begin{aligned} h_0 = 2120, \quad h_1 = -33.65, \\ b_0 = 1.573, \quad b_1 = 0.002764, \quad b_2 = 9.387 \times 10^{-4}, \end{aligned} \tag{2.2.20}$$

if P is in bars and u_2 in cm/sec. There is of course no reason why equation (2.2.18) should be the asymptotic form of u_2. The measured u_2 has been used [157, 158] with equation (2.1.37) for the entropy and equation (2.1.65) for C_P, to calculate ρ_s/ρ. These results are shown on logarithmic scales for three isobars in Figure 2.29 as a function of ε. This figure has insufficient resolution to reveal some of the details contained in the data, and serves only

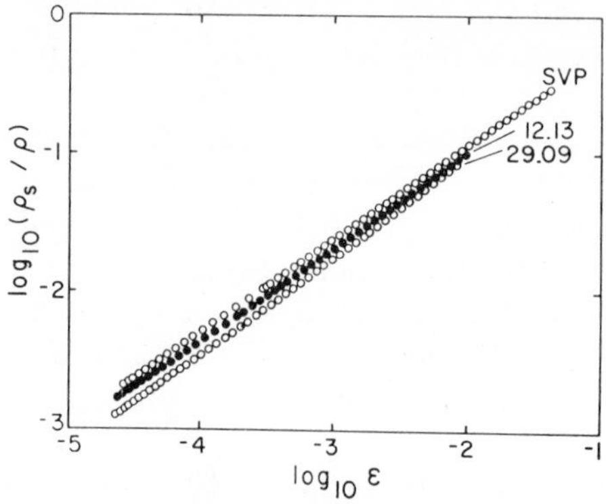

FIGURE 2.29. The superfluid density of He^4 on isobars as a function of ε on logarithmic scales. The numbers are the pressures in bars. After Reference [158].

the purpose of giving a general idea of the size of ρ_s/ρ at various ε and P. In order to reveal additional features in graphical form, ρ_s/ρ was multiplied by $\varepsilon^{-2/3}$, which, according to the vapor pressure measurements discussed in Sections 2.2.1(a) and 2.2.1(b) and equations (2.2.2) and (2.2.3), should yield approximately a constant. Figure 2.30 shows $\log_{10}[(\rho_s/\rho)\,\varepsilon^{-2/3}]$ as a function of $\log_{10}\varepsilon$. Also shown in Figure 2.30 as a solid line are equations (2.2.2) and (2.2.3), which are the best fit to the SVP results obtained by Tyson using the Andronikashvili technique [132, 133]. The two sets of results never differ by more than 2%, and any differences are within the permitted errors of the measurements [132, 133] and the thermodynamic parameters. The results for ρ_s/ρ under pressure, by Romer and Duffy [134], also obtained by the Andronikashvili technique, are shown with their permitted errors as a shaded area. Although these results are not very precise near T_λ, they are consistent at all pressures with the results derived from u_2.

Although the difference between the ρ_s/ρ derived from u_2 at SVP and ρ_s/ρ obtained by Tyson is never large, it does appear systematic, and one should enquire whether these systematic departures are permitted by the known uncertainties. This question is answered by fitting the results derived from u_2 to equation (2.2.2). For data with $2 \times 10^{-5} \le \varepsilon \le 10^{-2}$, this yields at SVP

$$k_{SVP} = 2.534, \quad \zeta_{SVP} = 0.674 \pm 0.001. \tag{2.2.21}$$

The value for ζ_{SVP} differs from that obtained by Tyson and quoted in equation (2.2.3) only by the sum of the standard errors, and the two sets of measurements are, therefore, consistent with each other. This comparison provides a rather sensitive test of the validity of equation (2.2.12).

The second-sound velocity under pressure has been measured recently by Brillouin light scattering from the very weak density variations which are

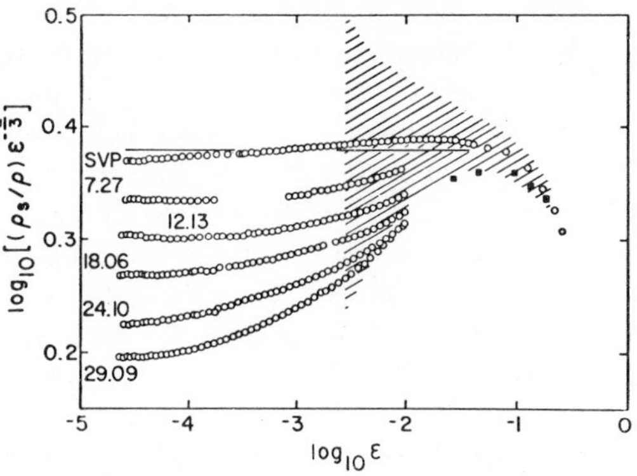

FIGURE 2.30. The superfluid density of He^4 on isobars multiplied by $\varepsilon^{-2/3}$, as a function of ε on logarithmic scales. The numbers are the pressures in bars. After Reference [157].

associated with the spontaneous second-sound waves if the thermal expansion coefficient is nonzero. These measurements were made at constant wave vector k. They typically have errors of the order of 1%. The results of both Winterling et al. [159a] for $\varepsilon \gtrsim 10^{-4}$ at 25.3 bars and of Vinen et al. [159b] for $\varepsilon \gtrsim 2 \times 10^{-4}$ at 20 bars agree with equations (2.2.18)–(2.2.20). This consistency with the low-frequency results indicates that even at the rather large k of $10^5\ \text{cm}^{-1}$, which is characteristic of the light-scattering experiments, there is no appreciable dispersion [see also Section 2.3.1(c)].

(*d*) *The Velocity of Fourth Sound.* When He (II) is contained in a porous medium, the velocities of the characteristic modes are strongly modified. Let us consider the case where the pore size d is much smaller than the viscous penetration length λ [130] [see equation (2.2.1)]. This limit has been considered first by Pellam [160], and then by Atkins [161]. It may be well approximated by assuming that the normal-fluid velocity v_n is zero because any $v_n > 0$ will result in a considerable viscous force acting on the normal fluid. If $v_n = 0$, the hydrodynamic equations yield one propagating mode, called fourth sound, with a velocity u_4 given by [162]*

$$u_4^2 = \left(\frac{\rho_s}{\rho}\right) u_{10}^2 + \left(\frac{\rho_n}{\rho}\right) c_2^2 \left(1 - \frac{2\alpha_P u_{10}^2}{\gamma S}\right)$$

$$= u_{10}^2 \left[\frac{\rho_s}{\rho} + f(T)\right], \qquad (2.2.22)$$

where α_P is the isobaric thermal expansion coefficient and S the entropy. At vapor pressure, $f(T) < 10^{-2}$ at all T [163]. Fourth sound was first observed

*A more complete discussion of the hydrodynamics of fourth sound has been given recently [D. J. Bergman, B. I. Halperin, and P. C. Hohenberg, *Phys. Rev.* B **11**, 4253 (1975)].

experimentally by Rudnick and Shapiro [164]. At low temperatures where $\rho_s/\rho \cong 1$, u_4 approaches the velocity of first sould. However, this velocity will be equal to u_1 for bulk helium at vapor pressure only if the pores are sufficiently large. For small pores, there will be size effects, and effects due to the vander Waals forces, which modify the velocity. Near T_λ, u_4 vanishes approximately as $\rho_s^{1/2}$. Thus the study of fourth sound may yield interesting information about the hydrodynamics near the transition. However equation (2.2.1) imposes serious limitations upon the detail in which information can be obtained about the transition in the bulk system. Near T_λ, equation (2.2.1) becomes $d \ll 2 \times 10^{-2}\ \omega^{-1/2}$ cm. In most experimental circumstances, $10^3 < \omega < 10^5$, yielding $d \ll 5 \times 10^{-4}$ cm. Measurements have been made [165, 166] on He (II) contained in packed powders with particle diameters of 10^{-4} cm. For $\omega \geq 10^3$, this size is about the largest that would satisfy equation (2.2.1). Although He (II) contained in geometries with characteristic dimensions of 10^{-4} cm will not differ strongly from the bulk, finite size effects are not absent. One can reasonably estimate [167] by extrapolation from smaller d that the transition temperature is depressed by $\Delta T_\lambda \cong 10^{-5}$ K (see also Figure 2.31 below). Therefore, it appears that fourth-sound measurements cannot be relied upon to yield quantitative information about the transition in the bulk system when $(T_\lambda - T) \lesssim 10^{-3}$ K. Measurements with smaller $T_\lambda - T$ or smaller d are, of course, extremely useful for the study of size effects [163, 165, 166] upon the transition, but that is not the topic of this chapter. If the pore size is made larger, the propagating mode becomes more attenuated. The dispersion relations have been examined for that case [167a, 167b], but it is difficult to see how systems with large pore sizes can be utilized for the study of the phase transition.

The fourth-sound results of Kriss and Rudnick [166] in a system with $d \cong 10^{-4}$ cm have been used by these authors to derive a very good approximation to the bulk ρ_s/ρ from equation (2.2.22) and the necessary thermodynamic parameters. Although it appears from equation (2.2.22) that absolute values of ρ_s/ρ can be determined, a temperature-independent correction to the measured resonance frequency that accounts for multiple scattering by the porous medium in which the helium is contained must be determined by normalizing the data to a known value of ρ_s/ρ at one temperature. Most reliably this can be done at low T, where $\rho_s/\rho \cong 1$. The measurements cover the range $10^{-3} \leq \varepsilon \leq 0.5$. When data for $\varepsilon \lesssim 10^{-1}$ were fitted to the power law equation (2.2.2), the parameters [166]

$$k = 2.38, \quad \zeta = 0.665 \pm 0.005 \tag{2.2.23}$$

were obtained. These results must be regarded as consistent with the values obtained by all other techniques, including the slightly different second-sound results given in equation (2.2.21), particularly since the quoted errors do not include any possible systematic contributions from finite size effects.

Yet another propagating mode, called third sound, exists in He (II) films. Here $v_n = 0$, just as it was in the case that yielded fourth sound, but in this system the thickness may be periodic, and mass exchange with the vapor is possible. This mode is not particularly useful for the study of the superfluid transition in bulk helium because He (II) films generally have a thickness $d \lesssim 3 \times 10^{-6}$ cm, and this results in a depression of the transition temperature $\Delta T_\lambda \gtrsim 2 \times 10^{-3}$ K. Third sound is, of course, useful for the study of size effects upon the transiton. A good, recent review [168] is available, and we shall not discuss this mode further here.

(e) The Superfluid Density and Critical Phenomena. It has not been possible so far to make quantitative measurements of the magnitude of the order parameter ψ for the superfluid transition. However, Josephson [128] has derived a scaling law that relates the superfluid density to $|\psi|^2$. If an exponent β is defined by the asymptotic proportionality $|\psi| \sim \varepsilon^\beta$, and if ζ is the asymptotic exponent of ρ_s/ρ, then the Josephson scaling law predicts

$$\zeta = 2\beta - \eta\nu'. \tag{2.2.24}$$

In this relation, ν' is an exponent given by the divergence of the length

$$\xi = \xi_0 \varepsilon^{-\nu'} \tag{2.2.25}$$

which describes the spatial range over which fluctuations in $|\psi|$ are correlated, and η describes the extent to which the order–parameter order–parameter correlation function departs from Ornstein–Zernike behavior [1]. It is expected that η is nearly zero, and perhaps about equal to 0.04 [92], although there is no experimental information on this point. If η is indeed sufficiently small, then measurements of ρ_s/ρ may be expected to reflect strongly the behavior of the square of the order parameter $|\psi|^2 \sim \varepsilon^{2\beta}$, but a quantitative test of equation (2.2.24) by itself is not possible with available information. However, equation (2.2.24) with other scaling laws previously derived by Widom [71] and Kadanoff [72] yields

$$\zeta = \frac{2-\alpha'}{3}, \tag{2.2.26}$$

and this equation may be compared with experiment.

It would be comparatively easy to compare equation (2.2.26) with experimental results if it could be assumed that C_P and ρ_s can be described by pure power laws for small $|\varepsilon|$. However, we discussed the determination of α' in some detail in Section 2.1.2(e), and saw that the measured C_P agrees with the predictions of scaling only if large higher-order contributions to C_P are permitted even for $|\varepsilon| \lesssim 10^{-3}$. The results for ρ_s/ρ in Figure 2.30 demonstrate even more explicitly the existence of singular, large corrections to the asymptotic behavior of this parameters. If the pure power law equation (2.2.2) were adequate to describe ρ_s/ρ, then the data in Figure 2.30 should fall on straight lines, and the slopes of these lines would be equal to $\zeta - \frac{2}{3}$. However, the results at the higher pressures in Figure 2.30 reveal that a straight line cannot represent the data within their precision over an appreciable range of ε even for $|\varepsilon| \lesssim 10^{-4}$. Therefore, a least-squares fit of the results to equation (2.2.2) will not yield the asymptotic exponent ζ to which equation (2.2.26) pertains. If such a fit is performed nonetheless, a pressure-dependent ζ_{eff} is obtained [158, 159]. In an attempt to determine the asymptotic behavior of ρ_s in spite of the presence of large corrections, the data were fitted to the equation

$$\frac{\rho_s}{\rho} = k(P)|\varepsilon|^{\zeta}[1 + a(P)\,|\varepsilon|^{y}]; \quad y > 0. \tag{2.2.27}$$

The greater number of parameters in equation (2.2.27) resulted in much larger uncertainties for ζ than those given in equation (2.2.21), and

$$\begin{aligned} 0.66 \leq \zeta \leq 0.68, \\ 0.4 \leq y \leq 0.6 \end{aligned} \tag{2.2.28}$$

was obtained at all pressures.* This result for ζ agrees well with the exponent equation (2.1.69) of C_P and the scaling prediction equation (2.2.26). The value of y has been calculated explicitly by Wegner [94]. It is equal to that of the correction exponent for C_P and has approximately the value 0.5. Equations (2.2.28) and (2.1.69) agree with this prediction. They are, of course, also consistent with universality [88], in that they reveal no pressure dependence of ζ or y. When it was *assumed* that ζ was independent of P and that $y \equiv \frac{1}{2}$, the result

$$\frac{\rho_s}{\rho} = k(P)\,|\varepsilon|^{0.6692}\,[1 + a(P)\,|\varepsilon|^{1/2}], \tag{2.2.29}$$

with

$$\begin{aligned} k &= 2.396 - 0.02883P_\lambda, \\ a &= 0.6514 - 0.04548P_\lambda + 5.265 \times 10^{-3}P_\lambda^2, \end{aligned} \tag{2.2.30}$$

was obtained. It is interesting to note that the best value of ζ in equation

*The measurements of u_2 and derivation of ρ_s are more accurate at vapor pressure than they are at higher pressures. In addition, the coefficient a in equation (2.2.27) happens to be small at SVP. For these reasons, a fit of the data at vapor pressure to equation (2.2.27) yields ζ with quite high accuracy. It gives $\zeta = 0.674 \pm 0.001$. This value, and the result $\alpha' = -0.026 \pm 0.004$ (see footnote on p. 117) obey the scaling law equation (2.2.26). This comparison of ζ and α provides the only test of any scaling law based on experimental results that were analyzed in terms of functions that include confluent singularities.

(2.2.29) and in equation (2.2.28) is greater than $\frac{2}{3}$. With equation (2.2.26), this yields $\alpha' < 0$ in agreement with the conclusions of Section 2.1.2(e).

Equations (2.2.29) and (2.2.30) provide a good description in closed form of the data, and departures of individual measurements generally do not exceed 1 %.†

(f) The "Superfluid Healing Length" and Finite Geometry Effects. We shall not deal extensively in this chapter with size effects upon the transition for their own sake. However, measurements of the effect of restricted geometries with a sufficiently small characteristic size d upon the hydrodynamic properties may be interpreted in terms of a length which is an intrinsic property of the bulk system, and which, therefore, concerns us here. A number of experiments, some of them similar to those discussed in Sections 2.2.1(b) and 2.2.1(d), have yielded the average superfluid density $\langle\rho_s\rangle_d$ in restricted geometries. For sufficiently small d, $\langle\rho_s\rangle_d$ is less than the superfluid density ρ_{sb} of bulk helium at the same temperature. This depletion of ρ_s is attributed to the presence of a large surface area, and to the existence of a length l perpendicular to the surface over which in some average sense the fluid does not participate in superfluidity. Henkel, Smith, and Reppy [169], in connection with their investigation of unsaturated helium films, defined the length l by equating the product of ρ_{sb} and l with the excess normal-fluid mass that corresponds to the difference between $\langle\rho_s\rangle_d$ and ρ_{sb}. Other authors [170] have used more detailed modified mean-field theoretical approaches towards defining l, but the resulting definitions appear largely equivalent, and the approaches based on the more detailed theory are not necessarily on a firmer footing [83]. The approach of Henkel et al. [169] is particularly appealing because of its conceptual simplicity, and because it can be directly related to experiment. They have demonstrated that their definition is an experimentally meaningful concept by measuring the angular momentum L of a persistent superfluid current in an unsaturated film, using the gyroscopic technique discussed in Section 2.2.1(b). Then, without disturbing the velocity field of the persistent current, they added successive increments of thickness to the film, and remeasured L. They found that L was a linear function of the film thickness, as would be expected if increments of added thickness contribute increments of superfluid mass corresponding to the bulk ρ_{sb}. However, at zero L the data extrapolate to a nonzero thickness that increases

†The values obtained for ρ_s/ρ depend on the measured u_2, and the entropy and C_p. In view of the inconsistency between the measurements of C_p [39] and α_p (see the footnote on p. 117), it is possible that the estimates for ρ_s/ρ are in error by as much as 5% at the highest pressures. Near vapor pressure, C_p and α_p are consistent with each other and ρ_s/ρ is given within rather small errors by equations (2.2.29) and (2.2.30). Any errors in C_p at the higher pressures are unlikely to have a substantial effect on the exponent of ρ_s, and will primarily influence the amplitude.

with decreasing $T_\lambda - T$. This thickness is equal to l defined above (except for a temperature-independent additive contribution which is believed to correspond to immobile helium atoms, about one atomic layer thick, and held firmly to the substrate by van der Waals forces).

The measurements on the films have not been made near T_λ, and could, of course, never be made very near T_λ because of the large depression of the transition temperature. The angular-momentum measurements described above serve primarily to demonstrate the validity of the concept of a "healing length". However, Henkel et al. [169] measured the depression of $\langle\rho_s\rangle_d$ below ρ_{sb} much nearer T_λ in other systems with larger d, and interpreted the results in an analogous fashion. The length deduced from all of these experiments for $2 \times 10^{-4} \lesssim \varepsilon \lesssim 0.2$ could be written as

$$l = (4.0 \pm 0.5) \times 10^{-8}\varepsilon^{-(0.67\pm0.04)}, \tag{2.2.31}$$

where l is in centimeters.

On the basis of scaling, it is expected that near a critical point there exists only one pertinent length ξ [see equation (2.2.25)], and any reasonable operational definition of a length should differ from ξ at most by a constant factor of order unity. Thus, we expect l to be equal to ξ in equation (2.2.25), except for possibly a small difference in the amplitude. We already saw that ν' is equal to ζ if the scaling law equation (2.2.26) is valid. Indeed, the exponent in equation (2.2.31), which should be equal to that of ξ, agrees with ζ as given by equation (2.2.28). The amplitude of l is about equal to an interatomic distance, which would also be a reasonable value for ξ_0. Thus, the experimental result equation (2.2.31) is entirely consistent with scaling and the measured ρ_{sb}.

The depression below T_λ of the superfluid onset temperature T_0 in restricted geometries can be looked upon as the temperature at which l(or ξ) becomes equal (within a numerical constant of order unity) to the characteristic length d of the geometry. Then the depletion of ρ_s would be complete and $\langle\rho_s\rangle_d = 0$. This qualitative argument yields

$$\varepsilon_0 \equiv \frac{T_\lambda - T_0}{T_\lambda} = \left(\frac{\xi_0}{d}\right)^{1.5} \tag{2.2.32}$$

within a factor of order unity. This is consistent with many experimental measurements, but usually d is not known well and most of the available information is not very definitive. Some of the experimental results on the depression of T_λ are summarized (167) in Figure 2.31, and are quite consistent with the functional form of equation (2.2.32).

Although the data in Figure 2.31 tend to support equation (2.2.32), most of them suffer from considerable uncertainties which are discussed in some detail elsewhere (167), and which are attributable primarily to the ill-defined

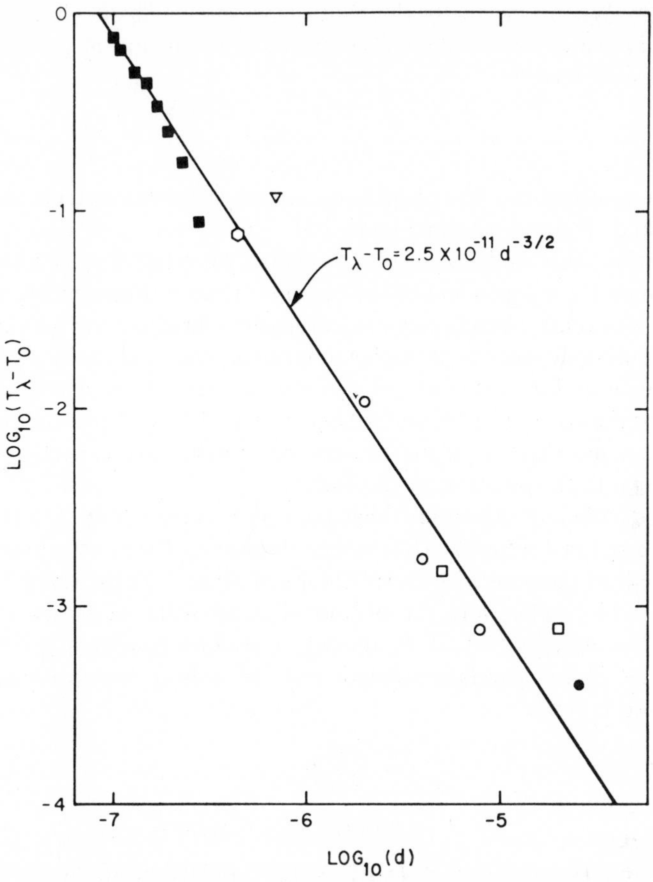

FIGURE 2.31. The depression $T_\lambda - T_0$ of the superfluid onset temperature in He^4 by restricted geometries of characteristic size d, with d in centimeters. The data are from various sources given in Reference [167]. After Reference [167].

geometries that had to be used, particularly for very small d. Considerably more uniform geometries have become available [170a] in recent years in the form of porous polycarbonate membranes with cylindrical pores of rather uniform diameter d, with d much less than the pore length. The onset of superfluidity for the ^4He contained in these pores was measured first by Williams et al. [155], and studied systematically by Ihas and Pobell [170b]. Both of these groups used the membranes as second-sound transducers [170c], and determined the value of $T_\lambda - T$ at which the second-sound amplitude generated by the vibrating membrane became negligible. At that temperature the superfluid density in the pores was expected to vanish.

Ihas and Pobell [170b] investigated nominal pore sizes of $d = 0.1, 0.2, 0.4,$

and 0.6 μm. Whereas virtually all previous measurements of ε_0 had been done at vapor pressure, they made measurements as a function of pressure up to 30 bar. They found

$$\varepsilon_0 = \left[\frac{d}{(5.7 \pm 0.6)} \times 10^{+8}\right]^{-1.54\pm0.05}, \tag{2.2.32a}$$

with d in centimeters. The result is consistent with, for instance, equation (2.2.31), and with the data in Figure 2.31.

In addition, Ihas and Pobell observed that, for fixed d, ε_0 is within their resolution of 2% independent of the pressure. If we assume that the onset of superfluidity occurs when the coherence length differs from the pore diameter by a pressure-independent multiplication factor, then the absence of a pressure dependence for ε_0 at constant d implies a pressure-independent amplitude ξ_0 for the coherence length [see equation (2.2.25)]. A pressure-independent ξ_0 is somewhat surprising because one might have expected ξ_0 to be proportional to the interatomic distance.

Ihas and Pobell observed also that the results for the bulk ρ_s/ρ [158] that were discussed in Section 2.2.1(e) yield values of ρ_s/T that at constant ε are independent of pressure within error limits of about 2%, provided $2 \times 10^{-5} \lesssim \varepsilon \lesssim 2 \times 10^{-4}$. This, and the pressure independence of ε_0, then implies that $\xi\rho_s/T$ is independent of P. Indeed, a relation has been given [170d] between the phase correlation length and the bulk ρ_s which leads to this latter result. It yields

$$\begin{aligned} \frac{\xi\rho_s}{T} &= \frac{m_4^2 k_B}{4\pi\hbar^2} \\ &\cong 0.044 \times 10^{-8}\ g/\text{cm}^2\text{K}. \end{aligned} \tag{2.2.32b}$$

The measurements result in $d\rho_s(\varepsilon_0)/T = (0.77 \pm 0.04) \times 10^{-8}\ g/\text{cm}^2\text{K}$ when ρ_s is evaluated for the bulk at ε_0. Since one would expect the superfluid density in the pores to vanish when the healing length is still less than the pore *radius*, the experimental result differs from the theoretical definition of ξ only by a numerical constant of order unity. Although a pressure-independent $\xi\rho_s/T$ is suggested by theory [170d], no particular reason seems to be known at this time for the pressure independence of ξ_0, ρ_s/T at constant ε, and ε_0 by themselves.

The relation between the superfluid healing length and ρ_s in ^{3}He–^{4}He mixtures will be discussed in Section 2.2.2(f).

(*g*) *Two-Parameter Universality*. Let us consider the parameter

$$u = G_s\xi^3/Vk_BT \tag{2.2.32c}$$

where G_s is the singular part of the Gibbs free energy per mole (see also Section 2.1.3) and ξ the correlation length given by equation (2.2.25). It has been suggested by Stauffer, Ferer, and Wortis [170e], and demonstrated on the basis of renormalization group theory [1] by Aharony [170f], that the parameter u is a universal parameter [88] near phase transitions, dependent only on such general properties of the system as the symmetry of the order parameter. One can obtain G_s from the singular part of the specific heat C_s [Section 2.1.2.(e)], and ξ from equation (2.2.32b) and the superfluid density [Section 2.2.1(e)]. Writing

$$C_s = \frac{A'}{\alpha'} \varepsilon^{-\alpha'}, \tag{2.2.32d}$$

one obtains

$$\alpha'\tilde{u} = k_B^2 \left(\frac{m_4}{\mathrm{h}}\right)^6 \frac{A'T^3}{k^3 \rho^3 V} \tag{2.2.32e}$$

which differs from αu only by a constant. Here k is the amplitude of ρ_s as given for instance by equation (2.2.30), ρ the density, and V the molar volume. In this equation, the dependence on ε has been eliminated by using the scaling law (2.2.26). It was shown by Ferer [170g] that the experimental results for A[39] and k[158] do indeed yield values of u which are, within the experimental uncertainty, independent of P along the λ line in pure ^{4}He. The most accurate value of u can be obtained at vapor pressure, and one has [117a]

$$\alpha\tilde{u} = 0.52 \pm 0.03. \tag{2.2.32f}$$

In mixtures of ^{3}He and ^{4}He, the experimental data do not define u very precisely; but within the uncertainties u is independent also of the ^{3}He concentration [117a].

2.2.2. ^{3}He –^{4}He Mixtures

(*a*) *Useful Thermodynamic Relations.* The results of the two-fluid hydrodynamics for superfluid mixtures of ^{3}He and ^{4}He have been presented in detail by Khalatnikov [127]. He found it convenient to consider the thermodynamic functions *per gram of solution*, and to use the mass concentration

$$c \equiv \frac{n_3 m_3}{n_3 m_3 + n_4 m_4}, \tag{2.2.33}$$

where n_3 and n_4 are the number of ^{3}He and ^{4}He atoms, and m_3 and m_4 the corresponding masses. The majority of the experimental results are, however, per *mole of solution*. Since the mass per mole M_s depends upon the concentration, the conversion from one to the other is not entirely trivial. This is particularly true when either the mass or the number of moles can be held constant when differentiating. Therefore we reproduce here some of the relations [107] of interest. For the concentrations one has

$$c = \frac{XM_3}{XM_3 + (1 - X)M_4}, \tag{2.2.34a}$$

where X is given by equation (2.1.74), and where M_3 and M_4 are the molar masses. For $(\partial c/\partial X)_{PT}$ one obtains

$$\left(\frac{\partial c}{\partial X}\right)_{PT} = \frac{M_4 c^2}{M_3 X^2}. \tag{2.2.34b}$$

In terms of the Gibbs free energy $\tilde{G}$ of the solution one can define the chemical potential of ^{3}He by

$$\mu_3^K = \frac{1}{m_3}\left(\frac{\partial \tilde{G}}{\partial n_3}\right)_{P,T,n_3m_3+n_4m_4}, \tag{2.2.35}$$

and similarly for the chemical potential μ_4^K of ^{4}He. These are the definitions used by Khalatnikov [127]. One also has

$$\mu_3^K = \frac{\partial g}{\partial c}, \tag{2.2.36}$$

with the same variables constant as in equation (2.2.35). Here g is the Gibbs free energy per gram of solution. The relation between μ_3 defined in equation (2.1.75) and μ_3^K is

$$\mu_3 = \left(\frac{c}{X}\right)M_4\mu_3^K - (M_4 - M_3)g, \tag{2.2.37}$$

where M_3 and M_4 are the molar masses of ^{3}He and ^{4}He. The difference in chemical potential

$$\Phi^K \equiv \mu_3^K - \mu_4^K \tag{2.2.38}$$

is related to Φ in equation (2.1.77) by

$$\Phi = \frac{c}{X}M_4\Phi^K - 2(M_4 - M_3)g. \tag{2.2.39}$$

The derivative of Φ with respect to X is related to that of Φ^K with respect to c by

$$\begin{aligned}\left(\frac{\partial \Phi}{\partial X}\right)_{TP} &= \left(\frac{M_4^2}{M_3}\right)\left(\frac{c^3}{X^3}\right)\left(\frac{\partial \Phi^K}{\partial c}\right)_{TP} \\ &= M_3^2 M_4^2\,[M_3 X + M_4(1 - X)]^{-3}\left(\frac{\partial \Phi^K}{\partial c}\right)_{TP}.\end{aligned} \tag{2.2.40}$$

For the derivative of the density ρ with respect to the concentration we have

$$\begin{aligned}\frac{c}{\rho}\left(\frac{\partial \rho}{\partial c}\right)_{TP} &= \left[\frac{1 - X}{1 - c}\right]\frac{X}{\rho}\left(\frac{\partial \rho}{\partial X}\right)_{TP} \\ &= \left[1 + \left(\frac{M_3}{M_4} - 1\right)X\right]\frac{X}{\rho}\left(\frac{\partial \rho}{\partial X}\right)_{TP}\end{aligned} \tag{2.2.41}$$

Although it is trivial, we give for completeness the relation

$$S = [XM_3 + (1 - X)M_4]s \tag{2.2.42a}$$

between the entropy S per mole of solution and the entropy s per gram of solution. Similarly, for the temperature derivatives

$$\left(\frac{\partial S}{\partial T}\right)_{PX} = \frac{C_{XP}}{T}$$

$$= [XM_3 + (1 - X)M_4]\left(\frac{\partial s}{\partial T}\right)_{cP} \tag{2.2.42b}$$

For the concentration derivative of the entropy one obtains

$$c\left(\frac{\partial s}{\partial c}\right)_{TP} = \frac{X}{M_4}\left[\frac{M_4 - M_3}{XM_3 + (1 - X)M_4}S + \left(\frac{\partial S}{\partial X}\right)_{TP}\right]. \tag{2.2.43}$$

(*b*) *The Andronikashvili Experiment.* The principle involved in this experiment has been described in Section 2.2.1(a). Measurements of the superfluid density by this technique have been made in ^{3}He–^{4}He mixtures over the entire concentration range [170h, 170i], but all the results available so far have been for $\varepsilon_\lambda \equiv 1 - T/T_\lambda(X) \gtrsim 10^{-2}$. The data are roughly consistent, however, with a value near 0.7 for the exponent ζ of ρ_s/ρ, and, therefore, are in agreement with expectations based upon universality arguments [88] [see Section 2.2.1(e)] and with the results for pure ^{4}He [equations (2.2.3) and (2.2.21)]. Although the data are not very detailed, they are of interest because they permit a comparison of ρ_s/ρ near T_λ (say at $\varepsilon_\lambda = 10^{-2}$) with results obtained from the second-sound velocity to be discussed in Sections 2.2.2(c) and 2.2.2(e). The values of ρ_s/ρ obtained by these very different techniques agree with each other within the experimental uncertainties [117a], and, therefore, tend to confirm the predictions of two-fluid hydrodynamics that will be discussed in Section 2.2.2(c).

No measurements have been made so far of the angular momentum of persistent currents in the mixtures near T_λ [see Section 2.2.1(b) for persistent currents in pure ^{4}He].

(*c*) *The Velocities of First and Second Sound.* In the superfluid mixtures, as in the pure system, there are two propagating modes, which are still called first sound and second sound. One would like to understand the measured velocities in He (II) in terms of the predictions based on two-fluid hydrodynamics, and in terms of independently measurable thermodynamic parameters and ρ_s/ρ. The hydrodynamic equations that determine the velocities of the propagating modes have been derived by Khalatnikov [127, 171a–171c]. They were solved originally only under the assumption that

$$\left(\frac{\partial \rho}{\partial T}\right)_{XP} \cong 0, \tag{2.2.44a}$$

which is equivalent to assuming that the following formula

$$\gamma_X \equiv \frac{C_{XP}}{C_{XV}} \tag{2.2.44b}$$

is about equal to unity. In that approximation Khalatnikov obtained $u_1^2 \cong c_1^2$ for the velocity of first sound where

$$c_1^2 \equiv \left(\frac{\partial P}{\partial \rho}\right)_{Tc}(1 + d), \tag{2.2.45}$$

and where

$$d \equiv \left(\frac{c}{\rho}\right)^2 \left(\frac{\partial \rho}{\partial c}\right)^2_{TP} \left(\frac{\rho_s}{\rho_n}\right). \tag{2.2.46}$$

Here c is the mass concentration given by equation (2.2.33). In terms of X, d can be obtained easily with the help of equation (2.2.41). For the velocity of second sound, the assumption $\gamma_X \cong 1$ yielded $u_2^2 \cong u_{20}^2$, with

$$u_{20}^2 = \frac{\rho_s}{\rho_n}\left[\bar{s}^2\left(\frac{\partial T}{\partial s}\right)_{cP} + c^2\left(\frac{\partial \Phi^K}{\partial c}\right)_{TP}\right](1 + d)^{-1}. \tag{2.2.47a}$$

Here

$$\begin{aligned} \bar{s} &= s - c\left(\frac{\partial s}{\partial c}\right)_{TP} \\ &= \frac{S}{M_4}\left[1 - \frac{X}{S}\left(\frac{\partial S}{\partial X}\right)_{TP}\right]. \end{aligned} \tag{2.2.47b}$$

In terms of molar quantities, we can write u_{20}^2 as

$$\begin{aligned} u_{20}^2 = {} & \frac{XM_3 + (1 - X)M_4}{M_4^2}\left(\frac{\rho_s}{\rho_n}\right)\left\{\left[S - X\left(\frac{\partial S}{\partial X}\right)_{PT}\right]^2 \frac{T}{C_{XP}} \right. \\ & \left. + X^2\left(\frac{\partial \Phi}{\partial X}\right)_{PT}\right\}(1 + d)^{-1}. \end{aligned} \tag{2.2.47c}$$

The quantity $1 + d$ in equations (2.2.45) and (2.2.47a) generally differs from unity by no more than 1% in the temperature range of interest near T_λ, and can be estimated from available thermodynamic data [3] and ρ_s/ρ. As X vanishes, d also vanishes, and equation (2.2.45) yields $u_1^2 \cong c_1^2 = (\partial P/\partial \rho)_{TX}$. This is equal to the result equation (2.2.14) for pure ^{4}He when $\gamma = 1$. Similarly, equation (2.2.47a) yields $u_2^2 \cong u_{20}^2 = (\rho_s/\rho_n)s^2(\partial T/\partial s)$ when $X = 0$, and for $\gamma = 1$ this agrees with equation (2.2.8b) for pure ^{4}He. As T_λ is approached for $X > 0$, d vanishes because it is proportional to ρ_s/ρ; therefore u_1 becomes equal to $(\partial P/\partial \rho)_{TX}$ also for $X > 0$ in this limit. For the case $\gamma_X = 1$, this velocity is continuous with that of ordinary sound in He (I), as one would expect since the λ transition is not of first order.

We saw in Section 2.1.2(e) that in the pure system γ [see equation (2.2.6)] becomes as large as 1.6 at experimentally resolvable values of $T_\lambda - T$ when the pressure is near the melting pressure. Thus, we must expect appreciable departures of γ_X from unity also for the mixtures under certain conditions, and, in that case, the velocities may differ from the results given above. In the general case (arbitrary γ), Khalatnikov's equations lead to the relation [171d]

$$u^4 - u^2 \left[u_{20}^2(1 + d)\gamma_X + u_{10}^2\right](1 + b) + u_{10}^2 u_{20}^2 = 0 \tag{2.2.48}$$

for the velocities. Here

$$u_{10}^2 = c_1^2 \gamma_X, \tag{2.2.49a}$$

and

$$b = \frac{\mathscr{A} \times \mathscr{B}}{[u_{10}^2/\gamma_X + u_{20}^2(1 + d)]}, \tag{2.2.49b}$$

with

$$\mathscr{A} = \frac{\rho_s}{\rho_n}\left(\frac{\partial T}{\partial s}\right)_{cP}\left(\frac{\partial P}{\partial \rho}\right)_{Tc}\frac{\bar{s}}{s}, \tag{2.2.49c}$$

and

$$\mathscr{B} = \frac{2cs}{\rho^2}\left(\frac{\partial \rho}{\partial c}\right)_{TP}\left(\frac{\partial \rho}{\partial T}\right)_{cP} - \frac{sc^2}{\bar{s}\rho^2}\left(\frac{\partial \Phi^K}{\partial c}\right)_{TP}\left(\frac{\partial \rho}{\partial T}\right)_{cP}^2. \tag{2.2.49d}$$

In terms of molar quantities, we can write $\mathscr{A}$ and $\mathscr{B}$ as

$$\mathscr{A} = \frac{[XM_3 + (1 - X)M_4]^2}{M_4}\left(\frac{T}{C_{XP}}\right)\left(\frac{\partial P}{\partial \rho}\right)_{TX} \times \left[1 - \frac{X}{S}\left(\frac{\partial S}{\partial X}\right)_{TP}\right]\left(\frac{\rho_s}{\rho_n}\right), \tag{2.2.49e}$$

and

$$\mathscr{B} = \frac{2SX}{M_4\rho^2}\left(\frac{\partial \rho}{\partial X}\right)_{TP}\left(\frac{\partial \rho}{\partial T}\right)_{XP} + \left(\frac{X^2}{M_4\rho^2}\right)\left(\frac{\partial \phi}{\partial X}\right)_{TP}\left(\frac{\partial \rho}{\partial T}\right)_{XP}^2\left[1 - \frac{X}{S}\left(\frac{\partial S}{\partial X}\right)_{PT}\right]^{-1}. \tag{2.2.49f}$$

Equation (2.2.48) can be compared with equation (2.2.7), which pertains to pure ^{4}He. Both b and d vanish as X vanishes; therefore equation (2.2.48) with equations (2.2.11a) and (2.2.11b) yields equation (2.2.7) in this limit. Equation (2.2.48) can be solved for the velocities u_1 and u_2 of first and second sound, and gives to first order in the small parameter [see also Section 2.2.1 (c)] u_{20}^2/u_{10}^2 [117a]

$$u_1^2 = (1 + b)u_{10}^2\left\{1 + [(1 + d)\gamma - (1 + b)^{-2}]\frac{u_{20}^2}{u_{10}^2} + \cdots\right\}, \tag{2.2.50}$$

and

$$u_2^2 = (1 + b)^{-1}u_{20}^2\left\{1 - [(1 + d)\gamma - (1 + b)^{-2}]\frac{u_{20}^2}{u_{10}^2} + \cdots\right\}. \tag{2.2.51}$$

The velocity u_1 of first sound near T_λ was measured over a wide concentration and temperature range with medium temperature resolution by Roberts and Sydoriak [103]. These results indicate that the singularity in u_1 at T_λ (see Figure 2.28) becomes much less pronounced as X is increased, and that the velocity itself decreases appreciably with increasing X. More detailed measurements which extend to within a few μK of T_λ were obtained for $X = 0.0506$ by Barmatz [171e], and for $X = 0.100$, 0.200, 0.395, and 0.580 by Thomlinson and Pobell [171f]. These authors compared their results for u_1 with the measured C_{XP}, using equation (2.1.36).* This PBF relation is valid only if the velocity of sound is given by

$$u^2 = \left(\frac{\partial P}{\partial \rho}\right)_{SX}. \tag{2.2.52}$$

Although we know from equation (2.2.50) that equation (2.2.52) does not give u_1 for $T < T_\lambda$, it probably is a sufficiently good approximation over the range of ε_λ involved in the analysis. In any event, it yields u_1 in the limit as ε_λ vanishes, because b and d vanish at T_λ. The plots of $u_1 - u_0$ (u_0 is the smallest measured velocity) vs. C_P^{-1} looked very similar [171*e*, 171f] to that in Figure 2.5 for pure ^{4}He. They were consistent with equation (2.1.36), in that the data above and below T_λ tended towards the same straight line as the transition was approached. The relatively weak dependence upon X of these graphs indicates that the observed reduction in the amplitude of the singularity [103] in the velocity is associated primarily with the reduction in the size of the singularity in C_{XP}. The values of $(\partial S/\partial P)_{\lambda X}$, which were deduced from the data and the assumption that equation (2.1.36) is valid, differ only slightly from that in pure ^{4}He, and permit at most a mild concentration dependence of $(\partial S/\partial P)_{\lambda X}$. Values of $(\partial S/\partial P)_{\lambda X}$ were obtained also from the velocities above T_λ, and agreed within experimental errors with those obtained from u_1 below T_λ. Above T_λ, one expects equation (2.2.52) to give the velocity of ordinary sound; thus the PBF relation equation (2.1.36) should be exact.

It is interesting to note that direct measurements in mixtures of T_λ as a function of pressure [172, 173] reveal that the addition of moderate amounts of ^{3}He has only a small effect upon $(\partial P/\partial T)_{\lambda X}$. It appears that the various derivatives along the λ line at constant X are affected only mildly by the addition of ^{3}He.

Thomlinson and Pobell [171f] fitted their minimum measurable velocity u_0 to a polynominal in X, and obtained the following formula:

*Equation (2.1.36), when applied to the mixtures. should be modified because C_{XP} is finite at T_λ. This modification is accomplished by replacing C_{XP}^{-1} with $C_{XP}^{-1} - [C_{XP}\,(T_\lambda)]^{-1}$. This change applies in principle also to pure ^{4}He since $\alpha < 0$. It does not affect the determination of $(\partial S/\partial P)_{\lambda X}$, because this derivative depends only the slope of the data in Figure 2.5

$$u_0 = 21783 - 3444X + 577X^2 \text{ cm/sec}, \tag{2.2.52a}$$

for $X \leq 0.58$. This minimum velocity is expected to be larger than the thermodynamic velocity at T_λ [see equation (2.1.31)] by 60–90 cm/sec, primarily because of the effect of gravity [44].

The velocity of second sound was measured in very dilute mixtures by Noble and Sandiford [174]. Their results are only for $T_\lambda - T \gtrsim 10^{-2}$ K, and are limited to $X \leq 0.0355$. When fitted to the power law

$$u_2 = u_0\left(\frac{T_\lambda - T}{T_\lambda}\right)^m, \tag{2.2.52b}$$

they yielded $m = 0.382 \pm 0.004$ and independent of X. Although equation (2.2.52b) with this m almost certainly does not represent the asymptotic behavior of u_2, it provides a good means of comparing different sets of data, and the result for m agrees well with equations (2.2.15) and (2.2.16), which pertain to $X = 0$. The amplitude u_0 was only slightly different for $X = 0.0355$ than it was for $X = 0$.

More recently, measurements were made by Terui and Ikushima, [174a] by Ahlers and Greywall [107, 174b], and by Thomlinson et al. [174c]. (see also reference [174d]). These data reveal that u_2 at constant ε is only very mildly dependent upon X for $X \lesssim 0.15$. For larger X, u_2 at constant ε decreases with increasing X and becomes quite small near the tricritical concentration [107]. A summary of all these data is given in Reference [170k].

Rather detailed measurements of u_2 have been made at large ^{3}He concentrations, along the coexistence curve near the tricritical temperature T_t, and along lines of constant X from T_λ to the phase-separation temperature T_σ [107]. Over a fairly wide range of T and X, the measurements are shown in Figure 2.32. Along the phase-separation curve, the data can be represented within 0.5% by the function

$$u_2 = u_{0\sigma}\varepsilon_t^{m_\sigma}[1 + A_\sigma\varepsilon_t], \tag{2.2.52c}$$

with the least-squares-adjusted parameters

$$u_{0\sigma} = 43.26 \text{m/sec}, \quad m_\sigma = 0.476,$$
$$A_\sigma = -0.979, \qquad T_t = 0.8672 \text{ K}, \tag{2.2.52d}$$

for $\varepsilon_t \equiv 1 - T/T_t \lesssim 0.1$. This result is consistent within about 2% with earlier measurements by Elliott and Fairbank [175] which covered the range $\varepsilon_t \gtrsim 0.05$. Along the phase-separation curve, the behavior of u_2 is different from that encountered at constant X in dilute solutions. Although $u_{0\sigma}$ is similar in size to u_0 for small X, we have $m_\sigma \cong \frac{1}{2}$, which differs from the exponent at small and constant X. Along the phase-separation curve the system reveals tricritical behavior, and not behavior characteristic of the

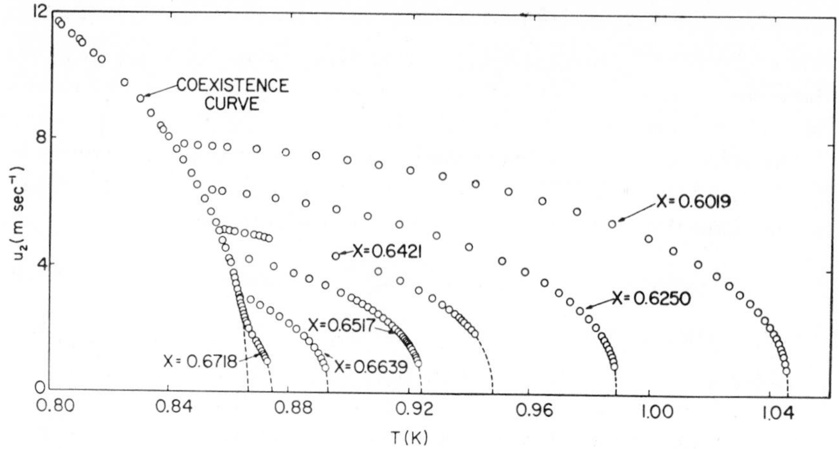

FIGURE 2.32. The velocity of second sound in ^3He–^4He mixtures as a function of T along the coexistence curve, and along several lines of constant X. After Reference [107].

ordinary λ points. This can be understood in terms of Figure 2.22, where the phase-separation curve is in region I, and where a line of constant X near T_λ is in region II. At constant X, even for X near X_t, the results near T_λ yield $m \cong 0.4$, which is similar to the result for small X. However, for X near X_t, the amplitude u_0 is much reduced, and strongly dependent upon X. Typically, u_0 is nearly an order of magnitude smaller than in the pure system at the same ε.

(*d*) *The Velocity of Fourth Sound.* The limitations and merits of fourth-sound measurements for the study of the phase transition have been discussed in Section 2.2.1(d). The relation between the velocity u_4, thermodynamic parameters, and ρ_s/ρ has been derived for the mixtures by Sanikidze and Chernikova [176]. For the case $\gamma_X = 1$, it can be written as

$$u_{40}^2 = \left(\frac{\rho_s}{\rho}\right)\frac{u_{10}^2[1 + (c/\rho)(\partial\rho/\partial c)_{TP}]^2}{(1 + d)} + (\rho_n/\rho)u_{20}^2(1 + d). \qquad (2.2.52e)$$

The parameters u_{10}, u_{20}, and d are given by equations (2.2.49a), (2.2.47a), and (2.2.46), respectively. Although the case $\gamma_X > 1$ has also been considered [176], we shall not repeat the result here.

Measurements of u_4 have been obtained over the entire concentration range [177, 178], but none of these results is sufficiently near T_λ to be very useful for the study of the phase transition. It may be of interest to note, however, that the measurements yield values of ρ_s/ρ that agree within experimental error with those obtained by the techniques described in Sections 2.2.2(b) and 2.2.2(c).

(*e*) *The Superfluid Density.* The results for u_2 given in Section 2.2.2(c) can be used to obtain fairly detailed values of the superfluid density by using equation (2.2.51). At vapor pressure, there is adequate thermodynamic information [179a] to extract ρ_s/ρ from the measured u_2 and equation (2.2.51) [179b]. For $\varepsilon_\lambda \lesssim 10^{-2}$, calculations can be based upon equations (2.1.89)–(2.1.92) and Table 2.2 for C_{XP}. With the help of the λ-line parameters in Table 2.1, $C_{\Phi P}$ can then be obtained from equation (2.1.96), and equation (2.1.80) yields $(\partial\Phi/\partial X)_{TP}$. The entropy is readily obtained by integrating equation (2.1.89). For $(\partial S/\partial X)_{TP}$ one has

$$\left(\frac{\partial S}{\partial X}\right)_{TP} = \left(\frac{\partial S}{\partial X}\right)_{tP} - \left(\frac{\partial T}{\partial X}\right)_{tP} \frac{C_{XP}}{T}. \tag{2.2.53a}$$

The parameter $\rho^{-1}(\partial\rho/\partial X)$ that occurs in equation (2.2.46) can be estimated from available data [3, 125e], and is about equal to 0.55, and only mildly dependent on concentration. In order to obtain the thermal expansion coefficient, it is necessary to use equation (2.1.10), whichpertains to C_{XP} and α_{XP} in the form

$$C_{XP} = T\left(\frac{\partial S}{\partial T}\right)_{tX} + VT\left(\frac{\partial P}{\partial T}\right)_{tX}\alpha_{XP}. \tag{2.2.53b}$$

Unfortunately, the concentration dependences of the parameters $(\partial S/\partial T)_{tX}$ and $(\partial P/\partial T)_{tX}$ are not known accurately from experiment, but the available data [171f, 172, 173] seem to indicate that α_{XP} may be estimated roughly by using the parameters for pure ^{4}He. This appears to be sufficiently accurate since α_{XP} contributes little to u_2. In fact, at vapor pressure only a very small error would be introduced by using equation (2.2.47c), which assumes $\alpha_{XP} = 0$, rather than equation (2.2.51). Near the tricritical point, ρ_s/ρ can be obtained easily over a wider range of ε_λ or ε_t. Here the contribution from $(\partial\Phi/\partial X)_{TP}$ becomes quite small, and the specific heat and entropy have only extremely mild singularities. Therefore, it becomes a great deal simpler to obtain the ε dependence of ρ_s/ρ. The necessary thermodynamic parameters have been tabulated over a wide concentration and temperature range [106, 109].

The measured velocities near the tricritical point [107], together with equations (2.1.115)–(2.1.117) for $(\partial\Phi/\partial X)_{PT}$ and the entropy and C_{XP} tabulated by Islander and Zimmermann [106], yield the data which are shown on logarithmic scales in Figure 2.33. Along the coexistence curve, and for sufficiently small ε_t, the date are consistent with

$$\frac{\rho_s}{\rho} = k_\sigma \varepsilon_t^{\zeta_\sigma} \tag{2.2.53c}$$

and $\zeta_\sigma = 1.0$.

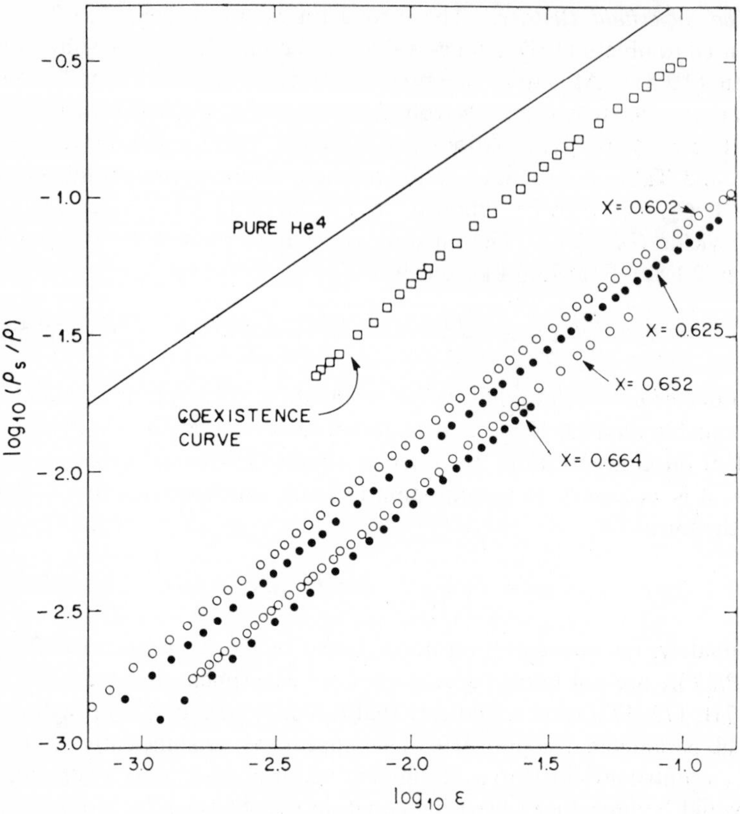

FIGURE 2.33. The superfluid density in He^3–He^4 mixtures along the phase separation curve and along several paths of constant X, as a function of $1 - T_\sigma/T_t$ or $1 - T/T_\lambda$, on logarithmic scales. After Reference [107].

It is evident that the dependence upon ε_t of ρ_s/ρ along the coexistence curve is very different from the dependence upon $\varepsilon_\lambda \equiv 1 - T/T_\lambda(X)$ along lines of constant concentration. Along the coexistence curve, the exponent of ρ_s/ρ is predicted by Riedel and Wegner [122]. They obtain $\beta_\sigma = \frac{1}{2}$ and $\eta_\sigma = 0$, where β_σ is the exponent of the superfluid order parameter along the phase-separation curve and η_σ describes the departure of the correlation function from Ornstein–Zernike behavior. Josephson's scaling law [128] $\zeta = 2\beta - \eta\nu'$ then yields $\zeta_\sigma = 2\beta_\sigma = 1$, in excellent agreement with the measurement. In addition to the asymptotically dominant term, Wegner and Riedel [123] have predicted the existence of logarithmic corrections along the coexistence curve. Available experimental results cannot distinguish between the absence or presence of these logarithms.*

It can also be shown on the basis of scaling and the exponents calculated

*A good fit to the data with $\varepsilon \leq 0.03$ is provided by $\rho_s/\rho = 1.826\,\varepsilon_t(-\ln \varepsilon_t)^{0.679}$. Logarithmic corrections to the leading power law behavior bave been predicted by Wegner and

by Riedel and Wegner [122] that ρ_s/ρ on the tricritical isotherm should be a linear function of $1 - X/X_t$. The data in Figure 2.31 permit an evaluation of ρ_s/ρ along this path, and are consistent with the expected behavior.

In spite of the very fine agreement between theory and experiment near the tricritical point, considerable caution is still in order, and the agreement should not yet be taken too seriously. There are no measurements of any parameter of the system for $\varepsilon_t \lesssim 3 \times 10^{-3}$, and it is possible that the behavior will be different for sufficiently small ε_t. Clearly, precise data for smaller ε_t would be most desirable to test the recent theoretical predictions, but it is exceedingly difficult to make measurements for smaller ε_t because of the effect of gravity [Section 2.1.3(g)]. For the superfluid density, there is the additional problem that ρ_s becomes rather small. Nonetheless, measurements for ε_t as small as, say, 3×10^{-4} seem feasible under ideal circumstances.

When ρ_s/ρ along lines of constant concentration was fitted to a pure power law in ε_λ, an effective exponent with a vaule near 0.8 was obtained for $X \gtrsim 0.55$. One would expect the exponents near T_λ to be universal, i.e., independent of X, just as they were expected to be independent of P in the pure system. Therefore, it is expected that the asymptotic value of ζ is rather close to 0.67, as found in pure ^4He [equation (2.2.28)]. Therefore, in order to obtain consistency between theory and experiment, singular correction terms must be invoked to the asymptotically dominant term. The problem is, in some sense, similar to that encountered for ρ_s/ρ in pure ^4He under pressure [157, 158]; as in the pure systme the results for the mixtures could be fitted to the equation [see equation (2.2.27)]

$$\frac{\rho_s}{\rho} = k(X)\,\varepsilon_\lambda^\zeta\,[1 + a(X)\,\varepsilon_\lambda^y]. \qquad (2.2.54)$$

They were consistent with $\zeta = 0.67$ and $y = 0.5$, but considerably different exponents are not ruled out by the data. However, ζ is clearly less than $\zeta_\sigma = 1$, and, consistent with Figure 2.22, the behavior at constant X and near T_λ is different from that near T_t. The consistency of the data with $\zeta = 0.67$ and $y = 0.5$, which also agrees with the results for pure ^4He [157, 158], tends to support the validity of the concept of universality [88].

Although equation (2.2.54), which was used in the data analysis, is similar to equation (2.2.27), the departure from power law behavior near T_λ for X near X_t may well have a different origin from that which yields higher-order

Riedel [123]. The power of the logarithm is expected to be 1/2. Although the measurements of ρ_s yield a slightly larger exponent, the uncertainties are large and the data should be regarded as consistent with the prediction. The results do not provide strong support for the theory, however, because there are alternate interpretations that do not involve logarithms. Thus the function $\rho_s/\rho = 5.01\ \varepsilon_t(1-0.60\varepsilon_t)$ fits the dat with $\varepsilon_t \leq 0.03$ equally well. More definitive experimental confirmation of similar predicted logarithmic terms, but for dipolar Ising systems, have been obtained recently from specific heat measurements for the uniaxial ferromagnet $LiTbF_4$ [G. Ahlers, A. Kornblit, and H. J. Guggenheim, *Phys. Rev. Lett.* **34**, 1227 (1975)].

terms in pure ^{4}He. We saw in Figure 2.22 that a line of constant X passes from region III where tricritical behavior dominates into region II where λ-point behavior dominates. Thus the departure from pure power law behavior may be caused by the proximity of the crossover line to the λ-line. A more thorough analysis of available data may shed additional light on these possible crossover effects.

Although the dependence of ρ_s/ρ upon $T_\lambda - T$ is not grossly altered when He^3 is added to the system, examination of Figure 2.33 reveals immediately that near X_t the amplitude k in equation (2.2.54) is much smaller than in pure ^{4}He. It appears that k is a decreasing function of X, and becomes very small as $1 - X/X_t$ vanishes. The amplitudes are shown as a function of $1 - X/X_t$ on logarithmic scales in Figure 2.34. On the basis of scaling and the exponents calculated by Riedel and Wegner [122] $k(X)$ is expected to vanish at X_t as $(1 - X/X_t)^{0.33}$. The data in Figure 2.34 are consistent with this prediction over about one decade in $1 - X/X_t$.

(*f*) *The "Superfluid Healing Length" and Finite Geometry Effects.* Finite size effects in the mixtures, as in pure ^{4}He, are expected to be determined by a healing length [see Section 2.2.1(f)] that differs from the phase correlation length [170d]

$$\xi = \frac{m^2 k_B T}{4\pi\hbar^2 \rho_s} \tag{2.2.55}$$

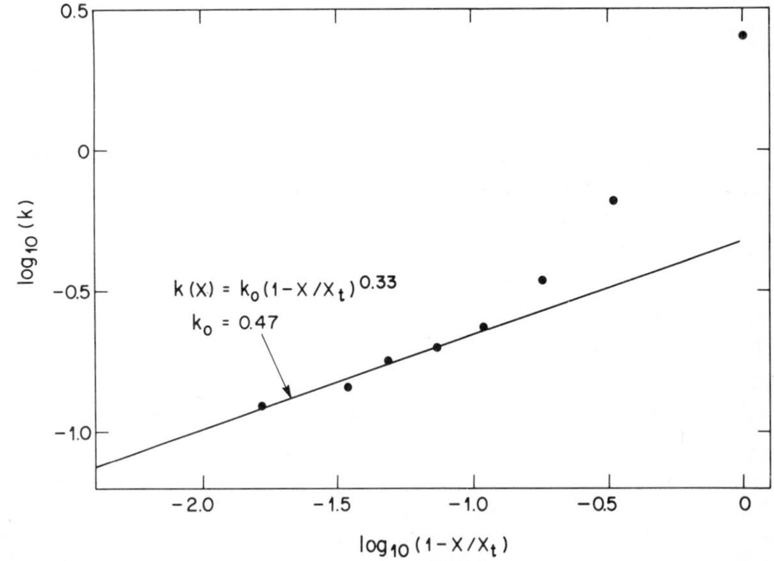

FIGURE 2.34. The amplitude near $T_\lambda(X)$ of power-law fits to ρ_s/ρ along lines of constant X in ^{3}He–^{4}He mixtures. The exponent was held fixed at 2/3. The solid circles are from Reference [107], and the open ones were calculated from the u_2 of Reference [174c].

only by a temperature- and concentration-independent multiplicative factor of order unity. It is, therefore, expected on the basis of the measured ρ_s [Section(e)] and equation (2.2.55) that the amplitude ξ_0 of ξ will become larger with increasing X. In agreement with this, it was found [107] that second sound could not be generated with porous microphone transducers [see Section 2.2.1(f)] near the tricritical concentration for $\varepsilon_\lambda \lesssim 10^{-3}$, indicating that ξ had become comparable to the 0.5-μm hole diameter of the transducer pores already at this rather large reduced temperature. Transducers with larger holes will have to be used if the tricritical point or λ point are to be approached more closely. Unfortunately the holes cannot be much larger because they must be smaller than the viscous penetration length for the normal fluid [see equation (2.2.1)]. For these reasons, it probably will not be possible to obtain u_2 with superleak transducers for ε_λ much less than 10^{-3} near X_t.

The onset of superfluidity in porous membranes was studied more systematically by Thomlinson et al. [174c]. They used the same technique employed previously [155, 170b] for pure ^4He [see Section 2.2.1(f)], and made measurements of the depression ε_0 of the onset temperature for $X \lesssim 0.40$. Consistent with the results near the tricritical point [107] discussed above, they found an increase by a factor of 6 for ε_0 at constant pore diameter d between $X = 0$ and $X = 0.40$. However, over the same concentration range, the amplitude of ρ_s decreases considerably, and the product $\rho_s(\varepsilon_0)/T$ remains constant within experimental error. Since $\xi(\varepsilon_0)$ is also a constant and determined solely by d, this result agrees with the prediction equation (2.2.55). It would, of course, be interesting to make these quantitative measurements nearer the tricritical point where the amplitude of ρ_s is even smaller.

(g) *Two-Parameter Universality.* As in pure ^4He, the parameter u (equation 2.2.32c) or $\alpha\tilde{u}$ (equation 2.2.32e) should be universal [88] also in the mixtures. In this case one would expect u and $\alpha\tilde{u}$ to be independent of X. However the amplitudes A and k are those of the heat capacity at constant chemical potential $C_{\Phi P}$ and of ρ_s/ρ, both along paths of constant Φ. These amplitudes, and $\alpha\tilde{u}$, have been calculated [117a] from C_{XP} (see equations 2.1.89–2.1.92) and the parameters in Tables 2.1 and 2.2, using the methods of Section 2.1.3(a). The uncertainty in $\alpha\tilde{u}$ increases rapidly with X; but within the estimated errors $\alpha\tilde{u}$ is independent of X as expected from theory.

2.3. TRANSPORT PROPERTIES

2.3.1. Pure ^4He

In Section 2.2 we discussed the experimental measurements that may be compared with some of the results of hydrodynamics that are obtained by considering the superfluid motion without dissipative terms in the equations.

Energy dissipation may be included in the theory by allowing terms in the fluxes that depend upon spacial derivatives of the velocities and the thermodynamic variables. The predictions for the various dissipative processes can then be stated in terms of several transport coefficients. For a single-component normal fluid [e.g., He (I)], there may be three independent coefficients [130]. They are the thermal conductivity κ, the first or shear viscosity η, and the second or bulk viscosity ζ. In the superfluid, it is possible to have a larger number of transport coefficients. There is still the thermal conductivity κ, which describes the irreversible contribution to the heat transport which is associated specifically with the presence of a temperature gradient. There is also a shear or first viscosity η, which is a property purely of the normal fluid. The bulk or second viscosity is replaced by four transport coefficients ζ_1, ζ_2, ζ_3, and ζ_4. Of these, only three are independent, because $\zeta_1 = \zeta^4$. All transport coefficients must be positive except for ζ_1, which must satisty the inequality $\zeta_1^2 \leq \zeta_2\zeta_3$. The transport coefficients may have singularities at T_λ, and some of these singularities are predicted by theory. We will examine the present status of experimental information about the transport coefficients, and, where possible, make comparisons with predictions.

Although in some cases, such as η, a single transport coefficient by itself has been obtained from experiment, direct measurements are not always feasible. Often only a combination of coefficients that is given by hydrodynamics can be inferred from such results as the damping of propagating modes or the heat transport due to a thermal gradient. Therefore, it is more reasonable to discuss the determination of certain combinations of transport coefficients, rather than to discuss the coefficients individually.

(*a*) *The Shear Viscosity* η. The shear viscosity η occurs in the hydrodynamics of both He (I) and He (II). Calculations indicate [180] that η remains finite both above and below T_λ, but the detailed behavior of η as a function of ε is not known from theory.

Particulary for $T < T_\lambda$ and at SVP, η has been inferred from a variety of experiments, and there is general agreement between different types of measurement that are related to each other through hydrodynamics. Many of the several types of results have been discussed elsewhere [3], and most of the data near T_λ [181–187] have been summarized recently [188]. By far the most precise measurements with the best temperature resolution are those by Webeler and Allen [186, 187].* These authors submerged a cylindrical quartz crystal in the liquid, and excited a torsional mode in the crystal at the re-

*Even more accurate results were obtained recently by R. Biskeborn and R. W. Guernsey, Jr. [*Phys. Rev. Lett.* **34**, 455 (1955)]. These authors find at saturated vapor pressure and for pure ^{4}He that a fit of their data with $|\varepsilon| \lesssim 0.005$ to equation (2.3.1) yields 0.65 ± 0.03 and 0.80 ± 0.05 for the exponent below and above T_λ, respectively. These results seem to

sonance frequency ω. They then measured the attenuation due to the liquid, which over a length $\lambda = (2\eta/\rho_n\omega)^{1/2}$ [see also equation (2.2.1)] from the crystal surface tends to participate in the oscillations. This measurement, like essentially all other viscosity measurements in helium, yields η/ρ_n, but ρ_n is known well enough (see Section 2.2.1) to calculate η without introducing serious errors. Webeler and Allen were able to obtain better results near the transition in dilute mixtures of ^3He in ^4He, because the depressed transition temperature permitted them to provide easily a thermally homogenous environment for their sample by surrounding it with superfluid ^4He. It is likely that their η for 0.5% ^3He in ^4He does not differ substantially from that of pure ^4He, and surely the dependence upon $T_\lambda - T$ should be essentially unaltered. They were able to show with rather high resolution (0.1% of η) that η is continous at T_λ. However, near T_λ, η varies extremely rapidly with T and it is likely that η is singular at T_λ [188]. All results from the various sources can be represented by

$$1 - \frac{\eta}{\eta_\lambda} = 5.19\,|\varepsilon|^{0.85} \text{ for } T < T_\lambda,$$
$$1 - \frac{\eta}{\eta_\lambda} = -\,1.82\,|\varepsilon|^{0.75} \text{ for } T > T_\lambda \qquad (2.3.1)$$

provided small adjustments are made in the values of η_λ for the several sets of data. This is demonstrated in Figure 2.35, where most of the available data are displayed. It is likely that an adjustment in η_λ will also largely take the difference between the dilute mixtures and the pure system into account. In pure ^4He, the best estimate of η_λ is

$$\eta_\lambda = (2.47 \pm 0.03) \times 10^{-5} \text{ poise.} \qquad (2.3.2)$$

There seem to be no measurements of η near T_λ in pure ^4He under pressure. One would expect the singularity in η to remain largely unchanged as P is increased. Precise measurements with which to test this universality would be most welcome. In the mixtures, it would be interesting to see in what fashion the singularity would be altered by the impurity. Measurements already exist for $X \leq 0.1$, and show that η_λ decreases linearly with $T_\lambda(X)$. Additional measurements for larger X would be desirable. Some results near the tricritical point have become available very recently [189].

To a large extent, the behavior of the shear viscosity is reflected also in the

rule out equal exponents above and below T_λ for the viscosity. However the data permit equal exponents if a confluent singularity is included in the data analysis. Confluent singularities have been shown to contribute to other properties near T_λ (see, for instance, Section 2.1.2e and equation (2.2.68) for a discussion of the confluent singularity in the specific heat and Section 2.2.1e and equation (2.2.27) for the superfluid density).

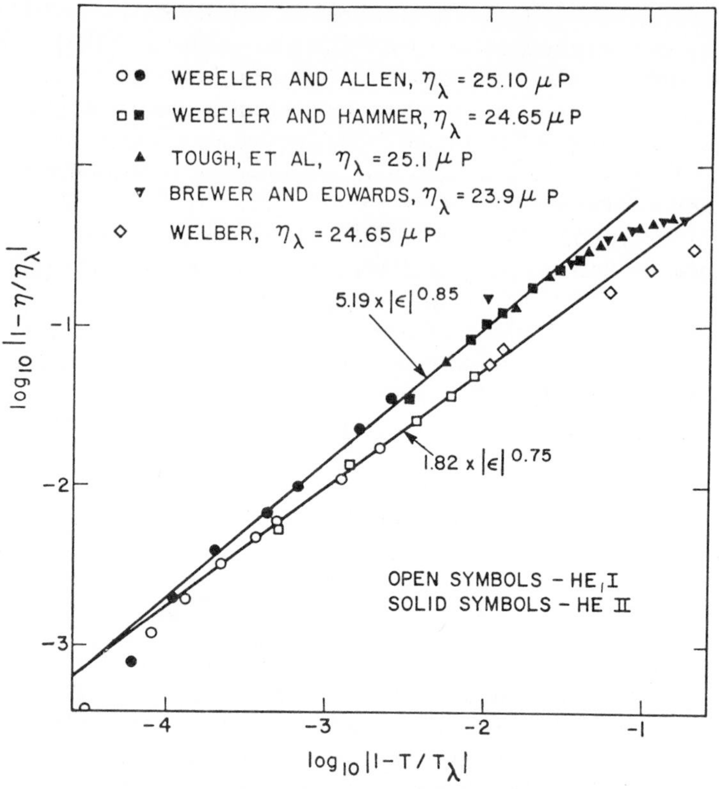

FIGURE 2.35. The dependence of the shear viscosity of ^{4}He at vapor pressure upon $1 - T/T_\lambda$. Circles, Reference [186, 187]; squares, Reference [185], triangles, Reference [184], inverted triangles, Reference [182]; diamonds, Reference [183]. After Reference [188].

mobility μ of positive or negative ions in helium near T_λ [190–194]. The mobility can apparently be represented by a power law [193] similar to equation (2.3.1) for η, but there seems to be some uncertainty at this time [193, 194] about the parameters. As a first approximation, one would expect $\mu \propto \eta^{-1}$, for this result should be valid if the motion of the charge carriers through the fluid can be described by hydrodynamics. However, the carrier radii R are only of the order of 10^{-6} cm, and there is no temperature range near T_λ where the radii are much larger than the coherence length. We therefore cannot expect the hydrodynamic result to apply quantitatively; indeed, a comparison of the measured μ with the measured η at vapor pressure has revealed departures from $\mu \propto \eta^{-1}$ [119]. There are no theoretical predictions about the detailed behavior of μ in the nonhydrodynamic regime where $R \gtrsim \xi$.

(*b*) *The Thermal Conductivity* κ. For $T > T_\lambda$, the thermal conductivity

κ of liquid helium can be measured by conventional means, and is given by the ratio between the time-independent heat current passing through a sample and the associated temperature gradient. Measurements near T_λ were made first by Kerrisk and Keller [195]. These authors determined κ on several isotherms as a function of the pressure P, and found that κ became rather large near P_λ. At about the same time, several theories were being developed which predicted a divergent κ at T_λ.

Let us consider in somewhat more detail the predictions of dynamic scaling [196–198] that pertain to κ. The theory predicts a characteristic frequency ω^* that describes the time dependence of fluctuations in the order parameter. Below T_λ, ω^* is equal to u_2/ξ^-, where ξ is a length which describes the spatial range over which fluctuations in the order parameter are correlated, and where the superscript on ξ indicates that $T < T_\lambda$. For $T > T_\lambda$, ω^* is given by $D_T/(\xi^+)^2$, where

$$D_T = \frac{\kappa}{\rho C_P} \tag{2.3.3}$$

is the thermal diffusivity. Dynamic scaling asserts that ω^* above T_λ will be equal to ω^* below T_λ for those pairs of temperatures for which ξ^+ is equal to ξ^- (except for temperature-independent multiplicative constants of other unity, which we shall neglect). This "matching" condition for the critical frequency leads to the prediction

$$\frac{D_T}{(\xi^+)^2} = \frac{u_2}{\xi^-} \tag{2.3.4}$$

Using equation (2.3.3) for D_T and equation (2.2.12) for u_2, one obtains

$$\kappa \sim \left[\frac{C_P^+}{(C_P^-)^{1/2}}\right]\frac{(\xi^+)^2}{(\xi^-)^{3/2}}, \tag{2.3.5}$$

where we used the static scaling result $\rho_s \sim (\xi^-)^{-1}$. The parameters in equation (2.3.5) should be evaluated at those temperatures for which $\xi^+ = \xi^-$. The coherence lengths can, of course, be written as $\xi^- = \xi_0^- |\varepsilon|^{-\nu'}$ and $\xi^+ = \xi_0^+ \varepsilon^{-\nu}$. Let us write $\kappa \sim \varepsilon^{-x}$. If, as implied by static scaling, ξ and C_p each have the same exponents above and below T_λ, then for positive α'

$$x = \tfrac{1}{2}(\nu' + \alpha'). \tag{2.3.6}$$

If α and α' are nearly equal to zero, then $x \cong \nu'/2$. However, since in any experimentally accessible temperature region C_P is dependent upon ε even if α and α' are equal to or near zero, it may be appropriate in the determination of $x = \alpha'/2$ from experimental data to include C_P in the analysis. In general the "logarithmic" (i.e., $\alpha \cong 0$) corrections to the asymptotic behavior are believed to be not necessarily given realiably by the theory, but the use of equation (2.3.4), with ξ^+ and ξ^- written as power laws and C_P taken from experiments, seems the most reasonable approach to the data analysis.*

When the results of Kerrisk and Keller were interpreted in terms of a power-law divergence for κ [195], they yielded values between 0.27 and 0.80 for x, with a tendency towards values larger than the expected $x \cong \nu'/2 \cong 0.34$. More precise results for κ soon became available at SVP [61]. These new measurements spanned the range of ε from 10^{-7} to 10^{-2}, and except at the smallest ε had a precision of a few percent. For $\varepsilon \lesssim 10^{-3}$, they could be fitted within their scatter by the prediction equation (2.3.4) and when the predicted corrections from C_p were included in the analysis, they yielded $\nu'/2 = 0.334 \pm 0.005$ for the asymptotic exponent of κ. This result appeared to agree exceedingly well with the dynamic scaling prediction and the measurements of ρ_s which were then available [132,133]. The results for ρ_s gave $\zeta/2 = \nu'/2 = 0.333 \pm 0.003$ [see equation (2.2.3)]. In particular, it was believed that this excellent agreement could not have occurred unless "logarithmic" factors from C_P in equation (2.3.5) were given correctly by the theory.

Unfortunately, it now appears from even more precise measurements, and particularly from results at higher pressure, that the good agreement at vapor pressure was fortuitous and did not pertain to the asymptotic behavior of κ [38]. In Figure 2.36 the new data are shown as a function of $\log_{10}(\varepsilon)$. Most of the results were obtained in a cell of length $L = 0.1$ cm. At saturated vapor pressure, the results with a cell length of 1 cm are those [61] that appeared to agree so well with theory. Data obtained with the two cell lengths agree within their experimental errors. However, with the greater length, extremely small power densities had to be used and thermal time constants were large. For these reasons, the precision was only a few percent for large ε, and perhaps 5% for $\varepsilon \lesssim 2 \times 10^{-5}$. For $L = 0.1$ cm, higher power densities could be used, and thermal time constants were shorter. Here the precision was limited by the temperature resolution, and varied between 1% at 10^{-7} W/cm^2 and 0.1% for power densities greater than about 10^{-6} W/cm^2. Unfortunately the shorter cell has the disadvantage that corrections for the thermal boundary resistance are larger, increasing considerably possible systematic errors in the derived exponents.

For κ, the asymptotic behavior proved to be extremely difficult to obtain. This is most apparent from the data for 22.3 bars. The lower dashed line in Figure 2.36 represents what now seems to be a good representation of equation (2.3.5) with the "logarithmic" corrections given by theory, but with the

*A more detailed treatment of the critical dynamics near T_λ, based on renormalization group theory, has been developed recently by B. I. Halperin, P. C. Hohenberg, and E. D. Siggia [*Phys. Rev. B*, (to be published)]. For $\alpha < 0$, these authors find that $x = \nu/2$, consistent with equation (2.3.5). However, the correction terms which contain C_p and depend upon $\varepsilon^{-\alpha}$, differ from those given by the matching condition of dynamic scaling.

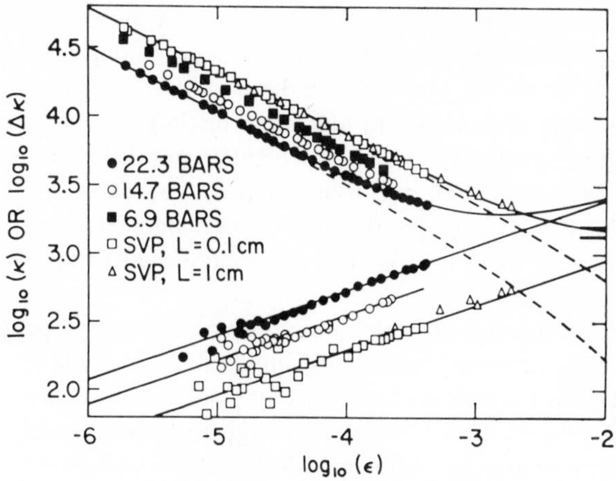

FIGURE 2.36. The thermal conductivity κ in erg/sec cm K of ^{4}He along several isobars as a function of $\varepsilon = 1 - T/T_\lambda$. The upper set of data is the total κ, and the lower set is an estimate of higher orders singular contributions to κ. After Reference [38].

asymptotic exponent chosen to yield agreement with the data at small ε. When $\varepsilon \gtrsim 10^{-5}$, there are appreciable contributions to κ from higher-order singular correction terms that are not included in the prediction equation (2.3.5). Therefore it was assumed that

$$\kappa = \kappa_1 + \Delta\kappa, \tag{2.3.7}$$

where κ_1 is given by equation (2.3.5), and where a power law

$$\Delta\kappa = a\varepsilon^{-y};\ y < x \tag{2.3.8}$$

adequately describes the higher order terms over an appreciable accessible range of ε. One might hope that the inclusion of $\Delta\kappa$ in the analysis would extend the range of ε over which the data can be fitted by at least a decade. Indeed, equations (2.3.7) and (2.3.8) fitted the results for ε as large as 10^{-3}. This can be seen from the good fit by a straight line of the lower three sets of points in Figure 2.36, which are $\Delta\kappa$ obtained by subtracting κ_1 with the best x from the data. However, the values of x varied from 0.39 to 0.41 as the pressure was changed from vapor pressure to 22 bars. Systematic errors in x, due primarily to uncertainties in the boundary resistance, probably are not larger than 0.02. Therefore x seems to differ significantly from the prediction $x = \nu'/2$ which pertains when $\alpha' < 0$. The previous apparently good agreement at V.P. was a result of nearly perfect cancellation between departures from theory and contributions from singular correction terms to the asymptotic behavior. It is now clear that such singular corrections must generally be included in the analysis of precise experimental results near critical points; this inclusion will usually substantially increase the uncertainties in the leading exponent. Singular corrections to the asymptotic behavior are known to occur also in the superfluid density [see 2.2.1(e)], and probably occur in C_P [see 2.1.2.(e)].*

Measurements of κ above T_λ also were made by Archibald et al. [199]. These authors measured the heat flow between parallel plates separated by a very thin layer of helium, and found that κ depended upon the spacing d between the plates for $d \lesssim 10^{-2}$ cm. No satisfactory explanation of this effect appears to exist at this time, and additional experiments might be desirable.

For $T < T_\lambda$, it is not possible to establish a nonzero temperature gradient with a small heat current, because any imposed $\nabla T > 0$ will result in $\nabla \mu > 0$ (μ is the chemical potential) and, therefore, accelerate the superfluid. When the relative velocity $v_s - v_n$ is large enough for the superfluid and normal fluid counterflow to carry all the heat, ∇T will again be zero, and a measurement will yield an effective thermal conductivity κ^{eff} which is infinite or exceedingly large. This κ^{eff}, however, is not related to the thermal kinetic coefficient κ because κ corresponds specifically to the irreversible heat transport associated with $\nabla T > 0$. For $T < T_\lambda$, κ can only be inferred from less direct results, such as the damping of second sound. There, however, it occurs in combination with other transport coefficients, which are also singular and which may dominate near T_λ. Therefore it is not known from experiment on pure ^{4}He whether κ is singular as T_λ is approached from below. However, in mixtures of ^{3}He and ^{4}He, it is possible to measure the finite effective thermal conductivity $\kappa_{\text{eff}} \equiv q/\nabla T$; κ_{eff} contains an additive contribution from the thermal conductivity κ in the absence of an impurity current (see Section 2.3.2). At a concentration of about 15% ^{3}He where κ_{eff} has been measured [200], the contribution from κ is appreciable. However, as T_λ is approached from below, any singularity in κ_{eff} is, at most, extremely weak and has not been detected experimentally. Although the addition of impurities may well have altered the behavior of κ, κ is unlikely to be strongly divergent in pure ^{4}He when the high-resolution measurements that are available for the mixture do not even reveal a mild cusp in κ_{eff} as T_λ is approached from below. However, this conjecture that κ in pure ^{4}He contains, at most, a mild singularity stands in contrast to theoretical predictions [180, 201] that give the same behavior for κ above and below T_λ and, therefore, yield $\kappa \sim |\varepsilon|^{-1/3}$ on both sides of the transition.†

*The experimental data have not yet been analyzed in terms of the new predictions for for the $\varepsilon^{-\alpha}$ terms (see the foot note on p. 180). It will be interesting to see whether proper inclusion of these terms in the analysis will yield better agreement between theory and experiment for the leading exponent.

†For $T > T_\lambda$, Kawasaki and Gunton [221] have predicted that the leading singularities of the transport coefficients which contribute to the thermal conductivity in the mixtures should

(c) The Damping of Second Sound. The damping constant of second sound D_2 can be defined in terms of the second-sound attenuation α_2 by the relation

$$\alpha_2(\omega, T) = \left(\frac{\omega^2}{2u_2^3}\right) D_2(T), \qquad (2.3.9)$$

where ω is the angular frequency of the second-sound wave and u_2 the velocity given by equation (2.2.10). The damping constant D_2 is related by two-fluid hydrodynamics to the transport coefficients. As with the velocity, it is readily given in closed form if it is assumed that $\gamma \equiv C_P/C_V = 1$. In that case, $D_2 = D_{20}$, with [127]

$$D_{20} \equiv \left(\frac{1}{\rho}\right)\left(\frac{\rho_s}{\rho_n}\right)\left[\zeta_2 + \frac{4}{3}\eta + \left(\frac{\rho_n}{\rho_s}\right)\left(\frac{\kappa}{C_P}\right) + \rho^2\zeta_3 - 2\rho\zeta_1\right] \qquad (2.3.10)$$

In the regime where hydrodynamics applies, D_2 is independent of ω and a function only of T. In this region, therefore, α_2 is proportional to ω^2.

Sufficiently near T_λ where $\gamma - 1$ is not negligible, we may write

$$D_2 = D_{20} - D'. \qquad (2.3.11)$$

The correction D' has been obtained by Hohenberg [202], and is given by

$$D' = \frac{\rho_s}{\rho_n}\beta(2 - \beta) D_{10} - 2\frac{\rho_s}{\rho_n}\beta\zeta_1 + \left(\frac{u_{20}^2}{u_{10}^2}\right)(\gamma - 1)(D_T + D_{20}). \qquad (2.3.12)$$

Here

$$D_{10} = \rho^{-1}(\zeta_2 + \tfrac{4}{3}\eta), \qquad (2.3.13)$$

$$\beta = \frac{\gamma - 1}{\gamma a}, \qquad (2.3.14)$$

$$a^2 = \frac{(\gamma - 1)\rho_s u_{10}^2}{\rho_n u_{20}^2 \gamma^2}, \qquad (2.3.15)$$

and u_{20}, u_{10}, D_T, and D_{20} are given by equations (2.2.11a), (2.2.11b), (2.3.3), and (2.3.10), respectively. We shall see that D_{10}, and probably ζ_1 diverge at T_λ as ε^{-1}, whereas D_2 diverges roughly as $\varepsilon^{-1/3}$. Therefore, from equation (2.3.12), D' has the same ε dependence as D_2, and $1 - D_2/D_{20}$ is of order β at all ε. Values of β [203] are avilable [202]; near melting, where the correction is largest, $1 - D_2/D_{20}$ becomes as large at 10^{-1}. Since the correction is mildly ε dependent, it should be considered in the comparison of precise secnd-sound damping measurements with other results for the transport coefficients.

We already saw that η is finite at T_λ. The contribution $\tfrac{4}{3}\eta\rho_s/\rho\rho_n$ to D_2,

cancel. If a similar cancellation occurs for $T < T_\lambda$, then κ_{eff} could remain finite although κ diverges.

therefore, vanishes at the transition. If, as we conjectured, κ is also finite, then the contribution $\kappa/\rho C_P$ will vanish likewise if C_P diverges. If, however, κ diverges below T_λ as predicted by theory, then this term will contribute a divergence to D_2 that will be roughly proportional to $|\varepsilon|^{-1/3}$. Although there are no measurements, it is expected from theory that ζ_1, ζ_2, and ζ_3 all diverge near T_λ as $|\varepsilon|^{-1}$. Therefore there may be contributions from these transport coefficients which are approximately proportional to $|\varepsilon|^{-1/3}$, and it is not possible to determine the behavior of κ alone or of ζ_1, ζ_2, and ζ_3 alone from a measurement of D_2. The behavior of D_2 near T_λ is predicted by dynamic scaling [196–198], and by this theory is given as

$$D_2 = u_2\xi \tag{2.3.16}$$

within a numerical constant of order unity that we shall ignore. With the aid of equations (2.2.12) and (2.2.55), we have

$$D_2 \propto (C_P\rho_s)_,^{-1/2} \sim \varepsilon^{-0.34} \tag{2.3.17}$$

if $\rho_s \sim \varepsilon^{0.67}$.

The attenuation of second sound was measured at saturated vapor pressure by Hanson and Pellam [204], and by Notarys [205], but their results were only for $\varepsilon \gtrsim 10^{-2}$. More recently, new results were obtained by Tyson [206] for $\varepsilon \gtrsim 10^{-4}$. He found that, even at his smallest ε, $\alpha_2 \propto \omega^2$, and his data largely confirm the prediction equation (2.3.16). They are shown in Figure 2.37 in the form of D_2 vs. $T_\lambda - T$ on logarithmic scales. When fitted to a pure power law, they yield the exponent 0.34 ± 0.06 for D_2. Although this value agrees well with the predicted value, one might have expected a somewhat lower result, say 0.28, from a fit to a pure power law if the predicted correction from C_P in equation (2.3.17) is indeed present. However the uncertainty of 0.06 is too large to rule out a contribution from C_P. More precise measurements of D_2, although very difficult, would be extremely desirable because they would provide a very direct test of the details of the dynamic scaling prediction. Such a test is particularly important in view of the apparent difference between the thermal conductivity of He (I) and the scaling prediction that was discussed in Section 2.3.1(b).

Very little information exists about the second-sound damping under pressure and near T_λ. We expect from equation (2.3.16) and the measured u_2 [157,158] that D_2 will diverge approximately as $\varepsilon^{-1/3}$ at all pressures, and that the size of D_2 at constant ε will depend only mildly upon P. The quality factor Q of resonances observed during velocity measurements under pressure [157,158], in a range of ε where the intrinsic absorption dominates, can be used to estimate D_2; the relation $D_2 = u_2^2/\omega Q$ yielded $D_2 \cong 1.2 \times 10^{-3}$ cm^2/sec at 29 bars and $\varepsilon = 3.8 \times 10^{-5}$. This, with equation (2.3.16), predicts $\xi_0 \cong 2.5 \times 10^{-8}$ cm at 29 bars if it is assumed that $\zeta = \frac{2}{3}$. At

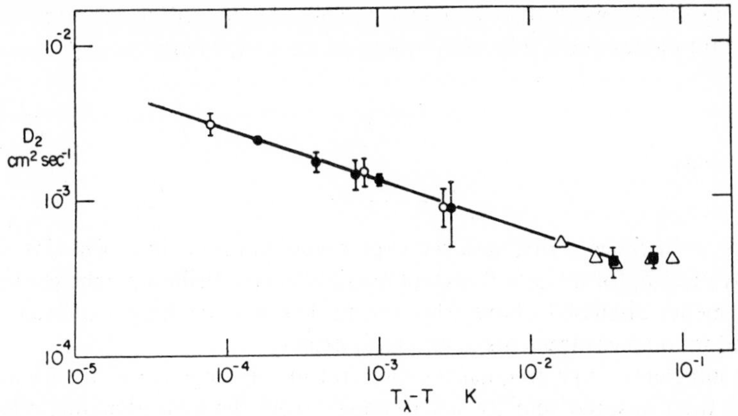

FIGURE 2.37. The second-sound damping constant of ^4He at vapor pressure. After Reference [206].

vapor pressure and a similar ε, this procedure resulted in $D_2 \cong 2.0 \times 10^{-3}$ cm^2/sec and $\xi_0 \cong 2.4 \times 10^{-8}$ cm. The values of ξ_0 are reasonable, and the absence of an appreciable pressure dependence of ξ_0 is consistent with our expectations. More detailed measurements of D_2 under pressure obviously are desirable.

Recently, second sound has been studied near T_λ under pressure by the scattering of light [207, 208]. The width $\Delta\omega_2$ of the Brillouin component [209] corresponding to second sound, although it is only about 2 MHz, has been resolved from the instrumental broadening. If hydrodynamics applies, it is expected that $\Delta\omega_2 = D_2 k^2$, where k is the wave vector. The experiment is done at constant $k \cong 2.3 \times 10^4$ cm^{-1}, and for such a large k departures from hydrodynamics would be expected to occur for experimentally accessible ε. The measurements [207] yield a width for $\varepsilon \gtrsim 10^{-2}$ which is of the same size as that predicted by hydrodynamics from the low-frequency results for D_2 at SVP which are shown in Figure 2.37. But for smaller ε, this width remains constant and shows no indication of the expected divergence. One might be tempted to attribute the absence of any obvious singularity to the large wave vector used in the measurements. However, even with the large k, a rather large amplitude ξ_0 for the coherence length would be required to give $k\xi \cong 1$ near $\varepsilon = 10^{-2}$. Such a large ξ_0 is particularly unlikely, in view of the fact that these same mesurements reveal no dispersion contribution to the velocity even for ε as small as 10^{-4}!

The hydrodynamic regime is expected to be limited at all frequencies to those values of $|\varepsilon|$ and ω at which the sound-wave length λ is much larger than any other relevant length. Since, according to static scaling, only one length ξ is important near the transition, the condition $\lambda \approx \xi$ serves as the dividing line between the two regimes. This can be written in terms of the sound frequency and a relaxation time τ as $\omega\tau \approx 1$, where

$$\tau \equiv \frac{\xi}{u_2}. \tag{2.3.18}$$

This yields

$$\tau \sim \varepsilon^{-1.0}, \tag{2.3.19}$$

if $\rho_s \sim \varepsilon^{0.67}$. Since τ diverges, we expect nonhydrodynamic behavior sufficiently near T_λ for all $\omega > 0$; except possibly for the Brillouin light-scattering experiments discussed above, this regime has not yet been explored with second sound measurements to any great extent.

At sufficiently high frequencies, but still in the range $\omega\tau < 1$, one might expect the measured velocity ω/k to depart from the hydrodynamic velocity u_2 given by equation (2.2.12). It has been suggested [209a] that this dispersion might be given by a relation of the form

$$\frac{\omega_2}{k} = u_2[1 + iA_1k\xi + A_2(k\xi)^2 + \cdots]. \tag{2.3.20}$$

Here the imaginary part corresponds to the damping that we just discussed, and the dispersive contribution is of order $(k\xi)^2$. So far, the only experiments that have been performed with sufficiently large k or sufficient resolution in ω_2/k to detect a contribution of the size predicted by equation (2.3.20) are the light-scattering measurements mentioned above. It is somewhat surprising that they have not revealed any dispersive contribution to the velocity near T_λ.

(*d*) *The Damping and Dispersion of First Sound.* The damping constant D_1 of first sound can be defined in terms of the first-sound attenuation α_1 by the relation

$$\alpha_1(\omega, T) = \left(\frac{\omega^2}{2u_1^3}\right) D_1(T). \tag{2.3.21}$$

Since u_1 is only mildly singular at T_λ, the behavior of α_1 near T_λ is almost the same as that of D_1. As was the case for the second-sound damping, D_1 is also given in terms of the transport coefficients by two-fluid hydrodynamics; when $\gamma = 1$ it is equal to D_{10}, which is given by equation (2.3.13). For $\gamma > 1$, there are additional contributions, and D_1 may be written as [202]

$$D_1 = D_{10} + (\gamma - 1)\, D_T + D', \tag{2.3.22}$$

where D_T and D' are defined by equations (2.3.3) and (2.3.12). This relation also pertains to ordinary sound above T_λ [130]. In that case, however, $D' = 0$ [see equation (2.3.12)], and ζ_2 in equation (2.3.13) is equal to the bulk viscosity ζ. Again, in the regime where hydrodynamics applies, $\alpha_1 \propto \omega^2$ and D_1 is independent of ω.

Both above and below T_λ, η makes a finite contribution to D_1. Near vapor pressure, the term $\kappa(\Upsilon - 1)/\rho C_P$ is likely to be small because $\Upsilon - 1$, although singular, remains small at all accessible ε. Under pressure, however, $(\Upsilon - 1)$ becomes appreciable at small ε [see Section 2.1.2(e)], and there could be a noticeable contribution from κ which, at least for $T > T_\lambda$, would vary approximately as $|\varepsilon|^{-1/3}$. For $T < T_\lambda$, this contribution is uncertain as it was for second sound [see Section 2.3.1(c)] because the behavior of κ is not known [see Section 2.3.1(b)]. Although ζ_2 by itself also is not known from independent experiments, it is expected from theory [180, 201] that ζ_2 diverges both above and below T_λ as $|\varepsilon|^{-1}$. The contribution from ζ_2, therefore, should dominate, and one expects $D_1 \sim |\varepsilon|^{-1}$ on both sides of the transition, but higher-order contributions, particularly under pressure, that diverge as $|\varepsilon|^{-1/3}$ might be expected from κ. Ferrell et al. [196] have also applied dynamic scaling arguments to the damping of first sound, although, in the strictest sense, one might expect this theory to apply only to the damping D_2 of the critical mode [197]. They also obtain [196] $D_1 \sim |\varepsilon|^{-1}$.

Measurements of the first-sound attenuation near T_λ were made first by Chase [210] at a frequency of 10^6 Hz. These results have been reviewed recently [146], and, for $T < T_\lambda$ and sufficiently large $|\varepsilon|$, they are consistent with $\alpha \propto \varepsilon^{-1}$ as predicted by equation (2.3.21) and (2.3.22). The measurements also reveal that, for sufficiently small $|\varepsilon|$, equation (2.3.21) is no longer valid. In this region, the attenuation is smaller than predicted by a pure power law, and has a maximum for $T < T_\lambda$. This effect is attributable to departures from the hydrodynamic prediction, because some relevant relaxation time τ increases with decreasing $|\varepsilon|$ and becomes comparable in size to the inverse of the sound frequency ω_1. If it is assumed that $\omega\tau \approx 1$ when α_1 reaches its maximum, the experimental data yield a τ of the same size as that predicted by equation (2.3.18) for second sound.

Although the early measurements [210] at 10^6 Hz revealed the existence of the nonhydrodynamic regime, they did not yield a great deal of detail about the behavior of α for $\omega\tau \gtrsim 1$, and, in particular, they did not provide definitive information about the value of ω at or extremely near T_λ, where one might expect $\omega\tau \gg 1$. They also were not quantitative for $T > T_\lambda$. However, more recent results by Barmatz and Rudnick [31] and especially by Williams and Rudnick [211] have provided much more detailed information about the frequency and ε dependence of α for He (II) and He (I). The data of Williams and Rudnick are shown in Figure 2.38. These measurements still do not yield the frequency and ε dependence in the hydrodynamic regime with great precision, but for sufficiently large ε they are consistent both in He (I) and He (II) with $\alpha \propto \omega^2$. In the region where this hydrodynamic frequency

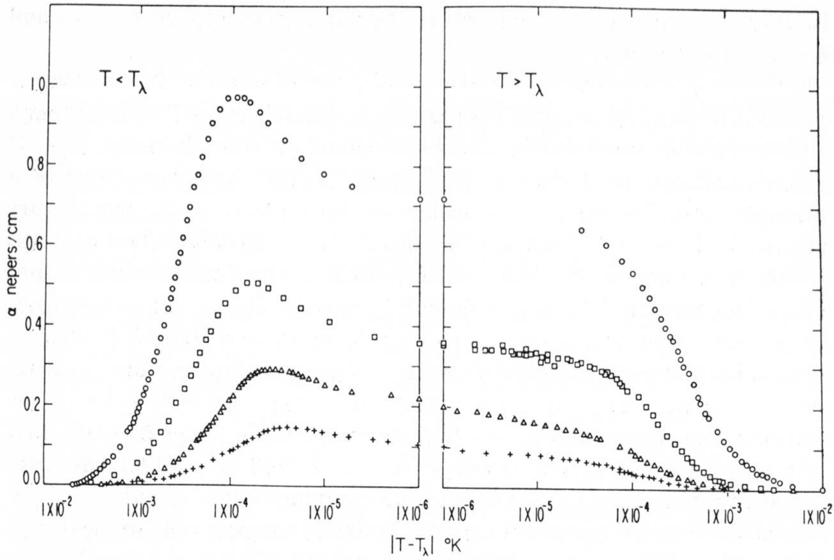

FIGURE 2.38. The attenuation of first sound in ^4He at vapor pressure at several frequencies. From top to bottom, the sets of data are at 3.17, 1.75, 1.00, and 0.60 MHz. After Reference [211].

dependence seems to prevail, the data also suggest $\alpha \propto \varepsilon^{-1}$. This is consistent with the earlier measurements below T_λ [31, 210], and agrees in both phases with expectations based on scaling arguments [196–198] and microscopic calculations [180, 201]. It suggests that ζ_2 diverges as ε^{-1} both above and below T_λ. The experiments do not yield very accurate exponents, however, because the range of ε over which hydrodynamics is applicable and over which α_1 is large enough to be measurable with reasonable precision is only perhaps a decade.

Most of the results of Figure 2.38 are not in the hydrodynamic regime. From the data near T_λ, where hydrodynamics does not apply, it is apparent that the attenuation in this region is *not* symmetric about T_λ. For $T < T_\lambda$, α_1 has a maximum, whereas for $T > T_\lambda$, α_1 increases monotonically with decreasing ε. Williams and Rudnick [211] analyzed their data by assuming that below T_λ the attenuation can be written as the sum of two terms. One of these they assumed to be symmetric about T_λ. The other contributes to α_1 only for $T < T_\lambda$ and is obtained by subtracting α_1 at the same $|\varepsilon|$ above T_λ from the total attenuation. The term that exists only below T_λ can be described reasonably well by the prediction which results if a single relaxation mechanism with a unique ε-dependent relaxation rate is assumed to be responsible for this contribution. Such a mechanism was invoked near T_λ first by Landau and Khalatnikov [212]. They considered the relaxation time τ_{LK} with which the order parameter relaxes to its equilibrium value after an initial disturbance. This classical mechanism for sound attenuation due to an internal relaxation exists only below T_λ, where the order parameter (or ρ_s/ρ) has a nonzero value. It yields [213]

$$\alpha_1 = \left(\frac{\Delta u}{u^2}\right)\omega^2 \frac{\tau_{LK}}{(1 + \omega^2\tau_{LK}^2)} \tag{2.3.23}$$

and should be associated with a dispersive contribution [212]

$$u_\omega - u = \frac{\Delta u \omega^2 \tau_{LK}^2}{(1 + \omega^2\tau_{LK}^2)} \tag{2.3.24}$$

to the velocity u_ω at frequency ω. The relaxation time τ_{LK} is discussed in detail in Chapter 1 of this book by Khalatnikov, and is expected to be equal (within a numerical constant of order unity) to the time τ given by equation (2.3.18). Williams and Rudnick demonstrated [211] that, after subtraction of the symmetric part, both the shape of their attenuation vs. ε curves below T_λ and the positions of their attenuation maxima are consistent with equation (2.3.23) and

$$\tau_{LK} = \frac{\xi}{u_2} \tag{2.3.25}$$

over about two decades in ω.

The contribution to α_1 which exists also above T_λ has been discussed by Kawasaki [214] and by Hohenberg [198]. It may be attributed to interaction of the sound wave with intrinsic fluctuations of the order parameter which occur spontaneously near T_λ. This mechanism yields, consistent with experiment, an attenuation that is approximately symmetric about T_λ. It is described by the same relaxation time ξ/u_2 that pertains also to the Landau–Khalatnikov mechanism below T_λ, but a unique τ is applicable only when $\omega\tau \ll 1$. For larger $\omega\tau$, equation (2.3.23) does not pertain; instead of vanishing at T_λ, the attenuation is predicted [198, 214] to reach the nonzero value

$$\alpha_\lambda = \frac{\Delta u}{u^2}\,\omega. \tag{2.3.26}$$

The measurements in He (I) are roughly consistent in their ε dependence and magnitude with the theory. At T_λ, however, α_λ seems to increase somewhat more rapidly with ω than predicted by equation (2.3.26), and empirically is found [211] to vary as $\omega^{1.2}$.

The attenuation near T_λ has also been measured [215–218, 207] at very high frequencies of about 10^9 Hz. This was done by conventional acoustic techniques by Imai and Rudnick [215], and by Commins and Rudnick [216]. In addition, this hypersonic attenuation has been inferred from both stimulated [217] and spontaneous [207, 218] Brillouin scattering of light. The results of Commins and Rudnick [216] are shown in Figure 2.39. An estimate of the contribution to α_1 from the order-parameter relaxation mechanism can be

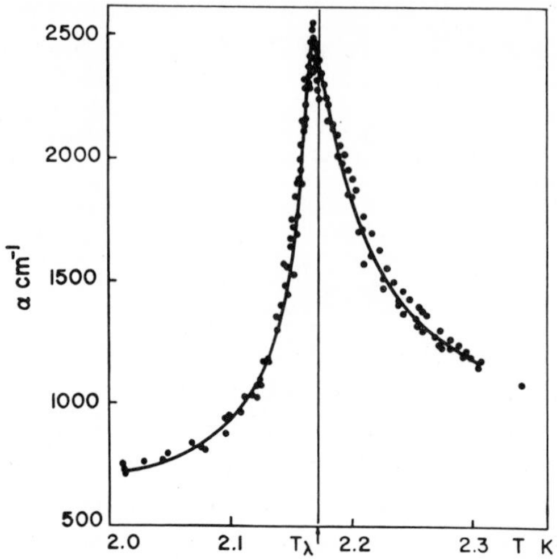

FIGURE 2.39. The attenuation of first sound in ^{4}He at vapor pressure at a frequency of 1 GHz. After Reference [216].

obtained from equation (2.3.23), with the parameters from the low-frequency measurements. This yields a maximum α_1 of about 180 cm^{-1}, which should occur 0.094 K below T_λ. At that temperature, the data in Figure 2.39 do not reveal a maximum, and α_1 is already larger than the estimate. There is a maximum in α much nearer T_λ, however. One might be inclined to associate the very large attenuation near T_λ with the scattering of sound by order-parameter fluctuations which was discussed above. However, this contribution is expected to have a maximum *at* T_λ, whereas the measurements yield a maximum near, but distinctly below, T_λ. According to equation (2.3.26), the attenuation is expected to vary as ω near its maximum. Since the relaxation mechanism described by equation (2.3.23) yields an attenuation maximum below T_λ that also varies as ω, the relative size of the two contributions would be expected to remain about the same. This is contrary to the hypersonic frequency results. The size of the attenuation near T_λ and at 10^9 Hz is consistent, however, with $\alpha_\lambda \propto \omega^{1.2}$ which was already suggested [211] by the lower-frequency data.

The only experiments so far on the attenuation near T_λ under pressure are the very high-frequency measurements based upon spontaneous Brillouin light scattering [207, 218]. These results indicate that, near 10^9 Hz, there is no qualitative change in the attenuation with pressure. Quantitative results as a function of pressure and ω would be quite interesting. It has been pointed out by Hohenberg [198] that the parameter Δu in equations (2.3.23) and (2.3.26) has a very characteristic pressure dependence, which can be calculated from thermodynamic data and the theory given by Khalatnikov in

Chapter 1 of this book. A measurement of Δu as a function of P would, therefore, be an interesting test of the details of that theory.

We already mentioned that the Landau–Khalatnikov relaxation mechanism which yielded the attenuation given by equation (2.3.23) should also result in a dispersive contribution to the velocity which is given by equation (2.3.24). In addition, one wuld expect dispersion from the mechanism that yielded the contribution to α_1 that is symmetric about T_λ. It is not known in any detail what from this latter contribution should take when $\omega\tau \gtrsim 1$. For $\omega\tau \ll 1$, one expects

$$u_\omega - u \sim \omega^2\tau^2 \sim \varepsilon^{-2} \tag{2.3.27}$$

for both mechanisms.

The experimental data at 2.2×10^4 Hz and 4.4×10^4 Hz by Barmatz and Rudnick [31] and those at 10^6 Hz by Chase [210] have been reviewed recently [146]. For $\omega\tau \ll 1$, the data are consistent with equation (2.3.27), but they are not precise enough to provide a quantitative confirmation. They also do not reveal whether there are two contributions from different mechanisms to the dispersion for $\omega\tau \gtrsim 1$.

Recently, the dispersion at vapor pressure was measured by Thomlinson and Pobell [218a] at frequencies up to 2×10^5 Hz. Even though these data have a velocity resolution of 1 part in 10^5, they are still not precise enough to provide quantitative confirmation for the behavior given by equation (2.3.27) for the hydrodynamic regime. They do, however, provide convincing qualitative confirmation of the expected dispersion for $\omega\tau \gtrsim 1$. The measurements yield a single, asymmetric peak near T_λ, with the dispersion in He (II) considerably larger than in He (I). When the expected contribution from the order-parameter relaxation is calculated from equation (2.3.24) and subtracted, a peak that is roughly symmetric about T_λ remains. This remaining contribution can reasonably be attributed to scattering of the sound wave by spontaneous fluctuations in the order parameter. The measurements yield a maximum value for the dispersion, which near 2×10^5 Hz is larger by a factor of 2 or 3 than the estimated maximum contribution from equation (2.3.24). This excess dispersion, due to fluctuations in the region $\omega\tau \gg 1$, increases monotonically with frequency, and is consistent with the even larger dispersion of 53 cm/sec deduced [44] from the measurements [210] at 10^6 Hz.

Recently, the velocity of sound has also been measured by Brillouin light scattering from spontaneous first sound waves [207, 218]. These results correspond to a frequency near 5×10^8 Hz. The dispersive contribution can be extracted from the measurement by comparison with the known thermodynamic ($\omega = 0$) velocity [39, 44]. This yields a maximum dispersion near T_λ of about 230 cm/sec. A dispersion of this size is much larger than any attributable to the relaxation process, and indicates that the fluctuation contribution continues to increase considerably with ω between 10^6 Hz and 5×10^8 Hz. There are no measurements over this wide gap in frequency, however.

Except for spontaneous Brillouin light scattering measurements at a frequency near 8×10^8 Hz and pressures of about 25 bar [207] and 20 bar [218], there are no results for the first-sound dispersion under pressure near T_λ. The measurements at 8×10^8 Hz reveal a maximum dispersion, located at or very near T_λ, of over 400 cm/sec. This value is even larger than that observed at vapor pressure at a similar frequency.

2.3.2. ^{3}He–^{4}He Mixtures

Experimental investigations of the transport properties of ^{3}He–^{4}He mixtures near T_λ are far from complete. Viscosity [186, 187], thermal conductivity [200], and mass diffusion [218b] have been measured, but the results do not nearly span the concentration range over which the transition exists, and the very interesting tricritical region has hardly been investigated at all. The damping constants of the hydrodynamic modes are not known from experiment, except for qualitative measurements [219] of second-sound absorption. Although theoretical predictions are also not complete, the hydrodynamics of the mixtures is well known [127], and the relations between the damping constants and the transport coefficients have been obtained explicitly [220]. The diffusivity corresponding to the critical mode is given by dynamic scaling [196, 197]. Mode–mode coupling calculations of the transport coefficients also have been performed [221, 221a] recently.

The experimental results for the shear viscosity were mentioned in Section 2.3.1, and are for $X \leq 0.1$. There seems to be no sizable change in the behavior of η at small concentrations from that of the pure system.

For the second-sound damping one would expect that equation (2.3.16) pertains to the mixtures as well as to pure ^{4}He, and that D_2 is determined by u_2 and ξ. There are no quantitative measurements, but qualitative observations based on the width of the resonances used in the measurements [107] of u_2 [Section 2.2.2(c)] were generally consistent with equation (2.3.16). For the same $u_2 T/\rho_s$, and for ε_λ or ε_t sufficiently small that the major contribution to the damping was intrinsic absorption in the fluid, the quality factor Q of the resonances was, within a factor of 2, independent of X at constant ω. This agrees with equations (2.3.16) and (2.2.55).*

*The attenuation and dispersion of first sound have been measured recently for mixtures with $X \leq 0.52$ by Buchal, Pobell, and Thomlinson [*Phys. Lett.* **A51**, 19 (1975)], and by Buchal and Pobell [*Proceedings of the fourteenth International Conference on Low Temperature Physics*, Helsinki, Finland, August, 1975 (to be published)]. As for pure ^{4}He (see Section 2.3.1 (d)], these authors found that the attenuation had a peak below T_λ. The value of $T_\lambda - T$ at this attenuation maximum increased by an order of magnitude as X increased from 0 to 0.5; but assuming that $\omega\tau \cong 1$ at the maximum [see equation (2.3.23)] yielded values of τ

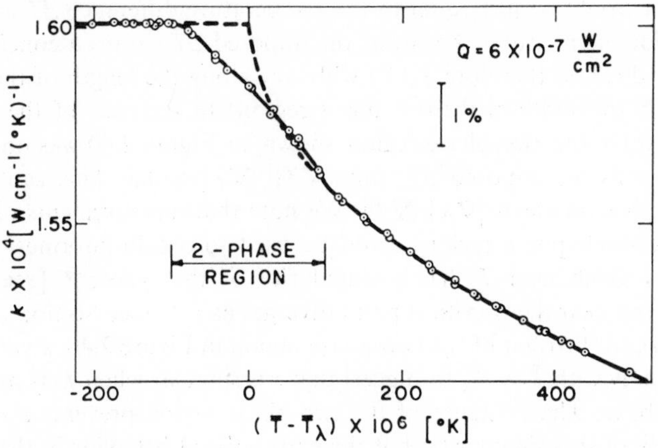

FIGURE 2.40. The effective thermal conductivity of a ^{3}He–^{4}He mixture with $X = 0.15$. After Reference [200].

Although the thermal conductivity κ has been measured near T_λ only at one concentration, very precise and detailed results have been obtained [200] for $X = 0.15$. These data are shown in Figure 2.40 as a function of the difference between the average sample temperature $\bar{T}$ and the average transition temperature $\bar{T}_\lambda$. At temperatures corresponding to the extreme left of the figure, the entire sample was superfluid. Here, the measured κ is an effective thermal conductivity κ_{eff}, and can be written [127] as a combination of the diffusion coefficient D, the coefficient of thermal diffusion $k_T D$, and the thermal conductivity κ in the absence of an impurity current. Figure 2.40 shows that there is no indication of a singularity in κ_{eff} upon approaching the transition from $T < T_\lambda$, and κ_{eff} is constant within a small fraction of 1% over the range of the data in the figure. Over a much wider temperature range in the superfluid, κ_{eff} can be represented by a regular function of T all the way up to T_λ [200]. We made reference to this regular behavior of κ_{eff} below T_λ near the end of Section 2.3.1(b). It would seem to imply that the relevant transport coefficients themselves are regular, unless there is an exact cancellation between the singularities in several of them. As we shall see below, a cancellation of divergent terms has indeed been suggested [221] above T_λ, but in that case there still remains a milder singularity, presumably due to higher-order terms that do not cancel each other.

Over the range $|\bar{T} - \bar{T}_\lambda| \lesssim 10^{-4}$ K, the sample consisted of two coexisting phases, one normal and one superfluid. In a homogeneous single component system where T_λ is a constant [see, however, Section 2.1.2(b)], the width of

which agreed with equation 2.3.25 if ξ is taken from equation (2.2.55) and the measured ρ_s. The overall size of the attenuation decreased rapidly with increasing X, indicating a strong concentration dependence of Δu in equation (2.3.23).

this region would be just equal to the temperature difference ΔT across the sample. For the mixtures, however, the imposed ΔT causes a concentration difference ΔX, and therefore $T_\lambda(X)$ will vary along the length of the sample and modify the width of the two-phase region. In the case of the solution with $X = 0.15$, the two-phase region shown in Figure 2.40 was only about half as wide as the imposed ΔT; thus $|\nabla T_\lambda|/|\nabla T| \approx 0.5$. This can be used to obtain an estimate of $|\nabla X|/|\nabla T|$. We note that measurements of ∇X vs. ∇T in the single-phase regions could be used to obtain information about $(\partial X/\partial \Phi)_{TP}$, which near T_λ has a singularity of the type $\varepsilon^{-\alpha'}$ [see Section 2.1.3(c)], and near the tricritical point diverges as $\varepsilon_t^{-\gamma'}$ [see Section 2.1.3(f)].

In He (I), to the right of the two-phase region in Figure 2.40, κ varies sufficiently rapidly with $\bar{T} - \bar{T}_\lambda$ to suggest that κ is singular when T_λ is approached from above. The existence of the two-phase region prevents a more detailed study of this singularity, but the extrapolated behavior in the limit of zero heat current ($\nabla T = \nabla X = 0$) is shown by the dashed line. For He (I), Kawasaki and Gunton [221] recently have carried out microscopic calculations that yield a finite κ. They found that the thermal conductivity in the absence of a concentration gradient (which does not seem to be easily accesible to experiment) and the thermal diffusion coefficient both diverge, but in the usual thermal conductivity these two transport coefficients occur in such a combination that the leading signularities cancel. The measured thermal conductivity, therefore, remains finite. The theory seems to make no prediction about higher-order singular contributions, which must evidentally be responsible for the remaining singularity that is observed experimentally in κ above T_λ.*

For $T > T_\lambda$, there are two normal modes in the mixtures that are combinations of mass diffusion and heat conduction. Their diffusivities have been given by Griffin [220] in terms of the diffusion coefficient D and the thermal diffusivity D_T. The diffusivity of one of the two modes is predicted by dynamic scaling [196–198] to diverge as $u_2\xi$, but the theory cannot determine which mode is the critical mode with the divergent diffusivity. However, if either one of the diffusivities diverges, then it can be shown [200] that, in the limit as $T - T_\lambda$ vanishes, the two modes become pure mass diffusion and pure thermal diffusion. Since we know from experiment that $D_T = \kappa/\rho C_{XP}$

*Measurements of the Rayleigh line width Γ_0 on the coexistence curve near the tricritical point of $^3He-^4He$ mixtures have been reported recently [P. Leiderer, D. R. Nelson, D. R. Watts, and W. W. Webb, *Phys. Rev. Lett.* **34**, 1080 (1975)]. From hydrodynamics, it is expected [220] that Γ_0 is determined essentially by the effective thermal conductivity which remains finite at the tricritical point [107], and by the inverse of the concentration susceptibility which vanishes at T_t as $T_t - T$. Indeed the experiment yielded $\Gamma_0 \propto T_t - T$, with an amplitude which agreed with that calculated from the thermal conductivity, measured thermodynamic parameters, and hydrodynamics.

remains finite at T_λ (see Figure 2.40), it follows that the predictions of dynamic scaling pertain to the mass-diffusion coefficient D; we have

$$D = u_2\xi, \tag{2.3.28}$$

which gives D above T_λ in terms of u_2 below T_λ. Here we have assumed numerical constants to be equal to unity.

One might have thought that a good upper bound for D could be provided by NMR measurements of the spin-diffusion coefficient D_s. Spin–spin relaxation time measurements [222] have indeed revealed an anomaly near T_λ, but they do not show the strong divergence suggested by equation (2.3.28). Furthermore, Eggington and Leggett [223] have presented arguments that indicate radically different behavior for D and D_s. They show that results for D_s depend upon the type of experiment that is performed, and that ordinary spin-echo measurements, which do not involve a gradient of the chemical potential, will yield a diffusivity without a strong singularity which has no particular relation to D.

For small X, the diffusion rate of ^{3}He into ^{4}He has been studied by measuring the absorption of neutrons as a function of time [224]. The neutron absorption is strongly dependent upon the ^{3}He concentration. These results have been interpreted to imply that D remains finite at T_λ [224], contrary to the prediction equation (2.3.28). It is not obvious that the measurements really have the resolution to permit that conclusion; in addition, it is not clear whether the experiment was performed under isothermal conditions.

More recently, the characteristic times τ associated with the relaxation of *isothermal* concentration gradients have been measured for $X = 0.10$, 0.21, and 0.40 [218b]. These results are shown in Figure 2.41 on logarithmic scales as a function of $T/T_\lambda - 1$. Under isothermal conditions, the mass diffusion coefficient D is related to τ by $D = L^2/\tau$, where L is a characteristic length of the experimental apparatus (for the geometry used in Reference 218b roughly equal to 0.7 cm). The data in Figure 2.41, therefore, imply that the mass diffusivity can be represented by

$$D = d\left(\frac{T - T_\lambda}{T_\lambda}\right)^{-z}. \tag{2.3.29}$$

For z, the measurements yield 0.40, 0.36, and 0.33, for $X = 0.10$, 0.21, and 0.40, respectively. The uncertainty for z in each case is about $\pm$ 0.06. If one assumes that the appropriate correlation length for equation (2.3.28) is given by a pure power law with a universal exponent of 0.67, and if one uses the measured u_2, one obtains a predicted effective exponent of about 0.27 at all three concentrations. This prediction is inconsistent with the experiment. However, equation (2.3.28) permits alternate interpretations. Thus, one might

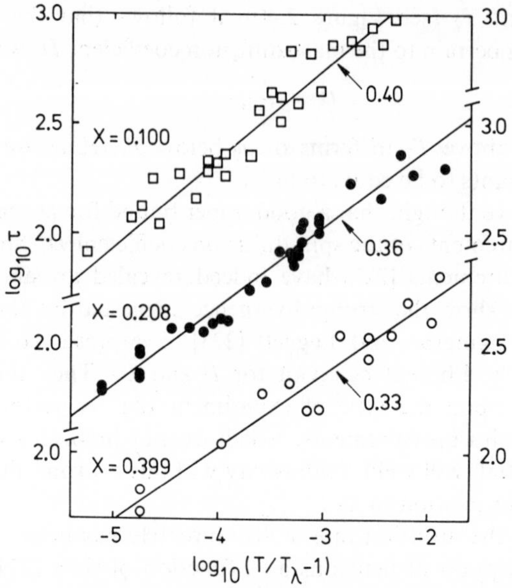

FIGURE 2.41. The relaxation times τ of isothermal concentration gradients for three ^3He–^4He mixtures. The mass diffusivity D is proportional to τ^{-1}. After Reference [218b].

assume that equation (2.3.28) is only asymptotically correct, and that the contributions to u_2 from the thermodynamic factors [see Section 2.2.2.(c) which have no influence upon the asymptotic behavior, are not given correctly. For pure He^4, one obtains $D \sim \varepsilon^{-0.34}$ from equation (2.3.28) and the measured ρ_s. If the pertinent exponents are universal, this relation is valid for all X. The value 0.34 for z cannot be excluded by the experimental result, although for $X = 0.1$ it appears somewhat low.

The mass diffusion coefficient is predicted also from the mode–mode coupling calculations [221, 221a] mentioned above in connection with the thermal conductivity. These theories yield a result equivalent to equation (2.3.28), but they too do not give singular correction terms such as those from the thermodynamic contributions to u_2. Both theoretical and experimental results, therefore, are subject to sufficiently large uncertainties to avoid a conflict. We can hope for more detailed predictions, based on renormalization group theory, in the future.

REFERENCES AND NOTES

1. For an introduction to the theory of phase transitions, see, for instance, H. E. Stanley, *Introduction to Phase Transitions and Critical Phenomena* (Oxford Univ. Press, 1971). For a review of the more recently developed renormalization group theory of critical phenomena, see K. G. Wilson and J. Kogut, *Physics Reports* **12C**, 76 (1974).

2. K. R. Atkins, *Liquid Helium* (Cambridge Univ. Press, 1959).
3. J. Wilks, *The Properties of Liquid and Solid Helium* (Clarendon, Oxford, 1967).
4. W. E. Keller, *Helium-3 and Helium-4* (Plenum Press, New York, 1969).
5. G. Kukich, R. P. Henkel and, J. D. Reppy, *Phys. Rev. Lett.* **21,** 197 (1968).
6. J. R. Clow and J. D. Reppy, *Phys. Rev. Lett.* **19,** 291 (1967).
7. J. R. Clow and J. D. Reppy, *Phys. Rev. A* **5,** 424 (1972).
8. J. S. Langer and J. D. Reppy, in *Progress in Low Temperature Physics*, edited by C. J. Gorter (North-Holland, Amsterdam, London, 1970) Vol. VI, Chap. 1.
9. K. D. Erben and F. Pobell, *Phys. Lett.* **26A,** 368 (1968).
10. S. M. Bhagat and B. M. Winer, *Phys. Lett.* **27A,** 537 (1968).
11. G. Ahlers, *Phys. Rev. Lett.* **22,** 54 (1969).
12. F. Pobell, W. Schoepe, and W. Veith, *Phys. Lett.* **25A,** 209 (1967).
13. W. Veith and F. Pobell, *Phys, Lett.* **27A,** 254 (1968).
14. P. Leiderer and F. Pobell, *Z. Physik* **223,** 378 (1969).
15. J. Andelin, F. Pobell, and F. Wagner, *J. Low Temp. Phys.* **1,** 417 (1969).
15a. M. J. Crooks and D. L. Johnson, *Can. J. Phys.* **49,** 1035 (1971).
16. G. Ahlers, *Phys. Rev.* **164,** 259 (1967).
17. P. Leiderer and F. Pobell, *J. Low Temp. Phys.* **3,** 577 (1970).
18. S. M. Bhagat and R. A. Lasken, *Phys. Rev. A* **3,** 340 (1971).
19. S. M. Bhagat and R. S. Davis, *Phys. Lett.* **34A,** 233 (1973); and *J. Low Temp. Phys.* **7,** 157 (1972).
20. S. M. Bhagat and R. A. Lasken, *Phys. Rev A* **5,** 2297 (1972).
21. H. J. Mikeska, *Phys. Rev.* **179,** 166 (1969).
22. H. J. Mikeska, *Z. Physik* **229,** 57 (1969).
23. G. Ahlers, A. Evenson, and A. Kornblit, *Phys. Rev. A* **4,** 804 (1971).
24. S. M. Bhagat, R. S. Davis, and R. A. Lasken, in *Proceedings of the Thirteenth International Conference on Low Temperature Physics,* edited by K. D. Timmerhaus, W. J. O'Sullivan, and E. F. Hammel (Plenum, New York, 1974), Vol. 1, p. 328.
25. O. K. Rice, J. *Chem. Phys.* **22,** 1535 (1954).
26. A. B. Pippard, *Phil. Mag.* **1,** 473 (1956).
27. A. B. Pippard, *The Elements of Classical Thermodynamics* (Cambridge Univ. Press, 1957), Chap. IX.
28. M. J. Buckingham and W. M. Fairbank, in *Progress in Low Temperature Physics*, edited by C. J. Gorter (North-Holland, Amsterdam, 1961), Vol. III, p. 80.
29. O. V. Lounasmaa, *Phys. Rev.* **130,** 847 (1963).
30. K. C. Lee and R. D. Puff, *Phys. Rev.* **158,** 170 (1967).
31. M. Barmatz and I. Rudnick, *Phys. Rev.* **170,** 224 (1968).
32. H. A. Kierstead, *Phys. Rev.* **162,** 153 (1967).
33. G. Ahlers, *J. Low Temp. Phys.* **7,** 361 (1972).
34. C. F. Kellers, Ph.D. Thesis, Duke Univ. (Durham N. C., 1960) (unpublished).
35. J. C. Wheeler and R. B. Griffiths, *Phys. Rev.* **170,** 249 (1968).
36. B. J. Lipa and M. J. Buckingham, *Phys. Lett.* **26A,** 643 (1968).
37. M. E. Fisher, *Phys. Rev.* **176,** 257 (1968).

38. G. Ahlers, in *Proceedings of the Twelfth International Conference on Low Temperature Physics*, edited by E. Kanda (Academic Press of Japan, 1971), p. 21.
39. G. Ahlers, *Phys. Rev. A* **8,** 530 (1973).
40. O. V. Lounasmaa and E. Kojo, *Ann. Acad. Sci. Fenn.* **AVI,** No. 36 (1959).
41. O. V. Lounasmaa, *Cryogenics* **1,** 1 (1961).
42. C. G. Waterfield, J. K. Hoffer, and N. E. Phillips (unpublished).
43. C. E. Chase, *Phys. Rev. Lett.* **2,** 197 (1959).
44. G. Ahlers, *Phys. Rev.* **182,** 352 (1969).
45. H. van Dijk, M. Durieux, J. R. Clement, and J. K. Logan, Natl. Bur. Std. (U.S.) Monograph No. 10 (U.S. GPO, Washington, D.C., 1960).
46. E. C. Kerr and R. Dean Taylor, *Ann. Phys. (N.Y.)* **26,** 292 (1964).
47. H. A. Kierstead, *Phys. Rev.* **138,** A1594 (1965).
48. J. H. Vignos and H. A. Fairbank, *Phys. Rev. Lett.* **6,** 265, 646 (1961); *Phys. Rev.* **147,** 185 (1966).
49. G. Ahlers, *Phys. Rev.* **135** A10 (1964); *Phys. Rev. Lett.* **10,** 439 (1963).
50. A. F. Schuch and R. L. Mills, *Phys. Rev. Lett.* **8,** 469 (1962).
51. D. L. Elwell and H. Meyer, *Phys. Rev.* **164,** 245 (1967).
51a. V. P. Peshkov and A. P. Borovikov *Zh. Eksp. Teor. Fiz.* **50,** 844 (1966) [English transl.: *Sov. Phys.—JETP* **23,** 559 (1966)].
51b. M. Barmatz and I. Rudnick, *Phys. Rev.* **173,** 275 (1968).
52. R. W. Hill and O. V. Lounasmaa, *Phil. Mag.* **2,** 143 (1957).
53. K. R. Atkins and R. A. Stasior, *Can. J. Phys.* **31,** 1156 (1953).
54. G. Ahlers, *Phys. Rev.* **171,** 275 (1968).
55. L. V. Kiknadze, Yu. G. Mamaladze, and O. D. Cheishvili, *Zh. Eksp. Teor. Fiz. Pis. Red.* **3,** 305 (1966) [English transl.: *Sov. Phys.—JETP Letters* **3,** 197 (1966)].
56. P. C. Hohenberg, as quoted in Ref. [54].
57. G. Ahlers (unpublished).
58. L. V. Kiknadze, Yu. G. Mamaladze, and O. D. Cheishvili, in *Proceedings of the Tenth International Conference on Low Temperature Physics,* edited by M. P. Malkov (Moscow, 1967), Vol. 1, p. 491; and private communication.
58a. V. A. Slyusarev and M. A. Strzhemechnyi, *Zh. Eksp. Teor. Fiz.* **58,** 1757 (1970) [English transl.: *Sov. Phys.—JETP* **31,** 941 (1970)].
58b. A. A. Sobyanin, *Zh. Eksp. Teor. Fiz.* **63,** 1780 (1972) [English transl.: *Sov. Phy.—JETP* **36,** 941 (1973)].
59. Yu. G. Mamaladze, *Zh. Eksp. Teor. Fiz.* **52,** 729 (1967). [English transl.: *Sov. Phys.—JETP* **25,** 479 (1967)].
60. V. L. Ginzburg and L. P. Pitaevskii, *Zh. Eksp. Teor. Fiz.* **34,** 1240 (1958). [English transl.: *Sov. Phys.—JETP* **34,** 858 (1958)].
61. G. Ahlers, *Phys. Rev. Lett.* **21,** 1159 (1968).
61a. V. L. Ginzburg and A. A. Sobyanin, *ZhETF Pis.Red.* **17,** 698 (1973) [English transl.: *JETP Letters* **17,** 483 (1973)].
62. W. M. Fairbank, M. J. Buckingham, and C. F. Kellers, in *Proceedings of the Fifth International Conference on Low Temperature Physics and Chemisty,* edited by J. R. Dillinger (Univ. of Wisconsin Press, Madison, 1958), p. 50.

63. W. M. Fairbank, in *Proceedings of the International School of Physics "Enrico Fermi,"* Course XXI, edited by G. Careri (Academic, New York, 1963), p. 293.

64. W. M. Fairbank and C. F. Kellers, in *Critical Phenomena, Proceedings of a Conference,* edited by M. S. Green and J. V. Sengers, Natl. Bur. Std. Misc. Pub. No. 273 (U.S. GPO, Washington, D.C., 1966), p. 71.

65. G. Ahlers, *Phys. Rev. Lett.* **23,** 464 (1969); **23,** 739 (1969) (E).

66. G. Ahlers, *Phys. Rev.* **A3,** 696 (1971).

67. F. Gasparini and M. R. Moldover, *Phys. Rev Lett.* **23,** 749 (1969); and *Phys. Rev. B,* to be published.

68. F. Gasparini, Ph.D. Thesis, Univ. of Minnesota (1970) (unpublished).

69. F. M. Gasparini and M. R. Moldover, in *Proceedings of the Thirteenth International Conference on Low Temperature Physics* edited by K. D. Timmerhaus, W. J. O'Sullivan, and E. F. Hammel (Plenum, New York, 1974), Vol. 1, p. 618.

70. H. C. Kramers, J. D. Wasscher, and C. J. Gorter, *Physica* **18,** 329 (1952).

71. B. Widom, *J. Chem. Phys.* **43,** 3892 (1965); **43,** 3898 (1965).

72. L. P. Kadanoff, *Physics* **2,** 263 (1966).

73. R. B. Griffiths, *Phys. Rev.* **158,** 176 (1967).

74. M. E. Fisher, *Rep. Progr. Phys.* **30,** 615 (1967).

75. L. P. Kadanoff, W. Götze, D. Hamblen, R. Hecht, E. A. S. Lewis, V. V. Palciauskas, M. Rayl, and J. Swift, *Rev. Mod. Phys.* **39,** 395 (1967).

76. K. R. Atkins and M. H. Edwards, *Phys. Rev.* **97,** 1429 (1955).

77. M. H. Edwards, *Can. J. Phys.* **36,** 884 (1958).

78. C. E. Chase, E. Maxwell, and W. E. Millett, *Physica* **27,** 1129 (1961).

79. R. F. Harris-Lowe and K. A. Smee, *Phys. Rev. A***2,** 158 (1970); *Phys. Lett.* **28A,** 246 (1968).

80. K. R. Atkins and Y. Narahara, *Phys. Rev.* **138,** A437 (1965).

80a. F. M. Gasparini, J. Eckardt, D. O. Edwards, and S.Y. Shen, *J. Low Temp. Phys.* **13,** 437 (1973).

81. A. A. Sobyanin, *Zh. Eksp. Teor. Fiz.* **61,** 433 (1971) [English transl.: *Sov. Phys.—JETP* **34,** 229 (1972)].

82. K. Binder and P. C. Hohenberg, *Phys. Rev. B* **6,** 3461 (1972).

83. P. C. Hohenberg, *J. Low Temp. Phys.* **13,** 433 (1973).

84. H. A. Kierstead, *Phys. Rev.* **153,** 258 (1967).

85. T. H. McCoy and E. H. Graf, *Phys. Lett.* **38A,** 287 (1972).

86. G. Ahlers, *Phys. Lett.* **39A,** 335 (1972).

87. E. R. Grilly, *Phys. Rev.* **149,** 97 (1966).

88. Early statements of the hypothesis of universality may be found in M. E. Fisher, *Phys. Rev. Lett.*, **16,** 11 (1966); P. G. Watson, *J. Phys. C: Proc. Phys. Soc. London* **2,** 1883, 2158 (1969); D. Jasnow and M. Wortis, *Phys. Rev.* **176,** 739 (1968). More recent references are L. P. Kadanoff, in *Critical Phenomena, Proceedings of the International School "Enrico Fermi"*, edited by M. S. Green (Academic, New York, 1971); R. B. Griffiths, *Phys. Rev. Lett.* **24,** 1479 (1970); D. Stauffer, M. Ferer, and M. Wortis, *Phys. Rev. Lett.* **29,** 345 (1972); K. Wilson and J. Kogut, *Physics Report* **12C,** 76 (1974).

89. R. B. Griffiths, in *Critical Phenomena in Alloys, Magnets, and Superconductors,* edited by R. E. Mills, E. Ascher, and R. I. Jaffee (McGraw-Hill, New York, 1971), p. 377.

90. R. B. Griffiths, *Phys. Rev. Lett.* **24,** 715 (1970).

91. K. G. Wilson and M. E. Fisher, *Phys. Rev. Lett.* **28,** 240 (1972); and reference 1.

92. K. G. Wilson, *Phys. Rev. Lett.* **28,** 548 (1972).

93. E. Brézin, D. J. Wallace, and K. G. Wilson, *Phys. Rev. Lett.* **29,** 591 (1972); *Phys. Rev. B* **7,** 232 (1973).

94. F. J. Wegner, *Phys. Rev. B*5,4529 (1972); and A. D. Bruce and A. Aharony, *Phys. Rev. B* **10**, 2078 (1974); and E. Brezin, J. C. LeGuillou, and J. Zinn-Justin, *Phys. Rev. D* **8,** 2418 (1973); and J. W. Swift and M. K. Grover, *Phys. Rev A* **9,** 2579 (1974).

95. A. Kornblit, G. Ahlers, and E. Buehler, *Phys. Lett. A* **43,** 531 (1973); and A. Kornblit and G. Ahlers. *Phys. Rev. B* **8,** 5163 (1973).

96. J. A. Lipa, C. Edwards, and M. J. Buckingham, *Phys. Rev. Lett.* **25,** 1086 (1970).

97. G. R. Brown and H. Meyer, *Phys. Rev. A* **6,** 364 (1972).

97a. A. V. Voronel, V. G. Gorbunova, V. A. Smirnov, N. G. Shamakov, and V. V. Shchekochikhina, *Zh. Eksp. Teor. Fiz.* **63,** 964 (1972) [English transl.: *Sov. Phys.—JETP* **36,** 505 (1973)].

97b. G. Ahlers, A. Kornblit, and M. B. Salamon, *Phys. Rev. B* **9,** 3932 (1974).

98. S. Ma, *Phys. Rev. Lett.* **29,** 1311 (1972).

99. E. Brézin and D. J. Wallace, *Phys. Rev. B* **7,** 1967 (1973).

99a. Equation (2.1.70a) is obtained from the $(4-d)$—expansion for the exponents (see Ref. [1]) γ and η given in Ref. [92], and from the scaling relation $\alpha = 2 - d\gamma/(2-\eta)$.

99b. E. Brezin, J.C. LeGuillou, and J. Zinn-Justin, *Phys. Lett.* **47 A,** 285 (1974).

100. E. A. Guggenheim, *Thermodynamics* (North-Holland, Amsterdam; John Wiley, New York) 5th Ed., 1967.

101. K. W. Taconis and R. de Bruyn Ouboter, in *Progress in Low Temperature Physics,* edited by C. J. Gorter (North-Holland Amsterdam, 1964) Vol. 4, p. 38.

102. S. G. Sydoriak and T. R. Roberts, *Phys. Rev.* **118,** 901 (1960).

103. T. R. Roberts and S. G. Sydoriak, *Phys. Fluids* **3,** 895 (1960).

104. E. H. Graf, D. M. Lee, and J. D. Reppy, *Phys. Rev. Lett.* **19,** 417 (1967).

105. T. Alvesalo, P. Berglund, S. Islander, G. R. Pickett, and W. Zimmermann, Jr., *Phys. Rev. Lett.* **22,** 1281 (1969); and *Phys. Rev. A* **4,** 2354 (1971).

106. S. T. Islander and W. Zimmermann, Jr., *Phys. Rev. A* **7,** 188 (1973).

107. G. Ahlers and D. S. Greywall, *Phys. Rev. Lett.* **29,** 849 (1972); and in *Proceedings of the Thirteenth International Conference on Low Temperature Physics* edited by K. D. Timmerhaus, W. J. O'Sullivan, and E. F. Hammel (Plenum, New York, 1974), Vol. 1, p. 586, and to be published.

108. G. J. Goellner, Ph.D. Thesis, Duke Univ. (Durham, N. C., 1972) (unpublished).

109. G. J. Goellner, R. Behringer, and H. Meyer, *J. Low Temp. Phys.* **13,** 113 (1973).

110. D. R. Watts and W. W. Webb, in *Proc. Thirteenth Int. Conf. Low Temp. Phys.* edited by K. D. Timmerhaus W. T. O'Sullivan, and E. F. Hammel (Plenum, New York, 1974), Vol. 1, p. 581.; and D. R. Watts, Ph.D. Thesis, Cornell Univ. (Ithaca, N.Y., 1973), and P. Leiderer, D. R. Watts, and W. W. Webb, *Phys. Rev. Lett.* **33,** 483 (1974).

111. G. K. Walters and W. M. Fairbank, *Phys. Rev.* **103,** 262 (1956).

112. D. O. Edwards, E. M. Ifft, and R. E. Sarwinski, *Phys. Rev.* **177,** 380 (1969); and references therein.

113. R. B. Griffiths, *Phys. Rev. Lett.* **24,** 715 (1970).

114. J. G. Daunt and C. V. Heer, *Phys. Rev.* **79,** 46 (1950).

115. R. de Bruyn Ouboter, K. W. Taconis, C. le Pair, and J. J. M. Beenakker, *Physica* **26,** 853 (1960).

116. D. O. Edwards, D. F. Brewer, P. Seligman, M. Skertic, and M. Yaqub, *Phys. Rev. Lett.* **15,** 773 (1965).

117. T. R. Roberts and B. V. Swartz, in *Proceedings of the Second Symposium on Liquid and Solid He*^{3} (Ohio State Univ. Press, Columbus, 1960).

117a. G. Ahlers, *Phys. Rev. A* **10,** 1670 (1974).

118. L. D. Landau, *Collected Papers,* edited by D. ter Haar (Gordon and Breach, New York, 1967), p. 193.

119. See, for instance, E. A. Guggenheim, *J. Chem. Phys.* **13,** 253 (1945).

119a. We will use an incomplete set of tricritical point parameters, and will not fully conform to the necessarily complicated notation which is being established in the current literature. For a more complete system of notation, see R. B. Griffiths, *Phys. Rev. B* **7** 545 (1973).

120. See Ref. [89].

121. E. K. Riedel, *Phys. Rev. Lett.* **28,** 675 (1972).

122. E. K. Riedel and F. J. Wegner, *Phys. Rev. Lett.* **29,** 349 (1972).

123. F. J. Wegner and E. K. Riedel, *Phys. Rev. B* **7,** 248 (1973).

123a. E. K. Riedel and F. J. Wegner, *Phys. Rev. B* **9,** 294 (1974); and D. R. Nelson, *Phys. Rev. B* **11,** 3504 (1975).

124. O. V. Lounasmaa, *J. Chem. Phys.* **33,** 443 (1960).

125. G. Goellner and H. Meyer, *Phys. Rev. Lett.* **26,** 1534 (1971).

125a. N. R. Brubaker and M. R. Moldover, in *Proceedings of the Thirteenth International Conference on Low Temperature Physics,* edited by K. D. Timmerhaus, W. J. O'Sullivan and E. F. Hammel (Plenum, New York, 1974), Vol. 1, p. 612.

125b. The results for $\Phi(X) - \Phi(X_t)$ originally reported in Ref. [125] were smaller than those given in Ref. [106] by up to a factor of two. This difference was largely removed in Ref. [109] however.

125c. M. Barmatz and P. C. Hohenberg, *Phys. Rev. Lett.* **24,** 1225 (1970); and P. C. Hohenberg and M. Barmatz, *Phys. Rev. A* **6,** 289 (1972).

125d. A. T. Berestov, A. V. Voronel, and M. Sh. Giterman, *Zh. ETF P 273is .Red.* **15,** (1972) [English transl.: *JETP Letters* **15,** 190 (1972)].

125e. E. C. Kerr, *Phys. Rev. Lett.* **12,** 185 (1964).

126. L. D. Landau, *Zh. Eksp. Teor. Fiz.* **11,** 592 (1941); *J. Phys. Moscow* **5,** 71 (1941); **8,** 1 (1944), see also E. M. Lifshitz and E. L. Andronikashvili, *A Supplement to Helium* (Consultants Bureau Inc., 1959).

127. I. M. Khalatnikov, *Introduction to the Theory of Superfluidity,* transl. by P. C. Hohenberg (W. A. Benjamin, New York and Amsterdam, 1965).

128. B. D. Josephson, *Phys. Lett.* **21,** 608 (1966).

129. E. L. Andronikashvili, *Zh. Eksp. Teor. Fiz.* **16,** 780 (1946); **18,** 424 (1948).

130. See for instance, L. D. Landau and E. M. Lifshitz, *Fluid Mechanics* (Pergamon Press, Long Island City, N.Y., 1959).

131. J. G. Dash and R. D. Taylor, *Phys. Rev.* **105,** 7 (1957).
132. J. A. Tyson and D. H. Douglass, Jr., *Phys. Rev. Lett.* **17,** 472 (1966).
133. J. A. Tyson, *Phys. Rev.* **166,** 166 (1968).
134. R. H. Romer and R. J. Duffy, *Phys. Rev.* **186,** 255 (1969).
135. H. E. Hall, *Phil. Trans. Roy. Soc. (London)* **A250,** 980 (1957).
136. W. F. Vinen, *Proc. Roy. Soc. (London)* **A260,** 218 (1961).
137. P. J. Bendt, *Phys. Rev.* **127,** 1441 (1962).
138. J. B. Mehl and W. Zimmermann, Jr., *Phys. Rev.* **167,** 214 (1968).
139. J. D. Reppy and D. Depatie, *Phys. Rev. Lett.* **12,** 187 (1964).
140. W. E. Keller and E. F. Hammel, *Physics* **2,** 221 (1966).
141. J. D. Reppy, *Phys. Rev. Lett.* **14,** 733 (1965).
142. J. B. Mehl and W. Zimmermann, Jr., *Phys. Rev. Lett.* **14,** 815 (1965).
143. J. R. Clow and J. D. Reppy, *Phys. Rev. Lett.* **16,** 887 (1966).
144. J. D. Reppy (private communication).
145. C. E. Chase, *Phys. Fluids* **1,** 193 (1958).
146. G. Ahlers, *J. Low Temp. Phys.* **1,** 609 (1969).
147. I. Rudnick and K. A. Shapiro, *Phys. Rev. Lett.* **15,** 386 (1965).
148. M. Revzen, A. Ron, and I. Rudnick, *Phys. Rev. Lett.* **15,** 384 (1965).
149. D. H. Douglass, Jr., *Phys. Rev. Lett.* **15,** 951 (1965).
150. H. A. Kierstead, *Phys. Rev. Lett.* **16,** 343 (1966).
151. C. J. Pearce, J. A. Lipa, and M. J. Buckingham, *Phys. Rev. Lett.* **20,** 1471 (1968).
152. J. A. Tyson and D. H. Douglass, Jr., *Phys. Rev. Lett.* **21,** 1308 (1968).
153. D. L. Johnson and M. J. Crooks, *Phys. Lett.* **27A,** 688 (1968).
154. D. L. Johnson and M. J. Crooks, *Phys. Rev.* **185,** 253 (1969).
155. R. Williams, S. E. A. Beaver, J. C. Fraser, R. S. Kagiwada, and I. Rudnick, *Phys. Lett.* **29A,** 279 (1969).
156. G. Ahlers, *Phys. Lett.* **28A,** 507 (1969).
157. D. S. Greywall and G. Ahlers, *Phys. Rev. Lett.* **28,** 1251 (1972).
158. D. S. Greywall and G. Ahlers, *Phys. Rev. A* **7,** 2145 (1973).
159. G. Terui and A. Ikushima, *Phys. Letters* **31A,** 161 (1972); A. Ikushima and G. Terui, *J. Low Temp. Phys.* **10,** 397 (1973).
159a. G. Winterling, W. S. Holmes, and T. J. Greytak, in *Proceedings of the Thirteenth International Conference on Low Temperature Physics*, edited by K. D. Timmerhaus, W. J. O'Sullivan, and E. F. Hammel (Plenum, New York, 1974), Vol. 1, p. 537.
159b. W. F. Vinen, C. J. Palin, and J. M. Vaughan, in *Proceedings of the Thirteenth International Conference Low Temperature Physics*, edited by K. D. Timmerhaus, W. J. O'Sullivan, and E. F. Hammel (Plenum, New York, 1974), Vol. 1, p. 524.
160. J. R. Pellam, *Phys. Rev.* **73,** 608 (1948).
161. K. R. Atkins, *Phys. Rev.* **113,** 962 (1959).
162. K. A. Shapiro and I. Rudnick, *Phys. Rev.* **137A,** 1383 (1965).
163. I Rudnick, E. Guyon, K. A. Shapiro, and S. A. Scott, *Phys. Rev. Lett.* **19,** 488 (1967).
164. I. Rudnick and K. A. Shapiro, *Phys. Rev. Lett.* **9,** 191 (1962).
165. E. Guyon and I. Rudnick, *J. Phys.* **29,** 1081 (1968).

166. M. Kriss and I. Rudnick, *J. Low Temp. Phys.* **3,** 339 (1970).

167. G. Ahlers, J. *Low Temp. Phys.* **1,** 159 (1969).

167a. H. Wiechert and Z. Meinhold-Heerlein, *J. Low Temp. Phys.* **4,** 273 (1971).

167b. Z. Rácz, *J. Low Temp. Phys.* **11,** 509 (1973).

168. K. R. Atkins and I. Rudnick, in *Progress in Low Temperature Physics,* edited by C. J. Gorter (North-Holland Amsterdam, London, 1970), Vol. VI, Chap. 2.

169. R. P. Henkel, E. N. Smith, and J. D. Reppy, *Phys. Rev. Lett.* **23,** 1276 (1969).

170. R. S. Kagiwada, J. C. Frasser, I. Rudnick, and D. Bergman, *Phys. Rev. Lett.* **22,** 338 (1969).

170a. *Nuclepore Filter Membranes,* (General Electric Co., Pleasanton, Calif.).

170b. G. G. Ihas and F. Pobell, *Phys. Rev. A* **9,** 1278 (1974).

170c. See also R. A. Sherlock and D. O. Edwards, *Rev. Sci. Instr.* **41,** 1603 (1970).

170d. R. A. Ferrell, N. Menyhárd, H. Schmidt, F. Schwabl, and P. Szépfalusy, *Ann. Phys. (N. Y.)* **47,** 565 (1968), B. I. Halperin and P. C. Hohenberg, *Phys. Rev.* **177,** 952 (1969); and M. E. Fisher, M. N. Barber, and D. Jasnow, *Phys. Rev. A* **8,** 1111 (1973).

170e. D. Stauffer, M. Ferer, and M. Wortis, *Phys. Rev. Lett.* **29,** 345 (1972); M. Ferer and M. Wortis, *Phys. Rev. B*6, 3426 (1972).

170f. A. Aharony, *Phys. Rev. B***9,** 2107 (1974).

170g. M. Ferer, *Phys. Rev. Lett.* **33,** 21 (1974).

170h. V. N. Grigor'ev, B. N. Esel'son, V. P. Mal'khanov, and V. I. Sobolev, *Zh. Eksp. Teor. Fiz.* **51,** 1059 (1966) [English transl.: *Sov. Phys.—JETP* **24,** 707 (1967)].

170i. V. I. Sobolev and B. N. Esel'son, *Zh. Eksp. Teor. Fiz.* **60,** 240 (1971) [English trnsl.: *Sov. Phys.—JETP* **33,** (1971)].

171a. I. M. Khalatnikov, *Zh. Eksp. Teor. Fiz.* **23,** 265 (1952).

171b. I. M. Khalatnikov, *Usp. Fiz. Nauk* **60,** 69 (1956).

171c. The pertinent equations, Eq. 24–68 of Ref. [127] and Eq. 8–18 of Ref. [171b], contain typographical errors. Equations 1–11 to 1–13 of Ref. [171a] are given correctly, however.

171d. J. Swift (private communication).

171e. M. Barmatz, *J. Low Temp. Phys.* **5,** 419 (1971).

171f. W. C. Thomlinson and F. Pobell, *Phys. Lett.* **44A,** 155 (1973).

172. J. H. Vignos and H. A. Fairbank, Phys. Rev. **147,** 185 (1966).

173. T. Satoh and A. Kakizaki, in *Proceedings of the Thirteenth International Confreence on Low Temperature Physics*, edited by K. D. Timmerhaus, W. J. O'Sullivan, and E. F. Hammel (Plenum, New York, 1974), Vol. 1, p. 627

174. S. M. Noble and D. J. Sandiford, *J. Phys. C* **3,** L123 (1970).

174a. G. Terui and A. Ikushima, *Phys. Lett.* **43A**. 255 (1973); and *J. Low Temp. Phys.* **16** 291 (1974), and A. Ikushima and G. Terui, *Phys. Lett.* **47A,** 387 (1974).

174b. G. Ahlers and D. S. Greywall (to be published).

174c. W. C. Thomlinson, G. G. Ihas and F. Pobell, *Phys. Rev. Lett.* **31,**1284 (1973); and *Phys. Rev. B* **11,**4292 (1975).

174d. F. Uehara, T. Kobayashi, A. Tominaga, and Y. Narahara, *Phys. Lett.* **50A,** 83 (1974). A. Tominaga, T. Kobayashi; F. Uehara, and T. Narahara, *J. Phys. C* **8,** 420 (1975).

175. S. D. Elliott, Jr., and H. A. Fairbank, in *Proceedings of the Fith International Con-*

ference on Low Temperature Physics and Chemistry, edited by J. R. Dillinger (Univ. of Wisconsin Press, Madison, 1958), p. 180.

176. D. G. Sanikidze and D. M. Chernikova, *Zh. Eksp. teor. Fiz.* **46,** 1123 (1964) [English transl.: *Sov. Phys.—JETP* **19,** 760 (1964)].

177. B. N. Esel'son, N. E. Dyumin, É. Ya. Rudavskii, and I. A. Serbin, *Zh. Eksp. Teor. Fiz.* **51,** 1064 (1966) [English transl.: *Sov. Phys.—JETP* **24,** 711 (1967)].

178. N. E. Dyumin, B. N. Esel'son, E. Ya. Rudavskii, and I. A. Serbin, *Zh. Eksp. Teor. Fiz.* **56,** 747 (1969) [English transl.: *Sov. Phys.—JETP* **29,** 406 (1969)].

179. G. Ahlers, *Phys. Lett.* **46A,** 89 (1973).

179a. For small X, ideal dilute solution theory has sometimes been used [174a] to evaluate the necessary thermodynamic parameters. This can lead to appreciable errors near the phase transition [179], even for moderately small X.

179b. It should be pointed out that the contribution $X\,(\partial S/\partial X)_{TP}$ to Eq. (2.2.47c), which enters into Eq. (2.2.51), has been erroneously neglected in some of the recent literature [3,107,174c,179]. Although this term vanishes for $X = 0$, its omission is not justified because it makes an appreciable contribution even for moderately small X. Neglect of this term is responsible for the incorrect conclusion [179] that the amplitude of ρ_s/ρ has an exceptionally strong concentration dependence [170k]. The omission has only a small effect, however, upon the exponents which have been quoted [107,174c] for ρ_s/ρ. The correct, full expression for u_2 [Eq. (2.2.51)] has been used in some of the most recent work [170k,174b].

180. A. M. Polyakov, *Zh. Eksp. Teor. Fiz.* **57,** 2144 (1969) [English transl.: *Sov. Phys.—JETP* **30,** 1164 (1970)].

181. R. Dean Taylor and J. G. Dash, *Phys. Rev.* **106,** 398 (1957).

182. D. F. Brewer and D. O. Edwards, *Proc. Roy. Soc. A* **251,** 247 (1959).

183. B. Welber, *Phys. Rev.* **119,** 1816 (1960).

184. J. T. Tough, W. D. McCormick, and J. G. Dash, *Phys. Rev.* **132,** 2373 (1963).

185. R. W. H. Webeler and D. C. Hammer, *Phys. Lett.* **15,** 233 (1965).

186. R. W. H. Webeler and G. Allen, *Phys. Lett.* **33A,** 213 (1970).

187. R. W. H. Webeler and G. Allen, *Phys. Rev. A* **5,** 1820 (1972).

188. G. Ahlers, Phys. Lett. **37A,** 151 (1971).

189. C. M. Lai and T. A. Kitchens, in *Proceedings of the Thirteenth International Conference on Low Temperature Physics,* edited by K. D. Timmerhaus, W. J. O'Sullivan, and E. F. Hammel (Plenum, New York, 1974), Vol. 1, p. 576.

190. D. T. Grimsrud and F. Scaramuzzi, in *Proceedings of the Tenth International Conference on Low Temperature Physics* edited by M. P. Malkov (Moscow Press, 1967), Vol. 1, p. 197.

191. G. Ahlers and G. Gamota, *Phys. Lett.* **38A,** 65 (1972).

192. K. W. Schwarz, *Phys. Rev. A* **6,** 837 (1972).

193. D. M. Sitton and F. Moss, *Phys. Lett.* **34A,** 159 (1971).

194. D. Goodstein, A. Savoia, and F. Scaramuzzi, *Bull. Am. Phys. Soc.* **18,** 475 (1973); *Phys. Rev. A* **9,** 2151 (1974).

195. J. Kerrisk and W. E. Keller, *Bull. Am. Phys. Soc.* **12,** 550 (1967); *Phys. Rev.* **177,** 341 (1969).

196. R. A. Ferrell, N. Menyha'rd, H. Schmidt, F. Schwabl. and P. Sze'pfalusy, *Phys. Rvv. Lett.* **18,** 891 (1967); *Phys. Lett.* **24A,** 493 (1967); *Ann. Phys. (N. Y.)* **47,** 565 (1968).

197. B. I. Halperin and P. C. Hohenberg, *Phys. Rev. Lett.* **19,** 700 (1967); *Phys. Rev.* **177,** 952 (1969).

198. For an extensive discussion of dynamic scaling near T_λ, see also P. C. Hohenberg, in *Critical Phenomena, Proceedings of the International School "Enrico Fermi,"* edited by M. E. Green (Academic, New York, 1971).

199. M. Archibald, J. M. Mochel, and L. Weaver, *Phys. Rev. Lett.* **21,** 1156 (1968).

200. G. Ahlers, *Phys. Rev. Lett.* **24,** 1333 (1970).

201. J. Swift and L. P. Kadanoff, *Ann. Phys. (N.Y.)* **50,** 312 (1968).

202. P. C. Hohenberg, *J. Low Temp. Phys.* **11,** 745 (1973).

203. G. Ahlers (unpublished).

204. W. B. Hanson and J. R. Pellam, *Phys. Rev.* **95,** 321 (1954).

205. N. A. Notarys, Ph.D. Thesis, California Institute of Technology (1964) (unpublished).

206. J. A. Tyson, *Phys. Rev. Lett.* **21,** 1235 (1968).

207. G. Winterling, F. S. Holmes, and T. J. Greytak, in *Proceedings of the Thirteenth International Conference on Low Temperature Physics,* edited by K. D. Timmerhaus, W. J. O'Sullivan, and E. F. Hammel (Plenum, New York, 1974), Vol. 1, p. 537; and *Phys. Rev. Letters* **30**, 427 (1973); and G. Winterling, J. Miller, and T. J. Greytak, *Phys. Letters* **48A**, 343 (1974).

208. W. F. Vinen, C. J. Palin, and J. M. Vaughan, in *Proceedings of the Thirteenth International Conference on Low Temperature Physics,* edited by K. D. Timmerhaus W. J. O'Sullivan, and E. F. Hammel (Planum, New York, 1974), Vol. 1, p. 524.

209. For a discussion of light scattering at critical points in ordinary fluids, see for instance Ref. [1, Ch. 13]. For a detailed discussion of the spectrum of light scattered by superfluid helium near T_λ, see W. E. Vinen, *Physics of Quantum Fluids,* edited by R. Kubo and F. Takano (Syokabo Publishing Co., Tokyo, Japan, 1971), p.1; and Ref. [202].

209a. P. C. Hohenberg, as quoted in Ref. [31].

210. C. E. Chase Phys. *Fluids* **1,** 193 (1958).

211. R. D. Williams and I. Rudnick, *Phys. Rev. Lett.* **25,** 276 (1970).

212. L. D. Landau and I. M. Khalatnikov, *Dokl. Akad. Nauk SSSR* **96,** 469 (1954) [English transl.: *Collected Papers of L. D. Landau* edited by D. ter Haar (Gordon and Breach, London, 1965), p. 626].

213. L. I. Mandel'shtam and M. A. Leontovich, as quoted in Ref. [130, p. 304].

214. K. Kawasaki, *Phys. Lett.* **31A,** 165 (1970).

215. J. S. Imai and I. Rudnick, *Phys. Rev. Lett.* **22,** 694 (1969).

216. D. E. Commins and I. Rudnick, in *Proceedings of the Thirteenth International Conference Low Temperature Physics,* edited by K. D. Timmerhaus, W. J. O'Sullivan, and E. F. Hammel (Plenum, New York, 1974), Vol. 1, p. 356.

217. W. Heinicke, G. Winterling, and K. Dransfeld, *Phys. Rev. Lett.* **22,** 170 (1969).

218. J. M. Vaughan, W. F. Vinen, and C. J. Palin, in *Proceedings of the Thirteenth International Conference on Low Temperature Physics,* edited by K. D. Timmerhaus, W. J. O'Sullivan, and E. F. Hammel (Plenum, New York, 1974), Vol. 1, p. 532.

218a. W. C. Thomlinson and F. Pobell, *Phys. Rev. Lett.* **31,** 283 (1973).

218b. G. Ahlers and F. Pobell, *Bull. Am. Phys. Soc.* **19,** 32 (1974); *Phys. Rev. Lett.* **32,** 144 (1974).

219. G. Ahlers and D. S. Greywall (unpublished).

220. A. Griffin, *Can. J. Phys.* **47,** 429 (1969).

221. K. Kawasaki and J. D. Gunton, *Phys. Rev. Lett.* **29,** 1661 (1972).

221a. M. K. Grover and J. Swift, *J. Low Temp. Phys.* **11,** 751 (1973).

222. S. Saito, G. Terui, and E. Kanda, in *Proceedings of the Twelveth International Conference Low Temperature Physics* edited by E. Kanda (Academic, Japan, 1971), p. 55; and S. Saito, *Physics Lett.* **43A,** 241 (1973).

223. M. A. Eggington and A. J. Leggett, *J. Low Temp. Phys.* **5,** 275 (1971).

224. G. M. Drabkin, V. A. Noskin, V. A. Trunov, A. F. Shchebetov, and A. Z. Yagud, *Zh. Tekh. Fiz.* **42,** 180 (1972) [English transl.: *Sov. Phys.—Tech. Phys.* **17,** 142 (1972)]; G. M. Drabkin, V. A. Noskin, fand A. Z. Yagud, *ZhETF Pis. Red.* **15,** 504 (1972) [English transl.: *Sov. Phys.—JETP Lett.* **15,** 357 (1972)]; and G. M. Drabkin, V. A. Noskin, E. G. Tarovik, and A. Z. Yagud, *Phys. Lett.* **43A,** 83 (1973).

CHAPTER 3

Vortices and Ions in Helium*

Alexander L. Fetter†

Institute of Theoretical Physics, Department of Physics
Stanford University, Stanford, California

*Research sponsored by the Air Force Office of Scientific Research, Office of Aerospace Research, U.S. Air Force, under AFOSR Contract F44620–71–C–0044.
†Alfred P. Sloan Foundation Research Fellow.

3.1. INTRODUCTION

Our present understanding of liquid helium [1–12] owes much to experimental and theoretical studies of vortices [13–17a] and ions [18–20]. Although such inherently nonuniform states are more complicated than those of the bulk fluid, they correspondingly can offer new insights. For example, quantized circulation [21, 22] in ^{4}He (II) implies the existence of macroscopic correlations, and spatially varying velocity fields have been crucial in verifying these long-range effects. Ions also have yielded valuable information on the properties of liquid helium. In particular, their internal structure characterizes the response of the helium to an external charge. Furthermore, the mobility of the resulting ion complex varies inversely with the net drag due to scattering by quasiparticles in the liquid. Experiments with the superfluid phase have been able to separate the effect of rotons, phonons, and ^{3}He impurities [23–25] thus providing a vivid confirmation of Landau's quasiparticle picture [26].

In addition to their value as probes of liquid helium, vortices and ions have an intrinsic interest and merit separate study. Moreover, their mutual interaction has been examined through the capture and escape of ions from vortex lines [27] and through the creation of vortex rings by fast ions [28, 29]. The existence of such interactions implies a close relationship between these apparently distinct entities.

Quantized circulation seems to require long-range order, and such states are thought to occur only in a superfluid phase; Section 3.2 examines the case of vortices in ^{4}He (II). On the other hand, the structure of ions in He depends primarily on *atomic* properties; hence quantum statistics play a less fundamental role in this phenomenon, which is considered in Section 3.3. The final topic is the interaction between ions and vortices, which again refers only to the superfluid phase; Section 3.4 analyzes the situation in ^{4}He (II) and in dilute mixtures.

3.2. VORTICES

Experiments with superfluid ^{4}He have amply confirmed the existence of vortices, which are very similar to those of classical hydrodynamics. They have one special feature, however, that betrays their quantum-mechanical origin. In a Bose system at low temperature, the particles tend to condense into the same quantum state. This behavior is most familiar for an ideal stationary Bose gas, but it also occurs in more general configurations. For example, if all the particles have angular momentum $l\hbar$ about the z-axis, the azimuthal velocity $l\hbar/mr$ varies as r^{-1}, which is typical of a rectilinear vortex [30–32]. Its strength is most simply characterized by the *circulation*

$$\kappa \equiv \oint d\mathbf{s}\cdot\mathbf{v} = \frac{lh}{m} \tag{3.2.1}$$

and we see that κ is quantized in units of $h/m \approx 10^{-3}$ cm^2/sec for ^{4}He. This result, which was first suggested by Onsager [21] and by Feynman [22], may also be inferred from the Bohr–Sommerfeld relation [33]

$$\oint d\mathbf{s}\cdot\mathbf{p} = m\oint d\mathbf{s}\cdot\mathbf{v} = lh, \tag{3.2.2}$$

but its applicability to an N-body system is not entirely obvious. Experiments have confirmed equation (3.2.1) both for circulating flow about fine wires [34, 35] and for vortex rings [28, 29]. Consequently, we shall first adopt a simple phenomenological model, in which the vortices satisfy the equations of classical incompressible hydrodynamics, but with quantized circulation. As seen below, this treatment adequately describes most of the present experiments.

3.2.1. Semiclassical Description

(*a*) *General Theory.* The velocity field of a classical incompressible fluid satisfies the condition

$$\nabla\cdot\mathbf{v} = 0 \tag{3.2.3}$$

throughout the medium. Such a fluid is said to contain one or more vortices if the velocity is everywhere irrotational, apart from certain singular lines, which constitute the vortex cores. As a result, the vorticity

$$\boldsymbol{\zeta} \equiv \nabla \times \mathbf{v} \tag{3.2.4}$$

is confined to the core regions and vanishes elsewhere. The vector identity $\nabla\cdot\boldsymbol{\zeta} = \nabla\cdot(\nabla \times \mathbf{v}) = 0$ then implies (i) that the circulation κ is independent of the path used in evaluating equation (3.2.1), as long as it never crosses the vortex core, and (ii) that the vortex axis either closes on itself or terminates on the boundary of the fluid.

When augmented by suitable boundary conditions, the hydrodynamic equations completely determine the velocity throughout the fluid. Although the solution is straightforward for an unbounded system, it is simpler to note a mathematical analogy with the magnetostatics of current-carrying wires. In particular, the magnetic field $\mathbf{B}$ in vacuum satisfies (in cgs units) [36]

$$\nabla \cdot \mathbf{B} = 0, \tag{3.2.5a}$$

$$\oint d\mathbf{s} \cdot \mathbf{B} = \frac{4\pi I}{c}, \tag{3.2.5b}$$

where I is the current enclosed by the line integral. Since the Biot–Savart law solves this magnetostatic problem, we immediately infer the desired velocity field [37]

$$\mathbf{v}(\mathbf{R}) = \frac{\kappa}{4\pi} \int \frac{d\mathbf{s}' \times (\mathbf{R} - \mathbf{R}')}{|\mathbf{R} - \mathbf{R}'|^3}, \tag{3.2.6}$$

where $\mathbf{R}$ and $\mathbf{R}'$ denote three-dimensional vectors, and the line integral follows the vortex axis in the positive sense.

The dynamical motion may be obtained from the Euler equation for an incompressible fluid [38]

$$\frac{d\mathbf{v}}{dt} \equiv \frac{\partial \mathbf{v}}{\partial t} + (\mathbf{v} \cdot \nabla)\,\mathbf{v} = -\rho^{-1} \nabla p, \tag{3.2.7}$$

where the left side defines the total time derivative seen by an observer moving with the fluid. The vector identity

$$(\mathbf{v} \cdot \nabla)\mathbf{v} = \tfrac{1}{2} \nabla(v^2) - \mathbf{v} \times (\nabla \times \mathbf{v}) \tag{3.2.8}$$

allows us to rewrite the curl of equation (3.2.7) as

$$\frac{\partial \boldsymbol{\zeta}}{\partial t} = \nabla \times (\mathbf{v} \times \boldsymbol{\zeta}). \tag{3.2.9a}$$

Furthermore, the solenoidal nature of $\mathbf{v}$ and $\boldsymbol{\zeta}$ permits the simplification to

$$\frac{d\boldsymbol{\zeta}}{dt} \equiv \frac{\partial \boldsymbol{\zeta}}{\partial t} + (\mathbf{v} \cdot \nabla)\boldsymbol{\zeta} = (\boldsymbol{\zeta} \cdot \nabla)\,\mathbf{v}. \tag{3.2.9b}$$

To interpret this equation, consider a vector $\boldsymbol{\delta}$ joining particles at two neighboring points $\mathbf{r}_1$ and $\mathbf{r}_2 = \mathbf{r}_1 + \boldsymbol{\delta}$. In a small time interval dt, these particles move to $\mathbf{r}_1 + \mathbf{v}(\mathbf{r}_1)\,dt$ and $\mathbf{r}_2 + \mathbf{v}(\mathbf{r}_2)\,dt \approx \mathbf{r}_2 + \mathbf{v}(\mathbf{r}_1)\,dt + [(\boldsymbol{\delta} \cdot \nabla)\mathbf{v}(\mathbf{r})]_{\mathbf{r}=\mathbf{r}_1}\,dt$. Thus the rate of change of the vector $\boldsymbol{\delta}$ is given by

$$\frac{d\boldsymbol{\delta}}{dt} = (\boldsymbol{\delta} \cdot \nabla)\mathbf{v}. \tag{3.2.10}$$

In particular, suppose that $\boldsymbol{\delta}$ initially lies in the vortex core along $\boldsymbol{\zeta}$. Equations (3.2.9b) and (3.2.10) then show that $\boldsymbol{\delta}$ and $\boldsymbol{\zeta}$ remain parallel and in the same ratio for all subsequent times. Consequently, the vortex core always contains the same physical particles, and each element of the core moves with the actual fluid velocity at that point. This result, which also may be obtained from Kelvin's circulation theorem [39], completely determines the dynamics of a system of vortices.

(*b*) *Rectilinear Vortices.* The simplest example is a rectilinear vortex, which may be taken to lie along the z-axis. Equation (3.2.6) is easily evaluated to give [compare the discussion preceding equation (3.2.1)]

$$\mathbf{v} = \frac{\kappa\hat{\theta}}{2\pi r}, \tag{3.2.11}$$

where we now use cylindrical polar coordinates (r, θ, z) and $\hat{\theta}$ is a unit azimuthal vector. Although $\mathbf{v}$ becomes singular as $r \to 0$, it is evident from symmetry that the core remains stationary. Thus a rectilinear vortex has no *self-induced* motion, which also follows directly from the stationary nature of the stream lines. More general vortex configurations have qualitatively similar velocity fields and stream lines, but the resulting motion can become quite complicated.

The energy of a vortex in an incompressible fluid is merely $\int d^3R\,\frac{1}{2}\rho v^2$, but the divergence for small r requires a specific model for the core. If a rectilinear vortex has a hollow core for $r < \xi$ and the velocity field (3.2.11) for $r > \xi$, the kinetic energy K' per unit length becomes

$$K' = \pi\rho \int_\xi^R r\,dr v^2 = \left(\frac{\rho\kappa^2}{4\pi}\right)\ln\left(\frac{R}{\xi}\right), \tag{3.2.12}$$

where R is a macroscopic cutoff that represents either the size of the container or the distance between vortices. More generally, K' takes the form

$$K' = \frac{\rho\kappa^2}{4\pi}\left[\ln\left(\frac{R}{\xi}\right) + \delta\right], \tag{3.2.13}$$

where δ is a dimensionless constant that depends on the structure of the core. As a concrete example, suppose that the core executes solid-body rotation with

$$\mathbf{v} = \left(\frac{\kappa r}{2\pi\xi^2}\right)\hat{\theta}; \quad r < \xi. \tag{3.2.14}$$

An elementary integration gives the core contribution $\rho\kappa^2/16\pi$ to K', so that $\delta = \frac{1}{4}$ in this case.

These considerations are readily extended to a system of parallel rectilinear vortices in unbounded fluid. If the jth vortex has circulation κ_j and intersects the x–y plane at $\mathbf{r}_j$, the integral in equation (3.2.6) now runs along each vortex axis and yields

$$\mathbf{v}(\mathbf{r}) = \sum_j \frac{\kappa_j}{2\pi}\frac{\hat{z} \times (\mathbf{r} - \mathbf{r}_j)}{|\mathbf{r} - \mathbf{r}_j|^2} \tag{3.2.15}$$

for points outside the core regions. The absence of self-induced motion implies that each vortex moves under the influence of all the other vortices, and we therefore obtain

$$\dot{\mathbf{r}}_i = \sum_{j \neq i} \frac{\kappa_j}{2\pi} \frac{\hat{z} \times (\mathbf{r}_i - \mathbf{r}_j)}{|\mathbf{r}_i - \mathbf{r}_j|^2}. \tag{3.2.15}$$

Equations (3.2.15) and (3.2.16) have the notable feature that the velocity field and translational motion are wholly determined by the instantaneous configuration of the vortices. Thus the dynamical equations are *first* order in the time derivatives, unlike the usual dynamics of point masses. An important example is a vortex pair, which consists of two rectilinear vortices with circulation $\pm\kappa$ separated by a distance $2d$. The resulting motion is easily seen to be a uniform translation with speed

$$v = \frac{\kappa}{4\pi d} \tag{3.2.17}$$

along the perpendicular bisector of the plane joining their axes.

Another interesting example is a large number N of identical vortices uniformly distributed throughout a circle of radius R. If $n_v = N/\pi R^2$ is the vortex density, the summation in equation (3.2.15) may be approximated by an integral [40]

$$\begin{aligned}\mathbf{v}(\mathbf{r}) &\approx \frac{n_v\kappa}{2\pi}\,\hat{z} \times \int d^2r' \,\frac{\mathbf{r} - \mathbf{r}'}{|\mathbf{r} - \mathbf{r}'|^2} \\ &= \frac{n_v\kappa}{2\pi}\,\hat{z} \times \hat{r} \int_0^R r'\,dr' \int_0^{2\pi} d\theta' \,\frac{r - r'\cos\theta'}{r^2 - 2rr'\cos\theta' + r'^2} \\ &= n_v\kappa\hat{z} \times \hat{r} r^{-1} \int_0^r r'\,dr' \\ &= \tfrac{1}{2}\, n_v\kappa\hat{z} \times \mathbf{r}.\end{aligned} \tag{3.2.18}$$

The resulting mean velocity is a uniform rotation, and the whole array rotates rigidly with an angular velocity

$$\Omega = \tfrac{1}{2}\, n_v\kappa. \tag{3.2.19}$$

Furthermore, the last line of (3.2.18) shows that the motion at a distance r_0 arises solely from the region $r < r_0$.

This calculation is directly relevant to a rotating vessel of ^{4}He (II), where the equilibrium vortex states necessarily rotate rigidly with the same angular velocity as the walls. Although this conclusion follows from the Hamiltonian dynamics of rectilinear vortices [41], it also is evident physically from the frictional drag on each vortex due to the normal fluid, which itself executes solid-body rotation with the walls. Hence equation (3.2.19) may be reinterpreted as fixing the equilibrium vortex density [22]

$$n_v = \frac{2\Omega}{\kappa}, \tag{3.2.20}$$

which varies linearly with Ω. For ^{4}He (II) rotating at 1 rad/sec, the vortex density is $\approx 2000\ \text{cm}^{-2}$. The existence of such arrays has been demonstrated conclusively by experiments on second-sound attenuation [42] and ion capture [27, 43] in rotating ^{4}He (II). Propagation perpendicular to $\hat{\Omega}$ is attenuated by an amount proportional to Ω, but propagation parallel to $\hat{\Omega}$ is essentially independent of Ω. These experiments will be discussed further in Sections 3.2.3 and 3.4.1.

(*c*) *Vortex Rings.* A vortex ring is one whose core forms a closed circle. This configuration is simply described in cylindrical polar coordinates, with the symmetry axis perpendicular to the plane of the ring. The general equation (3.2.6) may be rewritten exactly as

$$\mathbf{v}(\mathbf{R}) = \nabla \times \mathbf{A}(\mathbf{R}), \tag{3.2.21}$$

where

$$\mathbf{A}(\mathbf{R}) = \frac{\kappa}{4\pi}\int d\mathbf{s}' \frac{1}{|\mathbf{R}-\mathbf{R}'|} \tag{3.2.22}$$

is analogous to the vector potential of magnetostatics. For a vortex ring of radius r_0 in the x–y plane, equation (3.2.22) can be reduced to

$$\mathbf{A}(\mathbf{R}) = A_\theta(r,z)\hat{\theta} = \frac{\kappa r_0 \hat{\theta}}{4\pi}\int_0^{2\pi} d\theta' \frac{\cos\theta'}{(r^2 - 2rr_0\cos\theta' + r_0^2 + z^2)^{1/2}}, \tag{3.2.23}$$

where $\hat{\theta} = -\hat{x}\sin\theta + \hat{y}\cos\theta$ is the unit azimuthal vector at the point $\mathbf{R} = (r, \theta, z)$. This integral can be evaluated in terms of the complete elliptic integrals [44–47]

$$A_\theta(r, z) = \frac{\kappa}{2\pi}\left(\frac{r_0}{r}\right)^{1/2}\left[\left(\frac{2}{k} - k\right)K(k) - \frac{2}{k}E(k)\right], \tag{3.2.24}$$

where

$$k^2 = \frac{4rr_0}{(r+r_0)^2 + z^2}. \tag{3.2.25}$$

The resulting velocity field then follows by differentiation [see equation (3.2.21)]

$$v_r(r, z) = -\frac{\partial}{\partial z}A_\theta(r, z), \quad v_z(r, z) = \frac{1}{r}\frac{\partial}{\partial r}[rA_\theta(r, z)]. \tag{3.2.26}$$

For historical reasons, these relations are frequently expressed in terms of a (Stokes') stream function [47]

$$\Psi(r, z) \equiv -rA_\theta(r, z), \tag{3.2.27}$$

but this interpretation offers no special advantage here.

As for a rectilinear vortex, the energy of the ring is wholly kinetic, and equation (3.2.21) allows us to write

$$K = \tfrac{1}{2}\rho\int d^3Rv^2 = \tfrac{1}{2}\rho\int d^3R\mathbf{v}\cdot(\nabla\times\mathbf{A})$$
$$= \tfrac{1}{2}\rho\int d\mathbf{S}\cdot(\mathbf{A}\times\mathbf{v}) - \tfrac{1}{2}\rho\int d^3R\mathbf{A}\cdot(\nabla\times\mathbf{v}), \qquad (3.2.28)$$

where the second line follows from Gauss' theorem, with the surface integral extending over the boundaries of the fluid. If the ring has a hollow circular core with radius $\xi \ll r_0$, then $\boldsymbol{\zeta} = \nabla\times\mathbf{v}$ vanishes throughout the fluid. Moreover, the surface at infinity makes no contribution because $A \sim R^{-2}$ and $v \sim R^{-3}$ for $R \to \infty$. Hence equation (3.2.28) reduces to an integral over the surface of the core, where $r - r_0$ and z are both of order ξ. In this limit, equation (3.2.24) may be simplified to [45]

$$A_\theta(r, z) \approx \frac{\kappa}{2\pi}\left[\ln\left(\frac{8r_0}{[(r - r_0)^2 + z^2]^{1/2}}\right) - 2\right]; \quad (r - r_0)^2 + z^2 \ll r_0^2 \qquad (3.2.29)$$

and the resulting stream lines are approximately concentric circles enclosing the core with flow speed [compare equation (3.2.11)]

$$v(r, z) \approx \frac{\kappa}{2\pi[(r - r_0)^2 + z^2]^{1/2}}; \quad (r - r_0)^2 + z^2 \ll r_0^2. \qquad (3.2.30)$$

The remaining evaluation of equation (3.2.28) is elementary and gives

$$\left.\begin{aligned} K &= \frac{1}{2}\rho\kappa^2 r_0\left[\ln\left(\frac{8r_0}{\xi}\right) - 2\right] \\ &\approx 2\pi r_0\left(\frac{\rho\kappa^2}{4\pi}\right)\ln\left(\frac{1.083r_0}{\xi}\right)\end{aligned}\right\} \text{hollow core,} \qquad (3.2.31)$$

where the second form exhibits the relation to the energy of an isolated vortex line (3.2.12). If instead the core has uniform vorticity $\kappa/\pi\xi^2$, an analogous calculation yields [47]

$$K = \frac{1}{2}\rho\kappa^2 r_0\left[\ln\left(\frac{8r_0}{\xi}\right) - 2 + \delta\right], \qquad (3.2.32)$$

where again $\delta = \frac{1}{4}$. Alternatively, this latter result follows directly from the considerations used in deriving δ from equation (3.2.14).

The motion of a vortex ring is similar to that of a pair [equation (3.2.17)]; in both cases, a larger configuration moves more slowly because of the diminished velocity field induced at the core. In detail, however, the motion of a ring is more complicated than that of a pair, owing to the curvature of the ring's axis [48]. Indeed, most calculations of the translational velocity are both lengthy and devious. For this reason, we here present a relatively straightforward derivation for a hollow-cored ring, based merely on the motion of a particle at the surface of the core.

The velocity field of a ring of radius r_0 located in the plane $z = z_0$ may be obtained directly from equations (3.2.24) and (3.2.26). In the vicinity of the hollow core, the resulting expressions take the approximate form [45]

$$v_r(r, z) \approx \frac{\kappa(z - z_0)}{2\pi[(r - r_0)^2 + (z - z_0)^2]} - \frac{\kappa}{4\pi r_0} \frac{(r - r_0)(z - z_0)}{(r - r_0)^2 + (z - z_0)^2}, \quad (3.2.33a)$$

$$v_z(r, z) \approx \frac{-\kappa(r - r_0)}{2\pi[(r - r_0)^2 + (z - z_0)^2]} + \frac{\kappa}{4\pi r_0}\left[\ln\left(\frac{8r_0}{[(r - r_0)^2 + (z - z_0)^2]^{1/2}}\right) - \frac{(z - z_0)^2}{(r - r_0)^2 + (z - z_0)^2}\right], \quad (3.2.33b)$$

where the remaining corrections vanish as $(r - r_0)^2 + (z - z_0)^2 \to 0$. The dominant first terms represent circular flow about the core, and a particle on the surface of the core has an approximate angular velocity [compare equation (3.2.30)] $\omega = \kappa/2\pi\xi^2$. This simple motion is perturbed by the second terms, which are of order ξ/r_0 relative to the first ones, leading to two new effects: (i) The ring acquires a slow translational velocity $\mathbf{v} = v\hat{z}$ perpendicular to its plane. (ii) The surface of the core becomes slightly displaced. To exhibit these effects, it is convenient to transform to a (primed) coordinate system moving with the unknown velocity $\mathbf{v}$, where the motion is steady; the corresponding velocity fields become $v_r' = v_r$ and $v_z' = v_z - v$. Moreover, we shall introduce local plane polar coordinates (ρ, χ) about the vortex axis through the equations

$$r - r_0 = \rho \cos \chi,$$
$$z - z_0 = \rho \sin \chi, \quad (3.2.34)$$

and assume that the vortex-core surface is specified by the time-independent equation

$$\rho \equiv [(r - r_0)^2 + (z - z_0)^2]^{1/2} = \xi + \eta(\chi), \quad (3.2.35)$$

where $\eta \ll \xi$. In this moving frame, an expansion of (3.2.33) gives the approximate fluid velocity at the free surface

$$v_r' \approx \frac{\kappa \sin \chi}{2\pi\xi} - \frac{\kappa\eta \sin \chi}{2\pi\xi^2} - \frac{\kappa \sin \chi \cos \chi}{4\pi r_0}, \quad (3.2.36a)$$

$$v_z' \approx -\frac{\kappa \cos \chi}{2\pi\xi} + \frac{\kappa\eta \cos \chi}{2\pi\xi^2} + \frac{\kappa}{4\pi r_0}\left[\ln\left(\frac{8r_0}{\xi}\right) - \sin^2\chi\right] - v. \quad (3.2.36b)$$

The dynamical motion of the fluid obeys two boundary conditions. First, the stationary form of the free surface implies that $\mathbf{v}' \cdot \hat{n}$ vanishes all along the core, where $\hat{n} \approx \hat{\rho} - \hat{\chi}\, d\eta/d\chi$ is the normal to the free surface; a simple calculation then yields the relation

$$\frac{\kappa}{2\pi\xi^2}\frac{d\eta}{d\chi} = -\frac{\kappa \sin\chi}{4\pi r_0}\left[\ln\left(\frac{8r_0}{\xi}\right) - 1\right] + v \sin\chi. \tag{3.2.37}$$

Second, Bernoulli's theorem and hydrostatic equilibrium at the free surface require a constant $(v')^2 = (v_r')^2 + (v_z')^2$ along the core. A first-order expansion with (3.2.36) gives

$$\frac{\kappa\eta}{2\pi\xi^2} = -\frac{\kappa \cos\chi}{4\pi r_0}\ln\left(\frac{8r_0}{\xi}\right) + v\cos\chi + \text{const}, \tag{3.2.38}$$

where the constant may be absorbed in the overall pressure. Simultaneous solution of (3.2.37) and (3.2.38) provides the self-consistent expression [49, 50]

$$\eta \approx -\xi^2\frac{\cos\chi}{4r_0}, \tag{3.2.39}$$

$$v = \left(\frac{\kappa}{4\pi r_0}\right)\left[\ln\left(\frac{8r_0}{\xi}\right) - \frac{1}{2}\right] \text{hollow core.} \tag{3.2.40}$$

The first shows that the core remains circular with radius ξ, but its center is shifted to the point $r = r_0 - \xi^2/4r_0$, $z = z_0$; the second determines the translational velocity of a classical vortex ring with a hollow core. More generally, Lamb considers a large ring ($\xi \ll r_0$) with essentially arbitrary core and proves that [47, 51, 52]

$$\begin{aligned} v &= \frac{K}{2\pi\rho\kappa r_0^2} + \frac{3\kappa}{8\pi r_0} \\ &= \left(\frac{\kappa}{4\pi r_0}\right)\left[\ln\left(\frac{8r_0}{\xi}\right) - \frac{1}{2} + \delta\right], \end{aligned} \tag{3.2.41}$$

where the second line follows from (3.2.32). If the core has uniform vorticity (solid-body rotation with $\delta = \frac{1}{4}$), the particular expression was first stated by Kelvin [53].

These results have been applied to quantized vortex rings in ^{4}He (II) [28, 29], where $\kappa = h/m$ [see equation (3.2.1)]. Apart from logarithmic factors, K and v vary linearly and inversely with r_0; thus their product varies only slowly, and a graph of K *vs.* v should be essentially hyperbolic. Rayfield and Reif have been able to create such vortex rings with fast ions whose energy greatly exceeds k_BT. The ring and ion remain bound, so that the ring may be considered to have a charged core (see Section 3.4.4). This property allows an unambiguous determination of the ring's velocity v with a time-of-flight spectrometer, and the energy K is fixed by the original accelerating potential. At low temperature ($T \lesssim 0.7$ K), the ring experiences only a small frictional drag and moves like a classical vortex ring in unbounded inviscid fluid. Figure 3.1 shows the data along with the curve obtained by eliminating r_0

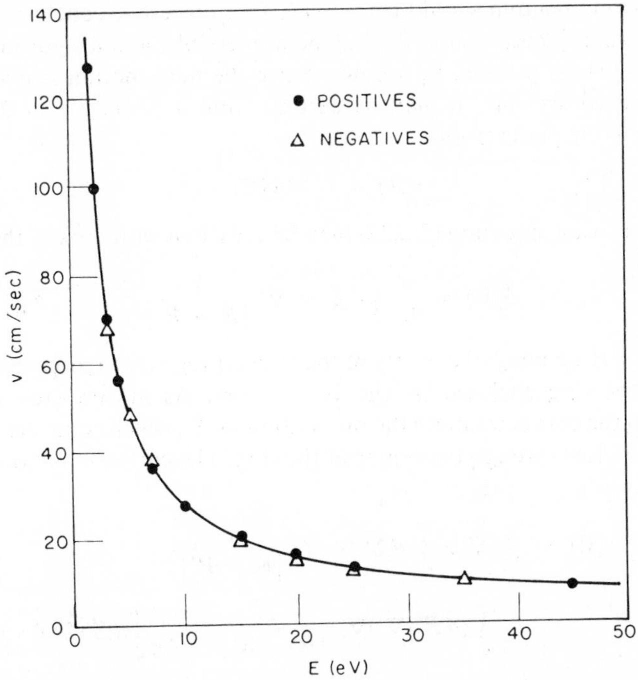

FIGURE 3.1. Dynamical properties of a singly quantized charged vortex ring; the curve is obtained by eliminating r_0 from equations (3.2.32) and (3.2.41), with $\delta = 1/4$. [From G. W. Rayfield and F. Reif, *Phys. Rev. A* **136**, 1194 (1964), Fig.2. Reprinted by permission of the authors and the American Institute of Physics.]

between equations (3.2.32) and (3.2.41). With $\delta = \frac{1}{4}$, the best fit yielded $\kappa = (1.00 \pm 0.03) \times 10^{-3}$ cm^2/sec and $\xi = (1.28 \pm 0.13)$Å. In all cases, the rings were singly quantized.

A related series of experiments used collision and transmission techniques to study the size of vortex rings. When a uniform beam of monoenergetic rings is directed at a grid [54], the transmission depends on the radius of the ring and the size of the aperture. For low energies (small rings), the transmission coefficient was indpendent of energy, but it fell sharply above a critical value. As might be expected, the corresponding critical diameter $2r_0$ [see equation (3.2.32)] was comparable with the size of the opening, thus providing a rather direct measure of the ring's geometric size. Similar experiments investigated the collision cross section between rings and other rings [55] or rectilinear vortices [56, 57]. Although the theoretical analysis becomes quite intricate, the experimental data generally confirm equation (3.2.32).

We have proved that a vortex ring moves through the fluid with a velocity of order $(\kappa/4\pi r_0)$ ln (r_0/ξ). If it has an effective mass of order ρr_0^3, the corresponding momentum should be $\approx \rho\kappa r_0^2$. This suggestive conclusion cannot be taken literally because the fluid is incompressible, and a rigorous discussion must instead proceed as follows: Since the fluid motion is irrotational outside the vortex core, it may be derived from a velocity potential $\Phi(\mathbf{R})$ according to the prescription

$$\mathbf{v}(\mathbf{R}) = -\nabla\Phi(\mathbf{R}). \tag{3.2.42}$$

For a vortex ring, equation (3.2.22) may be rewritten with Stokes' theorem

$$\mathbf{A}(\mathbf{R}) = \frac{\kappa}{4\pi}\int d\boldsymbol{\Sigma}' \times \nabla' \frac{1}{|\mathbf{R} - \mathbf{R}'|}, \tag{3.2.43}$$

where the surface integral consists of the circular region of radius $\approx r_0$ in the plane of the ring enclosed by the vortex core. As in Ampère's law, the vorticity in the core determines the orientation of $\boldsymbol{\Sigma}'$, which coincides with the direction of flow through the center of the ring. The curl of equation (3.2.43) gives [46]

$$\begin{aligned}\mathbf{v}(\mathbf{R}) &= \frac{\kappa}{4\pi}\nabla \times \int d\boldsymbol{\Sigma}' \times \nabla' \frac{1}{|\mathbf{R} - \mathbf{R}'|} \\ &= \frac{\kappa}{4\pi}\int (d\boldsymbol{\Sigma}'\cdot\nabla')\nabla' \frac{1}{|\mathbf{R} - \mathbf{R}'|} - \frac{\kappa}{4\pi}\int d\boldsymbol{\Sigma}'(\nabla')^2 \frac{1}{|\mathbf{R} - \mathbf{R}'|} \\ &= -\frac{\kappa}{4\pi}\nabla\int d\boldsymbol{\Sigma}'\cdot\frac{\mathbf{R} - \mathbf{R}'}{|\mathbf{R} - \mathbf{R}'|^3},\end{aligned} \tag{3.2.44}$$

where we have used the relations $\nabla'|\mathbf{R} - \mathbf{R}'|^{-1} = -\nabla|\mathbf{R} - \mathbf{R}'|^{-1}$ and $\nabla^2|\mathbf{R} - \mathbf{R}'|^{-1} = 0$ for $\mathbf{R} \neq \mathbf{R}'$. Comparison of equations (3.2.42) and (3.2.44) yields [58]

$$\Phi(\mathbf{R}) = \frac{\kappa}{4\pi}\int d\boldsymbol{\Sigma}'\cdot\frac{\mathbf{R} - \mathbf{R}'}{|\mathbf{R} - \mathbf{R}'|^3}, \tag{3.2.45}$$

which has a simple interpretation as $\kappa/4\pi$ times the solid angle subtended at $\mathbf{R}$ by the oriented surface $\boldsymbol{\Sigma}'$ enclosed by the core of the ring. In particular, Φ is everywhere continuous except in the plane of the ring, where it has the limiting value

$$\lim_{\eta\to 0}\Phi(r, \theta, z_0 \pm \eta) = \begin{cases} \pm\frac{1}{2}\kappa; & r < r_0 \\ 0; & r > r_0 \end{cases} \tag{3.2.46}$$

Draw an imaginary fixed surface S_0 around the instantaneous position of the ring, enclosing a disc-shaped region of radius just greater than r_0 and negligible thickness. Let V be the volume exterior to S_0 and bounded externally by a large spherical surface S_1 of radius R_1. The momentum balance in this region is expressed by the equation

$$\frac{\partial}{\partial t}\int_V d^3R\rho\mathbf{v} = -\int d\mathbf{S}\,p - \int (d\mathbf{S}\cdot\mathbf{v})\,\rho\mathbf{v}, \tag{3.2.47}$$

where the first term on the right is the hydrostatic force and the second represents the convection of momentum density $\rho\mathbf{v}$. Here the surface integral includes both S_0 and S_1, with the normal pointing outward from the volume V. Since the motion is solenoidal and irrotational in V, we may use Bernoulli's theorem [59]

$$p = \rho\frac{\partial\Phi}{\partial t} - \frac{1}{2}\rho v^2 + \text{const} \tag{3.2.48}$$

to rewrite equation (3.2.47) as

$$\frac{\partial}{\partial t}\int_V d^3R\rho\mathbf{v} = -\rho\int d\mathbf{S}\frac{\partial\Phi}{\partial t} + \frac{1}{2}\rho\int d\mathbf{S}v^2 - \rho\int (d\mathbf{S}\cdot\mathbf{v})\mathbf{v}. \tag{3.2.49}$$

It is easy to see that the surface S_1 makes negligible contribution as $R_1 \to \infty$, because $\mathbf{v}$ and $\partial\Phi/\partial t$ are each of order R_1^{-3}. (Φ itself is of order R_1^{-2}, but the time derivative introduces an additional factor R_1^{-1}) Equation (3.2.49) then reduces to an integral over the *fixed* surface S_0

$$\frac{\partial}{\partial t}\int_V d^3R\rho\mathbf{v} = -\frac{\partial}{\partial t}\int_{S_0} d\mathbf{S}\rho\Phi + \frac{1}{2}\rho\int_{S_0} d\mathbf{S}v^2 - \rho\int_{S_0} (d\mathbf{S}\cdot\mathbf{v})\mathbf{v}. \tag{3.2.50}$$

The last two terms vanish identically because v is continuous through the plane of the ring, leaving [60]

$$\frac{\partial}{\partial t}\int d^3R\rho\mathbf{v} = \frac{\partial}{\partial t}\mathbf{I}, \tag{3.2.51}$$

where

$$\mathbf{I} = -\int_{S_0} d\mathbf{S}\rho\Phi \tag{3.2.52}$$

is known as the *impulse* of the vortex ring. This equation asserts that the *rate of change of the impulse* exactly equals the *rate of change of the total momentum of the fluid.* In the present case of a single ring in unbounded fluid, an elementary calculation with (3.2.46) and (3.2.52) yields [47, 58]

$$\mathbf{I} = \pi\rho\kappa r_0^2\hat{v}, \tag{3.2.53}$$

where $\hat{v}$ is a unit vector parallel to the translational velocity of the ring. Hence any external force that alters either the radius r_0 or the orientation $\hat{v}$ of the ring necessarily produces a corresponding change in the total linear momentum of the fluid; conversely, each side of equation (3.2.51) vanishes identically for a *free* vortex ring in an infinite domain, where r_0 and $\hat{v}$ are constant.

It is essential to realize that equation (3.2.51) relates *changes* of impulse and momentum, rather than the quantities themselves. Indeed, any attempt to calculate the total linear momentum from the definition $\int d^3R\rho\mathbf{v}$ leads to indeterminate integrals with constant contributions that depend on the shape of the bounding surface at infinity [61–63]. In contrast, the impulse (3.2.52) or (3.2.53) is always well defined, which has sometimes led to the use of **I** as a measure of the corresponding momentum. Although such an identification can never be generally correct, it may be justified in certain cases [64–69]. For example, consider an infinitesimal charged ring that grows from a small initial radius r_i to a large final radius $r_f \gg r_i$ under the influence of an external electric field. Equations (3.2.51) and (3.2.53) show that the total momentum of the fluid increases by $\pi\rho\kappa(r_f^2 - r_i^2)$; to the extent that the initial impulse and momentum are negligible, we see that $\pi\rho\kappa r_f^2$ correctly represents the momentum associated with the presence of the final vortex ring of radius r_f. Alternatively, if a large ring shrinks to a tiny one under the influence of a weak frictional force, the decrease in impulse precisely equals the momentum removed from the fluid. Huggins [65] has also examined the more complicated situation in a tube of radius R, where an initial ring of radius r_i can disappear either by growing to the outer wall or by shrinking to the center. The change of momentum in the two processes is $\pi\rho\kappa(r_i^2 - R^2)$ and $\pi\rho\kappa r_i^2$, respectively; if the final vortex-free state is assigned zero momentum, the initial momentum of the ring cannot be uniquely defined.

This equality between changes in impulse and momentum has been demonstrated by Gamota and Barmatz [70], who directed a beam of charged vortex rings at a flexible diaphragm and measured the deflection for different values of the ring energy K. As might be expected from equations (3.2.32) and (3.2.53), the net force varied approximately as K^2. This result may be understood by assuming that each ring disintegrated near the wall, transferring its momentum and impulse, but the interpretation involves several subtle points [71–72a].

Roberts and Donnelly [73] have pointed out a relation between impulse, energy, and velocity of a vortex ring. If the radius of the ring changes by dr_0, the impulse of the ring and the momentum of the fluid each change by $dI = dP = 2\pi\rho\kappa r_0\, dr_0$. To compute the corresponding change in the total energy, they assumed a fixed core volume $\mathscr{V}_c = 2\pi^2\xi^2 r_0$, which eliminates the work involved in altering the core against an external pressure. In this way, ξ becomes a function of r_0, and equation (3.2.32) gives $dK = \frac{1}{2}\rho\kappa \times [\ln(8r_0/\xi) - \frac{1}{2} + \delta]$. Comparison with equation (3.2.41) yields

$$v = \frac{dK}{dI} = \frac{dK}{dP}, \tag{3.2.54}$$

so that K serves as a "quasi-Hamiltonian" for a vortex ring. This identification had previously been verified experimentally through the motion of vortex rings in electric [28, 29, 73a] and magnetic [74] fields. A similar result holds for other models such as a fixed core radius ξ, but K must then be augmented by the work done in increasing the core volume.

Recent studies of interacting charged vortex rings have provided further confirmation of this Hamiltonian analysis. The system is treated as a charged fluid or plasma [75, 76], each particle having the energy and impulse (momentum) of a vortex ring [(3.2.32) and (3.2.53)], with a corresponding negative effective mass, defined as dP/dv. This feature leads to two striking predictions: (i) An arbitrary initial bunch ultimately evolves to a universal final form; and (ii) an extended beam exhibits unstable collective modes for certain wavenumbers. Experiments verify both effects [75, 77].

(*d*) *Vortex Waves.* In one of the very early papers on vortices, Kelvin [78] investigated the stability of a rectilinear vortex with respect to oscillations about its equilibrium configuration. On dimensional grounds, the natural frequency is $\kappa/2\pi\xi^2$, which corresponds in ^{4}He (II) ($\xi \approx 1$ Å) to a rather high temperature $\hbar^2/m\xi^2 k_B \approx 10$ K. Consequently, only modes with frequencies considerably less than $\kappa/2\pi\xi^2$ will be thermally excited in practical situations. It is convenient to use cylindrical polar coordinates, with a harmonic perturbation proportional to $\exp i(kz + l\theta - \omega t)$. Kelvin then showed that the only low-lying mode had $l = \pm 1$ and $k\xi \ll 1$, with the frequency spectrum [17, 78–80]

$$\omega \approx \mp \left(\frac{k\kappa^2}{4\pi}\right)\left[\ln\left(\frac{2}{k\xi}\right) - \gamma + \delta\right], \tag{3.2.55}$$

where $\gamma \approx 0.577$ (Euler's constant) and δ again characterizes the core structure [see (3.2.40) and (3.2.41)]. These modes are circularly polarized with a helical deformation that propagates along the vortex axis; each element of the core performs a small circular orbit in the plane perpendicular to the undeformed axis. On the other hand, the short-wavelength behavior is more complicated, for the core structure affects even the qualitative form of the spectrum [17, 78, 81]. Since such classical models are unrealistic for ^{4}He, we shall not pursue this matter further.

In Kelvin's treatment, the vortex motion emerges only after a lengthy analysis of the linearized hydrodynamic equations. Fortunately, a more physical description follows directly from a *qualitative* analogy between a vortex line and a fine elastic filament. The basic observation is that a cylinder with circulation κ moving with velocity $\mathbf{v}_L$ through an ideal fluid experiences a Magnus force per unit length [82–84]

$$\mathbf{F}'_M = \rho_s \kappa \hat{\zeta} \times (\mathbf{v}_L - \mathbf{v}_s), \tag{3.2.56}$$

where $\hat{\zeta}$ is a unit vector along the axis of the cylinder and ρ_s and $\mathbf{v}_s$ are the density and velocity of the ideal fluid. This expression remains valid even for a massless cylinder, which serves as a simple model for a vortex line. Suppose that the axis is displaced from its equilibrium position by a two-dimensional vector $\mathbf{u}(z, t)$ perpendicular to $\hat{\zeta}$. If the vortex has an effective tension T, a restoring force $T\,\partial^2\mathbf{u}/\partial z^2$ acts on a unit length of the core. Since the vortex has negligible mass [29], the vanishing of the total force yields the dynamical equation [14, 17]

$$T\frac{\partial^2\mathbf{u}}{\partial z^2} + \rho_s\kappa\hat{\zeta} \times \frac{\partial\mathbf{u}}{\partial t} = 0 \tag{3.2.57}$$

because $\mathbf{v}_s = 0$. This equation predicts circularly polarized modes $\mathbf{u}(z, t) = u(\hat{x} \mp i\hat{y})e^{i(kz-\omega t)}$, with a frequency

$$\omega = \pm\frac{Tk^2}{\rho_s\kappa}. \tag{3.2.58}$$

Although T is frequently identified with the energy per unit length of line (3.2.12), a more detailed hydrodynamic analysis for a hollow core gives an effective long-wavelength tension [85]

$$T_{\text{eff}} = \left(\frac{\rho_s\kappa^2}{4\pi}\right)\left[\ln\left(\frac{2}{k\xi}\right) - \gamma\right]. \tag{3.2.59}$$

Equations (3.2.58) and (3.2.59) together reproduce equation (3.2.55) [86].

The Magnus force has been used to describe the vibration modes of a fine wire in superfluid helium. If the fluid is stationary, the degenerate modes may be resolved in any two perpendicular planes. The presence of a circulation κ about the wire lifts the degeneracy and splits the modes by an amount [34, 35, 85, 87]

$$\Delta\omega = \frac{\kappa\rho_s}{\pi\xi^2(\rho_s + \rho')}, \tag{3.2.60}$$

where ρ' is the mass density of the wire. Vinen [34, 35] used this expression to determine the stable values of circulation in ^{4}He (II). His data exhibited a marked tendency for the value h/m, providing the first clear evidence for quantized circulation. Subsequently, Whitmore and Zimmermann [87] observed the values $2h/m$ and $3h/m$ and also verified the temperature dependence of $\Delta\omega$ implicit in the factor $\rho_s(T)$.

These experiments relied on a small but macroscopic wire to render the fluid multiply connected, so that the resulting oscillations differ in principle from those on microscopic vortex lines. Although the vibration of a solitary vortex remains undetected, related effects occur in bulk rotating ^{4}He (II), where the fluid contains an array of rectilinear vortices with axes parallel to $\hat{\Omega}$ and a density $2\Omega/\kappa$ [see (3.2.20)]. The small-oscillation spectrum of such an array depends on the wave vector $\mathbf{k} = k_{\|}\hat{\Omega} + \mathbf{k}_{\perp}$. If the vortices form a regular two-dimensional lattice, then $\mathbf{k}_{\perp}$ must lie inside the first Brillouin zone with a maximum magnitude $\approx O(b^{-1})$, where $b = (n_v \pi)^{-1/2}$ characterizes the intervortex spacing; in contrast, $k_\| b$ can become large. The resulting dispersion relation depends on the precise lattice structure. For example, Tkachenko [88] proved that triangular and square lattices are respectively stable and unstable for motion without bending ($k_\| = 0$). The analysis is more difficult for general **k**, but the following approximation usually suffices [89]

$$\omega = 2\Omega\left(\frac{k_\|^2 + \frac{1}{16}k_\perp^4 b^2}{k_\|^2 + k_\perp^2}\right)^{1/2} + \frac{1}{4}k_\|^2 b^2 \ln\left(\frac{1}{k_\| \xi}\right). \tag{3.2.61}$$

If $k_\| = 0$, we recover Tkachenko's result [88]

$$\omega \approx \tfrac{1}{2}\Omega k_\perp b; \quad k_\| = 0, \quad k_\perp b \ll 1 \tag{3.2.62}$$

which represents an elastic wave in the vortex lattice. Conversely, if $k_\perp = 0$, then [86, 89, 90]

$$\omega \approx 2\Omega + \left(\frac{\kappa}{4\pi}\right)k_\|^2 \ln\left(\frac{1}{k_\| \xi}\right). \tag{3.2.63}$$

Hall used this relation in studying the oscillations of a disk suspended in rotating helium. If the resonant frequency ω of the disk was much smaller than the angular velocity Ω of the container, the imaginary wavenumber in (3.2.63) indicates a localized disturbance at the surface of the disk. For $\omega \gg \Omega$, on the other hand, typical resonance effects occurred whenever the distance between the free surface and the disk could accommodate simple discrete multiples of the wavelength $2\pi/k_\|$ [90, 91]. In this way, Hall verified equation (3.2.63) and even obtained an estimate $\xi \approx 7$ Å for the core size [92].

Hall's original analysis has been extended to include the normal component, [15, 16, 93–96] which also can exhibit its own circularly polarized vibration modes with a frequency [97] $\omega = 2\Omega k_\| (k_\|^2 + k_\perp^2)^{-1/2}$. Although this expression is formally identical with the classical limit of equation (3.2.61) as $b = (\hbar/m\Omega)^{1/2} \to 0$, the dynamical motion of a rotating viscous fluid differs considerably from that observed by Hall, thus permitting a clear distinction between normal and superfluid waves.

The thermal excitation of vortex waves means that rotating ^{4}He (II) has an additional entropy and heat capacity [89, 98], A simple Debye theory of this effect gives an extra contribution per unit volume

$$S_{\text{vor}} \approx 3\zeta\left(\frac{3}{2}\right)n_v k_B\left(\frac{k_B T}{2\hbar\kappa}\right)^{1/2}\left[\ln\left(\frac{\hbar\kappa}{\xi^2 k_B T}\right)\right]^{-1/2},$$

$$C_{\text{vor}} \approx \tfrac{1}{2} S_{\text{vor}}, \tag{3.2.64}$$

where $\zeta(\frac{3}{2}) \approx 2.612$ is the Riemann zeta function and n_v is the vortex density. Unfortunately, the phonon specific heat greatly exceeds C_{vor} except at very low temperatures and large n_v; even at $T = 10$ mK and $\Omega = 10^3$ rad/sec, for example, we find $C_{\text{vor}}/C_{ph} = 0.01$. The same vibration modes also induce zero-point oscillations of the core. For ^{4}He at zero temperature, the resulting root-mean-square displacement is $\approx (\pi n_4 \xi)^{-1/2} \approx 3.8$ Å, where n_4 is the bulk number density and we take $\xi \approx 1$ Å [89]. Alternatively, this expression may be used to determine a self-consistent core radius $\xi_{sc} = (\pi n_4)^{-1/3} \approx 2.4$ Å, comparable with the interparticle spacing [81].

Similar theoretical analyses predict the oscillation modes of a vortex ring. The toroidal symmetry complicates the problem considerably, however, and we merely refer to the original papers [99]. At present, such states of vortex rings in ^{4}He (II) have not been observed.

(*e*) *Effect of Boundaries.* Our previous treatment of vortices in infinite fluid will now be extended to include the walls, which perturb the velocity field and hence alter the vortex motion. Since the complete boundary-value problem is very difficult, most work has been restricted to two-dimensional arrays of rectilinear vortices. With this simplification, it becomes feasible to study the dynamics of vortices in several geometries, but the resulting orbits are usually irrelevant to experiments with equilibrium rotating helium. In particular, we noted above equation (3.2.20) that an equilibrium vortex array must rotate rigidly with the angular velocity Ω of the walls [41]. This condition immediately eliminates most textbook examples as equilibrium states because they almost invariably imply relative motion of the vortices [30].

We therefore consider the more physical problem of finding the equilibrium states of rotating helium at a given temperature T, volume V, and angular velocity Ω, which are obtained by minimizing the modified free energy

$$\bar{F}(T, V, \Omega) = F(T, V) - \boldsymbol{\Omega} \cdot \mathbf{L}(T, V). \tag{3.2.65}$$

Here $F(T, V)$ is the Helmholtz free energy and $\mathbf{L}$ is the total angular momentum. The validity of this expression follows most simply by noting that true equilibrium can exist only in the rotating frame, because otherwise the time-dependent boundary forces would do work on the system. The corresponding Hamiltonian becomes $\mathrm{H} - \boldsymbol{\Omega} \cdot \mathbf{L}$, and the usual techniques of statistical mechanics then give equation (3.2.65) [100]. Alternatively, if the total angular momentum is specified, then the equilibrium state maximizes the entropy (in the microcanonical ensemble) [101] or minimizes the Helmholtz free energy (in the canonical ensemble) [13, 17, 102], subject to the additional constraint of fixed $\mathbf{L}$. The associated Lagrange multiplier is readily identified with $\boldsymbol{\Omega}$, which again yields (3.2.65). For ^{4}He (II) at low temperatures, the variational quantity reduces to

$$\bar{F} = \int d^3r \left[\tfrac{1}{2}\rho_s v_s^2 - \boldsymbol{\Omega}\cdot\mathbf{r}\times\rho_s\mathbf{v}_s\right], \tag{3.2.66}$$

because the normal fluid automatically takes the desired form $\mathbf{v}_n = \boldsymbol{\Omega}\times\mathbf{r}$; equation (3.2.66) must be supplemented by the restriction of irrotational flow and quantized circulation.

If the walls of the container are invariant under translation along $\hat{\Omega}$, the motion becomes two dimensional, and the minimum principle (3.2.66) predicts a critical angular velocity Ω_{c1} for the creation of a singly quantized vortex line. The simplest example is a right circular cylinder of radius R rotating about its symmetry axis. In this case, the only *purely* irrotational state has $\mathbf{v}_s = 0$ so that the fluid remains stationary for low Ω. At the critical value [34, 103–104a]

$$\Omega_{c1} \approx \left(\frac{\kappa}{2\pi R^2}\right)\ln\left(\frac{R}{\xi}\right) = \left(\frac{\hbar}{mR^2}\right)\ln\left(\frac{R}{\xi}\right) \tag{3.2.67}$$

a single vortex first has a lower free energy; it lies on the symmetry axis, maximizing the angular momentum. More generally, Hess [41] calculated the free energy for a vortex a distance r_0 off axis and showed that $\bar{F}$ increases with increasing r_0 except very near the wall or for very low Ω (see also Reference [105]). Above Ω_{c1}, the system undergoes a sequence of discrete transitions that represent the addition of more vortex lines. This behavior was first studied experimentally by Hess and Fairbank [106], who measured the angular momentum of the superfluid as a function of the angular velocity of the container. Although they were unable to resolve the separate steps, they confirmed that the fluid remains stationary for $\Omega \lesssim \Omega_{c1}$ and verified the qualitative increase in angular momentum for $\Omega > \Omega_{c1}$. More recently Packard and Sanders [105] detected the presence of vortices through their tendency to trap free charges (ions) (see Section 3.4.1). The resulting step structure strikingly confirmed the vortex model; moreover, their data indicated appreciable hysteresis between increasing and decreasing Ω.

A similar analysis is possible for an annulus rotating about its symmetry axis. There is one new feature, however, because the multiply connected boundary allows purely irrotational states with quantized circulation about the inner cylinder. These states represent the true equilibrium for low Ω, where theory predicts that the actual quantum numbers increase in unit jumps with increasing Ω [107, 108]. Eventually, physical vortices appear at an angular velocity Ω_{c1} that depends on the geometry of the container. If the width d of the gap is small compared to the mean radius R, a detailed calculation [108, 109] gives $\Omega_{c1} \approx (\kappa/\pi d^2)\ln(2d/\pi\xi)$. To study these effects, Bendt and Donnelly [110, 111] measured the attenuation of radial second-sound modes in the gap of a narrow annulus. Since vortices attenuate second sound (see Section 3.2.3), the experiment was able to verify the predicted critical angular velocity. For reasons that remain obscure, hysteresis seems less evident in a narrow annulus [110, 111] than in a wide one or in a simply connected cylinder. Indeed, Vinen [35] was unable to create doubly quantized circulation about a wire, even though he far exceeded the theoretical critical angular velocity (see, however, Reference [87]).

It is also interesting to consider simply connected containers that are not invariant under rotation. The rotating walls now have a finite normal component of velocity, which induces *irrotational* (*circulation-free*) flow; nevertheless, the angular momentum is nonzero. This conclusion has sometimes caused difficulty, for if the fluid has no circulation about any path encircling the axis of rotation, how can it have angular momentum? To resolve this question, consider a circular path of radius r about the rotation axis and lying wholly in the fluid. By assumption, the associated circulation

$$\Gamma(r) = \int_0^{2\pi} r\, d\theta\, v_\theta(r,\theta) \tag{3.2.68}$$

vanishes for all suitable r, and the angular momentum per unit length becomes

$$L_z' = \int d^2r\, \mathbf{r} \times \mathbf{v} = \int r^2\, dr \int d\theta\, v_\theta(r, \theta). \tag{3.2.69}$$

A combination with equation (3.2.68) shows that the fluid enclosed in *any* circle inscribed by the boundary makes no contribution to L_z'; consequently, the whole angular momentum resides in the extremities of the container. This behavior is easily verified for an elliptic cylinder with semiaxes a and b ($a > b$) rotating about its symmetry axis with angular velocity Ω [62, 112]. The resulting velocity field is strictly irrotational, yet each particle acquires a mean angular velocity in the direction of rotation. Detailed calculations show that the total angular momentum is reduced from the value for solid-body rotation by a factor $[(a^2 - b^2)/(a^2 + b^2)]^2$, which correctly vanishes for a circular container. In the opposite limit ($a \gg b$), L_z' approaches that for a rotating solid plate, as expected from physical considerations.

The preceding discussion implies that an arbitrary multiply connected container of ^{4}He (II) can support three distinct types of superfluid velocity fields: circulation-free irrotational flow caused by the moving boundaries, quantized but irrotational circulation about the inner boundaries, and quantized vortex lines in the bulk fluid. Most studies of such flow measured the persistent current in a stationary container [113–123]. These experiments, which are discussed in more detail in Section 3.2.4, could detect only the latter two flow states. More recently, Kojima *et al.* [124] have also observed the first type with a Doppler-shifted fourth-sound resonance that measures the mean velocity field while the container is actually rotating. In this case, the external rotation speed Ω_{ex} induces a finite angular velocity Ω_{ind} in the fluid, precisely because the narrow powder-filled annulus is not invariant under rotation. Moreover, the observed Ω_{ind} varies linearly and reversibly with Ω_{ex} up to a critical angular velocity of order $\hbar/mRd$, above which quantized circulation or vortices enter the fluid. If Ω_{ex} is subsequently reduced below the critical value, Ω_{ind} exhibits hysteresis, very like that observed in highly irreversible type-II superconductors.

Most of the above discussion has considered arrays with few vortices, when symmetry makes the spatial arrangement evident. As Ω increases, however, the number of vortices becomes so large that an exact calculation becomes impractical. For this reason, analyses generally rely on statistical methods, in which some large number of vortices seeks its equilibrium from an initial (usually random) configuration. Such a program has been carried out in detail for a circular cylinder [125, 126], where the equilibrium states generally consist of vortices arranged in concentric circles with ≈ 6 nearest neighbors. Although the resulting configurations sometimes have the six-fold symmetry of a triangular lattice, this feature is by no means universal. At present, some indirect experimental evidence indicates that the vortex array in bulk rotating helium has the short-range order of a close-packed lattice [127], but no convincing demonstration of long-range triangular order has yet been given. One promising approach to this problem is direct photography [128], either of the induced dimples on a rotating free surface [129–131] or after decorating the vortex axes with neutral [132–135] or charged [105, 136, 137] impurities. This elusive and difficult experiment remains one of the most interesting challenges in low-temperature physics (see also Addendum).

When the angular velocity increases still further, the vortex configuration again becomes relatively simple, for it approaches a uniform array with density $2\Omega/\kappa$ [22, 138]. The first experiment on this regime was performed by Osborne [139] (see also References [140–142]), who actually tried to detect vortex-free states. Since Landau's original theory [26] predicted that the superfluid was everywhere irrotational, Osborne expected that $\mathbf{v}_s$ would vanish in a simply connected rotating cylinder, giving a reduced meniscus proportional to ρ_n/ρ. In fact, he observed a purely classical free surface, which is now interpreted as arising from the combined effect of many vortices, each contributing to the angular momentum [22]. The existence of this dense array is now well established through its attenuation of second sound and ion currents, as discussed in Sections 3.2.3 and 3.4.1.

Although the vortices effectively fill any rotating container, the presence of walls slightly alters their spatial distribution, giving a narrow vortex-free strip near the boundary [13, 17, 125, 143]. Such effects have been detected both with ions [144] and with second sound [145], in reasonable agreement with the theoretical predictions.

Another case of boundary effects occurs when a vortex ring approaches a wall. From the method of images, this system is equivalent to two equal and opposite rings moving toward each other. The induced velocity field acts to enlarge the ring, simultaneously slowing it down [32, 44, 71, 146]. At present, neutral vortex rings have not been detected directly [147], and it may be feasible to study them through the resulting (very small) pressure on the wall [71] (see also Addendum).

We shall close this section with a brief mention of rotating spherical boundaries, which are relevant to the (assumed) superfluid interior of a neutron star or pulsar [148]. The theoretical analysis is complicated by the mixed cylindrical and spherical symmetry, and, indeed, no one has calculated the equilibrium configuration of one vortex line displaced from the symmetry axis nor even proved that Ω_{c1} is nonzero in this geometry. For many vortices, the array presumably approximates a uniform rotation except for a thin surface region, but this question also remains unexplored. Furthermore, the continued slow deceleration of the pulsar crust implies interesting time-dependent and secular effects, which have only been discussed qualitatively [149–150a]. For these reasons, the spherical geometry merits further examination.

3.2.2. Quantum-Mechanical Description

The preceding section dealt exclusively with a semiclassical vortex theory, in which the usual equations of inviscid irrotational incompressible hydrodynamics were supplemented by the condition of quantized circulation. While such an approach has had notable success, it cannot clarify the origin of quantized circulation, nor can it describe the vortex core. Consequently, we now consider a more fundamental treatment that incorporates the quantum-mechanical properties from the start rather than as afterthoughts. It is simplest to begin with a weakly interacting Bose gas and then turn to the more difficult problem of ^{4}He (II).

(*a*) *Weakly Interacting Bose Gas.* The low-lying states of a perfect Bose gas have the characteristic feature that nearly all the particles occupy the same single-particle state. This qualitative behavior persists for sufficiently weak interactions, which motivates the Hartree approximation introduced by Gross [151, 152] and by Pitaevskii [153]. They write the N-body wave function as a product of normalized single-particle ones.

$$\Psi_H(\mathbf{r}_1, \cdots, \mathbf{r}_N) = \prod_{j=1}^{N} \psi(\mathbf{r}_j), \tag{3.2.70}$$

where ψ may be determined variationally by minimizing the expectation value of the total Hamiltonian. If the short-range repulsive potential is approximated by a delta function $g\delta(\mathbf{r})$, this procedure yields the corresponding

Hartree equation [154]

$$-\frac{\hbar^2\nabla^2}{2m}\psi(\mathbf{r}) + (N-1)g|\psi(\mathbf{r})|^2\psi(\mathbf{r}) = \varepsilon\psi(\mathbf{r}), \tag{3.2.71}$$

where the second term denotes the interaction with the remaining $N - 1$ particles and ε is a single-particle energy. It is often convenient to write ψ as

$$\psi(\mathbf{r}) = V^{-1/2}\, e^{iS(\mathbf{r})} f(\mathbf{r}), \tag{3.2.72}$$

where $S(\mathbf{r})$ and $f(\mathbf{r})$ are real functions of order unity and V is the volume of the system. In the thermodynamic limit ($N \to \infty$, $V \to \infty$, but $n = N/V$ fixed), a simple calculation yields the expectation value of the particle density and current

$$n(\mathbf{r}) = n[f(\mathbf{r})]^2, \tag{3.2.73a}$$

$$\mathbf{j}(\mathbf{r}) = n(\mathbf{r})\hbar m^{-1}\nabla S(\mathbf{r}), \tag{3.2.73b}$$

whose ratio defines the velocity field

$$\mathbf{v} \equiv \frac{\mathbf{j}}{n} = \hbar m^{-1}\nabla S(\mathbf{r}), \tag{3.2.74}$$

Moreover, equation (3.2.71) becomes the Gross–Pitaevskii [151–153, 155] equation

$$e^{-iS(\mathbf{r})}\left(-\frac{\hbar^2\nabla^2}{2m}\right)[e^{iS(\mathbf{r})}f(\mathbf{r})] + gn[f(\mathbf{r})]^3 = \varepsilon f(\mathbf{r}), \tag{3.2.75}$$

which can describe both uniform and nonuniform states of an imperfect Bose gas.

Consider a single-particle state with $l_z = \hbar$; the phase $S(r) = \theta$ increases by 2π on once encircling the polar axis, and the corresponding N-body wave function becomes

$$\Psi_H(\mathbf{r}_1, \cdots, \mathbf{r}_N) = \prod_{j=1}^{N} [V^{-1/2} e^{i\theta_j} f(\mathbf{r}_j)]. \tag{3.2.76}$$

Equations (3.2.73b) and (3.2.74) immediately give the circulating velocity

$$v_\theta(r) \equiv \frac{j_\theta(r)}{n(r)} = \frac{\hbar}{mr}, \tag{3.2.77}$$

which is precisely that of a singly quantized vortex [21, 22] [see equation (3.2.11)]. The radial functon $f(r)$ satisfies an ordinary nonlinear differential equation and must be obtained numerically [67, 155], but it has the simple variational approximation [17]

$$f(r) \approx \left(\frac{r^2}{r^2 + 2\xi_H^2}\right)^{1/2}, \tag{3.2.78}$$

where

$$\xi_H \equiv \left(\frac{\hbar^2}{2mng}\right)^{1/2} \tag{3.2.79}$$

is the Hartree healing length. The centrifugal barrier associated with the unit angular momentum depresses the density for $r \lesssim \xi_H$, which represents the vortex core region [151–153, 155–157]. Furthermore, the asymptotic normalization fixes the parameter $\varepsilon = ng$. As in the classical system, the energy diverges logarithmically at infinity and must be cut off at some large distance R; a numerical evaluation with the exact $f(r)$ then yields the energy (kinetic plus potential) per unit length [155, 158]

$$\begin{aligned} E' &= \left(\frac{\pi\hbar^2 n}{m}\right)\left[\ln\left(\frac{R}{\xi_H}\right) + 0.385\right] \\ &= \left(\frac{\rho\kappa^2}{4\pi}\right)\left[\ln\left(\frac{R}{\xi_H}\right) + 0.385\right]. \end{aligned} \tag{3.2.80}$$

If we somewhat arbitrarily take ξ_H as the core radius, comparison with equation (3.2.13) implies the Hartree value for the core energy

$$\delta_H = 0.385. \tag{3.2.81}$$

The variational approximation (3.2.78) gives the very similar value $\delta_v = 0.403$.

This Hartree description of a single vortex has been extended in several ways. If the axis of the vortex is bent into a circle of radius r_0, it becomes the quantum analog of the classical ring with $\kappa = h/m$. The total energy E is largely kinetic with a small contribution from the altered potential energy near the core; a detailed calculation yields [156, 158]

$$E = \frac{1}{2}\rho\kappa^2 r_0\left[\ln\left(\frac{8r_0}{\xi_H}\right) - 1.615\right] \tag{3.2.82}$$

in agreement with equations (3.2.32) and (3.2.81). Although a slowly moving ring no longer represents a strictly stationary state, a suitable Galilean transformation eliminates this difficulty, yielding the translational velocity of the ring [158] (see also Reference [159])

$$v = \left(\frac{\kappa}{4\pi r_0}\right)\left[\ln\left(\frac{8r_0}{\xi_H}\right) - 0.615\right]. \tag{3.2.83}$$

This expression satisfies the Hamiltonian relation (3.2.54), but it differs slightly from that expected from equations (3.2.41) and (3.2.61) (compare References [50] and [80]).

A different generalization describes a set of singly quantized rectilinear vortices with a single-particle wave function $\psi(\mathbf{r})$ in product form. If equation (3.2.71) is rewritten with $i\hbar\partial\psi/\partial t$ on the right side, the resulting time dependence of the initial wave function represents a rigid translation of each vortex with the classical velocity (3.2.16) [159]. The same approximate product wave function permits a study of the equilibrium states in a rotating cylinder; it reproduces the essential features of the previous semiclassical results [160]).

To this point, we have assigned every particle to the same single-particle state, whose structure incorporates the interactions only through a Hartree field. This description cannot be fully self-consistent, because the interparticle potential also excites particles to other single-particle states, thereby depleting the condensed mode. Such behavior is especially clear in a uniform system, where the wave vector **k** provides a natural label for these one-body states. The Hartree ground state has N particles in the mode with $\mathbf{k} = 0$, producing an eigenstate of linear momentum with eigenvalue zero. In perturbation theory, the interparticle potential excites pairs of particles to states with $\pm\, \mathbf{k}$, preserving the total (zero) momentum. Since the number of excited particles N' remains small in the weak-coupling limit, Bogoliubov [161] was able to solve the resulting equations completely, verifying that $N' \ll N$.

A similar description applies to a singly quantized vortex, where the Hartree state has all N particles in the lowest single-particle state with $l_z = \hbar$. As in the uniform system, the interparticle potential alters this distribution, but it preserves the total angular momentum ($L_z = N\hbar$) by exciting pairs of particles to the single-particle states with $l_z = (1 \pm p)\hbar$, where p is an integer. Hence a vortex in a weakly interacting Bose gas has N' particles in the excited modes, and $N_0(< N)$ remaining in the original condensed mode. The resulting vortex structure depends on the detailed wave functions, and we merely summarize the qualitative features.

The single-particle states have the natural quantum numbers l_z and k_z, denoting the component of angular momentum and wavenumber along the vortex axis. The radial dependence permits a further classification as bound or scattering states, according as the wave function vanishes exponentially or oscillates for $r \to \infty$. The most important bound state has $l_z = \pm\, \hbar$ relative to that of the condensate; it represents the quantum analog of Kelvin's classical vortex wave and has the long-wavelength dispersion relation [153, 162–163a]

$$\omega = \left(\frac{\kappa k^2}{4\pi}\right)\left[\ln\left(\frac{2}{k\xi_H}\right) - \gamma - 0.115\right], \qquad (3.2.84)$$

which again differs slightly from that expected from equations (3.2.55) and (3.2.81) [compare References [50] and [80], and the discussion below (3.2. 83)]. To interpret the scattering states, we recall that the uniform imperfect Bose gas has phononlike excitations [161], which are here modified by the vortex line. The resulting phase shifts and differential cross section [164, 165] are very similar to those for scattering of sound waves by a classical vortex [166, 167].

Far from the vortex axis, all possible excited modes contribute to the properties of the vortex; for small r, however, only the modes with $l_z = 0$ or $\pm\hbar$ are significant, because the other radial wave functions vanish too rapidly. Those particles with $l_z = 0$ yield a finite density at $r = 0$ [151], but they do not affect the circulating current or kinetic-energy density because their wave function has cylindrical symmetry. A detailed analysis shows that the resulting total density and circulating current may be approximated by [168]

$$n(r) \approx n_0 r^2(r^2 + \xi_H^2)^{-1} + n', \tag{3.2.85a}$$

$$j_\theta(r) \approx \frac{n\hbar r}{m(r^2 + \xi_H^2)}, \tag{3.2.85b}$$

where n_0 and n' are the condensate and noncondensate densities in a uniform system with the same particle density $n = N/V$ and coupling constant g. Note that the asymptotoic current is propertional to the *total* density n, independent of the depletion of the condensate.

(*b*) *Liquid Helium (II)*. The weak-coupling theory of a vortex in a Bose gas requires that $(n\xi_H^3)^{1/2} \gg 1$ [equivalently, that $n^{1/2}(mg)^{3/2}\,\hbar^{-3} \ll 1$, see equation (3.2.79)]. Liquid ^{4}He violates this inequality, for the interparticle spacing $n_4^{-1/3} \approx 3.6$ Å exceeds most plausible estimates of the core size. For example, the assumption of incompressible flow necessarily fails inside the distance $\xi_s = \hbar/ms \approx 0.7$ Å, where the circulating velocity field equals the speed of sound [169] s $\approx 2.4 \times 10^4$ cm/sec. Alternatively, if the core is taken as hollow with the macroscopic surface tension [170] $\sigma \approx 0.37$ erg/cm^2, the surface pressure [171] equals the Bernoulli pressure drop in the fluid at [22] $\xi_\sigma = \rho\kappa^2/8\pi^2\sigma \approx 0.5$ Å. A third possibility is $\xi_{sc} \approx 2.4$ Å, mentioned below (3.2.64). Although none of these estimates should be taken seriously, they agree qualitatively with the value $\xi \approx 1.3$ Å reported by Rayfield and Reif [28, 29].

The quantitative failure of the weak-coupling model has led to several phenomenological theories of vortex structures in liquid helium. Some of these introduce condensate wave functions, whose phase $S(\mathbf{r})$ determines the local velocity field through equation (3.2.74) or various generalizations. For example, Ginzburg and Pitaevskii [155] use a modified Ginzburg–Landau theory [172, 173] to study the vicinity of T_λ; their model predicts that ξ diverges near T_λ as $(T_\lambda\text{–T})^{-1/2}$, but there is no direct evidence for this increased core size (see also References [174] and [175]). In a related calculation Thouless [176] has proposed a low-temperature superfluid hydrodynamics that incorporates the effect of shear near the vortex core.

A very different phenomenological vortex theory relies on Landau's quasiparticle model [26] of phonons and rotons, suitably modified to incorporate the large circulating velocity field near the vortex core. Even in a uniform medium [26, 177], the normal-fluid density ρ_n increases with increasing relative velocity $|\mathbf{v}_s - \mathbf{v}_n|$, and a similar effect occurs in the vicinity of a vortex line [79, 178–182]. These calculations predict an essentially normal core with a radius of a few angstrom units, surrounded by a region of enhanced roton density. Although precise experimental confirmation is difficult [183–187], the general picture agrees with that inferred from the mobility of negative ions trapped on vortex lines in rotating ^{4}He (II) [79, 188] (see Section 3.4.2).

A more fundamental description of a vortex line uses an N-body trial wave function of the form suggested by Feynman [22]

$$\Psi_F(\mathbf{r}_1, \cdots, \mathbf{r}_N) = \Phi_0(\mathbf{r}_1, \cdots, \mathbf{r}_N) \prod_{j=1}^{N} [e^{i\theta_j} f(\mathbf{r}_j)], \tag{3.2.86}$$

where f is a variational function qualitatively like that in equation (3.2.78). Here Φ_0 denotes the exact ground state of uniform interacting ^{4}He; it vanishes whenever any two particles come close together and therefore incorporates the short-range correlations in the fluid. The second factor in (3.2.86) enforces the vortex character of the state by assigning unit angular momentum to every particle, with the radial function f eliminating the logarithmic divergence of the core energy. Detailed calculations [189] with Ψ_F are quite intricate (see also References [190] and [191]). The resulting particle density $n(r)$ vanishes at $r = 0$ with a core radius ≈ 1 Å; $n(r)$ then approaches the bulk value with decaying oscillations that reflect the short-range correlations.

In the case of a weakly interacting gas, we noted that the relative occupation of states with $l_z = 0, \pm \hbar$ determined the structure of the vortex core, with $n(0)$ arising solely from those particles with $l_z = 0$. Since Ψ_F assumes all particles have $l_z = \hbar$, however, it necessarily provides only a partial description of the core region [168, 192]. Unfortunately, it has not yet been possible to devise a tractable generalization that allows for the simultaneous presence of many excited pairs; a solution to this problem would be of great interest.

Although the detailed structure of the vortex core remains unsolved, the quantum theory of superfluid vortices has had considerable success in clarifying vortex dynamics. In this approach, the condensate wave function is identified with the off-diagonal matrix element of the second-quantized field operator between states of N and $N \pm 1$ particles [193, 194]. The resulting time-dependent phase increases at a rate

$$\frac{\partial S(\mathbf{r}, t)}{\partial t} = \hbar^{-1}\bar{\mu}(\mathbf{r}, t), \tag{3.2.87a}$$

where $\bar{\mu}$ is the chemical potential. Consider a contour C joining two points $\mathbf{r}_1$ and $\mathbf{r}_2$. At time t, the increase in phase $\Delta S_{21}(t)$ along C is $\int_1^2 d\mathbf{s}\cdot\nabla S(\mathbf{r}, t)$, which evidently depends on C. In particular, a singly quantized vortex represents a singularity with net phase 2π [151–153], so that $\Delta S_{21}(t)$ changes by 2π every time a vortex crosses C. More generally, if n vortices cross C in a time T, the mean vortex flow rate is given by [195–198]

$$\begin{aligned}\langle \dot{n}\rangle_{\text{av}} &= (2\pi T)^{-1}[\Delta S_{21}(T) - \Delta S_{21}(0)] \\ &= \frac{1}{2\pi T}\int_0^T dt\,\frac{\partial}{\partial t}\,\Delta S_{21}(t) \\ &= \frac{1}{2\pi T\hbar}\int_0^T dt[\bar{\mu}(\mathbf{r}_2, t) - \bar{\mu}(\mathbf{r}_1, t)] \\ &= h^{-1}\langle \bar{\mu}(\mathbf{r}_2) - \bar{\mu}(\mathbf{r}_1)\rangle_{\text{av}},\end{aligned} \tag{3.2.87b}$$

where the third line follows from (3.2.87a). This "phase-slippage" equation provides a simple explanation of various "ac-Josephson" effects [199] in ^{4}He (II), in which $\langle \dot{n}\rangle_{\text{av}}$ and hence the chemical potential difference seems synchronized to an external frequency [200–203] (see, however, Addendum).

3.2.3. Mutual Friction and Interaction with Quasiparticles

Our discussion of vortex dynamics in ^{4}He (II) has neglected coupling to the normal fluid, which is strictly correct only if the normal-fluid density vanishes. Otherwise, the normal fluid exerts a force on a moving vortex line, with components both parallel and perpendicular to the direction of relative motion. These forces have been treated on two distinct levels. In the simpler one, Hall [13, 90, 204], and Bekarevich and Khalatnikov [205] generalized Landau's two-fluid hydrodynamics to include a coarse-grain average vorticity $\boldsymbol{\zeta}$ in rotating ^{4}He (II), necessarily considering regions containing many vortices. In the other (more fundamental) description, Hall and Vinen [42] dealt directly with the scattering of quasiparticles by individual vortex lines. In the limit of many vortices, both descriptions agree, and the resulting average theory is frequently known as the HVBK equations. Since the derivations are somewhat involved [206], we shall merely present a brief summary.

(*a*) *Phenomenological Equations.* The original form of Landau's equations assumed that $\mathbf{v}_s$ was irrotational everywhere [26]. When the superfluid contains vortex lines, however, $\nabla \times \mathbf{v}_s$ becomes singular in the vortex cores. The resulting picture based on individual vortices is convenient for low vortex density, but is impractical for a dense array, when the mean vorticity

$$\boldsymbol{\zeta} \equiv \langle \nabla \times \mathbf{v}_s \rangle \tag{3.2.88}$$

provides a more convenient description. Here the angular brackets denote an average over regions containing many vortices. The theory makes the fundamental assumption that unit volumes of superfluid and normal fluid experience mutual forces $\mathbf{F}_s$ and $\mathbf{F}_n = -\mathbf{F}_s$, respectively, with

$$\mathbf{F}_s = \left(\frac{B\rho_s\rho_n}{2\rho}\right)\hat{\zeta} \times [\boldsymbol{\zeta} \times (\mathbf{v}_s - \mathbf{v}_n + \nu_s\nabla \times \hat{\zeta})] + \left(\frac{B'\rho_s\rho_n}{2\rho}\right)\boldsymbol{\zeta} \times (\mathbf{v}_s - \mathbf{v}_n + \nu_s\nabla \times \hat{\zeta}). \tag{3.2.89}$$

Here B and B' are phenomenological dimensionless constants characterizing the strength of the mutual friction, $\hat{\zeta}$ is a unit vector parallel to $\boldsymbol{\zeta}$, and ν_s has the dimensions of a kinematic viscosity with an approximate magnitude

$$\nu_s \approx \left(\frac{\kappa}{4\pi}\right)\ln\left(\frac{b}{\xi}\right) = \left(\frac{\hbar}{2m}\right)\ln\left(\frac{b}{\xi}\right), \tag{3.2.90}$$

where b is the intervortex spacing. In principle, equation (3.2.89) may contain a third contribution parallel to $\hat{\zeta}$, but experiments indicate that the corresponding coefficient B_z is either small or zero [207–209].

The phenomenological theory acquires dynamical content by incorporating the vortex-dependent forces $\mathbf{F}_s$ and $\mathbf{F}_M$ [see equation (3.2.56)] into Landau's two-fluid hydrodynamics. The resulting set of equations is very complicated, and only certain special cases have been investigated in detail [94, 95, 204–206]. First, it is evident that $\mathbf{F}_s$ vanishes for a uniform rotation, where $\boldsymbol{\zeta} = 2\boldsymbol{\Omega}$ and $\mathbf{v}_s = \mathbf{v}_n = \boldsymbol{\Omega} \times \mathbf{r}$. More important, the theory can describe the propagation of second sound through rotating ^{4}He (II), where $\mathbf{v}_s - \mathbf{v}_n$ oscillates at essentially constant density and pressure. As a first approximation, the terms proportional to $\nabla \times \hat{\zeta}$ may be neglected, and the linearized hydrodynamic equations predict the phenomena [42, 95, 206]:

i. When second sound propagates parallel to $\boldsymbol{\Omega}$, it suffers an attenuation proportional to ΩB_z. The experimental absence of such an effect implies that B_z is small.
ii. When second sound propagates perpendicular to $\boldsymbol{\Omega}$, the first term of equation (3.2.89) lies parallel to $\mathbf{v}_s - \mathbf{v}_n$ and acts like a viscous drag. Consequently, the wave amplitude is attenuated with a characteristic constant proportional to ΩB. The linear dependence on Ω agrees with equation (3.2.20), and the fit to the experiments determines B for various temperatures [42, 210–213]. In contrast, the second term of equation (3.2.89) lies perpendicular both to $\boldsymbol{\Omega}$ and to $\mathbf{v}_s - \mathbf{v}_n$. Instead of attenuating the wave, it combines with the Coriolis force to split the resonant second-sound modes

of a rotating cavity by an amount proportional to $\Omega(2-B')$ [214, 215]. Experiments verify this splitting and yield $B'(T)$ [213–216]. Such observations conclusively rule out an earlier quantum hydrodynamics, which, in effect, set $B' = 2$ and therefore predicted no mode splitting [69]. Lucas [213] has recently summarized the various measurements of B and B' (which he denotes B_{exp} and B'_{exp}).

The generalized two-fluid equations for rotating helium have an abundance of wavelike solutions in addition to the second-sound modes considered above [94, 95]. Some of these (first-sound and second-sound) are only slightly affected by the rotation, but the remaining ones (inertial waves in the normal fluid [97] and vortex waves in the superfluid) occur only in a rotating system. The vortex waves were first studied by Hall [13, 90] [see the discussion below equation (3.2.63)], who solved the equations for the geometries relevant to his experiments. These modes are especially interesting, for they involve the terms $\nu_s \nabla \times \hat{\zeta}$ in equation (3.2.89) and therefore measure the vortex parameter ν_s in equation (3.2.90). In this way, Hall obtained the estimate $\xi \approx 7$ Å mentioned below equation (3.2.63) [92]. More generally, Andronikashvili and his group have considered the full set of equations for both inertial and vortex waves, including mutual friction that couples the two modes [96]. The comparison with their extensive experiments is summarized in References [15, 16].

Although we have emphasized the special properties of parallel rectilinear vortices, similar but more complicated vortex configurations occur frequently in liquid ^{4}He (II). Indeed, Feynman [22] has suggested that superfluid turbulence represents a tangled isotropic mass of quantized vortex lines. This picture, which permits a further test of (3.2.89), has been extensively verified by Vinen [13, 14, 217] through measurements of mutual friction in heat currents [217a].

(*b*) *Quasiparticle Description.* Landau's quasiparticle model describes the normal fluid as a gas of elementary excitations consisting in pure ^{4}He of phonons and rotons, supplemented by ^{3}He atoms in dilute superfluid mixtures. From this viewpoint, mutual friction is merely the interaction between vortex lines in the superfluid and the quasiparticles. In particular, a single vortex line moving slowly through stationary normal fluid experiences a force perpendicular to its axis and proportional to its speed, but with components both parallel and perpendicular to its velocity [218]. The parallel component exerts a "viscous" drag and may be expressed in terms of the momentum-transfer cross section for scattering of quasiparticles by the vortex line. The perpendicular component is more unusual, for it orginates in the left–right asymmetry of the differential cross section. This effect is a direct consequence of the circulating velocity field about the vortex and would be absent, for example, for a cylinder with no circulation moving slowly through the normal fluid.

The force on a moving vortex line has been observed in two separate ways, either directly through its dynamical effects on single vortex rings or indirectly as the source of the mutual friction in (3.2.89). Rayfield and Reif [29] used the first approach to equate the measured energy loss of charged vortex rings with the work done against the parallel component of the force. In this way, they obtained the thermally averaged transport cross sections $\bar{\sigma}_r \approx 9.5$ Å and $\bar{\sigma}_3 \approx 18.3$ Å for rotons and ^{3}He impurities, respectively. They also detected a small phonon contribution, but did not attempt to extract a corresponding mean transport cross section $\bar{\sigma}_{\text{ph}}$. Unfortunately, their method cannot detect the perpendicular component of the force because of the circular symmetry of the ring.

The second way of studying the quasiparticle interaction with a moving vortex is to evaluate the mutual friction force $\mathbf{F}_s$ (3.2.89) in terms of the collision cross sections [42, 204]. The calculation is complicated because the force on a single line depends on the relative velocity $\mathbf{v}_L - \mathbf{v}_{qp}$ between the line and the quasiparticle gas, whereas the spatially averaged force $\mathbf{F}_s$ depends on the relative velocity $\mathbf{v}_s - \mathbf{v}_n$ between the superfluid and normal fluid. In general, these four velocities are all distinct, because of various hydrodynamic effects involving $\mathbf{v}_n - \mathbf{v}_{qp}$, which arises because the moving vortex drags the nearby normal fluid, $\mathbf{v}_L - \mathbf{v}_s$, which is related to the Magnus force, and $\mathbf{v}_L - \mathbf{v}_n$, which is an added contribution introduced by Iordanskii [219, 220] and studied recently by Hall [221, 221a], and Titus [222, 223] The final result of this analysis expresses the coefficients B and B' in (3.2.89) in terms of the elementary cross sections for quasiparticle-vortex scattering, which may therefore (in principle) be obtained from the experimental studies of second sound in rotating ^{4}He (II). At present, only the roton contribution has been detected, because phonon contributions are important only below 1 K, where the long mean free path makes second sound difficult to propagate. Lucas [213] has presented the most complete analysis of this work, but he does not express his results in terms of roton-vortex cross sections, thereby precluding a direct comparison with the data of Rayfield and Reif [29]. Such a comparison would be most valuable, as the derivation of B and B' from the cross sections is not entirely transparent. In addition, an experimental study of second sound in rotating superfluid ^{3}He–^{4}He mixtures would be desirable, for it could give an independent determination of Rayfield and Reif's $\bar{\sigma}_3$.

On an even more fundamental level, there have been several attempts to calculate the differential cross sections for phonon [165–167, 223], roton, [42, 223–226], and ^{3}He [223, 227] scattering by a vortex line. After suitable angular integrations and thermal averages, the resulting mean transport cross sections should be just those measured by Rayfield and Reif (assuming that the charged ion trapped on the large vortex ring has negligible effect). For phonons and ^{3}He impurities, the different calculations predict the same basic features and provide a reasonable fit to the data. For rotons, however, there are some discrepancies between the various calculations; this situation arises from the lack of a microscopic theory of rotons, which necessitates an undue reliance on roton models. Regrettably, the uncertainty in the roton-vortex cross sections further complicates the subsequent calculation of roton contributions to B and B'. Thus theoretical and experimental studies of the ^{3}He contributions are especially important. The present situation is well summarized in References [223] and [226].

3.2.4. Creation and Destruction of Vortices

Most of the previous discussion of vortices in ^{4}He (II) has considered only equilibrium states or those differing from equilibrium by infinitesimal perturbations. These states, which are characterized by temperature-independent circulation quantum numbers, presumably appear in experiments that cool a uniformly rotating sample through T_λ to a final temperature T well below the critical region; moreover, data obtained in this way for a given geometry and Ω should be quite reproducible. On the other hand, it is far less evident that accelerating a container from rest at fixed T will produce the same final state. Although thermodynamics would demand such an outcome for a system in equilibrium, the second procedure is much more likely to encounter metastability and hysteresis. This behavior, which has occurred in various experiments [105, 124], is particularly obvious in the persistent supercurrent remaining when a rotating container is decelerated to rest [35, 87, 113–121, 227a]. In this latter case, the system remains metastable with fixed velocity and circulation throughout a wide range of $T < T_\lambda$, instead of reaching the true equilibrium state of zero velocity ($\mathbf{v}_s = 0$). Since the normal fluid is stationary after the container is stopped, the total angular momentum in the persistent current should vary linearly with $\rho_s(T)$; experiments with superfluid gyroscopes fully confirm this prediction and now serve to determine the behavior of $\rho_s(T)$ near T_λ [119–121].

Theoretical analysis of these questions is extremely difficult, and most work has been confined to an idealized problem of determining the critical velocity v_c for the destruction of superflow in a long narrow channel with clamped normal fluid ($v_n = 0$). In general, v_c could depend both on the channel width d and on the temperature T. Experiments seem to indicate a somewhat simpler situation, however, with two quite distinct regimes. Very near T_λ, the critical velocity is nearly independent of channel size and has the temperature dependence [228–230]

$$v_c \approx u_0 \left(1 - \frac{T}{T_\lambda}\right)^{2/3}, \tag{3.2.91}$$

with $u_0 \approx 670$ cm/sec. At lower temperature, v_c is approximately independent of temperature, but it increases with decreasing d. The precise functional dependence is somewhat uncertain, but either of the functions [22, 231, 232]

$$\left.\begin{aligned} v_c &\approx \frac{h}{md} \\ v_c &\approx \left(\frac{\hbar}{md}\right)\ln\left(\frac{d}{\xi}\right) \end{aligned}\right\} T \ll T_\lambda, \tag{3.2.92}$$

with ξ an atomic dimension seems to give an order-of-magnitude fit to many experiments. For fixed d and decreasing T, the critical velocity is given by (3.2.91) until it reaches the value (3.2.92); below this temperature, v_c remains approximately constant, with the form shown in Figure 3.2. The experimental basis for these conclusions involves considerable selection among the available data; References [231–232a] contain complete and critical discussions.

All theoretical interpretations of this behavior have invoked the creation of vortex rings with their self-induced velocity opposite to that of the uniform stream $\mathbf{v}_s$ [98, 233–237]. The resulting statistical mechanics in a moving fluid involves the modified energy [238]

$$\bar{E} = E - Pv_s, \tag{3.2.93}$$

where E is the energy of an excitation and P is its corresponding momentum. Although P is indeterminate for a vortex ring in an incompressible fluid, it has been customary to identify P with the impulse [equation (3.2.53)], which does indeed equal the net increase in momentum when a ring grows from

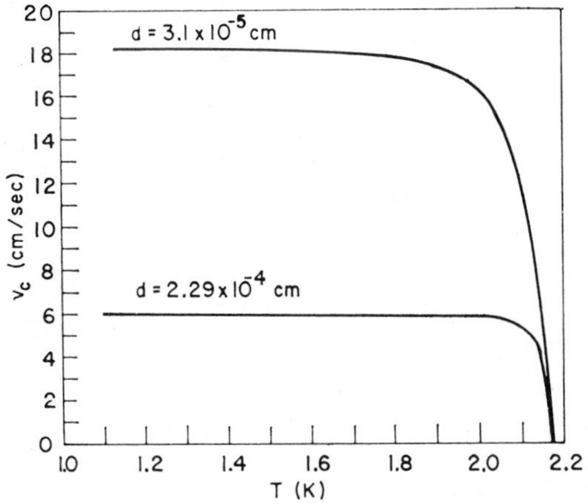

FIGURE 3.2. Temperature dependence of the critical velocity for the destruction of superflow in two different channels. [From E. F. Hammel and W. E. Keller, in *Superfluid Helium*, edited by J. F. Allen (Academic, London, 1966), Fig. 7, p. 130. Reprinted by permission.]

negligible size to its equilibrium radius r_0. As a result, equation (3.2.93) becomes [see equation (3.2.32) and (3.2.53)]

$$\bar{E}(r_0, v_s) = \tfrac{1}{2}\,\rho_s\kappa^2 r_0\eta - \pi\rho_s\kappa r_0^2 v_s, \qquad (3.2.94)$$

where the expression

$$\eta \equiv \ln\left(\frac{8r_0}{\xi}\right) - 2 + \delta \qquad (3.2.95)$$

will be treated as constant in the following discussion. If $v_s = 0$, then $\bar{E}$ increases approximately linearly with increasing r_0, becoming arbitrarily large and positive as $r_0 \to \infty$. For finite v_s, however, $\bar{E}$ reaches a maximum value

$$\bar{E}_{\max}(v_s) \approx \frac{\rho_s\kappa^3\eta^2}{16\pi v_s} \qquad (3.2.96)$$

at the critical radius

$$r_{0c}(v_s) \approx \frac{\kappa\eta}{4\pi v_s}, \qquad (3.2.97)$$

and decreases monotonically for $r_0 > r_{0c}$. This behavior is shown schematically in Figure 3.3.

Consider a system with uniform superflow in thermal equilibrium at

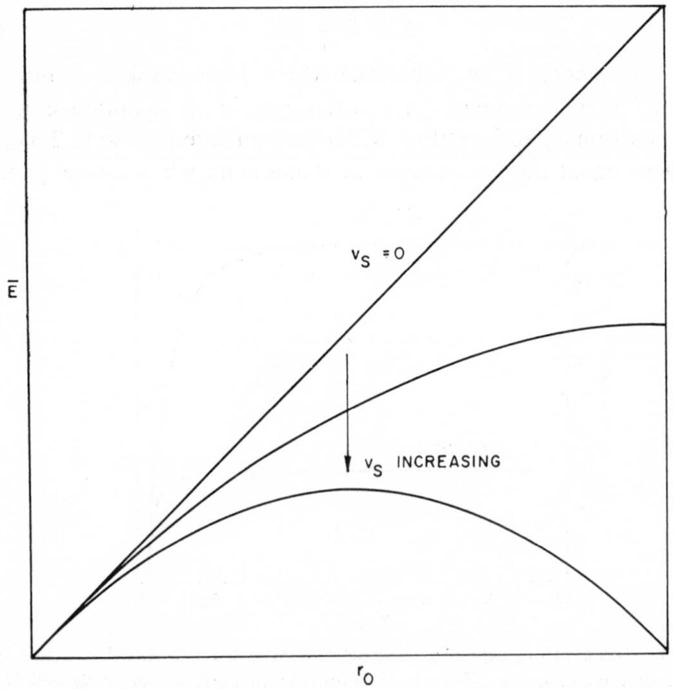

FIGURE 3.3. Transformed energy $\bar{E}$ (3.2.94) for a vortex ring of radius r_0 in a uniform superfluid flowing with velocity v_s.

temperature T. One possible way to reduce the flow is to create an oppositely directed vortex ring with radius r_0. If $r_0 < r_{0c}$, then the dissipative forces on the ring act to reduce $\bar{E}$, thereby shrinking the radius to zero. On the hand, if $r_0 > r_{0c}$, these forces expand the ring, which thus becomes stable. The relevant energy is that needed to get over the barrier $\bar{E}_{\max}(v_s)$, and the probability of creating such a ring is proportional to the Boltzmann factor

$$\text{Probability} \propto \exp\left[-\frac{\bar{E}_{\max}(v_s)}{k_B T}\right]. \tag{3.2.98}$$

Hence the critical velocity should have a temperature dependence given by

$$\frac{\bar{E}_{\max}(v_c)}{k_B T} = \text{const} \tag{3.2.99}$$

or, equivalently,

$$v_c = \text{const}\left(\frac{\rho_s \kappa^3 \eta^2}{16\pi k_B T}\right). \tag{3.2.100}$$

Since $\rho_s(T)$ varies near T_λ as $(1 - T/T_\lambda)^{2/3}$ [121], equation (3.2.100) has the correct functional form for $T_\lambda - T \ll T_\lambda$ [see equation (3.2.91)]. The calculation of the constant in equation (3.2.100) is difficult, however, and generally yields too large a value for v_c by an order of magnitude [230]. Various improvements to equation (3.2.100) have included the effect of walls and the entropy associated with the core [98, 236, 237], but the problem would likely benefit from further study.

Equation (3.2.100) has the notable defect that v_c is predicted to diverge as T^{-1} as $T \to 0$, in clear disagreement with the experimental data of Figure 3.2 (see, however, Addendum). Consequently, the low-temperature critical velocity cannot involve thermally activated vortex rings, but must instead depend on other processes [22, 239]. One suggestion [240–242] invokes an unspecified quantum transition probability that spontaneously creates a full-grown vortex ring with zero transformed energy $\bar{E}$. For a particular v_s, this condition fixes the radius r_0 through the implicit relation

$$v_s = \frac{E(r_0)}{p(r_0)} \approx \frac{\kappa\eta}{2\pi r_0}. \tag{3.2.101}$$

No rings appear for small v_s, because the corresponding diameter $2r_0$ exceeds the width of the channel d. As v_s increases, however, the creation process first becomes possible at a critical (Feynman) velocity [22]

$$v_c \approx \frac{\kappa\eta}{\pi d} \approx \left(\frac{2\hbar}{md}\right)\ln\left(\frac{d}{\xi}\right), \tag{3.2.102}$$

which agrees qualitatively with that observed for the destruction of superfluid flow [equation (3.2.92)]. In essence, this argument applies Landau's critical velocity to vortex rings with $2r_0 \approx d$ and ignores the inconsistency in using expressions for unbounded fluid to obtain a geometrical effect [65–67,

160, 242, 243]. Moreover, the possibility of an arbitrary additive constant in the impulse renders the approach even more dubious. For this reason, the temperature-independent critical velocity has sometimes been ascribed to the presence of previously existing vortices [244], but these arguments also suffer from conceptual difficulties. In summary, there is still no wholly satisfactory explanation for the experimental v_c well below T_λ.

3.3. IONS

In treating the structure of ions in liquid helium, it is necessary to distinguish between the two charge species, because the free positive molecular ion $(^4He)_2^+$ is stable with a binding energy ≈ 2.4 eV [245], whereas the free negative ion He^- is unstable [246]. As a result, the positive ion in the fluid is essentially a dressed molecular complex, but the negative ion turns out to be a single electron trapped in a small hollow cavity.

The dominant feature in determining the structure of an ion is the interaction between the bare charged entity and the individual He atoms. Consequently, the quantum statistics of the background fluid are rather unimportant, and the ion has essentially the same form in ^3He and in ^4He. As a first approximation, we therefore ignore the isotopic concentration; the small residual effects arising from the different atomic masses and equilibrium densities (see Table 3.1) may then be included rather simply.

TABLE 3.1. Characteristic properties of liquid helium at T = 0, p = 0 (unless noted otherwise)

	^4He	^3He
Atomic mass m^a	6.646×10^{-24}g	5.007×10^{-24}g
Mass density ρ^b	0.145 g/cm^3	0.0815 g/cm^3
Number density $n = \rho/m$	2.18×10^{22} cm^{-3}	1.63×10^{22} cm^{-3}
Interparticle separation $n^{-1/3}$	3.58 Å	3.95 Å
Equivalent hard-sphere radius $r_0 = (3/4\pi n)^{1/3}$	2.22 Å	2.45 Å
Dielectric constant^c $\varepsilon = 1 + 4\pi n\alpha$	1.0556	1.0415
Surface tension σ^d	0.37 dyne/cm	0.16 dyne/cm
Solidification pressure p_s at $T = 0$	25 atm^e	34 atm^f

[a]Reference [5, Table A 19, p. 679].
[b]Reference [5, Table A1, p. 666, and Table A14, p. 677].
[c]Reference [255].
[d]Reference [170].
[e]Reference [5, Section 2.4].
[f]Extrapolated value obtained from data in R. A. Scribner. M. F. Panczyk, and E. D. Adams, *J. Low Temp. Phys.* **1**, 313 (1969).

3.3.1. Structure of a Positive Ion

The $(^4\text{He})_2^+$ ion has an internuclear separation [245] of ≈ 1.08A, considerably smaller than the interparticle spacing $n_4^{-1/3} \approx 3.58$ Å (see Table 3.1). Since a comparable situation presumably exists for ^{3}He, where $n_3^{-1/3} \approx$ 3.95 Å, we consider the bare positive ion as a singly charged entity with twice the atomic mass [247–249]. In vacuum, such a configuration would produce an electrostatic field $(e/r^2)\hat{r}$, but this behavior is modified by the surrounding dielectric fluid. To understand the resulting structure, we first consider the simpler problem of a small neutral polarizable solid placed at a point $\mathbf{r}_0$ in an inhomogeneous external electrostatic field $\vec{\mathscr{E}}_0$. This field induces a charge density ρ_e ($\mathbf{r}$) in the solid, which thus experiences a net force

$$\mathbf{F} = \int d^3r\, \rho_e(\mathbf{r})\vec{\mathscr{E}}_0(\mathbf{r}), \tag{3.3.1}$$

with the integral extending over the volume of the solid. If $\vec{\mathscr{E}}_0$ varies slowly in space, it may be expanded in a Taylor series

$$\vec{\mathscr{E}}_0(\mathbf{r}) \approx \vec{\mathscr{E}}_0(\mathbf{r}_0) + [(\mathbf{r} - \mathbf{r}_0)\cdot\nabla_0]\vec{\mathscr{E}}_0(\mathbf{r}_0) + \cdots. \tag{3.3.2}$$

Substitution into equation (3.3.1) yields [250]

$$\mathbf{F} = \int d^3r\, \rho_e(\mathbf{r})\{\vec{\mathscr{E}}_0(\mathbf{r}_0) + [(\mathbf{r} - \mathbf{r}_0)\cdot\nabla_0]\vec{\mathscr{E}}_0(\mathbf{r}_0)\} = (\mathbf{p}\cdot\nabla_0)\vec{\mathscr{E}}_0(\mathbf{r}_0), \tag{3.3.3}$$

where the charge neutrality eliminates the first term and

$$\mathbf{p} = \int d^3r\, \rho_e(\mathbf{r})\,(\mathbf{r} - \mathbf{r}_0) \tag{3.3.4}$$

is the dipole moment of the solid. Equation (3.3.3) specifies the force on any small electric dipole $\mathbf{p}$; in the present case, $\mathbf{p}$ may be expressed in terms of the polarizability α of the solid

$$\mathbf{p} = \alpha\, \vec{\mathscr{E}}_0(\mathbf{r}_0), \tag{3.3.5}$$

and equation, (3.3.3) becomes

$$\mathbf{F} = \alpha\{[\vec{\mathscr{E}}_0(\mathbf{r})\cdot\nabla]\vec{\mathscr{E}}_0(\mathbf{r})\}_{\mathbf{r}=\mathbf{r}_0}. \tag{3.3.6}$$

Furthermore, the vector identity

$$(\mathbf{A}\cdot\nabla)\mathbf{A} = \tfrac{1}{2}\nabla(A^2) - \mathbf{A}\times(\nabla\times A) \tag{3.3.7}$$

and the irrotational character of $\vec{\mathscr{E}}_0$ together give

$$\mathbf{F} = \tfrac{1}{2}\alpha\{\nabla[\vec{\mathscr{E}}_0(\mathbf{r})]^2\}_{\mathbf{r}=\mathbf{r}_0}, \tag{3.3.8}$$

which identifies the potential energy

$$V(\mathbf{r}) = -\tfrac{1}{2}\alpha[\vec{\mathscr{E}}_0(\mathbf{r})]^2 \tag{3.3.9}$$

of the solid in the external field. Note that any polarizable object is attracted toward a region of high field strength.

This description forms the basis for Atkins' "snowball" model [251–254] of a positive ion in liquid helium where the atomic polarizability α is ≈ 0.203 Å^3 [255]. The inhomogeneous electric field of the bare ion attracts the atoms inward, enhancing the density near the origin. Static equilibrium obtains when the resulting increased pressure balances the inward electrostatic forces. A quantitative description [256] of this process starts from the actual electric field in the fluid

$$\vec{\mathscr{E}}_0(\mathbf{r}) = \left(\frac{e}{\varepsilon r^2}\right)\hat{r}, \tag{3.3.10}$$

where ε is the dielectric constant at the particular point r. The corresponding polarization potential

$$V_{\text{pol}} = -\frac{\alpha e^2}{2\varepsilon^2 r^4} \tag{3.3.11}$$

varies asymptotically as r^{-4}, but its short-range form is more complicated if $\varepsilon(\mathbf{r})$ varies significantly. Consider a small fixed volume element V in the fluid. The helium particles contained in V experience a net electrostatic force

$$\mathbf{F}_{\text{el}} = \int_V d^3r\, n(\mathbf{r})\, \tfrac{1}{2}\alpha\nabla[\vec{\mathscr{E}}_0(\mathbf{r})]^2, \tag{3.3.12}$$

where $n(\mathbf{r})$ is the local number density. In addition, the neighboring fluid elements exert a pressure p on the surface, giving a hydrostatic force

$$\mathbf{F}_h = -\int d\mathbf{S}\, p(\mathbf{r}) = -\int_V d^3r\, \nabla p(\mathbf{r}), \tag{3.3.13}$$

where $d\mathbf{S}$ points along the outward normal and the second form follows from Gauss' theorem. Static equilibrium implies $\mathbf{F}_{\text{el}} + \mathbf{F}_h = 0$, or

$$\tfrac{1}{2}\alpha\nabla[\vec{\mathscr{E}}_0(\mathbf{r})]^2 = [n(\mathbf{r})]^{-1}\nabla p(\mathbf{r}). \tag{3.3.14}$$

Integrate this equation from some fixed point $\mathbf{r}_0$ to a variable point $\mathbf{r}$:

$$\frac{1}{2}\alpha[\vec{\mathscr{E}}_0(\mathbf{r})]^2 - \frac{1}{2}\alpha[\vec{\mathscr{E}}_0(\mathbf{r}_0)]^2 = \int_{\mathbf{r}_0}^{\mathbf{r}} \frac{d\mathbf{s}'\cdot\nabla' p(\mathbf{r}')}{n(\mathbf{r}')}. \tag{3.3.15}$$

Since $d\mathbf{s}\cdot\nabla p$ is just the change in pressure dp along the path element $d\mathbf{s}$, we may change variables on the right side and obtain

$$\frac{1}{2}\alpha[\vec{\mathscr{E}}_0(\mathbf{r})]^2 - \frac{1}{2}\alpha[\vec{\mathscr{E}}_0(\mathbf{r}_0)]^2 = \int_{p(\mathbf{r}_0)}^{p(\mathbf{r})} \frac{dp'}{n(p')}, \tag{3.3.16}$$

where $n(p)$ is given by the equation of state. Equation (3.3.16) determines the pressure p throughout the fluid; it was first derived by Atkins with a thermodynamic argument [251, 252], but the present direct approach used by Dahm and Sanders [256] illuminates its physical basis.

Landau and Lifshitz [257] point out that, if $n(\mathbf{r})$ is constant in the absence of the electrostatic field, equation (3.3.16) may be replaced by the simpler equilibrium condition

$$\tfrac{1}{2}\, n(\mathbf{r})\alpha[\vec{\mathscr{E}}_0(\mathbf{r})]^2 - p[n(\mathbf{r})] = \text{const} \tag{3.3.17}$$

because the correction terms are of higher order than the linear approximation used in equation (3.3.5). A combination of (3.3.10) and (3.3.17) gives the spherically symmetric equation

$$\frac{\alpha n(r)e^2}{2r^4[\varepsilon(r)]^2} = p[n(r)] - p_\infty, \tag{3.3.18}$$

where p_∞ is the applied hydrostatic pressure. The only remaining unknown quantity is the dielectric constant, which we assume to depend on $\mathbf{r}$ only through the density. Since $n\alpha$ is very small for all densities in liquid helium (see Table 3.1), the internal field corrections are negligible, and equation (3.3.5) yields the usual relation between the electric displacement $\vec{\mathscr{D}}$, the polarization $\vec{\mathscr{P}}$, and $\vec{\mathscr{E}}_0$ [258]

$$\vec{\mathscr{D}} \approx \vec{\mathscr{E}}_0 + 4\pi\vec{\mathscr{P}} = \vec{\mathscr{E}}_0 + 4\pi n\alpha\vec{\mathscr{E}}_0. \tag{3.3.19}$$

Hence, the dielectric constant becomes

$$\varepsilon = 1 + 4\pi n\alpha, \tag{3.3.20}$$

which completes the set of equations. Figure 3.4 shows the density profile near a singly charged ion evaluated from equation (3.3.18) and (3.3.20) for ^{4}He at $\approx$ 1.25 K [259]. At $\mathbf{r} \approx 6.8$ Å, the pressure reaches the solidification pressure $\approx$ 25 atm, and the interior should be considered a quasisolid. It must be noted, however, that this model relies on a bulk continuum description and should not be interpreted literally on an atomic scale. Atkins obtained an essentially identical curve by integrating equation (3.3.16) numerically [251].

The simplest interpretation of Figure 3.4 is that the positive ion in ^{4}He has a solid core of radius $R_+ \approx$ 6–7 Å, surrounded by an enhanced density caused by the attractive polarization potential (3.3.11). Using this model, Atkins estimated the total excess mass of a static ion to be $\approx 40m_4$, arising both from the solid core ($r < \mathrm{R}_+$) and from the increased density for $r > R_+$. It is not obvious, however, that this quantity represents the dynamic effective mass associated with a moving positive ion. For example, a solid neutral

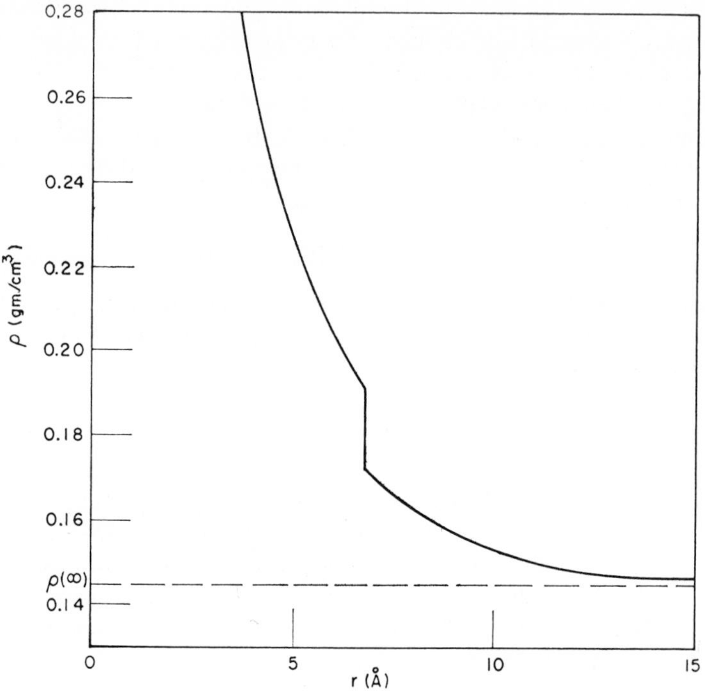

FIGURE 3.4. Density profile near a singly charged positive ion in ^{4}He at ≈ 1.25 K, evaluated with (3.3.18) and (3.3.20) and data from Reference [259].

sphere with radius R_+ and density ρ_4 equal to that of the bulk fluid would have an effective mass $2\pi\rho_4 R_+^3$, where the hydrodynamic flow field contributes one-half the mass of fluid displaced [260]. If we use the radius and unperturbed density appropriate for Atkins' model ($R_+ \approx 6.8$ Å, $\rho_4 \approx 0.145$g/cm^3), this fictitious neutral entity in ^{4}He would have an effective mass $\approx 43m_4$. Barrera and Baym [261] have analyzed the corresponding motion of a charged sphere in a dielectric fluid. They find that the hydrodynamic contribution is slightly less than the classical value $\frac{2}{3}\pi\rho_4 R_+^3$ because the increased density near the ionic surface reduces the local velocity and kinetic energy relative to that for a neutral sphere. Their final value $M_+^* \approx 40m_4$ is very close to Atkins', although the resulting physical picture is quite different from the original static model. At present, the only direct experimental data on M_+^* were obtained with a microwave-relaxation technique by Dham and Sanders [256], who give $M_+^* = (37 \pm 11)m_4$ and a corresponding radius $R_+ = (5.7 \pm 0.6)$ Å, about 1 Å smaller than the solid boundary in Atkins' model (see, however, Addendum). This radius may also be compared with the value (7.9 ± 0.1) Å derived by Parks and Donnelly [262] from the binding energy of positive ions to vortex rings (see Section 3.4.3) and with the various values (see Section 3.3.3) obtained from the mobility of positive ions in different regimes of temperature and pressure.

Atkins' original model implies a curious behavior when the external pressure approaches the solidification pressure p_s. In this limit, equation (3.3.18) predicts that the core radius should increase like $(p_s - p_\infty)^{-1/4}$. Pratt and Zimmermann [263] saw no trace of such an effect in their study of ion trapping by vortex lines in rotating ^4He (II) above 1.1 K; they therefore concluded that R_+ remained smaller than $\approx$ 12 Å, even when liquid and solid coexist. This apparent failure of equation (3.3.18) is not necessarily serious, for Atkins had previously noted that the analysis neglects the possibility of a liquid–solid surface tension σ_{ls} [251]. Indeed, recent experiments on positive-ion mobility can be fitted with $\sigma_{ls} \approx 0.1$ dyne/cm, which suffices to eliminate the effect completely [264–267]

The success of the snowball model indicates that it correctly incorporates the relevant physical phenomena. In particular, it provides a concrete description of the local compression due to the attractive polarization potential. Nevertheless, the reliance on a classical continuum is somewhat unsatisfactory, especially in the core region, where the resulting density and pressure vary appreciably in an atomic dimension. Unfortunately even bulk uniform ^4He has eluded a full quantum-mechanical description, so that any attempt at a microscopic theory of an ion in ^4He is impractical. The only progress in this program has been with simple models involving a charged impurity in an imperfect Bose gas [268–270]. In principle, such calculations provide a slightly more satisfactory description, but their final predictions are very similar to those given here.

The previous discussion of positive ions has considered only ^4He, but a qualitatively similar structure should occur in liquid ^3He. In this case, equation (3.3.18) implies a smaller solidification radius $R_{+3} \approx 6.2$ Å at zero applied pressure, arising from both the increased volume per particle and the increased solidification pressure (Table 3.1). Moreover, the effective mass should be correspondingly reduced; for example, a neutral sphere with radius 6.2 Å and density 0.0815 g/cm^3 would have a total effective mass $\approx 24m_3$. The only relevant experiments are studies of the mobility of positive ions in ^3He, where a combination of Stokes' law with the measured viscosity gives a crude value $R_{+3} \approx 4.5$ Å [271–272]. Since the theoretical analysis is particularly difficult in this case (see Section 3.3.3), such an estimate may not be reliable.

3.3.2. Structure of a Negative Ion

The short lifetime of a free negative He ion [246] eliminates the "snowball" as a possible negative ion in liquid helium. Among the various alternatives, the currently accepted configuration involves a cavity or "bubble." Careri and his co-workers [18, 253, 254] first suggested this form for the negative ion in helium (see also Reference [273]) although Ferrell [274] had previously proposed a similar structure for positronium to explain its anomalously long lifetime in liquid ^{4}He.

The physical basis for the cavity [275] is the electron-helium potential, which is attractive at large distances [see (3.3.11)] but repulsive at short distances because of the Pauli principle. The description becomes especially simple at low energy, where the s-wave scattering length a suffices to characterize the interaction. This quantity is positive, which implies a net repulsion, and has the value $a \approx 1.19\ a_0 \approx 0.63$ Å, where $a_0 = \hbar^2/m_e e^2 \approx 0.529$ Å is the Bohr radius [276–278]. As a result, it requires work to inject an electron into liquid helium. To estimate the magnitude of this energy barrier, we introduce the total electron-helium potential energy

$$V_{\text{e-He}} = \sum_{i=1}^{N} U(\mathbf{r} - \mathbf{r}_i) \tag{3.3.21}$$

obtained by summing the potential $U(\mathbf{r} - \mathbf{r}_i)$ between the single electron at $\mathbf{r}$ and all the helium atoms at $\{\mathbf{r}_i\}$. If U is temporarily taken as integrable, the background medium shifts the energy of an electron by a first-order amount

$$\begin{aligned} \Delta E &= \int d^3r\, d^3r_1 \cdots d^3r_N\, |\psi_{\text{el}}(\mathbf{r})|^2 |\Phi_0(\mathbf{r}_1, \cdots, \mathbf{r}_N)|^2 V_{\text{e-He}} \\ &= \sum_{i=1}^{N} \int d^3r\, d^3r_1 \cdots d^3r_N\, |\psi_{\text{el}}(\mathbf{r})|^2 |\Phi_0(\mathbf{r}_1, \cdots, \mathbf{r}_N)|^2\, U(\mathbf{r} - \mathbf{r}_i) \\ &= NV^{-1} \int d^3r\, d^3r_1\, |\psi_{\text{el}}(\mathbf{r})|^2 U(\mathbf{r} - \mathbf{r}_1) \end{aligned} \tag{3.3.22}$$

where $\psi_{\text{el}}(\mathbf{r})$ is the unperturbed electron wave function and Φ_0 is the exact helium ground-state wave function. Here the third line follows from the symmetry and normalization of Φ_0. In the low-energy limit, ψ_{el} reduces to a constant $V^{-1/2}$, and we find

$$\Delta E \approx NV^{-1} \int d^3r\ U(\mathbf{r}) = n \int d^3r\ U(\mathbf{r}) \tag{3.3.23}$$

The volume integral of U is merely $2\pi\hbar^2/m_{\text{red}}$ times the Born approximation a_B to the s-wave electron-helium atom scattering length, where $m_{\text{red}} \approx m_e$ is the reduced mass [279, 280]. We may therefore improve our estimate of ΔE by inserting the exact scattering length $a_B \to a$; the resulting expression [275]

$$\Delta E \approx \frac{2\pi n a \hbar^2}{m_e} \approx 0.66 \text{ eV} \tag{3.3.24}$$

remains valid even for a singular potential.

This first-order estimate for ΔE assumes that each helium atom is uniformly and independently distributed. It therefore neglects the He–He correlations in Φ_0, which tend to keep the atoms apart and thus increase their effective interaction with the electron. Such a "shadowing" correction [281] leads to an additional factor [282] $1 - 4\pi an \int_0^\infty dr\, rh(r)$, where $h(\mathbf{r} - \mathbf{r}') = g(\mathbf{r} - \mathbf{r}') - 1$ is the two-body correlation function in bulk helium [283, 284], As a simple model for the effect of the short-range repulsion, we may take a step function $g(r) = \theta(r - r_0)$, where r_0 is determined by the sum rule [282, 283]

$$\int d^3r\, h(r) = -n^{-1} \tag{3.3.25}$$

The value $r_0 \approx 2.22$ Å appropriate to ^{4}He (Table 3.1) yields an improved energy shift

$$\Delta E \approx \left(\frac{2\pi n a \hbar^2}{m_3}\right)(1 + 2\pi\, a n r_0^2) \approx 0.94\,\text{eV}, \tag{3.3.26}$$

in good qualitative agreement with more detailed analyses [285, 286], which give $\Delta E \approx 1.09$ eV and ≈ 1.04 eV. For comparison, experimental studies obtain $\Delta E = (1.3 \pm 0.3)$ eV and (1.02 ± 0.08) eV from transmission of electrons into liquid ^{4}He through the free surface [287] and from direct photoinjection [288], respectively.

The previous discussion indicates that bulk helium presents an energy barrier to an electron of ≈ 1 eV relative to vacuum. This rather large value favors the formation of a localized electron state, in which the electron avoids the helium atoms by carving out a spherical cavity of radius R. The surrounding fluid exerts a net inward pressure across the surface, which is balanced by the zero-point pressure of the electron. If the helium density is approximated by a step function, the effect of the neighboring helium atoms may be described with an effective surface tension σ. Moreover, the large repulsive energy ΔE permits us to treat the walls of the resulting square well as impenetrable, giving the ground-state electronic energy $\hbar^2\pi^2/2m_e R^2$. Consequently, the total energy of the ion becomes [274, 289]

$$E_T = \frac{\hbar^2\pi^2}{2m_e R^2} + 4\pi\sigma R^2 + \frac{4}{3}\pi R^3 p, \tag{3.3.27}$$

where the last term represents the effect of an external hydrostatic pressure p[290]. As expected on physical grounds, E_T has a minimum at an equilibrium radius R_-, determined by the condition

$$p = \frac{\hbar^2\pi}{4m_e R_-^5} - \frac{2\sigma}{R_-}. \tag{3.3.28}$$

Using the bulk ^{4}He surface tension (Table 3.1) [170], we find the zero-pressure values [263, 287, 289]

$$R_-(p = 0) = \left(\frac{\hbar^2 \pi}{8m_e \sigma}\right)^{1/4} \approx 18.9 \text{ A}, \tag{3.3.29a}$$

$$E_T(p = 0) = 2\left(\frac{2\sigma \hbar^2 \pi^3}{m_e}\right)^{1/2} \approx 0.21 \text{ eV}, \tag{3.3.29b}$$

It is evident that the cavity shrinks with increasing pressure; if σ is assumed constant, equation (3.3.28) predicts $R_- \approx 12.5$ Å at the ^{4}He solidification pressure of 25 atm [263].

The present model omits several interesting physical effects:

i. The polarizable medium surrounding the cavity alters the energy (3.3.27) by an amount [287] $-(\varepsilon - 1)e^2/2\varepsilon R$, obtained as the change in the total electrostatic energy $(8\pi)^{-1}\int \vec{\mathscr{D}}\cdot\vec{\mathscr{E}}\, d^3r$ due to the presence of the dielectric fluid for $r > R$. This small negative contribution (≈ -0.02 eV) by itself shifts the zero-pressure radius to ≈ 18.5 Å [263]. In contrast to the positive ion, the large radius here renders polarization effects rather unimportant.

ii. The square well has a finite depth ≈ 1 eV, which slightly decreases the localization of the electron [291] and therefore reduces its ground-state energy and zero-point pressure. By itself, this effect shifts the zero-pressure radius to ≈ 17.6 Å [263]. Since the well depth (3.3.24) or (3.3.26) increases with increasing density and pressure, the finite value of ΔE becomes less significant at elevated pressure.

iii. The concept of an effective surface tension may not be applicable on a microscopic scale, and Kuper [292] instead ascribes the compressional force on the bubble surface to the unbalanced pressure in the bulk fluid. His essentially macroscopic analysis yields a radius $R_-(p = 0) \approx 12$ Å, considerably smaller than (3.3.29a). Gross [293] (see also Reference [294, Section 2]) has compared the two approaches in some detail. As a compromise, it is also possible to evaluate the surface tension from first principles instead of using the bulk free-surface value. The resulting estimate [295] ≈ 0.53 dyne/cm tends to reduce the bubble radius. When combined with the finite well depth, this approach predicts [286, 295] $R_-(p = 0) \approx 12.4$ Å, but the pressure dependence of R_- is now complicated by the density dependence of σ [156, 296].

iv. The step function cannot be considered a realistic description of the helium density near the electron. Since an accurate solution would involve the difficult $N + 1$ body problem, most analyses have used a simpler Hartree model [275, 293, 297]. Such a calculation raises the same questions discussed above in (iii). Numerical estimates for an imperfect Bose gas yield [297] $R_-(p = 0) \approx 12.5$ Å and $E_T(p = 0) \approx 0.16$ eV, but these values are not reliable for the relatively high density of bulk helium.

The bare mass of the negative ion is negligible, with the effective mass

M_-^* arising solely from the hydrodynamic contribution of half the mass of fluid displaced [260]

$$M_-^* \approx \frac{2}{3} \pi R_-^3 \, m_4 n_4. \tag{3.3.30}$$

For fixed fluid density, M_-^* varies rapidly with R_-, from $\approx 315 m_4$ for $R_- =$ 19 Å to $\approx 80 m_4$ for $R_- = 12$ Å; in the latter case, polarization effects may become important. The only *direct* measurement of M_-^* is the microwave experiment of Dahm and Sanders [256], who conclude that M_-^* lies between $100 m_4$ and $200 m_4$ (see, however, Addendum). When combined with equation (3.3.30), they obtain the value $R_-(p = 0) = (31.5 + 3, - 1)$ Å, substantially smaller than predicted by the simplest bubble model, but in reasonable agreement with some of the improved versions.

In addition to the static ground-state configuration, the negative ion also can support a variety of interesting dynamical oscillations. These may be divided into mechanical vibrations of the cavity and electronic excitations. In the first case, the typical mechanical frequencies (10^{11}–10^{12} rad/sec) are small compared to the spacing between electronic energy levels ($\Delta E/\hbar \approx 2 \times 10^{14}$ rad/sec). Consequently, the electron remains in its ground state, which is only adiabatically perturbed by the slowly varying cavity. Several calculations [294, 298–300] predict the existence of a breathing mode and surface modes with deformation proportional to the lth spherical harmonic ($l \geqslant 2$). (The mode with $l = 1$ represents a pure translation.) The breathing mode is expected to have a frequency of order 10^{11} rad/sec, but it may be significantly broadened by phonon emission into the bulk fluid [298]. In contrast, the $l = 2$ quadrupole mode should be quite sharp with a somewhat higher frequency. The existence of these mechanical oscillations was first proposed by Cope and Gribbon [300] to explain certain discrete structure in the low-field mobility of negative ions [300–303]. They ignored the presence of the electron, however, which alters the frequency in a qualitative way [294, 298]. Moreover, the original effect itself remains controversial (see the discussion in Section 3.3.3). At present, we know of no convincing experimental evidence for such vibration modes, and their detection is an interesting challenge.

The other modes of the negative ion represent excited electron states in a square well of finite depth [304]. For each distinct electron state, the equilibrium radius is obtained by minimizing the analog of equation (3.3.27) [299]; the resulting electron wave functions have been used to analyze two distinct optical effects: photoabsorption [299, 305, 306] and inelastic (Raman) scattering, [305, 307, 308] which we shall now consider in sequence.

i. As a function of the photon energy, the theoretical photoabsorption cross section exhibits a few narrow peaks, corresponding to transitions from the $1s$ ground state to the various bound p states, and a much broader low peak representing transitions to the continuum [299, 306]. The $1s$–$1p$ peak at a photon frequency $\omega \approx 2 \times 10^{14}$ rad/sec dominates the spectrum, with nearly unit f value and a maximum absorption cross section of $\approx$ 20 Å^2. (Note that the lowest p state is conventionally denoted $1p$, as in nuclear physics, instead of $2p$, as in atomic physics.) The higher transition $1s$–$2p$ gives a much smaller peak at $\approx 8 \times 10^{14}$ rad/sec with a maximum absorption cross section $\approx$ 0.3 Å^2. Increasing pressure shifts the structure systematically to higher frequency, largely because the decreased bubble radius raises the electronic energy. These calculations were motivated by a series of photoabsorption experiments with negative ions in ^{4}He (II) by Northby, Sanders, and Zipfel [309, 310], in which the absorption of infrared photons with wavelength $\approx$ 1 micron was detected through the appearance of an electron photocurrent. Although the actual source of this current remains somewhat obscure, it seems likely that photoabsorption indeed took place. The original analysis relied on a simple square-well model [311] and ascribed the observed resonant photoabsorption cross section to transitions from the $1s$ state into the continuum. This approach yielded a radius $R_-(p = 0) \approx$ 16 Å, decreasing to $\approx$ 12 Å at 18 atm, but it implied a surprisingly small well depth of $\approx$ 0.6 eV [309, 310], The refined version of References [299] and [306] reanalyzes the data in terms of the discrete $1s$–np transitions ($n \gg 2$) and yields an improved well depth of $\approx$ 1 eV, in good agreement with direct experimental values [287, 288]. This latter theory also predicts a strong $1s$–$1p$ transition at longer wavelengths (10–25 micron) than those used previously [306]; an experimental search for such a feature would be most desirable.

ii. The other possible optical process is Raman scattering, in which the incident photon is scattered inelastically, exciting the electron from its original $1s$ ground state to another even-parity state, such as $2s$ or $1d$ [305, 307, 308]. The resulting scattering cross sections are of order 10^{-29} cm^2, much smaller than those for photoabsorption, because the Raman process proceeds through an intermediate state. Even with lasers, this effect is probably undetectable at presently available negative-ion densities ($\lesssim 10^8$ cm^{-3}), but it might become feasible at densities of order 10^{12} cm^{-3} [308].

We have already treated two distinct experimental studies of the electron bubble: The relaxation time determines the effective mass $M_\pm^*$ [256], and the photoabsorption yields the effective well depth and radius [306, 309, 310]. Other important approaches rely on the negative-ion mobility in ^{4}He (I) and ^{4}He (II) (Section 3.3.3), the escape time of electrons trapped in bubbles near the free fluid surface (Section 3.3.4), and negative-ion trapping by vortex lines and rings (Section 3.4.3). Although various experimental and theoretical values differ in detail, they generally confirm the basic bubble model with $R_-(p = 0) \approx$ 16 Å, $M_\pm^* \approx 190 m_4$, and a well depth of $\approx$ 1 eV [296]. Furthermore, the experimental pressure dependence agrees with well the empirical relation [312] (see also References [263, 313, 314])

$$\frac{R_-(p)}{R_-(p = 0)} \approx 1 - 0.077p^{1/2}, \tag{3.3.31}$$

where p is in atmospheres.

Our analysis has considered only a negative ion in liquid ^{4}He, but the same description applies to ^{3}He, to mixtures of ^{3}He– ^{4}He, to solid helium [315–318] and even to certain other liquids [319–321]. The detailed ionic structure in ^{3}He is somewhat altered, however, because of two separate effects. The decreased particle density (Table 3.1) reduces the well depth significantly; for example, equation (3.3.26) gives $\Delta E_3 \approx 0.68$ eV, and a more elaborate calculation predicts $\Delta E_3 = 0.74$ eV at 1 K [296]. In addition, the bulk surface tension is less than half the ^{4}He value (Table 3.1). For an infinite square well, the reduced σ by itself increases the equilibrium radius by a factor $\approx \frac{5}{4}$ and the effective mass (now measured in units of m_3) by a factor ≈ 1.9 [see equation (3.3.29a) and (3.3. 30)]. The true factors are somewhat less, because the smaller well depth decreases the zero-point pressure. These considerations suggest several interesting experiments. Photoemission [288] from a cathode immersed in bulk ^{3}He could determine ΔE_3 directly and thus test the theoretical estimates; moreover, various calculations predict the dependence of ΔE on n, which might be verified for both isotopes by photoemission experiments with the liquid at elevated pressure or with the vapor at reduced pressure [275]. The resulting square-well depth for ^{3}He could be combined with infrared photoabsorption by negative ions in ^{3}He to determine the corresponding bubble radius R_{-3} [306, 309, 310]. Although this last approach employs a specific model, it could in any case determine the ratio of the bubble radii in ^{4}He and ^{3}He (or, even more interesting, in mixtures, as discussed below). In principle, the relaxation-time technique [256] could also determine $M^*_\pm$ and R_- both in ^{3}He and in mixtures, but the greatly decreased ionic mobility might preclude such an experiment in pure ^{3}He. At present, mobility studies provide the only information on negative ions in ^{3}He, but the subsequent determination of R_{-3} is rather indirect (see Section 3.3.3).

The situation is more complicated in ^{3}He–^{4}He mixtures because the lighter ^{3}He atom exerts a higher zero-point pressure and occupies a larger volume than the ^{4}He atom it replaces [322, 323]. As a result, ^{3}He atoms tend to accumulate on a free surface, where they can reduce their localization energy [324–326]. For the same reason, the surface of a negative ion in a dilute superfluid mixture is likely to have an excess concentration of ^{3}He relative to the bulk fluid [327–329]. Such an effect was first suggested in connection with ionic critical velocities in mixtures (see Section 3.4.4) [327, 328], and it has subsequently been analyzed in some detail to explain the observed mobility of negative ions [329–332]. Since the atoms near the surface determine the general structure of the bubble, a small concentration of ^{3}He in ^{4}He should affect the bubble's radius and square-well depth more than might be expected from a linear interpolation between the values for the two pure fluids. Conversely, small concentrations of ^{4}He in ^{3}He should have relatively less effect because the ^{4}He atoms presumably avoid the surface region. Careful experiments [for example, photoemission of electrons from a cathode, photoabsorption by negative ions, or escape of electrons from bubbles near the surface (see Section 3.3.4)] in mixtures of varying concentrations could test these qualitative predictions.

We shall close this section with a brief summary of other localized excitations in liquid ^{4}He. Doake and Gribbon [333], and later Ihas and Sanders [334], have detected fast negative ions, which presumably are considerably smaller than the usual electron bubble. The latter authors [334] tentatively suggest a bubble structure containing an atomic He$^-$ ion whose lifetime could be much increased by the surrounding fluid [335]. They also report other charged entities, but the present sparse data preclude positive identification. In addition to these charged ions, liquid helium can support a variety to neutral excitations, which have been created both by radioactive sources [19, 146, 336–340] and by energetic electrons [341, 342]. These various entities lie somewhat outside the scope of the present chapter and will not be considered further.

3.3.3. Mobility of Ions

An ion or other impurity moving through liquid helium experiences a drag due to interactions with the particles of the liquid. At relatively high temperatures ($T \gtrsim 2$ K), the mean free path in the bulk fluid is comparable with the ionic radius, so that Stokes' law for viscous drag on a sphere provides an adequate description of the motion. As the temperature falls, however, the increasing mean free path necessitates a more fundamental approach, and the drag must be considered to arise from collisions with individual quasiparticles. Since the quasiparticle density in superfluid ^{4}He falls rapidly at low temperatures, the corresponding ionic mobility may be expected to increase dramatically [343]. In pure ^{3}He, on the other hand, the mobility remains relatively small and temperature independent because the impurity interacts with all particles inside a weakly temperature-dependent energy shell near the Fermi surface. In principle, this reduced scattering probability as $T \to 0$ implies that the low-temperature ionic mobility in pure ^{3}He ultimately should increase as T^{-2}, but there is no compelling evidence for such behavior above 17 mK. Each of these three cases (^{4}He and ^{3}He above ≈ 2 K, superfluid ^{4}He (II) below ≈ 1.5 K, and ^{3}He at low temperatures) requires a separate theoretical description, and we now consider them in order.

(a) *^{4}He and ^{3}He above 2 K.* If the atomic mean free path l in a fluid is short compared to other characteristic lengths, the system is able to achieve local thermodynamic equilibrium, and continuum hydrodynamics adequately describes the motion of an impurity. In the present case of a slow ion in liquid helium, the ionic radius R serves as the characteristic length, permitting a hydrodynamic description for $l \lesssim 10$ Å. To estimate l in the normal phase, we treat the fluid as a dilute gas and use the elementary kinetic expression for the viscosity [344]

$$\eta = \tfrac{1}{3}\rho l v, \tag{3.3.32}$$

where η is of order 25 μpoise both for liquid ^{3}He and for normal liquid ^{4}He [345]. If v is taken as the root-mean-square thermal velocity $(3k_BT/m)^{1/2}$, equation (3.3.32) gives $l \ll 6$ Å for $T \gtrsim 2$ K. In the superfluid phase, on the other hand, we may use Landau and Khalatnikov's estimate of the roton–roton cross section [346]

$$\sigma_{rr} \approx 5 \times 10^{-15}\, T^{-1/2}, \tag{3.3.33}$$

with T in degrees Kelvin and σ_{rr} in cm^2 to obtain the roton-roton mean free path [344]

$$l_{rr} = (\sqrt{2} n_r \sigma_{rr})^{-1}, \tag{3.3.34}$$

where n_r is the roton number density. With the typical values [26, 347] $T \approx 2$ K and $n_r(T) = 1.7 \times 10^{21}$ cm^{-3}, equation (3.3.34) yields $l_{rr} \approx 12$ Å. In all these cases, the relatively short mean free path justifies a qualitative hydrodynamic treatment of ionic mobility.

Consider a spherical ion with radius R and charge $\pm e$ moving with drift velocity $\mathbf{v}_D$ through a viscous fluid. Stokes' drag force [348] $\mathbf{F} = -6\pi\eta R\mathbf{v}_D f$ is balanced by the applied electric field

$$\pm e\vec{\mathscr{E}} + \mathbf{F} = \pm e\vec{\mathscr{E}} - 6\pi\eta R\mathbf{v}_D f = 0, \tag{3.3.35}$$

where f is a dimensionless constant with the value 1 for a hard sphere and $\frac{2}{3}$ for a bubble. Since the low-field mobility is defined by the relation

$$\mu = \lim_{\mathscr{E}\to 0} \frac{|v_D|}{|\mathscr{E}|}, \tag{3.3.36}$$

we immediately identify

$$\mu = \frac{e}{6\pi\eta R f}. \tag{3.3.37}$$

Traditionally, experimentalists measure μ in units of cm^2/V sec, but multiplication by 300 V/statV gives μ in the cgs units cm^2/statV sec. As an estimate, take $R = 10$ Å, $\eta = 25\mu$poise, and $f = 1$; equation (3.3.37) then predicts

$\mu = 3.4 \times 10^{-2}$ cm^2/V sec, in reasonable agreement with the experimental data. Indeed (3.3.37) determines an ionic Stokes' radius from the measured μ and η [271]:

$$R_+ = 4.3\ \text{Å},\quad R_- = 13\ \text{Å}\quad \text{in } ^4\text{He (I) at 3 K};$$
$$R_+ = 5\ \text{Å},\quad R_- = 21\ \text{Å}\quad \text{in } ^3\text{He at 3.2 K};$$

where we have multiplied the quoted negative radii by $\frac{3}{2}$ to account for the altered boundary condition. Although continuum hydrodynamics cannot be wholly correct, these values are surprisingly plausible (see Sections 3.3.1 and 3.3.2). More recent work has improved the description of the ionic structures [266] and considered the behavior through the phase transitions at T_c [266, 349–352] (for both isotopes) and at T_λ [353, 354]. For ^{4}He at the vapor pressure, μ_+ and μ_- have been measured accurately by Brody [265] (1.25K < T < 2.2 K) and by Schwarz [355] (0.27 K < T < 5.18 K); for ^{3}He, on the other hand, the data are much less complete, being largely restricted to $T < 1$ K [356–357a] (see, also References [271, 272]).

Throughout our discussion of negative ions, we have explicitly assumed a hollow cavity that eliminates the strong electron-helium repulsion. In his original theoretical treatment, however, Atkins [251, 252] also considered the possibility of an extended electronic state, which greatly reduces the kinetic of localization. The bubble state turns out to be favored at the liquid density but there is good evidence that the extended state actually occurs in the vapor phase at low density. Levine and Sanders [275] studied the mobility of negative ions in ^{4}He gas for $T < T_c$ and p below the saturated vapor pressure p_{sat}. At sufficiently low density, they found a high mobility ($\mu_- \approx 10^4$ cm^2/V sec for $p \approx 0.1\, p_{\text{sat}}$), which fell sharply to $\approx 10^{-1}$ cm^2/V sec as $p \to p_{\text{sat}}$. Since the low-density mobility is comparable with that expected for a classical free electron in a dilute gas [275], this dramatic decrease in μ is interpreted as a transition from an extended state at low density to a localized bubble at high density. Recent theoretical work has had considerable success in fitting the data throughout the experimental range of temperature and pressure [358–360].

(b) *Low-Temperature Mobility in Superfluid ^{4}He [360a].* For normal helium above 2 K, the short He–He mean free path permits an approximate hydrodynamic treatment of slow ionic motion through the fluid. In the superfluid phase of ^{4}He or of dilute ^{3}He–^{4}He mixtures, however, this picture becomes incorrect for $T \lesssim 2$ K, because the mean free path increases rapidly. An accurate description of the transition region is very difficult, but the situation again becomes simple at lower temperature, when the bulk fluid instead may be considered a gas of weakly interacting quasiparticles (rotons, phonons, and ^{3}He impurities) moving in an inert superfluid background. A fixed ion in stationary fluid would clearly experience no net force, because the quasiparticle distribution function is isotropic; a slowly moving ion, in contrast, suffers a net retardation from the asymmetry in the momentum transferred by the collisons. To calculate the resulting mobility, we first treat a single collision and then average over the relevant distribution functions [361]. In general, the kinematics is complicated by ionic recoil, which turns out to be especially significant in pure ^{3}He; for superfluid ^{4}He, on the other hand, we now show that recoil is relatively unimportant.

Assume that the ions constitute an ideal Boltzmann gas in thermodynamic equilibrium at temperature T. The mean ionic energy is $\frac{3}{2}k_BT$, and the mean ionic velocity has a random direction with approximate magnitude

$$\bar{v}_i = \left(\frac{3k_BT}{M^*}\right)^{1/2}, \tag{3.3.38}$$

where M^* is the ionic effective mass. Consider an ion that collides with a quasiparticle, acquiring an additional momentum $\hbar\mathbf{q}$. The corresponding final energy $(M^*\bar{\mathbf{v}}_i + \hbar\mathbf{q})^2/2M^*$ implies a mean recoil energy

$$\langle E_{\text{rec}}\rangle = (2M^*)^{-1}\langle(M^*\bar{\mathbf{v}}_i + \hbar\mathbf{q})^2 - (M^*\bar{\mathbf{v}}_i)^2\rangle = \frac{\hbar^2q^2}{2M^*} \tag{3.3.39}$$

because $\langle\bar{\mathbf{v}}_i\cdot\mathbf{q}\rangle$ vanishes for an isotropic distribution. Similarly, the spread in recoil energies may be characterized by

$$\begin{aligned}\langle \Delta E_{\text{rec}}\rangle &\equiv [\langle E_{\text{rec}}^2\rangle - \langle E_{\text{rec}}\rangle^2]^{1/2}\\ &= \frac{\bar{v}_i\hbar q}{\sqrt{3}} = \hbar q\left(\frac{k_BT}{M^*}\right)^{1/2}\\ &= [2k_BT\langle E_{\text{rec}}\rangle]^{1/2}.\end{aligned} \tag{3.3.40}$$

If $\langle E_{\text{rec}}\rangle$ (and hence $\langle \Delta E_{\text{rec}}\rangle$) is small compared to k_BT, recoil effects play a negligible role in the ionic motion, which may therefore be considered to arise from a series of elastic collisions.

We now apply these general ideas to ionic collisions with the three separate quasiparticles in superfluid ^{4}He and verify that each approximately satisfies the criteria for small recoil.

i. The typical roton momentum [362] $\hbar k_0 \approx \hbar(1.9\ \text{Å}^{-1})$ provides an estimate for the momentum transferred to an ion in a collision with a roton. Equation (3.3.39) then gives the mean recoil energy

$$\langle E_{\text{rec}}\rangle_{\text{r}} \approx \frac{\hbar^2k_0^2}{2M^*} \approx \frac{k_BT_{\text{r}}m_4}{M^*}, \tag{3.3.41}$$

where

$$T_{\text{r}} \equiv \frac{\hbar^2k_0^2}{2m_4k_B} \approx 22\ \text{K}, \tag{3.3.42}$$

defines a characteristic "roton" temperature. As a result, the important dimensionless ratio becomes

$$\frac{\langle E_{\text{rec}}\rangle_{\text{r}}}{k_B T} = \frac{T_{\text{r}}}{T}\frac{m_4}{M^*}, \tag{3.3.43}$$

implying that recoil effects are small for $T \gg T_{\text{r}} m_4/M^*$. Taking $M_+^* \approx 40m_4$ and $M_-^* \approx 190m_4$, we find the conditions $T \gg 0.55$ K and $T \gg 0.12$ K for positive and negative ions, respectively. Since rotons are important only above ≈ 0.8 K [362], ionic recoil in roton collisions is generally small for negative ions, but it may become significant for positive ions, especially below 1 K. As a first approximation, however, we shall neglect the recoil entirely, treating ion–roton collisions as elastic.

ii. Unlike the situation for rotons, the typical phonon wave number $k_B T/\hbar s$ varies linearly with temperature, where $s \approx 2.4 \times 10^4$ cm/sec is the speed of first sound in pure ^{4}He [169]. Correspondingly, equation (3.3.39) implies

$$\langle E_{\text{rec}}\rangle_{\text{ph}} \approx k_B T\left(\frac{T m_4}{2T_{\text{ph}} M^*}\right), \tag{3.3.44}$$

where

$$T_{\text{ph}} \equiv \frac{m_4 s^2}{k_B} \approx 28 \text{ K}. \tag{3.3.45}$$

It is evident from inspection that $\langle E_{\text{rec}}\rangle_{\text{ph}} \ll k_B T$, so that ion–phonon scattering is always elastic.

iii. Dilute ^{3}He impurities constitute a weakly interacting Fermi gas [363–365] with number density $\approx x_3 n_4$ and Fermi wave number

$$k_F = (3\pi^2 x_3 n_4)^{1/3}, \tag{3.3.46}$$

where x_3 is the small fractional impurity concentration, and we here neglect the increased zero-point volume of an ^{3}He atom. The corresponding Fermi temperature in degrees Kelvin is

$$T_F = \frac{\hbar^2 k_F^2}{2m_3^* k_B} \approx 2.5 x_3^{2/3}, \tag{3.3.47}$$

where $m_3^* \approx 2.4 m_3$ is the effective mass [365]; the typical value $x_3 = 0.05$ yields $T_F = 0.34$ K. For $T \gg T_F$, the impurities act like a Boltzmann gas, with a thermal velocity $(3k_B T/m_3^*)^{1/2}$; for $T \ll T_F$, the Pauli principle restricts the available states, and the mean velocity is comparable with $v_F \equiv \hbar k_F/m_3^*$. As a simple interpolation formula, we take $(3k_B T m_3^* + \hbar^2 k_F^2)^{1/2}$ as the mean momentum of a ^{3}He atom. When an ion collides with an impurity, the ionic recoil energy is

$$\langle E_{\text{rec}}\rangle_3 = \left(\frac{m_3^*}{M^*}\right)\left(\frac{3}{2}k_B T + k_B T_F\right), \tag{3.3.48}$$

which is smaller than $k_B T$ for $T > T_0$, where

$$T_0 \equiv \frac{T_F m_3^*}{M^*}. \tag{3.3.49}$$

Since the limiting low-temperature solubility of ^{3}He is $x_3 \approx 0.06$, recoil of positive ions in impurity scattering is negligible for $T \gg T_{0+} \approx 15$ mK, where we use $m_3^*/M_+^* = 0.045$ and $x_3 = 0.05$. The situation is even more favorable for negative ions because $M_-^*/M_+^* \approx 4.8$.

The foregoing considerations allow us to calculate the ionic mobility in superfluid helium with the following simple and elegant argument of Baym, Barrera, and Pethick [366]: Consider an ion moving slowly with drift velocity $\mathbf{v}_D$ through stationary fluid. In a particular collision, suppose that a quasiparticle is scattered from an incident momentum $\hbar\mathbf{k}$ to a final momentum $\hbar\mathbf{k}'$ with a transition probability $\Gamma_{\mathbf{v}_D}(\mathbf{k} \to \mathbf{k}')$ per unit time. The net rate for this process is given by $f(\varepsilon_k)\,[1 \pm f(\varepsilon_{k'})]\,\Gamma_{\mathbf{v}_D}(\mathbf{k} \to \mathbf{k}')$, where

$$f(\varepsilon) = [e^{\beta(\varepsilon - \bar{\mu})} \mp 1]^{-1} \tag{3.3.50}$$

is the equilibrium distribution function at temperature $T \equiv 1/k_B\beta$ for quasiparticles with energy ε and chemical potential $\bar{\mu}$, and the upper (lower) sign refers to bosons (fermions). Equation (3.3.50) simplifies somewhat for rotons and phonons because $\bar{\mu}$ vanishes for quasiparticles whose number is not conserved [367]. The second factor $(1 \pm f)$ in the net rate reflects the quantum statistics in the final state. As a result of all such scattering events, the total quasiparticle momentum $\mathbf{P}_{qp}$ increases at a rate

$$\frac{\mathbf{dP}_{qp}}{dt} = (2s+1)^2 \sum_{\mathbf{kk}'} \hbar(\mathbf{k} - \mathbf{k}')\, f(\varepsilon_k)\,[1 \pm f(\varepsilon_{k'})]\,\Gamma_{\mathbf{v}_D}(\mathbf{k} \to \mathbf{k}'), \tag{3.3.51}$$

where the factor $(2s + 1)^2$ incorporates the spin degeneracy for the case of ^{3}He impurities. This expression depends on $\mathbf{v}_D$ both through the transition rate Γ and through the Doppler shift of $\mathbf{k}'$ arising from the elastic scattering by a moving ion. Baym, Barrera, and Pethick now make the crucial observation that $d\mathbf{P}_{qp}/dt$ would obviously vanish identically if the quasiparticles were in thermal equilibrium with the moving ion; in this hypothetical case, $f(\varepsilon_k)$ would be replaced by [238]

$$\bar{f}(\varepsilon_k) \equiv f(\varepsilon_k - \hbar\mathbf{k}\cdot\mathbf{v}_D). \tag{3.3.52}$$

Consequently, equation (3.3.51) may be rewritten

$$\frac{d\mathbf{P}_{qp}}{dt} = (2s+1)^2 \sum_{\mathbf{kk}'} \hbar(\mathbf{k} - \mathbf{k}')\,\{f(\varepsilon_k)\,[1 \pm f(\varepsilon_{k'})] - \bar{f}(\varepsilon_k)\,[1 \pm \bar{f}(\varepsilon_{k'})]\}\,\Gamma_{\mathbf{v}_D}(\mathbf{k} \to \mathbf{k}'). \tag{3.3.53}$$

The quantity in braces is evidently of order $\mathbf{v}_D$, so that the first-order contribution to $d\mathbf{P}_{qp}/dt$ may be obtained by neglecting the Doppler correction to

$\mathbf{k}'$ and replacing $\Gamma_{\mathbf{v}_D}$ by its value Γ_0 for a stationary ion. Time-reversal invariance implies that Γ_0 is symmetric in $\mathbf{k}$ and $\mathbf{k}'$, and the antisymmetry of the first factor in (3.3.53) eliminates the term proportional to $f(\varepsilon_k)\,f(\varepsilon_{k'})$ and $\bar{f}(\varepsilon_k)\bar{f}(\varepsilon_{k'})$. Expanding the remaining terms to first order in $\mathbf{v}_D$, we obtain

$$\frac{d\mathbf{P}_{qp}}{dt} = (2s+1)^2 \sum_{\mathbf{k}} \hbar\mathbf{k}\cdot\mathbf{v}_D\left[-\frac{\partial f(\varepsilon_k)}{\partial \varepsilon_k}\right] \sum_{\mathbf{k}'} \hbar\,(\mathbf{k}-\mathbf{k}')\,\Gamma_0\,(\mathbf{k}\to\mathbf{k}'). \tag{3.3.54}$$

Consider a particular direction $\hat{n}$ described by the polar angles (θ_n, ϕ_n) with respect to the incident direction $\hat{k}$. The summation over $\mathbf{k}'$ will be performed first by requiring that $\hat{k}' = \hat{n}$, and then by integrating over $\hat{n}$ and the magnitude of k'. To ensure that $\hat{k}' = \hat{n}$, insert an angular δ function $\delta(\hat{n}-\hat{k}') \equiv \delta(\cos\theta_n - \cos\theta_{k'})\,\delta(\phi_n - \phi_{k'})$ into equation (3.3.54):

$$\frac{d\mathbf{P}_{qp}}{dt} = (2s+1)^2 \sum_{\mathbf{k}} \hbar\mathbf{k}\cdot\mathbf{v}_D\left[-\frac{\partial f(\varepsilon_k)}{\partial \varepsilon_k}\right] \int d\Omega_n \sum_{\mathbf{k}'} \delta(\hat{n}-\hat{k}') \times \hbar(\mathbf{k}-\mathbf{k}')\Gamma_0(\mathbf{k}\to\mathbf{k}'). \tag{3.3.55}$$

The equivalence of (3.3.55) and (3.3.54) follows immediately by carrying out the angular integration over Ω_n, but we now reverse the order of operations and evaluate the sum over $\mathbf{k}'$. In most cases of interest, the restriction to elastic scattering implies $k' = k$; this condition fails for the double-valued roton spectrum, but the approximation $k' \approx k \approx k_0$ suffices for all relevant temperatures. Consequently, equation (3.3.55) becomes

$$\frac{d\mathbf{P}_{qp}}{dt} = (2s+1) \sum_{\mathbf{k}} \hbar\mathbf{k}\cdot\mathbf{v}_D\left[-\frac{\partial f(\varepsilon_k)}{\partial \varepsilon_k}\right] \int d\Omega_n \hbar(\mathbf{k}-k\hat{n}) \times \left[(2s+1)\sum_{\mathbf{k}'} \delta(\hat{n}-\hat{k}')\Gamma_0(\mathbf{k}\to\mathbf{k}')\right]. \tag{3.3.56}$$

The quantity in square brackets is the total transition probability per unit time for a quasiparticle in an initial state $\mathbf{k}$ to scatter elastically to a final state with $\hat{k}' = \hat{n}$ (there are two such states for rotons); equivalently, it is the product of the differential cross section $d\sigma/d\Omega_n$ and the incident quasiparticle flux v_k/V [368], where V is the volume of the sample and

$$v_k = \hbar^{-1}\left|\frac{\partial \varepsilon_k}{\partial k}\right| \tag{3.3.57}$$

is the speed of the quasiparticle. Consequently, the momentum-transfer rate becomes

$$\frac{d\mathbf{P}_{qp}}{dt} = \frac{(2s+1)}{V} \sum_{\mathbf{k}} \hbar\mathbf{k}\cdot\mathbf{v}_D\left[-\frac{\partial f(\varepsilon_k)}{\partial \varepsilon_k}\right]\left|\frac{\partial \varepsilon_k}{\partial k}\right| \times \int d\Omega_n\,(\mathbf{k}-k\hat{n})\,\frac{d\sigma}{d\Omega_n}. \tag{3.3.58}$$

As a final step, note that the integral over $\hat{n}$ necessarily produces a vector parallel to $\hat{k}$ with magnitude

$$\hbar k \int d\Omega_n (1 - \hat{k}\cdot\hat{n})\left(\frac{d\sigma}{d\Omega_n}\right) \equiv \hbar k\, \sigma_{\text{tr}}(k), \tag{3.3.59}$$

where $\sigma_{\text{tr}}(k)$ is the transport cross section. When the sum over $\mathbf{k}$ is replaced by an integral, we obtain the result

$$\begin{aligned}\frac{d\mathbf{P}_{qp}}{dt} &= \frac{(2s+1)\hbar}{(2\pi)^3}\int d^3k\, \mathbf{k}(\mathbf{k}\cdot\mathbf{v}_D)\left[-\frac{\partial f(\varepsilon_k)}{\partial \varepsilon_k}\right]\left|\frac{\partial \varepsilon_k}{\partial k}\right|\sigma_{\text{tr}}(k)\\ &= \frac{(2s+1)\hbar}{6\pi^2}\mathbf{v}_D\int_0^\infty k^4\,dk\left[-\frac{\partial f(\varepsilon_k)}{\partial \varepsilon_k}\right]\left|\frac{\partial \varepsilon_k}{\partial k}\right|\sigma_{\text{tr}}(k).\end{aligned} \tag{3.3.60}$$

This force on the quasiparticles must precisely equal the applied electric force $\pm e\vec{\mathscr{E}}$, which thus gives the desired mobility [366]:

$$\frac{e}{\mu} = \frac{(2s+1)\hbar}{6\pi^2}\int_0^\infty k^4\,dk\left[-\frac{\partial f(\varepsilon_k)}{\partial \varepsilon_k}\right]\left|\frac{\partial \varepsilon_k}{\partial k}\right|\sigma_{\text{tr}}(k). \tag{3.3.61}$$

Other authors had previously obtained essentially the same expression, but the derivations all assumed that the scattering object moved with the superfluid [369, 370]; in contrast, the present approach of Baym, Barrera, and Pethick is completely general.

When rotons, phonons, and ^{3}He are present simultaneously, each species separately transfers momentum to the ion, and the corresponding mobilities add inversely

$$\mu^{-1} = \mu_r^{-1} + \mu_{\text{ph}}^{-1} + \mu_3^{-1} \tag{3.3.62}$$

To interpret the various terms, it is convenient to define a thermal-averaged transport cross section

$$\bar{\sigma} = \frac{\int_0^\infty k^4\,dk[-\partial f(\varepsilon_k)/\partial\varepsilon_k]\,|\partial\varepsilon_k/\partial k|\,\sigma_{\text{tr}}(k)}{\int_0^\infty k^4\,dk[-\partial f(\varepsilon_k)/\partial\varepsilon_k]\,|\partial\varepsilon_k/\partial k|} \tag{3.3.63}$$

and equation (3.3.61) then becomes

$$\frac{e}{\mu} = \frac{(2s+1)\hbar}{6\pi^2}\bar{\sigma}\int_0^\infty k^4\,dk\left[-\frac{\partial f(\varepsilon_k)}{\partial \varepsilon_k}\right]\left|\frac{\partial \varepsilon_k}{\partial k}\right|. \tag{3.3.64}$$

This useful form was first introduced by Rayfield and Reif [29]; since the integral is easily evaluated for any particular quasiparticle, (3.3.64) provides a direct relation between μ and $\bar{\sigma}$.

Rotons. The roton spectrum may be approximated as [362]

$$\varepsilon_k = \Delta + \frac{\hbar^2(k-k_0)^2}{2m_r}, \tag{3.3.65}$$

with $\Delta/k_B \approx 8.65$ K, $K_0 \approx 1.9$ Å^{-1}, and $m_r \approx 0.16 m_4$. The corresponding group velocity $\hbar^{-1}\,\partial\varepsilon_k/\partial k = \hbar(k-k_0)/m_r$ can take either sign, so that the absolute value in equations (3.3.63) and (3.3.64) (omitted in Reference [370]) becomes essential. The inequality $e^{-\beta\Delta} \ll 1$ allows the use of Boltzmann statistics, and a simple calculation gives the approximate expression

$$\frac{e}{\mu_r} = \left(\frac{\hbar k_0^4}{3\pi^2}\right) \bar{\sigma}_r \, e^{e-\beta\Delta} \cdot \tag{3.3.66}$$

The most striking feature here is the strong (approximately exponential) temperature dependence, which may be ascribed qualitatively to the equilibrium roton density [347]

$$n_r = \left(\frac{k_0^2}{\pi\hbar}\right)\left(\frac{m_r k_B T}{2\pi}\right)^{1/2} e^{-\beta\Delta}. \tag{3.3.67}$$

The roton-limited mobility was first studied by Meyer and Reif [23–25], who verified the near linear dependence of ln μ_r on T^{-1} for both species above ≈ 0.8 K. Their original analysis used elementary kinetic theory and predicted that μ_r varies inversely with the ionic mass M^* [25, 314, 371]. Such an effect is absent in the present description because of the explicit neglect of ionic recoil. Bowley [372], and Barrera and Baym [261] have evaluated the recoil corrections in detail; as expected from equation (3.3.43), the additional terms are of order $\hbar^2 k_0^2 / 2M^* k_B T$, which is small in most cases of interest.

Meyer and Reif [373] (see also References [374, 374a]) also measured e/μ_r at elevated pressure, where the roton gap parameter decreases from its zero-pressure value $\Delta/k_B \approx 8.65$ K to ≈ 7.0 K at 25 atm [362]. For negative ions, the mobility first increases with increasing pressure due to the decreased bubble radius, but it eventually decreases again because of the reduced gap Δ. For positive ions, only the second effect occurs and μ_{r+} decreases monotonically with increasing pressure. For $p \gtrsim 10$ atm, μ_{r+} and μ_{r-} become nearly equal [373], which has not been explained by present theories.

Recent work at low pressure has been summarized by Schwarz, whose data imply $\bar{\sigma}_{r+} \approx 2.8 \times 10^{-14}\, T^{-1/2}$ cm^2 for the positive ion [see equation (3.3.66)] [267, 375], where T is measured in degrees Kelvin. The data for the negative ion are less certain, but the cross section is of order [267] $\bar{\sigma}_{r-} \approx 4.7 \times 10^{-14}$ cm^2, with a weak temperature dependence lying between T^0 and $T^{-1/2}$. It is notable that the two cross sections are so similar in magnitude, considering that the bare radii differ by a factor $\approx$ 2–3. If $\bar{\sigma}_r$ is interpreted as arising from the short-wavelength transport cross section [376] for a hard sphere of radius $a \equiv (\bar{\sigma}_r/\pi)^{1/2}$, then the positive and negative ions have equivalent radii [377] $a_+ \approx 9.4$ Å and $a_- \approx 12$ Å at 1 K. This rather large value for a_+ implies considerable roton scattering by the compressed fluid surrounding the core [261, 267].

Phonons. The phonon excitation spectrum in ^{4}He (II) is $\varepsilon_k = \hbar s k$, with $s \approx 2.4 \times 10^4$ cm/sec [169], and a straighforward calculation with (3.3.64) gives

$$\frac{e}{\mu_{ph}} = \left(\frac{2\pi^2\hbar}{45}\right)\left(\frac{k_B T}{\hbar s}\right)^4 \bar{\sigma}_{ph}. \tag{3.3.68}$$

This quantity can be studied only below ≈ 1 K, because otherwise the roton density n_r exceeds the phonon density [347]

$$n_{ph} \approx 0.12\left(\frac{k_B T}{\hbar s}\right)^3. \tag{3.3.69}$$

Careful experiments with negative ions [355, 378] in this temperature range indicate $\mu_{ph-} \propto T^{-3}$. The corresponding theoretical treatment of $\bar{\sigma}_{ph-}$ [366] fits the data satisfactorily with classical acoustic scattering by a bubble with $R_- \approx 15.4$ Å. Schwarz's more recent analysis [267] suggests that $\bar{\sigma}_{ph-}$ is rather insensitive to R_-, however and he concludes merely that 16 Å $\lesssim R_- \lesssim$ 19 Å. The situation is more definite for positive ions [355, 375], where Schwarz [267] has evaluated $\bar{\sigma}_{ph+}$ for Atkins' snowball model [375a]. The compressibility of the core and the enhanced surrounding density each make a significant contribution, so that the cross section is considerably larger than that for acoustic scattering by a rigid sphere of radius R_+. In addition, the final best fit $R_+ \approx 5.1$ Å [267] included a liquid–solid interfacial energy (see the end of Section 3.3.1) $\sigma_{ls} \approx 0.1$ erg/cm^2, in reasonable agreement with values obtained from studies of μ_+ near the liquid–solid transition [264, 265] and in the vapor phase [266].

^{3}He impurities. The energy spectrum of a single ^{3}He impurity dissolved in ^{4}He is given by [363–365]

$$\varepsilon_k = \varepsilon_0 + \frac{\hbar^2 k^2}{2m_3^*}, \tag{3.3.70}$$

where $m_3^* \approx 2.4 m_3$ is the effective mass and $\varepsilon_0 \approx -0.23$ meV is the binding energy [379]. If $T \gtrsim T_F$ [see (3.3.47)], which holds in most cases of interest, then the impurities constitute a Boltzmann gas with [380]

$$f(\varepsilon_k) \approx e^{\beta\bar{\mu}} e^{-\beta\varepsilon_k} = \frac{1}{2} x_3 n_4 \left(\frac{2\pi\beta\hbar^2}{m_3^*}\right)^{3/2} e^{-\beta\varepsilon_k}. \tag{3.3.71}$$

A combination with equation (3.3.64) for $s = \frac{1}{2}$ yields

$$\frac{e}{\mu_3} = x_3 n_4 \bar{\sigma}_3 \frac{8}{3}\left(\frac{2m_3^* k_B T}{\pi}\right)^{1/2}, \tag{3.3.72}$$

implying that $\mu_3 x_3$ is independent of concentration and varies with temperature as $(\bar{\sigma}_3 T^{1/2})^{-1}$. The first prediction is well satisfied above $\approx \frac{1}{3}$ K, but it fails for negative ions at lower temperature [330–332, 381–383]. Moreover, μ_{3+} and μ_{3-} have distinct temperature dependences, and we therefore consider the two species separately (see Figure 3.5). For positive ions, μ_{3+} is approximately temperature independent for 20 mK < T < 700 mK with magnitude given by [331, 383] $\mu_{3+} x_3 \approx 5 \times 10^{-2}$ cm^2/V sec. Since a simple hard-sphere interaction between a positive ion and a ^{3}He atom would incorrectly predict a constant $\bar{\sigma}_{3+}$, Bowley and Lekner [384, 385] suggested augmenting the hard-core potential for $r < R_+$ by a *repulsive* ion–^{3}He polarization potential $\propto r^{-4}$ [386]. This effect may be understood by considering the total interaction energy obtained from (3.3.11), first for pure ^{4}He and then for ^{4}He containing a single ^{3}He at the point r. The increased zero-point volume v_3 diminishes the total attractive energy, leading to a net effective repulsion between the ion and the ^{3}He atom

$$V_{i3} \approx \frac{\alpha e^2}{2r^4}\left(\frac{v_3 - v_4}{v_4}\right), \tag{3.3.73}$$

where α is the polarizability and [323] $v_3 - v_4 \approx 0.28\, v_4$. The resulting transport cross section falls from $\approx 4\pi\alpha\, m_3^* e^2 (v_3 - v_4)/\hbar^2 v_4 = 1.8 \times 10^4$ Å^2 at $k \approx 0$ to $\approx \pi R_+^2 \approx 10^2$ Å^2 at short wavelengths. The corresponding thermal-average cross section $\bar{\sigma}_{3+}$ decreases as T increases, with $\bar{\sigma}_{3+}\, T^{1/2}$ remaining approximately constant near the experimental value [383–385].

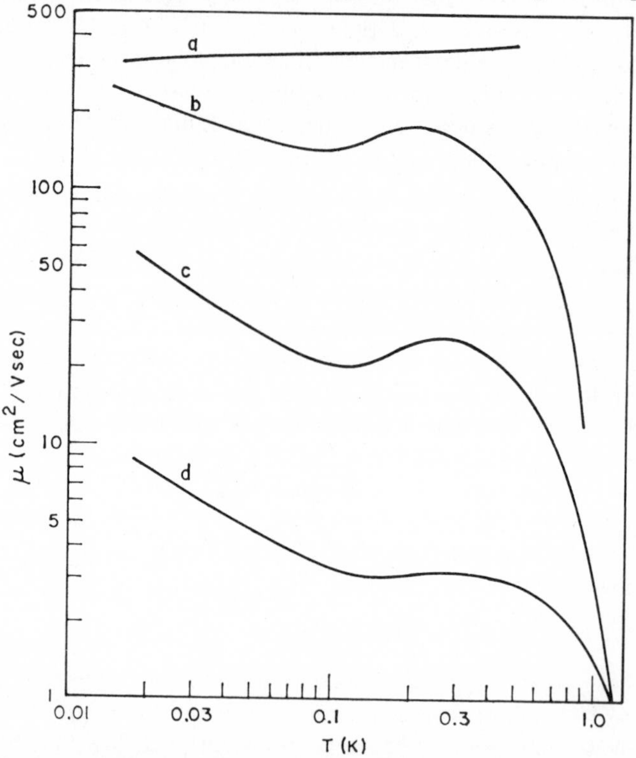

FIGURE 3.5. Ion mobility in dilute superfluid helium at low pressure: (*a*) positive ion, $x_3 = 1.7 \times 10^{-4}$; (*b*) negative ion, $x_3 = 7.7 \times 10^{-6}$; (*c*) negative ion, $x_3 = 5.8 \times 10^{-4}$; (*d*) negative ion, $x_3 = 3.4 \times 10^{-3}$. [From J. B. Ketterson, M. Kuchnir, and P. R. Roach, in *Proceedings of the Twelfth International Conference on Low Temperature Physics*, edited by E. Kanda (Keigaku Publishing Co., Ltd., Tokyo, 1971), Fig. 1, p. 105. Reprinted by permission.]

The situation is more complicated for negative ions below $\approx \frac{1}{2}$ K because the product $\mu_{3-}x_3$ varies with impurity concentration (see Figure 3.5). Moreover, μ_{3-} reaches a definite maximum as a function of temperature in the vicinity of $\frac{1}{3}$–$\frac{1}{4}$ K; it then decreases with decreasing temperature down to $\approx$ 100 mK, where it again starts to increase. As a qualitative explanation of this behavior, Kramer [329] has invoked the condensation of ^{3}He atoms on the bubble surface (see the discussion at the end of Section 3.3.2). At high temperature ($T \gtrsim \frac{1}{2}$ K), thermal excitation keeps the number of bound ^{3}He atoms small, with the resulting ^{3}He contribution to the mobility μ_{3-} equal to that for a hard sphere with $R_- \approx 21$ Å [383]. As the temperature falls, ^{3}He atoms eventually occupy the available surface states, with a correspondingly reduced surface tension and increased bubble radius nearly equal to that for a negative ion in pure ^{3}He. In the transition region, which should occur at higher temperatures in more concentrated solutions, the mobility can actually fall with falling temperature because of the increasing geometric cross section. Although these qualitative predictions agree with the experimental trends [329, 331], attempts at a quantitative theory have not been wholly successful. Kramer also considered a correction to the "Knudsen" limit [329], proportional to the ratio of R_- to the mean free path for ^{3}He–^{3}He interactions and acting to reduce the drag force (that is, to increase the mobility). This effect predicts a slight increase in $\mu_{3-}x_3$ with increasing concentration; it is more important in pure ^{3}He, however, and will be considered further in that context [387]. At lower temperatures ($\lesssim$ 50 mK), the bubble structure becomes independent of temperature and concentration for reasonable x_3; in this limit, a hard-sphere potential should again describe μ_{3-} in (3.3.72), but with R_- suitably increased to account for the adsorbed ^{3}He. The low-temperature data (see Figure 3.5) confirm these expectations, with $\mu_{3-}x_3T^{1/2}$ approximately constant. The experimental value $\mu_{3-}x_3 \approx 25 \times 10^{-5}$ cm^2/V sec at 20 mK implies $\bar{\sigma}_{3-} \approx 24 \times 10^{-14}$ cm^2, comfortably between the two extremes πR_-^2 and $4\pi R_-^2$, for $R_- \approx$ 20–25 Å (see the discussion at the end of Section 3.3.2). A more precise theoretical prediction requires numerical evaluation of (3.3.63), because the thermal wave number $(3k_BTm_3^*/\hbar^2)^{1/2}$ for ^{3}He impurities at 20 mK is of order 10^7 cm^{-1}, necessitating the inclusion of several partial waves [331].

In closing this description of low-field mobilities in superfluid ^{4}He, we shall mention the elusive velocity discontinuities [300–303, 388–391], where the mobility sometimes appears to decrease by small discrete amounts at approximately integral multiples of a critical velocity. This behavior, which also can appear in other simple liquids [392, 393], depends markedly on the precise technique used to measure μ [394]. In particular, no discontinuities are seen when the drift velocity is measured directly from the time of flight across a fixed distance [395, 396]; hence the phenomenon is probably unrelated to any fundamental property of superfluid helium.

(*c*) *Low-Temperature Mobility in Pure* ^{3}*He*. The most notable feature of ionic motion in pure ^{3}He is that the mobility [356, 357] remains below 0.1 cm^2/V sec throughout the observed temperature range 20m K $< T < 4$ K(see Figure 3.6). This upper limit is well below typical values in pure superfluid ^{4}He, where, for example $\mu_+ \approx 3.4 \times 10^4$ cm^2/V sec and $\mu_- \approx 4.2 \times 10^2$ cm^2/V sec at 0.4 K [355]. Moreover, $\mu_\pm$ in pure ^{3}He is relatively constant, whereas $\mu_\pm$ in ^{4}He increases greatly at low temperature. In part, these features reflect the contrast between the constant quasiparticle density in ^{3}He and the rapidly varying n_r and n_{ph} in ^{4}He (see the beginning of Section 3.3.3). The detailed explanation of $\mu_\pm$ in ^{3}He has proved very difficult, however, and it will only be possible to summarize the essential physical ideas [397].

We already noted that the short mean free path l_{33} for ^{3}He–^{3}He interactions permits a Stokes' law description of the ionic mobility in pure ^{3}He above ≈ 2 K. As T falls below the ^{3}He Fermi temperature $T_F = \hbar^2 k_F^2/2m_3^* k_B \approx 1.62$ K [365], the Pauli restriction on the possible atomic collisions increases l_{33} [398], which ultimately varies at low temperature as $(T_F/T)^2$. (Here $m_3^* \approx 3m_3$ is the effective mass of pure ^{3}He; it slightly exceeds that for dilute solutions [365].) In this degenerate temperature regime, viscous drag forces no longer can account for the mobility, and we instead use the quasiparticle

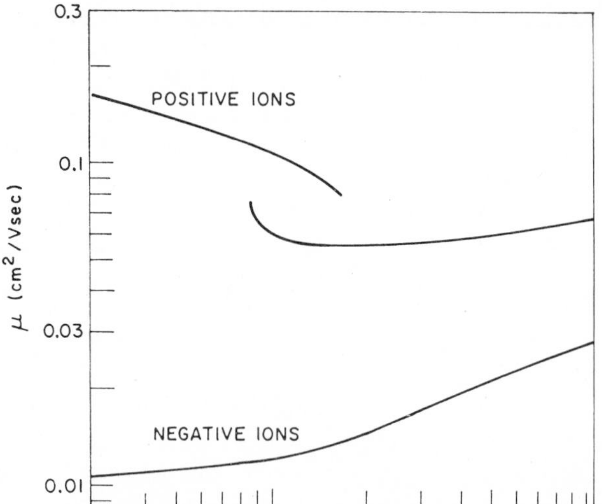

FIGURE 3.6. Ion mobility in pure ^{3}He at low pressure. [From A. C. Anderson, M. Kuchnir, and J. C. Wheatley, *Phys. Rev.* **168**, 261 (1968), Figs. 6 and 9. Reprinted by permission of the authors and the American Institute of Physics.]

picture introduced previously in connection with superfluid ^{4}He. The simple criteria used in equation (3.3.49) for recoil effects in dilute mixtures also apply here [384]; consequently, the scattering is certainly elastic for $T \gg T_0 \equiv T_F m_3^*/M^*$. Using the above values for T_F and m_3^* and the plausible estimates $M_+^* = 30m_3$ and $M_-^* = 300m_3$ for the ionic effective masses in pure ^{3}He, we obtain $T_{0+} \approx 160$ mK and $T_{0-} \approx 16$ mK. Of these two quantities, T_{0+} is unlikely to vary much with pressure, but T_{0-} should increase considerably with increasing pressure, because of the reduced R_- and M_-^*. There is also a slow variation of T_F and m_3^* with pressure [399], but this effect is less significant than the change in M_-^*.

These considerations imply that equation (3.3.64) should give an adequate account of $\mu_\pm$ for $T_0 \lesssim T \lesssim T_F$, where $f(\varepsilon)$ must now be the full Fermi–Dirac distribition function from (3.3.50).

$$\frac{e}{\mu} = \frac{\hbar}{3\pi^2}\,\bar{\sigma}\left(\frac{2m_3^*}{\hbar^2}\right)^2 \int_0^\infty d\varepsilon\, \varepsilon^2 \left[-\frac{\partial}{\partial\varepsilon}\frac{1}{\exp\beta(\varepsilon - \bar{\mu}) + 1}\right] \tag{3.3.74}$$

The low-temperature asymptotic expansion of (3.3.74) for $T \ll T_F$ is a standard calculation [400, 401] and yields

$$\frac{e}{\mu} \approx \frac{\hbar}{3\pi^2}\,\bar{\sigma}\left(\frac{2m_3^*\bar{\mu}}{\hbar^2}\right)^2\left[1 + \frac{\pi^2}{3}\left(\frac{k_B T}{\bar{\mu}}\right)^2\right], \tag{3.3.75}$$

apart from corrections that vanish exponentially as $T/T_F \to 0$. A similar calculation for $\bar{\mu}(T)$ gives [401]

$$\bar{\mu} \approx \varepsilon_F\left[1 - \frac{1}{12}\pi^2\left(\frac{k_B T}{\varepsilon_F}\right)^2 + \cdots\right], \tag{3.3.76}$$

and a combination with (3.3.75) leads to [329, 387, 402–404]

$$\frac{e}{\mu} \approx \frac{\hbar k_F^4}{3\pi^2}\,\bar{\sigma}\left[1 + \frac{\pi^2}{6}\left(\frac{k_B T}{\varepsilon_F}\right)^2\right], \tag{3.3.77}$$

where $\varepsilon_F = k_B T_F$ is the zero-temperature Fermi energy. Since $k_F = (3\pi^2 n_3)^{1/3}$, this expression may be rewritten in the intuitive form [405]

$$\frac{e}{\mu} = \hbar k_F n_3\, \bar{\sigma}\left[1 + \frac{1}{6}\pi^2\left(\frac{k_B T}{\varepsilon_F}\right)^2\right], \tag{3.3.78}$$

valid for $T_0 \lesssim T \lesssim T_F$.

The experimental data for μ_+ and μ_- in pure ^{3}He at low pressure are shown in Figure 3.6; similar behavior persists at elevated pressure. The relatively constant values $\mu_+ \approx 0.065$ cm^2/V sec and $\mu_- \approx 0.011$ cm^2/V sec for $T \lesssim T_0$ are readily explicable with the leading temperature-independent term of equation (3.3.78). Using the value [365] $k_F \approx 7.9 \times 10^7$ cm^{-1} and n_3 from Table 3.1, we find the equivalent hard-sphere radii $a_+ \equiv (\bar{\sigma}_+/\pi)^{1/2} \approx 7.6$ Å and $a_- \equiv (\bar{\sigma}_-/\pi)^{1/2} \approx 18.5$ Å, in qualitative agreement with other determinations. If $\bar{\sigma}$ were temperature independent, equation (3.3.78) predicts that μ should decrease slightly with increasing temperature, in contradiction with the data in Figure 3.6. Since (3.3.78) describes positive ions only in a very restricted range, the negligible observed temperature dependence of μ_+ for $T \gtrsim T_{0+} \approx 160$ mK is not very worrisome. On the other hand, (3.3.78) might be expected to describe negative ions quite accurately between (say) 30 mK and 1 K, so that the discrepancy between theory and experiment for μ_- must be taken seriously. In principle, it could be ascribed to a temperature-dependent $\bar{\sigma}$ [385], giving an additional correction of order $(T/T_F)^2$, but this explanation is unlikely because the large value ($\gtrsim 10$) of $k_F R_-$ implies that $\sigma_{tr}(k)$ varies only slowly with k. Kramer [387] has proposed an alternative and more attractive explanation based on the local change in the quasiparticle distribution caused by the moving ion. This effect tends to increase the mobility, with a correction of order $R_-/l_{33} \propto (T/T_F)^2$ [see the paragraph following (3.3.73) and the discussion preceding (3.3.74)]; it yields a plausible fit to the data for negative ions in the range $T_{0-} \ll T \ll T_F$. Since the correction is proportional to R, it also accounts for the much weaker temperature dependence of μ_+ above T_{0+}. Still another explanation of the same data invokes Friedel oscillations near the impurity [403], but the theory apparently requires the stringent and impractical condition $T_{0-} \ll T \ll T_F(k_F R_-)^{-1}$.

We now consider the low-temperature regime ($T \lesssim T_0 \equiv T_F m_3^*/M^*$), which has caused considerable confusion and has even led to qualitatively different theoretical predictions. To understand the discrepancies, we recall the theory of linear response, which evaluates a transport coefficient in the unperturbed but interacting ensemble [406]. For example, equation (3.3.61) expresses the low-field mobility μ [see (3.3.36)] in terms of the quasiparticle distribution function in the absence of the electric field. This approach evidently requires that the perturbed distribution functions for the quasiparticles and the impurity differ only slightly from those in thermal equilibrium; in particular, the ion's drift velocity v_D must be smaller than its thermal velocity $\bar{v}_i$. This condition has presented no difficulty even for a negative ion ($M_-^* \approx 300 m_3$) at 17 mK [357], where $\bar{v}_i$ still exceeds 50 cm/sec, but it becomes increasingly restrictive as $T \to 0$. For this reason, Kramer [329, 387] has also considered the alternative possibility $v_D \gg \bar{v}_i$, where the ion is assumed to move uniformly without recoil. In such a limit, μ takes the temperature-independent value calculated previously in equation (3.3.78), but it no longer represents the low-field value defined by (3.3.36).

In the more typical experimental situation ($v_D \ll \bar{v}_i$), the mean ionic energy remains of order $k_B T$. For $T \ll T_0$, the Pauli principle then allows only scattering events that involve initial ^{3}He quasiparticles within an energy shell of thickness $\approx (T/T_0)\varepsilon_F$ near the Fermi surface; the resulting reduced drag force might be expected to increase the mobility in pure ^{3}He for $T \ll T_0$. Although this conclusion probably becomes correct at sufficiently low temperature, it requires a careful analysis of ionic recoil, because the Fermi momentum $\hbar k_F$ now exceeds the momentum width of the ion's Maxwell distribution.

To study the recoil effects [384, 397] consider an ensemble of ions of mass M^*, each of which receives a momentum $\hbar\mathbf{q}$ at time $t = 0$. For an ideal gas at zero temperature, the ion would move forever with velocity $\hbar\mathbf{q}/M^*$, and its energy could be measured with arbitrary accuracy. In reality, each recoiling ion undergoes thermal motion and simultaneously experiences a random fluctuating collision force, both of which tend to destroy the information contained in the original ensemble. This degradation occurs in a characteristic time Δt associated with the loss of phase coherence among the initial plane waves for the different members of the ensemble. Equivalently, Δt is approximately the time needed for an ion to move a distance of order q^{-1} in its initial isotropic distribution. Since any measurement of the energy must be completed within the time Δt, the corresponding energy uncertainty will be of order $\hbar/\Delta t$. If this width is smaller than $k_B T$, then the scattering may be considered elastic; otherwise, recoil effects play an important role in calculating the mobility μ.

We may make a quantitative estimate of Δt with the Langevin equation [407, 408], which determines the motion of heavy impurity subject simultaneously to a random fluctuating force and a viscous drag with friction constant e/μ. The resulting mean-square displacement $\langle r^2(t)\rangle$ depends on the ratio t/τ, where

$$\tau \equiv \frac{M^*\mu}{e} \tag{3.3.79}$$

may be interpreted as the time required for the ion to attain its equilibrium drift speed $\mu\mathscr{E}$ following the sudden application of a dc electric field $\mathscr{E}$. For $t \ll \tau$, the ion behaves like a free particle with

$$\langle r^2(t)\rangle \approx (\bar{v}_i t)^2 = \frac{3k_B T t^2}{M^*} \; ; \; t \ll \tau \tag{3.3.80a}$$

for $t \gg \tau$, on the other hand, the ion executes a random walk with

$$\langle r^2(t)\rangle \approx \frac{6k_B T t\tau}{M^*} \; ; \; t \gg \tau \tag{3.3.80b}$$

in accordance with the "Einstein relation" $D/\mu = k_B T/e$ between the mobility μ and the diffusion constant D [407, 408]. If Δt is defined by the condition

$$q^2 \langle r^2(\Delta t)\rangle \approx 1, \tag{3.3.81}$$

a simple calculation gives

$$\Delta t \approx \begin{cases} \left(\dfrac{M^*}{3k_BTq^2}\right)^{1/2}, & \left(\dfrac{M^*}{3k_BTq^2}\right)^{1/2} \ll \tau, \\ \dfrac{M^*}{6k_BTq^2\tau}, & \left(\dfrac{M^*}{6k_BTq^2}\right)^{1/2} \gg \tau, \end{cases} \tag{3.3.82}$$

with the corresponding energy widths

$$\Delta E \approx \frac{\hbar}{\Delta t} \approx \begin{cases} \left(\dfrac{3k_BT\hbar^2q^2}{M^*}\right)^{1/2}, & \left(\dfrac{M^*}{3k_BTq^2}\right)^{1/2} \ll \tau, & (3.3.83a) \\ \dfrac{6k_BT\hbar q^2\tau}{M^*}, & \left(\dfrac{M^*}{6k_BTq^2}\right)^{1/2} \gg \tau. & (3.3.83b) \end{cases}$$

In the first limit, the ion recoils with the kinematics of a free particle, and equation (3.3.83a) reproduces (3.3.40). In the second case, the motion becomes a collision-dominated diffusion process, and ΔE is smaller than that for a free particle by a factor of order $\tau(k_BTq^2/M^*)^{1/2} \ll 1$. Since ionic recoil is negligible whenever $\Delta E \ll k_BT$, the additional factor can prove decisive [397].

These ideas may be applied to pure ^{3}He, for which the typical momentum transfer is $\hbar k_F$. As noted above (3.3.49), recoil is certainly negligible for $T \gtrsim T_0$, and this conclusion can only be strengthened if (3.3.83a) is replaced by the smaller quantity (3.3.83b). To investigate the behavior below T_0, we first use the experimental values (Figure 3.6) $\mu_+(T_{0+}) \approx 0.06$ cm^2/V sec, $\mu_-(T_{0-}) \approx 0.01$ cm^2/V sec to estimate

$$\tau_+ \approx 6 \times 10^{-12} \text{ sec}, \quad \tau_- \approx 9 \times 10^{-12} \text{ sec} \tag{3.3.84}$$

[see equation (3.3.79) and the discussion preceding (3.3.74)]. Since these characteristic times are each small compared to $(M_+^*/6k_BT_{0+}k_F^2)^{1/2} \approx 1.3 \times 10^{-11}$ sec and $(M_-^*/6k_BT_{0-}k_F^2)^{1/2} \approx 1.3 \times 10^{-10}$ sec, both species act like *diffusing* particles at their respective $T_{0\pm}$, with energy widths characterized by the dimensionless parameter [384, 397]

$$\frac{\Delta E}{k_BT_{0\pm}} \approx \frac{6\hbar k_F^2\mu_\pm(T_{0\pm})}{e} \tag{3.3.85}$$

[see (3.3.83b)]. For the negative ion, this ratio is $\approx \frac{1}{4}$, which indicates that the scattering is approximately elastic; the resulting mobility μ_- should then be given by (3.3.78) and remain nearly constant as the temperature falls somewhat below T_{0-}. For positive ions, however, (3.3.85) is ≈ 1.5, and any calculation of μ_+ for $T \lesssim T_{0+}$ must include inelastic effects. Note that this description is self-consistent, because μ itself determines the particular approximation to be used in calculating μ.

These considerations were first introduced by Josephson and Lekner [397], who derived the following expression for the mobility in pure ^{3}He:

$$\frac{e}{\mu} = \frac{\pi n_3 \hbar k_F}{4} \int d\Omega\,(1 - \cos\theta)\,\frac{d\sigma}{d\Omega}$$

$$\times \int_{-\infty}^{\infty} d\zeta\, \operatorname{sech}^2\left(\frac{1}{2}\pi\zeta\right) F\left(2k_F \sin\frac{1}{2}\theta, \frac{\hbar\zeta}{k_B T}\right). \quad (3.3.86)$$

Here $d\sigma/d\Omega$ is the differential cross section for scattering of quasiparticles with wavenumber k_F by a stationary ion, and $F(q, t)$ is an ensemble average

$$F(q, t) \equiv \langle e^{-i\mathbf{q}\cdot\mathbf{r}(0)}\, e^{i\mathbf{q}\cdot\mathbf{r}(t)}\rangle\,, \quad (3.3.87)$$

with $\mathbf{r}(t)$ the Heisenberg operator for the ion's position. This expression constitutes a precise formulation of the previous discussion because F decreases rapidly from 1 as soon as t exceeds the characteristic time Δt [see (3.3.81)]. Consequently, the second integral over ζ is of order 1 for $\hbar/k_B T \ll \Delta t$ and of order $k_B T\,\Delta t/\hbar$ for $\hbar/k_B T \gg \Delta t$. In the first case, the scattering can be considered elastic [compare equation (3.3.83)] and equation (3.3.86) is easily shown to reproduce the leading term of equation (3.3.78). In the second case, inelastic effects are important, and e/μ falls below its low-temperature elastic value $\sigma_{\mathrm{tr}}(k_F) n_3 \hbar k_F$. When the ion can be treated as a diffusing particle, Josephson and Lekner conclude that μ should rise only very slowly with decreasing T below T_0 [397]. At sufficiently low temperature, however, this increased mobility may eventually make τ so large [see (3.3.79)] that the ion must instead be considered a free particle. Since $\mathbf{r}(t)$ then reduces to $\mathbf{r}(0) + \mathbf{p}t/M^*$, the corresponding F in (3.3.87) is readily evaluated explicitly. A lengthy calculation for this case predicts that μ increases as T^{-2} as $T \to 0$, which was first derived by Abe and Aizu [385, 409–413]. Although the detailed derivations cannot be included, Josephson and Lekner's work is significant as the first explanation of the weak temperature dependence of $\mu_\pm$ even for $T \lesssim T_{0\pm}$. Their model has been extended by Bowley [385], who also suggested [414] that the apparently double-valued μ_+ shown in Figure 3.6 may represent an instability arising from the simultaneous existence of two self-consistent solutions to equation (3.3.86).

3.3.4. Ionic Behavior near Surfaces

The preceding analysis of ionic structure and mobility assumed a uniform background medium filling all space. Although this model is useful for bulk properties, it evidently fails near a surface separating liquid He from its vapor [414a] or an interface between the two phases in isotopic mixtures. Indeed, the presence of such surfaces leads to several interesting physical effects, most of which are readily explained with elementary electrostatics. Consider first a point charge q placed in an infinite linear dielectric with dielectric constant ε. The resulting electrostatic energy density [415] is given by $\vec{\mathscr{E}}\cdot\vec{\mathscr{D}}/8\pi = q^2/8\pi\varepsilon r^4$, where $\vec{\mathscr{D}} = \varepsilon\vec{\mathscr{E}} = q\hat{r}/r^2$ is the displacement and r is the distance from the charge. Since this energy varies inversely with ε, a charge in an *inhomogeneous* dielectric would tend to move in the direction of increasing ε, thereby lowering the total energy. In particular, when placed on either side of a plane boundary separating two uniform dielectrics with ε_1 and ε_2, such a charge experiences a force toward the region of larger ε. This qualitative conclusion is easily verified explicitly with the method of images. If the charge is located in region 1, a distance x to the right of the interface, the net (inverse-square) force to the right is [416]

$$F_x = \frac{q^2}{4x^2}\frac{\varepsilon_1 - \varepsilon_2}{\varepsilon_1(\varepsilon_1 + \varepsilon_2)}, \tag{3.3.88}$$

which is positive or negative according as $\varepsilon_1 > \varepsilon_2$ or $\varepsilon_1 < \varepsilon_2$. Equivalently, the charge experiences a net potential energy

$$V(x) = \frac{q^2}{4x}\frac{\varepsilon_1 - \varepsilon_2}{\varepsilon_1(\varepsilon_1 + \varepsilon_2)}. \tag{3.3.89}$$

(a) *Ions in the Vapor Phase.* The preceding calculation applies directly to charges near the surface of liquid He, and we first consider singly charged ions in the vapor phase, where $\varepsilon_1 \approx 1$ and ε_2 is that for bulk helium (see Table 3.1). The net attractive potential energy becomes

$$V(x) = -Ze^2x^{-1}, \tag{3.3.90}$$

where

$$Z = \frac{1}{4}\frac{\varepsilon - 1}{\varepsilon + 1} \approx \begin{cases} 6.8 \times 10^{-3} & \text{for } ^4\text{He}. \\ 5.1 \times 10^{-3} & \text{for } ^3\text{He}, \end{cases} \tag{3.3.91}$$

If the charged particle is a positive $(\text{He})_2^+$ ion, it approaches the surface and presumably passes into the fluid, where the free surface now repels it [see equation (3.3.88) and the associated discussion]. The ion thus moves into the bulk fluid, acquiring the snowball structure studied in Section 3.3.1.

On the other hand, an electron behaves very differently, for bulk helium presents a repulsive barrier with height ≈ 1 eV. Hence the electron is likely to remain trapped in the exterior region ($x > 0$) by the attractive image potential (3.3.90) [417–420]. The Schrödinger equation for this case is similar to that for s-wave hydrogenic orbitals around a nucleus with charge Ze, and the corresponding "Bohr radius" $\hbar^2/m_e e^2 Z \approx 78$ Å and "Rydberg" $Z^2e^4m_e/2\hbar^2 \approx 0.63$ meV characterize the perpendicular dimension and binding energy of the electron surface state. Since this energy corresponds to a temperature ≈ 7.3 K, the less tightly bound excited states are unimportant except for $T \lesssim 1$ *K*. More refined calculations incorporating the finite barrier height do not alter these qualitative conclusions [417–420]. When the interelectron potentials are included, Crandall and Williams [421, 421a] have suggested the formation of a two-dimensional Wigner [422] surface lattice, but it is unclear that this ordered array would be stable [423–423b]. Several experiments have sought direct evidence for these bound surface electrons [424–427]; the results are contradictory, however, and even the existence of such states remains uncertain (see, however, Addendum). Further experimental and theoretical work would be valuable.

Although the electrons are predicted to be bound in the x direction, they can still move parallel to the surface. Such an electron should feel a drag force from scattering both by surface waves on the fluid and by He atoms in the vapor. Cole [418] and Shikin [420] have independently studied the resulting surface mobility, but their qualitatively similar predictions are not borne out by recent experiments [426, 427]. Cole [428] has also considered electrons outside other insulating fluids and solids such as bulk ^{3}He and layered dielectrics.

(b) *Ions in the Liquid Phase.* We now study the alternative possibility of a singly charged ion located inside liquid helium, where the image potential becomes repulsive. If the bulk fluid is taken to occupy the region $x > 0$, equation (3.3.89) reduces to

$$V(x) = Z'e^2x^{-1}\,, \tag{3.3.92}$$

where $Z' = Z/\varepsilon$ is slightly less than Z. This repulsion means that an external electric field is required to keep an ion near the surface. For the positive species, experiments have been unable to extract and measure a current through the surface, but additional work would be desirable [254, 429, 430]. For the negative species, however, the surface transmission current has been studied rather extensively [254, 429–434]. When an electric field $\mathscr{E}$ is applied perpendicular to the surface, the total potential energy becomes

$$V_T(x) = \frac{Z'e^2}{x} + e\mathscr{E}x\,, \tag{3.3.93}$$

which has a minimum value $V_{\min} = 2(\mathscr{E}Z'e^3)^{1/2}$ at the point $x_{\min} = (Z'e/\mathscr{E})^{1/2}$ (see, also, Addendum). Increasing $\mathscr{E}$ shifts this equilibrium position toward the surface; it thus raises the probability for the electron to escape from the bubble into the vapor, which reduces its zero-point kinetic energy. As a result, the trapping time τ_{tr} for an electron near the inside surface should fall rapidly as $\mathscr{E}$ increases.

The escape process has been studied both by monitoring the total current extracted through the surface [254, 429, 431] and by measuring the trapping time τ_{tr} [430, 432, 433]. Since the former data may also be analyzed to give τ_{tr} [434, 435], we need consider only the latter method. The most complete study is by Rayfield and Schoepe [430, 433], who treat the massive negative ion as a diffusing Brownian particle in the potential well (3.3.93). The actual escape seems to occur by tunneling, but thermal activation may sometimes be important. Their fit [436] to the data on ^{4}He yields $R_{-4} \approx 24.8$ Å and an energy $\Delta E_4 - E_0 \approx 0.70$ eV, where ΔE is the repulsive barrier between bulk ^{4}He and an electron [see the discussion following (3.3.26)] and E_0 is the zero-point energy of the trapped electron in the bubble. Similar data [436] for ^{3}He yield $R_{-3} \approx 30.1$ Å and $\Delta E_3 - E_0 \approx 0.56$ eV. Although both radii are too large, the ratio R_{-3}/R_{-4} is very close to the value $\frac{5}{4}$ predicted by the simple bubble model (see the end of Section 3.3.2). Furthermore, as shown below, the binding energies agree well with other estimates.

These data have been supplemented by studies of electron escape from negatively charged vortex rings in ^{4}He [437, 438], where the electrons detected in the vapor phase have a maximum energy *independent* of the radius or energy of the ring [437–441]. Mitchell and Rayfield [437] identify this maximum with the zero-point energy E_0 of the trapped electron, and the resulting barrier $\Delta E_4 = 0.84$ eV is close to the more direct determinations and to the theoretical values. The absence of superfluidity in ^{3}He precludes a similar approach there, but the estimate $\Delta E_3 \approx 0.65$ eV [436] is again close to the theoretical values discussed at the end of Section 3.3.2.

(*c*) *Transmission through* ^{4}*He*–^{3}*He Phase Boundary.* We close this section by considering the transmission of negative ions through the phase boundary in ^{4}He–^{3}He mixtures, which has been studied by Kuchnir, Roach, and Ketterson [442]. Since the bubble radius is larger in pure ^{3}He than in ^{4}He, its total energy is lower on the ^{3}He-rich side by ≈ 0.025 eV, which includes the slightly decreased polarization energy. Although a negative ion on the ^{4}He-rich side experiences a weak repulsive image potential (3.3.92) with $Z'_{43} = (\varepsilon_4 - \varepsilon_3)/4\varepsilon_4(\varepsilon_4 + \varepsilon_3) = 1.6 \times 10^{-3}$, it can lower its energy by moving through the interface to the ^{3}He-rich side. The transmitted ion current exhibits the expected temperature dependence, but its magnitude is larger than that expected for a discontinuous density profile. A finite boundary width acts to reduce the reflection probability, and a surface thickness ≈ 5.5 Å yields a satisfactory fit to the data. Williams and Crandall [434] have reanalyzed the same data to obtain the trapping time τ_{tr} at the surface; since τ_{tr} appears to be very long (up to 10^5 sec in some cases), direct measurements would be very interesting.

3.4. INTERACTION BETWEEN IONS AND VORTICES

The previous sections have treated vortices and ions as basically unrelated. This approximation is frequently adequate, but it omits several interesting effects arising from their mutual interactions. For example, the anisotropic attenuation of negative ions in rotating ^{4}He (II) [43] helped confirm the Onsager–Feynman [21, 22] model (compare Section 3.2.3), and the creation of vortex rings by fast ions [28, 29] has served as the basis for nearly all experiments with quantized vortex rings [147]. We first study the capture of ions by vortex lines (Section 3.4.1) and the subsequent axial motion along the vortex axes (Section 3.4.2). Such phenomena can occur only if the ion remains trapped, and the corresponding escape probability is considered in Section 3.4.3. The final section (3.4.4) treats the creation of vortex rings and the dynamics of the ion–vortex complex.

3.4.1. Capture of Ions by Vortex Lines

Consider a small particle of volume V placed in an inhomogeneous irrotational velocity field $\mathbf{v}(\mathbf{r})$. It experiences a hydrodynamic force (3.3.13) owing to the nonuniform pressure distribution, and, as a first approximation, we assume that the flow is steady and unaffected by the particle's presence. In this case, equations (3.2.48) and (3.3.13) give

$$\begin{aligned} \mathbf{F}_h &= \int d\mathbf{S}\,\tfrac{1}{2}\,\rho v^2 \\ &\approx \int d^3r\,\nabla(\tfrac{1}{2}\,\rho v^2) \\ &\approx V\nabla(\tfrac{1}{2}\rho v^2), \end{aligned} \tag{3.4.1}$$

where the last form assumes that ρv^2 varies slowly. Alternatively, the force may be obtained from a hydrodynamic potential

$$V_h(\mathbf{r}) \approx -\tfrac{1}{2}\,\rho v^2 V, \tag{3.4.2}$$

showing that the particle is attracted toward a region of higher velocity as expected from Bernoulli's theorem [59]. This expression may be interpreted as the kinetic energy removed by the presence of the particle. If the flow represents a singly quantized vortex in ^{4}He (II), then $\mathbf{v}$ is given by (3.2.11) and ρ may be identified with the superfluid density ρ_s; furthermore,

$$V_h(r) = -\frac{\rho_s V \kappa^2}{8\pi^2 r^2} = -\frac{1}{2}\,\rho_s V\left(\frac{\hbar}{mr}\right)^2 \tag{3.4.3}$$

is purely radial, with a corresponding attractive inverse-cube force [27, 222, 227, 443–445]. For large distances, the actual potential is more attractive by a factor $\frac{3}{2}$ because the velocity field is locally perturbed [263, 444, 445], but the qualitative behavior is unchanged.

In a typical experiment, a weak electric field $\vec{\mathscr{E}}$ perpendicular to the symmetry axis draws a beam of ions through rotating ^{4}He (II). In addition the ions are attracted toward the vortex cores and experience a frictional drag

$$\mathbf{F}_d = -\frac{M^*\mathbf{v}_i}{\tau} = -\frac{\mu\mathbf{v}_i}{e} \tag{3.4.4}$$

from collisions with the fluid particles [see equation (3.3.79)]. Since the ion's thermal velocity generally exceeds its mean drift velocity, the ion may be considered a massive Brownian particle diffusing through the fluid under the combined electric, hydrodynamic, and frictional forces. The resulting motion has been studied by Donnelly and Roberts [27, 443, 444], who find that each vortex captures ions with a cross section

$$\sigma = O\left(\frac{r_{\mathscr{E}}}{\ln r_{\mathscr{E}}}\right), \tag{3.4.5a}$$

where

$$r_{\mathscr{E}} \equiv \frac{2k_BT}{e\mathscr{E}} \tag{3.4.5b}$$

is typically several hundred angstrom units, much larger than the vortex-core radius ξ or the ion radius R. Consequently, the capture process provides little information on the intrinsic properties of either the vortex or the ion. The approximate inverse dependence of σ on $\mathscr{E}$ reflects the linear mobility relation (3.3.36), because the long-range hydrodynamic potential is more likely to capture a slow ion than a fast one.

Consider an ion current i passing through ^{4}He (II) rotating with angular velocity Ω. In a path length dx, the capture process changes i by an amount

$$di = -\, n_v \sigma i\, dx, \tag{3.4.6}$$

where $n_v = 2\Omega/\kappa = 2m\Omega/h$ is the vortex density (3.2.20). Hence the current is exponentially attenuated [27, 443, 444]

$$i(x) = i(0)\exp\left(\frac{-2m\Omega\sigma x}{h}\right), \tag{3.4.7}$$

with a characteristic length

$$x_0 = \frac{h}{2m\Omega\sigma} = O\left(\frac{he\mathscr{E}}{4mk_BT\Omega}\right). \tag{3.4.8}$$

Several experiments have verified that x_0 is proportional to $\mathscr{E}/\Omega$ [446–449], but the temperature dependence is somewhat more complicated than that predicted here. The discrepancy has several sources. At higher temperatures ($T \gtrsim 1.7$ K), the negative ions escape from the vortex lines with nearly unit probability, so that the effective capture cross section falls to zero (see Section 3.4.3) [27, 443, 444]. For 1.2 K $\lesssim T \lesssim$ 1.7 K, σ is indeed of order $r_{\mathscr{E}}$ and falls slowly with falling temperature [449], but it seems to decrease more rapidly below $\approx$ 1.2 K [449]. Since the detailed theory is not expected to hold below 1 K [444], this altered temperature dependence may merely reflect the failure of one or more of the theoretical assumptions, but a more general analysis has not yet been performed.

For positive ions, the calculations of Section 3.4.3 indicate nearly unit escape probability for $T \gtrsim 0.6$ K. This prediction is confirmed by experiments above 1 K, where only negative ions are attenuated by vortex lines [43, 446–450]. At low temperatures ($\lesssim 0.7$ K), on the other hand, vortex rings can trap ions of both signs [29]; moreover, Schwarz and Donnelly [56, 57] have observed the transfer of positive charges from vortex rings to vortex lines below 0.5 K, showing that rectilinear vortices also can trap positive ions at low temperature.

Two other experiments are relevant to this question. In the first, vortex lines seem to scatter positive ions above 1 K through capture and immediate escape [450], but this effect requires low electric fields to make σ large and high ion density. The resulting extra drag reduces the positive-ion mobility, with an observed magnitude in qualitative agreement with theoretical estimates [443, 444]. In the second experiment, fast positive ions have created vortex rings up to 1.5 K [451], but the ion does not appear to remain trapped on the ring for very long. This rather complicated phenomenon is discussed in Section 3.4.4.

3.4.2. Mobility of Trapped Ions

If an ion remains trapped on a vortex line for a sufficiently long time, its axial mobility along the vortex core may be measured with suitable electric fields. As noted in the preceding section, positive ions have a large escape probability above 1 K, so that only negative ions have been studied. The experiments [79, 180, 188, 452] consist in building up a space charge of free and trapped ions at the bottom of a rotating container of ^{4}He (II). The fields are then switched to draw the ions along the rotation axis to a collector near the top. Two separate pulses are seen, representing the faster free ions and the slower trapped ions.

Two specific models have been proposed to account for the additional drag force on the trapped ions. At higher temperatures ($1.0\text{ K} \lesssim T \lesssim 1.7$ K), the predominant mechanism seems to be rotons trapped near the vortex core, with the roton momentum directed opposite to the circulating superfluid [178, 180]. The qualitative basis for this suggestion is the shifted distribution function $\bar{f}(\varepsilon_k) = f[\varepsilon_k - \hbar\mathbf{k}\cdot(\mathbf{v}_n - \mathbf{v}_s)]$ seen by an observer moving with velocity $\mathbf{v}_n - \mathbf{v}_s$ relative to the center of mass of the excitations [see equation (3.3.52)]. If $\mathbf{v}_n = 0$, then $\mathbf{k}\cdot\mathbf{v}_s$ has a negative average value. To the extent that $\mathbf{v}_s$ is locally uniform, a simple calculation gives the excess roton density [26, 79, 177]

$$\frac{n_r(v_s)}{n_r(0)} = \frac{\sinh(\hbar k_0 v_s/k_B T)}{\hbar k_0 v_s/k_B T}, \tag{3.4.9}$$

assuming $\hbar k_0 v_s \ll \Delta$. Such an approximation evidently fails for small r, where a more careful analysis would be desirable, especially with regard to the existence and structure of bound roton states. Nevertheless, this rather simple model has had considerable success in fitting the mobility of trapped negative ions as a function of both T and p [178, 180, 444].

The other proposal for excess drag on a trapped ion involves the reflection of vortex waves (Sections 3.2.1 and 3.2.2) [79, 453–455], in close analogy with the calculations of mobility in bulk superfluid ^{4}He (II) (Section 3.3.3). When a slow ion moves along the vortex axis, the corresponding vortex-wave contribution to its inverse mobility is [compare equation (3.3.61)]

$$\frac{e}{\mu_{vw}} = \frac{2\hbar}{\pi}\int_0^\infty k^2 dk\left[-\frac{\partial f(\varepsilon_k)}{\partial k}\right]|R(k)|^2, \qquad (3.4.10)$$

where ε_k is the energy spectrum for the vortex waves and $|R|^2$ is the reflection probability for an ion in the core. The various calculations of $|R|^2$ all rely on classical models; although different in detail, they provide acceptable fits to the mobility of trapped negative ions for $0.75\ \mathrm{K} \lesssim T \lesssim 1\ \mathrm{K}$ [79, 453, 455]. At higher temperatures, however, μ_{vw} is considerably larger than the observed value, indicating the importance of trapped rotons.

Electrons trapped on vortex lines have also been extracted with negligible energy loss through the free surface into the vapor phase [456]. This behavior is central to the elegant experiments of Packard and Sanders [105], discussed in Section 3.2.1.

In principle, it should be possible to study various resonance oscillations of the trapped ions by augmenting the axial dc electric field with a transverse rf one [457, 458]. Unfortunately, the low ion density probably renders such an experiment marginal at best, but further theoretical and experimental studies would be helpful.

3.4.3. Escape of Ions from Vortex Lines

The preceding sections (3.4.1 and 3.4.2) dealt with negative ions trapped on vortex lines, neglecting the possibility of thermally activated escape. This effect merits careful study, however, as it becomes predominant above $\approx$ 1.7 K for negative ions and above $\approx$ 0.6 K for positive ions. Donnelly and Roberts [27, 443, 444] developed a theory of such processes from a stochastic model, and we now briefly discuss its physical basis.

A vortex attracts an ion or other impurity with a hydrodynamic potential (3.4.2) approximately equal to the kinetic energy eliminated by the particle. If the ion is located at the vortex core, then this quantity will be of order $2R\,(\rho_s\kappa^2/4\pi)\ln(R/\xi)$ [compare equation (3.2.12)], where the argument of the logarithm is model dependent. In particular, Parks and Donnelly [262] have used (3.2.11) for all r, with the logarithmic divergence removed by a superfluid density of the form [compare equation (3.2.78)]

$$\rho_s(r) = \rho_s(\infty)r^2(r^2 + \xi^2)^{-1}, \qquad (3.4.11)$$

and they find [27, 262, 444]

$$|V_h(0)| = \frac{\rho_s \kappa^2 R}{2\pi}\left[\ln\left(\frac{2R}{\xi}\right) - 1\right], \tag{3.4.12}$$

apart from corrections that vanish as $\xi/R \to 0$. Since $Vh(r)$ tends to zero as $r \to \infty$, the binding energy of an ion is $|V_h(0)|$. Taking $R_+ \approx 6$ Å, $R_- \approx 16$ Å, and $\rho_s \approx \rho$, we obtain $|V_{h+}(0)|/k_B \approx 15$ K and $|V_{h-}(0)|/k_B \approx 66$ K.

In the presence of a transverse electric field $\vec{\mathscr{E}} \perp \hat{\Omega}$, the ion experiences a total potential

$$V_T(\mathbf{r}) = V_h(r) \mp e\vec{\mathscr{E}}\cdot\mathbf{r}, \tag{3.4.13}$$

where the upper (lower) sign refers to positive (negative) ions. This function has a local minimum M near the vortex axis and a saddle point S in the direction of $\pm\vec{\mathscr{E}}$ [262] (see Figure 3.7). The escape probability P of a particle trapped in this well is proportional to $\exp(-\Delta F/k_B T)$, where ΔF is the difference in free energy between S and M [459]. Donnelly and Roberts neglected any change in the vortex and therefore set $\Delta F = \Delta V$ (see Figure 3.7). A rather intricate calculation then gives the final form as in the following:

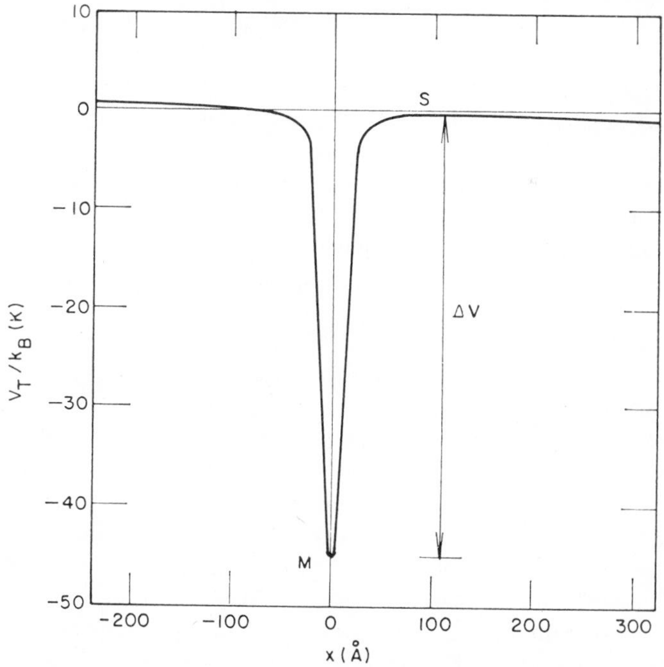

FIGURE 3.7. Total potential energy $V_T(x, 0, 0)/k_B$ of a negative ion for $\vec{\mathscr{E}} = -\mathscr{E}\hat{x}$, $R_- = 16$ Å, $T = 1.64$ K. [From P. E. Parks and R. J. Donnelly, *Phys. Rev. Lett.* **16**, 45 (1966), Fig. 2b. Reprinted by permission of the authors and the American Institute of Physics.]

$$P = P_0 \exp\left(\frac{-\Delta V}{k_B T}\right), \tag{3.4.14}$$

where the prefactor P_0 has the dimensions $\sec^{-1}$; it varies linearly with the freeion mobility and depends on the precise shape of V_T near M and S. Alternatively, the liftetime $\tau_{tr} \equiv P^{-1}$ becomes

$$\tau_{tr} = \tau_0 \exp\left(\frac{\Delta V}{k_B T}\right), \tag{3.4.15}$$

where τ_0 and ΔV are known functions of R and $\mathscr{E}$[27, 263, 444, 460].

The trapping time for negative ions was first determined by Douglass [452] with the apparatus used to study the trapped-ion mobility (Section 3.4.2). After allowing the charge to accumulate for a standard length of time, he then measured the trapped charge Q_{tr} remaining after a variable time delay Δt. The observed Q_{tr} decayed exponentially with Δt, and the resulting mean trapping time indeed exhibited the predicted linear dependence of $\ln \tau_{tr}$ on T^{-1} [461]. Parks and Donnelly [262] have fitted these data to equation (3.4.15) and infer $R_- \approx (16.0 \pm 0.4)$ Å at the vapor pressure. This procedure is not entirely straightforward, however, for the slope $(= \Delta V/k_B)$ and intercept $(= \ln \tau_0)$ of $\ln \tau_{tr}$ vs. T^{-1} should yield consistent values for R_-. In fact, Parks and Donnelly [262] chose R_- to fit the value of $\ln \tau_{tr}$ at ≈ 1.65 K, but the resulting theoretical slope was somewhat too small. In contrast, Pratt and Zimmermann [263] used the alternative procedure for their measurements of τ_{tr} from 1.09 to 1.68 K and from zero pressure to the solidification pressure; their preferred value $R_-(p = 0) \approx (19.5 \pm 0.6)$ Å fits the observed slope, but it overestimates τ_0 by a factor $\approx 10^4$. Although this radius is somewhat larger than the generally accepted value (see Section 3.3.2), their data on the relative pressure dependence $R_-(p)/R_-(0)$ agree well with the simple bubble model and also with (3.3.31), obtained by quite a different method (see below [312–314]).

This discrepancy has led to two distinct explanations.

i. McCauley and Onsager [81] note that the vortex-wave contribution to the total free energy changes when the ion moves from infinity to the vortex core; this effect should therefore be included in the barrier ΔF. Their numerical estimate increases the theoretical slope of $\ln \tau_{tr}$ vs. T^{-1}, and the revised theory agrees reasonably with the experimental data.

ii. Padmore [462] has suggested a different mechanism arising from the local deformation of the vortex axis caused by the presence of the ion. The net effect is to multiply ΔV in equation (3.4.15) by a temperature-dependent factor that also increases the slope of $\ln \tau_{tr}$ vs. T^{-1}. This calculation apparently ignores the vector character of the Magnus force (3.2.56), however, so that it must be considered suggestive rather than rigorous.

Since both mechanisms may be significant, a combined analysis would be useful.

The trapping time decreases rapidly with increasing temperature; for negative ions at zero pressure, it reaches $\approx$ 1 sec at $\approx$ 1.7 K, above which trapping soon becomes unobservable [263, 452]. Since the relevant ratio is $\approx \Delta V/k_B T$, the corresponding "critical" temperature for positive ions should be of order $\frac{1}{2}$ K. Below 0.6 K, Cade [463] has studied the binding of positive ions to vortex rings through a field-assisted escape process. Since he used high electric fields (10–20 kV/cm), the $\mathscr{E}$ dependence in equation (3.4.15) becomes important. A theoretical fit [262] to these data yielded an equivalent positive-ion radius (7.9 $\pm$ 0.1) Å, which probably exceeds the hard-core radius R_+ because the locally enhanced density (see Figure 3.4) increases $|V_{h+}(0)|$.

The temperature variation of τ_{tr} affects the measurement of the capture cross section (Section 3.4.1) because the attenuation of the ion beam becomes negligible as soon as τ_{tr} falls below a time t characterizing the transit along the vortex axis to the collector. If equation (3.4.7) is used to determine the effective cross section σ_{eff}, it will, therefore, have the approximate form [443]

$$\sigma_{eff} \approx \sigma \exp\left(\frac{-t}{\tau_{tr}}\right), \tag{3.4.16}$$

with τ_{tr} given by (3.4.15). This sharp decrease in σ_{eff} with increasing T has allowed a determination of R_-, not only at the vapor pressure [27, 447, 449] but also at elevated pressure [312–314], where the reduced binding energy $|V_{h-}(0)|$ yields a corresponding reduction in τ_{tr}. Equation (3.4.16) also accounts for the absence of positive-ion trapping above $\approx$ 1 K.

Although (3.4.15) predicts that τ_{tr} increases monotonically at low temperature, Douglass [79, 464] has instead observed an anomalous decrease of τ_{tr} for negative ions ($\tau_{tr} \lesssim$ 1–100 sec) below 1 K. Moreover, the addition of ^{3}He impurities increases the trapping time in this regime, whereas it lowers τ_{tr} at higher temperature [465] through the reduced ρ_s in equation (3.4.12). Douglass [464] speculated that these data might indicate a disruption of the vortex lines; more specifically, Packard [466] has suggested that the greatly reduced ρ_s [and hence mutual friction; see equation (3.2.89)] below 1 K might allow the vortex lines to execute large amplitude oscillations, thus "shaking off" the trapped ions. This proposal explains the qualitative effect of the ^{3}He impurities, but it apparently contradicts Douglass' tentative observation [464] of positive-ion trapping below 0.65 K with a lifetime greater than that for negative ions. The question remains unresolved and requires considerably more experimental data (see, however, Addendum).

We close this section with a short comment on the behavior of ^{3}He impurities near a vortex line. Equations (3.4.12) and (3.4.15) imply that the corresponding τ_{tr} will be very small at reasonable temperatures, but this macroscopic classical model is suspect for an atomic impurity. A more satisfactory approach to a single ^{3}He atom [467] derives the bound states near a vortex axis from a Schrödinger equation, in direct analogy with the treatment of scattering states [223] (see Section 3.2.3). This calculation suggests a maximum binding energy of order 10^{-17} erg (≈ 0.1 K in thermal units); consequently, ^{3}He impurities in a dilute rotating superfluid are unlikely to collect near the vortex axes except below ≈ 100 mK. Unfortunately, these estimates are quite sensitive to the vortex-core structure and may not be reliable. Another treatment [468] based on a weakly interacting Bose gas implies a binding energy ≈ 3.5 K, but this theory involves a drastic extrapolation to describe physical helium. Finally, the behavior of rotating mixtures has also been analyzed with thermodynamic arguments [467, 469], based on the phase diagram of stationary mixtures. Such calculations can only describe distances larger than the mean free path of the ^{3}He quasiparticles, which generally exceeds 10–20 Å, even for a several percent solution [467]; thus they cannot provide insight into the core region. In principle, studies of ion mobilities in rotating mixtures might indicate whether ^{3}He actually accumulates on the vortex cores [188], but Douglass' [464] observation of reduced trapping time might preclude such studies in the most interesting region below 1 K.

3.4.4. Ion Critical Velocity and Vortex-Ring Creation

When an ion moves uniformly through ^{4}He (II) under the influence of an applied electric field $\mathscr{E}$, the resulting drift velocity v_D exhibits three distinct regions (Figure 3.8). For small fields, the drift velocity varies linearly with $\mathscr{E}$, defining the mobility regime of Section 3.3.3. As $\mathscr{E}$ increases, nonlinear effects become important, and the curve of v_D vs. $\mathscr{E}$ generally bends downward. Eventually, the ion reaches a maximum critical velocity v_{c1}, which signals the transition to a new regime [470–472]. Although the corresponding critical field $\mathscr{E}_{c1}$ varies greatly with temperature T, pressure p, and ^{3}He concentration x_3, v_{c1} itself is rather insensitive to changes in these parameters, remaining between 25 and 50 m/sec. The characteristic feature of the second regime is a rapid fall in v_D for $\mathscr{E} \gtrsim \mathscr{E}_{c1}$. This behavior is ascribed to the formation of a quantized vortex ring at v_{c1} and reflects its peculiar dynamics (see Figure 3.1). For each $\mathscr{E}$, the ring has a characteristic radius r_v and speed that varies approximately inversely with r_v. As $\mathscr{E}$ increases, v_D falls more slowly and ultimately reaches a minimum at a second critical field $\mathscr{E}_{c2}$; the system then enters a third regime [470–472], where v_D increases slowly with increasing $\mathscr{E}$ because the ion escapes from the ring through a field-assisted stochastic process (Section 3.4.3) [451]. Since a free ion tends to accelerate to

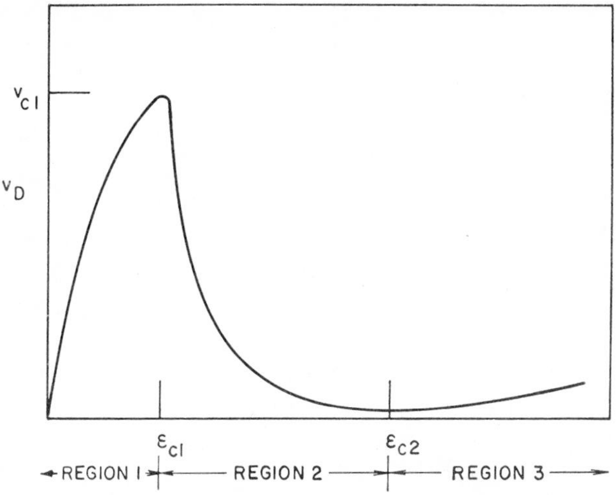

FIGURE 3.8. Sketch of ion drift velocity v_D at various applied electric fields $\mathcal{E}$. [From T. C. Padmore, *Phys. Rev. A* **5**, 356 (1972), Fig. 1. Reprinted by permission of the author and the American Institute of Physics.]

v_{c1} and nucleate a new ring, the average drift speed exceeds that expected from an extrapolation of the curve for region 2. We now examine the various regimes in more detail.

(*a*) *Bare Ions.* In region 1, the ion propagates like a free particle of mass M^* under the influence of an electric field $\vec{\mathcal{E}}$ and a frictional drag $\mathbf{F}_i(\mathbf{v}_i)$, which depends on the instantaneous ion velocity $\mathbf{v}_i$. In the usual situation, the ion accelerates rapidly from rest and may be considered to move uniformly at the drift velocity $\mathbf{v}_D$, determined by the implicit equation

$$\pm e\vec{\mathcal{E}} + \mathbf{F}_i(\mathbf{v}_D) = 0. \tag{3.4.17}$$

For small $\mathbf{v}_i$, the friction varies linearly with the velocity (see Section 3.3.3)

$$\mathbf{F}_i \approx \pm\left(\frac{e}{\mu}\right)\mathbf{v}_i, \tag{3.4.18}$$

and the behavior for all $\mathbf{v}_i$ may be inferred from Figure 3.8. The form of $\mathbf{F}_i(\mathbf{v}_i)$ has been studied by Strayer, Donnelly, and Roberts [237, 473, 474], who suggest that a fast ion acquires an enhanced roton density near its equator, where $|\mathbf{v}_s - \mathbf{v}_n|$ becomes large. This effect is similar to that near a vortex core (Section 3.4.2); it yields an additional drag, so that $\mathbf{F}_i(\mathbf{v}_i)$ grows more rapidly with increasing $\mathbf{v}_i$. The detailed theory agrees well with experimental data [474] throughout region 1 for positive and negative ions at the vapor pressure in the range $0.6\ \mathrm{K} \lesssim T \lesssim 1.0\ \mathrm{K}$; it would be desirable to extend the comparison to a wider range of T, p, and x_3.

(*b*) *The Critical Velocity* v_{c1}. When the drift velocity reaches v_{c1}, the bare ion apparently becomes a charged vortex ring. Early calculations of this event assumed a sudden creation process, with the vortex ring then capturing the ion [237, 389, 475, 476]. Subsequently, however, a series of experiments by Rayfield [477] indicated that the vortex ring appears continuously by "peeling off" the moving ion, which therefore remains attached to the ring. This qualitative picture has been extended by Donnelly and Roberts [237], who have constructed a stochastic theory of the transition at $\mathscr{E}_{c1}$. In their model, the enhanced roton density provides the vorticity that ultimately grows through thermal fluctuations into a macroscopic vortex ring (compare the discussion in Section 3.2.4). Although they explicitly identify the rotons as elementary microscopic vortex rings, they also observe that the calculation is unchanged if a small vortex ring itself has a minimum energy for a radius r_{v0} comparable with atomic dimensions. From a quantum-mechanical viewpoint, the increasing curvature of the wave function for small r_v would probably produce this type of energy spectrum. Furthermore, such a small ring at its energy minimum would have zero group velocity and therefore act very much like a roton [22].

These calculations predict the critical velocity v_{c1} for positive and negative ions as a function of T, p, and x_3 [237]. Since the initial vortex ring is expected to have a radius comparable with that of the ion, the resulting critical velocity v_{c1} should decrease with increasing R [see equation (3.2.41)], and, indeed v_{c1+} typically exceeds v_{c1-}. The actual theory is considerably more complicated, however, and provides a good fit to the value and temperature dependence of $v_{c1\pm}$ for pure ^{4}He (II) at the vapor pressure [451, 478–480]; it also describes the detailed evolution of an ion pulse for $\mathscr{E}$ near $\mathscr{E}_{c1}$ [481, 481a]. At elevated pressure, the two species behave distinctly, because of their different internal structure (see Sections 3.3.1 and 3.3.2). The relative constancy expected of R_+ implies that v_{c1+} should not vary much with p, in reasonable agreement with the experiments [478]. On the other hand, v_{c1-} should increase with increasing p, because of the corresponding decrease in R_-. Experiments verify this qualitative prediction [482], but the observed change is considerably greater. In fact v_{c1-} increases so rapidly with p that it reaches the (pressure-dependent) Landau critical velocity [26] $v_c(p) \approx \Delta(p)/\hbar k_0(p)$ for roton production at about 12–15 atm [373, 482–484]. For still higher pressures, the ions cease to create rings and instead move at the velocity v_c, leaving a wake of rotons. Rayfield [327], and Neeper and Meyer [330] also studied the effect of ^{3}He impurities, which depressed v_{c1-} but left v_{c1+} unchanged, as expected from the previous discussion of ionic structure in dilute mixtures (Sections 3.3.1 and 3.3.2). Although the theory predicts the correct qualitative trends, it cannot fit the observed magnitudes; since the data are somewhat fregmentary, a systematic study of v_{c1} for various T, p, and x_3 would be very useful.

(*c*) *Charged Vortex Rings.* Once the charged ring appears, it rapidly gains energy from the field, growing to its equilibrium radius r_v and simultaneously decelerating to a final velocity [see equation (3.2.41)]

$$v = \left(\frac{\kappa}{4\pi r_v}\right) \ln\left(\frac{c r_v}{\xi}\right), \tag{3.4.19}$$

where

$$c = 8 \exp\left(\delta - \tfrac{1}{2}\right) \tag{3.4.20}$$

is a dimensionless constant. This process generally occurs in a short distance, so that v may be equated with the drift velocity v_D; the condition of eventual equilibrium becomes

$$\pm e\vec{\mathscr{E}} + \mathbf{F}_{\text{ir}}(\mathbf{v}_D) = 0, \tag{3.4.21}$$

where $\mathbf{F}_{\text{ir}}$ is the drag force on the ion–ring complex. The situation is particularly simple for large rings, where Rayfield and Reif [29] have shown that the mutual friction (Section 3.2.3) yields a frictional force per unit length

$$\mathbf{F}'_{\text{ir}}(\mathbf{v}_D) = -2\alpha\kappa^{-1}\mathbf{v}_D, \tag{3.4.22}$$

with α a temperature-dependent parameter proportional to the mean transport cross section $\bar{\sigma}$ for quasiparticle scattering by a straight vortex. The total frictional force is then given by

$$\mathbf{F}_{\text{ir}}(\mathbf{v}_D) = 2\pi r_v \mathbf{F}'_{\text{ir}}(\mathbf{v}_D) = -4\pi\alpha r_v\kappa^{-1}\mathbf{v}_D, \tag{3.4.23}$$

and a combination with equation (3.4.19) yields the simple expression

$$F_{\text{ir}} = -\alpha \ln\left(\frac{c r_v}{\xi}\right), \tag{3.4.24}$$

which has, in fact, been used to determine α and ξ as a function of T, p, and x_3 [29, 183, 185, 478]. Comparison with (3.4.21) fixes the equilibrium radius

$$r_v = c^{-1}\xi \exp\left(\frac{e\mathscr{E}}{\alpha}\right), \tag{3.4.25}$$

and the corresponding drift velocity reduces to [470, 471]

$$v_D = \frac{\kappa c}{4\pi\xi}\frac{e\mathscr{E}}{\alpha} \exp\left(-\frac{e\mathscr{E}}{\alpha}\right). \tag{3.4.26}$$

Since $e\mathscr{E}_{c1}/\alpha$ is usually large, equation (3.4.26) predicts a sharp fall in v_D with increasing $\mathscr{E} > \mathscr{E}_{c1}$, as seen in Figure 3.8.

This simple analysis has been generalized to include the drag on the trapped ion [389, 485, 486] and the finite length d of drift space [486, 487], which occasionally affects the interpretation of v_D. For sufficiently large $\mathscr{E}$, the latter effect can invalidate equation (3.4.25), because the total energy of the ring $\approx \frac{1}{2} \rho_s \kappa^2 r_v \ln (r_v/\xi)$ cannot exceed $e\mathscr{E}d$; in this case, the ion never attains its equilibrium velocity (3.4.26), and r_v instead varies approximately linearly with $\mathscr{E}$ [486, 487], in qualitative agreement with the high-field data.

We shall briefly mention a few additional observations.

i. In some experiments, the radioactive source emits ions with sufficient speed to nucleate vortex rings with no additional electric field [330, 477, 478, 488], Such a phenomenon is particularly evident at low temperature, where α is small; it probably accounts [237] for the second (lower) critical velocity observed by Cunsolo, Maraviglia, and Ricci [479, 480].

ii. In dilute mixtures at low temperature, the drift velocity v_D actually bends up just below $\mathscr{E}_{c1}$ [332, 489], in contrast to the more common behavior shown in Figure 3.8. Padmore [486] has explained this effect by invoking a low-field dynamical instability of the ion–ring complex, first suggested by Huang and Olinto [389].

iii. For $\mathscr{E} \gtrsim \mathscr{E}_{c1}$ in dilute mixtures, the curve of v_D vs. $\mathscr{E}$ can exhibit a second break [332, 489, 490]. This behavior has led to the suggestion of a second species of vortex ring, but such a drastic conclusion seems unwarranted without further evidence.

(*d*) *Field-Assisted Ion Escape.* It is evident from Figure 3.7 that the lifetime for an ion trapped on a vortex line decreases with increasing $\mathscr{E}$. Thus it is not surprising that a sufficiently large electric field can lead to field-assisted disintegration of charged vortex rings. After becoming free, the ion quickly accelerates to its critical velocity v_{c1}, when it once again nucleates a quantized vortex ring. Since the resulting time-averaged velocity exceeds that of the equilibrium ring, v_D falls less rapidly and even increases above a critical field $\mathscr{E}_{c2}$. The presence of two stochastic processes (nucleation of rings and subsequent escape from them) complicates the theory considerably. Nevertheless, Padmore [486, 487] can fit the data [451] throughout the whole range $\mathscr{E} \gtrsim \mathscr{E}_{c1}$.

Above $\mathscr{E}_{c2}$, the ion beam is observed to penetrate rather long field-free regions ("persistence current" [491, 492]), where it can also produce a large attenuation of second sound [491]. Early discussions of these effects incorrectly interpreted the relatively high drift velocity as evidence for small rings, which should neither penetrate sufficiently far nor attenuate second sound sufficiently strongly to account for the data. The suggestion of field-assisted escape [451, 486, 487] and the consequent rise in v_D removes these discrepancies; the relevant vortex rings are large, energetic, and slow—well able to travel long distances and to attenuate second sound.

ADDENDUM (*added January 1974*)

This article was completed in August 1972, and the present Addendum discusses a few significant developments since that time. A complete literature search proved impractical, however, and this brief account is not exhaustive.

1. *Detection of Vortex Lattice.* Williams and Packard [493] have refined earlier attempts to detect the vortex lattice in uniformly rotating ^{4}He (II) through ion trapping. Although their optical system should suffice to resolve single vortex lines, no discrete structure was visible. They speculate that either the vortices move rapidly or the trapped charge does not emerge at the center of the meniscus. Further experiments would be desirable.*

2. *Vortex Rings.* Gamota [494] has produced neutral vortex rings by critical flow through orifices a few microns in size. The presence of neutral rings was detected through their ability to capture charge and transport it to a collector. Measurements of both the velocity and radius of the rings fixed the circulation as h/m. In another interesting experiment [495], charged vortex rings incident on an orifice induce a level difference between two adjacent chambers. Although the data are analyzed in terms of phase slippage [195] [compare equation (3.2.87b)], a classical explanation based on the increased pressure in front of a vortex ring [71] might also be possible.

3. *AC Josephson Effects.* Recent studies [496, 497] have cast doubt on the previous interpretation of the ac Josephson effects in ^{4}He (II). In particular, two baths connected by a small orifice respond differently to a temperature difference than to a level difference, although both quantities contribute to the chemical potential. Moreover, detailed studies with an external acoustic transducer indicate that the induced stable steps of zero flow likely represent acoustic standing waves rather than a true ac Josephson state.†

4. *Critical Flow Velocity and Creation of Vorticity.* Several recent experiments [498–500] support the fluctuation model [230] for the onset of dissipation in ^{4}He (II). In particular, Hess [500–502] has studied the critical

*Williams and Packard [508] have successfully photographed the vortex array in rotating ^{4}He (II) The mean number of lines increases linearly with Ω, but they apparently execute continuous motion insted of assuming a fixed lattice structure.

†Gamota [509] apparently has observed a macroscopic interference effect in superfluid ^{4}He, analogous to one involving two superconducting Josephson junctions. See also Reference 510 for additional comments on the ac Josephson effect in ^{4}He (II).

velocity through channels or orifices guarded by superleak filters on each side. With the filters present, the observed critical velocity has the "intrinsic" form (3.2.100), but removal of the filters reduces v_c to values comparable with the "Feynman" criterion (3.2.92). Additional evidence comes from studies of isothermal film oscillations [503] and saturation effects in thermally driven film flow [504]. It now seems probable that intrinsic (fluctuation-induced) dissipation could be observed at any temperature if it were possible to suppress extrinsic (accidental) mechanisms that destroy superfluidity.

5. *Effective Mass of Ions in* ^{4}*He (II)*. Poitrenaud and Williams [505] have used a simple but elegant technique to determine the effective mass of positive and negative ions near the surface of liquid ^{4}He. In the presence of an external field, the ions are held in a potential well (3.3.93), which has a natural oscillation frequency fixed both by their effective mass and by the curvature of $V_T(x)$ near the minimum. When this frequency coincides with that of an applied radio-frequency field, the increased energy absorption is readily observed with standard microwave techniques. The resulting effective masses $M_+/m_4 = 43.6 \pm 2$ and $M^*_-/m_4 = 243 \pm 5$ agree well with those obtained by other methods (see Sections 3.3.1 and 3.3.2), as do the corresponding effective radii $R_+ = 6.1 \pm 0.1$ Å, $R_- = 17.2 \pm 0.15$ Å.

6. *Electrons on the Outer Surface of* ^{4}*He (II)*. Brown and Grimes [506] verified the existence of an electron surface layer in the vapor phase with cyclotron resonance in an applied magnetic field **B**. The observed resonance frequency depends only on the normal component of **B**, as expected for a planar array. Moreover, the width of the resonance determines the electron mobility, which (for $T \gtrsim 1.2$ K) seems limited solely by collisions with ^{4}He atoms in the vapor. At these temperatures, thermally excited surface waves do not seem to contribute. Grimes and Brown [507] also observed the excitation of the bound surface electrons from the ground state to the next higher bound level. The experimental value of the energy difference confirms both the simple hydrogenic form of the potential (3.3.90) and its detailed numerical parameters.

7. *Ion Trapping on Vortex Lines.* Williams, DeConde, and Packard [511] have observed trapping of postive ions on rectilinear vortex lines in rotating superfluid ^{4}He below 0.7 K. They also studied the trapping of negative ions below 1.5 K [512] in great detail, confirming the earlier suggestions [464, 466] that the vortex lines perform additional motions as the temperature falls.

ACKNOWLEDGEMENT

I am grateful to R. M. Bowley, J. Grant, H. E. Hall, J. L. McCauley, Jr., S. J. Putterman, G. W. Rayfield, L. L. Tankersley, and W. Zimmermann, Jr., for correspondence and suggestions. Part of the final draft was prepared at The Research Institute for Theoretical Physics, University of Helsinki, and I thank M. Vuorio for his hospitality.

REFERENCES AND NOTES

1. In addition to the other chapters in the book, general descriptions of helium may be found in Refs. [2–12]; the special problems of rotating ^{4}He (II) and of ions are considered in Refs. [13–17a] and [18–20], respectively.
2. K. R. Atkins, *Liquid Helium* (Cambridge Univ. Press, Cambridge, 1959).
3. F. London, *Superfluids* (Dover, New York, 1964), Vol. II.
4. R. J. Donnelly, *Experimental Superfluidity* (Univ. of Chicago Press, Chicago, 1967).
5. J. Wilks, *The Properties of Liquid and Solid Helium* (Oxford Univ. Press, Oxford, 1967).
6. W. E. Keller, *Helium-3 and Helium-4* (Plenum Press, New York, 1969).
7. *Proceedings of the VIIth International Conference on Low Temperature Physics,* edited by G. M. Graham and A. C. H. Hallet (University of Toronto Press, Toronto, 1961).
8. *Proceedings of the Eighth International Conference on Low Temperature Physics,* edited by R. O. Davies (Butterworths, London, 1963).
9. *Low Temperature Physics LT9,* edited by J. G. Daunt, D. O. Edwards, F. J. Milford, and M. Yaqub (Plenum Press, New York, 1965), Part A.
10. *Proceedings of the Xth International Conference on Low Temperature Physics,* edited by M. P. Malkov (Moscow, Viniti, 1967), Vol. I.
11. *Proceedings of the Eleventh International Conference on Low Temperature Physics,* edited by J. F. Allen, D. M. Finlayson, and D. M. McCall (Univ. of St. Andrews, Scotland, 1968), Vol. I.
12. *Proceedings of the Twelfth International Conference on Low Temperature Physics,* edited by E. Kanda (Keigaku Publishing Co., Tokyo, 1971).
13. H. E. Hall, *Advan. Phys.* **9,** 89 (1960).
14. W. F. Vinen, in *Progress in Low Temperature Physics,* edited by C. J. Gorter (North-Holland, Amsterdam, 1961), Vol. III, p. 1.
15. E. L. Andronikashivili and Yu. G. Mamaladze, *Rev. Mod. Phys.* **38,** 567 (1966); **39,** 494 (1967).
16. E. L. Andronikashvili and Yu. G. Mamaladze, in *Progress in Low Temperature Physics,* edited by C. J. Gorter (North-Holland, Amsterdam, 1967), Vol. V, p. 79.
17. A. L. Fetter, in *Lectures in Theoretical Physics,* edited by K. T. Mahanthappa and W. E. Brittin (Gordon and Breach, New York, 1969), Vol. XI-B, p. 321.

17a. S. J. Putterman *Phys. Rep.* **4C,** 67 (1972).

18. G. Careri, in *Progress in Low Temperature Physics,* edited by C. J. Gorter (North-Holland, Amsterdam, 1961), Vol. III, p. 58.

19. F. Reif, in *Quantum Fluids,* edited by N. Wiser and D. J. Amit (Gordon and Breach, London, 1970), p. 165.
20. G. Gamota, *J. Phys. (Paris)* **31,** Suppl. C **3,** 39 (1970).
21. L. Onsager, *Nuovo Cimento* **6,** Suppl. 2, 249 (1949).
22. R. P. Feynman, in *Progress in Low Temperature Physics,* edited by C. J. Gorter (North-Holland, Amsterdam, 1955), Vol. I, p. 17.
23. L. Meyer and F. Reif, *Phys. Rev.* **110,** 279 (1958).
24. L. Meyer and F. Reif, *Phys. Rev. Lett.* **5,** 1 (1960).
25. F. Reif and L. Meyer, *Phys. Rev.* **119,** 1164 (1960).
26. L. D. Landau, *J. Phys. (USSR)* **5,** 71 (1941); **11,** 91 (1947).
27. Reference [4, Secs. 24–26].
28. G. W. Rayfield and F. Reif, *Phys. Rev. Lett.* **11,** 305 (1963).
29. G. W. Rayfield and F. Reif, *Phys. Rev.* **136,** A1194 (1964).
30. H. Lamb, *Hydrodynamics* (Dover, New York, 1945), 6th ed., Chap. VII.
31. L. M. Milne-Thomson, *Theoretical Hydrodynamics* (Macmillan, New York, 1962), 4th ed., Chaps. XIII and XVIII.
32. A. Sommerfeld, *Mechanics of Deformable Bodies* (Academic, New York, 1950), Chap. IV.
33. See, for example, M. Born, *Atomic Physics* (Blackie, London, 1960), 6th ed., Chap. V.
34. W. F. Vinen, *Nature* **181,** 1524 (1958).
35. W. F. Vinen, *Proc. Roy. Soc. (London)* **A260,** 218 (1961).
36. See, for example, J. D. Jackson, *Classical Electrodynamics* (Wiley, New York, 1962), Chap. 5.
37. In the present section (3.2.1), we use the cumbersome but explicit notation **R** for a three-dimensional vector and **r** for a two-dimensional vector in the x-y plane. For the remainder of the chapter, however, **r** denotes a general three-dimensional vector.
38. See, for example, L. D. Landau and E. M. Lifshitz, *Fluid Mechanics* (Addison-Wesley, Reading, Mass., 1959), Sec. 2.
39. Reference [30, Secs. 145 and 146]; Ref. [31, pp. 81–85]; Ref. [38, Sec. 8].
40. The angular integral in the second line is given in H. B. Dwight, *Tables of Integrals and Other Mathematical Data* (Macmillan, New York, 1961), 4th ed., p. 228, 859.124.
41. G. B. Hess, *Phys. Rev.* **161,** 189 (1967).
42. H. E. Hall and W. F. Vinen, *Proc. Roy. Soc. (London)* **A238,** 204, 215 (1956).
43. G. Careri, W. D. McCormick, and F. Scaramuzzi, *Phys. Lett.* **1,** 61 (1962).
44. A. Walraven, *Phys. Rev. A* **1,** 145 (1970).
45. See, for example, Ref. [40, Chap. 9].
46. L. D. Landau and E. M. Lifshitz, *Electrodynamics of Continuous Media* (Addison-Wesley, Reading, Mass., 1960), pp. 119–125.
47. Reference [30, pp. 236–243].
48. R. J. Arms and F. R. Hama, *Phys. Fluids* **8,** 553 (1965).
49. W. M. Hicks, *Phil. Trans. Roy. Soc. (London)* **A175,** 183 (1884); **A176,** 725 (1885).
50. J. J. Thomson [*A Treatise on the Motion of Vortex Rings* (Macmillan, London, 1883) p. 33] also considered a vortex ring with a hollow core but obtained an expression

equivalent to taking $\delta = -\frac{1}{2}$ in equation (3.2.41). The discrepancy arises from his use of a purely kinematic boundary condition on the free surface (see also Ref. [44]), in contrast to Hicks' more physical dynamic method. In the present context, Thomson set $d\eta/d\chi = 0$ in equation (3.2.37), which (erroneously) places the free surface at $\rho = \xi$. [I am grateful to J. Grant (private communication) for elucidating this point.]

51. L. E. Fraenkel, *Proc. Roy. Soc. (London)* **A316,** 29 (1970).
52. P. G. Saffman, *Stud. Appl. Math.* **XLIX,** 371 (1970).
53. W. Thomson (Lord Kelvin), *Phil. Mag.* **33,** 511 (1867).
54 G. Gamota and T. M. Sanders, Jr., *Phys. Rev. Lett.* **15,** 949 (1965); *Phys. Rev. A* **4,** 1092 (1971).
55. G. Gamota and T. M. Sanders, Jr., *Phys. Rev. Lett.* **21,** 200 (1968).
56. K. W. Schwarz and R. J. Donnelly, *Phys. Rev. Lett.* **17,** 1088 (1966).
57. K. W. Schwarz, *Phys. Rev.* **165,** 323 (1968).
58. Reference [31, pp. 548–550].
59. See, for example, Ref. [38, Secs. 9 and 10].
60. Alternative derivations of these results may be found in Ref. [30, pp. 214–219] and Ref. [31, pp. 522–528].
61. Reference [38, p. 33, footnote 2].
62. C. Darwin [*Proc. Camb. Phil. Soc.* **49,** 342 (1953)] shows that both the physical interpretation and the numerical value of the integral change when the integration sequence is altered.
63. A. G. van Vijfeijken, A. Walraven, and F. A. Staas, *Physica* **44,** 415 (1969).
64. In fact, the impulse itself can take different values in a bounded system, depending on the choice of cuts used to render the velocity potential single valued; in all cases, however, (3.2.51) remains valid. This question is discussed in Refs. [65–69].
65. E. R. Huggins, *Phys. Rev. Lett.* **17,** 1284 (1966).
66. W. F. Vinen, in *Quantum Fluids,* edited by D. F. Brewer (North-Holland, Amsterdam, 1966), pp. 104–105.
67. M. P. Kawatra and R. K. Pathria, *Phys. Rev.* **151,** 132 (1966).
68. L. J. Campbell, *J. Low Temp. Phys.* **8,** 105 (1972).
69. C. C. Lin, in *Liquid Helium,* edited by G. Careri (Academic Press, New York, 1963), p. 93. See, especially, pp. 99–103.
70. G. Gamota and M. Barmatz, *Phys. Rev. Lett.* **22,** 874 (1969).
71. A. L. Fetter, *Phys. Rev. A* **6,** 402 (1972).
72. E. R. Huggins, *Phys. Rev. Lett.* **29,** 1067 (1972).
72a. M. C. Cross, *Phys. Rev. A* **10,** 1442 (1974).
73. P. H. Roberts and R. J. Donnelly, *Phys. Lett.* **31A,** 137 (1970).
73a. G. B. Hess, *Phys. Lett.* **41A,** 275 (1972), has used this Hamiltonian description to study the focal properties of electrostatic lenses on a beam of charged vortex rings.
74. L. Meyer, *Phys. Rev.* **148,** 145 (1966).
75. G. Gamota, A. Hasegawa, and C. M. Varma, *Phys. Rev. Lett.* **26,** 960 (1971).
76. A. Hasegawa and C. M. Varma, *Phys. Rev. Lett.* **28,** 1689 (1972).
77. G. Gamota *Phys. Rev. Lett.* **28,** 1691 (1972).

78. W. Thomson (Lord Kelvin), *Phil. Mag.* **10,** 155 (1880).

79. R. L. Douglass, *Phys. Rev.* **174,** 255 (1968).

80. J. Grant (private communication) has verified that J. J. Thomson's kinematic approach (Ref. [50]) again predicts $\delta = -\frac{1}{2}$ for a hollow core.

81. J. L. McCauley, Jr., in *Low Temperature Physics—LT 13*, edited by K. D. Timmerhaus W. J. O'Sullivan, and E. F. Hammel (Plenum Press, New York, 1974), Vol. 1, p. 421; J. L. McCauley, Jr., and L. Onsager, *J. Phys. A* **8,** 203 (1975).

82. See, for example, Ref. [31, pp. 247–248].

83. E. R. Huggins, *Phys. Rev. A* **1,** 327 (1970); *Phys. Rev. Lett.* **26,** 1291 (1971).

84. L. Turski, *Bull. Acad. Polon. Sci. Ser. Math. Astron. Phys.* **15,** 197 (1967); **18,** 103 (1970).

85. A. L. Fetter and K. Harvey, *Phys. Rev. A* **4,** 2305 (1971).

86. See also E. S. Raja Gopal, *Phys. Lett.* **9,** 230 (1964); *Ann. Phys.* (*N.Y.*) **29,** 350 (1964).

87. S. C. Whitmore and W. Zimmermann, Jr., *Phys. Rev. Lett.* **15,** 389 (1965); *Phys. Rev.* **166,** 181 (1968).

88. V. K. Tkachenko, *Zh. Eksp. Teor. Fiz.* **50,** 1573 (1966) [English transl.: *Sov. Phys.—JETP* **23,** 1049 (1966)]; see also V. K. Tkachenko, *Zh. Eksp. Teor. Fiz.* **49,** 1875 (1965) and **56,** 1763 (1969) [English transl.: *Sov. Phys.—JETP* **22,** 1282 (1966); **29,** 945 (1969)].

89. A. L. Fetter, *Phys. Rev.* **162,** 143 (1967); *B* **11,** 2049 (1975).

90. H. E. Hall, *Proc. Roy. Soc. (London)* **A 245,** 546 (1958).

91. See also E. L. Andronikashvili and D. S. Tsakadze, *Zh. Eksp. Teor. Fiz.* **37,** 322 (1959) [English transl.: *Sov. Phys.—JETP* **10,** 227 (1960)].

92. In fact, Hall (Ref. [13]) used the quantity ln $(1.046/k_{\parallel}\xi)$ instead of Kelvin's value ln $(1.12/k_{\parallel}\xi)$ for a hollow-cored vortex in (3.2.55), but the practical difference is negligible.

93. Yu. G. Mamaladze and S. G. Matinyan, *Zh. Eksp. Teor. Fiz.* **38,** 184, 656 (1960) [English transl.: *Sov. Phys.—JETP* **11,** 134, 471 (1960)].

94. E. L. Andronikashvili, Yu. G. Mamaladze, S. G. Matinyan, and D. S. Tsakadze, *Usp. Fiz. Nauk* **73,** 3 (1961) [English transl.: *Sov. Phys.—Usp.* **4,** 1 (1961)].

95. H. A. Snyder and P. J. Westervelt, *Ann. Phys. (N.Y.)* **43,** 158 (1967).

96. D. Stauffer, *Phys. Lett.* **24A,** 72 (1967).

97. The classical theory is lucidly derived by S. Chandrasekhar, *Hydrodynamic and Hydromagnetic Stability* (Oxford Univ. Press, Oxford, 1961), pp. 85–86.

98. S. V. Iordanskii, *Zh. Eksp. Teor. Fiz.* **48,** 708 (1965) [English transl.: *Sov. Phys.—JETP* **21,** 467 (1965)].

99. Reference [50, pp. 35–36]; see also, Ref. [30, p. 246, footnotes 3 and 4].

100. L. D. Landau and E. M. Lifshitz, *Statistical Physics* (Addison-Wesley, Reading, Mass., 1958), Secs. 26 and 34.

101. S. Putterman and G. E. Uhlenbeck, *Phys. Fluids* **12,** 2229 (1969).

102. A. J. Bennett and L. M. Falicov, *Phys. Rev.* **144,** 162 (1966); **152,** 494 (1966).

103. This qualitative behavior was first suggested by H. London, *Rep. Int. Conf. Low Temp.* (The Physical Society, London, 1947), Vol. II, p. 45.

104. R. G. Arkhipov, *Zh. Eksp. Teor. Fiz.* **33,** 116 (1957) [English transl.: *Sov. Phys.—JETP* **6,** 90 (1958)].

104a. In an *ideal* Bose gas, ξ must be replaced by a length of order R, as discussed by J. M. Blatt and S. T. Butler, *Phys. Rev.* **100,** 476 (1956), and by S. J. Putterman, M. Kac, and G. E. Uhlenbeck, *Phys. Rev. Lett.* **29,** 546 (1972).

105. R. E. Packard and T. M. Sanders, Jr., *Phys. Rev. Lett.* **22,** 823 (1969); *Phys. Rev. A* **6,** 799 (1972).

106. G. B. Hess and W. M. Fairbank, *Phys. Rev. Lett.* **19,** 216 (1967).

107. P. J. Bendt and T. A. Oliphant, *Phys. Rev. Lett.* **6,** 213 (1961).

108. A. L. Fetter, *Phys. Rev.* **153,** 285 (1967).

109. R. J. Donnelly and A. L. Fetter, *Phys. Rev. Lett.* **17,** 747 (1966).

110. P. J. Bendt and R. J. Donnelly, *Phys. Rev. Lett.* **19,** 214 (1967).

111. P. J. Bendt, *Phys. Rev.* **164,** 262 (1967).

112. Reference [31, pp. 254–255].

113. H. E. Hall, *Phil. Trans. Roy. Soc. (London) A* **250,** 359 (1957).

114. R. H. Walmsley and C. T. Lane, *Phys. Rev.* **112,** 1041 (1958).

115. P. J. Bendt, *Phys. Rev.* **127,** 1441 (1962).

116. D. Depatie, J. D. Reppy, and C. T. Lane, in Ref. [8, p. 75].

117. J. D. Reppy and D. Depatie, *Phys. Rev. Lett.* **12,** 187 (1964).

118. J. D. Reppy and C. T. Lane, *Phys. Rev.* **140,** A106 (1965).

119. J. D. Reppy, *Phys. Rev. Lett.* **14,** 733 (1965).

120. J. B. Mehl and W. Zimmermann, Jr., *Phys. Rev. Lett.* **14,** 815 (1965); *Phys. Rev.* **167,** 214 (1968).

121. J. R. Clow and J. D. Reppy, *Phys. Rev. Lett.* **16,** 887 (1966); *Phys. Rev. A* **5,** 424 (1972).

122. I. Rudnick, H. Kojima, W. Veith, and R. S. Kagiwada, *Phys. Rev. Lett.* **23,** 1220 (1969).

123. J. C. Weaver, *Phys. Lett.* **31A,** 97 (1970); *Phys. Rev. A* **6,** 378 (1972).

124. H. Kojima, W. Veith, S. J. Putterman, E. Guyon, and I. Rudnick, *Phys. Rev. Lett.* **27,** 714 (1971).

125. D. Stauffer and A. L. Fetter, *Phys. Rev.* **168,** 156 (1968).

126. L. J. Campbell, in Ref. [12, p. 77].

127. M. Le Ray, J. Bataille, M. François, and D. Lhuillier, *Phys. Lett.* **31A,** 249 (1970); in Ref. [12, p. 75].

128. See, for example, S. J. Putterman and I. Rudnick, *Phys. Today* **24**(8), 39 (1971).

129. H. E. Hall, in Ref. [7, p. 580].

130. A. A. Zatovskaya and L. S. Reut, *Ukr. Fiz. Zh.* **13,** 697 (1968) [English transl.: *Ukr. Phys. J.* **13,** 496 (1968)].

131. K. C. Harvey and A. L. Fetter, *J. Low Temp. Phys.* **11,** 473 (1973).

132. K. L. Chopra and J. B. Brown, *Phys. Rev.* **108,** 157 (1957).

133. D. Y. Chung and P. R. Critchlow, *Phys. Rev. Lett.* **14,** 892 (1965).

134. T. A. Kitchens, W. A. Steyert, R. D. Taylor, and P. P. Craig, *Phys. Rev. Lett.* **14** 942 (1965).

135. W. A. Steyert, R. D. Taylor, and T. A. Kitchens, *Phys. Rev. Lett.* **15,** 546 (1965).

136. L. S. Reut and I. Z. Fisher, *Ukr. Fiz. Zh.* **13,** 1910 (1968) [English transl.: *Ukr. Phys. J.* **13,** 1360 (1969)].

137. A. M. Finkel'shtein, *Zh. Eksp. Teor. Fiz.* **58,** 341 (1970) [English transl.: *Sov. Phys.—JETP* **31,** 183 (1970)].

138. A. L. Fetter, *Phys. Rev.* **152,** 183 (1966).

139. D. V. Osborne, *Proc. Phys. Soc.* **A63,** 909 (1950).

140. E. L. Andronikashvili and I. P. Kaverkin, *Zh. Eksp. Teor. Fiz.* **28,** 126 (1955) [English transl.: *Sov. Phys.—JETP* **1,** 174 (1955)].

141. J. D. Reppy, D. Depatie, and C. T. Lane, *Phys. Rev. Lett.* **5,** 541 (1960).

142. Osborne (Ref. [139]) also predicted a small temperature gradient, but R. Meservey [*Phys. Rev.* **133,** A1471 (1964)] subsequently showed that this prediction arose from an incorrect treatment of quadratic terms in the relative velocity $\mathbf{v}_s - \mathbf{v}_n$.

143. I. M. Khalatnikov, *An Introduction to the Theory of Superfluidity* (Benjamin, New York, 1965), pp. 97–98.

144. J. A. Northby and R. J. Donnelly, *Phys. Rev. Lett.* **25,** 214 (1970).

145. D. S. Shenk and J. B. Mehl, *Phys. Rev. Lett.* **27,** 1703 (1971).

146. C. M. Surko and F. Reif, *Phys. Rev.* **175,** 229 (1968); see, especially, the Appendix.

147. See, however, W. J. Trela and W. M. Fairbank, *Phys. Rev. Lett.* **19,** 822 (1967).

148. See, for example, D. Pines, in Ref. [12, p. 7].

149. G. Greenstein, *Nature* **227,** 791 (1970).

150. R. E. Packard, *Phys. Rev. Lett.* **28,** 1080 (1972).

150a. J. S. Tsakadze and S. J. Tsakadze, *Phys. Lett.* **41A,** 197 (1972).

151. E. P. Gross, *Nuovo Cimento* **20,** 454 (1961).

152. E. P. Gross, *J. Math. Phys.* **4,** 195 (1963).

153. L. P. Pitaevskii, *Zh. Eksp. Teor. Fiz.* **40,** 646 (1961) [English transl.: *Sov. Phys.—JETP* **13,** 451 (1961)].

154. See, for example, L. I. Schiff, *Quantum Machanics* (McGraw-Hill, New York, 1968), 3rd ed., pp. 431–433.

155. V. L. Ginzburg and L. P. Pitaevskii, *Zh. Eksp. Teor. Fiz.* **34,** 1240 (1958) [English transl.: *Sov. Phys.—JETP* **7,** 858 (1958)].

156. D. Amit and E. P. Gross, *Phys. Rev.* **145,** 130 (1966).

157. D. J. Amit [*Phys. Lett.* **23,** 665 (1966)] has extended the Hartree model to estimate the pressure dependence of the vortex core.

158. P. H. Roberts and J. Grant, *J. Phys. A* **4,** 55 (1971).

159. A. L. Fetter, *Phys. Rev.* **151,** 100 (1966).

160. A. L. Fetter, *Phys. Rev.* **138,** A429 (1965).

161. N. Bogoliubov, *J. Phys. (USSR)* **11,** 23 (1947).

162. A. L. Fetter, *Phys. Rev.* **138,** A709 (1965).

163. J. Grant, *J. Phys. A* **4,** 695 (1971).

163a. See, however, G. Rowlands, *J. Phys. A* **6,** 322 (1973), who instead finds a strictly quadratic spectrum with no logarithmic factor.

164. S. V. Iordanskii, *Zh. Eksp. Teor. Fiz.* **49,** 225 (1965) [English transl.: *Sov. Phys.—JETP* **22,** 160 (1960)].

165. A. L. Fetter, *Phys. Rev.* **140,** A452 (1965).
166. L. P. Pitaevskii, *Zh. Eksp. Teor. Fiz.* **35,** 1271 (1958) [English transl.: *Sov. Phys.—JETP* **8,** 888 (1959)].
167. A. L. Fetter, *Phys. Rev.* **136,** A1488 (1964).
168. A. L. Fetter, *Phys. Rev. Lett.* **27,** 986 (1971); *Ann. Phys. (N.Y.)* **70,** 67 (1972).
169. Reference [5, pp. 193–198 and Table A6, p. 670].
170. Reference [5, pp. 421–422].
171. Reference [38, Sec. 60].
172. V. L. Ginzburg and L. D. Landau, *Zh. Eksp. Teor. Fiz.* **20,** 1064 (1950).
173. Reference [100, Chap. XIV].
174. Yu. G. Mamaladze, *Zh. Eksp. Teor. Fiz.* **52,** 729 (1967) [English transl.: *Sov. Phys.—JETP* **25,** 479 (1967)].
175. I. Rudnick and J. C. Fraser, *J. Low Temp. Phys.* **3,** 225 (1970).
176. D. J. Thouless, *Ann. Phys. (N.Y.)* **52,** 403 (1969).
177. Reference [143, pp. 12–15].
178. W. I. Glaberson, D. M. Strayer, and R. J. Donnelly, *Phys. Rev. Lett.* **21,** 1740 (1968).
179. F. Pollock, *J. Low Temp. Phys.* **1,** 123 (1969).
180. W. I. Glaberson, *J. Low Temp. Phys.* **1,** 289 (1969).
181. L. S. Reut, *Ukrayin. Fiz. Zh.* **14,** 158 (1969).
182. I. Iguchi, *Phys. Rev. A* **6,** 1087 (1972).
183. W. I. Glaberson and M. Steingart, *Phys. Rev. Lett.* **26,** 1423 (1971).
184. F. Pollock, *Phys. Rev. Lett.* **27,** 303 (1971).
185. W. I. Glaberson, *Phys. Lett.* **38A,** 183 (1972).
186. M. Steingart and W. I. Glaberson, *Phys. Rev. A* **5,** 985 (1972).
187. M. Steingart and W. I. Glaberson, *J. Low Temp. Phys.* **8,** 61 (1972).
188. W. I. Glaberson, D. M. Strayer, and R. J. Donnelly, *Phys. Rev. Lett.* **20,** 1428 (1968).
189. G. V. Chester, R. Metz, and L. Reatto, *Phys. Rev.* **175,** 275 (1968).
190. V. Chiu and G. V. Chester, *Phys. Rev. A* **1,** 1549 (1970).
191. D. J. Amit, *Phys. Rev. A* **3,** 1198 (1971).
192. G. V. Chester, in *Lectures in Theoretical Physics,* edited by K. T. Mahanthappa and W. E. Brittin (Gordon and Breach, New York, 1969), Vol XI-B, pp. 271–279.
193. O. Penrose and L. Onsager, *Phys. Rev.* **104,** 576 (1956).
194. S. T. Beliaev, *Zh. Eksp. Teor. Fiz.* **34,** 417 (1958) [English transl.: *Sov. Phys.—JETP* **7,** 289 (1958)].
195. P. W. Anderson, *Rev. Mod. Phys.* **38,** 298 (1966).
196. R. J. Donnelly, *Phys. Rev. Lett.* **14,** 939 (1965).
197. W. Zimmermann, Jr., *Phys. Rev. Lett.* **14,** 976 (1965).
198. E. R. Huggins, *Phys. Rev. A* **1,** 332 (1970).
199. B. D. Josephson, *Phys. Lett.* **1,** 251 (1962).
200. P. L. Richards and P. W. Anderson, *Phys. Rev. Lett.* **14,** 540 (1965).
201. B. M. Khorana and B. S. Chandrasekhar, *Phys. Rev. Lett.* **18,** 230 (1967).
202. B. M. Khorana, *Phys. Rev.* **185,** 299 (1969).
203. P. L. Richards, *Phys. Rev. A* **2,** 1532 (1970).

204. H. E. Hall, in *Liquid Helium,* edited by G. Careri (Academic, New York, 1963), p. 326.

205. I. L. Bekarevich and I. M. Khalatnikov, *Zh. Eksp. Teor. Fiz.* **40,** 920 (1961) [English transl.: *Sov. Phys.—JETP* **13,** 643 (1961)]. See also Ref. [143, pp. 99–104].

206. See, for example, Ref. [4, Chap. 4].

207. D. S. Tsakadze, *Zh. Eksp. Teor. Fiz.* **42,** 985 (1962); **44,** 103 (1963) [English transl.: *Sov. Phys.—JETP* **15,** 681 (1962); **17,** 70 (1963)].

208. H. A. Snyder, *Phys. Fluids* **6,** 755 (1963).

209. R. H. Bruce, in Ref. [9, p. 174].

210. H. A. Snyder and Z. Putney, *Phys. Rev.* **150,** 110 (1966).

211. P. J. Bendt, *Phys. Rev.* **153,** 280 (1967).

212. J. A. Lipa, C. J. Pearce, and P. D. Jarman, *Phys. Rev.* **155,** 75 (1967).

213. P. Lucas, *J. Phys. C* **3,** 1180 (1970).

214. H. A. Snyder and P. J. Westervelt, *Phys. Rev. Lett.* **15,** 748 (1965).

215. P. Lucas, *Phys. Rev. Lett.* **15,** 750 (1965).

216. H. A. Snyder and D. M. Linekin, *Phys. Rev.* **147,** 131 (1966).

217. W. F. Vinen, *Proc. Roy. Soc. (London)* **A240,** 114, 128 (1957); **A242,** 493 (1957); **A243,** 400 (1958).

217a. Recent experiments on ion trapping in turbulent superfluid ^{4}He (II) (see Sec. 3.4.1) by D. M. Sitton and F. Moss, *Phys. Rev. Lett.* **29,** 542 (1972), and by R. A. Ashton and J. A. Northby, *Phys. Rev. Lett.* **30,** 1119 (1973) have further confirmed this picture.

218. L. J. Campbell, *J. Low Temp. Phys.* **3,** 175 (1970).

219. S. V. Iordanskii, *Ann. Phys. (N.Y.)* **29,** 335 (1964).

220. S. V. Iordanskii, *Phys. Lett.* **15,** 34 (1965).

221. H. E. Hall, *J. Phys. C* **3,** 1166 (1970).

221a. Hall's theory implies a coupling between first and second sound in rotating ^{4}He (II) [P. Lucas and A. J. Hillel, *J. Phys. C* **5,** 721 (1972)], but an experimental search for this effect [P. Lucas, *J. Phys. C* **6,** 3372 (1973)] has been unsuccessful.

222. W. J. Titus, *J. Low Temp. Phys.* **2,** 291 (1970).

223. W. J. Titus, *Phys. Rev. A* **2,** 206 (1970).

224. E. M. Lifshitz and L. P. Pitaevskii, *Zh. Eksp. Teor. Fiz.* **33,** 535 (1957) [English transl.: *Sov. Phys.—JETP* **6,** 418 (1958)].

225. I. Iguchi, *J. Low Temp. Phys.* **5,** 353 (1971).

226. S. E. Goodman, *Phys. Fluids* **14,** 1293 (1971).

227. A. L. Fetter, *Phys. Rev.* **151,** 112 (1966).

227a. A theoretical analysis of this effect has been given by F. Bloch, *Phys. Rev. A* **7,** 2187 (1973).

228. J. R. Clow and J. D. Reppy, *Phys. Rev. Lett.* **19,** 291 (1967).

229. G. Kukich, R. P. Henkel, and J. D. Reppy, *Phys. Rev. Lett.* **21,** 197 (1968); in Ref. [11, p. 140].

230. J. S. Langer and J. D. Reppy, in *Progress in Low Temperature Physics,* edited by C. J. Gorter (North-Holland, Amsterdam, 1970), Vol. VI, p. 1.

231. Reference [2, pp. 198–201].

232. Reference [6, Sec. 8.1].

232a. L. J. Campbell, in *The Helium Liquids,* Scottish Universities' Summer School 1974 (to be published).

233. W. F. Vinen, in *Liquid Helium,* edited by G. Careri (Academic, New York, 1963), p. 336.

234. J. S. Langer and M. E. Fisher, *Phys. Rev. Lett.* **19,** 560 (1967).

235. M. E. Fisher, in *Proc. Conf. Fluctuations in Superconductors,* edited by W. S. Goree and F. Chilton (Stanford Research Institute, Menlo Park, California, 1968), p. 357.

236. P. H. Roberts and R. J. Donnelly, *Phys. Rev. Lett.* **24,** 367 (1970).

237. R. J. Donnelly and P. H. Roberts, *Phil. Trans. Roy. Soc. (London)* **A271,** 41 (1971).

238. See, for example, Ref. [3, pp. 95–97].

239. G. E. Volovik, *Zh. Eksp. Teor. Fiz. Pis'ma* **15,** 116 (1972) [English transl.: *Sov. Phys.—JETP Lett.* **15,** 81 (1972)].

240. B. T. Geilikman, *Zh. Eksp. Teor. Fiz.* **37,** 891 (1959) [English transl.: *Sov. Phys.—JETP* **10,** 635 (1960)].

241. V. P. Peshkov, in *Progress in Low Temperature Physics,* edited by C. J. Gorter (North-Holland, Amsterdam, 1964), Vol. IV, p. 1.

242. J. C. Fineman and C. E. Chase, *Phys. Rev.* **129,** 1 (1963).

243. E. S. Raja Gopal, *Ann. Phys. (N. Y.)* **25,** 196 (1963).

244. W. I. Glaberson and R. J. Donnelly, *Phys. Rev.* **141,** 208 (1966).

245. L. Pauling, *The Chemical Bond* (Cornell University Press, Ithaca, N.Y., 1967), p. 189.

246. E. Haløien and J. Midtdal, *Proc. Phys. Soc. A* **68,** 815 (1955).

247. In fact, the bare mass is not very relevant, and similar phenomena occur for other positive ions in ^{4}He (II); see, for example, Refs. [248, 249].

248. G. G. Ihas and T. M. Sanders, Jr., *Phys. Lett.* **31A,** 502 (1970).

249. W. W. Johnson and W. I. Glaberson, *Phys. Rev. Lett.* **29,** 214 (1972).

250. B. Bleaney and B. I. Bleaney, *Electricity and Magnetism* (Oxford Univ. Press, Oxford, 1965), 2nd ed., Secs. 1.4. and 8.7.

251. K. R. Atkins, *Phys. Rev.* **116,** 1339 (1959).

252. K. R. Atkins, in *Liquid Helium,* edited by G. Careri (Academic Press, New York, 1963), p. 403.

253. G. Careri, F. Scaramuzzi, and J. O. Thomson, *Nuovo Cimento* **13,** 186 (1959).

254. G. Careri, U. Fasoli, and F. S. Gaeta, *Nuovo Cimento* **15,** 774 (1960).

255. J. O. Hirschfelder, C. F. Curtis, and R. B. Bird, *Molecular Theory of Gases and Liquids* (Wiley, New York, 1954), p. 946.

256. A. J. Dahm and T. M. Sanders, Jr., *Phys. Rev. Lett.* **17,** 126 (1966); *J. Low Temp. Phys.* **2,** 199 (1970).

257. Reference [46, pp. 68–69].

258. Reference [36, Sec. 4.6].

259. Reference [5, Table 3, p. 600; Table A2, p. 667; and Table A16, p. 678].

260. Reference [38, Sec. 11 and p. 36].

261. R. Barrera and G. Baym, *Phys. Rev. A* **6,** 1558 (1972).

262. P. E. Parks and R. J. Donnelly, *Phys. Rev. Lett.* **16,** 45 (1966).

263. W. P. Pratt, Jr. and W. Zimmermann, Jr., *Phys. Rev.* **177,** 412 (1969).

264. K. O. Keshishev. Yu. Z. Kovdrya, L. P. Mezhov-Deglin, and A. I. Shal'nikov, *Zh. Eksp. Teor. Fiz.* **56,** 94 (1969) [English transl.: *Sov. Phys.—JETP* **29,** 53 (1969)].

265. B. A. Brody, Ph.D. Thesis, Univ. of Michigan (1970); *Phys. Rev. B* **11,** 170 (1975).

266. R. M. Ostermeier and K. W. Schwarz, *Phys. Rev. A* **5,** 2510 (1972).

267. K. W. Schwarz, *Phys. Rev. A* **6,** 1958 (1972).

268. E. P. Gross, *Ann. Phys. (N.Y.)* **19,** 234 (1962).

269. E. P. Gross, in *Quantum Fluids,* edited by D. F. Brewer (North-Holland, Amsterdam, 1966), p. 275.

270. T. C. Padmore and A. L. Fetter, *Ann. Phys. (N.Y.)* **62,** 293 (1971).

271. L. Meyer, H. T. Davis, S. A. Rice, and R. J. Donnelly, *Phys. Rev.* **126,** 1927 (1962).

272. P. de Magistris, I. Modena, and F. Scaramuzzi, in Ref. [9, p. 349].

273. R. L. Williams, *Canad. J. Phys.* **35,** 134 (1957).

274. R. A. Ferrell, *Phys. Rev.* **108,** 167 (1957).

275. J. L. Levine and T. M. Sanders, Jr., *Phys. Rev. Lett.* **8,** 159 (1962); *Phys. Rev.* **154,** 138 (1967).

276. T. F. O'Malley, *Phys. Rev.* **130,** 1020 (1963), footnote 25.

277. D. E. Golden, *Phys. Rev.* **151,** 48 (1966).

278. B. Bederson and L. J. Kieffer, *Rev. Mod. Phys.* **43,** 601 (1971), Table II.

279. Reference [154, pp. 119, 324, 354–355].

280. A. L. Fetter and J. D. Walecka, *Quantum Theory of Many-Particle Systems* (McGraw-Hill, New York, 1971), Secs. 11 and 35.

281. R. J. Glauber, in *Lectures in Theoretical Physics,* edited by W. E. Brittin and L. G. Dunham (Interscience, New York, 1959), Vol. I, p. 315.

282. L. L. Foldy and J. D. Walecka, *Ann. Phys. (N.Y.)* **54,** 447 (1969). Strictly speaking, Foldy and Walecka define a modified $h(r)$ that is of order N^{-1} at infinity, but this change has no effect on our calculation.

283. E. Feenberg, *Theory of Quantum Fluids* (Academic, New York, 1969), Secs. 1.1. 1.2, and 2.1.

284. Alternatively, this factor may be rewritten $1 + 2a\pi^{-1}\int_0^\infty dk[1 - S(k)]$, where $S(k)$ is the *measured* structure factor; for a complete derivation, see L. L. Tankersley, *J. Low Temp. Phys.* **11,** 451 (1973).

285. B. Burdick, *Phys. Rev. Lett.* **14,** 11 (1965).

286. J. Jortner, N. R. Kestner, S. A. Rice, and M. H. Cohen, *J. Chem. Phys.* **43,** 2614 (1965).

287. W. T. Sommer, *Phys. Rev. Lett.* **12,** 271 (1964).

288. M. A. Woolf and G. W. Rayfield, *Phys. Rev. Lett.* **15,** 235 (1965).

289. A. Dahm, J. Levine, J. Penley, and T. M. Sanders, Jr., in Ref. [7, p. 495].

290. Reference [100, Sec. 147].

291. Reference [4, Sec. 23].

292. C. G. Kuper, *Phys. Rev.* **122,** 1007 (1961); in *Liquid Helium,* edited by G. Careri (Academic, New York, 1963), p. 414.

293. Reference [269, Sec. 5].

294. E. P. Gross and H. Tung-Li, *Phys. Rev.* **170,** 190 (1968).

295. K. Hiroike, N. R. Kestner, S. A. Rice, and J. Jortner, *J. Chem. Phys.* **43,** 2625 (1965).
296. B. E. Springett, M. H. Cohen, and J. Jortner, *Phys. Rev.* **159,** 183 (1967).
297. R. C. Clark, *Phys. Lett.* **16,** 42 (1965); in *Superfluid Helium,* edited by J. F. Allen (Academic, London, 1966), p. 93.
298. V. Celli, M. H. Cohen, and M. J. Zuckermann, *Phys. Rev.* **173,** 253 (1968).
299. W. B. Fowler and D. L. Dexter, *Phys. Rev.* **176,** 337 (1968).
300. J. A. Cope and P. W. F. Gribbon, *Phys. Lett.* **16,** 128 (1965); in *Superfluid Helium,* edited by J. F. Allen (Academic, London, 1966), p. 83.
301. G. Careri, S. Cunsolo, and P. Mazzoldi, *Phys. Rev. Lett.* **7,** 151 (1961); *Phys. Rev.* **136,** A303 (1964).
302. G. Careri, S. Cunsolo, and M. Vincentini-Missoni, *Phys. Rev.* **136,** A311 (1964).
303. L. Bruschi, P. Mazzoldi, and M. Santini, *Phys. Rev. Lett.* **17,** 292 (1966); *Phys. Rev.* **167,** 203 (1968).
304. D. L. Dexter and W. B. Fowler [*Phys. Rev.* **183,** 307 (1969)] have also considered the stability of doubly charged negative ions containing two electrons.
305. B. DuVall and V. Celli, *Phys. Rev.* **180,** 276 (1969).
306. T. Miyakawa and D. L. Dexter, *Phys. Rev. A* **1,** 513 (1970).
307. I. Fomin, *Zh. Eksp. Teor. Fiz. Pis.* **6,** 715 (1967) [English transl.: *Sov. Phys.—JETP Lett.* **6,** 196 (1967)].
308. B. DuVall and V. Celli, *Phys. Lett.* **26A,** 524 (1968).
309. J. A. Northby and T. M. Sanders, Jr., *Phys. Rev. Lett.* **18,** 1184 (1967).
310. C. Zipfel and T. M. Sanders, Jr., in Ref. [11, p. 296].
311. S. -Y. Wang, Ph. D. Thesis, Univ. of Michigan (1967).
312. B. E. Springett, in Ref. [11, p. 291].
313. B. E. Springett and R. J. Donnelly, *Phys. Rev. Lett.* **17,** 364 (1966).
314. B. E. Springett, *Phys. Rev.* **155,** 139 (1967).
315. M. H. Cohen and J. Jortner, *Phys. Rev.* **180,** 238 (1969).
316. K. O. Keshishev, L. P. Mezhov-Deglin, and A. I. Shal'nikov, *Zh. Eksp. Teor. Fiz. Pis.* **12,** 234 (1970) [English transl.: *Sov. Phys.—JETP Lett.* **12,** 160 (1971)].
317. V. E. Dionne, R. A. Young, and C. T. Tomizuka, *Phys. Rev. A* **5,** 1403 (1972).
318. G. A. Sai-Halasz and A. J. Dahm, *Phys. Rev. Lett.* **28,** 1244 (1972).
319. B. E. Springett, J. Jortner, and M. H. Cohen, *J. Chem. Phys.* **48,** 2720 (1968).
320. T. Miyakawa and D. L. Dexter, *Phys. Rev.* **184,** 166 (1969).
321. L. Bruschi, G. Mazzi, and M. Santini, *Phys. Rev. Lett.* **28,** 1504 (1972).
322. G. Baym, *Phys. Rev. Lett.* **17,** 952 (1966).
323. J. Bardeen, G. Baym, and D. Pines, *Phys. Rev.* **156,** 207 (1967).
324. A. F. Andreev, *Zh. Eksp. Teor. Fiz.* **50,** 1415 (1966) [English transl.: *Sov. Phys.—JETP* **23,** 939 (1966)].
325. J. Lekner, *Phil. Mag.* **22,** 669 (1970).
326. W. F. Saam, *Phys. Rev. A* **4,** 1278 (1971).
327. G. W. Rayfield, *Phys. Rev. Lett.* **20,** 1467 (1968).
328. A. J. Dahm, *Phys. Rev.* **180,** 259 (1969).

329. L. Kramer, *Phys. Rev. A* **1,** 1517 (1970).

330. D. A. Neeper and L. Meyer, *Phys. Rev.* **182,** 223 (1969).

331. J. B. Ketterson, M. Kuchnir, and P. R. Roach, in Ref. [12, p. 105].

332. M. Kuchnir, J. B. Ketterson, and P. R. Roach, *Phys. Rev. A* **6,** 341 (1972).

333. C. S. M. Doake and P. W. F. Gribbon, *Phys. Lett.* **30A,** 251 (1969).

334. G. G. Ihas and T. M. Sanders, Jr., *Phys. Rev. Lett.* **27,** 383 (1971).

335. Similar but neutral entities have been studied by A. P. Hickman and N. F. Lane, *Phys. Rev. Lett.* **26,** 1216 (1971).

336. C. M. Surko and F. Reif, *Phys. Rev. Lett.* **20,** 582 (1968).

337. M. Stockton, J. W. Keto, and W. A. Fitzsimmons, *Phys. Rev. Lett.* **24,** 654 (1970); *Phys. Rev. A* **5,** 372 (1972).

338. C. M. Surko, R. E. Packard, G. J. Dick, and F. Reif, *Phys. Rev. Lett.* **24,** 657 (1970).

339. G. W. Rayfield, *Phys. Rev. Lett.* **23,** 687 (1969).

340. R. P. Mitchell and G. W. Rayfield, *Phys. Lett.* **37A,** 231 (1971).

341. W. S. Dennis, E. Durbin, Jr., W. A. Fitzsimmons, O. Heybey, and G. K. Walters, *Phys. Rev. Lett.* **23,** 1083 (1969).

342. J. C. Hill, O. Heybey, and G. K. Walters, *Phys. Rev. Lett.* **26,** 1213 (1971).

343. Eventually, the quasiparticle density becomes so small that the ionic drag force instead arises from the creation of new quasiparticles; in this limit, S. V. Iordanskii {*Zh. Eksp. Teor. Fiz.* **54,** 1479 (1968) [English transl.: *Sov. Phys.—JETP* **27,** 793 (1968)]} finds that the drift velocity varies as $\mathscr{E}^{1/3}$, implying an effectively infinite mobility [see equation (3.3. 36)].

344. See, for example, F. Reif, *Fundamentals of Statistical and Thermal Physics* (McGraw-Hill, New York, 1965), Secs. 12.2 and 12.3.

345. Reference [5, pp. 18, 59, 431, 446].

346. Reference [5, p. 156].

347. Reference [4, Sec. 11 and Appendix, Table 17].

348. Reference [38, Sec. 20 and pp. 69–70].

349. I. Modena and F. P. Ricci, *Phys. Rev. Lett.* **19,** 347 (1967).

350. R. Cantelli, I. Modena, and F. P. Ricci, *Phys. Rev.* **171,** 236 (1968).

351. E. Achter, *Phys. Lett.* **27A,** 687 (1968).

352. B. E. Springett, *Phys. Rev.* **184,** 229 (1969).

353. D. T. Grimsrud and F. Scaramuzzi, Ref. [10, p. 197].

354. G. Ahlers and G. Gamota, *Phys. Lett.* **38A,** 65 (1972).

355. K. W. Schwarz, *Phys. Rev. A* **6,** 837 (1972).

356. A. C. Anderson, M. Kuchnir, and J. C. Wheatley, *Phys. Rev.* **168,** 261 (1968).

357. M. Kuchnir, P. R. Roach, and J. B. Ketterson, *Phys. Rev. A* **2,** 262 (1970).

357a. See also P. V. E. McClintock, *J. Low Temp. Phys.* **11,** 277 (1973), who confirms the previous measurements with a quite different technique.

358. H. E. Neustadter and M. H. Coopersmith, *Phys. Rev. Lett.* **23,** 585 (1969).

359. T. P. Eggarter and M. H. Cohen, *Phys. Rev. Lett.* **25,** 807 (1970); **27,** 129 (1971).

360. T. P. Eggarter, *Phys. Rev. A* **5,** 2496 (1972).

360a. K. W. Schwarz, in *Advances in Chemical Physics* (Interscience, New York, to be published), has reviewed this subject in great detail.

361. This approach neglects ion–ion interactions and thus ignores any concentration dependence in μ. Such effects are considered by M. Bailyn and R. Lobo, *Phys. Rev.* **176,** 222 (1968), and R. Lobo, *Phys. Lett.* **29A,** 33 (1969).

362. Reference [5, Sec. 5.4–5.7].

363. L. D. Landau and I. Pomeranchuk, *Dokl. Akad. Nauk SSR* **59,** 669 (1948).

364. I. Pomeranchuk, *Zh. Eksp. Teor. Fiz.* **19,** 42 (1949).

365. A good summary may be found in J. C. Wheatley, *Am. J. Phys.* **36,** 181 (1968).

366. G. Baym, R. G. Barrera, and C. J. Pethick, *Phys. Rev. Lett.* **22,** 20 (1969).

367. Reference [100, p. 172].

368. Reference [154, Sec. 37].

369. See Ref. [29] for a two-dimensional analog.

370. R. G. Arkhipov, *Usp. Fiz. Nauk* **88,** 185 (1966) [English transl.: *Sov. Phys.—Usp.* **9,** 174 (1966)].

371. Reference [4, Sec. 14].

372. R. M. Bowley, *J. Phys. C* **4,** 1645 (1971).

373. L. Meyer and F. Reif, *Phys. Rev.* **123,** 727 (1961).

374. S. Cunsolo and P. Mazzoldi, *Nuovo Cimento* **20,** 949 (1961).

374a. R. M. Ostermeier, *Phys. Rev. A* **8,** 514 (1973).

375. K. W. Schwarz and R. W. Stark, *Phys. Rev. Lett.* **22,** 1278 (1969).

375a. See also T. C. Padmore, *Ann. Phys. (N.Y.)* **70,** 102 (1972), who studied quasiparticle scattering by a hard sphere in an imperfect Bose gas.

376. Reference [154, p. 126].

377. I. Iguchi, *J. Low Temp. Phys.* **4,** 637 (1971).

378. K. W. Schwarz and R. W. Stark, *Phys. Rev. Lett.* **21,** 967 (1968).

379. Reference [5, p. 238].

380. Reference [280, Sec. 5].

381. Yu. Z. Kovdrya and B. N. Esel'son, *Zh. Eksp. Teor. Fiz. Pis.* **6,** 581 (1967) [English transl.: *Sov. Phys.—JETP Lett.* **6,** 90 (1967)]; *Zh. Eksp. Teor. Fiz.* **55,** 86 (1968) [English transl.: *Sov. Phys.—JETP* **28,** 46 (1969)].

382. B. N. Esel'son, Yu. Z. Kovdrya, and V. B. Shikin, *Zh. Eksp. Teor. Fiz.* **59,** 64 (1970) [English transl.: *Sov. Phys.—JETP* **32,** 37 (1971)].

383. K. W. Schwarz, *Phys. Rev. A* **6,** 1947 (1972).

384. R. M. Bowley and J. Lekner, *J. Phys. C* **3,** L127 (1970).

385. R. M. Bowley, *J. Phys. C* **4,** 853 (1971).

386. Alternatively, B. E. Springett [*Phys. Rev. A* **1,** 209 (1970)] has suggested an *attractive* polarization potential, but his model omits the effect of the ^{4}He background.

387. L. Kramer, *Phys. Rev. A* **2,** 2063 (1970).

388. J. A. Cope and P. W. F. Gribbon, *J. Phys. C* **3,** 460 (1970).

389. K. Huang and A. C. Olinto, *Phys. Rev.* **139,** A1441 (1965).

390. D. L. Goodstein, U. Buontempo, and M. Cerdonio, *Phys. Rev.* **171,** 181 (1968).

391. C. S. M. Doake and P. W. F. Gribbon, *J. Phys. C* **4,** 2457 (1971).

392. B. L. Henson, *Phys. Rev.* **135,** A1002 (1964); *Phys. Rev. Lett.* **24,** 1327 (1970).

393. L. Bruschi, G. Mazzi, and M. Santini, *Phys. Rev. Lett.* **25,** 330 (1970).

394. Most positive observations have used the Cunsolo method [S. Cunsolo, *Nuovo Cimento* **21,** 76 (1961)].

395. K. W. Schwarz, *Phys. Rev. Lett.* **24,** 648 (1970).

396. M. Steingart and W. I. Glaberson, *Phys. Rev. A* **2,** 1480 (1970).

397. B. D. Josephson and J. Lekner, *Phys. Rev. Lett.* **23,** 111 (1969).

398. For an elementary discussion of this result, see C. Kittel, *Introduction to Solid State Physics* (Wiley, New York, 1971), 4th ed., pp. 281–284.

399. J. C. Wheatley, in *Quantum Fluids,* edited by D. F. Brewer (North-Holland, Amsterdam, 1966), p. 205.

400. Reference [100, Sec. 57].

401. Reference [344, Sec. 9.17].

402. H. T. Davis and R. Dagonnier, *J. Chem. Phys.* **44,** 4030 (1966).

403. H. Gould and S. -K. Ma, *Phys. Rev. Lett.* **21,** 1379 (1968); *Phys. Rev.* **183,** 338 (1969).

404. Strictly, all these authors omitted the correction in (3.3. 76) and therefore merely obtained (3.3. 75).

405. J. Lekner and N. F. Mott (unpublished), quoted in Ref. [357].

406. See, for example, Ref. [280, Chaps. 5, 9].

407. S. Chandrasekhar, *Rev. Mod. Phys.* **15,** 1 (1943), Secs. II.1, II.2.

408. Reference [344, Secs. 15.5, 15.6].

409. R. Abe and K. Aizu, *Phys. Rev.* **123,** 10 (1961).

410. R. C. Clark, *Proc. Phys. Soc.* **82,** 785 (1963).

411. G. T. Schappert, *Phys. Rev.* **168,** 162 (1968).

412. H. Winter, *Z. Physik* **221,** 186 (1969).

413. R. Lobo and A. Massambani, *Phys. Rev. A* **4,** 426 (1971).

414. R. M. Bowley, *J. Phys. C* **4,** L207 (1971).

414a. M. W. Cole, *Rev. Mod. Phys.* **46,** 451 (1974), has recently reviewed this subject in great detail.

415. Reference [46, Sec. 10].

416. Reference [46, p. 40].

417. M. W. Cole and M. H. Cohen, *Phys. Rev. Lett.* **23,** 1238 (1969).

418. M. W. Cole, *Phys. Rev. B* **2,** 4239 (1970).

419. V. B. Shikin, *Zh. Eksp. Teor. Fiz.* **58,** 1748 (1970) [English transl.: *Sov. Phys.—JETP* **31,** 936 (1970)].

420. V. B. Shikin, *Zh. Eksp. Teor. Fiz.* **60,** 713 (1971) [English transl.: *Sov. Phys.—JETP* **33,** 387 (1971)].

421. R. S. Crandall and R. Williams, *Phys. Lett.* **34A,** 404 (1971).

421a. R. S. Crandall, *Phys. Rev. A* **8,** 2136 (1973).

422. E.P. Wigner, *Trans. Farad. Soc.* **34,** 678 (1938).

423. See, for example, R. E. Peierls, *Quantum Theory of Solids* (Oxford University Press, Oxford, 1956), pp. 66–67.

423a. N. D. Mermin, *Phys. Rev.* **176,** 250 (1968).

423b. J. F. Fernández, *Phys. Rev. B* **1,** 1345 (1970).

424. R. Williams, R. S. Crandall, and A. H. Willis, *Phys. Rev. Lett.* **26,** 7 (1971).

425. R. S. Crandall and R. Williams, *Phys. Rev. A* **5,** 2183 (1972).

426. W. T. Sommer and D. J. Tanner, *Phys. Rev. Lett.* **27,** 1345 (1971).

427. R. M. Ostermeier and K. W. Schwarz, *Phys. Rev. Lett.* **29,** 25 (1972).

428. M. W. Cole, *Phys. Rev. B* **3,** 4418 (1971).

429. L. Bruschi, B. Maraviglia, and F. E. Moss, *Phys. Rev. Lett.* **17,** 682 (1966).

430. W. Schoepe and G. W. Rayfield, *Z. Naturforsch.* **26a,** 1392 (1971), footnote 10.

431. W. Schoepe and C. Probst, *Phys. Lett.* **31A,** 490 (1970).

432. G. W. Rayfield and W. Schoepe, *Phys. Lett.* **34A,** 133 (1971).

433. G. W. Rayfield and W. Schoepe, *Phys. Lett.* **37A,** 417 (1971).

434. R. Williams and R. S. Crandall, *Phys. Rev. A* **4,** 2024 (1971).

435. B. E. Springett, *Phys. Rev. A* **5,** 2666 (1972).

436. W. Schoepe and G. W. Rayfield, *Phys. Rev. A* **7,** 2111 (1973).

437. R. P. Mitchell and G. W. Rayfield, *Phys. Lett.* **42A,** 267 (1972).

438. C. Hostika and B. E. Springett, *Phys. Rev. A* **6,** 492 (1972).

439. These observations are consistent with those of Surko and Reif, Ref. [146], but they disagree with those of Refs. [440, 441].

440. S. Cunsolo, G. Dall'Oglio, B. Maraviglia, and M. V. Ricci, *Phys. Rev. Lett.* **21,** 74 (1968).

441. S. Cunsolo and B. Maraviglia, in Ref. [11, p. 265].

442. M. Kuchnir, P. R. Roach, and J. B. Ketterson, *J. Low Temp. Phys.* **3,** 183 (1970); in Ref. [12, p. 103].

443. R. J. Donnelly, *Phys. Rev. Lett.* **14,** 39 (1965); in *Superfluid Helium,* edited by J. F. Allen (Academic, London, 1966), p. 103.

444. R. J. Donnelly and P. H. Roberts, *Proc. Roy. Soc.* (*London*) **A312,** 519 (1969).

445. W. Thomson (Lord Kelvin), *Phil. Mag.* **45,** 321 (1873).

446. D. J. Tanner, B. E. Springett, and R. J. Donnelly, in Ref. [9, p. 346].

447. B. E. Springett, D. J. Tanner, and R. J. Donnelly, *Phys. Rev. Lett.* **14,** 585 (1965).

448. R. L. Douglass, *Phys. Rev.* **141,** 192 (1966).

449. D. J. Tanner, *Phys. Rev.* **152,** 121 (1966).

450. See, however, I. Modena, A. Savoia, and F. Scaramuzzi, in Ref. [9, p. 342].

451. L. Bruschi, P. Mazzoldi, and M. Santini, *Phys. Rev. Lett.* **21,** 1738 (1968).

452. R. L. Douglass, *Phys. Rev. Lett.* **13,** 791 (1964).

453. A. L. Fetter and I. Iguchi, *Phys. Rev. A* **2,** 2067 (1970).

454. I. Iguchi, *Phys. Rev. A* **4,** 2410 (1971).

455. T. Ohmi and T. Usui, *Progr. Theoret. Phys. (Kyoto)* **45,** 1717 (1971).

456. W. Schoepe and K. Dransfeld, *Phys. Lett.* **29A,** 165 (1969).

457. J. W. Halley and A. Cheung, *Phys. Rev.* **168,** 209 (1968).

458. I. Iguchi, *Progr. Theoret. Phys. (Kyoto)* **47,** 45 (1972).

459. Reference [100, Sec. 111].

460. See also Reference [81], where the treatment is extended for small $\mathscr{E}$.
461. See also D. M. Sitton and F. Moss, *Phys. Rev. Lett.* **23,** 1090 (1969).
462. T. C. Padmore, *Phys. Rev. Lett.* **28,** 469 (1972).
463. A. G. Cade, *Phys. Rev. Lett.* **15,** 238 (1965).
464. R. L. Douglass, *Phys. Lett.* **28A,** 560(1969).
465. R. L. Douglass and C. le Pair, Ref. [10, p. 239].
466. R. E. Packard (private communication).
467. L. S. Reut and I. Z. Fisher, *Zh. Eksp. Teor. Fiz.* **55,** 722 (1968) [English transl.: *Sov. Phys.—JETP* **28,** 375 (1969)].
468. T. Ohmi and T. Usui, *Progr. Theoret. Phys. (Kyoto)* **41,** 1401 (1969).
469. T. Ohmi, T. Tsuneto, and T. Usui, *Progr. Theoret. Phys. (Kyoto)* **41,** 1395 (1969).
470. G. Careri, S. Cunsolo, P. Mazzoldi, and M. Santini, in Ref. [9, p. 335].
471. G. Careri, S. Cunsolo, P. Mazzoldi, and M. Santini, *Phys. Rev. Lett.* **15,** 392 (1965).
472. G. Careri, F. Dupré, and P. Mazzoldi, in *Quantum Fluids,* edited by D. F. Brewer (North-Holland, Amsterdam, 1966), p. 305.
473. R. J. Donnelly and P. H. Roberts, *Phys. Rev. Lett.* **23,** 1491 (1969).
474. D. M. Strayer, R. J. Donnelly, and P. H. Roberts, *Phys. Rev. Lett.* **26,** 165 (1971).
475. D. Pines, in *Quantum Fluids,* edited by D. F. Brewer (North-Holland, Amsterdam, 1966), p. 328.
476. B. K. Jones, *Phys. Rev.* **186,** 168 (1969).
477. G. W. Rayfield, *Phys. Rev. Lett.* **19,** 1371 (1967).
478. G. W. Rayfield, *Phys. Rev.* **168,** 222 (1968).
479. S. Cunsolo, B. Maraviglia, and M. V. Ricci, *Phys. Lett.* **26A,** 605 (1968).
480. S. Cunsolo and B. Maraviglia, *Phys. Rev.* **187,** 292 (1969).
481. D. M. Strayer and R. J. Donnelly, *Phys. Rev. Lett.* **26,** 1420 (1971).
481a. J. A. Titus and J. S. Rosenshein, *Phys. Rev. Lett.* **31,** 146 (1973).
482. G. W. Rayfield, *Phys. Rev. Lett.* **16,** 934 (1966).
483. D. A. Neeper, *Phys. Rev. Lett.* **21,** 274 (1968).
484. D. A. Neeper and L. Meyer, in Ref. [11, p. 270].
485. M. Steingart and W. I. Glaberson, *Phys. Lett.* **35A,** 311 (1971).
486. T. C. Padmore, *Phys. Rev. A* **5,** 356 (1972).
487. T. C. Padmore, *Phys. Rev. Lett.* **26,** 63 (1971).
488. G. Gamota and T. M. Sanders, Jr., in Ref. [11, p. 261].
489. M. Kuchnir, J. B. Ketterson, and P. R. Roach, *Phys. Rev. Lett.* **26,** 879 (1971).
490. M. Kuchnir, J. B. Ketterson, and P. R. Roach, *Phys. Lett.* **36A,** 287 (1971).
491. L. Bruschi, B. Maraviglia, and P. Mazzoldi, *Phys. Rev.* **143,** 84 (1966).
492. L. Bruschi and M. Santini, *Rev. Sci. Instr.* **41,** 102 (1970).
493. G. A. Williams and R. E. Packard, *Bull. Am. Phys. Soc.* **18,** 1598 (1973).
494. G. Gamota, *Phys. Rev. Lett.* **31,** 517 (1973).
495. R. Carey, B. S. Chandrasekhar, and A. J. Dahm, *Phys. Rev. Lett.* **31,** 873 (1973).
496. D. L. Musinski and D. H. Douglass, *Phys. Rev. Lett.* **29,** 1541 (1972).
497. P. Leiderer and F. Pobell, *Phys. Rev. A* **7,** 1130 (1973).

498. H. A. Notarys, *Phys. Rev. Lett.* **22,** 1240 (1969).

499. D. H. Liebenberg, *Phys. Rev. Lett.* **26,** 744 (1971).

500. G. B. Hess, *Phys. Rev. Lett.* **27,** 977 (1971).

501. G. B. Hess, *Phys. Rev. Lett.* **29,** 96 (1972).

502. G. B. Hess, *Phys. Rev. A* **8,** 899 (1973).

503. J. K. Hoffer, J. C. Faser, E. F. Hammel, L. J. Campbell, W. E. Keller, and R. H. Sherman, in *Low Temperature Physics—LT* 13, edited by K. D. Timmerhaus, W. J. O'Sullivan, and E. F. Hammel (Plenum Press, New York, 1974), Vol. 1, p. 253.

504. L. J. Campbell and D. H. Liebenberg, *Phys. Rev. Lett.* **29,** 1065 (1972).

505. J. Poitrenaud and F. I. B. Williams, *Phys. Rev. Lett.* **29,** 1230 (1972); **32,** 1213(E) (1974).

506. T. R. Brown and C. C. Grimes, *Phys. Rev. Lett.* **29,** 1233 (1972).

507. T. R. Brown and C. C. Grimes (to be published).

507. C. C. Grimes and T. R. Brown, *Phys. Rev. Lett.* **32,** 280 (1974).

508. G. A. Williams and R. E. Packard, *Phys. Rev. Lett.* **33,** 280 (1974).

509. G. Gamota, *Phys. Rev. Lett.* **33,** 1428 (1974).

510. P. W. Anderson and P. L. Richards, *Phys. Rev. B***11,** 2702 (1975).

511. G. A. Williams, K. DeConde, and R. E. Packard, *Phys. Rev. Lett.* **34,** 924 (1975).

512. K. DeConde, G. A. Williams, and R. E. Packard, *Phys. Rev. Lett.* **33,** 683 (1974).

CHAPTER 4

Brillouin and Raman Scattering in Helium*

Michael J. Stephen

Physics Department, Rutgers University,
New Brunswick, New Jersey

4.1. INTRODUCTION

It was first suggested in 1943 by Ginzburg [1] that light scattering would be a useful probe for investigating some of the properties of liquid He. During the past five years a number of experiments have been reported for the first time in which high-resolution studies have been made of the light scattered spontaneously by liquid He. These experiments were made possible by the advent of laser light sources and recent developments in optical techniques. Undoubtedly further experiments of this kind will be carried out in the future. In this chapter we attempt to give an elementary review of the principles of light scattering from liquid He and indicate what information can be obtained from the intensity and spectrum of the scattered light. In many ways the information obtained from Brillouin and Raman scattering in He is complementary to that obtained from neutron scattering.

Throughout this chapter we will be concerned with light scattering by liquid He when the liquid is close to thermal equilibrium. This scattering is called thermal or spontaneous scattering. It is also possible to scatter light

*Supported in part by the National Science Foundation.

when these conditions do not hold: Van den Berg et al. [2] and Woolf et al. [3] have observed light scattered from injected first-sound waves in superfluid ^{4}He, and Jacucci and Signorelli [4] from injected second-sound waves in superfluid ^{4}He. A number of authors [5–7] have observed stimulated Brillouin scattering in superfluid ^{4}He in which the sound wave is generated by the incident light itself. These experiments do not appear to be as informative as those in which spontaneous or thermal scattering is observed. We will also confine our attention to the superfluid phase of He and will not discuss Brillouin scattering [8] or Raman scattering [9] from the normal He (I) phase.

The scattering of light by liquid He is produced by the thermal fluctuations in the dielectric constant of the liquid. The polarizability α of a He atom is small ($\alpha \simeq 0.2$ Å^3), and the amount of light scattered is consequently small. The light scattering in liquid He is comparable with that produced by dry air. A typical scattering process can be represented as in Figure 4. 1*a*. An incident light beam of frequency (circular) Ω_i and wave vector K_i in the liquid is scattered through an angle θ giving rise to a scattered beam of frequency Ω_s and wave vector K_s. An excitation of frequency ω and wave vector $\mathbf{q}$ is either created or destroyed in the liquid. In superfluid ^{4}He at low temperatures these excitations would be single phonons, while at higher temperatures, in the hydrodynamic region, they would be first- or second-sound waves. The conservation of momentum and energy require that

$$\mathbf{q} = \mathbf{K}_i - \mathbf{K}_s, \tag{4.1.1a}$$

$$\omega = \Omega_i - \Omega_s. \tag{4.1.1b}$$

The frequency shift is $\omega = \pm uq$, where u is the sound velocity and is very much smaller than either Ω_i or Ω_s for visible light. Then to a good approximation from (4.1.1b) $|\mathbf{K}_i| = |\mathbf{K}_s|$ and the wave vector of the phonon is deter-

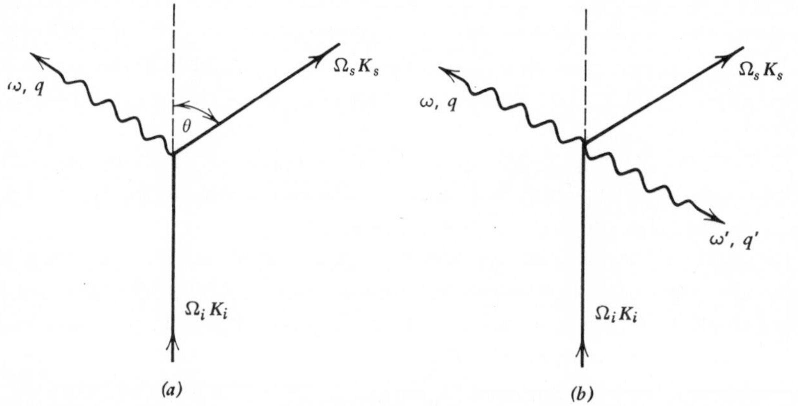

FIGURE 4.1. Representation of light scattering by excitations. (*a*) Creation of a single excitation; (*b*) Creation of two excitations.

mined by the angle of scattering (for a fixed value of $|\mathbf{K}_i|$). From (4.1.1a) one finds

$$q = 2K_i \sin \frac{\theta}{2}. \tag{4.1.2}$$

The largest wave vector excitation that can be produced is thus $2K_i$ when the light is scattered in the backward direction. For an argon ion laser $\Omega_i \simeq 3.6 \times 10^{15}$ sec^{-1}, $K_i = \Omega_i/c \simeq 1.2 \times 10^5$ cm/s. Thus, owing to the very small wave vector of the light, only excitations of small wave vector can be produced ($q \sim 10^5$ cm^{-1}). This is the case if only the first-order process in Figure 4.1a is considered and this process accounts for most of the intensity (99.9%) of the light scattering. In the above discussion we have only considered light scattered by sound waves. However, any excitation which leads to a fluctuation in the dielectric constant can be investigated by light scattering.

It is also possible for higher-order processes to occur in light scattering in which more than one excitation is produced in the scattering of a single photon. A second-order process in which two excitations are produced is shown in Figure 4. 1*b*. In this case the conservation of momentum and energy require

$$\mathbf{q} + \mathbf{q}' = \mathbf{K}_i - \mathbf{K}_s, \tag{4.1.3a}$$

$$\omega + \omega' = \Omega_i - \Omega_s. \tag{4.1.3b}$$

As $\mathbf{K}_i$ and $\mathbf{K}_s$ are small, for $q, q' > |\mathbf{K}_i|$, momentum conservation requires that the excitations have practically equal and oppsite momenta but puts no restriction on their magnitude. The frequency shift in (4.1.3b) is the sum of the frequencies of the two excitations and can be very much larger than the frequency shift in first-order Brillouin scattering. For this reason, this process is generally referred to as Raman scattering. It was first suggested in superfluid ^{4}He by Halley [10]. Observation of this process enables us to investigate the properties of pairs of excitations of almost equal and opposite momenta in ^{4}He. The intensity in this process is considerably weaker than in Brillouin scattering (by a factor of almost 10^{-4}).

With this brief introduction, we now turn to a more detailed discussion of some topics in spontaneous light scattering in the superfluid phase of He. In Section 4.2 we express the intensity and spectrum of the scattered light in terms of certain density correlation functions. In Section 4.3 we discuss light scattering from the first-and second-sound modes in pure superfluid ^{4}He. This discussion is extended in Section 4.4 to light scattering near T_λ and in Section 4.5 to superfluid ^{3}He–^{4}He mixtures. Finally, in Section 4.6 we discuss Raman scattering in He. Several review articles have recently appeared on the subject of light scattering from He [11–13] in which some of these subjects are also covered.

4.2. LIGHT SCATTERING BY DENSITY FLUCTUATIONS

In a monatomic liquid like He at low temperatures the dielectric constant of the liquid is determined by the local density. This enables us to express the light scattering spectrum in terms of certain density correlation functions. The density $\rho(\mathbf{r}, t)$ at point $\mathbf{r}$ and at time t in the liquid can be written $\rho(\mathbf{r}, t) = \rho + \delta\rho(\mathbf{r}, t)$, where ρ is the average density and $\delta\rho(\mathbf{r}, t)$ is the fluctuating part. We define density correlation functions

$$S(\mathbf{r}_{12}, t_{12}) = \langle \delta\rho(\mathbf{r}_1 t_1)\, \delta\rho(\mathbf{r}_2 t_2)\rangle, \tag{4.2.1}$$

$$S_3(\mathbf{r}_{12}, \mathbf{r}_{23}, t_{12}, t_{23}) = \langle \delta\rho(\mathbf{r}_1 t_1)\, \delta\rho(\mathbf{r}_2 t_2)\, \delta\rho(\mathbf{r}_3 t_3)\rangle, \tag{4.2.2}$$

$$S_4(\mathbf{r}_{12}, \mathbf{r}_{23}, \mathbf{r}_{34}, t_{12}, t_{23}, t_{34}) = \langle \delta\rho(\mathbf{r}_1 t_1)\, \delta\rho(\mathbf{r}_2 t_2)\, \delta\rho(\mathbf{r}_3 t_3)\, \delta\rho(\mathbf{r}_4 t_4)\rangle, \tag{4.2.3}$$

where $\mathbf{r}_{12} = \mathbf{r}_1 - \mathbf{r}_2$, $t_{12} = t_1 - t_2$, etc. The angular brackets indicate an ensemble average over the atomic motions in the liquid. The Fourier transform of (4.2.1) is defined by

$$S(\mathbf{q}, \omega) = \int d\mathbf{r}_{12}\, dt_{12}\, e^{-i\mathbf{q}\cdot\mathbf{r}_{12} + i\omega t_{12}}\, S(\mathbf{r}_{12}, t_{12}). \tag{4.2.4}$$

In what follows, we will require only the cubic correlation function (4.2.2) in the case $t_{23} = 0$ and the quartic correlation function (4.2.3) in the case $t_{12} = 0$ and $t_{34} = 0$. We define the appropriate Fourier transforms:

$$S_3(\mathbf{q}_1, \mathbf{q}_2, \omega) = \int d\mathbf{r}_{12}\, d\mathbf{r}_{23}\, dt_{12}\, e^{-i\mathbf{q}_1\cdot\mathbf{r}_{12} - i\mathbf{q}_2\cdot\mathbf{r}_{23} + i\omega t_{12}}\, S_3(\mathbf{r}_{12}, \mathbf{r}_{23}, t_{12}, 0). \tag{4.2.5}$$

$$S_4(\mathbf{q}_1, \mathbf{q}_2, \mathbf{q}_3, \omega) = \int d\mathbf{r}_{12}\, d\mathbf{r}_{23}\, d\mathbf{r}_{34}\, dt_{23}\, e^{-i\mathbf{q}_1\cdot\mathbf{r}_{12} - i\mathbf{q}_2\cdot\mathbf{r}_{23} - i\mathbf{q}_3\cdot\mathbf{r}_{34} + i\omega t_{23}}$$
$$S_4(\mathbf{r}_{12}, \mathbf{r}_{23}, \mathbf{r}_{34}, 0, t_{23}, 0). \tag{4.2.6}$$

Expressions for the light-scattering spectrum in terms of the correlation functions (4.2.4)–(4.2.6) have been given in a number of places [14–17] and we will give only a brief discussion before quoting the results.

We suppose that a plane polarized light wave with electric field $\mathbf{E}_0(\mathbf{r},t) = \mathbf{E}_0\, e^{i\mathbf{K}_i\cdot\mathbf{r} - i\Omega_i t}$ is incident in the liquid. The electric field $\mathbf{E}(\mathbf{r},t)$ at point $\mathbf{r}$ consists of the incident field and the field from all the induced dipoles on the atoms of the liquid. From Maxwell's equations it follows that

$$\mathbf{E}(\mathbf{r}, t) = \mathbf{E}_0(\mathbf{r}, t) + \int_\sigma d\mathbf{r}' \int dt' \mathbf{G}(\mathbf{r} - \mathbf{r}', t - t')\cdot\mathbf{P}(\mathbf{r}', t'), \tag{4.2.7}$$

where $\mathbf{P}(\mathbf{r}', t)$ is the polarization at $\mathbf{r}'$ and t' in the liquid. The integration in (4.2.7) is over the volume of the liquid. If the point $\mathbf{r}$ lies within the liquid it is necessary to exclude a small region σ surrounding the point $\mathbf{r}$. The dyadic

Green's function in (4.2.7) is given by

$$\mathbf{G}(\mathbf{R},t) = \frac{c}{2\pi}\int_{-\infty}^{\infty} dk\, e^{-ikct}\, \mathbf{G}_k(\mathbf{R}) \tag{4.2.8}$$

with

$$\mathbf{G}_k(\mathbf{R}) = (\nabla\nabla + k^2\mathbf{I})\left(\frac{e^{ikR}}{R}\right). \tag{4.2.9}$$

Of particular interest are the following limiting forms of $\mathbf{G}_k$:

$$\mathbf{G}_k(\mathbf{R}) = \frac{k^2 e^{ikR}}{R}\left(\mathbf{I} - \frac{\mathbf{RR}}{R^2}\right); \quad kR \gg 1 \tag{4.2.10a}$$

$$\mathbf{G}_k(R) = \frac{3\mathbf{RR} - \mathbf{I}R^2}{R^5}; \quad kR \ll 1. \tag{4.2.10b}$$

When (4.2.10a) is substituted in (4.2.7), we see that the field a long way from the liquid consists of the incident field and the field arising from the induced polarization that is a transverse spherical wave. On the other hand, if we examine the field within the liquid, it is appropriate to use (4.2.10b), and the field is the sum of the incident field and the field of all the induced atomic dipoles.

In a monatomic liquid like He it is reasonable to suppose that the polarization $\mathbf{P}$ is proportional to the local density and the local field [18]

$$\mathbf{P}(\mathbf{r},t) = \frac{\alpha}{m}\,\rho(\mathbf{r},t)\,\mathbf{E}(\mathbf{r},t), \tag{4.2.11}$$

where α is the polarizability of the He atom at the frequency of the incident light and m is the mass of a ^{4}He atom. When (4.2.11) is substituted in (4.2.7), we obtain an integral equation for the field $\mathbf{E}$

$$\mathbf{E}(1) = \mathbf{E}_0(1) + \frac{\alpha}{m}\int_\sigma d(2)\,\mathbf{G}(1-2)\cdot\rho(2)\,\mathbf{E}(2), \tag{4.2.12}$$

where we use the notation $1 \equiv (\mathbf{r}, t)$, $2 \equiv (\mathbf{r}', t')$. This equation has been solved approximately in Reference [16] for the scattered field $\delta\mathbf{E}$ as a power series in the density fluctuations $\delta\rho\,(\mathbf{r}, t)$. The first two terms in the series for the scattered field are

$$\delta\mathbf{E}_1(1) = \frac{\alpha}{m}\int d(2)\,\mathbf{G}(1-2)\cdot\delta\rho(2)\,\mathbf{E}_0(2) \tag{4.2.13}$$

and

$$\delta\mathbf{E}_2(1) = \frac{\alpha^2}{m^2}\int_\sigma d(2)\int d(3)\,\mathbf{G}(1-2)\cdot\delta\rho(2)\,\mathbf{G}(2-3)\cdot\delta\rho(3)\,\mathbf{E}_0(3). \tag{4.2.14}$$

In equation (4.2.13) the incident light is scattered by a single density fluctuation to give the scattered light. Equation (4.2.14) describes a second-order process in which the incident light polarizes a density fluctuation $\delta\rho(3)$ which then leads to a polarization at a neighboring density fluctuation $\delta\rho(2)$. This density fluctuation produces the scattered light. This second-order process will be considerably weaker than the first-order process in (4.2.13) because it is of second order in the small polarizability α. Higher-order processes are even weaker and not observable with present techniques, and will be neglected.

We suppose that the scattered light with polarization along a unit vector **p** is observed. A grating spectrometer or Fabry–Perot spectrometer measures the power spectrum of the scattered light. This is the Fourier transform with respect to the time of the correlation function of the scattered field. From (4.2.13) and (4.2.14) we can define the three correlation functions:

$$H_p^{(1)}(\tau) = \langle \mathbf{p}\cdot\delta\mathbf{E}_1^*(\mathbf{R}, t)\, \mathbf{p}\cdot\delta\mathbf{E}_1(\mathbf{R}, t+\tau)\rangle, \tag{4.2.15}$$

$$H_p^{(12)}(\tau) = \langle \mathbf{p}\cdot\delta\mathbf{E}_1^*(\mathbf{R}, t)\, \mathbf{p}\cdot\delta\mathbf{E}_2(\mathbf{R}, t+\tau)\rangle, \tag{4.2.16}$$

and

$$H_p^{(2)}(\tau) = \langle \mathbf{p}\cdot\delta\mathbf{E}_2^*(\mathbf{R}, t)\, \mathbf{p}\cdot\delta\mathbf{E}_2(\mathbf{R}, t+\tau)\rangle, \tag{4.2.17}$$

where **R** is the point of observation of the light a long way from the liquid and the angular brackets indicate, as in equations (4.2.1)–(4.2.3), an ensemble average over the atomic motions in the liquid. In each case the power spectrum is defined by

$$H_p(\Omega_s) = \int_{-\infty}^{\infty} d\tau\, e^{i\Omega_s\tau}\, H_p(\tau). \tag{4.2.18}$$

The differential extinction coefficient for light scattering, $w_p(\omega)\, d\Omega_s\, dS$ is the number of photons scattered into a solid angle dS with polarization **p** and frequency Ω_s per incident quantum and per unit path length in the liquid. This extinction coefficient is related to the power spectrum (4.2.18) by

$$w_p(\Omega_s) = \left(\frac{H_p(\Omega_s)R^2}{2\pi V E_0^2}\right)\text{sec/cm. sec.}, \tag{4.2.19}$$

where V is the volume of the liquid.

The power spectra of (4.2.15)–(4.2.17) can be expressed in terms of the Fourier transforms of the correlation functions (4.2.1)–(4.2.3). Substituting (4.2.13) in (4.2.15) and using (4.2.8), one finds for the first-order scattering

$$H_p^{(1)}(\Omega_s) = V\left(\frac{\alpha^2}{m^2R^2}\right)\left(\frac{\Omega_s}{c}\right)^4 (\mathbf{p}\cdot\mathbf{E}_0)^2 S(q,\omega), \tag{4.2.20}$$

where $\mathbf{q} = \mathbf{K}_i - \mathbf{K}_s$ and $\omega = \Omega_t - \Omega_s$ are the wave vector and frequency transfer in the scattering ($\mathbf{K}_s$ and Ω_s are the wave vector and frequency of the scattered light, see Figure 1a). The interesting information about the spectrum of the scattered light is contained in the density correlation function $S(q,\omega)$. The same is true in neutron scattering. The important difference between neutron and light scattering is in the magnitude of the wave vectors and frequency shifts. In the first-order light scattering we are confined to very small wave vectors and frequency shifts. The dependence of the intensity of the scattered light on the angle of scattering and the direction of polarization is contained in the factor $(\mathbf{p}\cdot\mathbf{E}_0)^2$ in (4.2.20). If the incident light is unpolarized and we sum over the two directions of polarization of the scattered light, this factor is replaced by $E_0^2\,(1 + \cos^2\theta)$, where θ is the angle of scattering. Equation (4.2.20) was derived with the approximation (4.2.11), which assumes that the dielectric constant is a linear function of the density. In a more accurate derivation [15] the factor α/m in (4.2.20) is replaced by $1/4\pi\,(\partial\varepsilon/\partial\rho)$, where ε is the dielectric constant at the incident light frequency.

Also of interest is the total intensity of the Brillouin scattering, $w_p^{(1)}$, in unit solid angle with a given polarization, per incident photon and per unit path length in the liquid. This is obtained by integrating (4.2.19) over all frequencies:

$$w_p^{(1)} = \int_{-\infty}^{\infty} d\Omega_s\, w_p^{(1)}(\Omega_s) = \frac{\alpha^2}{m^2}\left(\frac{\Omega_i}{c}\right)^4 (\mathbf{p}\cdot\hat{\mathbf{E}}_0)^2 S(q)/\text{cm. sec.}, \qquad (4.2.21)$$

where $\hat{\mathbf{E}}_0$ is a unit vector along $\mathbf{E}_0$ and $S(q)$ is the mean square fluctuations in density with wave vector q:

$$S(q) = \frac{1}{2\pi}\int_{-\infty}^{\infty} d\omega\, S(q,\omega) = \langle|\delta\rho_q|^2\rangle. \qquad (4.2.22)$$

In evaluating the integral in (4.2.21), we have neglected the frequency dependence of α and replaced the factor $(\Omega_s/c)^4$ by $(\Omega_i/c)^4$.

The higher-order correlation functions, (4.2.16) and (4.2.17), may be expressed in terms of the density correlation functions (4.2.5) and (4.2.6) [16]:

$$H_p^{(12)}(\Omega_s) = V\left(\frac{\alpha^3}{m^3R^2}\right)\left(\frac{\Omega_s}{c}\right)^4 (\mathbf{p}\cdot\mathbf{E}_0)[3(\mathbf{p}\cdot\mathbf{n})(\mathbf{E}_0\cdot\mathbf{n}) - \mathbf{p}\cdot\mathbf{E}_0]$$

$$\left(\frac{1}{2\pi}\right)^3 \int d\mathbf{q}\, g(q) P_2(\cos\theta_{qn}) S_3(\mathbf{K}_i - \mathbf{K}_s, \mathbf{q}, \omega), \qquad (4.2.23)$$

$$H_p^{(2)}(\Omega_s) = V\left(\frac{\alpha^4}{m^4R^2}\right)\left(\frac{\Omega_s}{c}\right)^4 \frac{3}{5} E_0^2[1 + \frac{1}{3}(\mathbf{p}\cdot\mathbf{E}_0)^2]$$

$$\frac{1}{(2\pi)^6}\int d\mathbf{q}\, d\mathbf{q}'\, g(q) g(q') P_2(\cos\theta_{qq'}) S_4(\mathbf{q}, 0, \mathbf{q}', \omega). \qquad (4.2.24)$$

Here $\mathbf{n}$ is a unit vector along the direction of $\mathbf{K}_i - \mathbf{K}_s$, P_2 is the second Legendre polynomial, θ_{qn} is the angle between $\mathbf{q}$ and $\mathbf{n}$ and $\theta_{qq'}$ is the angle between $\mathbf{q}$ and $\mathbf{q}'$. Owing to the isotropy of the fluid, the correlation functions in (4.2.23) and (4.2.24) only depend on these angles, respectively, and on the magnitudes of the wave vector and frequency appearing in them. In each case we have neglected $\mathbf{K}_i$ and $\mathbf{K}_s$ in comparison with $\mathbf{q}$ and or $\mathbf{q}'$. This is well justified, as the main contribution to the integrals in (4.2.23) and (4.2.24) come from regions of q space where the density of states is large, i.e., near the roton minumum and the maxima in the dispersion curve. The factor $g(q)$ arises because we have excluded a small region σ from the integration in (4.2.12). In the paper of Stephen [16], this was taken to be a spherical region of radius a equal to half the interparticle spacing in ^{4}He. This gives

$$g(q) = 4\pi\left(\frac{\sin qa}{(qa)^3} - \frac{\cos qa}{(qa)^2}\right). \tag{4.2.25}$$

A better treatment of this factor was given by Baeriswyl [19], who introduced the pair-distribution function $g(r)$ for the liquid, and showed that (4.2.25) should be replaced by

$$g(q) = \int_0^\infty \frac{dr\, g(r)\, j_2(qr)}{r}, \tag{4.2.26}$$

where j_2 is the second spherical Bessel function.

The three scattering processes given in (4.2.20), (4.2.23), and (4.2.24) have a very different dependence on scattering angle and direction of polarization $\mathbf{p}$ of the scattered light. This allows us to obtain some information about the processes separately. Thus both (4.2.20) and (4.2.23) vanish when $\mathbf{p}\cdot\mathbf{E}_0 = 0$. This can be achieved by collecting the light scattered along the direction of the electric vector of the incident light. The scattering in (4.2.24) is almost isotropic, whereas (4.2.20) and (4.2.23) depend strongly on the angle of scattering. This will be discussed in greater detail when we consider Raman scattering in Section 4.6.

It is also instructive to make a connection between the simple representation of light scattering given in Figure 4.1 and the results of equations (4.2.20), (4.2.23), and (4.2.24). In a linear approximation, we would associate a single phonon or excitation with a density fluctuation $\delta\rho$. Then equation (4.2.20) would represent the process shown in Figure 4. la in which a single excitation is created (or destroyed) in the scattering of a photon. Equation (4.2.24) represents the process shown in Figure 4.1b in which two excitations of equal and opposite momenta are created (or destroyed) in the scattering of a single photon. In the initial state the excitations have momenta $\mathbf{q}$ and $-\mathbf{q}$, and in the final state, $\mathbf{q}'$ and $-\mathbf{q}'$. Owing to the factor $P_2(\cos\theta_{qq'})$ in (4.2.24) the two excitations are in a state of angular momentum $l = 2$, i.e., a d state. The correlation function in (4.2.23) is identically zero in the linear approximation.

The fact that the hydrodynamic equations are nonlinear complicates the picture slightly. If these equations are solved by successive approximation, then $\delta\rho$ will contain terms involving one phonon amplitude, two phonon amplitudes, etc. Thus it is possible to get a contribution to scattering in which two phonons are created from equation (4.2.20). This will be discussed further in Section 4.6 but these nonlinear effects appear to be quite small.

4.3. BRILLOUIN SCATTERING FROM SUPERFLUID ^{4}HE

In the previous section we derived the basic formula for Brillouin scattering in a monatomic fluid like ^{4}He [see equation (4.2.20)]. The interesting spectral information is contained in the density correlation function, $S(q, \omega)$. In this section we will discuss the form of this correlation function for superfluid ^{4}He. Once this function is known, the power spectrum of the scattered light is given by equation (4.2.20), and the scattering cross section by (4.2.14). Also of interest is the total intensity, $w_p^{(1)}$ of the Brillouin light scattered, which is proportional to the mean square fluctuations in density, $S(q)$ [see equation (4.2.21)].

We begin by considering the total intensity of the Brillouin scattering. The mean square density fluctuations are easily evaluated by thermodynamics. The probability of having spontaneous fluctuations in density $\delta\rho(\mathbf{r})$ and temperature $\delta T(\mathbf{r})$ is proportional to $e^{-W/kT}$, where

$$W = \frac{1}{2}\int d\mathbf{r}\left[\frac{C_V}{T}\,\delta T(\mathbf{r})^2 + \frac{1}{\rho^2 K_T}\,\delta\rho(\mathbf{r})^2\right]. \tag{4.3.1}$$

Here C_V is the specific heat/unit volume at constant volume, and K_T is the isothermal compressibility $(K_T = 1/\rho(\partial\rho/\partial P)_T)$ and the integration is over the volume of the liquid. We have chosen to use the variables ρ and T because the fluctuations in these variables are independent. Introducing Fourier transforms in (4.3.1), we have

$$W = \frac{1}{2}\sum_{\mathbf{q}}\left[\frac{C_V}{T}\,|\delta T_{\mathbf{q}}|^2 + \frac{1}{\rho^2 K_T}\,|\delta\rho_{\mathbf{q}}|^2\right]. \tag{4.3.2}$$

The equipartition theorem then gives

$$S(q) = kT\rho^2 K_T. \tag{4.3.3}$$

The total scattering cross section from (4.2.21) is then

$$w_p^{(1)} = \frac{\alpha^2}{m^2}\left(\frac{\Omega_i}{c}\right)^4 (\mathbf{q}\cdot\hat{\mathbf{E}}_0)^2\, kT\rho^2 K_T \simeq 0.8\ 10^{-8}(\mathbf{q}\cdot\hat{\mathbf{E}}_0)^2/\text{cm sr}, \tag{4.3.4}$$

where we have used $\alpha = 0.2$ Å^3, $\Omega_i/c = 1.2\ 10^5$ cm^{-1}, $T = 1.2$ K, $\rho = 0.145$ g/cm^3, and $\rho K_T \simeq u_1^{-2}$ [where u_1 is the first-sound speed ($u_1 \sim 240$ m/sec)]. Thus the total amount of light scattered by superfluid ^{4}He is very small. The Brillouin scattering from first sound, which accounts for most of the light scattered for $T < T_\lambda$, was first observed by St. Peters et al. [20]. These experiments are discussed further below.

We now turn to a discussion of the spectrum of the light scattered from superfluid ^{4}He. This requires a knowledge of the correlation function, $S(q, \omega)$, which is determined once we know how the spontaneous density fluctuations evolve in time. In superfluid ^{4}He, the time evolution of the density fluctuations is governed by the linearized two-fluid hydrodynamic equations of Landau [21,22]. The linearized hydrodynamic equations possess two normal-mode solutions, first and second sound. First sound is predominantly a propagating pressure wave and second sound is predominantly a propagating entropy wave. The light-scattering spectrum will, thus, consist of two Brillouin doublets—a first-sound doublet centered at frequency shifts $\omega = \pm u_1 q$, where u_1 is the first-sound velocity, and a second doublet centered at frequency shifts $\omega = \pm u_2 q$, where u_2 is the second-sound velocity. The lines are broadened by the dissipative effects, viscosity and thermal conductivity, in the two-fluid equations. When the two doublets are well separated, we can then determine approximately the light scattered by each mode using a simple argument. The light scattered in the first-sound doublet will be due to density fluctuations which take place approximately at constant entropy. Thus, for first sound, we must replace the isothermal compressibility in (4.3.3) by the adiabatic compressibility K_σ. Using the thermodynamic relation $K_\sigma = K_T/\gamma$, where $\gamma = C_P/C_V$ is the ratio of the specific heats, we obtain for the mean-square density fluctuations associated with first sound

$$S^{(1)}(q) = \frac{kT\rho^2 K_T}{\gamma}. \qquad \text{(first sound).} \qquad (4.3.5)$$

The remainder of the scattered light will be due to density fluctuations at constant pressure and will be scattered in the second-sound doublet. From (4.3.3) and (4.3.5) we obtain for the mean square density fluctuations associated with second sound

$$S^{(11)}(q) = kT\rho^2 K_T\left(\frac{\gamma - 1}{\gamma}\right). \qquad \text{(second sound)} \qquad (4.3.6)$$

A physical interpretation of this result can be obtained by using the thermodynamic identity $(\gamma - 1)/\gamma = T\alpha^2/K_T C_P$, where α is the thermal expansion coefficient ($\alpha = -1/\rho\,(\partial\rho/\partial T)_P$).
Then

$$S^{(11)}(q) = \frac{kT^2\rho^2\alpha^2}{C_P}, \qquad \text{(second sound),} \qquad (4.3.7)$$

which shows that the second-sound modes couple to the density fluctuations through the thermal expansion coefficient α. The ratio of the intensity of light scattered in the second-sound doublet to that in the first-sound doublet is, from (4.3.5) and (4.3.6),

$$\frac{S^{(11)}(q)}{S^{(1)}(q)} = \gamma - 1. \tag{4.3.8}$$

More accurate expressions for this ratio have been given by Ginzburg [1] and Vinen [23]. Vinen has estimated that equation (4.3.8) is accurate to better than 2% in the temperature range 1 K to the λ point and at all pressures. At temperatures well below the λ point, $\gamma - 1 \ll 1$ and the light scattering by the second-sound modes is too weak to be observed. It was first pointed out by Ferrell et al. [24] that $\gamma - 1$ increases markedly with pressure and diverges as T_λ is approached. The light scattering by the second sound modes is thus more readily observed at elevated pressure and close to T_λ. This is discussed in more detail in Section 4.4.

We now turn to the complete spectrum of light scattered by the first and second-sound modes. The correlation function, $S(q, \omega)$, has been obtained by several authors [24,25] by solving the equations of two-fluid hydrodynamics. Under the conditions of most light-scattering experiments, the spectrum, to a good approximation, is the superposition of two Brillouin doublets. The first-sound contribution can be represented by a pair of normalized Lorentzian lines centered at $\omega = \pm u_1 q$:

$$S^{(1)}(q, \omega) = \frac{2(u_1^2 q^2)(D_1 q^2)}{(\omega^2 - u_1^2 q^2)^2 + (D_1 q^2 \omega)^2} S^{(1)}(q). \tag{4.3.9}$$

The second-sound contribution, at much smaller frequency shifts, is

$$S^{(11)}(q, \omega) = \frac{2q^2[D_\zeta u_2^2 q^2 + D_\kappa \omega^2]}{(\omega^2 - u_2^2 q^2)^2 + (D_\zeta + D_\kappa)^2 q^4 \omega^2} S^{(11)}(q). \tag{4.3.10}$$

In the derivation of these results an expansion in the small parameter $(u_2^2/u_1^2)(\gamma - 1)$ has been made and terms of this order have been neglected. To a first approximation the velocities and damping constants appearing in (4.3.9) and (4.3.10) are given by

$$u_1^2 = \left(\frac{\partial P}{\partial \rho}\right)_\sigma, \quad u_2^2 = \frac{\rho_s}{\rho_n}\frac{T\sigma^2}{C_P},$$

$$D_1 = \frac{1}{\rho}\left(\zeta_2 + \frac{4}{3}\eta\right), \quad D_\kappa = \frac{\kappa}{C_P}, \tag{4.3.11}$$

and

$$D_\zeta = \frac{\rho_s}{\rho_n}[\rho\zeta_3 - \zeta_1 - \zeta_4 + D_1],$$

where κ is the thermal conductivity, and the conventional notation [22] for the viscosity coefficients has been used. It should be noted from (4.3.10) that the scattering does not vanish at $\omega = 0$, but that

$$S^{(11)}(q, 0) = \frac{2D_\zeta}{u_2^2} S^{(11)}(q). \tag{4.3.12}$$

The line width of the second-sound mode contains contributions from viscosity and thermal conductivity. Experimentally it is found for $T < T_\lambda$ that the contribution of the thermal conductivity is about four times greater than the viscosity contribution.

The spontaneous Brillouin scattering from first sound has been observed by two experimental groups [20,26,27] using a Fabry–Perot etalon and laser light sources. We refer the reader to the original papers for details. A spectral trace which clearly shows the first- and second-sound modes is shown in Figure 4.2. This trace was taken close to T_λ where the second-sound intensity is not negligible. In the early experiments of St. Peters et al. [20], the first-sound velocity at ~ 700 MHz was measured with an accuracy of $\pm$ 0.1% and agreed with that obtained by more conventional methods at lower frequencies. Close to T_λ, some dispersion in the sound velocity was observed. St. Peters et al. were also able to measure the first-sound linewidths with an accuracy of about $\pm$ 10%. Again the results were approximately in accord with those expected from theory and more conventional experiments. We defer a discussion of the second-sound modes to the next section.

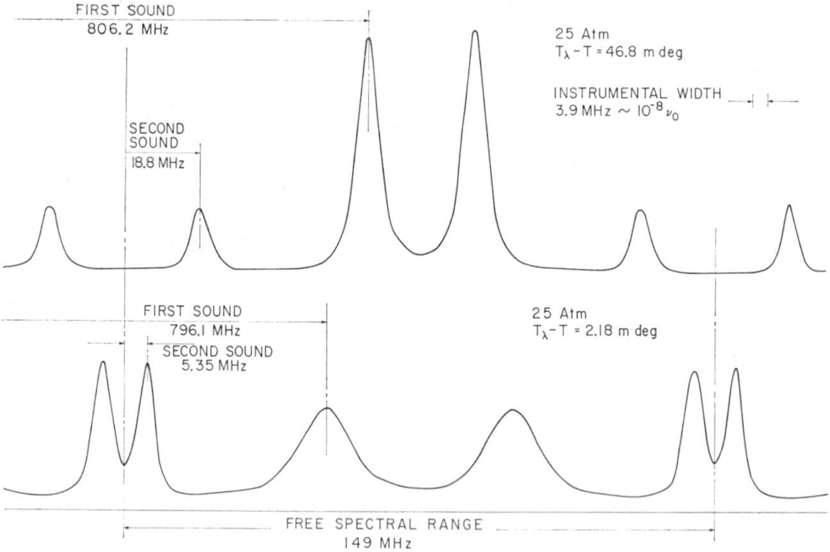

FIGURE 4.2. Light scattering spectrum from superfluid ^{4}He close to T_λ. Both the first-and second-sound modes are clearly shown. This trace was taken by Winterling, Holmes, and Greytak [27].

4.4. BRILLOUIN SCATTERING IN ^{4}HE NEAR T_λ

It was shown in the previous section on the basis of the two-fluid model that the ratio of the Brillouin scattering in the second-sound modes to that in the first-sound modes is determined by $\gamma - 1$, where $\gamma = C_P/C_V$. At SVP and $T < T_\lambda$, $\gamma - 1$ is very small. Korenman [28] gives the value 2.9×10^{-2} at $T_\lambda - T = 10^{-5}$ K, which leads to only about 3% of the total Brillouin scattering arising from the second-sound modes. At higher pressures $\gamma - 1$ is much larger; for example, at 29 atm and $T_\lambda - T = 10^{-5}$ K, $\gamma - 1$ is of the order of 1. Also, along the λ line in pure ^{4}He, $C_P \to \infty$, whereas it can be shown that C_V remains finite [29]. The isothermal compressibility, K_T, can also be shown to be divergent. The divergences of C_P and K_T are approximately logarithmic, i.e., $\sim \log (T_\lambda/(T_\lambda - T))$, and thus according to the two-fluid model the total Brillouin scattering intensity and the intensity in the second modes should be logarithmically divergent.*

Recently two experimental groups [27,30] have studied the Brillouin scattering from the second-sound modes in pure ^{4}He very close to T_λ at elevated pressures. Vinen et al. [30] have measured the total intensity of the scattered light at 18.5 atm. Winterling et al. [27], with greater resolution, have measured the intensity and line widths associated with the second-sound modes. A spectral trace taken by Winterling et al., showing the first and second modes, is given in Figure 4.2.

The ratio of the intensities in the second-sound and first-sound modes measured by Winterling et al. [27] is shown in Figure 4.3 and compared with the thermodynamic measurements of $C_P/C_V - 1$. For $T_\lambda - T > 1$ mK, the measured intensity ratio agrees well with the thermodynamic data of Grilly but less well with the more recent measurements of Ahlers.† The reason for this discrepancy is not known. For $T_\lambda - T < 1$ mK the measured intensity falls below the thermodynamic value of $\gamma - 1$. The measurements of the total intensity of Vinen et al. [30] show a similar departure. This behavior is well known in the Brillouin scattering from normal liquids near the liquid–vapor critical point and is explained by the Ornstein–Zernike theory [31]. This theory takes into account the fact that the correlation length ξ for the order-parameter fluctuations becomes comparable with the wavelength of the fluctuation which scatters the light, i.e., $q\, \xi \sim 1$, where q is the wave vector of the fluctuation.

We can use thermodynamics to discuss the effects of the finite correlation

*More recently it has been shown (G. Ahlers, this volume) that the specific heat is not divergent but shows a cusp.

†More recent measurement by O'Connor, Palin and Vinen (*J. Phys. C: Solid State Phys.* **8**, 101 (1975)) of this intensity ratio are in good agreement with the thermodynamic measurements of Ahlers (this volume).

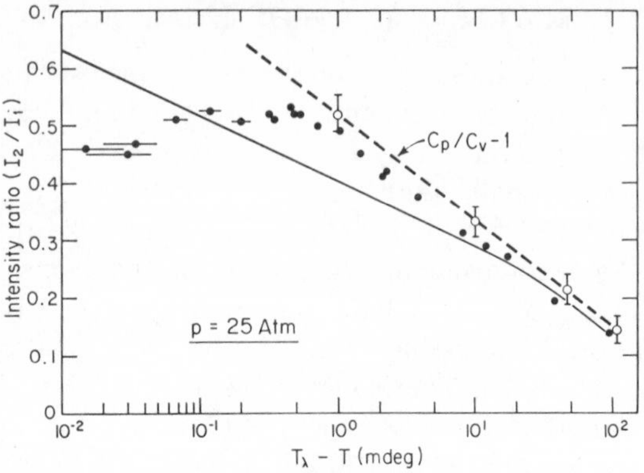

FIGURE 4.3. The ratio of the intensities of light scattered by second-sound to first-sound modes in[4] He close to T_λ. ●, Experimental results of Winterling et al. [27]; ○, thermodynamic values of $\gamma - 1$ [E. R. Grilly, *Phys. Rev.* **149**, 97 (1966); solid line, thermodyamic values of $\gamma - 1$ (G. Ahlers, Chapter 2, this volume).

length for the order-parameter fluctuations on the total Brillouin scattering. We suppose that the free energy density F of the ^{4}He depends explicitly on the order parameter, and we expand it in powers of $\nabla\psi$ [22]:

$$F = F_0\,(T,\rho,\,|\psi|^2) + \frac{\hbar^2}{2m}\,(\nabla\psi)^2. \tag{4.4.1}$$

We will assume that the superfluid density $\rho_s = m\,|\psi|^2$, thus setting the critical exponent $\eta = 0$ [32]. The free energy (4.4.1) is further expanded in the small fluctuations δT, $\delta\rho$, and $\delta\rho_s$ in the temperature, density, and superfluid density. The probability of these fluctuations is proportional to $e^{-W/kT}$, where

$$W = \frac{1}{2}\sum_q\Bigg[\frac{1}{T}(C_V - B\delta_2^2)\,(\delta T_q)^2 + \frac{1}{\rho^2}(K_T^{-1} + B\rho^2\delta_1^2)\,|\delta\rho_q|^2 + B\,(1 + \xi^2q^2)\,|\delta\rho_{s\mathbf{q}}|^2 - B\delta_1\,\delta\rho_{s\mathbf{q}}\,\delta\rho_{-\mathbf{q}}\Bigg]. \tag{4.4.2}$$

Here $\delta_1 = (\partial\rho_s/\partial\rho)_T$, $\delta_2 = (\partial\rho_s/\partial T)_\rho$, and $B = (\partial^2F/\partial\rho_s^2)_{T,\rho}$. The correlation length ξ is related to B by $\xi^2 = \hbar^2/4m^2\rho_s B$. The mean square density fluctuations with wave vector q from (4.4.2) are [33]

$$\langle|\delta\rho_q|^2\rangle = kTK_q\rho^2, \tag{4.4.3}$$

where K_q is the isothermal compressibility at finite q. K_q can be written in the two forms

$$K_q^{-1} = \rho\left(\frac{\partial P}{\partial\rho}\right)_T + B\rho^2\delta_1^2\left(\frac{\xi^2q^2}{1 + \xi^2q^2}\right) \tag{4.4.4}$$

and

$$K_q^{-1} = \rho\left(\frac{\partial P}{\partial \rho}\right)_{T,\,\rho_s} - \frac{B\rho^2\delta_1^2}{(1+\xi^2 q^2)}. \tag{4.4.5}$$

Thus for $q\xi > 1$, K_q is the isothermal compressibility at constant ρ_s; for $q\xi < 1$, it assumes its equilibrium or long-wavelength value. It is expected that any thermodynamic derivative at constant ρ_s is only weakly temperature dependent near T_λ as the fluctuations in ρ_s are suppressed. Thus, very close to T_λ the Brillouin scattering intensity should reach the constant value K_∞.

The various quantities appearing in (4.4.2) and (4.4.4) can be determined as follows:

$$\delta_1 = \left(\frac{\partial \rho_s}{\partial \rho}\right)_T \simeq \frac{2}{3}\frac{\rho_s}{\varepsilon}\frac{1}{T_\lambda}\left(\frac{\partial T_\lambda}{\partial \rho}\right), \tag{4.4.6}$$

$$\delta_2 = \left(\frac{\partial \rho_s}{\partial T}\right) \simeq \frac{-2}{3}\frac{\rho_s}{\varepsilon}, \tag{4.4.7}$$

$$\xi = \xi_0\, \varepsilon^{-(2-\alpha)/3}, \quad B = \frac{\hbar^2}{4m^2 \rho_s \xi^2}, \tag{4.4.8}$$

where $\varepsilon = (T_\lambda - T)/T_\lambda$, α is the specific-heat exponent and we have taken the temperature dependence of $\rho_s \sim \varepsilon^{(2-\alpha)/3}$. The zero-temperature correlation length ξ_0 is left as a free parameter. The superfluid density ρ_s has been measured at different pressures by Greywall and Ahlers [34] and the slope of λ line by Kierstead [35]. The total Brillouin intensity, given by (4.4.3) and (4.4.4), is plotted in Figure 4.4*a* assuming $\xi_0 = 2$ Å. It does show the expected departure from the $q = 0$ value but the overall decrease seems to be somewhat less than that observed by Vinen et al. [30].

It is also instructive to consider the coupling between density and superfluid-density fluctuations. The fluctuations in ρ_s are not directly observable in light scattering because the dielectric constant does not depend on ρ_s. The dynamics of the density and superfluid density fluctuations are coupled and the contribution of the fluctuations in ρ_s to the light scattering is determined in part by the correlation function

$$\langle \delta\rho_{s\mathbf{q}}\, \delta\rho_{-\mathbf{q}} \rangle = \frac{kTK_q\delta_1}{1+\xi^2 q^2}. \tag{4.4.9}$$

This function is small when $\xi q > 1$ and the fluctuations in ρ_s should not be observable in light scattering very close to T_λ (for $T < T_\lambda$ these fluctuations will be rapidly damped and hence also not observable). This decoupling of the fluctuations is expected on physical grounds; at T_λ the fluctuations in ρ_s are extremely large. If they were coupled to the density, the energy required would be too large.

We now turn to a discussion of the spectrum of the Brillouin scattering

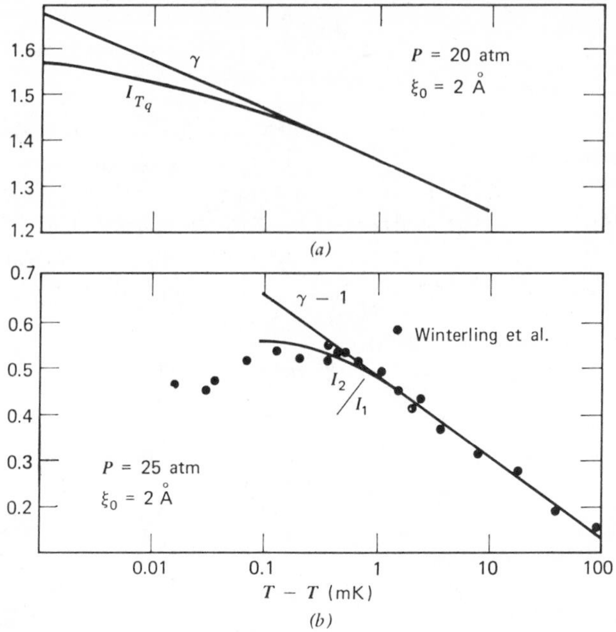

FIGURE 4.4. Ornstein–Zernike corrections to the intensity of light scattering in ^{4}He near T_λ. (*a*) The total intensity, equation (4.4.3) (normalized to γ). (*b*) The ratio of the intensities in the second-sound and first-sound modes, equation (4.4.12).

associated with the second-sound modes and the fluctuations in ρ_s. In order to determine the intensities and the line widths of the light scattering by these modes, information about the dynamics of the density fluctuations is needed. At the present time dynamical equations, analogous to the two-fluid hydrodynamic equations, when $q\xi \sim 1$, are not established.

Dynamical scaling [24, 36] assumes that the density–density correlation function takes the form

$$S(q, \omega) = 2\pi\, \omega_\xi^{-1}(q)\, f_{q\xi}\left(\frac{\omega}{\omega_\xi(q)}\right) S(q), \qquad (4.4.10)$$

where $S(q)$ is the static correlation function and $\omega_\xi(q)$ is a characteristic frequency that is taken in the form

$$\omega_\xi(q) = q^z \Omega(q\xi), \qquad (4.4.11)$$

where the exponent $z \simeq \frac{3}{2}$ in ^{4}He. The form of the function f is not established. From equation (4.4.11) it follows that for $q\xi < 1$ the damping of second sound should vary with temperatures approximately as $\varepsilon^{-1/3}$. This behavior was observed by Tyson [37] at SVP and low frequencies in experiments on the propagation of second-sound waves in the temperature range $10^{-4} < T_\lambda - T < 10^{-1}$ K. Winterling et al. [27], using light-scattering techniques, have measured the second-sound linewidths at 25 atm and found an almost temperature-independent linewidth in the temperature range $10^{-4} < T_\lambda - T < 10^{-2}$ K. The widths for $T_\lambda - T < 10^{-3}$ K are also much less than those obtained by scaling Tyson's values to the higher frequencies used in the light-scattering experiments. These linewidths can not be explained by invoking a $q\xi$ effect, as they are measured in the region where $q\xi < 1$ when $q\xi$ is determined from the intensity measurements. The thermal conductivity of ^{4}He above T_λ has been measured by Ahlers [40] and found to diverge as $\varepsilon^{-1/3}$ in agreement with the predictions of dynamical scaling.

For $q\xi > 1$, dynamical scaling predicts that the characteristic frequency $\omega_\xi(q) \sim q^{3/2}$. Schmidt and Szepfalusy [38] have suggested that the spectrum in this region should consist of two peaks (excluding the first-sound doublet) with both the splitting and linewidth varying as $q^{3/2}$. This has not yet been tested in ^{4}He but is observed in the antiferromagnet $RbMnF_3$ [39].

Khalatnikov [41] has introduced an additional equation in the two-fluid hydrodynamic equations that describes the relaxation of the order parameter in ^{4}He. He has argued that the relaxation time τ of the order parameter should be of order of magnitude ξ/u_2, where u_2 is the second-sound velocity. τ thus varies with temperature as ε^{-1} and becomes long close to T_λ. The attenuation of first sound due to order-parameter fluctuations has been determined by Khalatnikov and found to vary as $\varepsilon^{-1+\alpha}$, in agreement with experiment [42]. McCauley and Stephen [33] have calculated the spectrum of light scattered from the second sound and ρ_s fluctuations using the Khalatnikov model. The intensity associated with the second-sound modes is of the same form as in equation (4.3.7) except that the all thermodynamic derivatives are replaced [as in equation (4.4.3)] by their q-dependent values. The ratio of the second-sound intensity to the first-sound intensity is given by

$$\frac{S^{(11)}(q)}{S^{(1)}(q)} = \frac{K_q \beta_q^2 T}{C_{Vq}}, \tag{4.4.12}$$

where

$$\beta_q = \left(\frac{\partial P}{\partial T}\right)_\rho + B\rho\delta_1\delta_2\left(\frac{\xi^2 q^2}{1 + \xi^2 q^2}\right)$$

and (4.4.13)

$$C_{Vq} = C_V - BT\delta_2^2\left(\frac{\xi^2 q^2}{1 + \xi^2 q^2}\right).$$

Equation (4.4.12) reduces to (4.3.8) when $\xi q = 0$. This ratio is plotted in Figure 4.4*b* and compared with the experiments of Winterling et al. using the value $\xi_0 = 2$ Å. This value for ξ_0 agrees with that obtained by other methods [43]. Again the Ornstein–Zernike corrections reduce the intensity below the $q = 0$ value. The intensity in the unshifted fluctuations in ρ_s is small and vanishes as $(1 + \xi^2 q^2)^{-1}$, as found in equation (4.4.9). A different result has been derived by Kretzen and Mikeska [44] who predict that the intensity in the central component is proportional to $(\xi q)^4$.

The linewidths of the second-sound and central modes rising from the relaxation of the order parameter are also given in Reference [33]. It was found that the second-sound linewidth has an appreciable "background" contribution and only shows the expected $\varepsilon^{-1/3}$ temperature dependence for $T_\lambda - T < 10$ mK. at SVP and for $T_\lambda - T < 1$ mK at 25 atm. The Khalatnikov model makes no prediction about the temperature dependence of the thermal conductivity which also contributes to the second-sound linewidth. So this model does not provide a complete understanding of the observed linewidths. The Khalatnikov model predicts a spectrum for $q\xi > 1$ in which the linewidth is proportional to $q^{3/2}$ and is temperature independent, and the splitting decreases as $(q\xi)^{-1}$.

4.5. BRILLOUIN SCATTERING FROM SUPERFLUID ^{3}HE–^{4}HE MIXTURES

In Section 4.3. we discussed Brillouin scattering from superfluid ^{4}He and showed that the spectrum consists of two doublets arising from the first- and second-sound modes. If ^{3}He is now dissolved in superfluid ^{4}He, the ^{3}He impurity atoms become part of the normal fluid and move with the normal-fluid velocity. The ^{3}He atoms do not participate in the superfluid flow. It was first shown by Gorkov and Pitaevskii [45] that the light-scattering spectrum of a superfluid ^{3}He–^{4}He mixture should contain five lines—two Brillouin doublets arising from the first- and second-sound modes and a central or undisplaced line. This last line arises from a diffusive mode which is a combination of ^{3}He concentration and entropy fluctuations. Also as the ^{3}He atoms participate in the normal-fluid motion, the ^{3}He concentration fluctuations appear in an important manner in the second-sound mode. The two fluid hydrodynamic equations for superfluid ^{4}He have been generalized by Khalatnikov [22, 46] to ^{3}He–^{4}He mixtures and most of the discussion below on light scattering from these mixtures is based on these equations. The ^{3}He concentration fluctuations are readily observable by light scattering for the following reason: The ^{3}He and ^{4}He atoms are, of course, optically identical as they have the same polarizability. However, a ^{3}He atom dissolved in ^{4}He, owing to its smaller mass, has a larger amplutide of vibration than a ^{4}He atom. A ^{3}He atom, consequently, occupies a larger average volume, v_3, than that occupied by a ^{4}He atom (v_4 say). At SVP and low ^{3}He concentrations

$v_3/v_4 \simeq 1.3$ [47]. The addition of ^{3}He to ^{4}He thus leads to an important change in the dielectric constant of the solution, and fluctuations in the ^{3}He concentration consequently give rise to fluctuations in the dielectric constant.

The composition of a mixture is conveniently described in terms of the following variables (which are those used by Khalatnikov) [22]: the mass density $\rho = n_3 m_3 + n_4 m_4$ and the mass concentration $c = n_3 m_3/(n_3 m_3 + n_4 m_4)$.* n_3 and n_4 are number densities of ^{3}He and ^{4}He, respectively, and m_3 and m_4 are the mass of a ^{3}He atom and ^{4}He atom, respectively. The dielectric constant of a mixture $\varepsilon(\rho,c)$ will depend on both these variables. We again assume that ε is independent of the temperature. The fluctuations in the dielectric constant are then related to the fluctuations in ρ and c by

$$\delta\varepsilon = \left(\frac{\partial \varepsilon}{\partial \rho}\right)_c \delta\rho + \left(\frac{\partial \varepsilon}{\partial c}\right)_\rho \delta c. \tag{4.5.1}$$

In the present case, it is more convenient to use the variables P and c and the above equation can be written (neglecting the small thermal expansion coefficient)

$$\delta\varepsilon = \left(\frac{\partial \varepsilon}{\partial \rho}\right)_c \left(\frac{\partial \rho}{\partial P}\right)_{T,c} \delta P + \left(\frac{\partial \varepsilon}{\partial c}\right)_P \delta c. \tag{4.5.2}$$

It is generally an adequate approximation to regard the polarizabilities of the mixture as the sum of the polarizabilities of the individual atoms. Then

$$\varepsilon = 1 + 4\pi\alpha\,(n_3 + n_4), \tag{4.5.3}$$

and it is easy to show that the two derivatives in (4.5.1) are related by [11]

$$\frac{c}{\rho}\left(\frac{\partial \varepsilon}{\partial c}\right)_\rho = \left(\frac{\partial \varepsilon}{\partial \rho}\right)_c \frac{c}{c+3} = \frac{4\pi\alpha c}{3m_4}. \tag{4.5.4}$$

It is also possible to express the fluctuations in ε arising from fluctuations in the concentration in terms of the difference in the average volumes, $v_3 - v_4$, occupied by ^{3}He and ^{4}He atoms. Assuming that v_3 and v_4 depend on the pressure only and are independent of the concentration, it is easy to show that

$$\left(\frac{\partial \varepsilon}{\partial c}\right)_P = \frac{4\pi\rho^2}{m_3 m_4}(v_4 - v_3). \tag{4.5.5}$$

This equation shows directly the physical origin of the fluctuations in the dielectric constant arising from the ^{3}He concentration fluctuations.

The power spectrum for Brillouin scattering from pure ^{4}He, equation (4.2.20), is readily generalized to the case of a mixture of ^{3}He and ^{4}He. It is most easily expressed in terms of the fluctuations in the dielectric constant:

*The molar concentration $x = n_3/n_3 + n_4$ is related to c by $1/x - 1 = (m_3/m_4)(1/c - 1)$.

$$H_{\rho}^{(1)}(\Omega_s) = \frac{V}{16\pi^2 R^2}\left(\frac{\Omega_s}{c}\right)^4 (\mathbf{p}\cdot\mathbf{E}_0)^2\, S_\varepsilon\,(q,\,\omega), \tag{4.5.6}$$

where

$$S_\varepsilon\,(q,\,\omega) = \int_{-\infty}^{\infty} dt\; e^{i\omega t}\,\langle\delta\varepsilon_{\mathrm{q}}(t)\;\delta\varepsilon_{-\mathrm{q}}(0)\rangle, \tag{4.5.7}$$

$\delta\varepsilon_{\mathrm{q}}(t)$ is the fluctuation in ε with wave vector $\mathbf{q}$ at time t. It is related to the fluctuations in pressure and concentration by equation (4.5.2). The cross section for light scattering is again given by equation (4.2.19) and the total cross section for scattering is now given by [see equation (4.2.21)].

$$w_p^{(1)} = \frac{1}{16\pi^2}\left(\frac{\Omega_i}{c}\right)^4 (\mathbf{p}\cdot\mathbf{E}_0)^2 S_\varepsilon(q)/\text{cm, sr}, \tag{4.5.8}$$

where

$$S_\varepsilon\,(q) = \frac{1}{2\pi}\int_{-\infty}^{\infty} d\omega\; S_\varepsilon\,(q,\,\omega) = \langle|\delta\varepsilon_q|^2\rangle. \tag{4.5.9}$$

In the following discussion, we obtain results for the total fluctuations in the dielectric constant $S_\varepsilon(q)$ and the dielectric fluctuations associated with the central mode $S_\varepsilon^{(0)}\,(q)$, the second-sound mode $S_\varepsilon^{(11)}\,(q)$, and the first-sound mode $S_\varepsilon^{(1)}(q)$. The total intensity of the scattered light and those parts of the scattered light associated with the three modes is then given by (4.5.8), with the appropriate value for $S_\varepsilon(q)$.

We first consider the total intensity. The probability of fluctuations $\delta P(\mathbf{r})$, $\delta T(\mathbf{r})$, and $\delta c(\mathbf{r})$ in pressure, temperature, and concentration, respectively, is proportional to $e^{-W/kT}$, where

$$W = \frac{1}{2}\int d\mathbf{r}\left[\frac{C_P}{T}\,\delta T^2(\mathbf{r}) + K_T\,\delta P^2(\mathbf{r}) - 2\,\alpha\delta P(\mathbf{r})\,\delta T(\mathbf{r}) + \rho\beta\delta c^2(\mathbf{r})\right], \tag{4.5.10}$$

$K_T = 1/\rho(\partial\rho/\partial P)_{T,c}$ is the isothermal compressibility, $\alpha = -\,1/\rho(\partial\rho/\partial T)_{P,c}$ is the thermal expansion coefficient, $\beta = ((\partial/\partial c)\,Z/\rho)_{P,T}$ is the osomotic pressure coefficient, and $Z/\rho = \mu_3 - \mu_4$ is the difference in the chemical potentials per unit mass of the ^{3}He and ^{4}He. In what follows we will consistently neglect the small thermal expansion coefficient α. Introducing Fourier transforms in (4.5.10) and using (4.5.2) to relate the fluctuations in pressure and concentration to those in ε, we find

$$S_\varepsilon(q) = kT\left[\left(\frac{\partial\varepsilon}{\partial\rho}\right)_c^2 \rho^2\, K_T + \left(\frac{\partial\varepsilon}{\partial c}\right)_P^2 (\rho\beta)^{-1}\right]. \tag{4.5.11}$$

At low concentration of ^{3}He in ^{4}He the solution is ideal, and

$$\beta = \left(\frac{\partial}{\partial c}\,\frac{Z}{\rho}\right)_{P,T} = \frac{kT}{m_3 c}. \tag{4.5.12}$$

When this result is substituted in (4.5.11) and we use (4.5.5), we see that the scattering due to the ^{3}He concentration fluctuations for low concentrations is equivalent to that of an ideal gas whose atoms have an effective polarisability $\alpha((v_3 - v_4)/v_4)$.

The light scattering near the tricritical point of the mixture is of considerable interest. At this point there appears to be a conjunction of two lines of phase transitions—the normal–superfluid transition along the λ line and the two-phase coexistence curve for the miscibility of the solution. This tricritical point is at a molar concentration of about 0.67 and a temperature $T_c \simeq 0.87K$. The osmotic coefficient β in (4.5.11) is expected to approach zero as $(T-T_c)^\gamma$. Scaling arguments [52] predict that the critical exponent $\gamma \simeq 1$ when T_c is approached along either branch of the coexistence curve or in the single-phase region $T > T_c$. The total light-scattering intensity has been measured by Watts et al. [49] close to the tricritical point of a ^{3}He–^{4}He mixture. In general, their data for the osmotic coefficient is in good agreement with that obtained by vapor pressure measurements by Goellner and Meyer [50] and by Alvesalo et al. [51] (see Figure 4.5) The exponent γ is close to 1 for $T > T_c$ in the single-phase region and for $T < T_c$ in the normal phase. In the superfluid phase with $T < T_c$, $\gamma \simeq 1.6$. This latter result may indicate that the true asymptotic critical behavior has not been observed. More recent results [49] show that this is the case.

In order to determine the light-scattering intensities associated with the three modes, it is necessary to study the dynamics of the density fluctuations. This dynamics is described by the generalized two-fluid hydrodynamic equations of Khalatnikov [22]. The velocities of first and second sound in a mixture have been obtained by Khalatnikov from the solution of these equations The results are (for $\alpha = 0$ and assuming $u_2^2 \ll u_1^2$).

$$u_1^2 = (\rho K_T)^{-1}\left[1 + \frac{\rho_s}{\rho_n}\left(\frac{c}{\rho}\frac{\partial \rho}{\partial c}\right)^2\right] \tag{4.5.13}$$

and

$$u_2^2 = \frac{\rho_s}{\rho_n}\left[\frac{\bar{\sigma}^2 T\rho/C_P + c^2\beta}{1 + (\rho_s/\rho_n)\,[(c/\rho)\,(\partial\rho/\partial c)]^2}\right], \tag{4.5.14}$$

where $\bar{\sigma} = \sigma - c(\partial\sigma/\partial c)_{P,T}$, σ is the entropy/g, and all thermodynamic derivatives are at constant P,T. It is interesting to note that the properties of the mixtures become extremely simple in the case of low concentrations when the solutions are ideal and at low temperatures when the thermal properties of the mixtures is dominated by the impurities. (We assume that the temperature is higher than the Fermi degeneracy temperature of the mixture.) The thermodynamic functions in (4.5.14) are then [22]

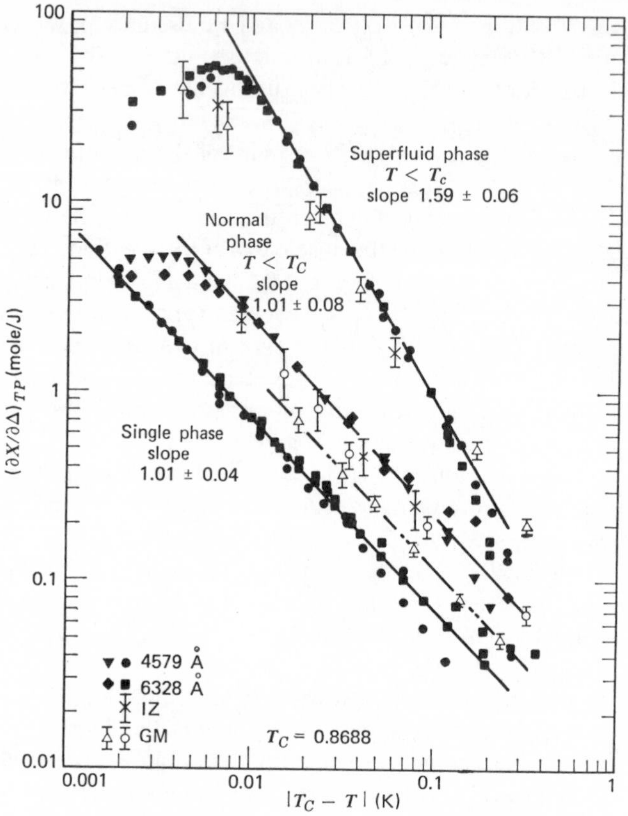

FIGURE 4.5. The osmotic coefficient $\beta^{-1} \sim \partial x/\partial \Delta$ near the tricritical point in a ^{3}He–^{4}He mixture. The light-scattering measurements at wavelengths 4579 and 6328 Å were made by Watts et al. [49]. The thermodynamic measurements are those of (IZ) Alvesalo et al. [51] and (GM) Goellner and Meyer [50]. In each case the slope gives the value of the exponent γ.

$$\rho_n \simeq \frac{\rho c m_3^*}{m_3}, \quad \bar{\sigma} \simeq \frac{kc}{m_4}, \quad C_P \simeq \left(\frac{3}{2}\right) k \left(\frac{\rho c}{m_3}\right), \tag{4.5.15}$$

where m_3^* is the effective mass of a ^{3}He atom in superfluid ^{4}He. Substituting these results together with (4.5.12), we find that at low frequencies

$$u_2^2 \simeq \frac{5}{3} \frac{kT}{m_3^*},$$

i.e., the adiabatic sound velocity is a nearly ideal gas of particles with effective mass m_3^*.

We now turn to a discussion of the intensities associated with the three modes of a ^{3}He–^{4}He mixture. We use the assumption that the three modes are well separated in the light-scattering spectrum. This is well justified for the first-sound modes, but not always true for the second-sound and diffusive modes. For this reason, the complete spectrum of the second sound and diffusive modes is given in Appendix A.

In the second-sound and diffusive modes the pressure fluctuations are small, and we will neglect them. These modes are then a combination of entropy and concentration fluctuations and the light scattering arises only from the second term in (4.5.2), i.e., the concentration fluctuations. The relation between the entropy and concentration fluctuations in the second-sound mode follows simply from the conservation relations for entropy and concentration

$$\frac{\partial}{\partial t}(\rho\sigma) + \nabla\cdot\rho\sigma\mathbf{v}_n = 0, \tag{4.5.16a}$$

$$\frac{\partial}{\partial t}(\rho c) + \nabla\cdot\rho c\mathbf{v}_n = 0. \tag{4.5.16b}$$

We have neglected dissipative terms in these equations. Eliminating $\mathbf{v}_n$ from the linearized form of those equations gives

$$\rho c\frac{\partial\sigma}{\partial t} - \rho\sigma\frac{\partial c}{\partial t} = 0. \tag{4.5.17}$$

Introducing the temperature as variable gives for the fluctuations in the second-sound mode

$$cC_P\delta T = \rho\bar{\sigma}\delta c,\ \delta P = 0. \tag{4.5.18}$$

Substituting these results in (4.5.10), we can calculate the mean square fluctuations in concentration arising from the second-sound mode. From (4.5.2), we then obtain the mean square fluctuations in dielectric constant associated with the second-sound modes:

$$S_\varepsilon^{(11)}(q) = \left(\frac{\partial\varepsilon}{\partial c}\right)_P^2 \frac{kTc^2\rho_s}{\rho\rho_n}\frac{(\rho K_T)^{-1}}{u_1^2u_2^2}. \tag{4.5.19}$$

In the diffusive central mode, the fluctuations in entropy and concentration are connected by the constraint μ_4 = constant. This is the condition that the superfluid is not accelerated. When this constraint is expressed in terms of the variables T and c, we obtain for the fluctuations in the central mode

$$\bar{\sigma}\delta T = -c\rho\delta c,\ \delta P = 0. \tag{4.5.20}$$

Substituting these results in (4.5.10), we can calculate the mean-square fluctuations in concentration arising from the central mode. Again using

(4.5.2), we find for the central mode

$$S_{\varepsilon}^{(0)}(q) = \left(\frac{\partial \varepsilon}{\partial c}\right)_P^2 \frac{kT^2\bar{\sigma}^2\rho_s}{\beta\rho_n C_P} \frac{(\rho K_T)^{-1}}{u_1^2 u_2^2}. \tag{4.5.21}$$

The results (4.5.19) and (4.5.21) for the mean square fluctuations in ε associated with the second mode and central mode were first given by Gorkov and Pitaevskii [45]. More accurate results* have been given by Vinen [11], with the approximation of zero thermal expension. Expressions for these intensities have also been derived by Ganguly and Griffin [53], but their results appear to be in error [11]. In the present approximation, (4.5.19) and (4.5.21) exhaust the concentration fluctuations at constant pressure. This is only approximately true in the more accurate results of Vinen. We can give a simple physical interpretation to the results (4.5.19) and (4.5.21) at low temperatures when the impurities dominate the thermodynamic properties of the mixture. In this situation the impurities behave like a nearly ideal gas. The light scattering in the second-sound and central modes is equivalent to the Brillouin scattering in a nearly ideal gas. The second-sound modes correspond to the Brillouin doublet, and the central mode to the Rayleigh component. This model also implies that the linewidth of the second sound should grow as the concentration decreases and at very low concentrations the central component and the second-sound doublet should merge into a single broad Gaussian line resulting from the Doppler shifts from the Maxwellian distribution of individual quasiparticle velocities [54].

To determine the intensity associated with the first-sound modes we use a more accurate approximation relating the pressure and concentration fluctuations in a first-sound wave. It can be shown from the two-fluid hydrodynamic equations that in the first-sound mode these fluctuations are connected by

$$\delta c = -\frac{(u_1^2\rho K_T - 1)}{u_1^2 (\partial\rho/\partial c)_{P,T}}\delta P. \tag{4.5.22}$$

Substituting this result in (4.5.2) and using the approximate result $\langle|\delta P_q|^2\rangle = kTK_T^{-1}$, we find for the fluctuations in ε associated with first sound [45]

$$S_{\varepsilon}^{(1)}(q) = kT\rho^2 K_T\left[\left(\frac{\partial\varepsilon}{\partial\rho}\right)_c - \left(\frac{\partial\varepsilon}{\partial c}\right)_P \frac{\rho_s c^2}{\rho_n\rho^2}\left(\frac{\partial\rho}{\partial c}\right)_{P,T}\frac{(\rho K_T)^{-1}}{n_1^2}\right]^2. \tag{4.5.23}$$

The velocities of first and second sound in ^{3}He–^{4}He mixtures have been measured using light-scattering techniques by Palin et al. [55] and by Benjamin et al. [54]. The results agree well with those obtained at low frequencies by conventional methods. These two experimental groups have also measured

*Expressions for the intensities associated with the these modes have been derived from the hydrodynamic equations without the neglect of thermal expanision by Rockwell, Benjamin and Greytak (*J. Low Temp. Phys.* **18**, 389 (1975)). They estimate that the Gorkov-Pitaevski results are accurate to $\pm 5\%$.

the intensity ratios $S_\varepsilon^{(0)}/S_\varepsilon^{(1)}$ and $S_\varepsilon^{(11)}/S_\varepsilon^{(1)}$ for the central and second-sound modes to the first-sound mode. The results are in reasonable agreement with the theoretical predictions of the Gorkov and Pitaevskii equations (4.5.19), (4.5.21), and (4.5.23).

The line widths of the five components of the spectrum are determined by the damping of the second-sound, diffusive, and first-sound modes. We again assume that the various components of the spectrum are well separated in frequency. These damping rates have been determined by Khalatnikov [46] and Gorkov and Pitaevskii [45]. Similar results have been obtained by Griffin [56]. For the second-sound mode the linewidth (full width at half maximum) is $D_2^{(3)}q^2$, where

$$D_2^{(3)} = D_\zeta + \frac{\rho_s\bar{\sigma}^2\rho T}{2\rho_n C_P^2 u_2^2}\left[\kappa + \frac{\beta\rho D}{T}\left\{\frac{c}{\bar{\sigma}}\frac{C_P}{\rho} + k_T\right\}^2\right], \tag{4.5.24}$$

where D_ζ is given in equation (4.3.11), κ is the termal conductivity, D is the diffusion constant, and k_T is the thermal diffusion ratio. The linewidth of the central mode is $D_0^{(3)}q^2$, where

$$D_0^{(3)} = \frac{\rho_s\beta c^2}{\rho_n C_P u_2^2}\left[\kappa + \frac{\beta\rho D}{T}\left\{\frac{\bar{\sigma}T}{\beta c} - k_T\right\}^2\right]. \tag{4.5.25}$$

It should be noted that these linewidths satisfy the simple relation

$$D_0^{(3)} + 2D_2^{(3)} = \frac{\kappa}{C_P} + D\left(1 + \frac{\beta\rho k_T}{TC_P}\right) + 2D_\zeta. \tag{4.5.26}$$

The linewidth of the first-sound mode is $D_1^{(3)}q^2$, where

$$D_1^{(3)} = D_1 + \frac{D}{\beta}\left(\frac{c}{\rho}\frac{\partial\rho}{\partial c}\right)^2, \tag{4.5.27}$$

and D_1 is given in equation (4.3.11). In the results (4.5.24)–(4.5.27) the thermal expansion coefficient has been neglected, which is justified except very close to T_λ. Terms in $\rho_s/\rho_n((c/\rho)/\partial\rho/\partial c)^2$ have also been neglected. This quantity does not exceed 4.10^{-2} at all concentrations. The second-sound linewidth contains contributions from viscosity, thermal conductivity, and diffusion. As in the case of second sound in pure ^{4}He the linewidth is dominated by the thermal conductivity (and diffusion) terms and the viscosity terms can be neglected in a first approximation. The second-sound linewidth has been measured in the mixtures by Benjamin et al. [54] by light scattering, but values for κ and D have not yet been obtained.

4.6. RAMAN SCATTERING IN SUPERFLUID ^{4}HE

We now turn to a study of light-scattering processes in which two excitations are involved. As discussed in Section 4.1, momentum conservation requires that the two excitations have almost equal and opposite momenta. Such a process is illustrated in Figure 4.6. Halley [10] first suggested that it would be interesting to observe these processes in superfluid ^{4}He. He pointed out that the intensity of the light scattering would be proportional to the joint density of states for the two excitations. In superfluid ^{4}He, this joint density of states (for noninteracting excitations) has two pronounced maxima: (i) at an energy $2\varepsilon_0$, where ε_0 is the roton energy; (ii) at an energy $2\varepsilon_1$, where ε_1 is the energy of the first maximum in the excitation dispersion curve

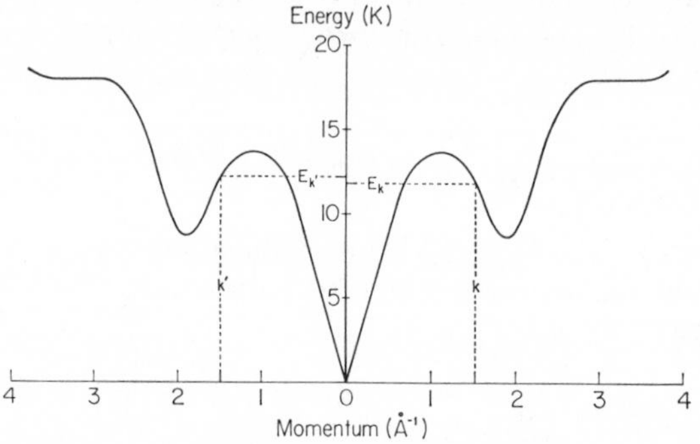

FIGURE 4.6. Raman scattering process in ^{4}He leading to the creation of two excitations of almost equal and opposite momenta.

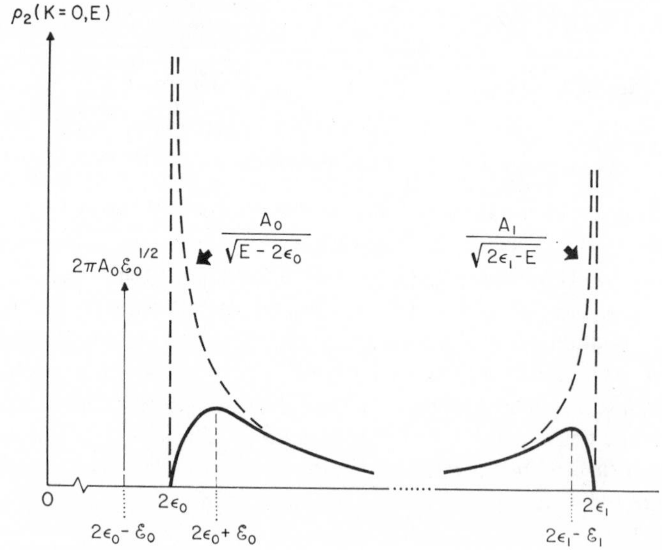

FIGURE 4.7. Density of states in ^{4}He. Dotted curve is the density of states for two non-interacting excitations, and solid lines for an attractive interaction between excitations. The total momentum of the excitations $K = 0$.

shown in Figure 4.6. For convenience we refer to excitations near ε_1 as "maxons." The joint density of states is shown in Figure 4.7. A third, less pronounced maximum arises from the cutoff point in the excitation curve where one excitation can decay into two rotons. The energy of this maximum in the joint density of states is appromimately $4\varepsilon_0$. On this simple picture, the Raman scattering spectrum would be expected to show three maxima corresponding to the three maxima in the joint density of states. The situation is actually more interesting and complicated by the interactions between the two excitations and gives valuable information concerning the interaction between the excitations (see below). The Raman scattering spectrum of superfluid ^{4}He was first observed by Greytak and Yan [57] and is shown in Figure 4.8. The observed spectrum shows a pronounced maximum at an energy shift $\sim 2\varepsilon_0$, which is identified with the creation of two rotons, and a broad maximum at an energy shift $\sim 4\varepsilon_0$, presumably arising from the cutoff point in the excitation curve. The feature expected at $2\varepsilon_1$ on the basis of the joint density of states argument is almost totally absent.

We showed in Section 4.2. that light scattering processes involving two excitations can arise in two ways: (a) through multiple scattering (or local field correction to the scattering), which led to equation (4.2.24), (b) through a density fluctuation $\delta\rho$, which by the nonlinearity of the hydrodynamic equations can contain terms quadratic in phonon amplitudes. These latter terms can, in the Born approximation to the scattering equation (4.2.20), lead to production of two excitations in the scattering of a single photon. The first process was considered by Stephen [16], and the second was included by Iwamoto [58] and by Fetter [59]. We will refer to these as the *d* and *s* processes, respectively, as these symbols describe their symmetry properties.

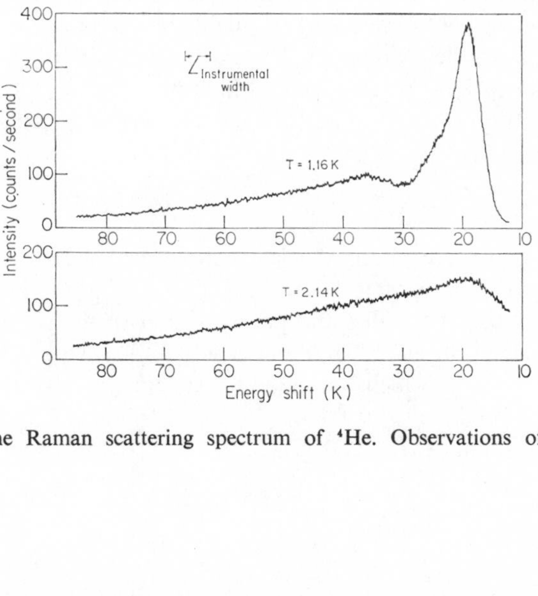

FIGURE 4.8. The Raman scattering spectrum of ^{4}He. Observations of Greytak and Yan [57].

These two processes have a very different dependence on the angle of scattering and give different polarization ratios for the scattered light. Thus information on the relative importance of these two processes can be inferred from a measurement of the polarization ratios. We also note that the s process can be entirely eliminated by observing the scattered light in such a geometry that $\mathbf{p}\cdot\hat{\mathbf{E}}_0 = 0$.

The polarization and angular dependence of the intensity of the scattered light is determined in the d process by the factor $1 + \frac{1}{3}(\mathbf{p}\cdot\hat{\mathbf{E}}_0)^2$, and in the s process by the factor $(\mathbf{p}\cdot\hat{\mathbf{E}}_0)^2$. Suppose that plane-polarized light is incident on the ^{4}He in the y direction with the electric vector of the light in the x direction. We use the notation $w_i(j)$ to denote the intensity of the light scattered in the direction j with polarization along direction i. With the above geometry, the light scattered along the z-axis is collected in a conventional light-scattering experiment. The polarization ratios in this direction for the two processes are

$$\frac{w_y(z)}{w_x(z)} = \frac{3}{4}, \qquad (d \text{ process}) \qquad (4.6.1)$$

$$= 0. \qquad (s \text{ process})$$

This ratio has recently been measured by Woerner and Greytak [60] with the result 0.74 ± 0.01. Fetter's [59] calculation, which included the s process, gave 0.69 for this ratio. The reason for this discrepancy is not understood. The experimental result does indicate, however, that the d process is dominant and we will confine our attention to this process only, i.e., equation (4.2.24). Woerner and Greytak [60] have also measured the polarization ratio, in the above geometry, for the light scattered along x, i.e., parallel to the electric vector of the incident light. This geometry has the advantage that there is no Brillouin scattering in this direction as $\mathbf{p}\cdot\hat{\mathbf{E}}_0 = 0$. The polarization ratio from the d process is 1; experimentally it is 1.0 ± 0.01.

If both polarizations of the scattered light are observed, the factor $1 + \frac{1}{3}(\mathbf{p}\cdot\hat{\mathbf{E}}_0)^2$ in equation (4.2.24) must be summed over both polarization directions:

$$\sum_p \left(1 + \tfrac{1}{3}(\mathbf{p}\cdot\hat{\mathbf{E}}_0)^2\right) = \tfrac{7}{3}\left(1 - \tfrac{1}{7}(\hat{\mathbf{K}}_s\cdot\hat{\mathbf{E}}_0)^2\right), \qquad (4.6.2)$$

where $\hat{\mathbf{K}}_s$ is a unit vector along the scattering direction. This shows that the Raman scattered light is almost isotropic. The ratio of the intensities of the Raman scattered light parallel and perpendicular to the polarization of the laser $(\hat{\mathbf{E}}_0)$ has been measured in Reference [60]. The experimental intensity ratio is 0.845 ± 0.015, which is in good agreement with the theoretical value from equation (4.6.2) of $\frac{6}{7} \simeq 0.857$. The result obtained by Fetter [59] including the s Raman process is 0.82.

We now turn to a discussion of the spectrum of the Raman scattered light. This is determined by the four-particle correlation function in equation (4.2.24). It is convenient to define

$$h(\omega) = \frac{1}{(2\pi)^6}\int d\mathbf{q}\, d\mathbf{q}' g(q)g(q')P_2(\cos\theta qq')S_4(\mathbf{q},0,\mathbf{q}',\omega). \tag{4.6.3}$$

The Raman scattering cross section from (4.2.24) and (4.2.19), including both polarizations, is

$$w_p^{(2)}(\Omega_s) = \frac{7\alpha^4}{10\pi m^4}\left(\frac{\Omega_s}{c}\right)^4\left[1 - \frac{1}{7}(\hat{\mathbf{K}}_s\cdot\hat{\mathbf{E}}_0)^2\right]h(\omega). \tag{4.6.4}$$

In order to determine the correlation function S_4, we first assume that the density fluctuation $\delta\rho_q$ can be expanded in the creation and annihilation operators a_q^+ and a_q for the elementary excitations of momentum q in ^{4}He. We only require the first term in this expansion

$$\delta\rho_q = iZ_q^{1/2}(a_q - a_{-q}^+), \tag{4.6.5}$$

where $Z_q^{1/2}$ is a measure of the amplitude of the single excitation contribution to a density fluctuation. Information about this factor is provided by neutron scattering and also by the f sum rule. We first consider the f sum rule, which states

$$\frac{1}{2\pi}\int_{-\infty}^{\infty} d\omega\, \omega\, S(q,\omega) = \frac{\hbar\rho q^2}{2}. \tag{4.6.6}$$

The single-particle contribution to $S(q,\omega)$ is form (4.6.5)

$$S'(q,\omega) = 2\pi Z_q[f_q\delta(\omega + \omega_q) + (1 + f_q)\,\delta(\omega - \omega_q)], \tag{4.6.7}$$

where $f_q = (e^{h\omega_q/kT} - 1)^{-1}$ is the Planck distribution function and ω_q is the frequency of an excitation of wave vector q. If it is assumed that the f sum rule is exhausted by the single-particle excitations from (4.6.7) and (4.6. 6), we find $Z_q = (\hbar\rho q^2/2\omega_q)$. This is only true as $q \to 0$, and from the neutron-scattering data we know that only a fraction α_q of the sum rule is used up by the single particle excitations. We assume, therefore,

$$Z_q = \frac{\hbar\rho q^2}{2\omega_q}\alpha_q. \tag{4.6.8}$$

From the neutron-scattering data [61] near the roton minimum $q = q_0$, $Z_{q_0} = 0.710^{-24}$ cgs and near the maximum $q = q_1$, $Z_{q_1} = 0.01510^{-24}$ cg.

The simplest approximation for the four-particle correlation function in (4.6.3) is to assume that it factorizes into a product of two-particle correlation functions. This is exact if the excitations do not interact with each other.

Thus,

$$S_4(\mathbf{q},0,\mathbf{q}',t) = \langle \delta\rho_{\mathbf{q}}(t)\, \delta\rho_{-\mathbf{q}}(t)\, \delta\rho_{\mathbf{q}'}(0)\, \delta\rho_{-\mathbf{q}'}(0)\rangle$$
$$= S'(q,t)\, S'(-q,t)\, (\delta_{\mathbf{q},\mathbf{q}'} + \delta_{\mathbf{q},-\mathbf{q}'}). \tag{4.6.9}$$

We have omitted a term that involves equal time correlation functions and only contributes to the scattering at zero frequency shift. Taking the Fourier transform of (4.6.9) and noticing that all functions in (4.6.3) are even functions of q and q' we have

$$S_4(\mathbf{q},0,\mathbf{q}',\omega) = \delta_{\mathbf{q},\mathbf{q}'} \frac{1}{\pi} \int d\omega'\, S'(q,\omega')\, S'(q,\omega - \omega'). \tag{4.6.10}$$

Finally we make use of (4.6.7) for the density–density correlation function with $f_q = 0$. This last approximation is justified as near the roton minimum or maximum $\hbar\omega_q \gg kT$ and $f_q \ll 1$. We then find

$$S_4(\mathbf{q},0,\mathbf{q}',\omega) = 4\pi\delta_{\mathbf{q},\mathbf{q}'}\, Z_q^2 \delta(\omega - 2\omega_q). \tag{4.6.11}$$

This result is then substituted in (4.6.3) to determine the scattering cross section. We are mainly interested in frequencies corresponding to twice the roton energy $2\varepsilon_0$ or twice the "maxon" energy $2\varepsilon_1$. Close to these energies, we may neglect the momentum dependence of $g(q)$ and Z_q in (4.6.3) and express the results in terms of the joint density of states for noninteracting excitations

$$\rho_0^{(2)} = \frac{1}{(2\pi)^3} \int dq\delta(\omega - 2\omega_q).$$

Near the roton minimum and the maximum ω_q is given by

$$\omega_q = \frac{\varepsilon_0}{\hbar} + \frac{\hbar(q - q_0)^2}{2\mu_0} \qquad \text{(rotons)}$$

and by (4.6.12)

$$\omega_q = \frac{\varepsilon_1}{\hbar} - \frac{\hbar(q - q_1)^2}{2\mu_1} \qquad \text{(maxons)}.$$

The joint density of states is

$$\rho_{20}^{(0)}(\omega) = \frac{q_0^2}{2\pi^2} \left(\frac{\mu_0}{\hbar\omega - 2\varepsilon_0} \right)^{1/2} \qquad \text{(rotons)}$$

and (4.6.13)

$$\rho_{21}^{(0)}(\omega) = \frac{q_1^2}{2\pi^2} \left(\frac{\mu}{2\varepsilon_1 - \hbar\omega} \right)^{1/2} \qquad \text{(maxons)}.$$

These densities of states for the rotons and "maxons" are shown in Figure 4.7. The function $h(\omega)$ for the rotons ($\omega \sim 2\varepsilon_0$) and the "maxons" ($\omega \sim 2\varepsilon_1$) is then given by

$$h_0(\omega) = 4\pi Z_{q_0}^2 g^2(q_0) \rho_{20}^{(0)}(\omega) \qquad \text{(rotons)}$$

and (4.6.14)

$$h_1(\omega) = 4\pi Z_{q_1}^2 g^2(q_1)\rho_{21}^{(0)}(\omega) \qquad \text{(maxons)}.$$

To obtain the total light scattered by the rotons, equation (4.6.4) is integrated over an interval

$$\frac{2\varepsilon_0}{\hbar} < \omega < \left(\frac{2\varepsilon_0}{\hbar} + \Delta\Omega\right),$$

where $\Delta\Omega$ is the spectral width (assumed rectangular) of the observing instrument. The total cross section for scattering by the rotons is

$$w^{(2)}(\text{rotons}) = \frac{14}{5}\frac{\alpha^4}{m^4\pi^2}\left(\frac{\Omega_i}{c}\right)^4\left(1 - \frac{1}{7}(\hat{\mathbf{K}}_s\cdot\hat{\mathbf{E}}_0)^2\right)(Z_{q_0}g(q_0)q_0)^2\left(\frac{\mu_0\Delta\Omega}{\hbar}\right)^{1/2}. \tag{4.6.15}$$

Using $Z_{q_0} = 0.710^{-24}$cgs, $g(q_0) = 1.1$ (which assumes a $= 1.8$ Å in equation (4.2.25)), $\Delta\Omega = 4.5\ k/\hbar = 6.3\ 10^{11}\ \text{sec}^{-1}$ (which is appropriate in the experiments of Woerner and Greytak [60]), $\hat{\mathbf{K}}\cdot\hat{\mathbf{E}}_0 = 0$, and other known constants appropriate for rotons we find $w^{(2)}$ (rotons) $\simeq 4.10^{-11}$/cm sr. The ratio of the two-roton Raman scattering to the Brillouin scattering in ^{4}He (equation (4.3.4) is thus $w^{(2)}$ (rotons)/$w^{(1)}$ (Brillouin) $= 4.5 \times 10^{-3}$ at $T = 1.2$K. This is considerably larger than the recent experimental value of $(3.4 \pm 0.08)\ 10^{-4}$ measured by Woerner and Greytak [60].

Two suggestions have been made for this discrepancy. Fetter [59] has made a better calculation of the four particle correlation function S_4 than the simple factorization in equation (4.6.9). His calculation is based on a convolution approximation and he has shown that it reduces the above ratio of two-roton Raman scattering to Brillouin scattering to 5.10^{-4}, much closer to the experimental value. Baeriswyl [19] has calculated the above intensity ratio using the more precise form that he devised for the factor $g(q_0)$ which is given in equation (4.2.26). He showed that this factor reduces the two-roton Raman intensity. This reflects the fact that $g(q_0)$ is quite sensitive to the value of a chosen in equation (4.2.25). Presumably the best theoretical estimate is obtained by including both these corrections, and the resulting intensity ratio (which neglects the interaction between the two rotons) is closer to the observed value. We show in the next section that including the interaction between the rotons increases the theoretical Raman intensity by about 1.5.

The ratio of the two-roton Raman intensity to the two-"maxon" Raman intensity is given by

$$\frac{w^{(2)}(\text{maxons})}{w^{(2)}(\text{rotons})} = \left(\frac{q_1}{q_0}\right)^6\left(\frac{g(q_1)\varepsilon_1}{g(q_0)\varepsilon_0}\right)^2\left(\frac{\mu_1}{\mu_0}\right)^{1/2} \simeq \frac{1}{6}. \tag{4.6.16}$$

This ratio is larger than that observed by Greytak and Yan [57], who do not see any special feature in the Raman scattering at frequency shifts $\omega \sim 2\varepsilon_1/\hbar$ corresponding to the creation of two "maxons." This discrepancy has been very elegantly explained independently by Ruvalds and Zawadowski [62] and Iwamoto [63]. They include an interaction between the excitations, which we now consider.

4.6.1. Effects of Interaction on the Raman Spectrum

It was pointed out independently by Ruvalds and Zawadowski [62] and by Iwamoto [63] that the interaction between the excitations can have a profound effect on the Raman spectrum. This can be simply understood as follows: The joint density of states for noninteracting excitations diverges as $(\hbar\omega - 2\varepsilon_0)^{-1/2}$ near the roton minimum (see equation (4.6.13)). It is interesting to first compare this joint density of states with the density of states for particles near zero energy in a system of dimension d. This density of states is proportional to $N(\omega_q) \sim q^{d-1}(dq/d\omega_q)$, and using $\omega_q = \hbar q^2/2m$, we find $N(\omega) \sim \omega^{(d-2)/2}$. Thus, near the roton minimum the joint density of states has the same energy dependence as in a one-dimensional system. It is well known from elementary quantum mechanics that if two particles, which attract each other, are confined to move in one dimension that a bound state is always obtained, however weak the attractive potential. This bound state results from the divergence in the density of states. This suggests that the same may occur for two rotons if their potential is attractive. A simple calculation, which follows, confirms this result.

We regard rotons as quasiparticles with an energy–momentum relation $\hbar\omega_q = \varepsilon_0 + \hbar^2(q - q_0)^2/2\mu_0$ and interacting via a potential $V(\mathbf{r}_1 - \mathbf{r}_2)$. The wave function for two rotons at $\mathbf{r}_1$ and $\mathbf{r}_2$ is $\psi(\mathbf{r}_1 - \mathbf{r}_2)$. When expanded in a Fourier series one has

$$\psi(\mathbf{r}_1 - \mathbf{r}_2) = \frac{1}{V} \sum_{\mathbf{q}} h_{\mathbf{q}} e^{i\mathbf{q}\cdot(\mathbf{r}_1 - \mathbf{r}_2)} \tag{4.6.17}$$

which is a superposition of plane wave roton states with equal and opposite momenta. The internal structure of a roton is neglected. The Schrödinger equation for the two rotons in momentum space is

$$2\hbar\omega_q h_{\mathbf{q}} + \frac{1}{V} \sum_{q'} V(\mathbf{q} - \mathbf{q}')h_{\mathbf{q}'} = Eh_q, \tag{4.6.18}$$

where $V(\mathbf{q})$ is the Fourier transform of the potential. This equation is easily solved if we assume that the real potential may be replaced by a simple pseudopotential $V(\mathbf{r}) = -g^{(0)}\delta(\mathbf{r})$. The Schrödinger equation is now

$$(2\hbar\omega_q - E)h_q = g^{(0)}C, \tag{4.6.19}$$

where $C = 1/V \sum_q h_q$. From this we find

$$h_q = \frac{g^{(0)} C}{2\hbar\omega q - E}, \tag{4.6.20}$$

and summing both sides on **q** gives the eigenvalue equation

$$1 = \frac{g^{(0)}}{V} \sum_q \frac{1}{2\hbar\omega_q - E}. \tag{4.6.21}$$

Substituting the roton energy for ω_q on the right-hand side and and assuming $E < 2\varepsilon_0$, we obtain for the binding energy (measured from $2\varepsilon_0$)

$$\mathscr{E}_0 = \frac{g^{(0)2}\mu_0 q_0^4}{4\pi^2\hbar^2}. \tag{4.6.22}$$

This calculation is easily generalized to a bound state of angular momentum L. In this case we assume a pseudopotential $V(\mathbf{q} - \mathbf{q}') = -g^{(L)}(2L + 1) P_L(\cos\theta_{qq'})$. This pseudopotential has the special feature that the angular momentum of the bound state is determined by the angular dependence of the potential. This potential then leads to a bound state with angular momentum L and binding energy as in equation (4.6.22) with $g^{(0)}$ replaced by $g^{(L)}$. We can also consider two excitations near the maximum in the dispersion curve. In this case, because of the negative effective mass of these excitations, a "bound" state with energy $\mathscr{E}_0 = g^{(L)2}\mu_1 q_1^4/4\pi^2\hbar^2$ *above* $2\varepsilon_1$ is obtained for a *repulsive* potential.

The simple calculation of the bound states given above shows that any roton–roton interaction will have a profound effect on the joint density of states and the Raman scattering spectrum. We turn to a more detailed calculation of the four-particle correlation function in equation (4.6.3). Analytic expressions can be obtained if we again replace the potential by a pseudopotential. The P_2 Legendre polynomial in (4.6.3) requires that the two rotons be in a relative d state, so we take a potential $V(\mathbf{q} - \mathbf{q}') = -5g^{(2)} P_2(\cos\theta_{qq'})$. Then a "ladder" approximation [62] to the four-particle correlation function gives for energies near $2\varepsilon_0$.

$$h_0(\omega) = 4\pi Z_{q_0}^2 g^2(q_0)\rho_2(\omega), \tag{4.6.23}$$

where

$$\rho_2(\omega) = \frac{\hbar}{\pi}\,\mathrm{Re}\frac{F(\omega)}{1 + ig^{(2)}F(\omega)}, \tag{4.6.24}$$

$$F(\omega) = \frac{-i}{(2\pi)^3\hbar}\int d\mathbf{q}\,\frac{1}{\omega - 2\omega_q - i\delta}, \tag{4.6.25}$$

and δ is a small positive quantity. A similar result is obtained for energies near twice the maximum energy except that q_0 is everywhere replaced by q_1.

The integral in (4.6.25) is easily evaluated, and we find for $\hbar\omega \sim 2\varepsilon_0$

$$\rho_2(\omega) = 2\hbar g^{(2)} A_0^2 \delta(2\varepsilon_0 - \mathscr{E}_0 - \hbar\omega);\ \hbar\omega < 2\varepsilon_0$$

and (4.6.26)

$$\rho_2(\omega) = \frac{\hbar}{\pi}\,\frac{A_0(\hbar\omega - 2\varepsilon_0)^{1/2}}{\hbar\omega - 2\varepsilon_0 + \mathscr{E}_0};\ \hbar\omega > 2\varepsilon_0$$

where $\mathscr{E}_0 = g^{(2)2} q_0^4 \mu_0 / 4\pi^2\hbar^2$ and $A_0 = (q_0^2/2\pi\hbar)\mu_0^{1/2}$. These results are for an attractive potential. For a repulsive potential the first term, resulting from the bound state, is absent, and $\rho_2(\omega)$ is just given by the second term. This latter term goes over into the noninteracting joint density of states when $g^{(2)} = 0$. In the attractive case the total integral of the density of states near the roton minimum is increased by

$$\int_{2\varepsilon_0 - \Delta\Omega}^{2\varepsilon_0 + \Delta\Omega} d\omega(\rho_2(\omega) - \rho_2^{(0)}(\omega)) = A_0 \mathscr{E}_0^{1/2}, \tag{4.6.27}$$

when $\hbar\Delta\Omega > \mathscr{E}_0$. This leads to an increase in the intensity of the two-roton scattering in equation (4.6.15) by a factor $1 + \pi/2(\mathscr{E}_0/\hbar\Delta\Omega)^{1/2} \simeq 1.5$, where we have used the experimental values [64] $\mathscr{E}_0 = 0.37\text{K}$ and $\hbar\Delta\Omega = 4.8$ K [60]. As mentioned above, this factor brings the calculated two-roton scattering intensity closer to the experimental value of Woerner and Greytak[60].

For two-"maxon" excitations the same expressions, equation (4.6.26) holds with q_0 and μ_0 replaced by q_1 and μ_1 and $\hbar\omega - 2\varepsilon_0$ replaced by $2\varepsilon_1 - \hbar\omega$. The "bound"-state contribution is only included if the potential is repulsive. The function $\rho_2(E)$ is shown in Figure 4.7 for an attractive potential and compared with the joint density of states for noninteracting excitations. The treatment given here of the effects of the interaction of two rotons has been greatly simplified by introducing a pseudopotential and neglecting the interaction of rotons with other excitations. A much more complete discussion has been given by Iwamoto [63] and Iwamoto et al. [65].

The assumption of an attractive interaction between excitations immediately explains the absence of any strong feature at energy shifts $\hbar\omega \sim 2\varepsilon_1$ corresponding to scattering from two maxons. In order to make a detailed comparison of the theoretical and experimental spectra, it is necessary to introduce the finite lifetime of the rotons. It was noted by Greytak et al. [64] that this can be done by convoluting the function $h(\omega)$ with an appropriate Lorentzian of unit area and width $2\delta\varepsilon_0$ where $\delta\varepsilon_0$ is the inverse lifetime or width of a single roton. This width for $T > 1$ K is primarily due to collisions with thermally excited rotons. Another source of broadening of the spectrum is the instrumental profile. Both the broadening effects have been taken into account by Greytak et al. in comparing the experimental and the theoretical spectra. This comparison is shown in Figure 4.9 and provides good evidence for the existence of the bound two-roton state. The binding energy is esti-

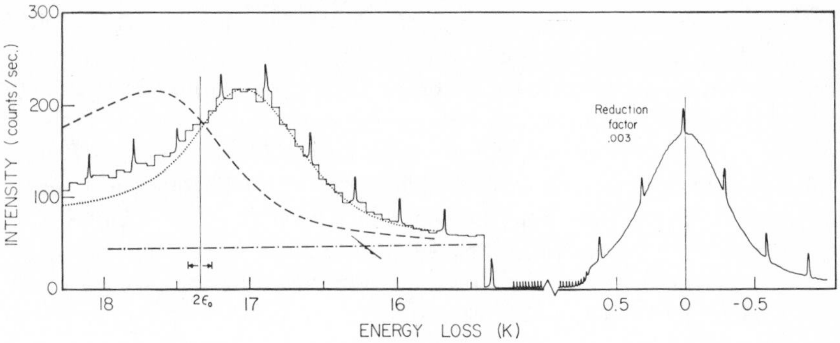

FIGURE 4.9. The two-roton bound state. The dashed line is the calculated spectrum assuming no interaction and the dotted line the calculated spectrum assuming an attractive interaction between rotons. The solid line is the observation of Greytak et al [64].

mated to be $\mathscr{E}_0 = 0.37$ K and corresponds to a value of $g^{(2)} = 1.2\ 10^{-39}$ erg cm^3.* We can also estimate the root mean square radius R of the bound state from the binding energy. It is $R \sim \hbar/(\mathscr{E}_0\mu_0)^{1/2} = 13$ Å, which is much larger than the atomic spacing in ^{4}He of 3.5 Å.

The existence of a two-roton bound state has been questioned by Cowley [66]. He compared the spectra of Greytak et al. [64] with theoretical predictions based on noninteracting and interacting rotons. In the opinion of the present author, the evidence for the bound state can be summarized as follows: (a) The peak in the Raman spectrum definitely falls below $2\varepsilon_0$. The largest error in determining the binding energy comes from our imprecise knowledge of ε_0 from neutron scattering. (b) The absence of any pronounced feature in the spectrum at an energy shift of $2\varepsilon_1$. This is easily accounted for if there is an attractive interaction between excitations.

Raman scattering in ^{4}He above T_λ has been observed by Pike and Vaughan [67]. They made the interesting observation that the "two-roton peak" moves to higher energies as the temperature is increased through T_λ. Above T_λ the Raman spectrum shows a single (very broad) peak at an energy closer to $2\varepsilon_1$ (i.e., two maxons) than to $2\varepsilon_0$. They made the suggestion that the sign of the interaction between excitations may change sign at T_λ, becoming repulsive for $T > T_\lambda$. According to the above arguments, this would have the effect of suppressing the two-roton peak and enhancing the two-maxon peak.

The linewidths of the two-roton Raman spectrum has been measured for $1.3\ \text{K} < T < 2$ K by Greytak and Yan [68]. They have interpreted the linewidth

*Recently a high resolution study of the two-roton scattering at $T = 0.6$ K and SVP has been made by Murray, Woerner and Greytak [J. Phys. C: *Solid State Phys.* **8**, L90 (1975)] The two-roton binding energy was estimated from the spectral shape to be 0.22 ± 0.07K and the low temperature line width to be 0.11 ± 0.01K. This residual line width at low temperatures presumably arises from the spontaneous decay of two rotons into two phonons.

in this temperature range as arising from collisions with thermally excited rotons. Their value for the roton–roton collision time is in good agreement with that estimated from the roton contribution to the normal-fluid viscosity [69]. The temperature dependence of the linewidth is determined mainly by the temperature dependence of the roton number density $n_r \sim e^{-\varepsilon_0}/kT$. This collision time also gives information about roton–roton interactions. In order to calculate the collision time, Landau and Khalatnikov [70] assumed that the interaction between two rotons is repulsive and could be represented by a pseudopotential $V_0\delta(\mathbf{r})$. The above linewidth measurements, together with the Landau–Khalatnikov [70] value for the collision frequency in Born approximation

$$\tau^{-1} = \left(\frac{\mu_0 q_0 v_0^2}{\hbar^3}\right) n_r,$$

gives a value of $V_0 = 1.73\ 10^{-38}$ erg cm^3. This is an order of magnitude larger than $g^{(2)}$ which was found from the bound state energy.

This leads us to a brief discussion of roton–roton interactions. Information on this interaction is provided by the following experimental observations:

a. The bound state with $L = 2$ measured by Raman scattering.
b. The collision time for two rotons which is obtained from (i) the line width in the Raman spectrum, (ii) the roton contribution to the normal viscosity (iii) line widths in neutron spectroscopy. All three experiments give collision times in good agreement.
c. The variation of the roton energy ε_0 with temperature. It is observed that ε_0 decreases with increasing temperature [61].

It is not possible to explain all these observations by a single pseudopotential. The observed bound state requires a weakly attractive pseudopotential in a *d* state. The collision time requires for its explanation a pseudopotential that is stronger by an order of magnitude, if Born approximation is used to describe the scattering. Calculations going beyond Born approximation of the roton–roton collision time have been made by several authors [71–74]. They showed that there exists an upper limit to the collision frequency if a single pseudopotential is used to represent the roton–roton interaction. This upper limit is smaller than the observed collision frequency by a factor of 4. The bound state does not affect the collision frequency appreciably as $\mathscr{E}_0 \ll kT$ for $T > 1$ K. It was suggested that it is important to take into account the range of the roton–roton interaction, and the roton–roton scattering will take place in several angular-momentum channels. Solana et al. [74] have suggested that hybridization of a single roton with bound roton pairs should also be taken into account in roton–roton scattering.

Then there is the decrease of the roton energy ε_0 with temperature. This requires an attractive roton–roton interaction for its explanation. A simple Hartree calculation shows that to get agreement with experiment an attractive interaction an order of magnitude larger than that found from the bound state is required. Tutto [75] and Nagai [76] have shown that there is also an upper limit to the variation in ε_0 with temperature if a single attractive pseudopotential is used, and that again it is necessary to take account of the interaction in several angular-momentum channels.

The conclusion is that it is at present not possible to provide a simple representation of roton–roton interactions in superfluid ^{4}He. The best description of a roton has been provided by Feynman [77] and Feynman and Cohen [78]. It can be pictured roughly as a vortex ring of small radius, comparable with the interparticle spacing in ^{4}He. Outside the ring there is a dipolar backflow field, which is taken at large distances to be of the hydrodynamic form. The energy of a roton is not dependent on the detailed nature of the core. At large separations of two rotons, a hydrodynamic picture is reliable, and the interaction of the two rotons results from the interaction of their backflow fields. This leads to an interaction energy [72, 79]

$$V(\mathbf{r}) = \frac{A_0}{m} \frac{3(\mathbf{p}_1\cdot\mathbf{r})(\mathbf{p}_2\cdot\mathbf{r}) - (\mathbf{p}_1\cdot\mathbf{p}_2)r^2}{r^5} \tag{4.6.28}$$

where $A_0 \simeq 3.8$ Å^3 and $\mathbf{p}_1$ and $\mathbf{p}_2$ are the momenta of the two rotons. This interaction is attractive in certain orientations and repulsive in others. A variational calculation by Yau and Stephen [72] shows that this interaction is sufficient to explain the bound d state of two rotons if the interaction is cut off for $r<5$ Å. The binding energy of two rotons in various states of angular momentum has been determined by Roberts and Pardee [80] by numerical solution of the Schrodinger equation using the potential (4. 6. 28). They found that the most tightly bound state has $L = 3$ and that the $L = 2$ state has a binding energy of about 0.29 K.

The scattering of two rotons has been calculated using Born approximation for the interaction (4.6.28) by Toigo [79]. He obtained a cross section that was too small to explain the observed scattering of two rotons. The classical calculation of Roberts and Donnelly [80] shows that the dipole interaction (4.6.28) can lead to strong back scattering of two rotons at small impact parameters $b \sim (\mu/4\pi\rho)(p_0^2/2\mu kT))^{1/3}$ which is approximately 4 Å at $T = 1$ K. This leads to a cross section $\sigma \sim \pi b^2$, which is of the right order of magnitude. It should be remarked that the interaction (4.6.28) may not be reliable at such short distances. The determination of the interaction of rotons at short distances from a microscopic model remains an interesting unsolved problem.

Finally, we mention some interesting recent data on Raman scattering in

^{4}He under pressure [81], in ^{3}He–^{4}He mixtures [81, 82, 83], and in pure ^{3}He [81]. In ^{4}He under pressure the two-roton peak is observed to shift to lower frequencies and decrease in intensity. The shift appears to be consistent with the variation in ε_0 with pressure as measured by neutron scattering [61]. In the ^{3}He–^{4}He mixtures the two-roton peak has been observed as a function of ^{3}He concentration and is found to be almost concentration independent up to a^3 He molar concentration of 31%. Then results show that the roton gap ε_0 is almost independent of concentration up to the highest concentration used. The roton gap has also been obtained from an analysis of the temperature dependence of the measured normal fluid density in mixtures [84]. If this analysis is based on the assumption of a parabolic dispersion relation for the ^{3}He quasiparticles in a mixture, the roton gap apparently decreases markedly with increasing ^{3}He concentration. It has been argued by Stephen and Mittag [85] that the ^{3}He quasiparticle dispersion relation is unlikely to be of the free particle form at large momenta but may exhibit a minimum or a plateau at momenta ~ 1.8 Å^{-1} and energy less than the roton gap. With this form for the ^{3}He dispersion relation it is possible to explain the Raman and normal fluid density measurements. In pure liquid ^{3}He a broad peak in seen in the Raman spectrum at an energy shift of ~ 21 K. This peak resembles that found in solid helium.

REFERENCES

1. V. L. Ginzburg, *J. Exp. Theor. Phys. (USSR)* **13,** 243 (1943).
2. G. J. Van den Berg, A. Van Itterbeck, G. M. V. Van Aardenne, and J. H. J. Herfkens, *Physica* **21,** 860 (1955).
3. M. A. Woolf, P. M. Platzman, and M. G. Cohen, *Phys. Rev. Lett.* **17,** 294 (1966).
4. G. Jacucci and G. Signorelli, *Phys. Lett.* **26A,** 5 (1967).
5. G. Winterling, G. Walda, and W. Heinicke, *Phys. Lett.* **26A,** 301 (1968).
6. I. I. Abrikosova and D. M. Bochkova, *JETP Lett.* **9,** 167 (1969).
7. W. Heinicke, G. Winterling, and K. Dransfeld, *Phys. Rev. Lett.* **22,** 170 (1969).
8. E. R. Pike, J. M. Vaughan, and W. F. Vinen, *J. Phys. C.* **3,** L37 (1970).
9. E. R. Pike and J. M. Vaughan, *J. Phys. C* **4,** L362 (1971).
10. J. W. Halley, *Phys. Rev.* **181,** 338 (1969).
11. W. F. Vinen, in *Physics of Quantum Fluids,* edited by R. Kubo and F. Takano (Syokabo Publ., Tokyo, 1971).
12. T. J. Greytak, ibid.
13. T. J. Greytak, in *Proceedings of the 13th International Conference on Low Temperature* Physics (Boulder Colo., 1972) (to be published).
14. E. L. Fabelinskii, *The Molecular Scattering of Light* (Plenum Press, New York, 1968).
15. L. D. Landau and E. M. Lifshitz, *Electrodynamics of Continuous Media* (Addison-Wesley, Mass., 1960), Chap. XIV.
16. M. J. Stephen, *Phys. Rev.* **187,** 279 (1969).

17. S. Nakajima, *Progr. Theor, Phys. (Japan)* **45,** 353 (1971).
18. E. C. Kerr and R. D. Taylor, *Ann. Phys.* **26,** 292 (1964).
19. D. Baeriswyl, *Phys. Lett.* **41A** 297 (1972).
20. R. S. St. Peters, T. J. Greytak, and G. B. Benedek, *Optics Commun.* **1,** 412 (1970).
21. L. D. Landau and E. M. Lifshitz, *Fluid Mechanics* (Addison—Wesley, Mass., 1960).
22. I. M. Khalatnikov, *Introduction to the Theory of Superfluidity* (W. A. Benjamin Inc., New York, 1965).
23. W. F. Vinen, *J. Phys. C.* **4,** L287 (1971).
24. R. A. Ferrell, N. Menyhard, H. Schmidt, F. Schwabl, and P. Szepfalusy, *Ann. Phys. (N.Y.)* **47,** 565 (1968).
25. P. C. Hohenberg and P. C. Martin, *Ann. Phys. (N. Y.)* **34,** 291 (1965).
26. E. R. Pike, J. M. Vaughan, and W. F. Vinen, *J. Phys. C.* **3,** L37 (1970).
27. G. Winterling, F. S. Holmes, and T. J. Greytak, in *Proceedings of the 13th International Conference on Low Temperature Physics* (Boulder, Colo., 1972), *Phys. Rev. Lett.* **30,** 427 (1973).
28. V. Korenman, Dept. of Phys. and Astron. Tech. Rep. (Univ. of Maryland, June, 1967) (unpublished). See also Ref. [24].
29. M. J. Buckingham and W. M. Fairbank in *Progress in Low Temperature Physics,* edited by C. J. Gorter (North-Holland, Amsterdam, 1961), Vol. 3, p. 80.
30. W. F. Vinen, C. J. Palin, and J. M. Vaughan, *Proc. of the 13th Int. Conf. on Low Temp. Phys.* (Boulder, Colo., 1972) and J. Phys. C: Solid State Phys. **5,** L139 (1972).
31. L. S. Ornstein and F. Zernike, *Proc. Acad. Sci. (Amsterdam).* **17,** 793 (1914).
32. B. D. Josephson, *Phys. Lett.* **21,** 608 (1966).
33. J. McCauley and M. J. Stephen (unpublished).
34. D. Greywall and G. Ahlers, *Phys. Rev. Lett.* **28,** 1251 (1972).
35. H. A. Kierstead, *Phys. Rev.* **162,** 153 (1967).
36. B. I. Halperin and P. C. Hohenberg, *Phys. Rev.* **177,** 952 (1969).
37. J. A. Tyson, *Phys. Rev. Lett.* **21,** 1235 (1968).
38. H. Schmidt and P. Szepfalusy, *Phys. Lett.* **32A,** 326 (1970).
39. H. Y. Lau, L. M. Corlin, A. Delapalme, J. M. Hastings, R. Nathans, and A. Tucciarone, *Phys. Rev. Lett.* **23,** 1125 (1969).
40. G. Ahlers, *Phys. Rev. Lett.* **21,** 1159 (1968).
41. I. M. Khalatnikov, *Zh. Eksp. Teor. Fiz.* **57,** 489 (1969) [English transl.: *Sov. Phys.—JETP* **30,** 268 (1970)].
42. M. Barmatz and I. Rudnick, *Phys. Rev.* **170,** 224 (1968).
43. R. P. Henkel, E. N. Smith, and J. D. Reppy, *Phys. Rev. Lett.* **23,** 1216 (1969).
44. H. H. Kretzen and H. F. Mikeska Phys. Lett. **43A,** 525 (1973).
45. L. P. Gorkov and L. P. Pitaevskii, *J. Exptl. Theoret. Phys. (USSR)* **33,** 634 (1957) [English transl.: *Sov. Phys.—JETP* **6,** 486 (1958)].
46. I. M. Khalatnikov, *Zh. Eksp. Teor. Fiz.* **23,** 265 (1952).
47. J. Wilks, *Liquid and Solid Helium* (Clarendon Press, Oxford, 1967).
48. D. R. Watts, W. I. Goldburg, L. D. Jackel, and W. W. Webb, *J. Phys.* **33,** Cl-155 (1972).

49. D. R. Watts and W. W. Webb. in *Proceedings of the 13th International Conference on Low Temperature Physics* (Boulder, Colo,. 1972) (to be published). See also P. Leiderer, D. R. Watts, and W. W. Webb, Phys. Rev. Lett, **33**, 483 (1974).
50. G. Goellner and H. Meyer, *Phys. Rev. Lett.* **26,** 1534 (1971).
51. T. Alvesalo, P. Berglund, S. Islander, G. R. Pickett, and W. Zimmermann, *Proceedings of the 12th International Conference on Low Temperature Physics* (Kyoto, Japan, 1970).
52. E. K. Riedel, *Phys. Rev. Lett.* **29,** 675 (1972).
53. G. N. Ganguly and A. Griffin, *Can J. Phys.* **46,** 1895 (1968).
54. R. F. Benjamin, D. A. Rockwell, and T. J. Greytak, in *Proceedings of the 13th International Conference on Low Temperature Physics* (Boulder, Colo., 1972) (to be published). See also Rockwell, Benjamin and Greytak *J. Low Temp. Phys*, **18**, 389 (1975).
55. C. J. Palin, W. F. Vinen, E. R. Pike, and J. M. Vaughan, *J. Phys. C Solid State Phys.* **4,** 1225 (1971).
56. A. Griffin, *Can. J. Phys.* **47,** 429 (1969).
57. T. J. Greytak and J. Yan, *Phys. Rev. Lett.* **22,** 987 (1969).
58. F. Iwamoto, *Progr. Theor.-Phys. (Japan)* **44,** 1135 (1970).
59. A. L. Fetter, *J. Low Temp. Phys.* **6,** 487 (1972).
60. R. L. Woerner and T. J. Greytak, *J. Low Temp. Phys.* October (1973).
61. R. A. Cowley and A. D. B. Woods, *Can. J. Phys.* **69,** 172 (1971). Note that their definition of $Z(Q)$ is related to the present one by $Z_q = \rho m Z(Q)$.
62. J. Ruvalds and A. Zawadowski, *Phys. Rev. Lett.* **25,** 333 (1970); also A. Zawadowski, J. Ruvalds, and J. Solana, *Phys. Rev. A* **5,** 399 (1972).
63. F. Iwamoto, *Progr. Theor. Phys.* **44,** 1121 (1970).
64. T. J. Greytak, R. Woerner, J. Yan, and R. Benjamin, *Phys. Rev. Lett.* **25,** 1547 (1970); for a review see T. J. Greytak and R. Woerner, *J. de Phys.* **33,** Cl-269 (1972).
65. F. Iwamoto, K. Nagai, and K. Nojima, in *Proceedings of the 12th International Conference on Low Temperature Physics* (Kyoto, Japan, 1970).
66. R. A. Cowley, *J, Phys.* **5,** L287 (1972).
67. E. R. Pike and J. M. Vaughan, *J. Phys. C* **4,** L362 (1971).
68. T. J. Greytak and J. Yan, in *Proceedings of the 12th International Conference on Low Temperature Physics* (Kyoto, Japan, 1970).
69. D. F. Brewer and D. O. Edwards, *Proceedings of the 8th International Conference on Low Temperature Physics* (London, 1962), p. 96.
70. L. D. Landau and I. M. Khalatnikov, *Zh. Eksp. Teor. Fiz.* **19,** 637 (1949). We differ from their result by a factor 2.
71. I. A. Fomin, *Zh. Eksp. Teor. Fiz.* **60,** 1178 (1971) [English transl.: *Sov. Phys.—JETP*].
72. J. Yau and M. J. Stephen, *Phys. Rev. Lett.* **27,** 482 (1971).
73. K. Nagai, K. Nojima, and A.-Hatano, *Progr. Theor. Phys. (Japan)* **47,** 355 (1972).
74. J. Solana, V. Celli, J. Ruvalds, I. Tutto, and A. Zawadowski, *Phys. Rev. A* **6,** 1665 (1972).
75. I. Tutto, *J. Low Temp. Phys.* **11,** 77 (1973)
76. K. Nagai, Progr. Theor. Phys. (Japan) **49,** 46 (1973).
77. R. P. Feyman, *Phys. Rev.* **94,** 262 (1959).

78. R. P. Feynman and M. Cohen, *Phys. Rev.* **102,** 1189 (1956).
79. F. Toigo, *Nuovo Cimento* **62B,** 103 (1969).
80. P. H. Roberts and W. J. Pardee, J. Phys. *A: Math. Phys.* **7**, 572 (1974).
81. R. E. Slusher and C. M. Surko, *Phys. Rev. Lett.* **27,** 1699 (1971); and private communication.
82. C. M. Surko and R. E. Slusher, *Phys. Rev. Lett.* **30,** 1111 (1973).
83. R. L. Woerner, D. A. Rockwell, and T. J. Greytak, *Phys. Rev. Lett.* **30,** 1114 (1973).
84. V. I. Sobolev and B. N. Esel'son, *Sov. Phys.—JETP* **33,** 132 (1971).
85. M. J. Stephen and L. Mittag, *Phys. Rev. Lett.* **31,** 923 (1973).

ACKNOWLEDGEMENT

I thank Dr. G. Winterling for Figure 4.2 and 4.3, Dr. W. Webb for Figure 4.5 and Dr. T. J. Greytak for Figures 4.6–4.9. I am grateful to Dr. R. Woerner and Dr. R. Benjamin for checking the numerical calculations.

Appendix A

In this section we calculate the correlation function of the concentration fluctuation in a superfluid ^{3}He–^{4}He mixture. This function is given by

$$S_{cc}(q,\omega) = \int_{-\infty}^{\infty} dt\, e^{i\omega t} \langle \delta c_q(t)\, \delta c_{-q}(o) \rangle. \qquad \text{(A.1)}$$

We begin with the two-fluid hydrodynamic equations of Khalatnikov and make the following simplifying approximations. The thermal expansion coefficient, terms in $\rho_s/\rho_n(c/\rho(\partial\rho/\partial c))^2 \simeq 0.02$, and the viscosity coefficients are neglected. These approximations are justified if the temperature is not very close to T_λ, and the concentration is not too low, so that the damping of the second sound is dominated by the diffusion and thermal conductivity. These approximations lead to the simplification that the pressure fluctuations are not coupled to the concentration and temperature fluctuations. After elimination of the velocities from the two-fluid equations, we obtain the two equations for the concentration and temperature fluctuations

$$\left[\rho\bar{\sigma}\frac{\partial}{\partial t} - \rho D\left(\bar{\sigma} - \frac{k_T}{T}\beta c\right)\nabla^2\right]\delta c$$
$$-\left[\frac{cC_p}{T}\frac{\partial}{\partial t} - \left(\frac{\kappa c}{T} - \frac{\rho D k_T}{T}\left(\bar{\sigma} - \frac{k_T\beta c}{T}\right)\right)\nabla^2\right]\delta T = 0, \qquad \text{(A.2)}$$

and

$$\left[\frac{T\rho}{C_p}\left(\frac{\partial\sigma}{\partial c}\right)\frac{\partial^2}{\partial t^2} - \frac{\beta c}{\sigma}u_{20}^2\nabla^2 + \frac{\rho D}{C_p}\left(k_T\beta + T\frac{\partial\sigma}{\partial c}\right)\frac{\partial}{\partial t}\nabla^2\right]\delta c$$

$$+\left[\frac{\partial^2}{\partial t^2} - \frac{\bar{\sigma}}{\sigma}u_{20}^2\nabla^2 - \frac{1}{C_p}\left(\kappa + \frac{\rho D k_T}{T}\left(k_T\beta + T\frac{\partial\sigma}{\partial c}\right)\right)\frac{\partial}{\partial t}\nabla^2\right]\delta T = 0, \quad \text{(A.3)}$$

where $u_{20}^2 = \rho_s\rho\sigma^2 T/\rho_n C_p$. The spatial Fourier transform and Laplace transform in time of these equations is taken. After solving for δc, we find for the correlation function (A.1)

$$S_{cc}(q,\omega) = 2\mathrm{Re}\left(\frac{N(s)}{D(s)}\right)_{s=i\omega}\langle|\delta c_q|^2\rangle, \quad \text{(A.4)}$$

where

$$N(s) = s^2 + sq^2\left[D_0 + 2D_2 - D\left(\frac{\bar{\sigma}}{\sigma} - \frac{k_T\beta c}{T\sigma}\right)\right]$$

$$+ q^2\left[\frac{\bar{\sigma}^2}{\sigma^2}u_{20}^2 + \frac{q^2 D}{C_p}\left(\frac{k_T\beta c}{T\sigma} + \frac{c}{\sigma}\frac{\partial\sigma}{\partial c}\right)\left(\kappa - \frac{\rho D k_T}{c}\left(\bar{\sigma} - \frac{k_T\beta c}{T}\right)\right)\right], \quad \text{(A.5)}$$

and

$$D(s) = s^3 + s^2q^2(D_0^{(3)} + 2D_2^{(3)}) + sq^2\left(u_2^2 + \frac{D\kappa}{C_p}q^2\right) + q^4u_2^2D_0^{(3)}, \quad \text{(A.6)}$$

where u_2^2 is given in (4.5.14), $D_0^{(3)}$ in (4.5.25), and $D_2^{(3)}$ in (4.5.24) (the viscosity contribution has been neglected). The function (A.4) is peaked at $\omega = 0$, $\omega \simeq \pm u_2 q$ with strengths proportional to (4.5.21) and (4.5.19). The roots of the denominator (A.6) are to a first approximation

$$\omega = iD_0^{(3)}q^2, \quad \omega = \pm u_2 q + iD_2^{(3)}q^2.$$

CHAPTER **5**

Microscopic Calculations for Condensed Phases of Helium

Chia-Wei Woo

Department of Physics
Northwestern University
Evanston, Illinois

5.1. INTRODUCTION

5.1.1. The Quantum Many-Body Problem

Consider a normalization volume Ω enclosing N particles, each of mass m and spin σ. By way of a limiting process, the number density $\rho = N/\Omega$ is held fixed, while N and Ω are caused to expand without limit, at least to the extent that surface effects become negligible. ρ is the only external parameter: an input to the problem.

Between each pair of particles i and j, a central spin-independent potential $v(r_{ij})$ is prescribed. This potential contains all the dynamical information available. The Hamiltonian of the system is then given by

$$H = \frac{-\hbar^2}{2m} \sum_{i=1}^{N} \nabla_i^2 + \sum_{i<j=1}^{N} v(r_{ij}). \tag{5.1.1}$$

The above provides a complete description of a specific, nonrelativistic quantum mechanical system. We have here a particularly simple model system: a pure, bulk substance characterized by a pairwise central potential. Obviously it cannot pretend to be a suitable model for all the systems to be discussed in this chapter. I have merely chosen a most convenient example for the purpose of defining and formulating the quantum many-body problem.

Among the standard theoretical tools available to a physicist are quantum mechanics and statistical mechanics. In principle one wishes to solve the Schrödinger equation

$$H\Psi_\alpha(1,2,\cdots,N) = E_\alpha\Psi_\alpha(1,2,\cdots,\mathrm{N}), \tag{5.1.2}$$

using the Hamiltonian (5.1.1), for all the eigenstates consistent with whatever exchange symmetry and boundary conditions prescribed. (Each argument i of Ψ_α is a shorthand notation for the position and spin variables $\mathbf{r}_i, s_i$; α stands for a complete set of quantum numbers.) Statistical mechanics is then applied to obtain equilibrium and transport properties of the system. This two-stage procedure offers one the maximum amount of macroscipic information extractable from the input microscopic data. In practice one must be content with much less: in almost all cases, present techniques yield little more than properties of the ground state and certain low-lying excited states, with the latter approximated by weakly coupled elementary excitations or "quasiparticles." It will be in these terms that we now approach the solution of equation (5.1.2) and discuss the microscopic theory of realistic helium systems.

Assume first the Hamiltonian (5.1.1) to be resolvable into two parts:

$$H = H_0 + H_1, \tag{5.1.3}$$

where H_0 is an unperturbed Hamiltonian, and H_1 the perturbation. In order to carry out a valid perturbation expansion, it is necessary for H_0 to contain a dominant fraction of the physical information, and for the unperturbed Schrödinger equation

$$H_0\Phi_\alpha(1, 2, \cdots, N) = \mathscr{E}_\alpha\Phi_\alpha(1, 2, \cdots, N) \tag{5.1.4}$$

to possess readily obtainable solutions. Often such demands cause one to choose H_0 as a sum of single-particle operators. For example, one might extract from the interaction term an average single-particle field $V(\mathbf{r}_i)$ and write

$$H_0 = \sum_{i=1}^{N} h(i), \tag{5.1.5}$$

with

$$h(i) = \frac{-\hbar^2}{2m}\nabla_i^2 + V(\mathbf{r}_i), \tag{5.1.6}$$

leaving as the perturbation

$$H_1 = \sum_{i<j=1}^{N} v(r_{ij}) - \sum_{i=1}^{N} V(\mathbf{r}_i). \tag{5.1.7}$$

In such cases, the solutions of equation (5.1.4) are immediately known:

$$\Phi_\alpha(1, 2, \cdots, N) = \prod_{i=1}^{N} \varphi_{\alpha_i}(i), \tag{5.1.8}$$

$$\mathscr{E}_\alpha = \sum_{i=1}^{N} e_{\alpha_i}, \tag{5.1.9}$$

where the single-particle functions, or "orbitals," satisfy the equation

$$h(i)\varphi_{\alpha_i}(i) = e_{\alpha_i}\varphi_{\alpha_i}(i). \tag{5.1.10}$$

To take into account statistics, Φ_α must then be symmetrized for bosons or antisymmetrized for fermions. To this end one adopts for convenience the second quantized representation

$$H_0 = \sum_{\alpha_1\alpha_2} \langle \alpha_1 \,|h|\, \alpha_2\rangle \, a^\dagger_{\alpha_1} a_{\alpha_2} = \sum_{\alpha} e_\alpha a^\dagger_\alpha a_\alpha, \tag{5.1.11}$$

$$H_1 = \tfrac{1}{2} \sum_{\alpha_1\alpha_2\alpha_3\alpha_4} \langle \alpha_1\alpha_2 \,|v|\, \alpha_3\alpha_4\rangle \, a^\dagger_{\alpha_1} a^\dagger_{\alpha_2} a_{\alpha_4} a_{\alpha_3} - \sum_{\alpha_1\alpha_2} \langle \alpha_1 \,|V|\, \alpha_2\rangle \, a^\dagger_{\alpha_1} a_{\alpha_2}, \tag{5.1.12}$$

where

$$\langle \alpha_1\alpha_2 \,|v|\, \alpha_3\alpha_4\rangle = \sum_{s_i s_j} \int d\mathbf{r}_i \int d\mathbf{r}_j \varphi^*_{\alpha_1}(i)\varphi^*_{\alpha_2}(j) v(r_{ij}) \varphi_{\alpha_3}(i)\varphi_{\alpha_4}(j), \tag{5.1.13}$$

and

$$\langle \alpha_1 \,|V|\, \alpha_2\rangle = \sum_{s_i} \int d\mathbf{r}_i \varphi^*_{\alpha_1}(i) V(\mathbf{r}_i)\, \varphi_{\alpha_2}(i), \tag{5.1.14}$$

and the creation and annihilation operators $a^\dagger$ and a obey usual commutation rules.

For liquids and gases, translational invariance reduces the average single-particle field to a constant. H_0 and H_1 are then, respectively, the kinetic and potential energy operators. One finds each orbital to be a product of a plane wave and a spin function:

$$\begin{cases} \varphi_{\mathbf{k}_i\sigma_i}(\mathbf{r}_i s_i) = \dfrac{1}{\sqrt{\Omega}} e^{i\mathbf{k}_i\cdot\mathbf{r}_i} S_{\sigma_i}(s_i) = \begin{cases} \dfrac{1}{\sqrt{\Omega}} e^{i\mathbf{k}_i\cdot\mathbf{r}_i} \delta_{\sigma_i}(s_i) & \text{for fermions,} \quad (5.1.15a) \\ \dfrac{1}{\sqrt{\Omega}} e^{i\mathbf{k}_i\cdot\mathbf{r}_i} & \text{for bosons,} \quad (5.1.15b) \end{cases} \\ e_{\mathbf{k}_i\sigma_i} = e_{\mathbf{k}_i} = \dfrac{\hbar^2 k_i^2}{2m}. \quad (5.1.16) \end{cases}$$

Additionally,

$$\langle \mathbf{k}_1\sigma_1, \mathbf{k}_2\sigma_2 \,|v|\, \mathbf{k}_3\sigma_3, \mathbf{k}_4\sigma_4\rangle$$

$$= \begin{cases} \dfrac{1}{\Omega} \delta_{\mathbf{k}_1+\mathbf{k}_2, \mathbf{k}_3+\mathbf{k}_4} v(|\mathbf{k}-\mathbf{p}|)\, \delta_{\sigma_1,\sigma_3}\delta_{\sigma_2,\sigma_4} & \text{for fermions,} \quad (5.1.17a) \\ \dfrac{1}{\Omega} \delta_{\mathbf{k}_1+\mathbf{k}_2, \mathbf{k}_3+\mathbf{k}_4} v(|\mathbf{k}-\mathbf{p}|) & \text{for bosons,} \quad (5.1.17b) \end{cases}$$

where the definition

$$v(k) = \int d\mathbf{r} e^{i\mathbf{k}\cdot\mathbf{r}}\, v(r) \tag{5.1.18}$$

requires $v(r)$ to be Fourier transformable. The Kronecker delta $\delta_{\mathbf{k}_1+\mathbf{k}_2, \mathbf{k}_3+\mathbf{k}_4}$ implies momentum conservation, while the Kronecker deltas $\delta_{\sigma_i,\sigma_j}$ simply reiterate the spin independence of the interaction.

The resolution of the Hamiltonian in the manner described here allows one to attempt solving equation (5.1.2) by means of conventional perturbation theory, which I shall review presently. For systems with strong interparticle correlations, however, conventional perturbation expansions fail to converge. In other words, we are unable to regard a strongly correlated many-particle system simply as a collection of bare or skimpily veiled entities each moving essentially independently under the perturbing influence of its peers. The difficulties that arise advance quantum mechanics beyond its basic independent particle formulation into the subject matter known as the quantum many-body theory.

5.1.2. Conventional Perturbation Theory [1,2]

Let the chosen state of interest be nondegenerate, denoted by Ψ_γ, and normalized as follows:

$$\langle \Psi_\gamma | \Phi_\gamma \rangle = 1. \tag{5.1.19}$$

Define the projection operators P_γ and Q_γ,

$$P_\gamma \Phi_\alpha = \delta_{\gamma\alpha} \Phi_\alpha, \tag{5.1.20}$$

and

$$Q_\gamma = 1 - P_\gamma. \tag{5.1.21}$$

Operating with Q_γ from the left on the Schrödinger equation (5.1.2), which is now rewritten to read

$$(E_\gamma - H_0)\Psi_\gamma = H_1 \Psi_\gamma, \tag{5.1.22}$$

one finds with the aid of equation (5.1.19), which implies $P_\gamma \Psi_\gamma = \Phi_\gamma$, that

$$(E_\gamma - H_0 - Q_\gamma H_1 Q_\gamma)(\Psi_\gamma - \Phi_\gamma) = Q_\gamma H_1 (1 - Q_\gamma)\Psi_\gamma = Q_\gamma H_1 \Phi_\gamma. \tag{5.1.23}$$

Consequently,

$$\Psi_\gamma = \Phi_\gamma + (E_\gamma - H_0 - Q_\gamma H_1 Q_\gamma)^{-1} Q_\gamma H_1 \Phi_\gamma. \tag{5.1.24}$$

Also, taking the inner product of Φ_γ with equation (5.1.22) yields

$$E_\gamma = \mathscr{E}_\gamma + \langle \Phi_\gamma | H_1 | \Psi_\gamma \rangle. \tag{5.1.25}$$

Combining equation (5.1.24) and (5.1.25), one finds

$$E_\gamma = \mathscr{E}_\gamma + \langle \Phi_\gamma | H_1 + H_1 (E_\gamma - H_0 - Q_\gamma H_1 Q_\gamma)^{-1} Q_\gamma H_1 | \Phi_\gamma \rangle, \tag{5.1.26}$$

which is the common starting point for all perturbation expansions.

The binomial expansion of $(E_\gamma - H_0 - Q_\gamma H_1 Q_\gamma)^{-1}$ in powers of $(Q_\gamma H_1 Q_\gamma)/(E_\gamma - H_0)$ now leads to the Brillouin–Wigner (BW) perturbation theory:

$$E_\gamma = \sum_{n \geq 0} E_\gamma^{[n]}, \tag{5.1.27a}$$

$$E_\gamma^{[0]} = \mathscr{E}_\gamma, \tag{5.1.27b}$$

$$E_\gamma^{[1]} = \langle \Phi_\gamma | H_1 | \Phi_\gamma \rangle, \tag{5.1.27c}$$

$$E_\gamma^{[2]} = \langle \Phi_\gamma | H_1 \frac{Q_\gamma}{\mathscr{E}_\gamma - H_0} H_1 | \Phi_\gamma \rangle, \tag{5.1.27d}$$

$$E_\gamma^{[3]} = \langle \Phi_\gamma | H_1 \frac{Q_\gamma}{E_\gamma - H_0} H_1 \frac{Q_\gamma}{E_\gamma - H_0} H_1 | \Phi_\gamma \rangle, \tag{5.1.27e}$$

$$\cdots.$$

Note that starting with the second order, each term contains the entire series E_γ on the right-hand side. The employment of the BW formalism thus requires the solution of implicit equations. A more tractable form can be obtained if one further expands E_γ, which results in the Rayleigh–Schrödinger (RS) perturbation theory:

$$E_\gamma = \sum_{n \geq 0} E_\gamma^{(n)}, \tag{5.1.28a}$$

$$E_\gamma^{(0)} = \mathscr{E}_\gamma, \tag{5.1.28b}$$

$$E_\gamma^{(1)} = \langle \Phi_\gamma | H_1 | \Phi_\gamma \rangle, \tag{5.1.28c}$$

$$E_\gamma^{(2)} = \langle \Phi_\gamma | H_1 \frac{Q_\gamma}{\mathscr{E}_\gamma - H_0} H_1 | \Phi_\gamma \rangle, \tag{5.1.28d}$$

$$E_\gamma^{(3)} = \langle \Phi_\gamma | H_1 \frac{Q_\gamma}{\mathscr{E}_\gamma - H_0} H_1 \frac{Q_\gamma}{\mathscr{E}_\gamma - H_0} H_1 | \Phi_\gamma \rangle$$

$$- E_\gamma^{(1)} \langle \Phi_\gamma | H_1 \frac{Q_\gamma}{(\mathscr{E}_\gamma - H_0)^2} H_1 | {}_\gamma \Phi \rangle, \tag{5.1.28e}$$

We next move on to investigate how these conventional perturbation formalisms work when applied to infinitely extended systems.

The independent-particle picture of a fluid, as noted in Section 5.1.1, resolves the Hamiltonian (5.1.1) into a kinetic-energy operator H_0 and a potential-energy operator H_1. The unperturbed wave functions are the eigenfunctions of H_0, namely, properly symmetrized products of single-particle orbitals $\varphi_{\mathbf{k}_i\sigma_i}(\mathbf{r}_i s_i)$, equation (5.1.15). In this basis it is easy to see that for the ground state (i.e., $\gamma = 0$), $\mathscr{E}_0$ and $\langle \Phi_0 | H_1 | \Phi_0 \rangle$ are proportional to N as they should be, since energy is necessarily extensive. Beginning with the second order, one must analyze both the BW and the RS energy series with some care. It turns out that $E_0^{[2]}$ is of order N^0, while $E_0^{(2)}$ remains proportional to N, the difference being in the energy denominator. For an arbitrary intermediate state β, the RS energy denominator is $(\mathscr{E}_0 - \mathscr{E}_\beta)$, which is of order N^0, while for BW one has $(\mathscr{E}_0 + E_0^{[1]} + E_0^{[2]} + \cdots - \mathscr{E}_\beta) = E_0^{[1]} + E_0^{[2]} + \cdots + O(N^0) = O(N)$. A term-by-term analysis reveals that the BW energy corrections $E_0^{[n]}$ beyond $n=2$ are all of order N^0, which offhand suggests rapid convergence of the BW expansion. Actually the convergence is quite superficial; contributions from each set of about N terms accumulate to form a significant sum. However, in practice it is only possible to evaluate contributions from a few terms; hence the BW energy series, at least for the ground state of an extended system, is totally useless. On the other hand, the same term-by-term analysis applied to the RS expansion shows that in each order beyond $n = 2$, $E_0^{(n)}$ contains expressions proportional to N^ν, $\nu \geq 1$. At first sight, divergence of the series is clearly indicated. However, carefully constructed algebraic identities can be found to demonstrate that all divergent terms cancel exactly within each order of the perturbation expansion: Certain choices of the intermediate states give rise to the unlinked terms which diverge; such terms are, in turn, cancelled by corresponding unlinked parts of the renormalization corrections like the last term in equation (5.1.28e). It was suggested by Brueckner [1] and proved by Goldstone [3] that as in the S-matrix expansion of quantum field theory these unlinked terms bear no physical reality and *must* cancel to all orders. Therefore, provided that $E_0^{(n)}$ numerically diminish with n, the RS expansion is a valid perturbation theory for extended systems. Unfortunately, for systems characterized by strong interparticle correlations, this last provision fails to materialize.

There are three types of interparticle correlations to be considered. The first type originates from quantum degeneracy. For a system of bosons, Bose–Einstein condensation causes a macroscopic fraction of particles to be deposited in the zero-momentum orbital. As intermediate states are constructed for evaluating perturbation corrections, some of these particles are removed from the condensate. A typical energy denominator that results is

$$\mathscr{E}_0 - \mathscr{E}_{\mathbf{p},-\mathbf{p}} = \mathscr{E}_0 - (\mathscr{E}_0 + 2e_p) = -\frac{\hbar^2 p^2}{m}, \tag{5.1.29}$$

where $\mathscr{E}_{\mathbf{p},-\mathbf{p}}$ represents the unperturbed energy of an intermediate state with paired excitations of momenta $\mathbf{p}$ and $-\mathbf{p}$. Since there is an inexhaustible supply of condensate particles, such energy denominators appear all too frequently. Furthermore the condensate particles can emerge singly to make an encounter with an excitation of momentum $\mathbf{p}$ before returning to the condensate. Passive encounters of this nature also contribute factors of p^2 to the energy denominator. The subsequent summation over all intermediate states, which corresponds to integration over $\mathbf{p}$, thus diverges at the lower limit $p = 0$.

The second type is dynamical in nature and arises from long-range correlations due to, for example, the Coulomb interaction. In this case, even for fermions there is a piling up of p^2 in the denominator, where $\mathbf{p}$ now stands for the momentum transfer. I shall refrain from describing Coulomb divergences since they lie outside the scope of present discussion. The interested reader is referred to Reference [1].

The third and most common type is that due to "hard cores"—strong interparticle repulsions at short range. Such repulsive cores are present between each pair of molecules, and between each pair of nucleons. They are characteristic of all multimolecular and nuclear systems. If the molecules rarely find themselves in close proximity of one another, such as in the case of a gas in which the density is sufficiently low, or in the case of an ordinary crystal in which molecules are permitted only small excursions beyond their lattice sites, the omission of these repulsive cores will hardly matter. In liquids and quantum crystals, however, particles constantly find themselves nearly overlapping one another. The effect of the hard cores becomes devastating.

For helium, the pairwise interaction at long range dies off in accordance with van der Waals' r^{-6} law. At close approach, Pauli's exclusion principle imposed on the electronic configuration gives rise to strong repulsion approximately representable by a power law $r^{-\alpha}$, $\alpha \gtrsim 10$. A most popular form of the interaction is the Lennard-Jones 6–12 potential

$$v_{LJ}(r) = 4\varepsilon\left[\left(\frac{\sigma}{r}\right)^{12} - \left(\frac{\sigma}{r}\right)^{6}\right], \tag{5.1.30a}$$

$$= \varepsilon\left[\left(\frac{r_0}{r}\right)^{12} - 2\left(\frac{r_0}{r}\right)^{6}\right], \tag{5.1.30b}$$

where ε and σ, or ε and r_0, are numerical parameters that can be determined empirically. Early work by de Boer and Michels [4] gives

$$\varepsilon = 10.22 \text{ K}, \quad \sigma = 2.556 \text{ Å}, \tag{5.1.31a}$$

or

$$\varepsilon = 10.22 \text{ K}, \quad r_0 = 2.869 \text{ Å}, \tag{5.1.31b}$$

where the energy unit has a Boltzmann constant k_B omitted for convenience (see Figure 5.1.1). It is questionable that these values are the best available. In fact, even the form $v_{LJ}(r)$ may not be all that desirable. Later I shall return briefly to this topic, but for now equation (5.1.30) suffices for illustrating the source of difficulty. The matrix elements of H_1 diverge along with the Fourier transform of $v_{LJ}(r)$, rendering the energy series divergent in every order. Even if we argue that r^{-12} is too arbitrary and could be replaced by a softer, Fourier-transformable core, the latter would still be so large that the energy series rapidly diverges. The fault does not lie with the perturbation expansion, but with the independent-particle picture and the unrealistic resolution of the Hamiltonian; the interparticle forces are just too important to be treated as mere perturbation.

Over the years a large number of many-body methods have been developed for treating strongly correlated systems. They vary widely in approach and in underlying philosophy. As yet no unified formalism has arisen. Field theoretic techniques, which began with Brueckner's rearrangement of the RS

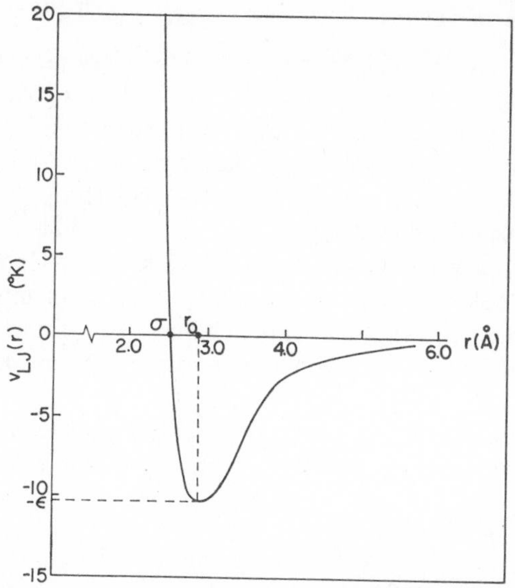

FIGURE 5.1.1. Lennard-Jones potential with parameters determined by deBoer and Michels.

series and culminated in the elegance of Green's functions, have as their common aim the removal of the spurious divergences. Variational calculations such as Jastrow's theory (first proposed by Bijl) and its generalization, and perturbation methods in the correlated representation (invented by Feenberg and Clark), seek to circumvent the difficulty by discarding the independent-particle resolution. Other methods are somewhat less popular, and are usually restricted to treating model systems. Within the limited space available, I intend to concentrate on microscopic theories which are directly applicable to realistic helium systems. The other methods, if not totally left out, will be mentioned in passing only for pedagogical reasons and for the sake of continuity.

5.2. GROUND STATE OF LIQUID ^{4}HE

Of the isotopes of helium, only ^{4}He and ^{3}He are stable. ^{4}He is by far the more abundant. For over half a century, various macroscopic quantum phenomena associated with liquid helium have alternately challenged and enlightened the physicist. In its capacity to help unravel the intricacies of quantum principles, liquid He^4 has played and continues to play a virtually unsurpassable role. In these two sections I shall review various microscopic calculations for realistic bulk He^4, first in regard to the ground state, then the elementary excitations.

5.2.1 Properties of Interest

Helium atoms have filled electronic shells. Their excitation energies exceed 19 eV. At liquid helium temperatures they can be safely regarded as spherically symmetric and practically structureless. The interaction between a pair of helium atoms must, therefore, be in the nature of induced multipoles, and is exceedingly weak. Expressed in the Lennard-Jones form, the well-depth is only 10.22 K. This should be compared to the interaction between other rare gas atoms such as argon; the well-depth in which case is 119.6 K. This very weak attraction coupled to the effect of large zero point motion, resulting from small atomic masses, causes bulk helium to be extraordinarily fluidic. It remains a liquid under its own vapor pressure even in the limit of zero temperature.

At zero temperature, by calculating the energy E_0 of the system as a function of the number density ρ, one obtains the equation of state

$$p = -\left(\frac{\partial F}{\partial \Omega}\right)_T$$

$$\to -\frac{\partial E_0}{\partial \Omega} = \rho^2 \frac{d(E_0/N)}{d\rho}, \quad T \to 0 \tag{5.2.1}$$

and from which the isothermal compressibility κ_T and the velocity of sound s:

$$\kappa_T = -\left(\Omega \frac{dp}{d\Omega}\right)_T^{-1} = \left(\rho \frac{dp}{d\rho}\right)_T^{-1}, \tag{5.2.2}$$

$$s = \frac{1}{\sqrt{m\rho\kappa_T}}. \tag{5.2.3}$$

If the calculation of $E_0(\rho)$ is sufficiently accurate, one finds it possible to determine the Grüneisen constant as well:

$$u = \frac{\rho}{s} \frac{ds}{d\rho}. \tag{5.2.4}$$

At equilibrium, $p = 0$. Solving equation (5.2.1) with $p = 0$ yields the equilibrium density ρ_0 and the chemical potential

$$\mu = \frac{E_0(\rho_0)}{N}. \tag{5.2.5}$$

These are some of the equilibrium thermodynamic properties associated with an isotropic, infinitely extended liquid at zero temperature.

Turning away from macroscopic quantities, it is also possible to inquire about the microsopic structure of the liquid. One defines in the configuration

TABLE 5.2.1. Parameters for Bruch–McGee's hybrid He–He potentials[a]

		ε'	R	c	d
Reference [29] (1967)	v_M	12.14	2.975	6.22697	3.59754
	v_R	13.04	2.97	9.05533	3.46696
	v_{FM}	12.54	2.98	8.00877	3.51078
Reference [30] (1970)	v_M	10.754	3.0238	6.12777	3.68280

[a]Energies are in degree Kelvin and distances in angstrom units.

space the zero-temperature n-particle distribution function

$$P_n(\mathbf{r}_1, \mathbf{r}_2, \cdots, \mathbf{r}_n) \equiv \rho^n g_n(\mathbf{r}_1, \mathbf{r}_2, \cdots, \mathbf{r}_N) = \frac{N!}{(N-n)!} \frac{\int |\Psi_0|^2 \, d\tau_{(1,2,\cdots,n)}}{\int |\Psi_0|^2 \, d\tau}, \qquad (5.2.6)$$

where Ψ_0 denotes the ground-state eigenfunction, $d\tau$ represents the $3N$-dimensional differential element $d\mathbf{r}_1 d\mathbf{r}_2 \cdots d\mathbf{r}_N$, and $d\tau_{(1,2,\cdots,n)}$ stands for $d\tau$ with $d\mathbf{r}_1 d\mathbf{r}_2 \cdots d\mathbf{r}_n$ omitted. Given Ψ_0, one should be able to evaluate $P_n(\mathbf{r}_1, \mathbf{r}_2, \cdots, \mathbf{r}_n)$, in particular the radial distribution function

$$g(r_{12}) \equiv g_2(r_{12}) \equiv \frac{P_2(\mathbf{r}_1, \mathbf{r}_2),}{\rho^2}, \qquad (5.2.7)$$

which is related to the zero-temperature liquid structure function $S(k)$

$$S(k) = 1 + \rho \int [g(r) - 1] e^{i\mathbf{k}\cdot\mathbf{r}} \, d\mathbf{r}. \qquad (5.2.8)$$

More generally, at any given temperature T the liquid structure function $S(k,T)$ is defined as the thermal average

$$S(k, T) = \left\langle \frac{1}{N} \rho_{\mathbf{k}}^{+} \rho_{\mathbf{k}} \right\rangle_T, \qquad (5.2.9)$$

where $\rho_{\mathbf{k}}$ is the density fluctuation operator

$$\rho_{\mathbf{k}} = \int \rho(\mathbf{r}) e^{i\mathbf{k}\cdot\mathbf{r}} \, d\mathbf{r} = \int \sum_{i=1}^{N} \delta(\mathbf{r} - \mathbf{r}_i) e^{i\mathbf{k}\cdot\mathbf{r}} \, d\mathbf{r} = \sum_{i=1}^{N} e^{i\mathbf{k}\cdot\mathbf{r}_i}. \qquad (5.2.10)$$

$S(k,T)$ can be measured over a wide range of wave vectors and temperature by X-ray and neutron-scattering experiments, and provides a sensitive tset for every microscopic theoretical calculation. In the hydrodynamic regime,

$$S(k,T) = \frac{\hbar k}{2ms} \coth \frac{\hbar s k}{2k_B T} \to \frac{k_B T}{ms^2} + \frac{\hbar^2 k^2}{8mk_B T}, \quad k \to 0 \qquad (5.2.11)$$

while

$$S(k) = \lim_{T \to 0} S(k, T) = \frac{\hbar k}{2ms}; \; k \to 0. \qquad (5.2.12)$$

Equation (5.2.11) is the well-known Ornstein–Zernike relation.

The distribution in the momentum space $N_{\mathbf{k}}$ is also of interest. In particular, one may wish to compute the fraction of atoms in the zero-momentum orbital: the Bose–Einstein condensate fraction $f_0 \equiv N_0/N$. Unlike the case of a noninteracting Bose system, the condensate is much depleted in liquid ^{4}He; yet its occupation, in all likelihood, remains macroscopic. By expressing the one-particle density matrix

$$\rho_1(\mathbf{r}-\mathbf{r}') = N\frac{\int \Psi_0^*(\mathbf{r}', \mathbf{r}_2, \cdots, \mathbf{r}_N)\, \Psi_0(\mathbf{r}, \mathbf{r}_2, \cdots, \mathbf{r}_N)\, d\tau_{(1)}}{\int |\Psi_0(\mathbf{r}_1, \mathbf{r}_2, \cdots, \mathbf{r}_N)|^2\, d\tau} \tag{5.2.13}$$

in the occupation number representation, thus,

$$\begin{aligned}\rho_1(\mathbf{r}-\mathbf{r}') &= \left\langle 0 \left| \frac{1}{\sqrt{\Omega}} \sum_{\mathbf{k}'} a_{\mathbf{k}'}^\dagger e^{-i\mathbf{k}'\cdot\mathbf{r}'} \frac{1}{\sqrt{\Omega}} \sum_{\mathbf{k}} a_{\mathbf{k}}\, e^{i\mathbf{k}\cdot\mathbf{r}} \right| 0 \right\rangle \\ &= \sum_{\mathbf{k}\mathbf{k}'} e^{-i\mathbf{k}'\cdot\mathbf{r}'+i\mathbf{k}\cdot\mathbf{r}} \langle 0 | a_{\mathbf{k}'}^\dagger a_{\mathbf{k}} | 0\rangle\, \langle 0|a_{\mathbf{k}}^\dagger \cdot a_{\mathbf{k}}|0\rangle/\Omega \\ &= \sum_{\mathbf{k}} e^{i\mathbf{k}\cdot(\mathbf{r}-\mathbf{r}')} N_{\mathbf{k}}/\Omega \\ &= \frac{N_0}{\Omega} + \frac{1}{(2\pi)^3} \int N_{\mathbf{k}} e^{i\mathbf{k}\cdot(\mathbf{r}-\mathbf{r}')}\, d\mathbf{k},\end{aligned} \tag{5.2.14}$$

one recognizes f_0 through the relation

$$f_0 = \frac{N_0}{\Omega}\frac{\Omega}{N} = \frac{1}{\rho} \lim_{r\to\infty} \rho_1(r). \tag{5.2.15}$$

Unfortunately, available knowledge concerning f_0 is meager.

Central to the calculation of all the properties discussed above is the ground-state wave function Ψ_0, which is the solution of equation (5.1.2) corresponding to the lowest eigenvalue E_0. For boundary condition one might employ the periodic condition:

$$\Psi_0(\mathbf{r}_1 + \mathbf{L}, \mathbf{r}_2 + \mathbf{L}, \cdots, \mathbf{r}_N + \mathbf{L}) = \Psi_0(\mathbf{r}_1, \mathbf{r}_2, \cdots, \mathbf{r}_N). \tag{5.2.16}$$

To account for exchange symmetry, one further requires

$$\Psi_0(\mathbf{r}_1, \mathbf{r}_2, \mathbf{r}_3, \cdots, \mathbf{r}_N) = \Psi_0(\mathbf{r}_2, \mathbf{r}_1, \mathbf{r}_3, \cdots, \mathbf{r}_N). \tag{5.2.17}$$

5.2.2. Jastrow's Theory

A time-honored method for obtaining the ground-state solution of a Schrödinger equation is the variational principle. The idea is to construct a sensible many-body trial function, evaluate the energy expectation value, and then minimize the latter with respect to the variational function. This yields an upper bound to the ground-state energy. By extending the trial wave function space, the true ground-state energy is approached from above, in principle, to whatever accuracy one desires.

The simplest many-body wave function that can be constructed is a properly symmetrized product of single-particle functions

$$\psi^{(1)}(1, 2, \cdots, N) = P\{\prod_{i=1}^{N} \varphi(i)\}, \tag{5.2.18}$$

where P denotes the symmetrization operator. Unlike equations (5.1.8)–(5.1.10), $\varphi(i)$ is not restricted to eigenstates of some known single-particle portion of the complete many-body Hamiltonian. It is totally arbitrary except for boundary conditions, and may be parameterized in any way deemed sensible. For fluids, translational invariance requires that momentum be a good quantum number, hence the only sensible single-particle functions are plane waves. In the presence of hard cores, then, the expectation value of the Hamiltonian diverges, rendering the wave function $\psi^{(1)}$ useless.

The next simplest variational function is a properly symmetrized product of two-particle functions:

$$\psi^{(2)}(1, 2, \cdots, \mathrm{N}) = P\{\prod_{i<j=1}^{N} \chi(i, j)\}. \tag{5.2.19}$$

It is possible to factor our the single-particle elements without loss of generality. Thus one writes

$$\psi^{(2)}(1, 2, \cdots, N) = P\{\prod_{l=1}^{N} \varphi(l) \prod_{i<j=1}^{N} f(i, j)\}. \tag{5.2.20}$$

In the case of liquid ^{4}He, there are no spin variables and the forces are central. The two-particle function $f(i, j)$, or $f(\mathbf{r}_i,\mathbf{r}_j)$, will therefore appear with spherical symmetry, taking on the form $f(r_{ij})$. Since the product of such functions is by nature symmetrical with respect to exchange, it may be taken outside the symmetrization operation, leaving the exchange property to be accounted for by the single-particle product. Thus,

$$\psi^{(2)}(1, 2, \cdots, N)= \prod_{i<j=1}^{N} f(r_{ij})\cdot P\{\prod_{l=1}^{N} \varphi(l)\}. \tag{5.2.21}$$

For the ground state, each of the plane waves reduces to unity. What remains is just a product of pair functions. Such a function is known as the Bijl–Dingle–Jastrow wave function [5–6], or simply the Jastrow function

$$\psi_J(1, 2, \cdots, N) \equiv \psi_0^{(2)}(1, 2, \cdots, N) = \prod_{i<j=1}^{N} f(r_{ij}) \equiv \prod_{i<j=1}^{N} \exp[\tfrac{1}{2}u(r_{ij})]. \tag{5.2.22}$$

By choosing $f(r)$ to vanish as the hard cores overlap, the divergence of the energy expectation value

$$E_J = \frac{\langle\psi_J|H|\psi_J\rangle}{\langle\psi_J|\psi_J\rangle} \tag{5.2.23}$$

can be prevented. The Jastrow function turns out to be the simplest usable trial function that arises under our systemmatic extension of the cluster-function space.

The "correlation factor" $f(r)$ possesses certain asymptotic properties. At short range, it should reduce to the solution of a two-particle Schrödinger equation. Assume for the sake of illustration a pair interaction of the Lennard-Jones 6–12 form. One finds the potential dominated by the r^{-12} term as $r \to 0$. The two-particle Schrödinger equation has a solution only if $f(r)$ takes on the form $\exp(-A/r^5)$. At long range, with the interparticle correlation rapidly diminishing, the correlation factor approaches a constant. An analysis of its asymptotic behavior can be carried out, based on the quantization of classical sound field, as summarized below [8].

Let $\varepsilon[\rho(\mathbf{r},t)]$ represent the energy density. The instantaneous, local fluid density $\rho(\mathbf{r},t)$ is permitted to deviate by small amounts from the average density ρ. Taylor-expanding the total energy about equilibrium, one finds

$$E[\rho(\mathbf{r}, t)] - E[\rho] \approx \frac{1}{2}\int\left[\frac{\delta^2\varepsilon}{\delta\rho(\mathbf{r}, t)^2}\right]_\rho \delta^2\rho(\mathbf{r}, t)\, d\mathbf{r}, \tag{5.2.24}$$

which can be identified to the elastic or deformation potential energy U. The thermodynamic relations (5.2.2)–(5.2.3) enable one to express it in the form

$$U = \frac{1}{2\rho^2\kappa_T}\int[\rho(\mathbf{r}, t) - \rho]^2\, d\mathbf{r}. \tag{5.2.25}$$

Along with the kinetic energy

$$K = \tfrac{1}{2}m\rho\int v^2(\mathbf{r}, t)\, d\mathbf{r}, \tag{5.2.26}$$

where $v(\mathbf{r},t)$ is the velocity field, a hydrodynamic Hamiltonian is obtained:

$$\begin{aligned} H &= K + U \\ &= \frac{1}{2_\rho\kappa_T N}\sum_{\mathbf{k}} q_{\mathbf{k}}^* q_{\mathbf{k}} + \frac{m\rho}{2\Omega}\sum_{\mathbf{k}} v_{\mathbf{k}}^* v_{\mathbf{k}} \\ &= \frac{ms^2}{2N}\sum_{\mathbf{k}} q_{\mathbf{k}}^* q_{\mathbf{k}} + \frac{m}{2N}\sum_{\mathbf{k}}\frac{1}{k^2}\dot{q}_{\mathbf{k}}^*\dot{q}_{\mathbf{k}}, \end{aligned} \tag{5.2.27}$$

where $q_{\mathbf{k}}(t)$ and $v_{\mathbf{k}}(t)$ are, respectively, the Fourier transforms of $[\rho(\mathbf{r}, t) - \rho]$ and $v(\mathbf{r}, t)$. The last line of equation (5.2.27) has been derived with the help of the equation of continuity

$$\dot{q}_{\mathbf{k}} = i\,\rho\, k v_{\mathbf{k}}. \tag{5.2.28}$$

Define next the generalized coordinate $Q^{\mathbf{k}}$ so that

$$Q_{\mathbf{k}} = \frac{1}{\sqrt{2}}(q_{\mathbf{k}} + q_{-\mathbf{k}}) \tag{5.2.29a}$$

in the upper half of **k** space,

$$Q_{\mathbf{k}} = \frac{i}{\sqrt{2}}(q_{\mathbf{k}} - q_{-\mathbf{k}}) \tag{5.2.29b}$$

in the lower half of **k** space, and the generalized momentum

$$P_{\mathbf{k}} = \frac{m}{Nk^2}\dot{Q}_{\mathbf{k}}. \tag{5.2.30}$$

The Hamiltonian emerges in the familiar harmonic form

$$H = \frac{ms^2}{2N}\sum_{\mathbf{k}} Q_{\mathbf{k}}^2 + \frac{N}{2ms^2}\sum_{\mathbf{k}}(sk)^2 P_{\mathbf{k}}^2. \tag{5.2.31}$$

The equation of motion

$$\begin{cases} \dot{P}_{\mathbf{k}} = \dfrac{-\partial H}{\partial Q_{\mathbf{k}}} = \dfrac{-ms^2}{N} Q_{\mathbf{k}} & (5.2.32a) \\[2ex] \dot{Q}_{\mathbf{k}} = \dfrac{\partial H}{\partial P_{\mathbf{k}}} = \dfrac{Nk^2}{m} P_{\mathbf{k}} & (5.2.32b) \end{cases}$$

then combine to yield

$$\dot{Q}_{\mathrm{k}} + (sk)^2 Q_{\mathrm{k}} = 0, \tag{5.2.33}$$

from which one finds the longitudinal phonon frequency

$$2\pi\nu(k) = sk. \tag{5.2.34}$$

Equation (5.2.31) suggests that at long range the wave function takes on the asymptotic form

$$\psi_\infty = \prod_{\mathbf{k}} \psi_{\mathbf{k}}, \tag{5.2.35}$$

where

$$\psi_{\mathbf{k}} = \exp\left[-\frac{ms}{2N\hbar k} Q_{\mathbf{k}}^2\right] = \exp\left[-\frac{ms}{2N\hbar k}\left|\rho_{\mathbf{k}}\right|^2\right]. \tag{5.2.36}$$

Chester and Reatto [9] pointed out that the cutoff limit on k beyond which the hydrodynamic approximation fails may be introduced through a cutoff parameter k_c, thus,

$$\psi_\infty = \prod_{\mathbf{k}} \exp\left[-\frac{ms\exp(-k/k_c)}{2N\hbar k}\left|\rho_{\mathbf{k}}\right|^2\right]; \tag{5.2.37}$$

furthermore, the exponents in equation (5.2.37) may be Fourier-transformed to configuration space to convert ψ_∞ into Jastrow's form. In this manner, the asymptotic behavior of $f(r)$ is established:

$$f(r) \to \exp\left[-\frac{mc}{4\pi^2\rho\hbar}\frac{1}{r^2 + k_c^{-2}}\right], r \to \infty. \tag{5.2.38}$$

In principle, the parameter k_c can be determined variationally. Francis, Chester, and Reatto [10] reported that a numerical calculation led to $k_c \approx 0.5$ Å^{-1}. Independent work by Miller [11], however, failed to confirm this value. It was found that the energy of the system does not depend sensitively upon the detailed long-range behavior of the wave function. This conclusion should not be surprising since the contribution to the energy comes mainly from the intermediate range in which the potential-well resides. Consequently the variational principle which relies on the energy as the sole criterion has very little to say about the asymptotic behavior of the wave function. Also, Miller [11] showed that most of the available schemes for evaluating the energy expectation value are inconsistent with the hydrodynamic behavior if one employs equation (5.2.38) with a finite k_c.

Having chosen $f(r)$, the problem reduces to a purely technical one: How does one evaluate the many-body integrals that occur in the expectation value of the energy?

First of all, E_J of equation (5.2.23) can be cast in a simpler form. One applies an integration by parts (frequently referred to as the Jackson–Feenberg transformation [12]) to the kinetic energy. For real ψ_J,

$$\sum_{i=1}^{N} \frac{-\hbar^2}{2m} \int \psi_J \nabla_i^2 \psi_J \, d\tau = \sum_{i=1}^{N} \frac{-\hbar^2}{4m} \left[\int \psi_J \nabla_i^2 \psi_J \, d\tau - \int \nabla_i \psi_J \cdot \nabla_i \psi_J \, d\tau \right]$$

$$= \sum_{i=1}^{N} \frac{-\hbar^2}{4m} \int \psi_J^2 \nabla_i^2 \ln \psi_J \, d\tau. \qquad (5.2.39)$$

Consequently,

$$E_J = \frac{N\rho}{2} \int \tilde{v}(r) g_J(r) \, d\mathbf{r} = N(2\pi\rho) \int_0^\infty \tilde{v}(r) g_J(r) r^2 \, dr, \qquad (5.2.40)$$

where

$$\tilde{v}(r) = v(r) - \frac{\hbar^2}{4m} \nabla^2 u(r), \qquad (5.2.41)$$

and $g_J(r)$ is as defined in equations (5.2.6)–(5.2.7) with ψ_J approximating ψ_0. Offhand, it seems that we have merely traded off the $3N$-fold integration for E_J in return for an equally unpleasant $(3N - 6)$-fold integration for $g_J(r)$. Actually, it proves to be a crucial move: $g_J(r)$ has a well-known counterpart in the theory of classical liquids. For each liquid characterized by a pairwise intermolecular potential $v(r_{ij})$, a radial distribution function at temperature T can be defined:

$$g_{cl}(r_{ij}, T) = \frac{N(N-1)}{\rho^2} \frac{\int \exp\left[\sum_{k<l} - v(r_{kl})/k_B T\right] d\tau_{(i,j)}}{\int \exp\left[\sum_{k<l} - v(r_{kl})/k_B T\right] d\tau}. \qquad (5.2.42)$$

The replacement of $[-v(r_{kl})/k_B T]$ by $u(r_{kl})$ leads immediately to $g_J(r_{ij})$. Thus the zero-temperature Bose liquid problem in the Jastrow formulation can be neatly mapped onto a finite-temperature classical liquid problem. In one sweep the entire arsenal of theoretical tools developed for classical liquids over the years is acquired. Among the techniques that can be borrowed are cluster expansion procedures, integral equations for summing cluster series, molecular dynamics, and Monte Carlo methods. All are approximation schemes that have withstood the rigorous test of time.

Let us first look at cluster-expansion procedures. Many such procedures are found in the literature, all of them being variations of the Ursell–Meyer expansion. A most direct derivation of the latter was given by van Kampen [13]. One begins with an auxiliary two-particle operator

$$X = \sum_{i<j=1}^{N} x(r_{ij}). \tag{5.2.43}$$

The expectation value of X with respect to the Jastrow function is given by

$$\langle X\rangle = \frac{\sum_{i<j}\int x(r_{ij})\exp[\sum_{k<l}u(r_{kl})]d\tau}{\int\exp[\sum_{k<l}u(r_{kl})]\,d\tau} = \frac{\rho^2}{2}\int x(r_{ij})g_J(r_{ij})d\mathbf{r}_i d\mathbf{r}_j. \tag{5.2.44}$$

By calculating $\langle X\rangle$ via an expansion procedure, $g_J(r)$ can be extracted from the resulting expression.

Define a generalized normalization integral

$$W(\beta) = \int\prod_{i<j}\exp[\beta x(r_{ij}) + u(r_{ij})]\,d\tau \equiv \int\prod_{i<j} q_{ij}\,d\tau = \overline{\prod_{i<j} q_{ij}}. \tag{5.2.45}$$

q_{ij} is a shorthand notation for $\exp[\beta x(r_{ij}) + u(r_{ij})]$, and the average is taken thus:

$$\overline{A(1,2,\cdots,n)} = \int A(1,2,\cdots,n)d\tau_{(n+1,\cdots,N)} \equiv \int A(1,2,\cdots,n)\,d\mathbf{r}_1 d\mathbf{r}_2\cdots d\mathbf{r}_n, \tag{5.2.46}$$

for any n-particle function $A(1, 2, \cdots, n)$. We note that

$$\langle X\rangle = \left[\frac{d}{d\beta}\ln W(\beta)\right]_{\beta=0} \tag{5.2.47}$$

It is therefore necessary and sufficient to develop $W(\beta)$. In the lowest cumulant approximation,

$$W_2(\beta) = \prod_{i<j}\overline{q_{ij}}. \tag{5.2.48}$$

Next,

$$W_3(\beta) = W_2(\beta)\prod_{i<j<k}\frac{\overline{q_{ij}q_{jk}q_{ki}}}{\overline{q_{ij}}\;\overline{q_{jk}}\;\overline{q_{ki}}} \equiv W_2(\beta)\prod_{i<j<k}\frac{\overline{q_{ij}q_{jk}q_{ki}}}{W_2^{ijk}(\beta)}, \tag{5.2.49}$$

where the notation $W_n^{i_1, i_2, \cdots, i_m}(\beta)$ has been introduced to mean $W_n(\beta)$ defined for a system of m particles: $i_1, i_2, \cdots, i_m$. Next, the following obtains:

$$W_4(\beta) = W_3(\beta) \prod_{i<j<k<l} \frac{\overline{q_{ij}q_{ik}q_{il}q_{jk}q_{jl}q_{kl}}}{W_3^{ijkl}(\beta)}$$

$$= W_3(\beta) \prod_{i<j<k<l} \frac{\overline{q_{ij}q_{ik}q_{il}q_{jk}q_{jl}q_{kl}}\,\overline{q_{ij}}\,\overline{q_{ik}}\,\overline{q_{il}}\,\overline{q_{jk}}\,\overline{q_{jl}}\,\overline{q_{kl}}}{\overline{q_{ij}q_{jk}q_{ki}}\;\overline{q_{ij}q_{jl}q_{li}}\;\overline{q_{ik}q_{kl}q_{li}}\;\overline{q_{jk}q_{kl}q_{lj}}}, \tag{5.2.50}$$

etc. And finally,

$$W_N(\beta) = W_{N-1}(\beta) \prod_{1\cdots N} \frac{\overline{\prod_{i<j} q_{ij}}}{W_{N-1}^{N}(\beta)} = W(\beta). \tag{5.2.51}$$

By truncating the expansion of $W(\beta)$ at the nth order, $W_n(\beta)$, one finds $\langle X \rangle$ approximated by $\langle X \rangle_n$:

$$\langle X \rangle_n \equiv \left[\frac{d}{d\beta} \ln W_n(\beta). \right]_{\beta=0}. \tag{5.2.52}$$

In the present case

$$\langle X \rangle^2 = \sum_{i<j} \left[\frac{d}{d\beta} \ln \overline{q_{ij}} \right]_{\beta=0} = \frac{1}{2} \sum_{ij}' \frac{\int x(r_{ij}) f^2(r_{ij})\, d\mathbf{r}_i\, d\mathbf{r}_j}{\int f^2(r_{ij})\, d\mathbf{r}_i\, d\mathbf{r}_j}, \tag{5.2.53}$$

$$\langle X \rangle_3 = \langle X \rangle_2 + \sum_{i<j<k} \left[\frac{d}{d\beta} (\ln \overline{q_{ij}q_{jk}q_{ki}} - \ln \overline{q_{ij}} - \ln \overline{q_{jk}} - \ln \overline{q_{ki}}) \right]_{\beta=0}$$

$$= \langle X \rangle_2 + \frac{1}{2} \sum_{ijk}' \left[\frac{\int x(r_{ij}) f^2(r_{ij}) f^2(r_{jk}) f^2(r_{ki})\, d\mathbf{r}_i\, d\mathbf{r}_j\, d\mathbf{r}_k}{\int f^2(r_{ij}) f^2(r_{jk}) f^2(r_{ki})\, d\mathbf{r}_i\, d\mathbf{r}_j\, d\mathbf{r}_k} - \frac{\int x(r_{ij}) f^2(r_{ij})\, d\mathbf{r}_i\, d\mathbf{r}_j}{\int f^2(r_{ij})\, d\mathbf{r}_i\, d\mathbf{r}_j} \right], \tag{5.2.54}$$

etc., where the primes on the summation signs indicate that no two indices are equal. From equations (5.2.52)–(5.2.54), one obtains $g_J(r_{ij})$. Note that the quantity

$$\eta \equiv \int [f^2(r) - 1]\, d\mathbf{r} \tag{5.2.55}$$

is of order unity (Ω^0). By retaining only terms of order Ω^0 in the expansion of $g_J(r_{ij})$, one obtains, after some algebra,

$$g_J(r_{ij}) = f^2(r_{ij}) + \rho f^2(r_{ij}) \int [f^2(r_{jk}) - 1]\,[f^2(r_{ki}) - 1]\, d\mathbf{r}_k + \cdots, \tag{5.2.56}$$

the first term on the right having arisen from $\langle X \rangle_2$, the second term from $\langle X \rangle_3 - \langle X \rangle_2$, etc.

The cluster expansion is an expansion in the "hole function" $[f^2(r) - 1]$ as well as the density ρ. There is no way to define a *numerical* expansion parameter, but a measure of magnitudes, if there is one, is clearly given by $\rho\eta$. Since the purpose of introducing $f(r)$ into the wave function is to prevent the overlap of hard cores, it is expected that at small r, $f(r)$ must quickly drop

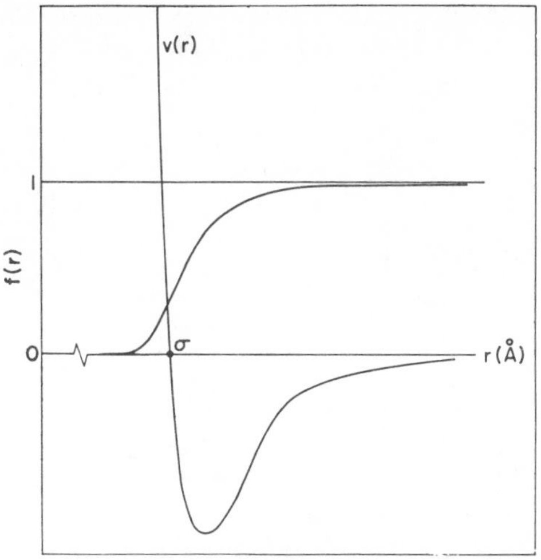

FIGURE 5.2.1. Two-particle correlation factor $f(r)$ for liquid helium.

off to zero as the repulsion builds up, which means that in the case of helium the hole function approaches -1 as r approaches σ from the right, as shown in Figure 5.2.1. This gives us an estimate of η:

$$\eta \approx \frac{4\pi}{3}(0.8\sigma)^3 = 36 \text{ Å}^3. \tag{5.2.57}$$

With $\rho = 0.0218$ Å^{-3} at equilibrium, $\rho\eta \approx 0.8$. The cluster expansion is, therefore, expected to converge very slowly at best.

The theory of classical liquids guides us toward a number of summation techniques. Integral equations relating $f(r)$ and $g_J(r)$ can be derived [14], which upon iteration reproduce the cluster series equation (5.2.56). By using approximation schemes, these integral equations can be put in forms that are solvable at least by numerical means. The most reliable and perhaps best-understood equation is that of Bogoliubov, Born, Green, Kirkwood, and Yvon (BBGKY). It is obtained simply by applying the gradient operator ∇_1 to the definition of $g_J(r_{12})$ and introducing in the resulting expression the three-particle distribution function

$$g_{3J}(\mathbf{r}_1, \mathbf{r}_2, \mathbf{r}_3) \equiv g_{3J}(r_{12}, r_{23}, r_{31}) = \rho^{-3}P_3(\mathbf{r}_1, \mathbf{r}_2, \mathbf{r}_3). \tag{5.2.58}$$

One finds

$$\nabla_1 g_J(r_{12}) = g_J(r_{12})\nabla_1 u(r_{12}) + \rho\int g_{3J}(r_{12}, r_{23}, r_{31})\nabla_1 u(r_{13})\, d\mathbf{r}_3. \tag{5.2.59}$$

This is really just the first of an hierarchy of integral equations. By applying ∇_1 to g_{3J} and introducing g_{4J} result, in the one obtains the second in the hierarchy, and so on. As they stand, these equations are exact but useless. For actual application, the hierarchy must be truncated by means of some approximate closure relation. Abe [15] showed that g_{3J} can be expressed as a functional of $g_J(r)$, thus,

$$g_{3J}(r_{12}, r_{23}, r_{31}) = g_J(r_{12})g_J(r_{23})g_J(r_{31}) \exp A(1, 2, 3), \qquad (5.2.60)$$

where

$$A(1, 2, 3) = \rho \int h_J(r_{14})h_J(r_{24})h_J(r_{34})d\mathbf{r}_4 + \cdots, \qquad (5.2.61)$$

and

$$h_J(r_{ij}) \equiv g_J(r_{ij}) - 1. \qquad (5.2.62)$$

$A(1, 2, 3)$ is an infinite series: an expansion in $\rho h_J(r)$. If one chooses to neglect A,

$$g_{3J}(r_{12}, r_{23}, r_{31}) \approx g_J(r_{12})g_J(r_{23})g_J(r_{31}) \qquad (5.2.63)$$

gives rise to what is known as the Kirkwood superposition approximation (KSA). On the other hand, if one neglects $h_J(r_{12})h_J(r_{23})h_J(r_{31})$ but includes the first term in A,

$$\begin{aligned} g_{3J}(r_{12}, r_{23}, r_{31}) \approx{} & 1 + h_J(r_{12}) + h_J(r_{23}) + h_J(r_{31}) \\ & + h_J(r_{12})h_J(r_{23}) + h_J(r_{23})h_J(r_{31}) + h_J(r_{31})h_J(r_{12}) \\ & + \rho \int h_J(r_{14})h_J(r_{24})h_J(r_{34})d\mathbf{r}_4 \end{aligned} \qquad (5.2.64)$$

gives rise to what is known as the convolution approximation (CA). The choice of terms to be retained is somewhat arbitrary. One hopes, however, to follow as many of the guidelines listed below as possible and still come up with a manageable form for g_{3J}. First of all, like $g_J(r)$,g_{3J} must vanish whenever two or more of the three particles overlap. Next, if one of the three particles is very far away from the remaining pair, g_{3J} must reduce to g_J for the pair. When all three particles are far apart, g_{3J} must reduce to unity. Also, from the definitions, there is the sequential relation

$$\int g_{3J}(r_{12}, r_{23}, r_{31})d\mathbf{r}_3 = \frac{N-2}{\rho} g_J(r_{12}) \qquad (5.2.65)$$

to be satisfied. Finally, g_{3J} should be symmetric in its three arguments. Neither the KSA, nor the CA, nor any other scheme comes close to satisfying all of these conditions. Such is the price for computational expediency.

Substituting for example the KSA in equation (5.2.59) and formally iterating, bearing in mind at every step the normalization of $g_J(r)$,

$$\frac{\rho^2}{N(N-1)} \int g_J(r_{12})\, d\mathbf{r}_1\, d\mathbf{r}_2 = 1, \tag{5.2.66}$$

one recovers the first two terms in the cluster series for $g_J(r)$, namely equation (5.2.56). The reader can demonstrate that beginning with four-particle clusters certain terms will be missing. A pattern of selection can indeed be established for every approximation scheme.

Other popular integral equations include the Percus–Yevick equation (PY) and the hypernetted chain (HNC):

$$\ln g_J(r) = u(r) + \ln\,[1 + p(r)], \qquad \text{(PY)} \quad (5.2.67)$$

$$\ln g_J(r) = u(r) + p(r), \qquad \text{(HNC)} \quad (5.2.68)$$

where

$$p(r) = \frac{1}{(2\pi)^3 \rho} \int \frac{[S_J(k) - 1]^2}{S_J(k)} e^{-i\mathbf{k}\cdot\mathbf{r}}\, d\mathbf{k}, \tag{5.2.69}$$

with $S_J(k)$ given by equation (5.2.8) for $g_J(r)$. These equations have been shown [16] less accurate than the BBGKY equation, but are more convenient to use, especially if only crude estimates are desired, for the following reason. Assuming a one-to-one correspondence between $u(r)$ and $g_J(r)$, one might vary $g_J(r)$ in the variational calculation instead of $u(r)$. In that case the BBGKY equation remains an integral equation for $u(r)$, but the PY and the HNC yield $u(r)$ without the need of solving integral equations. In terms of mathematical efforts and computer time, tremendous savings may be reaped with some sacrifice of accuracy. There is however no guarantee that a one-to-one correspondence exists between $u(r)$ and $g_J(r)$. Certain conditions must be met in order for a $g_J(r)$ to be realizable from a wave function [17, 18], and even then the conditions are merely necessary, not sufficient. (From our experience it appears that the conditions are not very restrictive. Of the many sensible-looking $g_J(r)$ constructed by various authors, few $g_J(r)$ which violate one of these conditions have so far been identified.)

Other forms of equations relating $u(r)$ and $g_J(r)$ are available but have not achieved a broad base of support. The interested reader is referred to References [19–22]. It must be borne in mind that good quantitative agreement with experiment does not suffice to guarantee the validity of a new equation, nor does a mere lowering of the energy expectation value. After all, the variational principle breaks down as soon as an approximation scheme is employed. It is always advisable to analyze the approximation against the complete cluster series and to show that the omitted terms do not contribute significantly. A most helpful and, indeed, indispensable tool for such an analysis is the use of diagrams, as discussed in standard texts [14] on statistical mechanics and classical liquids.

In order to avoid the type of errors consciously committed when terms from the cluster series are selectively summed, one must resort to stochastic methods, among which we find molecular dynamics and Monte Carlo procedures.

The analogy between the ground-state Jastrow problem and the classical liquid problem once again plays a prominent role. In the latter, the evaluation of $g_{cl}(r_{ij}, T)$ or the partition function makes use of the canonical ensemble. Recognizing the equivalence of the canonical ensemble with the microcanonical ensemble and the validity of the ergodic theorem, one replaces the ensemble average by time average and keeps track of the evolution of a classical system of particles toward equilibrium. This constitutes the basis for molecular dynamics. In more detail, one begins by setting a system of N particles at coordinates $(\mathbf{r}_1, \mathbf{r}_2, \cdots, \mathbf{r}_N)$ in volume Ω and giving each a momentum $\mathbf{p}_i$. The total classical energy E_{cl} is immediately determined, as the sum of the kinetic energies $p_i^2/2m$ and the total potential energy $\sum_{i<j} v(r_{ij})$. The motion of the particles will henceforth be regulated by the N coupled equations of motion

$$m \frac{d^2\mathbf{r}_i}{dt^2} = - \sum_{j(\neq i)} \nabla_i v(r_{ij}); \quad i = 1, 2, \cdots, N. \tag{5.2.70}$$

As the system progresses toward equilibrium, the total energy E_{cl} does not change, but a redistribution between kinetic and potential energies continuously takes place. After the elapse of a sufficient length of time, depending on the initial conditions and the strength of the interparticle forces, equilibrium is achieved. At that time one collects whatever information desired from the resulting configuration. In particular, the n-particle distribution functions are readily available. The statistical error can be determined by allowing the configuration to fluctuate about equilibrium.

The $g_{cl}(r_{ij}, T)$ thus determined represents the classical radial distribution function at temperature T. However, T cannot be predetermined. At the start of the "experiment," the system was not at equilibrium and T was not defined. One must await the conclusion of the experiment to discover T by setting

$$E_{cl} = \tfrac{3}{2} N k_B T + \langle \sum_{i<j} v(r_{ij}) \rangle_T. \tag{5.2.71}$$

This causes a bit of embarassment to the liquid-helium theorist who wishes to exploit the classical analogy for variational purpose; the correspondence $[- v(r_{ij})/k_B T] \leftrightarrow u(r_{ij})$ cannot be established until the calculation nears completion [23]. Moreover, the assumption of ergodicity requires the approach to the thermodynamic limit. While such a limit is central to the formulation of the many-body problem, the computer-experimenter has at his disposal but a finite amount of time and memory. He takes a small number of particles while employing periodic boundary conditions at the walls of a normalizing cell, making certain that the range of interparticle force does not exceed one-half the side of his cell. The statistical uncertainty can be embarassingly large if the number of particles is too small. In recent work the number has exceeded 800. As in all experimental work, the molecular dynamicist relies heavily upon past experience and intutition, and is, in his own right, an artist.

The Monte Carlo method is in philsophy quite different from molecular dynamics. One actually constructs a canonical ensemble of systems and computes ensemble averages. Since time plays no part in the calculation, momentum variables do not enter. The temperature T is fixed from the outset, but not the classical energy E_{cl}, for again one has equation (5.2.71), but each configuration of particles commands a different potential energy. It is a fundamental principle of statistical mechanics that in the complete set of configurations, or in a statistically significant set of random configurations, the distribution peaks sharply around the equilibrium. The fluctuation in E_{cl} is amall. This being the case, the statistical uncertainty decreases with increasing number of trial configurations.

If the trial configurations are chosen completely at random, the Boltzmann factor in equation (5.2.42) will yield extremely small values most of the time. Only those which are very close to equilibrium will contribute. Nearly all the time the computer will be made to generate garbage information. Clearly the usual Monte Carlo integration technique must give way to some intelligently designed biased selection procedure. The most widely accepted is the biased random walk [24].

Let the statistical weight $\exp[\sum_{i<j} - v(r_{ij})/k_BT]$ used in computing ensemble average be denoted by the probability distribution $W(R)$, where R denotes a configuration $(\mathbf{r}_1, \mathbf{r}_2, \cdots, \mathbf{r}_N)$. It is then possible to replace the ensemble average by a weighted sum, as long as the distribution of configurations within the sum is proportional to $W(R)$. One begins with an arbitrary configuation $R^{(1)}$, and constructs a new configuration R' by moving one of the N particles from $\mathbf{r}_i$ to

$$\mathbf{r}_i' = \mathbf{r}_i + \vec{\xi}d, \tag{5.2.72}$$

where $\vec{\xi}$ is a random vector with components sampled from the interval $(-1, 1)$. One now computes and compares $W(R^{(1)})$ and $W(R')$, and decides whether to accept R' as the second configuration $R^{(2)}$ in his ensemble. The decision is rendered in accordance with the following strategy:

$W(R')/W(R^{(1)}) \geq 1?$ yes $\longrightarrow R^{(2)} = R'$

no $\downarrow$

Choose η at random from the interval (0, 1) $\longrightarrow \left[W(R')/W(R^{(1)})\right] \geq \eta?$ yes $\rightarrow R^{(2)} = R'$; no $\rightarrow R^{(2)} = R^{(1)}$

The consequence is that, if the new configuration is a more probable one, it is accepted; if the new configuration is less probable than the starting configuration, it is accepted only with the probability $W(R')/W(R^{(1)})$. As this step is repeated, one advances toward regions of higher probability.

The proof that a long walk in this fashion will lead to the proper distribution of configurations is straightforward. Consider a configuration $R^{(i)}$, and a second configuration $R^{(j)}$ that is within one step's reach from $R^{(i)}$. Let there be, respectively, ν_i and ν_j configurations in the ensemble. One then wishes to show that

$$\nu_i/\nu_j = \frac{W(R^{(i)})}{W(R^{(j)})}. \tag{5.2.73}$$

Let the probability of moving from $R^{(i)}$ to $R^{(j)}$ be denoted by ω_{ij} and let $W(R^{(i)})/W(R^{(j)}) \geq 1$. According to the prescription given above, the number of systems in the ensemble which actually move from $R^{(i)}$ to $R^{(j)}$ equals $\nu_i\omega_{ij}\ W(R^{(j)})/W(R^{(i)})$, while the number of systems taking the opposite path is $\nu_j\omega_{ji}$. At equilibrium these numbers must be equal. And since $\omega_{ij} = \omega_{ji}$, we have equation (5.2.73) as desired.

The biased random walk method has a practical difficulty: At high densities it is possible to trap a small system in a configurational subspace far from equilibrium. One must, therefore, experiment with several starting configurations.

An alternate and exceedingly efficient method has been developed by Coldwell [25]. The essential idea is that in constructing each configuration, the particles are located one at a time. As each particle is located, the region around it that falls within its hard core diameter is excluded from the view of the next particle to be located. Thus the hard cores will never overlap, and each configuration that can be constructed will contribute. What remains to be carried out is a minimal amount of probability calculation to account for the bias. This method is capable of reaching much higher densities than both the usual Monte Carlo integration procedure and the biased random walk, and is consequently suitable for studying phase changes in classical systems. Also, it can compute with high accuracy small shifts in $g_{cl}(r_{ij},T)$, E_{cl}, or other ensemble averages, as $v(r)$ undergoes a slight change [26].

Let us now discuss several practical aspects of the Jastrow calculations before turning to a brief review of theoretical results.

The interaction potential between a pair of He atoms is central, and is rather easy to determine from second virial coefficient available over a wide range of temperatures. Since the Lennard-Jones 6–12 potential [equation (5.1.30)] with parameters determined by de Boer and Michels [equation(5.1.31)] seems to work reasonably well, and since the determination of interatomic potentials is generally regarded as in the realm of atomic and chemical physics, the helium theorist has concentrated his efforts strictly on the many-body aspects of the problem. Unlike the nuclear many-body theorist, he has grown all too complacent about the reliability of the only microscopic information available to him. An overwhelming majority of the variational calculations for liquid helium have been carried out with the deBoer–Michels potential, partly for the reasons given above, and partly for the more legitimate reason that a comparison of results is meaningful only if a common starting point is adopted. As will be seen shortly (in Table 5.2.2), there is general consistency in the results obtained, which in turn agree reasonably well with experiment. Some small discrepancies remain unacounted for, and may well be due to the quality of the input information.

There are two aspects to reconsider. First, does the deBoer-Michels potential provide the best fit to virial coefficients data? Secondly, what about the nonadditivity of multiatomic forces?

As for the best fit to the virial coefficients, we list without comment several other choices which are known to fit the data equally well.

i. Haberlandt [27]: Lennard–Jones 6–12 form, with $\varepsilon = 10.30$ K, $\sigma = 2.610$ Å$(r_0 = 2.929$ Å$)$.

ii Yntema–Schneider [28]: $v(r) = b\exp(-r/d) - (c_1/r)^6 - (c_2/r)^8$,

$$b = 8.7 \times 10^6 \text{ K},\ d = 0.212 \text{ Å},\ c_1 = 4.56 \text{ Å},\ c_2 = 3.29 \text{ Å}. \qquad (5.2.74)$$

TABLE 5.2.2. Typical results from Jastrow theory based on the deBoer–Michels potential

Method	Reference	$\frac{E_0(\rho_0)}{N}$ [K]	ρ_0[Å^{-3}]	s[m/sec] at $\rho = \rho_0^{\text{expt}}$	u
Integral equation	Massey–Woo [40, 44] (BBGKY)	–6.06	0.0229	257	2.96
	Francis–Chester–Reatto [10] (PY2XS, SR)	–6.29	0.0211	—	—
Molecular dynamics	Schiff–Verlet [41, 44]	–5.95	0.0196	248	2.26
Monte Carlo	McMillan [24]	–5.9	0.0194	267	—
	Murphy–Watts [42]	–5.96	0.0195	—	—
"Euler–Lagrange equation"	Campbell–Feenberg [38] (based on Massey–Woo)	–6.74	Fixed at 0.0229	—	—
	Pokrant [37]	–6.63	0.0205	258	—
Experiment	Abraham et al. [43]	–7.14	0.0218	238	2.84

iii. Bruch–McGee [29, 30] two-parameter hybrid potentials:

$$v(r) = v_{\text{emp}}(r), \quad r \geq R$$
$$= v_{l.r.}(r), \quad r > R. \tag{5.2.75}$$

$v_{\text{emp}}(r)$ is a three-parameter empirical potential that takes on one of three forms:

Morse: $$v_M(r) = \varepsilon'\left\{\exp\left[2c\left(1 - \frac{r}{d}\right)\right] - 2\exp\left[c\left(1 - \frac{r}{d}\right)\right]\right\}, \tag{5.2.76a}$$

Rydberg: $$v_R(r) = -\varepsilon'\left[1 + c\left(\frac{r}{d} - 1\right)\right]\exp\left[-c\left(\frac{r}{d} - 1\right)\right], \tag{5.2.76b}$$

Frost–Musulin: $$v_{FM}(r) = -\varepsilon'\left[1 + c\left(1 - \frac{d}{r}\right)\right]\exp\left[-c\left(\frac{r}{d} - 1\right)\right], \tag{5.2.76c}$$

and $v_{l.r.}(r)$ is the long-range interaction given by dipole–dipole and dipole–quadrupole terms in the multipole expansion

$$v_{l.r.}(r) = \left[-\left(\frac{c_1}{r}\right)^6 - \left(\frac{c_2}{r}\right)^8\right]\text{K}, \quad c_1 = 4.66\ \text{Å}, \quad c_2 = 3.59\ \text{Å}. \tag{5.2.77}$$

Two (c and d) of the four parameters (c, d, ε', and R) are determined by matching $v_{\text{emp}}(r)$ to $v_{l.r.}(r)$ continuously at R:

$$v_{\text{emp}}(R) = v_{l.r.}(R), \tag{5.2.78a}$$

$$\left[\frac{dv_{\text{emp}}}{dr}\right]_{r=R} = \left[\frac{dv_{l.r.}}{dr}\right]_{r=R}. \tag{5.2.78b}$$

Table 5.2.1 lists the values of parameters ε' and R that best fit the second virial coefficients.

iv. Beck [31]: $v(r) = A\exp(-\alpha r - \beta r^6) - \dfrac{0.869}{(r^2 + a^2)^3}\left(1 + \dfrac{2.709 + 3a^2}{r^2 + a^2}\right)$,

$$A = 46.39\ \text{K}, \alpha = 4.390\ \text{Å}^{-1}, \beta = 3.746 \times 10^{-4}\ \text{Å}^{-6}, a = 0.675\ \text{Å}. \tag{5.2.79}$$

v. Sposito [32]: Equations (5.2.75), (5.2.76a) and (5.2.77), with

$$R = 3.6\ \text{Å}, \varepsilon' = 9.25\ \text{K}, c = 6.2059, d = 2.948\ \text{Å}, c_1 = 4.36\ \text{Å},$$
$$c_2 = 3.58\ \text{Å}. \tag{5.2.80}$$

As for the nonadditivity of interatomic forces, the only calculations done were for the long-range triple-dipole dispersive interaction [33, 34]:

$$v_{\text{t.d.}}(r_{12}, r_{23}, r_{31}) = \nu\,\frac{1 + 3\cos(\mathbf{r}_{12}, \mathbf{r}_{13})\cos(\mathbf{r}_{21}, \mathbf{r}_{23})\cos(\mathbf{r}_{31}, \mathbf{r}_{32})}{(r_{12}r_{23}r_{31})^3}, \tag{5.2.81}$$

with

$$\nu = 1.5 \times 10^3 \text{ K Å}^9.$$

I discussed earlier the asymptotic properties of the correlation factor $f(r)$, or alternately $u(r)$. Let us now turn our attention to the actual parameterization and optimization of $f(r)$. As r_{ij} approaches zero, the pair i and j become removed from the rest of the system, and as a consequence $g_J(r_{ij})$ reduces to $f^2(r_{ij})$. This is clear from equation (5.2.56). Figure 5.2.1 and equation (5.2.40) then tell us that the region inside the core ($r \lesssim \sigma$) contributes little to the energy integral. Also, since both $v(r)$ and $\nabla^2 u(r)$, rapidly diminish outside $r = 6$ Å, the long-range part contributes little to E_J. Consequently, as far as variational calculations are concerned, the sensitive region is from 2 to 6 Å, in which $f(r)$ must be accurately determined. The asymptotic behavior of $f(r)$ is of mere academic interest, and need not be overly concerned with in the process of parameterization [35].

The ideal and rigorous way to determine $f(r)$ is by deriving and solving the Euler–Lagrange equation

$$\frac{\delta E_J}{\delta f(r)} = 0,$$

which unfortunately is coupled to the defining equation of $g_J(r)$, (5.2.6)–(5.2.7). One could replace the latter by one of the approximate integral equations (BBGKY, PY, HNC, etc.) and thereby obtain an optimization equation in closed form. Such an equation is highly nonlinear. Its iterative solution is perilously unstable, and tends to exact extravagant demands on the computer. Nevertheless several authors, in particular those from Broyles' group at Florida [36, 37], have led some valiant assaults in this direction. Their results are included in Table 5.2.2. Jackson, Feenberg and Campbell [10, 38] have taken what eventually turned out to be an alternate path toward optimizing the Jastrow function. Their method, known as the paired phonon analysis, has a perturbation theoretic origin, and will therefore be reviewed in Section 5.2.3. Their results are also included in Table 5.2.2. for comparison.

All the other approaches begin with parameterized forms of $f(r)$, or occasionally $g_J(r)$. The most popular form used is [39]:

$$f(r) = \exp\left[-\left(\frac{a\sigma}{r}\right)^m\right], \tag{5.2.82}$$

where a and m are variational parameters. Invariably the best value of m turns out to be between 4 and 5, with a ranging between 1.1 and 1.2. One advantage of the form (5.2.82) is, as pointed out by McMillan [24], that it permits scaling, so that with one strategic sweep of the parameter space at an arbitrary density, the energy minima for a wide range of densities can be located.

Table 5.2.2 summarizes some of the typical thermodynamic results obtained with Jastrow's theory. Among the entries are results of integral equations, molecular dynamics and Monte Carlo calculations. All are based on the deBoer–Michels potential. Also shown for the purpose of comparison are experimental data. Note that the velocity of sound and the Grüneisen constant quoted in rows 1 and 3 (MW and SV) were obtained after the energy and the equilibrium density had been fitted to experiment [44, 45]. This was necessary in order to make the comparison with experiment meaningful.

As for the contribution from three-body forces—the triple-dipole interaction—Davison and Feenberg [46] made an estimate by mapping liquid ^{4}He onto classical fcc and hcp lattices. In the latter cases, Axelrod [33] found the potential energy due to $v_{\text{t.d.}}$ to be

$$V_3 = \frac{1}{3!} \sum_{ijk}{}' v_{\text{t.d.}}(r_{ij}, r_{jk}, r_{ki}) = N \frac{2.89 \times 10^4}{r_{\text{n.n.}}^9} \text{ K}, \qquad (5.2.83)$$

where $r_{\text{n.n.}}$ denotes the nearest neighbor distance, which Davison and Feenberg (DF) took as 3.2 Å: the location of the first maximum of $g(r)$. The result is then 0.85 K per particle, or approximately 12% of the total energy. Murphy and Barker [47] (MB), however, found by two different methods that the three-body energy V_3 is nearly an order of magnitude lower. A quick estimate was first made via the use of a Monte Carlo program for classical hard spheres. Setting the hard-sphere diameter to $\sigma = 2.556$ Å, V_3 was found to be 0.17 K per particle. Next, a Monte Carlo calculation was carried out

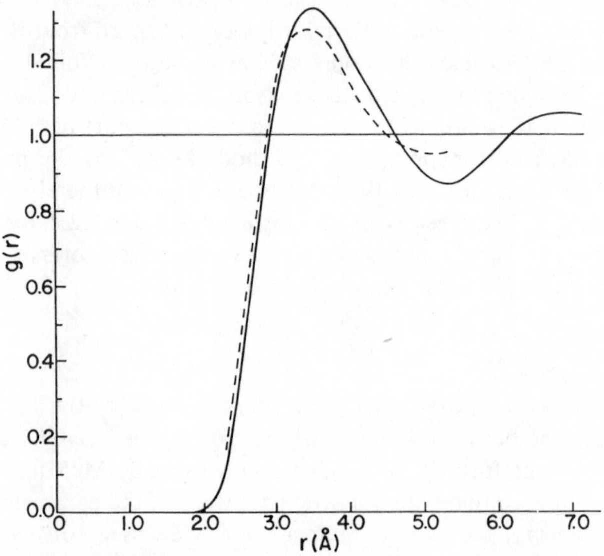

FIGURE 5.2.2. Radial distribution function of liquid ^{4}He.———, Experiment; ---------, optimum Jastrow-theory results.

directly for liquid ^{4}He with the use of a Jastrow function. V_3 turned out to be 0.14 K per particle. The reason for the discrepancy between DF and MB is as follows: For a fcc lattice, $r_{\text{n.n.}} = (1/\sqrt{2})\, a$, a being the lattice constant. Since there are four atoms to a unit cell, the number density ρ is given by $4/a^3$. If one wishes to map liquid ^{4}He onto such a lattice, ρ should be set equal to ρ_0, the equilibrium liquid ^{4}He density. This yields $r_{\text{n.n.}} \approx 4.03$ Å, which is quite different from the location of the first maximum of $g(r)$. Using equation (5.2.83), one then finds, $V_3 \approx 0.1$ K per particle. It should be emphasized that V_3 depends sensitively on $r_{\text{n.n.}}$ in the DF approximation, and on the hard sphere radius in MB's first method of estimation. Actual detailed calculations are laborious, but may well be unavoidable. In any case one can safely concede that the three-body forces do not affect the thermodynamic properties in any significant way.

Figure 5.2.2 shows $g_J(r)$ in comparison with experimental results. The latter came from $S(k)$ measurements by neutron and X-ray scattering experiments. Only one $g_J(r)$ is shown since all Jastrow calculations yield similar results. Note that the maximum is too low and too broad, indicating that the true liquid has more structure than what the Jastrow function is capable of generating. The same story is told by Figure 5.2.3, which compares $S_J(k)$ to the experimental $S(k)$. At small k, $S_J(k)$ should be linear, but no calculation has yet appeared which can account for the *manner* in which $S(k)$ approaches linear behavior. The $S_J(k)$ obtained by Francis, Chester, and Reatto [10] lies much lower than $S(k)$ in this region. One of the earliest triumphs of Jastrow's theory is the demonstration that a shoulder must exist in $S(k)$ in the region 0.4–0.8 Å^{-1}. Soon after Miller, Pines, and Nozieres [48] predicted the occurrence of such a shoulder, Massey [49] reported that every one of the

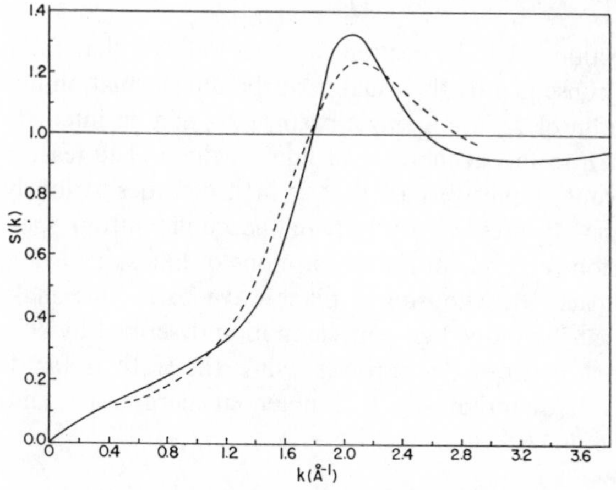

FIGURE 5.2.3. Static structure function of liquid ^{4}He.———, Experiment; - - - - - -, optimum Jastrow-theory results.

$g_J(r)$ that he obtained near the optimum of his variational results upon Fourier transformation to $S_J(k)$ yielded a shoulder in that region, even though $S_J(k)$ did not approach zero linearly. In Section 5.3 I shall present a more thorough discussion of $S(k)$.

The condensate fraction f_0 is difficult to calculate accurately since it depends on the asymptotic behavior of the one-particle density matrix. Estimates [8, 24, 41, 50, 51] based on Jastrow-type calculations range from 8 to 40%. No reliable experimental determination is as yet available. The reader is referred to Section 5.3 and to recent high momentum neutron-scattering work by Harling [52] and by Mook et al. [53]. This is undoubtedly one of the most crucial, unsolved problems of liquid helium. One might even venture to say that it is one of the fundamental problems in physics.

5.2.3. Feenberg's Perturbation Method

The usual formulation of a perturbation theory begins with the resolution of the Hamiltonian H into two parts, as in equation (5.1.3). The eigenfunctions of the unperturbed Hamiltonian H_0 serve as a complete set of basis functions, with respect to which the matrix elements of the perturbation H_1 are calculated. In the independent particle picture of an infinitely extended system, one takes H_0 to be a sum of single-particle operators. The eignefunctions are then products of plane waves. (The quantum numbers consist of momentum vectors, since in view of translational invariance momentum is a good single-particle quantum number.) The evaluation of the matrix elements of H_1 then requires knowledge of the Fourier transform of the two-particle potential $v(r)$:

$$v(k) = \int d\mathbf{r}\, e^{i\mathbf{k}\cdot\mathbf{r}} v(r), \tag{5.2.84}$$

as seen in Section 5.1.1. In Section 5.1.2 we showed that for helium, $v(k)$ diverges, and consequently the usual perturbation calculation fails.

The resolution of H into a single-particle H_0 and an interaction term H_1 can be *expected* to fail in the case of liquid helium. The reason is that $v(r)$ contains a strong, repulsive part that at high densities seriously affects the particle motion. It prevents overlap in the configuration space, much as Pauli's exclusion principle prohibits fermions of like spins to overlap in the momentum space. By choosing a plane-wave basis, one makes the tacit assumption that the individual particle motion described by H_0 overwhelms the correlations between the particles, while the truth is far different. We can also see this quantitatively. The potential energy for liquid helium [see equation (5.2.40)]

$$U = N(2\pi\rho) \int_0^\infty v(r) g(r) r^2\, dr, \tag{5.2.85}$$

evaluated with the aid of, say, a Lennard-Jones 6–12 $v(r)$ and the experimental $g(r)$, equals about -21 K per particle at equilibrium, while the total energy is about -7 K. This leaves 14 K for the kinetic energy K. It is neither like in the gaseous phase, in which the kinetic energy dominates, nor like a usual solid, in which the potential energy dominates. There exists strong cancellation between K and U; neither is capable of overcoming the other.

This does not mean that the ordinary perturbation theory is invalid. What is at fault is the independent particle resolution. A new set of basis functions is needed for handling the strong interparticle correlations. Quantum mechanics harbors no particular favoritism towards the plane-wave basis. In fact, if there is any discrimination at all, it ought to show preference for basis functions that contain as much dynamical information as possible. In the present case, the most relevant dynamical information concerns interparticle correlations. Consequently, it is highly desirable to incorporate such correlations into the basis. Any basis constructed with this purpose in mind is known as a *correlated basis*. The use of a set of correlated basis functions (CBF) to carry out perturbation expansions was first suggested by Feenberg and Clark [54], and has since become a powerful tool in the workshop of the quantum many-body theorist.

The application of CBF to the ground state of liquid ^{4}He is frequently mistaken as an extension of Jastrow's theory. The reason is that a natural way to introduce two-particle correlations into a set of basis functions, and therefore the method employed most often in early attempts, is to modify the independent-particle basis with Jastrow correlation factors. Let the plane-wave basis functions be denoted by

$$\phi_\alpha(1, 2, \cdots, N) \equiv \phi_{\mathbf{k}_1, \mathbf{k}_2, \cdots, \mathbf{k}_N}(1, 2, \cdots, N) \equiv P\left\{\prod_{i=1}^{N} \varphi_{\mathbf{k}_i}(\mathbf{r}_i)\right\}, \qquad (5.2.86)$$

and be used as "model functions". The correlated basis functions can be defined as

$$\Phi_\alpha(1, 2, \cdots, N) = F(1, 2, \cdots, N)\phi_\alpha(1, 2, \cdots, N), \qquad (5.2.87)$$

where in this example

$$F(1, 2, \cdots, N) = \prod_{i<j=1}^{N} f(r_{ij}). \qquad (5.2.88)$$

If the function $f(r)$ is predetermined via a variational calculation for the ground state, F becomes the Jastrow wave function ψ_J (1, 2, $\cdots$, N).

Let us emphasize immediately that the correlation factor F need not be of the Jastrow form ψ_J. In certain calculations it is much more convenient and accurate to employ a correlation factor which is of a different form, as will be seen in, e.g., Section 5.3. Moreover, since ψ_J is a product of two-particle functions in the configuration space, it is capable of handling nothing more then simple, central, two-particle dynamical correlations. While it is quite sufficient for liquid ^{4}He, it is not universally satisfactory. For instance, if there are important three-particle forces, or if the two-particle interaction is spin dependent as in nuclear systems, or momentum dependent, the correlation factor must be completely revamped to reflect these facts. Finally, the correlated basis functions need not even be of the multiplicative form (5.2.87). In short, the better the choice of the basis, the quicker the convergence of the perturbation expansion, the ultimate choice being the true eigenfunctions of the system. There is nothing mechanical about the construction of a set of CBF. In view of these statements, it is clear that in whatever way CBF might have originally been conceived, the philosophy underlying CBF as it stands today will be sadly misinterpreted and its utility lost if one refers to it as an extension or a generalization of Jastrow's theory.

Returning to the ground state of liquid ^{4}He, the prescription outlined in equations (5.2.86)–(5.2.88) does provide us with a usable set of CBF. First, note that the ground-state model function ϕ_0 is obtained by setting all the $\mathbf{k}_i$ to zero. Apart from a normalization constant,

$$\phi_0 = 1. \tag{5.2.89}$$

Next, one constructs a class of low-lying excited states by setting $\mathbf{k}_1$ to $\mathbf{k}$, and all other $\mathbf{k}_i$ to zero. On account of symmetrization, one obtains apart from normalization

$$\phi_{\mathbf{k}} = \sum_i e^{i\mathbf{k}\cdot\mathbf{r}_i} \equiv \rho_{\mathbf{k}}. \tag{5.2.90}$$

Next, with $\mathbf{k}_1 = \mathbf{k}$, $\mathbf{k}_2 = \mathbf{l}$, and $\mathbf{k}_i = 0$ for i > 2, one has the class

$$\phi_{\mathbf{k},\mathbf{l}} = \sum_{i,j} e^{i\mathbf{k}\cdot\mathbf{r}_i} e^{i\mathbf{l}\cdot\mathbf{r}_j} - \sum_i e^{i(\mathbf{k}+\mathbf{l})\cdot\mathbf{r}_i} \equiv \rho_{\mathbf{k}}\rho_{\mathbf{l}} - \rho_{\mathbf{k}+\mathbf{l}}, \tag{5.2.91}$$

and so on. The symmetrization operation caused the independent particle products to become expressible in terms of density fluctuations! Since the second term in $\phi_{\mathbf{k},\mathbf{l}}$, namely $\rho_{\mathbf{k}+\mathbf{l}}$, has already been included in the class of functions $\phi_{\mathbf{k}}$, one loses no generality in omitting it from the definition of the model functions. Consequently,

$$\Phi_0 = \psi_J, \tag{5.2.92a}$$

$$\Phi_{\mathbf{k}} = \rho_{\mathbf{k}}\psi_J \equiv \rho_{\mathbf{k}}\Phi_0, \tag{5.2.92b}$$

$$\Phi_{\mathbf{k},\mathbf{l}} = \rho_{\mathbf{k}}\rho_{\mathbf{l}}\psi_J \equiv \rho_{\mathbf{k}}\rho_{\mathbf{l}}\Phi_0, \tag{5.2.92c}$$

$$\Phi_{\mathbf{k},\mathbf{l},\mathbf{p}} = \rho_{\mathbf{k}}\rho_{\mathbf{l}}\rho_{\mathbf{p}}\psi_J \equiv \rho_{\mathbf{k}}\rho_{\mathbf{l}}\rho_{\mathbf{p}}\Phi_0, \tag{5.2.92d}$$

etc.

are formed. For reasons to become clear in Section 5.3, they will be known as free phonon states or Feynman phonon functions, or simply though inaccurately as "phonons".

We now turn to the evaluation of matrix elements with respect to the free phonon states. In constructing the CBF, we paid no attention to the orthogonality property. Unlike in textbook cases, the basis functions here are in general not orthogonal to one another. ($\Phi_{\mathbf{k}}$, $\Phi_{\mathbf{k},\mathbf{l}}$, with $\mathbf{k} + \mathbf{l} \neq 0$, $\Phi_{\mathbf{k},\mathbf{l},\mathbf{p}}$, with $\mathbf{k} + \mathbf{l} + \mathbf{p} \neq 0$, etc., are all orthogonal to Φ_0 on account of momentum conservation, but are not necessarily orthogonal to one another. For example, states with the same number of phonons are not orthogonal if their momenta are equal, e.g., $\langle \Phi_{\mathbf{k},\mathbf{l}} | \Phi_{\mathbf{p},\mathbf{q}} \rangle \neq 0$ for $\mathbf{k} + \mathbf{l} = \mathbf{p} + \mathbf{q}$.) The eigenstates which we seek must eventually diagonalize both H and 1. It might have been more convenient to begin with a set of functions which are orthogonal. On the other hand, there is no reason why the diagonalization of 1 is any more urgent than H. One might argue that H and 1 should be diagonalized simultaneously, indeed to the same level of accuracy in every approximation. In any case, our first concern is to weave as much physical information into the basis functions as possible. The pursuit of mathematical convenience as can be offered by orthogonality has to be relegated to a secondary role.

Each of the matrix elements is a many-body integral, not unlike the expectation value of H with respect to the Jastrow function ψ_J. In fact, by equation (5.2.92a) the latter *is* a normalized diagonal matrix element of H. Since the radial distribution function $g(r)$ is already known, either from experiment or from a variational calculation, it will suffice to express all matrix elements in terms of $g(r)$ or $S(k)$. The diagonal elements are relatively simple to obtain. For example, for $\langle \Phi_{\mathbf{k}} | H | \Phi_{\mathbf{k}} \rangle$ one uses integration by parts

$$\begin{aligned}
\langle \rho_{\mathbf{k}}\Phi_0 | \nabla_i^2 | \rho_{\mathbf{k}}\Phi_0 \rangle &= \int \rho_{\mathbf{k}}^* \Phi_0 \nabla_i^2 \rho_{\mathbf{k}} \Phi_0 \, d\tau \\
&= \int |\rho_{\mathbf{k}}|^2 \Phi_0 \nabla_i^2 \Phi_0 \, d\tau + \int \Phi_0^2 \rho_{\mathbf{k}}^* \nabla_i^2 \rho_{\mathbf{k}} \, d\tau + \int (\rho_{\mathbf{k}}^* \nabla_i \rho_{\mathbf{k}}) \cdot \nabla_i \Phi_0^2 d\tau \\
&= \int |\rho_{\mathbf{k}}|^2 \Phi_0 \nabla_i^2 \Phi_0 \, d\tau - \int (\nabla_i \rho_{\mathbf{k}}^* . \nabla_i \rho_{\mathbf{k}}) \Phi_0^2 \, d\tau \\
&= \int |\rho_{\mathbf{k}}|^2 \Phi_0 \nabla_i^2 \Phi_0 \, d\tau - k^2 \int \Phi_0^2 \, d\tau,
\end{aligned}$$

to obtain

$$\begin{aligned}
\langle \Phi_{\mathbf{k}} | H | \Phi_{\mathbf{k}} \rangle &= \int |\rho_{\mathbf{k}}|^2 \Phi_0 \left[\sum_i \frac{-\hbar^2}{2m} \nabla_i^2 + \sum_{i<j} v(r_{ij}) \right] \Phi_0 \, d\tau + \sum_i \frac{\hbar^2 k^2}{2m} \int \Phi_0^2 \, d\tau \\
&= E_0 \langle \Phi_{\mathbf{k}} | \Phi_{\mathbf{k}} \rangle + N \frac{\hbar^2 k^2}{2m} \langle \Phi_0 | \Phi_0 \rangle, \qquad (5.2.93)
\end{aligned}$$

provided that Φ_0 is the true ground-state eigenfunction of H. It turns out that even though ψ_J is not the true ground-state wave function, equation (5.2.93) remains valid [55] for the optimum ψ_J, and one finds

$$\langle \Phi_{\mathbf{k}} | H | \Phi_{\mathbf{k}} \rangle = E_J \langle \Phi_{\mathbf{k}} | \Phi_{\mathbf{k}} \rangle + N \frac{\hbar^2 k^2}{2m} \langle \psi_J | \psi_J \rangle. \tag{5.2.94}$$

As for the matrix element of 1, $\langle \Phi_{\mathbf{k}} | \Phi_{\mathbf{k}} \rangle$, one finds

$$\begin{aligned} \langle \Phi_{\mathbf{k}} | \Phi_{\mathbf{k}} \rangle &= \int \rho_{\mathbf{k}}^* \rho_{\mathbf{k}} \Phi_0^2 \, d\tau = N \int \Phi_0^2 \, d\tau + \sum_{i \neq j} \int \Phi_0^2 \, e^{i\mathbf{k}\cdot\mathbf{r}_{ji}} \, d\tau \\ &= N \int \Phi_0^2 \, d\tau \left\{ 1 + \frac{\rho^2}{N} \int [g(r_{ij}) - 1] e^{i\mathbf{k}\cdot\mathbf{r}_{ij}} \, d\mathbf{r}_i \, d\mathbf{r}_j \right\} \\ &= NS(k) \langle \Phi_0 | \Phi_0 \rangle. \end{aligned} \tag{5.2.95}$$

From equations (5.2.94) and (5.2.95) emerges the normalized diagonal matrix element

$$\frac{\langle \Phi_{\mathbf{k}} | H | \Phi_{\mathbf{k}} \rangle}{\langle \Phi_{\mathbf{k}} | \Phi_{\mathbf{k}} \rangle} = E_0 + \frac{\hbar^2 k^2}{2mS(k)}. \tag{5.2.96}$$

The diagonal matrix element in $\Phi_{\mathbf{k},\mathbf{l},\cdots,\mathbf{p}}$ for a finite number of free phonons is given by [55]

$$\frac{\langle \Phi_{\mathbf{k},\mathbf{l},\cdots,\mathbf{p}} | H | \Phi_{\mathbf{k},\mathbf{l},\cdots,\mathbf{p}} \rangle}{\langle \Phi_{\mathbf{k},\mathbf{l},\cdots,\mathbf{p}} | \Phi_{\mathbf{k},\mathbf{l},\cdots,\mathbf{p}} \rangle} = E_0 + \frac{\hbar^2 k^2}{2mS(k)} + \frac{\hbar^2 l^2}{2mS(l)} + \cdots + \frac{\hbar^2 p^2}{2mS(p)} + O\left(\frac{1}{N}\right). \tag{5.2.97}$$

The nondiagonal matrix elements cannot be obtained simply as functionals of $g(r)$ or $S(k)$, except in the case of matrix elements connecting the paired phonon states $\Phi_{\mathbf{k},-\mathbf{k}}$ to Φ_0. But if Φ_0 is the true ground state, the calculation proceeds with little difficulty.

In the evaluation [56] of matrix elements connecting one-phonon states $\Phi_{\mathbf{k}}$ to two-phonon states $\Phi_{\mathbf{k}-\mathbf{l},\mathbf{l}}$, the following products of density fluctuation operators frequently appear, and have the same action as the expressions on the right:

$$\rho_{\mathbf{k}-\mathbf{l}}^\dagger \rho_{\mathbf{l}}^\dagger \to \rho_{\mathbf{k}}^\dagger + N(N-1) \exp(-i\mathbf{k}\cdot\mathbf{r}_1 + i\mathbf{l}\cdot\mathbf{r}_{12}) \tag{5.2.98a}$$

$$\rho_{\mathbf{k}-\mathbf{l}}^\dagger \rho_{\mathbf{l}}^\dagger \rho_{\mathbf{k}} \to |\rho_{\mathbf{k}}|^2 + |\rho_{\mathbf{l}}|^2 + |\rho_{\mathbf{k}-\mathbf{l}}|^2 - 2N + N(N-1)(N-2) \exp(i\mathbf{k}\cdot\mathbf{r}_{31} + i\mathbf{l}\cdot\mathbf{r}_{12}). \tag{5.2.98b}$$

Consequently, two- and three-particle distribution functions appear in the integrals. The three-particle distribution functions must be approximated by one of the standard schemes such as the KSA, equation (5.2.63), or the CA, equation (5.2.64). In either case one finds [56]

$$\langle \Phi_{\mathbf{k}-\mathbf{l},\mathbf{l}} | \Phi_{\mathbf{k}} \rangle = N[1 - S(k) - S(l) - S(|\mathbf{k}-\mathbf{l}|) + S(k)S(l) +$$

$$S(l)\,S(|\mathbf{k}-\mathbf{l}|)+S(|\mathbf{k}-\mathbf{l}|)S(k)]+\mathcal{N}(\mathbf{k},\mathbf{l}). \tag{5.2.99}$$

where

$$\text{in KSA: } \mathcal{N}(\mathbf{k},\mathbf{l})=\frac{\Omega}{(2\pi)^3}\int[S(p)-1][S(|\mathbf{k}-\mathbf{p}|)-1] \cdot[S(|\mathbf{l}-\mathbf{p}|)-1]d\mathbf{p}, \tag{5.2.100a}$$

$$\text{in CA: } \mathcal{N}(\mathbf{k},\mathbf{l})=N[S(k)-1][S(l)-1][S(|\mathbf{k}-\mathbf{l}|)-1]. \tag{5.2.100b}$$

Also,

$$\langle\Phi_{\mathbf{k}-\mathbf{l},\mathbf{l}}|H|\Phi_{\mathbf{k}}\rangle=\left[E_0+\frac{\hbar^2k^2}{2mS(k)}\right]\langle\Phi_{\mathbf{k}-\mathbf{l},\mathbf{l}}|\Phi_{\mathbf{k}}\rangle+N\frac{\hbar^2}{2m}[\mathbf{k}\cdot\mathbf{l}S(|\mathbf{k}-\mathbf{l}|) + \mathbf{k}\cdot(\mathbf{k}-\mathbf{l})S(l)-k^2S(l)S(|\mathbf{k}-\mathbf{l}|)]+\mathcal{H}(\mathbf{k},\mathbf{l}), \tag{5.2.101}$$

where

$$\text{in KSA: } \mathcal{H}(\mathbf{k},\mathbf{l})=N\frac{\hbar^2k^2}{2mS(k)}\Big\{[S(k)-1][S(l)-1][S(|\mathbf{k}-\mathbf{l}|)-1] - \frac{1}{(2\pi)^3\rho}\int[S(p)-1][S(|\mathbf{k}-\mathbf{p}|)-1][S(|\mathbf{l}-\mathbf{p}|) - 1]d\mathbf{p}\Big\}, \tag{5.2.102a}$$

$$\text{in CA: } \mathcal{H}(\mathbf{k},\mathbf{l})=0. \tag{5.2.102b}$$

From experience we find the convolution form more accurate. The convolution approximation need not be limited to just the three-particle distribution function. Wu generalized the CA to all n-particle distribution functions [57], and subsequently obtained [58, 55] the matrix elements of 1 and H between all n- and n'-phonon states using the CA. These general formulas will prove to be of immense help to theorists employing CBF for quantum liquids. To save space, we shall refrain from showing them unless the situation calls for explicit display. The reader should be warned that simple expressions like equations (5.2.99)–(5.2.102) are correct only when Φ_0 is the *exact* ground state wave function. For $\Phi_0=\psi_J$, as in equation (5.2.92), the formulas are much more complicated. Certain reduction of complexity is made possible by having the Jastrow function optimized, but even then there will be additional terms in most matrix elements.

The knowledge of the matrix elements of 1 in principle allows one to orthogonalize the entire basis. In practice, it is only easy to orthogonalize each subclass of the basis functions with respect to other subclasses. This can be accomplished by means of Schmidt's procedure as in the following:

$$|0\rangle = \frac{\Phi_0}{\langle\Phi_0|\Phi_0\rangle} \tag{5.2.103a}$$

$$|\mathbf{k}, -\mathbf{k}\rangle = \frac{(\Phi_{\mathbf{k},-\mathbf{k}} - \Phi_0\langle\Phi_0|\Phi_{\mathbf{k},-\mathbf{k}}\rangle)}{I_1}, \tag{5.2.103b}$$

$$|\mathbf{k}, \mathbf{l}, -\mathbf{k} - \mathbf{l}\rangle = \frac{(\Phi_{\mathbf{k},\mathbf{l},-\mathbf{k}-\mathbf{l}} - \Phi_0\langle\Phi_0|\Phi_{\mathbf{k},\mathbf{l},-\mathbf{k}-\mathbf{l}}\rangle - \sum_{\mathbf{p}\neq 0}\Phi_{\mathbf{p},-\mathbf{p}}\langle\Phi_{\mathbf{p},-\mathbf{p}}|\Phi_{\mathbf{k},\mathbf{l},-\mathbf{k}-\mathbf{l}}\rangle)}{I_2} \tag{5.103c}$$

etc., where $I_1, I_2, \cdots$ represent normalization factors. The states within each subclass remain generally nonorthogonal. This does not matter as long as one stops at second-order perturbation theory. The knowledge of the matrix element $\langle 0|H|\mathbf{k}, \mathbf{l}, \cdots\rangle$ then permits one to proceed toward calculating the second-order correction to E_J.

Let us inspect the *RS* series, equation (5.1.28):

$$E_0 = \mathscr{E}_0 + \langle\Phi_0|H_1|\Phi_0\rangle + \langle\Phi_0|H_1\frac{Q_0}{\mathscr{E}_0 - H_0}H_1|\Phi_0\rangle + \cdots$$

$$= \langle\Phi_0|H_0|\Phi_0\rangle + \langle\Phi_0|H_1|\Phi_0\rangle + \sum_{\gamma\neq 0}\frac{\langle\Phi_0|H_1|\Phi_\gamma\rangle\langle\Phi_\gamma|H_1|\Phi_0\rangle}{\langle\Phi_0|H_0|\Phi_0\rangle - \langle\Phi_\gamma|H_0|\Phi_\gamma\rangle} + \cdots. \tag{5.2.104}$$

In the present case, one does not exactly know how to resolve H into H_0 and H_1 in the configuration representation. One, however, has on hand a complete set of basis functions and all the matrix elements of H. It is possible to choose H_0 and H_1 such that H_0 has only diagonal elements and H_1 only nondiagonal elements .Thus,

$$\langle\Phi_\gamma|H_0|\Phi_\beta\rangle = \langle\Phi_\gamma|H|\Phi_\gamma\rangle\,\delta_{\gamma\beta}, \tag{5.2.105a}$$

$$\langle\Phi_\gamma|H_1|\Phi_\beta\rangle = \langle\Phi_\gamma|H|\Phi_\beta\rangle(1 - \delta_{\gamma\beta}). \tag{5.2.105b}$$

Equation (5.2.104) then reads

$$E_0 = \langle\Phi_0|H|\Phi_0\rangle + \sum_{\gamma\neq 0}\frac{\langle\Phi_0|H|\Phi_\gamma\rangle\langle\Phi_\gamma|H|\Phi_0\rangle}{\langle\Phi_0|H|\Phi_0\rangle - \langle\Phi_\gamma|H|\Phi_\gamma\rangle} + \cdots, \tag{5.2.106}$$

which is just the *RS* perturbation series in matrix representation. If the nondiagonal elements are small, as should be the case when the basis functions contain enough physical information to resemble the true eigenstates of the system, the perturbation series should converge rapidly. In second order, using equation (5.2.92), the ground-state energy of liquid ^{4}He is then given by

$$E_0 = E_J + \frac{1}{2}\sum_{\mathbf{k}}{}'\frac{|\langle 0|H|\mathbf{k}, -\mathbf{k}\rangle|^2}{E_J - \langle\mathbf{k}, -\mathbf{k}|H|\mathbf{k}, -\mathbf{k}\rangle}$$

$$+\frac{1}{6}\sum_{\mathbf{k},\mathbf{l}}{}' \frac{|\langle 0|H|\mathbf{k},\mathbf{l},-\mathbf{k}-\mathbf{l}\rangle|^2}{E_J-\langle \mathbf{k},\mathbf{l},-\mathbf{k}-\mathbf{l}|H|\mathbf{k},\mathbf{l},-\mathbf{k}-\mathbf{l}\rangle}+\cdots, \quad (5.2.107)$$

where the prime on the summation indicates that none of the momentum labels equals zero.

For taking into account perturbative corrections involving paired excitations like the second term on the right hand side of equation (2.5.107), Jackson and Feenberg [12] invented a procedure known as the "paired phonon analysis." In this procedure, one does not make use of the complete set of correlated basis functions, but only those of the form

$$|n_1, n_2) \equiv \rho_{\mathbf{k}}^{n_1}\rho_{-\mathbf{k}}^{n_2}\psi_J, \quad (5.2.108)$$

where ψ_J is a Jastrow function, not necessarily optimized, and n_1 and n_2 are nonnegative integers. In the class described by equation (5.2.108), the momenta of the states considered are $(n_1 - n_2)\hbar\mathbf{k}$. The paired phonon analysis is a procedure for diagonalizing H in the space spanned by such classes of wave functions, to *all* orders of the perturbation theory.

To calculate the matrix elements of H and 1 in this space for an arbitrary Jastrow function, one defines a generalized normalization integral

$$I(\beta)=\int\psi_J^2\exp\left[\beta\sum_{i<j}\tilde{v}(r_{ij})\right]d\tau=\int\exp\left\{\sum_{i<j}[u(r_{ij})+\beta\tilde{v}(r_{ij})]\right\}d\tau, \quad (5.2.109)$$

a generalized distribution function

$$g(r_{12}|\beta)=\frac{N(N-1)}{\rho^2}\int\exp\left\{\sum_{i<j}[u(r_{ij})+\beta\tilde{v}(r_{ij})]\right\}d\tau_{(1,2)}/I(\beta), \quad (5.2.110)$$

and a generalized structure function

$$S(k|\beta)=1+\rho\int[g(r|\beta)-1]\,e^{i\mathbf{k}\cdot\mathbf{r}}\,d\mathbf{r}. \quad (5.2.111)$$

It can be shown [12] that

$$(n_1n_2|\exp\left[\sum_{i<j}\beta\tilde{v}(r_{ij})\right]|n_3, n_4)=N^{1/2(n_1+n_2+n_3+n_4)}\,\delta_{n_1+n_4-n_3-n_2}(n_1+n_4)!$$
$$\cdot I(\beta)[S(k|\beta)]^{n_1+n_4}, \quad (5.2.112)$$

from which one obtains the matrix element $(n_1, n_2|\sum_{i<j}v(r_{ij})|n_3, n_4)$ as the derivative of equation (5.2.112) with respect to β, evaluated at $\beta = 0$. Algebraic manipulations much like those used for deriving equations (5.2.99) and (5.2.101) lead to

$$(n_1, n_2|H-\sum_{i<j}\tilde{v}(r_{ij})|n_3, n_4)=\frac{\hbar^2k^2}{2m}\left\{\frac{n_1n_3+n_2n_4}{(n_1+n_4)S(k)}+\frac{n_1+n_4}{2}\right.$$
$$\left.\left[1-\frac{1}{S(k)}\right]\right\}(n_1, n_2|1|n_3, n_4). \quad (5.2.113)$$

Combining these results, one finds

$$(n_1, n_2|H|n_3, n_4) = \left\{E_J + \frac{n_1n_3 + n_2n_4}{n_1 + n_4}\frac{\hbar^2k^2}{2mS(k)} + (n_1 + n_4)\Omega(k)\right\} \cdot (n_1, n_2|1|n_3, n_4), \tag{5.2.114}$$

where $S(k)$ is identical to $S(k|0)$,

$$\Omega(k) = \frac{S'(k|0)}{S(k)} + \frac{\hbar^2k^2}{4m}\left[1 - \frac{1}{S(k)}\right], \tag{5.2.115}$$

and $S'(k|0)$ denotes $[(d/d\beta)S(k|\beta)]_{\beta=0}$.

The essential step that follows is a linear transformation to orthonormalize the basis $|n_1, n_2)$. Let

$$|s, \nu\rangle = \sum_{n=0}^{\nu} a_{s,\nu;n}|n + s, n), \tag{5.2.116}$$

so that

$$\langle s, \nu|1|s, \nu'\rangle = \delta_{\nu\nu'}. \tag{5.2.117}$$

In this new representation, the matrix elements of H vanish except when the quantum numbers s and ν in the initial and final states are equal, or when ν changes by 1. Under this circumstance, the most convenient way to express the Hamiltonian is in terms of creation and annihilation operators $(c_{\mathbf{k}}^\dagger, c_{\mathbf{k}})$ that obey boson commutation relations, as in the usual treatment of quantum oscillators. Extending such an analysis to all $\mathbf{k}$, one finds

$$H = E_J + \sum_{\mathbf{k}}\left\{\left[\frac{\hbar^2k^2}{2mS(k)} + \Omega(k)\right](c_{\mathbf{k}}^\dagger c_{\mathbf{k}} + c_{-\mathbf{k}}^\dagger c_{-\mathbf{k}}) + \Omega(k)(c_{\mathbf{k}}c_{-\mathbf{k}} + c_{\mathbf{k}}^\dagger c_{-\mathbf{k}}^\dagger)\right\}. \tag{5.2.118}$$

Equation (5.2.118) is in the form of a Bogoliubov Hamiltonian, and can be diagonalized by means of a linear transformation to be discussed in Section 5.3. One of the results is then the correction of the constant term E_J by a term

$$\Delta E = \frac{1}{2}\sum_{\mathbf{k}}\left\{\left[\left(\frac{\hbar^2k^2}{2mS(k)}\right)^2 + \frac{\hbar^2k^2\Omega(k)}{mS(k)}\right]^{1/2} - \frac{\hbar^2k^2}{2mS(k)} - \Omega(k)\right\}, \tag{5.2.119}$$

and an improvement in the ground-state wave function. Campbell and Feenberg [38] showed that the new wave function is still of the Jastrow form. By repeating this procedure many times using the improved wave function as ψ_J, eventually ΔE diminishes toward zero, and the Jastrow function becomes optimized.

Quantitatively, the difficulty of applying the paired-phonon analysis lies with the evaluation of $S(k|\beta)$ and its derivative $S'(k|0)$. Campbell and Feenberg resorted to approximation schemes like the HNC, PY, and BBGKY procedures described earlier in connection with the function $S(k)$, and found

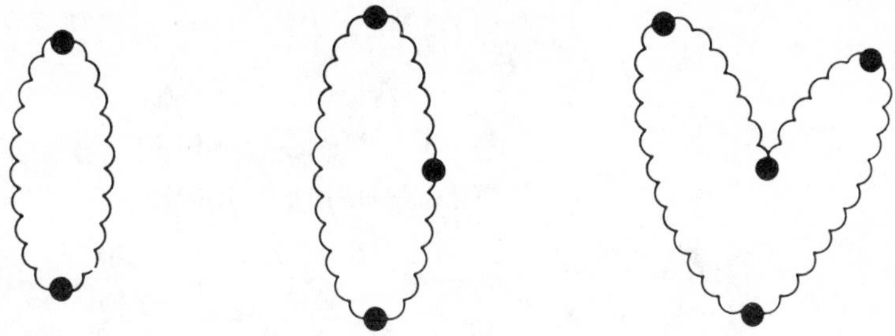

FIGURE 5.2.4. Some typical paired-phonon correction terms to the ground-state energy of liquid ^{4}He.

ΔE diminishing rapidly with each iteration. The total correction amounted to about -0.68 K per particle, which when applied to the variational result by Massey and Woo [40] led to the -6.74 K listed in Table 5.2.2. For the purpose of improving the accuracy, Feenberg and Rahman [59] proposed that a molecular dynamics calculations be carried out for $S'(k\,|\,0)$. The attempt was unsuccessful on account of technical difficulties.

The paired-phonon analysis amounts to summing all contributions to the ground state arising from pair excitations. There remain large classes of terms outside the sum. The leading contribution is given by the third term in equation (5.2.107): a second-order correction involving three-phonon intermediate states. It is convenient for bookkeeping purposes to depict perturbative corrections in diagrams. Using wavy lines to denote free phonons, Figure 5.2.4 shows some typical members of the infinite series summed by the paired-phonon analysis. The term that we shall next consider is shown in Figure 5.2.5.

FIGURE 5.2.5. Three-phonon correction to the ground-state energy of liquid ^{4}He.

The matrix elements required are $\langle 0\,|\,H\,|\,\mathbf{k}, \mathbf{l}, -\mathbf{k}-\mathbf{l}\rangle$ and $\langle \mathbf{k}, \mathbf{l}, -\mathbf{k}-\mathbf{l}\,|\,H\,|\,\mathbf{k}, \mathbf{l}, -\mathbf{k}-\mathbf{l}\rangle$, and can be evaluated [55] in a similar way as the derivation of equations (5.2.99) and (5.2.101). For an optimized ψ_J, Davison and Feenberg [46] found

$$\langle 0\,|\,H\,|\,\mathbf{k}, \mathbf{l}, -\mathbf{k}-\mathbf{l}\rangle = \frac{\hbar^2}{8m}\{-[k^2 + l^2 + |\mathbf{k}+\mathbf{l}|^2]S_J(k)S_J(l)S_J(|\mathbf{k}+\mathbf{l}|)$$

$$+ 2k^2S_J(l)S_J(|\mathbf{k}+\mathbf{l}|) + 2l^2S_J(k)S_J(|\mathbf{k}+\mathbf{l}|)$$
$$+ 2|\mathbf{k}+\mathbf{l}|^2S_J(k)S_J(l) - 2\mathbf{k}\cdot(\mathbf{k}+\mathbf{l})S_J(l)$$
$$- 2\mathbf{l}\cdot(\mathbf{k}+\mathbf{l})S_J(k) + 2\mathbf{k}\cdot\mathbf{l}S_J(|\mathbf{k}+\mathbf{l}|)\}$$
$$\cdot [NS_J(k)S_J(l)S_J(|\mathbf{k}+\mathbf{l}|)]^{1/2}, \tag{5.2.120}$$

$$\langle \mathbf{k}, \mathbf{l}, -\mathbf{k}-\mathbf{l}|H|\mathbf{k}, \mathbf{l}, -\mathbf{k}-\mathbf{l}\rangle = E_J + \frac{\hbar^2k^2}{2mS_J(k)} + \frac{\hbar^2l^2}{2mS_J(l)} + \frac{\hbar^2|\mathbf{k}+\mathbf{l}|^2}{2mS_J(|\mathbf{k}+\mathbf{l}|)}, \tag{5.2.121}$$

and an energy correction of

$$\frac{1}{6N}\sum_{\mathbf{k},\mathbf{l}}{}' \frac{|\langle 0|H|\mathbf{k}, \mathbf{l}, -\mathbf{k}-\mathbf{l}\rangle|^2}{E_J - \langle \mathbf{k}, \mathbf{l}, -\mathbf{k}-\mathbf{l}|H|\mathbf{k}, \mathbf{l}, -\mathbf{k}-\mathbf{l}\rangle} \approx -0.76\text{ K}. \tag{5.2.122}$$

Combining the results of Massey and Woo [40], Campbell and Feenberg [38], and Davison and Feenberg [46], the ground-state energy becomes

$$-6.06 - 0.68 - 0.76\text{ K} = -7.50\text{ K} \tag{5.2.123}$$

per particle, lower than the experimental value of -7.14 K. Even though it appears that one need not worry about violating the variational principle here, since equation (5.2.122) has emerged as a result of a perturbation calculation, in Section 5.2.5 it will be seen that there may well be a contradiction after all. At that point, one must return to investigate several sources of uncertainties in the theory.

At the conclusion of this brief review of ground-state calculations, I must confess that a number of important contributions to the field have been left out. In particular, there is the integral equation for $S(k)$ derived by Mihara and Puff [60], an extension of the Bohm–Salt method by Sposito [61], reaction-matrix calculations by Østgaard [62, 63], a diagrammatic perturbation theory by Brandow [64, 65], a two-particle density matrix approach by Tan and Nosanow [66], etc. Some of these methods [60, 61] require He–He potentials that are Fourier transformable or of certain "soft" forms to assure convergence. Some [60, 62, 63] give rise to numbers that deviate significantly from experiment, and others [64–66] have not yet produced quotable quantitative results. Brandow's theory of interacting bosons promises to play an important part both in the microscopic calculation of liquid helium properties and in unifying Jastrow's theory and CBF with diagrammatic perturbation methods. It will be reviewed briefly in the next section. The rest must be ruled, somewhat arbitrarily, as outside the present scope of discussion which deals strictly with "realistic" helium systems.

5.2.4. Diagrammatic Analysis of Jastrow's and Feenberg's Methods

Jastrow's theory and the method of correlated basis functions have by now been applied to all conceivable quantum liquids, fermion as well as boson, and have in every case achieved considerable quantitative success. They have continued to mystify the many-body theorist trained in the conventional field-theoretic perturbation language. What diagrams are being summed? And to what order are they summed accurately?

Liquid helium is not a weakly interacting Bose gas. Yet, in order to establish connections between various methods in a formal way, one has to work with *some* system for which all methods render solutions. The weakly interacting Bose gas serves such a function in the following comparative analysis.

Consider N bosons in a volume Ω. Denote the number density by ρ. The bosons are each of mass m, and interact via a Fourier-transformable pairwise potential $v(r)$. As given in Section 5.1.1, the Hamiltonian of the system in the second quantized form is

$$H = \sum_{\mathbf{k}} \frac{\hbar^2 k^2}{2m} a_{\mathbf{k}}^{\dagger} a_{\mathbf{k}} + \frac{1}{4\Omega} \sum_{\mathbf{klpq}} [v(|\mathbf{k} - \mathbf{p}|) + v(|\mathbf{k} - \mathbf{q}|)]\delta_{\mathbf{k+l,p+q}} a_{\mathbf{k}}^{\dagger} a_{\mathbf{l}}^{\dagger} a_{\mathbf{p}} a_{\mathbf{q}}, \tag{5.2.124}$$

where $v(k) = \int d\mathbf{r}\, e^{i\mathbf{k}\cdot\mathbf{r}}\, v(r)$. (5.2.125)

The creation and annihilation operators obey boson commutation laws. The only obstacle to an immediate application of the diagrammatic perturbation theory, namely the Brueckner–Bethe–Goldstone linked-cluster analysis, is Bose condensation. The ground state does not form a vacuum with respect to zero-momentum annihilation operators a_0, since, as we know, even in the presence of strong interactions between particles, the occupation in the zero-momentum orbital remains macroscopic. Field-theoretic treatment of the Bose gas by Beliaev [67] and by Hugenholtz and Pines [68] (HP) begins with consideration of the condensate. I shall briefly state as follows the HP prescription.

Following Bogoliubov [69], HP replace $a_0^{\dagger}$ and a_0 by the c number $\sqrt{N_0}$. The condensate occupation number N_0 is restricted by the obvious fact that it may not exceed N, but is otherwise a free parameter. It is to be determined by minimizing the ground-state energy E_0. The replacement of $a_0^{\dagger}$ and a_0 by c-numbers leads to another problem: The Hamiltonian no longer conserves the number of particles. To overcome this difficulty, HP introduce a Lagrange multiplier, the chemical potential μ, to accompany the constraint

$$\langle \sum_{\mathbf{k}\neq 0} a_{\mathbf{k}}^{\dagger} a_{\mathbf{k}} \rangle = N - N_0, \tag{5.2.126}$$

which implies conservation of particle number on the average. The way is now clear for a valid definition of the vacuum, and the linked-cluster theorem follows.

Three kinds of Green's functions are defined, corresponding, respectively, to ground-state expectation values of time-ordered products of one creation and one annihilation operator, two annihilation operators, and two creation operators, denoted, respectively, by G, $\tilde{G}$, and $\bar{G}$. Likewise three types of proper self-energy parts, Σ_{11}, Σ_{02}, and Σ_{20}, are defined. Coupled Dyson-like equations are derived and solved to relate these functions. In terms of the Green's function G, one has

$$\frac{E_0}{\Omega} = \frac{1}{2}\rho\mu + \frac{i}{(2\pi)^4}\int d\mathbf{k}\int_c d\varepsilon \frac{1}{2}\left(\varepsilon + \frac{\hbar^2 k^2}{2m}\right) G(\mathbf{k},\varepsilon), \quad (5.2.127)$$

where

$$\rho = \rho_0 + \frac{i}{(2\pi)^4}\int d\mathbf{k}\int_c d\varepsilon G(\mathbf{k}, \varepsilon), \quad (5.2.128)$$

with the contour c closing in the upper half-plane. The chemical potential μ can be determined by solving the differential equation self-consistently:

$$\mu = \frac{d}{d\rho}\left(\frac{E_0}{\Omega}\right), \quad (5.2.129)$$

or by the theorem

$$\mu = \Sigma_{11}(0,0) - \Sigma_{02}(0,0). \quad (5.2.130)$$

We wish to compare here energy formulas obtained by different methods. It is, therefore, necessary to introduce some identifiable ordering parameter. To this end, a strength parameter λ is attached to $v(r)$, or $v(k)$. All relevant quantities will be expanded in powers of λ. We begin by computing the Green's functions and the proper self-energy parts by perturbation theory, in orders of the number of vertices contained in each diagram, and then carry out a second expansion to rearrange all contributions in orders of λ. First, diagrams are constructed using the seven kinds of vertices obtained from replacing four or less $a_0^\dagger$ and a_0 by $\sqrt{N_0}$ (Figure 5.2.6). These vertices are strung together with free-particle propagators:

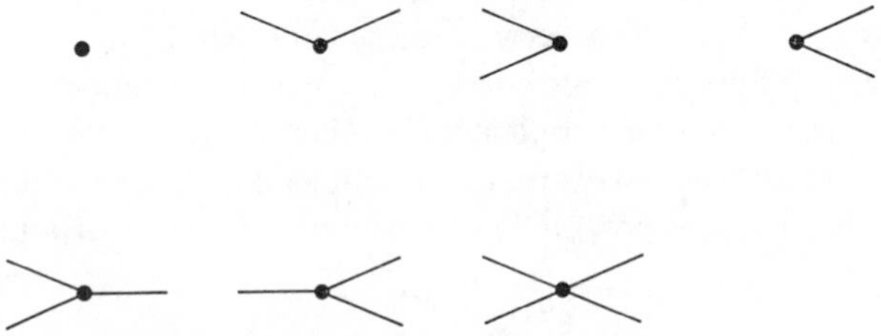

FIGURE 5.2.6. Vertices in the theory of weakly interacting Bose gas.

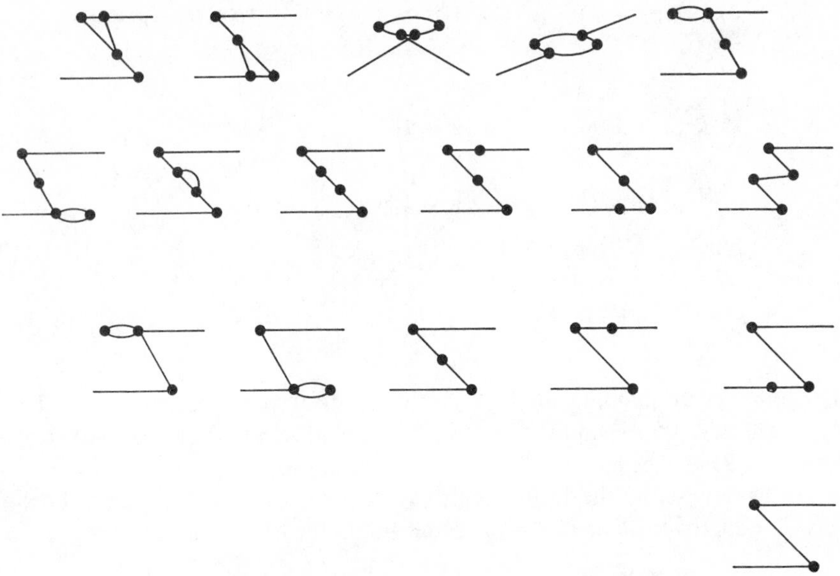

FIGURE 5.2.7. Contributions to $G(\mathbf{k}, \varepsilon)$ in the theory of weakly interacting Bose gas.

$$G^{(0)}(\mathbf{k}, \varepsilon) = \left(\varepsilon - \frac{\hbar^2 k^2}{2m} + \mu + i\delta\right)^{-1} \tag{5.2.131}$$

The rules for evaluating the contribution from each diagram are straightforward, and will not be cited here. (See Reference [68].) Figure 5.2.7 shows all contributions to $G(\mathbf{k}, \varepsilon)$ through four orders of interaction. The diagrams for Σ_{11} and Σ_{02} are not shown. Substitution of G, Σ_{11}, and Σ_{02} into equations (5.2.128) and (5.2.130) then yields N_0 and μ. Finally, equation (5.2.127) gives E_0, which is arranged in powers of λ:

$$E_0 = \sum_{m>0} \lambda^m E_0^{(m)}, \tag{5.2.132}$$

with

$$\frac{E_0^{(1)}}{\Omega} = \frac{1}{2}\rho^2 v(0) \equiv e_a, \tag{5.2.133a}$$

$$\frac{E_0^{(2)}}{\Omega} = \frac{-\rho^2 m}{2\hbar^2} \sum_{\mathbf{k}} \frac{v^2(k)}{k^2} \equiv e_b, \tag{5.2.133b}$$

$$\frac{E_0^{(3)}}{\Omega} = \frac{\rho^3 m^2}{\hbar^4} \sum_{\mathbf{k}} \frac{v^3(k)}{k^4} + \frac{\rho^2 m^2}{2\hbar^4} \sum_{\mathbf{kl}} \frac{v(k)v(l)v(|\mathbf{k}+\mathbf{l}|)}{k^2 l^2} \equiv e_c + e_d, \tag{5.2.133c}$$

$$\frac{E_0^{(4)}}{\Omega} = \frac{-5\rho^4 m^3}{2\hbar^6} \sum_{\mathbf{k}} \frac{v^4(k)}{k^6} + \frac{\rho^3 m^3}{\hbar^6} \sum_{\mathbf{kl}} \frac{v^2(k)v^2(l)}{k^4 l^2}$$

$$- \frac{2\rho^3 m^3}{\hbar^6} \sum_{\mathbf{kl}} \frac{v^2(k)v(l)v(|\mathbf{k}+\mathbf{l}|)}{k^4 l^2} - \frac{2\rho^3 m^3}{\hbar^6} \left[\sum_{\mathbf{kl}} \frac{v^2(k)v(l)v(|\mathbf{k}+\mathbf{l}|)}{k^2 l^2 |\mathbf{k}+\mathbf{l}|^2} \right.$$

$$\left. + \sum_{\mathbf{kl}} \frac{v^2(k)v^2(|\mathbf{k}+\mathbf{l}|)}{k^2|\mathbf{k}+\mathbf{l}|^2(k^2+l^2+|\mathbf{k}+\mathbf{l}|^2)} \right]$$

$$- \frac{2\rho^3 m^3}{\hbar^6} \left[\sum_{\mathbf{kl}} \frac{v^2(k)v(l)v(|\mathbf{k}+\mathbf{l}|)}{k^4(k^2+l^2+|\mathbf{k}+\mathbf{l}|^2)} + \sum_{\mathbf{kl}} \frac{v^2(k)v^2(|\mathbf{k}+\mathbf{l}|)}{k^4(k^2+l^2+|\mathbf{k}+\mathbf{l}|^2)} \right]$$

$$- \frac{\rho^2 m^3}{2\hbar^6} \sum_{\mathbf{klp}} \frac{v(l)v(p)v(|\mathbf{k}+\mathbf{l}|)v(|\mathbf{k}+\mathbf{p}|)}{k^2 l^2 p^2} \equiv e_e + e_f + e_g + [e_h] + [e_i]$$

$$+ e_j, \tag{5.2.133d}$$

etc.

Diagrams corresponding to these terms are displayed in Figure 5.2.8. To $0(\lambda^4)$, we have thus established a framework against which Jastrow's theory and the CBF will be measured.

Let us now set up the Jastrow calculation for the weakly interacting Bose gas. The energy formula (5.2.40) is better put in the following form:

$$\frac{E_J}{\Omega} = \frac{1}{2}\rho^2 \int v(r)g_J(r)\,d\mathbf{r} + \rho^2 \frac{\hbar^2}{8m} \int u'(r)g'_J(r)\,d\mathbf{r}. \tag{5.2.134}$$

Or, in terms of the expansions

$$g_J(r) = \sum_{m\geq 0} \lambda^m g_J^{(m)}(r) = \sum_{m\geq 0} \lambda^m \sum_{\mathbf{k}} e^{-i\mathbf{k}\cdot\mathbf{r}} g_J^{(m)}(k), \tag{5.2.135}$$

$$u(r) = \sum_{m\geq 0} \lambda^m u^{(m)}(r) = \sum_{m\geq 0} \lambda^m \sum_{\mathbf{k}} e^{-i\mathbf{k}\cdot\mathbf{r}} u^{(m)}(k): \tag{5.2.136}$$

$$E_J = \sum_{m>0} \lambda^m E_J^{(m)}, \tag{5.2.137}$$

with

$$\frac{E_J^{(1)}}{\Omega} = \frac{1}{2}\rho^2 v(0), \tag{5.2.138a}$$

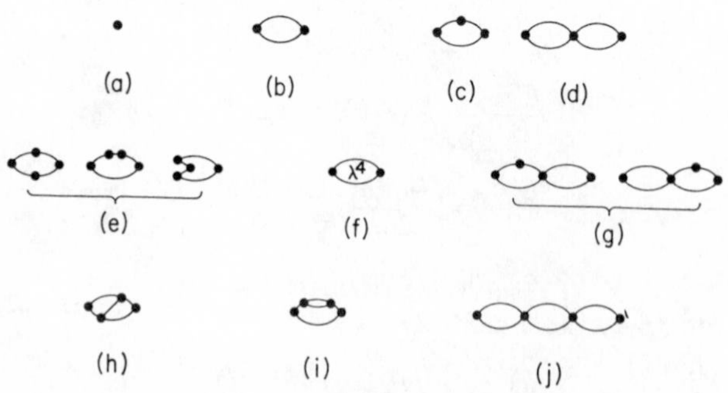

FIGURE 5.2.8. Contributions to the ground-state energy of a weakly interacting Bose gas.

$$\frac{E_J^{(2)}}{\Omega} = \frac{1}{2}\rho^2 \sum_{\mathbf{k}} v(k) g_J^{(1)}(k) + \rho^2 \frac{\hbar^2}{8m} \sum_{\mathbf{k}} k^2 u^{(1)}(k) g_J^{(1)}(k), \qquad (5.2.138b)$$

$$\frac{E_J^{(3)}}{\Omega} = \frac{1}{2}\rho^2 \sum_{\mathbf{k}} v(k) g_J^{(2)}(k) + \rho^2 \frac{\hbar^2}{8m} \sum_{\mathbf{k}} k^2 [u^{(1)}(k) g_J^{(2)}(k) + u^{(2)}(k) g_J^{(1)}(k)], \qquad (5.2.138c)$$

$$\frac{E_J^{(4)}}{\Omega} = \frac{1}{2}\rho^2 \sum_{\mathbf{k}} v(k) g_J^{(3)}(k) + \rho^2 \frac{\hbar^2}{8m} \cdot \sum_{\mathbf{k}} k^2 [u^{(1)}(k) g_J^{(3)}(k) + u^{(2)}(k) g_J^{(2)}(k) + u^{(3)}(k) g_J^{(1)}(k)]. \qquad (5.2.138d)$$

Since the functions $u^{(m)}(k)$ and $g_J^{(m)}(k)$ are related by the BBGKY equation, which in its exact form is given by equations (5.2.59)–(5.2.62), each $u^{(m)}(k)$ can be solved in terms of $g_J^{(m)}(k)$ and substituted back into equations (5.2.138). E_J is then expressed as a functional in $g_J^{(m)}(k)$. Thereafter, either by solving the Euler–Lagrange equations

$$\frac{\delta E_J}{\delta g_J^{(m)}(k)} = 0, \qquad (5.2.139)$$

or by beginning with some reasonable set of $g^{(m)}(k)$ and applying the paired-phonon analysis, it becomes possible to optimize the variational function and find the best energy Jastrow's theory has to offer. In References [70, 71], Sim, Woo and Buchler started with what was thought to be the exact $g^{(m)}(k)$, and watched the paired-phonon analysis distort it in its effort to optimize $u^{(m)}(k)$. This was not surprising, since the exact ground-state wave function was not expected to be of the Jastrow form anyway. Later, Bhattacharyya and Woo [72] found an (unimportant) error in Reference [71]: The exact $g^{(m)}(k)$ to order λ^2 was actually generated by the optimum Jastrow function. So to $O(\lambda^2)$ in the energy, the paired-phonon analysis was really not necessary, once the exact $g^{(m)}(k)$ was calculated to $O(\lambda^2)$ to provide a starting point. The latter can be obtained by way of HP's prescription [68]

$$\begin{aligned} S_J(k) &= 1 + \rho \int [g_J(r) - 1]\, e^{i\mathbf{k}\cdot\mathbf{r}}\, d\mathbf{r} \\ &= 1 + \rho \sum_{m>0} \lambda^m g_J^{(m)}(k) \\ &= \frac{1}{\rho} \int_{c'} d\omega \frac{i}{2\pi} F_k(\omega), \end{aligned} \qquad (5.2.140)$$

where c' closes either above or below the real axis and

$$F_k(\omega) = \int_{-\infty}^{\infty} dt\, e^{i\omega t} F_k(t), \qquad (5.2.141)$$

$$F_k(t) = -i\langle 0 | T\{\rho_{\mathbf{k}}(t)\rho_{-\mathbf{k}}(0)\} | 0\rangle, \qquad (5.2.142)$$

$$\rho_{\mathbf{k}}(t) = \sum_{\mathbf{l}} a_{\mathbf{l}}^\dagger(t) a_{\mathbf{k}+\mathbf{l}}(0). \qquad (5.2.143)$$

Depending on the number of $a_0^\dagger$ and a_0 contained, $F_k(t)$ appears with 2, 3, or 4 $a_{\mathbf{k}}^\dagger(t)$ and $a_{\mathbf{k}}(t)$, and is related in each case to the Green's functions. Lengthy algebraic work leads to

$$\begin{cases} g_J^{(0)}(k) = 1, & (5.2.144a) \\ g_J^{(1)}(k) = -\dfrac{2mv(k)}{\hbar^2 k^2}, & (5.2.144b) \\ g_J^{(2)}(k) = \dfrac{6\rho m^2 v^2(k)}{\hbar^4 k^4} - \dfrac{2m^2}{\hbar^4}\sum_{\mathbf{l}} \dfrac{v(l)v(|\mathbf{k}+\mathbf{l}|)}{k^2 l^2} + \dfrac{m^2}{\hbar^4}\sum_{\mathbf{l}} \dfrac{v(l)v(|\mathbf{k}+\mathbf{l}|)}{l^2|\mathbf{k}+\mathbf{l}|^2}, & \\ \text{etc.,} & (5.2.144c) \end{cases}$$

and

$$\begin{cases} \dfrac{E_J^{(m)}}{\Omega} = \dfrac{E_0^{(m)}}{\Omega}, \quad m \le 3 & (5.2.145a) \\ \dfrac{E_J^{(4)}}{\Omega} = \dfrac{E_0^{(4)}}{\Omega} - \dfrac{\delta E_0^{(4)}}{\Omega}, & (5.2.145b) \\ \text{etc.,} & \end{cases}$$

where the discrepancy $\delta E_0^{(4)}/\Omega$ from the exact formula is given by

$$\begin{aligned} \frac{\delta E_0^{(4)}}{\Omega} = {} & \frac{\rho^3 m^3}{\hbar^6}\sum_{\mathbf{kl}} \frac{v^2(k)v^2(l)}{k^4 l^2} - \frac{2\rho^3 m^3}{\hbar^6}\sum_{\mathbf{kl}} \frac{v^2(k)v^2(l)}{k^4[k^2 + l^2 + |\mathbf{k}+\mathbf{l}|]} \\ & - \frac{2\rho^3 m^3}{\hbar^6}\sum_{\mathbf{kl}} \frac{v^2(k)v^2(l)}{k^2 l^2[k^2 + l^2 + |\mathbf{k}+\mathbf{l}|^2]} \\ & - \frac{\rho^3 m^3}{2\hbar^6}\sum_{\mathbf{kl}} \frac{v^2(k)v(l)v(|\mathbf{k}+\mathbf{l}|)}{k^2 l^2 |\mathbf{k}+\mathbf{l}|^2} + \frac{\rho^3 m^3}{\hbar^6}\sum_{\mathbf{kl}} \frac{v^2(k)v(l)v(|\mathbf{k}+\mathbf{l}|)}{k^4 l^2} \\ & - \frac{2\rho^3 m^3}{\hbar^6}\sum_{\mathbf{kl}} \frac{v^2(k)v(l)v(|\mathbf{k}+\mathbf{l}|)}{k^4[k^2 + l^2 + |\mathbf{k}+\mathbf{l}|^2]}. \end{aligned} \qquad (5.2.146)$$

Note that terms with denominators involving $[k^2 + l^2 + |\mathbf{k}+\mathbf{l}|^2]$ can never emerge from Jastrow calculations.

Inspect equation (5.2.145) and (5.2.146) against the diagrams in Figure 5.2.8. It is clear that the diagrams most susceptible to summation by the Jastrow theory are the single rings and the ladders. In $O(\lambda^4)$, the ring–ladder combination (e_g), the vertex-renormalizing "cross-bar" diagram (e_h), and the propagator-renormalizing "side-loop" diagram (e_i) are only partially included.

As we turn to CBF, the first correction arises from the diagram depicting the excitation and subsequent de-excitation of three virtual phonons, Figure 5.2.5. The prospect that such a correction may complete the energy series to $O(\lambda^4)$ is quite good, since equations (5.2.121) and (5.2.122) suggest that the energy denominator will indeed assume the form $[k^2 + l^2 + |\mathbf{k} + \mathbf{l}|^2]$. An actual calculation [73] showed that the correction yields precisely $\delta E_0^{(4)}/\Omega$ of equation (5.2.146). In Reference [73] the structure of certain higher-order contributions was also analyzed. It was concluded that Jastrow's theory picks up the ring diagrams and the ladder diagrams to at least $O(\lambda^5)$, and the simplest (second-order) CBF theory in effect sums an additional class of selected diagrams

The same kind of analysis has by now been carried out for the charged Bose gas as well [74], and for the excitation spectra of both the weakly interacting Bose gas [72] and the charged Bose gas [74, 75]. Even though no general theorem has been proven, it is at least strongly suggestive that the strength of Jastrow's theory and the CBF resides in their ability to sum large classes of important diagrams without the use of sophisticated field-theoretic apparatuses.

An alternate and complementary approach toward relating Jastrow's method to diagrammatic perturbation theory is to work from the latter: Collect the diagrams that are expected to be important, apply whatever approximation schemes deemed necessary or convenient, and then match the resulting perturbed wave function against the Jastrow form. It is not yet clear how such an analysis might be generalized for comparison with the CBF, but what has already been established by Brandow and Day for the Jastrow theory provides much cause for enthusiasm and hope.

It is not possible to even highlight, let alone outline, the highly developed technology of diagrammatic perturbation theory. The reader must refer to standard texts [77–79] and the cross-referenced publications by Brandow [64, 65] and Day [80]. The papers by Brandow are beautifully detailed and amply footnoted. Those by Day are concise and lucid.

Brandow's theory is based on the Brueckner–Bethe–Goldstone expansion for fermions. By considering a system of N fermions with spin degeneracy greater than N and interacting via a spin-independent, pairwise potential, he finds it possible to describe the ground-state wave function as consisting of a completely symmetric spatial factor and an antisymmetric product of spin functions. Once the fermion problem is formulated, the perturbation series can be partially summed in a particle-conserving manner, provided that to each zero-momentum particle a chemical potential μ and a depletion factor $(1 - N_0/N)$ is assigned. At the conclusion of the calculation, he sheds the spin functions and returns to the many-*boson* problem along with its correct wave function. The actual calculation as carried out by Day begins with the construction of contributing diagrams to the wave function—diagrams that have external lines only at the top, and have reaction matrix vertices in place of bare He–He interactions to account for the hard cores. The hole lines represent the annihilation of particles in the "boson sea," and have momenta zero. When only a subclass of diagrams—those with noninteracting hole lines—are summed, and approximations are made regarding (i) the mode of operation of reaction matrices on many-body wave functions and (ii) the handling of off-energy-shell effects, it becomes possible to demonstrate that the wave function always reduces to a product of two-particle functions, i.e., the Jastrow form. It is, of course, not clear whether more diagrams can be absorbed into the partial summation scheme without sacrificing the Jastrow form, and whether a "natural" extension of the present summation scheme will take us beyond the Jastrow regime into the space spanned by CBF. Finally, it remains to be seen whether "hybrid" calculations mingling diagrammatic techniques with Jastrow and CBF methods could be made to work to our advantage. These are by no means easy questions to answer, but it is indeed fortunate and encouraging that instead of an impasse we have discovered fertile grounds between two major camps of many-body theory for further exploration and eventual exploitation.

5.2.5. Variational Calculation beyond Jastrow's Theory [81, 82]

The establishment of minimal contacts between theories based on the correlated representation and the independent-particle representation has already borne certain fruits, as we shall see in this section.

As is clear from the preceding sections, Jastrow's theory and the method of CBF have been rather successful in calculating ground-state properties of liquid ^{4}He, considering that these calculations do not require *any* adjustable parameter. The energy obtained by Jastrow-type calculations is around -6 K per particle, about 20% higher than the experimental value. But since the total energy results from strong cancellations between potential and kinetic parts, a 20% discrepancy must be regarded as tolerable. The correction by the paired-phonon analysis brings the energy much closer to experiment; and the CBF three-phonon correction further improves it to -7.50 K, a value that is now a little too low. One must not forget to include contributions from real three-body forces such as the triple-dipole interaction, which amounts to about $+0.15$ K. All in all, each of these corrections seems to nudge the theoretical value yet a little closer toward experiment. Except for some cautionary remarks to be stated near the end of this section, we have no reason to be dissatisfied with the state of the matter as far as energy is concerned.

The same cannot be said of the wave function. While there is little interest in the wave function itself, we are entitled to getting good results for distribution functions—probability amplitudes that can usually be deduced from the ground-state wave function. This is particularly important because they are experimentally measurable. There we find significant discrepancy between theory and experiment, as can be seen from Figure 5.2.2, in which the radial distribution functions $g(r)$ and $g_J(r)$ are compared [83], or from Figure 5.2.3. in which the liquid-structure functions [84, 85] $S(k)$ and $S_J(k)$ are compared. $g_J(r)$, or $S_J(k)$, shows that theory predicts much less structure than experiment, even though the latter is obtained at higher temperatures.

There are two sources to which we can attribute this discrepancy. One concerns the accuracy of the He–He potential. The other concerns the variational wave function. Surely the liquid structure can be sharpened if the potential is made more attractive, e.g., by deepening the well. While there are good reasons why the deBoer–Michels potential used in most Jastrow calculations should be modified, we cannot arbitrarily reset the potential to force an inaccurate quantum mechanical calculation into agreement with experiment. The first-order business is to improve the theoretical calculation until we have gained an irreproachable degree of reliability. This is one of the motivations for our earlier reversion to the perturbation theory—CBF. The Jastrow function *explicitly* takes into account only two-particle correlations. For the system to lower its energy, every triplet of particles should be permitted to choose a configuration which allows them to sample more of one another's attraction. This calls for explicit three-particle correlations not contained in Jastrow's theory. In the CBF calculation of Davison and Feenberg [46], three-particle correlations enter through intermediate three-phonon states, which accounts for the sizable lowering of the ground-state energy. In principle, then, one knows how to handle three-particle correlations, and should be able to study their effects on the liquid structure. In practice, however, the method fails, since it is well known that for an infinite system a finite-order perturbation theory does not produce realistic many-body wave functions. (If only two-particle potentials are present, a second-order theory, for example, limits the perturbed wave function to the subspace spanned by states with two or less excited particles.) In Section 5.3.3 we shall present a simple variation of the CBF theory to compute an improved (perturbed) $S(k)$, but first we must explore an obvious alternative: the extension of the variational function space.

Return to the introductory remarks in Section 5.2.2. Equation (5.2.19) is a natural extension of equation (5.2.18). Beyond equation (5.2.19), the next simplest variational function is a properly symmetrized product of three-particle functions

$$\psi^{(3)}(1,2,\cdots,N) = P\{\prod_{\substack{i<j<k\\=1}}^{N} \xi(i,j,k)\}. \tag{5.2.147}$$

By factoring out the one- and two-particle parts and realizing that in an isotropic, homogeneous liquid the wave function depends only on *relative* coordinates, we can rewrite equation (5.2.147) in the following form:

$$\psi^{(3)}(1,2,\cdots,N) = P\{\prod_{\substack{i<j<k\\=1}}^{N} b(i,j,k) \prod_{\substack{l<m\\=1}}^{N} f(l,m) \prod_{n=1}^{N} \varphi(n)\}$$

$$= \prod_{\substack{i<j<k\\=1}}^{N} b(r_{ij},r_{jk},r_{ki}) \prod_{\substack{l<m\\=1}}^{N} f(r_{lm}) P\{\prod_{n=1}^{N} \varphi(n)\}. \tag{5.2.148}$$

For the ground state of a Bose liquid, equation (5.2.148) further reduces to

$$\psi_t(1,2,\cdots,N) = \prod_{\substack{i<j<k\\=1}}^{N} b(r_{ij},r_{jk},r_{ki}) \prod_{\substack{l<m\\=1}}^{N} f(r_{lm})$$

$$\equiv \exp\{\sum_{\substack{i<j<k\\=1}}^{N} \tfrac{1}{2} w(r_{ij},r_{jk},r_{ki}) + \sum_{\substack{l<m\\=1}}^{N} \tfrac{1}{2} u(r_{lm})\}. \tag{5.2.149}$$

To determine u and w, the expectation value

$$E_t \equiv \frac{\langle\psi_t|H|\psi_t\rangle}{\langle\psi_t|\psi_t\rangle} \tag{5.2.150}$$

should be minimized with respect to u and w independently. In practical calculations, however, one probably loses little by setting $u(r)$ to the best Jastrow result:

$$\psi_t(1,2,\cdots,N) \approx \psi_J(1,2,\cdots,N) \exp\{\tfrac{1}{2} \sum_{\substack{i<j<k\\=1}}^{N} w(r_{ij},r_{jk},r_{ki})\}, \tag{5.2.151}$$

with ψ_J optimized.

That equation (5.2.149) or (5.2.151) is capable of improving $S_J(k)$ toward the experimental $S(k)$ was demonstrated in a recent publication [26]. A biased selection Monte Carlo procedure developed by Coldwell [25] (Section 5.2.2) was employed. The Jastrow function used was first optimized in a preliminary calculation, and was checked out against earlier work using molecular dynamics [86] and a variety of integral equations [87]. The three-particle factor was chosen to be of the convenient form

$$w(r,s,|\mathbf{r} - \mathbf{s}|) = -\left(\frac{C}{R}\right)^n \equiv w(R), \tag{5.2.152}$$

where

$$R^2 = r^2 + s^2 + |\mathbf{r} - \mathbf{s}|^2. \tag{5.2.153}$$

The choice of $w(R)$ was based on the fact that, for positive n, equilateral triangles are favored over a large class of other less-structured configurations. This is necessary because by placing the particles at the vertices of equilateral triangles, they become best located to sample the attractive wells of their neighbors, thus helping to lower the energy. A side effect is that the distribution at nearest-neighbor distance is enhanced, causing sharper configurational structure. Such effects are ignored by the Jastrow function whose main function is to prevent the overlap of hard cores. It is recognized that equation (5.2.152) is not likely to be the best form available, but one expects it to do well enough in demonstrating the existence of an effect. The actual calculation was done in two dimensions to save computer time. $g(r)$ and the energy, which in two dimensions reads

$$\frac{E}{N} = \frac{\rho}{2}\int g(r)\left\{v(r) - \frac{\hbar^2}{4m}\left[u''(r) + \frac{u'(r)}{r}\right]\right\} d\mathbf{r}$$
$$+ \frac{\rho^2}{6}\int g_3(r,s,|\mathbf{r}-\mathbf{s}|)\left\{\frac{3\hbar^2}{8m}\left[w''(R) + \frac{3w'(R)}{R}\right]\right\} d\mathbf{r}\, d\mathbf{s}, \tag{5.2.154}$$

where $\rho = 0.0358$ Å^{-2}(equilibrium areal density), were not evaluated directly. Instead, the change $\Delta g(r) \equiv g_t(r) - g_J(r)$ was calculated in the minimization of $\Delta E \equiv E_t - E_J$. The results are shown in Fig. 5.2.9. The change in the liquid structure function $\Delta S(k) \equiv S_t(k) - S_J(k)$ is shown in Table 5.2.3. The three-particle correlation effects obtained in this manner are admittedly small, but they do exist and are in the right direction.

TABLE 5.2.3. Three-particle correlation effects on the two-dimensional liquid structure function at $\sigma = 0.0358$ Å^{-2}

k(Å^{-1})	$S_J(k)$	$\Delta S(k)$
0.5	0.749	−0.027
0.7	0.796	−0.007
0.9	0.888	+0.010
1.1	1.024	+0.019
1.3	1.159	+0.020
1.5	1.225	+0.015
1.7	1.194	+0.006
1.9	1.098	−0.005
2.1	0.999	−0.013
2.3	0.934	−0.015
2.5	0.907	−0.010
2.7	0.906	−0.001
2.9	0.920	+0.007
3.1	0.948	+0.011
3.3	0.981	+0.007
3.5	1.011	+0.001
3.7	1.031	−0.005
3.9	1.042	−0.007
4.1	1.046	−0.004
4.3	1.041	−0.002

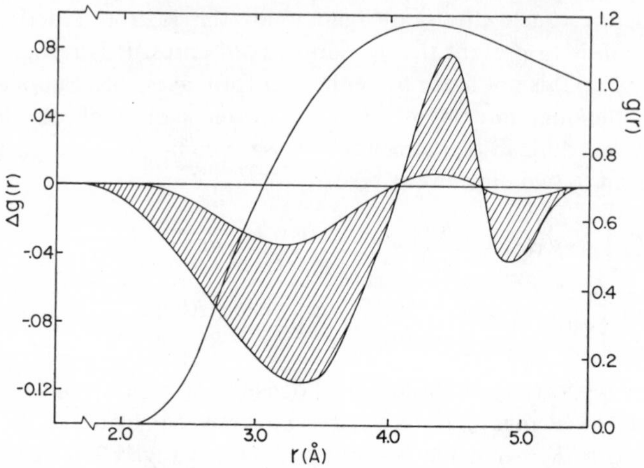

FIGURE 5.2.9. Improvement, $\Delta g(r)$, in the radial distribution function of two-dimensional helium liquid brought about by the inclusion of three-particle correlation factors. The solid curve is the optimum Jastrow result. The spread represents uncertainties in $\Delta g(r)$.

Judging from the smallness of the effects, one must conclude that the form of w is inadequate. In the Jastrow-type calculation, much is known about the two-particle factor $f(r)$, or $u(r)$. We discussed, for example, its asymptotic behavior in Section 5.2.2. The experimental $g(r)$ is available to provide certain guidance toward constructing $u(r)$. In fact, sometimes $g(r)$ itself is parameterized and varied. On the other hand, little is known about the three-particle factor $w(R)$. Presumably a major part of the discrepancy that exists between $g_J(r)$ and the experimental $g(r)$ must be accountable to w; yet such information is unreliable: Neither the experimental nor the Jastrow $g(r)$ is sufficiently accurate to assure a precise determination of their difference. This is where the diagrammatic analysis of Jastrow theory comes in.

In the case of the weakly interacting Bose gas, to $O(\lambda^3)$ Jastrow's theory reproduces exactly the energy series E_0 of equations (5.2.132)–(5.2.133). In $O(\lambda^4)$, the terms grouped together in equation (5.2.146) as $\delta E_0^{(4)}$ are missing, only to be picked up by the second-order three-phonon diagram in CBF perturbation theory. Now, since the latter diagram accounts for three-particle correlations, its effect must be reproducible by a variational function which includes explicit three-particle factors, namely, ψ_t of equation (5.2.149) or (5.2.151). One has then a basic requirement for w: If properly chosen, it must yield the energy series exactly to $O(\lambda^4)$.

Assuming ψ_t, let us now redo the variational calculation [81–82]. The energy expectation value, equation (5.2.134), is replaced by

$$\frac{E_t}{\Omega} = \frac{1}{2}\frac{\rho^2}{\Omega}\int \tilde{v}(ij)g_t(ij)\, d\mathbf{r}_i\, d\mathbf{r}_j - \frac{\rho^3}{16\Omega}\int g_3(ijk)\nabla_i^2 w(ijk) d\mathbf{r}_i\, d\mathbf{r}_j\, d\mathbf{r}_k, \quad (5.2.155)$$

where the argument (ij) denotes r_{ij}, and (ijk) denotes (r_{ij}, r_{jk}, r_{ki}). For the moment, let us assume that ψ_J is the optimum Jastrow function. We Fourier-transform and expand $u(r)$ and $g_t(r)$ as in equation (5.2.136) and (5.2.135). Also, let

$$
\begin{aligned}
w(ijk) &= \sum_{m\geq 0} \lambda^m w^{(m)}(ijk) \\
&= \sum_{m\geq 0} \lambda^m \sum_{\mathbf{kl}}{}' \exp[-i\mathbf{k}\cdot\mathbf{r}_i - i\mathbf{l}\cdot\mathbf{r}_j + i(\mathbf{k}+\mathbf{l})\cdot\mathbf{r}_k] w^{(m)}(\mathbf{k}, \mathbf{l}, -\mathbf{k}-\mathbf{l}),
\end{aligned} \tag{5.2.156}
$$

and

$$E_t = \sum_{m>0} \lambda^m E_t^{(m)}. \tag{5.2.157}$$

The relations between $u^{(m)}(k)$ and $g_t^{(m)}(k)$ are not the same as those between $u^{(m)}(k)$ and $g_J^{(m)}(k)$. Generalized BBGKY equations must be derived and solved. One finds

$$
\begin{aligned}
\nabla_1 g_t(12) &= g_t(12)\nabla_1 u(12) + \rho \int g_{3,t}(123)\nabla_1 u(13)\, d\mathbf{r}_3 \\
&\quad + \rho \int g_{3,t}(123)\nabla_1 w(123)\, d\mathbf{r}_3 + \tfrac{1}{2}\rho^2 \int g_{4,t}(1234)\nabla_1 w(134)\, d\mathbf{r}_3\, d\mathbf{r}_4,
\end{aligned} \tag{5.2.158a}
$$

$$
\begin{aligned}
\nabla_1 g_{3,t}(123) &= g_{3,t}(123)[\nabla_1 u(12) + \nabla_1 u(13) + \nabla_1 w(123)] \\
&\quad + \rho \int g_{4,t}(1234)[\nabla_1 u(14) + \nabla_1 w(124) + \nabla_1 w(134)]\, d\mathbf{r}_4 \\
&\quad + \tfrac{1}{2}\rho^2 \int g_{5,t}(12345)\nabla_1 w(145)\, d\mathbf{r}_4\, d\mathbf{r}_5,
\end{aligned} \tag{5.2.158b}
$$

$$
\begin{aligned}
\nabla_1 g_{4,t}(1234) &= g_{4,t}(1234)[\nabla_1 u(12) + \nabla_1 u(13) + \nabla_1 u(14) + \nabla_1 w(123) \\
&\quad + \nabla_1 w(124) + \nabla_1 w(134)] + \rho \int g_{5,t}(12345)[\nabla_1 u(15) + \nabla_1 w(125) \\
&\quad + \nabla_1 w(135) + \nabla_1 w(145)] d\mathbf{r}_5 + \tfrac{1}{2}\rho^2 \int g_{6,t}(123456)\nabla_1 w(156)\, d\mathbf{r}_5\, d\mathbf{r}_6,
\end{aligned} \tag{5.2.158c}
$$

etc.

Define for convenience $x(123)$ and $y(1234)$:

$$g_{3,t}(123) \equiv g_t(12)g_t(23)g_t(31)[1 + x(123)], \tag{5.2.159}$$

$$g_{4,t}(123) \equiv g_t(12)g_t(13)g_t(14)g_t(23)g_t(24)g_t(34)[1 + y(1234)]; \tag{5.2.160}$$

one finds, for example, that $x(123)$ is no longer give by Abe's correction; i.e.

$$x(123) \neq \rho \int [g_t(14) - 1][g_t(24) - 1][g_t(34) - 1] d\mathbf{r}_4 + \cdots.$$

Instead, it leads off with $w(123)$, as one might expect by considering classical liquids with three-particle interactions. Solving the coupled equations (5.2.158), one finds their Fourier transforms $x^{(m)}(\mathbf{k}, \mathbf{l}, -\mathbf{k}-\mathbf{l})$ and $y^{(m)}(\mathbf{k}, \mathbf{l}, \mathbf{p}, -\mathbf{k}-\mathbf{l}-\mathbf{p})$ along with $g_t^{(m)}(k)$. It turns out that

$$g_t^{(m)}(k) = g_J^{(m)}(k), \quad m < 3. \tag{5.2.161}$$

But beginning with $g_t^{(3)}(k)$, $x^{(m)}$, $y^{(m)}$, and $w^{(m)}$ enter. Substituting the results into equation (5.2.155) and writing E_t in the form (5.2.157), one obtains

$$E_t^{(m)} = E_J^{(m)}, \quad m < 4 \tag{5.2.162}$$

and

$$E_t^{(4)} = E_J^{(4)} + \Delta E^{(4)}, \tag{5.2.163}$$

where

$$\begin{aligned}\frac{\Delta E^{(4)}}{\Omega} = &-\frac{1}{4}\rho^3 \sum_{\mathbf{kl}}{}' k^2 g_t^{(1)}(k) g_t^{(1)}(l) w^{(2)}(\mathbf{k}, \mathbf{l}, -\mathbf{k}-\mathbf{l}) \\ &+ \frac{1}{16}\rho^3 \sum_{\mathbf{kl}}{}' k^2 w^{(2)}(\mathbf{k}, \mathbf{l}, -\mathbf{k}-\mathbf{l}) w^{(2)}(-\mathbf{k}, -\mathbf{l}, \mathbf{k}+\mathbf{l}) \\ &+ \frac{1}{16}\rho^3 \sum_{\mathbf{kl}}{}' k^2 [g_t^{(1)}(k) g_t^{(1)}(l) + g_t^{(1)}(k) g_t^{(1)}(|\mathbf{k}+\mathbf{l}|) \\ &+ g_t^{(1)}(l) g_t^{(1)}(|\mathbf{k}+\mathbf{l}|)] w^{(2)}(\mathbf{k}, \mathbf{l}, -\mathbf{k}-\mathbf{l}).\end{aligned} \tag{5.2.164}$$

To $O(\lambda^4)$, this expression is what one wishes to minimize with respect to the undetermined function $w^{(2)}$. The Euler–Lagrange equation

$$\frac{\delta E_t}{\delta w^{(2)}} = 0 \tag{5.2.165}$$

is then solved readily to yield

$$\begin{aligned} w^{(2)}(\mathbf{k}, \mathbf{l}, -\mathbf{k}-\mathbf{l}) = \\ &- \frac{\mathbf{k}\cdot\mathbf{l}\, g^{(1)}(k) g^{(1)}(l) + \mathbf{k}\cdot(-\mathbf{k}-\mathbf{l}) g^{(1)}(k) g^{(1)}(|\mathbf{k}+\mathbf{l}|)}{k^2 + l^2 + |\mathbf{k}+\mathbf{l}|^2} \\ &+ \frac{\mathbf{l}\cdot(-\mathbf{k}-\mathbf{l}) g^{(1)}(l) g^{(1)}(|\mathbf{k}+\mathbf{l}|)}{k^2 + l^2 + 1\mathbf{k}+\mathbf{l}|^2},\end{aligned} \tag{5.2.166}$$

and

$$\Delta E^{(4)} = \delta E_0^{(4)}, \tag{5.2.167}$$

which is precisely what one needs to complete $O(\lambda^4)$.

The same results obtain when $u(ij)$ is allowed to vary along with $w(ijk)$, as one can easily verify by solving $\delta E_t/\delta u = 0$ along with equation (5.2.165).

We note that the explicit inclusion of three-particle factors enables the variational theory to sum a totally new class of diagrams, the leading ones being those in $O(\lambda^4)$ left out by Jastrow's theory. This statement is actually stronger than a similar one made earlier in connection with the CBF calculation. Being variational, the theory here retains the upper-bound property of the energy.

In analogy to the paired-phonon analysis, the present work suggests a "triplet-phonon analysis": It should be possible to show that the matrix element connecting the ground-state wave function ψ_t and the three-phonon state $\rho_{\mathbf{k}}\rho_{\mathbf{l}}\rho_{-\mathbf{k}-\mathbf{l}}\,\psi_t$ vanishes when ψ_t is optimized [88].

For improving the ground-state calculation of liquid ^{4}He, the above analysis suggests a wave function of the form (5.2.151), with $w(r_{ij}, r_{jk}, r_{ki})$ given by its Fourier transform

$$w(\mathbf{k}, \mathbf{l}, -\mathbf{k}-\mathbf{l}) = -\frac{\mathbf{k}\cdot\mathbf{l}h(k)h(l) + \mathbf{k}\cdot(-\mathbf{k}-\mathbf{l})h(k)h(|\mathbf{k}+\mathbf{l}|) + \mathbf{l}\cdot(-\mathbf{k}-\mathbf{l})h(l)h(|\mathbf{k}+\mathbf{l}|)}{k^2 + l^2 + |\mathbf{k}+\mathbf{l}|^2}, \tag{5.2.168}$$

where

$$h(k) = g(k) - 1. \tag{5.2.169}$$

I predict a significant improvement in the liquid-structure function, and an energy close to the -7.50 K per particle obtained by Davison and Feenberg [46] using the CBF, but here it represents an upper bound. One may protest that -7.50 K is *already* lower than the experimental energy of -7.14 K. However, the discrepancy is reduced somewhat by the positive contribution from the triple-dipole energy. Also, there may well be some error in the -0.68 K contribution from the paired-phonon analysis. The uncertainty in the He–He potential should finally be held responsible for whatever discrepancy that remains.

5.3. ELEMENTARY EXCITATIONS IN LIQUID ^{4}He

5.3.1. Phonon–Roton Spectrum

The elementary excitations in liquid ^{4}He are discussed in Chapter 1 and will not be covered in full detail here. What I wish to discuss in this chapter deals strictly with the microscopic derivation of the excitation spectrum. First, a few words about the development of the elementary excitation concept.

As we move away from zero temperature, thermodynamic functions other than those associated with the ground state make their appearance, the most important being the Helmholtz free energy F, the internal energy U, the entropy S, and the specific heat c_Ω. In order to calculate these quantities and to relate them to experiment, one must be able to evaluate the partition function Z, which in turn requires full knowledge concerning the energy levels of the system—in other words, the solutions of the Schrödinger equation (5.1.2). On the other hand, if one's interest is in the transport properties, such as thermal conductivity, viscosity, and diffusion coefficients, it is important to identify the modes of excitation and the coupling between these modes. Again, one is driven to carefully exploring the excitation spectrum of the system. By first applying quantum mechanics, then statistical mechanics, the physicist divides his complete, microscopic investigation of the physical properties of a many-particle system neatly into two stages.

At sufficiently low temperatures where quantum degeneracy reigns, only low-lying energy levels contribute significantly toward the partition function and take part in the collision processes to determine transport properties. We shall, therefore, concern ourselves only with low-lying states.

In 1941, Landau [89] postulated the existence of two branches of excitation in liquid ^{4}He at low tempratures. One corresponds to quantized sound waves, and forms an exact analogy to phonons in crystals, with the exception that only a longitudinal mode can be supported—as in all liquids. The other, Landau felt, is related to the rotational motion of clusters of atoms. This he proceeded to name "rotons". At every wave vector **k** both branches exist, with a linear phonon spectrum intersecting the origin

$$\omega^{\mathrm{ph}}(k) = \hbar s k, \tag{5.3.1}$$

and a quadratic roton branch perching precariously above:

$$\omega^{\mathrm{rot}}(k) = \Delta + \frac{\hbar^2 k^2}{2\mu}. \tag{5.3.2}$$

Statistical mechanics was applied to compute the specific heat, and comparison to experiment was made for determining the parameters s(sound velocity), Δ, and μ. Guided by second-sound data that subsequently became available, Landau revised [90] his form of the roton spectrum to

$$\omega^{\mathrm{rot}}(k) = \Delta + \frac{\hbar^2 (k - k_0)^2}{2\mu}. \tag{5.3.3}$$

Using equations (5.3.1), and (5.3.3) the contributions to the specific heat per particle can be evaluated:

$$c^{\mathrm{ph}} = \frac{2\pi^2 k_B^4}{15\rho\hbar^3 s^3} T^3, \tag{5.3.4}$$

$$c^{\mathrm{rot}} = \frac{2\mu^{1/2} k_0^2 \Delta^2}{(2\pi)^{3/2} \rho \hbar k_B^{1/2}} \left[1 + \frac{k_B T}{\Delta} + \frac{3}{4}\left(\frac{k_B T}{\Delta}\right)^2\right] T^{-3/2} \exp\left[\frac{-\Delta}{k_B T}\right]. \tag{5.3.5}$$

The fit to experiment then yields (energy as usual measured in units of k_B)

$$s = 238 \text{ m/sec}, \tag{5.3.6a}$$

$$\Delta = 8.65 \text{ K}, \tag{5.3.6b}$$

$$k_0 = 1.91 \text{ Å}^{-1}, \tag{5.3.6c}$$

$$\mu = 0.16m. \tag{5.3.6d}$$

Since the 1940s, the phonon–roton spectrum shown in Figure 5.3.1 has continued to serve as the universally accepted way of describing excitations in liquid ^{4}He.

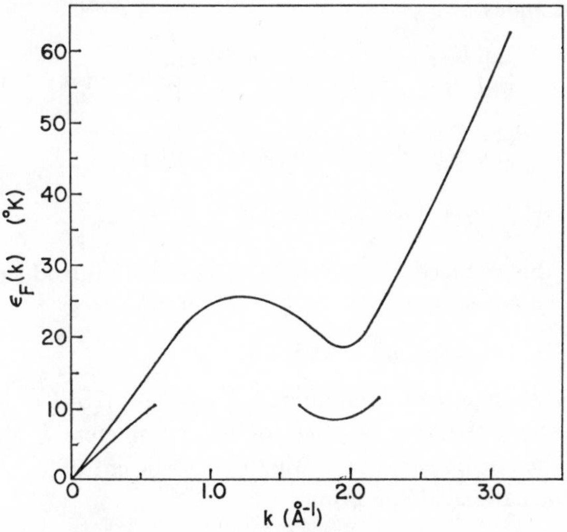

FIGURE 5.3.1. The phonon–roton spectrum in liquid ^{4}He. The continuous curve is the theoretical result obtained by Feynman [92].

The phenomenological theory of Landau was given a firm microscopic basis in 1947 by Bogoliubov [69]. The model considered by Bogoliubov was actually far too idealized for liquid ^{4}He; it was a weakly interacting Bose system at low density. The interaction was assumed Fourier transformable.

The Hamiltonian of such a system is best described in the second-quantized form (5.1.11)–(5.1.18), or equivalently equation (5.2.124),

$$H = \sum_{\mathbf{k}} \frac{\hbar^2 k^2}{2m} a_{\mathbf{k}}^\dagger a_{\mathbf{k}} + \frac{1}{2\Omega} \sum_{\mathbf{klp}} v(p) a_{\mathbf{k}}^\dagger a_{\mathbf{l}}^\dagger a_{\mathbf{k}+\mathbf{p}} a_{\mathbf{l}-\mathbf{p}}. \tag{5.3.7}$$

Since for the ground state of a Bose gas the occupation of the zero-momentum orbital is macroscopic, the commutator between $a_0^\dagger$ and a_0 can be neglected. Consequently, $a_0^\dagger$ and a_0 can be regarded as c numbers and replaced by $\sqrt{N_0}$, square root of the condensate occupation number. Moreover, if the interaction is weak, the depletion of the condensate will not be significant, so that N_0 can be approximated by N. From the second part of equation (5.3.7), terms containing four, two, and one $a_0^\dagger$ or a_0 can be separated out. The term with four $a_0^\dagger$ or a_0 is a constant and is of interest only for the ground state. There are no terms with three condensate operators; they are ruled out by the Kronecker δ. The terms with two $a_0^\dagger$ or a_0 are proportional to N while the terms with one are proportional to $\sqrt{N}$. In a first approximation, terms with zero or one condensate operator can be neglected. The

Hamiltonian becomes

$$H \approx \text{const} + \sum_{\mathbf{k}} \left[\frac{\hbar^2 k^2}{2m} + \rho v(k) \right] a_{\mathbf{k}}^\dagger a_{\mathbf{k}} + \tfrac{1}{2}\rho \sum_{\mathbf{k}} v(k)[a_{\mathbf{k}}^\dagger a_{\mathbf{k}}^\dagger + a_{\mathbf{k}} a_{-\mathbf{k}}]$$

$$= \text{const} + \sum_{\mathbf{k}} \left[\frac{\hbar^2 k^2}{2m} + \rho v(k) \right] [a_{\mathbf{k}}^\dagger a_{\mathbf{k}} + a_{-\mathbf{k}}^\dagger a_{-\mathbf{k}}]$$

$$+ \sum_{\mathbf{k}} \rho v(k) \, [a_{\mathbf{k}}^\dagger a_{-\mathbf{k}}^\dagger + a_{\mathbf{k}} a_{-\mathbf{k}}]. \tag{5.3.8}$$

To diagonalize this reduced Hamiltonian, Bogoliubov introduced his now-famous linear transformation

$$a_{\mathbf{k}}^\dagger = \mu \alpha_{\mathbf{k}}^\dagger + \nu \alpha_{-\mathbf{k}}, \tag{5.3.9}$$

which expresses H in a form containing $\alpha_{\mathbf{k}}^\dagger \; \alpha_{\mathbf{k}}^\dagger$ and $[\alpha_{\mathbf{k}}^\dagger \alpha_{-\mathbf{k}}^\dagger + \alpha_{\mathbf{k}} \alpha_{-\mathbf{k}}]$, each carrying a coefficient that is a function of the parameters μ and ν. These parameters are then determined by setting the coefficient of $[\alpha_{\mathbf{k}}^\dagger \alpha_{-\mathbf{k}}^\dagger + \alpha_{\mathbf{k}} \alpha_{-\mathbf{k}}]$ to zero under the canonical condition

$$\mu^2 - \nu^2 = 1. \tag{5.3.10}$$

The transformed Hamiltonian has the form

$$H = E_0 + \sum_{\mathbf{k}} \omega(k) \, \alpha_{\mathbf{k}}^\dagger \alpha_{\mathbf{k}}, \tag{5.3.11}$$

where

$$\omega(k) = \left[\left(\frac{\hbar^2 k^2}{2m} \right)^2 + 2 \left(\frac{\hbar^2 k^2}{2m} \right) \rho v(k) \right]^{1/2} \tag{5.3.12}$$

defines the single excitation energy. The $\alpha_{\mathbf{k}}$'s are identified to quasiparticle operators, and $\omega(k)$ Landau's excitation spectrum.

As $k \to 0$, provided that $v(k)$ approaches a positive constant $v(0)$,

$$\omega(k) \to \left[\frac{\rho v(0)}{m} \right]^{1/2} \hbar k, \quad k \to 0 \tag{5.3.13}$$

which possesses Landau's phonon property, the sound velocity being $[\rho v(0)/m]^{1/2}$. As $k \to \infty$, provided that $v(k)$ does not diverge as fast as k^2,

$$\omega(k) \to \frac{\hbar^2 k^2}{2m}, \quad k \to \infty \tag{5.3.14}$$

the free helium-atom energy. In between these limits, the behavior of $\omega(k)$ depends sensitively upon the detailed form of $v(k)$. For example, if $v(k)$ equals a constant $v(0)$ throughout, the linear small k portion will join smoothly onto the quadratic tail, since $\omega(k)$'s first and second derivatives will both be deprived of zeros. If $v(k)$ decreases with k from a positive $v(0)$ toward some negative constant, $\omega(k)$ will have an inflection point. If $v(k)$ descreases as $k^{-2}e^{-rk}$ at large k, there will be a mximum and a minimum. It was shown by Brueckner and Sawada [91] that for $v(k) \propto \sin(kb)/(kb)$, Landau's phonon–

roton spectrum can be reproduced qualitatively. If we are willing to forget momentarily that liquid helium is, in reality, dense and strongly interacting, Bogoliubov's approximation must certainly be regarded as extremely attractive. After all, the noninteracting Bose gas model yields simply a free-particle spectrum $\hbar^2k^2/2m$, which upon the introduction of a small repulsive interaction was shown to convert readily to a (collective) phonon spectrum. Also, the two branches postulated by Landau seem to have turned out to be two sections of one continuous curve. Henceforth we shall often refer to the entire curve as the phonon spectrum.

To be more realistic, at high densities and strong interactions one must introduce a few refinements. The theory of elementary excitations in strongly interacting Bose liquids is a thoroughly worked-over field. We cite here several references mentioned earlier: Beliaev [67], Hugennholtz and Pines [68], Brueckner and Sawada [91], and Brandow [65]. The minimal ingredients in a successful theory must include proper treatment for (i) depletion of the condensate, (ii) repeated interactions between a pair of bosons, and (iii) the perturbing influence of terms with three and four operators.

Barring the availability of a sound formal analysis, one resorts to taking a more practical path. As in the case of the ground-state calculation, the variational principle becomes nearly indispensable. What is needed is a physically motivated trial wave function for an excited state with momentum $\hbar k$. Such a wave function is automatically orthogonal to the ground-state wave function. Therefore, the minimization of the expectation value of H with respect to the wave function will yield an upper bound to the energy of the excited state. In 1954, Feynman [92] published a paper of long-lasting influence in which he considered several possible models for an excitation in liquid ^{4}He. Each led to a wave function of the form

$$\Phi_{\mathbf{k}} = \sum_{i=1}^{N} f(\mathbf{r}_i)\,\Psi_0, \tag{5.3.15}$$

where Ψ_0 is the ground-state eigenfunction of the system and $f(\mathbf{r}_i)$ takes on the form of a plane wave. In other words,

$$\Phi_{\mathbf{k}} = \rho_{\mathbf{k}}\Psi_0. \tag{5.3.16}$$

The expectation value of H with respect to such a wave function was derived in Section 5.2.3, and was given by equation (5.2.96). Consequently, Feynman's excitation spectrum can be simply expressed as

$$\begin{aligned}\omega_F(k) &= E_{\mathbf{k}} - E_0\\ &\le \frac{\langle\Phi_{\mathbf{k}}|H|\Phi_{\mathbf{k}}\rangle}{\langle\Phi_{\mathbf{k}}|\Phi_{\mathbf{k}}\rangle} - E_0 = \frac{\hbar^2k^2}{2mS(k)}.\end{aligned} \tag{5.3.17}$$

Taking the zero-temperature liquid structure function $S(k)$ from low-temperature X-ray scattering experiment, he found the spectrum shown in

Figure 5.3.1. The roton gap was about twice as big as Landau's value, but several important features became uncovered. The phonon and roton excitations indeed fall on the same curve. Also, the excitation results from density fluctuation, and is intimately related to the liquid structure function. It can, therefore, be measured. And all this for *real* helium! It was accomplished with a great deal of physical insight (best exposed by Feynman himself in Reference [92]), very little algebra, and *no adjustable parameter.*

Two years later, in a sequel to Reference [92], Feynman and Cohen [93] proposed an improved trial wave function which took into account backflow. This resulted in an energy spectrum that lay only slightly above Landau's curve.

In 1957, Cohen and Feynman [94] proposed that inelastic neutron scattering may be used for probing the excitation spectrum directly. The idea is simple. Let a neutron of momentum $\hbar\mathbf{p}$ scatter once inside liquid helium and emerge with momentum $\hbar\mathbf{q}$. The momentum transfer is

$$\hbar\mathbf{k} \equiv \hbar(\mathbf{p} - \mathbf{q}). \tag{5.3.18}$$

In the inelastic process, liquid helium becomes excited, causing the neutron to suffer an energy loss ω. By assuming a δ-function potential between the neutron ($\mathbf{R}$) and the ith helium nucleus

$$v_{n\text{-He}} = a\delta(\mathbf{R} - \mathbf{r}_i), \tag{5.3.19}$$

one finds the total interaction operator to be

$$V = a \sum_{i=1}^{N} \delta(\mathbf{R} - \mathbf{r}_i). \tag{5.3.20}$$

The golden rule states that the scattering cross section is proportional to the matrix element of V, and energy is conserved. Using Born approximation and denoting the final states of helium by $|n\rangle$, one obtains after summing over all final states

$$\begin{aligned} \sigma &\propto \sum_n \left| a \sum_i \langle n | e^{i(\mathbf{p}-\mathbf{q})\cdot \mathbf{r}_i} | 0\rangle \right|^2 \delta(E_n - E_0 - \omega) \\ &\propto \sum_n |\langle n | \rho_{\mathbf{k}} | 0 \rangle|^2 \delta(E_n - E_0 - \omega) \equiv NS(k, \omega). \end{aligned} \tag{5.3.21}$$

Thus by locating the peak in the scattering amplitude for fixed k, one can find the excitation energy.

Over the last decade, much work has been done along these lines, in particular by Woods and co-workers [95–97]. We shall return often to compare results of microscopic calculations with the experimental data thus obtained. In the meantime, let us state a few useful sum rules which govern the scattering amplitude $S(k, \omega)$, alternately known as the dynamic structure function.

By integrating $S(k, \omega)$ over ω, applying the completeness condition, and comparing with equation (5.2.9), we find

$$\int_0^\infty S(k, \omega)\, d\omega = S(k). \qquad [S(k) \text{ sum rule}] \quad (5.3.22)$$

Thus inelastic neutron scattering also yields $S(k)$. In fact, neutron scattering [95] and X-ray scattering [84, 85] produce generally compatible data. By manipulating commutators between $\rho_{\mathbf{k}}^\dagger$, $\rho_{\mathbf{k}}$ and H, and by other shortcut methods, four other sum rules can be derived [98]:

$$\int_0^\infty \omega S(k, \omega)\, d\omega = \frac{\hbar^2 k^2}{2m}, \qquad [f \text{ sum rule}] \quad (5.3.23)$$

$$\lim_{k\to 0} \int_0^\infty \frac{1}{\omega} S(k, \omega)\, d\omega = \frac{1}{2ms^2}, \qquad [\text{compressibility sum rule}] \quad (5.3.24)$$

$$\int_0^\infty \omega^2 S(k, \omega)\, d\omega = \frac{\hbar^4 k^4}{2m^2 S(k)} + \hbar s k\left[2 - S(k) - \frac{1}{S(k)}\right] + \frac{4\hbar s}{k} D(k), \quad (5.3.25)$$

and
$$\int_0^\infty \omega^3 S(k, \omega)\, d\omega = \frac{\hbar^4 k^4}{4m^2}\left\{\frac{\hbar^2 k^2}{2m} + \frac{4\langle 0|\text{K.E.}|0\rangle}{N} + \frac{2\rho}{k^4}\int (1 - \cos \mathbf{k}\cdot\mathbf{r}) g(r)(\mathbf{k}\cdot\nabla)^2 v(r)\, d\mathbf{r}\right\}, \quad (5.3.26)$$

where

$$D(k) = \frac{(N-1)}{k^2}\int (\cos \mathbf{k}\cdot\mathbf{r}_{12} - 1)\,(\mathbf{k}\cdot\nabla_1\Psi_0)\,(\mathbf{k}\cdot\nabla_2\Psi_0)\, d\tau. \quad (5.327)$$

We have presented here briefly the historical facts, the basic concepts, and some rigorous theoretical results. These form the necessary background for the discussion that follows.

5.3.2. Tamm–Dancoff Approximation

Feynman [92] pointed out that the wave function that corresponds to two excitations in liquid ^{4}He should have the form

$$\Phi_{\mathbf{k},\mathbf{l}} = \rho_{\mathbf{k}}\rho_{\mathbf{l}}\Psi_0, \quad (5.3.28)$$

a natural generalization of equation (5.3.16). The expectation value of H with respect to $\Phi_{\mathbf{k},\mathbf{l}}$ is

$$\frac{\langle \Phi_{\mathbf{k},\mathbf{l}}|H|\Phi_{\mathbf{k},\mathbf{l}}\rangle}{\langle \Phi_{\mathbf{k},\mathbf{l}}|\Phi_{\mathbf{k},\mathbf{l}}\rangle} = E_0 + \omega_F(k) + \omega_F(l) + 0\left(\frac{1}{N}\right), \quad (5.3.29)$$

as was seen earlier in equation (5.2.97). The overlap integral $\langle \Phi_{\mathbf{k},\mathbf{l}} | \Phi_{\mathbf{k},\mathbf{l}}\rangle$ equals $N^2 S(k)S(l)$.

To estimate self-energy corrections for an excitation, one could follow Tamm [102] and Dancoff [103] and consider the dressing of Feynman's phonon by intermediate states involving two phonons. The relevant matrix elements include those of 1 and H connecting the one-phonon state $\Phi_{\mathbf{k}}$ to two-phonon states $\Phi_{\mathbf{k-l,l}}$. They were given earlier in equations (5.2.99)–(5.2.102), and were schematically represented by the vertices shown in Figure 5.3.2. Kuper [104] carried out a second-order Rayleigh–Schrödinger perturbation calculation, finding

$$E_{\mathbf{k}} \approx E_0 + \omega_F(k) + \frac{1}{2}\sum_{\mathbf{l}}{}' \frac{|\langle\Phi_{\mathbf{k}}|H - E_0 - \omega_F(k)|\Phi_{\mathbf{k-l,l}}\rangle|^2/[N^3S(k)S(l)S\,|(\mathbf{k}-\mathbf{l}|)]}{\omega_F(k) - \omega_F(l) - \omega_F(|\mathbf{k}-\mathbf{l}|),}, \tag{5.3.30}$$

and a lowering of Feynman's roton energy to the vicinity of Landau's empirical value. In certain regions of k, for example, for small k and assuming no dispersion, the energy denominator in equation (5.3.30) has zeros, which give rise to nonphysical divergences in the energy correction. Kuper further cautioned that the size of the correction (about $\frac{1}{2}\,\omega_F(k)$) indicates that the three-phonon interaction is not small, and the perturbation theory may well not converge.

Both of these difficulties were subsequently resolved by Jackson and Feenberg [105]. First, of all, to avoid singularities in the energy correction, one can use the Brillouin–Wigner perturbation formula

$$E_{\mathbf{k}} \approx E_0 + \omega_F(k) + \frac{1}{2}\sum_{\mathbf{l}}{}' \frac{|\langle\Phi_{\mathbf{k}}|H - E_0 - \omega_F(k)|\Phi_{\mathbf{k-l,l}}\rangle|^2/[N^3S(k)S(l)S(|\mathbf{k}-\mathbf{l}|)]}{E_{\mathbf{k}} - [E_0 + \omega_F(l) + \omega_F\,(|\mathbf{k}-\mathbf{l}|)]} \tag{5.3.31}$$

which differs from equation (5.3.30). in that the renormalized energy $(E_{\mathbf{k}} - E_0)$ itself appears in the energy denominator. The solution of equation

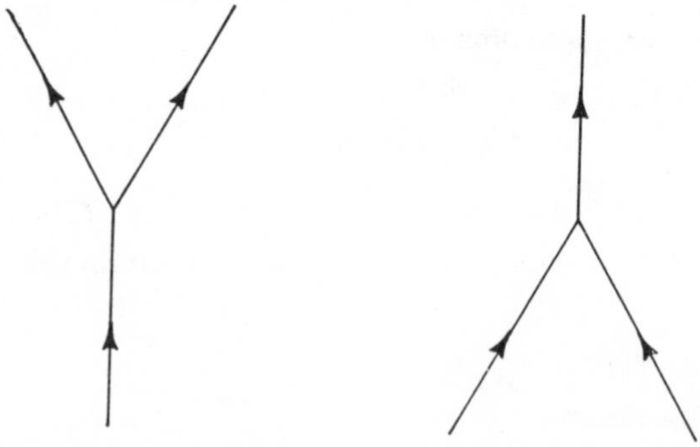

FIGURE 5.3.2. Virtual decay and coalescence of Feynman phonons.

(5.3.31) becomes now somewhat complicated for the same reason. One now has an "integral equation," which may be solved by iteration. For starting values of $E_{\mathbf{k}}$ one could, for example, employ the experimental curve $E_0 + \omega(k)$. On the other hand, note that equation (5.3.31) is not really an integral equation in the usual sense. For given $\mathbf{k}$, $(E_{\mathbf{k}} - E_0)$ is a parameter which appears in the integration; but there is no integration over $\mathbf{k}$. It is therefore possible to solve equation (5.3.31) as an implicit algebraic equation. Jackson and Feenberg's results appear in Figure 5.3.3 along with the Feynman–Cohen sepctrum [93], Kuper's roton energy [104], and the experimental curve obtained by neutron scattering [95]. As for the reliability of the perturbation expansion, Jackson and Feenberg computed the weight $P(k)$ of the two phonon component in the perturbed wave function, thus,

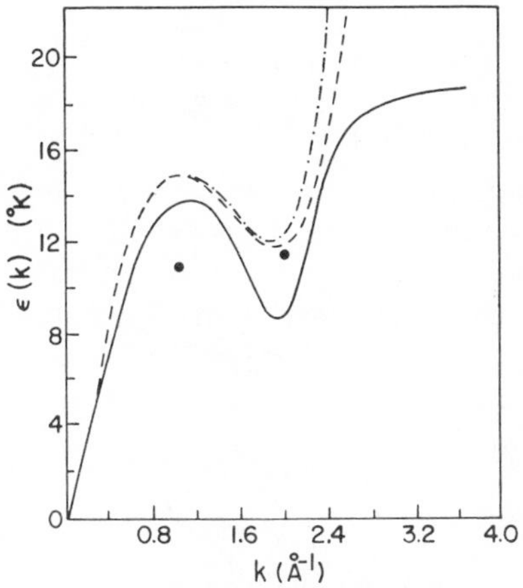

FIGURE 5.3.3. The phonon–roton spectrum.---------, Jackson–Feenberg [105]; -·-·-· Feynman–Cohen [93]; · ·, [104];———, experiment [95].

$$\frac{P(k)}{1-P(k)} = \frac{1}{2}\sum_{\mathbf{l}}{}' \frac{|\langle\Phi_{\mathbf{k}}|H - E_0 - \omega_F(k)|\Phi_{\mathbf{k-l,l}}\rangle|^2/[N^3S(k)S(l)S(|\mathbf{k}-\mathbf{l}|)]}{\{E_{\mathbf{k}} - [E_0 + \omega_F(l) + \omega_F(|\mathbf{k}-\mathbf{l}|)]\}^2} \tag{5.3.32}$$

and found $P(k)$ consistently small compared to 1. At the maximum of the spectrum ("maxon"), $P(1\ \text{Å}^{-1}) \approx 0.20$; and at the roton minimum, $P(2\ \text{Å}^{-1}) \approx 0.12$. Therefore, third- and higher-order contributions are not expected to be important.

Lee [106] considered the third-order contribution more accurately for small k by taking into account the four-phonon vertex connecting different two-phonon states. Matrix elements of 1 and H were computed and entered into the Rayleigh–Schrödinger correction formula

$$\Delta\omega_e(k) = \sum_{\mathbf{l},\mathbf{p}}{}' \frac{\langle \Phi_{\mathbf{k}} | H - E_0 - \omega_F(k) | \Phi_{\mathbf{k}-\mathbf{l},\mathbf{l}}\rangle \langle \Phi_{\mathbf{k}-\mathbf{l},\mathbf{l}} | H - E_0 - \omega_F(k) | \Phi_{\mathbf{k}-\mathbf{p},\mathbf{p}}\rangle}{[\omega_F(k) - \omega_F(l) - \omega_F(|\mathbf{k}-\mathbf{l}|)]}$$
$$\cdot \frac{\langle \Phi_{\mathbf{k}-\mathbf{p},\mathbf{p}} | H - E_0 - \omega_F(k) | \Phi_{\mathbf{k}}\rangle}{[\omega_F(k) - \omega_F(p) - \omega_F(|\mathbf{k}-\mathbf{p}|)]}$$
$$\times [N^5 S(k) S(l) S(|\mathbf{k}-\mathbf{l}|) S(p) S(|\mathbf{k}-\mathbf{p}|)]^{-1}. \tag{5.3.33}$$

The results are given in a power series of k:

$$\omega(k) = \omega_F(k)[1 + c_1' k + c_2' k^2 + \cdots], \tag{5.3.34}$$

with

$$c_1' = 0, \tag{5.3.35a}$$

and [107]

$$c_2' \approx c_{2a}' + c_{2e}' = -0.462 - 0.0678 = -0.530, \tag{5.3.35b}$$

where c_{2a}' comes from the second-order correction given in equation (5.3.30) and c_{2e}' comes from the third-order correction $\Delta\omega_e(k)$ of equation (5.3.33). (The subscripts will be explained in the next section.) This number can be improved slightly if one uses the variational formula of Goldhammer and Feenberg [108, 109]:

$$c_2' \approx \frac{(c_{2a}')^2}{(c_{2a}' - c_{2e}')}, \tag{5.3.36}$$

to -0.542.

A more formal way to present the Tamm–Dancoff approximation is as follows:

Take as a trial wave function a linear combination of one- and two-phonon states

$$\psi_{\mathbf{k}} = \zeta_{\mathbf{k}} |\mathbf{k}) + \tfrac{1}{2} \sum_{\mathbf{l}}{}' \xi_{\mathbf{k}}(\mathbf{l}) |\mathbf{k}-\mathbf{l}, \mathbf{l}), \tag{5.3.37}$$

where

$$|\mathbf{k}) = \frac{\rho_{\mathbf{k}} \Psi_0}{\sqrt{N S(k)}} \tag{5.3.38}$$

and

$$|\mathbf{k}-\mathbf{l}, \mathbf{l}) = \frac{\rho_{\mathbf{k}-\mathbf{l}} \rho_{\mathbf{l}} \psi_0}{\sqrt{N^2 S(|\mathbf{k}-\mathbf{l}|) S(l)}} \tag{5.3.39}$$

are normalized, and the coefficients $\zeta_{\mathbf{k}}$ and $\xi_{\mathbf{k}}$ (l) measure the relative weights of the one- and two-phonon components. The best linear combination requires the energy to be stationary, and can be determined variationally. To evaluate

$$E_{\mathbf{k}} \equiv E_0 + \omega(k) \approx \frac{\langle \psi_{\mathbf{k}} | H | \psi_{\mathbf{k}} \rangle}{\langle \psi_{\mathbf{k}} | \psi_{\mathbf{k}} \rangle}, \tag{5.3.40}$$

one needs

$$(\mathbf{k}|H|\mathbf{k}-\mathbf{l},\mathbf{l}) \equiv E^0(\mathbf{k}|\mathbf{k}-\mathbf{l}) + (\mathbf{k}|W|\mathbf{k}-\mathbf{l},\mathbf{l}), \tag{5.3.41}$$

and

$$(\mathbf{k}-\mathbf{h},\mathbf{h}|H|\mathbf{k}-\mathbf{l},\mathbf{l}) \equiv E_0(\mathbf{k}-\mathbf{h},\mathbf{h}|\mathbf{k}-\mathbf{l},\mathbf{l}) + (\mathbf{k}-\mathbf{h},\mathbf{h}|W|\mathbf{k}-\mathbf{l},\mathbf{l}), \tag{5.3.42}$$

where in convolution approximation,

$$(\mathbf{k}|W|\mathbf{k}-\mathbf{l},\mathbf{l}) = \frac{\hbar^2}{2m}[\mathbf{k}\cdot(\mathbf{k}-\mathbf{l})S(l) + \mathbf{k}\cdot\mathbf{l}S(|\mathbf{k}-\mathbf{l}|)]\cdot \cdot[NS(k)S(l)S(|\mathbf{k}-\mathbf{l}|)]^{-1/2}, \tag{5.3.43}$$

$$\begin{aligned}(\mathbf{k}-\mathbf{h},\mathbf{h}|W|\mathbf{k}-\mathbf{l},\mathbf{l}) = \frac{\hbar^2}{2m}[&(\mathbf{k}-\mathbf{h})\cdot(\mathbf{k}-\mathbf{l})S(h)S(l)S(|\mathbf{h}-\mathbf{l}|)\\ &+ \mathbf{h}\cdot\mathbf{l}S(|\mathbf{k}-\mathbf{h}|)S(|\mathbf{k}-\mathbf{l}|)S(|\mathbf{h}-\mathbf{l}|)\\ &+ \mathbf{h}\cdot(\mathbf{k}-\mathbf{l})S(|\mathbf{k}-\mathbf{h}|)S(l)S(|\mathbf{k}-\mathbf{h}-\mathbf{l}|)\\ &+ \mathbf{l}\cdot(\mathbf{k}-\mathbf{h})S(h)S(|\mathbf{k}-\mathbf{l}|)S(|\mathbf{k}-\mathbf{h}-\mathbf{l}|)]\\ &\times [N^2S(h)S(|\mathbf{k}-\mathbf{h}|)S(l)S(|\mathbf{k}-\mathbf{l}|)]^{-1/2},\end{aligned} \tag{5.3.44}$$

$$(\mathbf{k}|\mathbf{k}-\mathbf{l},\mathbf{l}) = N^{-1/2}[S(k)S(l)S(|\mathbf{k}-\mathbf{l}|)]^{1/2}, \tag{5.3.45}$$

and

$$\begin{aligned}(\mathbf{k}-\mathbf{h},\mathbf{h}|\mathbf{k}-\mathbf{l},\mathbf{l}) = N^{-1}[&-2 + S(k) + S(|\mathbf{h}-\mathbf{l}|)\\ &+ S(|\mathbf{k}-\mathbf{h}-\mathbf{l}|)][S(h)S(|\mathbf{k}-\mathbf{h}|)S(l)S(|\mathbf{k}-\mathbf{l}|)]^{1/2}.\end{aligned} \tag{5.3.46}$$

The minimization of $E_{\mathbf{k}}$ with respect to $\psi_{\mathbf{k}}$ leads to two Euler–Lagrange equations

$$\frac{\delta E_{\mathbf{k}}}{\delta \zeta_{\mathbf{k}}} = 0, \quad \text{and} \quad \frac{\delta E_{\mathbf{k}}}{\delta \xi_{\mathbf{k}}(\mathbf{l})} = 0,$$

or explicitly,

$$\begin{cases}[\omega(k) - \omega_F(k)]\,\zeta_{\mathbf{k}} = \frac{1}{2}\sum_{\mathbf{l}}{}'\,\xi_{\mathbf{k}}(\mathbf{l})\,[(\mathbf{k}|W|\mathbf{k}-\mathbf{l},\mathbf{l}) - \omega(k)\,(\mathbf{k}|\mathbf{k}-\mathbf{l},\mathbf{l})],\\ [\omega(k) - \omega_F(h) - \omega_F(|\mathbf{k}-\mathbf{h}|)]\,\xi_{\mathbf{k}}(\mathbf{h}) = \zeta_{\mathbf{k}}[(\mathbf{k}|W|\mathbf{k}-\mathbf{h},\mathbf{h})\end{cases} \tag{5.3.47}$$

$$\begin{aligned}&- \omega(\mathbf{k})(\mathbf{k}|\mathbf{k}-\mathbf{h},\mathbf{h})]\\ &+ \tfrac{1}{2}\sum_{\mathbf{l}(\neq\mathbf{h})}{}'\,\xi_{\mathbf{k}}(\mathbf{l})[(\mathbf{k}-\mathbf{h},\mathbf{h}|W|\mathbf{k}-\mathbf{l},\mathbf{l}) - \omega(k)(\mathbf{k}-\mathbf{h},\mathbf{h}|\mathbf{k}-\mathbf{l},\mathbf{l})],\end{aligned} \tag{5.3.48}$$

which can be solved simulatneously for $\omega(k)$.

If one neglects the four-phonon vertices $(\mathbf{k}-\mathbf{h},\ \mathbf{h}\,|\,\mathbf{k}-\mathbf{l},\ \mathbf{l})$ and $(\mathbf{k}-\mathbf{h}, \mathbf{h}\,|\,W\,|\,\mathbf{k}-\mathbf{l},\ \mathbf{l})$, the second equation (5.3.48) can be solved readily for $\xi_{\mathbf{k}}(\mathbf{h})$ and substituted into the first equation (5.3.47). As $\zeta_{\mathbf{k}}$ cancels out, one regains the second-order Brillouin–Wigner formula used by Jackson and Feenberg. By retaining the four-phonon vertices and iterating equations (5.3.47)–(5.3.48) formally, one obtains an infinite series that depicts virtual phonon–phonon scattering to all orders, as shown schematically in Figure 5.3.4. While it is expected that the series converges rapidly, a numerical solution of these coupled integral equations is the best way to assure that no surprise is in store. Such a calculation was recently carried out by Lin-Liu and Woo [110]. At long wavelengths, by expanding $\omega(k)$ in a power series of k as in equation (5.3.34), or alternately,

$$\omega(k) = \hbar sk(1 + \omega_1' k + \omega_2' k^2 + \cdots), \quad k \to 0, \tag{5.3.49}$$

one finds

$$(\mathbf{k}\,|\,W\,|\,\mathbf{k}-\mathbf{l},\mathbf{l}) - \omega(k)\,(\mathbf{k}\,|\,\mathbf{k}-\mathbf{l},\mathbf{l}) \to k^{3/2}\sqrt{\frac{\hbar^3 s}{2mN}}\left[1 - S(l) - \frac{l\cos^2(\mathbf{k},\mathbf{l})S'(l)}{S(l)}\right], \tag{5.3.50}$$

and

$$(\mathbf{k}-\mathbf{h},\mathbf{h}\,|\,W\,|\,\mathbf{k}-\mathbf{l},\mathbf{l}) - \omega(k)\,(\mathbf{k}-\mathbf{h},\mathbf{h}\,|\,\mathbf{k}-\mathbf{l},\mathbf{l})$$
$$\to \frac{\hbar^2}{mN}\,\mathbf{h}\cdot\mathbf{l}\,[S(|\mathbf{h}-\mathbf{l}|) - S(|\mathbf{h}+\mathbf{l}|)]; \tag{5.3.51}$$

and equations (5.3.47)–(5.3.48) reduce to

$$\xi_{\mathbf{k}\to 0}(\mathbf{h}) = \frac{-1}{N}\sum_{\mathbf{l}(\neq\mathbf{h})}' \frac{S(h)}{4h^2}$$

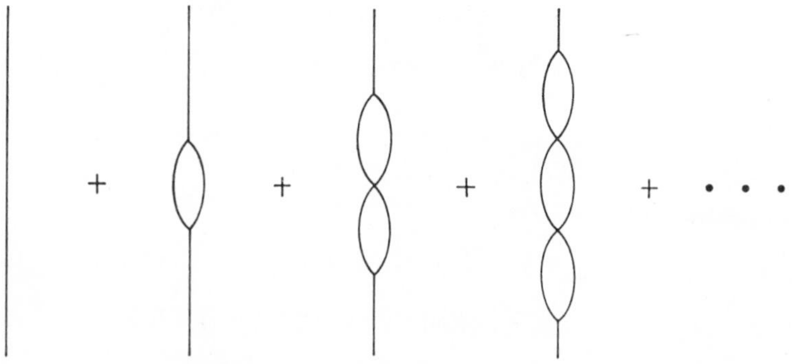

FIGURE 5.3.4. Renormalization of the phonon-roton spectrum by repeated scattering processes.

$$\cdot \left\{ \frac{[1 - S(l) - l\cos^2(\mathbf{k}, \mathbf{l})S'(l)/S(l)][1 - S(h) - h\cos^2(\mathbf{k}, \mathbf{h})S'(h)/S(h)]}{c_2'} \right.$$

$$\left. + 2\mathbf{h}\cdot\mathbf{l}\,[S(|\mathbf{h} - \mathbf{l}|) - S(|\mathbf{h} + \mathbf{l}|)] \right\} \xi_{\mathbf{k}\to 0}(\mathbf{l}). \qquad (5.3.52)$$

As shown in Reference [110], the eigenvalue c_2' turns out to be not very different from that given in equation (5.3.35b).

This procedure permits one to simultaneously seek a second branch of excitation: the two-roton branch [111–113]. By setting

$$\omega(k) \to 2\Delta - \delta, \quad k \to 0 \qquad (5.3.53)$$

where Δ is the roton gap, a second eigenvalue δ should be located at about the two-roton binding energy [111, 112]. In this picture, the phonon–roton branch and the two-roton branch appear naturally as the result of hybridizing one and two Feynman phonon states.

Yet another alternate way of renormalizing Feynman's phonon is available. The method was first attempted by Tan and is perhaps the most systematic. For convenience of illustration we shall deal only with the three-phonon vertex. Since it is possible to orthogonalize all two-phonon states with respect to the one-phonon state by Schmidt's process, one can consider the orthogonalization carried out and the resulting three-phonon matrix elements of H represented by $\langle \mathbf{k} | W | \mathbf{k} - \mathbf{l}, \mathbf{l} \rangle$. The Hamiltonian can then be approximated by the second-quantized form

$$H = E_0 + \sum_{\mathbf{k}} \omega_F(k) a_{\mathbf{k}}^\dagger a_{\mathbf{k}} + \tfrac{1}{2} \sum_{\mathbf{kl}}{}' \langle \mathbf{k} | W | \mathbf{k} - \mathbf{l}, \mathbf{l} \rangle\, a_{\mathbf{k}}^\dagger a_{\mathbf{k}-\mathbf{l}} a_{\mathbf{l}} + \text{H.c.} \qquad (5.3.54)$$

Introduce now the canonical transformation

$$a_{\mathbf{k}}^\dagger = e^{-B} \alpha_{\mathbf{k}}^\dagger e^{B}, \qquad (5.3.55)$$

where

$$B = B_3 + B_4, \qquad (5.3.56)$$

$$B_3 = \sum_{\mathbf{p},\mathbf{q}} (\mu_{\mathbf{pq}} \alpha_{\mathbf{p}}^\dagger \alpha_{\mathbf{q}}^\dagger \alpha_{\mathbf{p}+\mathbf{q}} - \mu_{\mathbf{pq}}^* \alpha_{\mathbf{p}+\mathbf{q}}^\dagger \alpha_{\mathbf{q}} \alpha_{\mathbf{p}}), \qquad (5.3.57)$$

$$B_4 = \sum_{\mathbf{l},\mathbf{p},\mathbf{q}} \nu_{\mathbf{lpq}} \alpha_{\mathbf{l}}^\dagger \alpha_{\mathbf{p}}^\dagger \alpha_{\mathbf{l}+\mathbf{q}} \alpha_{\mathbf{p}-\mathbf{q}}. \qquad (5.3.58)$$

The condition $\nu_{\mathbf{lpq}}^* = -\nu_{\mathbf{pl},-\mathbf{q}}$ is imposed to ensure that $B^\dagger = -B$, so that the transformation is canonical. One now makes a choice on ν, and then carry out the transformation on H. Collecting the coefficients of $\alpha^\dagger\alpha^\dagger\alpha$ or $\alpha^\dagger\alpha\alpha$ and setting the result to zero, μ is determined and H becomes approximately diagonal. If the exponentials e^{-B} and e^{B} are expanded in powers of the transformation parameters μ and ν, one finds as expected a perturbation expansion. In particular, the choice $\nu_{\mathbf{lpq}} = 0$ yields

$$\mu_{\mathbf{pq}} = \frac{1}{2} \frac{\langle \mathbf{p} + \mathbf{q} \,|\, W \,|\, \mathbf{p}, \mathbf{q} \rangle}{\omega_F(|\mathbf{p} + \mathbf{q}|) - \omega_F(p) - \omega_F(q)}: \tag{5.3.59}$$

the Rayleigh–Schrödinger result; the choice

$$-\tfrac{1}{2} \langle \mathbf{p} + \mathbf{q} | W | \mathbf{p}, \mathbf{q} \rangle + \mu_{\mathbf{pq}} [\omega_F(|\mathbf{p} + \mathbf{q}|) - \omega_F(p) - \omega_F(q)] = \tfrac{1}{2} \sum_{\mathbf{l}} \langle \mathbf{p} + \mathbf{q} | W | \mathbf{p} + \mathbf{l}, \mathbf{q} - \mathbf{l} \rangle \nu_{\mathbf{q-l,p+l,l}} \tag{5.3.60}$$

yields

$$\mu_{\mathbf{pq}} = \frac{1}{2} \frac{\langle \mathbf{p} + \mathbf{q} | W | \mathbf{p}, \mathbf{q} \rangle}{\omega(|\mathbf{p} + \mathbf{q}|) - \omega_F(p) - \omega_F(q)}: \tag{5.3.61}$$

the Brillouin–Wigner result. Clearly, this procedure can be extended to include two and more α terms in the transformation equation (5.3.55) in order to restore the truncated part of H.

The advantage of this method becomes clear when one wishes to transform a Hamiltonian that mixes phonon operators with operators representing other kinds of excitations present in the system. I shall come back to this later in discussing the microscopic theory of ^{3}He–^{4}He solutions.

5.3.3. Correlated Basis Functions

The Tamm–Dancoff approximation, presented in the way of the preceding section, may be regarded as the first stage in a CBF perturbation treatment. There are two kinds of expansions involved in the CBF. One is the expansion in powers of phonon interactions. The other is the expansion of the Hamiltonian itself—the decomposition of the Hamiltonian into various orders of "anharmonic" terms. When the Hamiltonian is truncated at the three- and four-phonon vertices of Figures 5.3.2 and 5.3.5, the Tamm–Dancoff approximation was shown to be adequate for summing the first type of expansion. We now turn out attention to the second type.

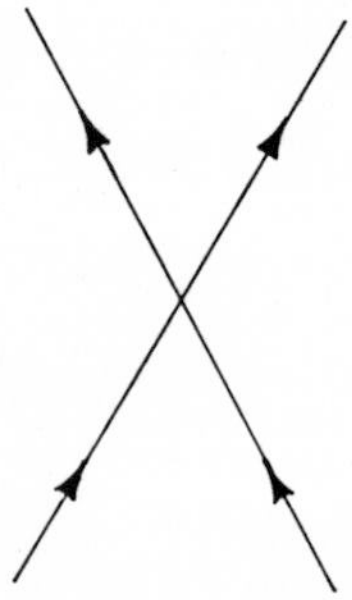

FIGURE 5.3.5. Scattering between Feynman phonons: the four-phonon vertex.

Let us construct a complete set of correlated basis functions, in much the same way as was done for the ground state [Section 5.2.3, equations (5.2.92) and (5.2.103)], or as in the last section with the inclusion of three and more phonons:

$$|\mathbf{k}-\mathbf{l}-\mathbf{h},\mathbf{l},\mathbf{h}) = \frac{\rho_{\mathbf{k}-\mathbf{l}-\mathbf{h}}\rho_{\mathbf{l}}\rho_{\mathbf{h}}\Psi_0}{\sqrt{N^3S(|\mathbf{k}-\mathbf{l}-\mathbf{h}|)S(l)S(h)}}, \tag{5.3.62}$$

etc.

Apply Schmidt's procedure to orthogonalize each class of n-phonon states with respect to all other classes, and let the resulting partially orthogonalized functions be represented by $|\mathbf{k}\rangle, |\mathbf{k}-\mathbf{l},\mathbf{l}\rangle, |\mathbf{k}-\mathbf{l}-\mathbf{h},\mathbf{l},\mathbf{h}\rangle$, etc. Evaluate all the matrix elements between these functions. Matrix elements of 1 will of course vanish between states with different numbers of phonons, but will remain (though modified) if the states connected belong to the same class. New matrix elements emerge: those connecting one- and three-phonon states (Figure 5.3.6), etc. In the limit $k \to 0$, $\langle\mathbf{k}|H|\mathbf{k}-\mathbf{l},\mathbf{l}\rangle$ and $\langle\mathbf{k}-\mathbf{h},\mathbf{h}|H|\mathbf{k}-\mathbf{l},\mathbf{l}\rangle$ are given by equations (5.3.50) and (5.3.51), while

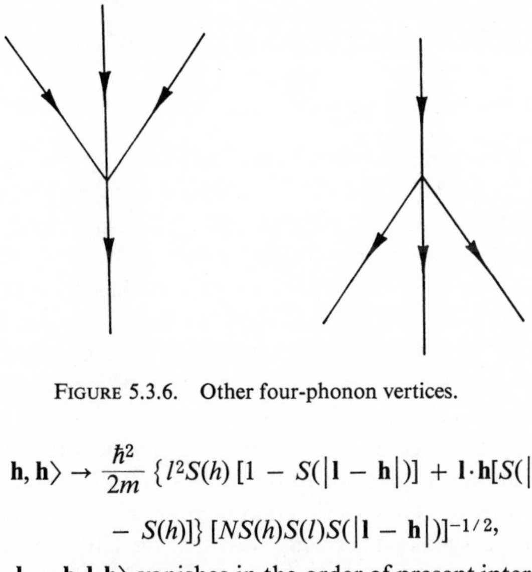

FIGURE 5.3.6. Other four-phonon vertices.

$$\langle\mathbf{l}|H|\mathbf{l}-\mathbf{h},\mathbf{h}\rangle \to \frac{\hbar^2}{2m}\{l^2S(h)[1-S(|\mathbf{l}-\mathbf{h}|)] + \mathbf{l}\cdot\mathbf{h}[S(|\mathbf{l}-\mathbf{h}|) - S(h)]\}[NS(h)S(l)S(|\mathbf{l}-\mathbf{h}|)]^{-1/2}, \tag{5.3.63}$$

and $\langle\mathbf{k}|H|\mathbf{k}-\mathbf{l}-\mathbf{h},\mathbf{l},\mathbf{h}\rangle$ vanishes in the order of present interest. Using the Rayleigh–Schrödinger expansion and classifying the contributions by the number of phonon momenta integrated over (so that the expansion parameter is roughly defined as the small quantity $\rho[S(l)-1]$ or $\rho S'(l)$), we [114] find the corrections to the energy spectrum represented by the diagrams in Figure 5.3.7, in which (a) and (e) are accounted for by the Tamm–Dancoff approximation, and were shown earlier to give rise to c'_{2a} and c'_{2e}. The rest

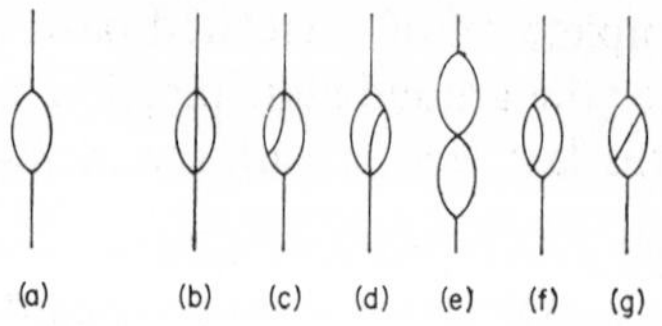

FIGURE 5.3.7. Perturbative corrections to Feynman's phonon spectrum.

of the diagrams are new. In the convolution approximation, diagrams (*b*)–(*d*) do not contribute, while

$$c'_{2f} = -0.0640, \tag{5.3.64a}$$

and

$$c'_{2g} = -0.0096. \tag{5.3.64b}$$

In Reference [114] an improved $S(h)$ was used, with the consequence that c'_{2e} now becomes -0.0591. Finally,

$$c'_2 \approx -0.462 + (-0.0591 - 0.0640 - 0.0096) = -0.595. \tag{5.3.65}$$

Or by using the Goldhammer–Feenberg [108, 109] formula,

$$c'_2 \approx \frac{(c'_{2a})^2}{[c'_{2a} - (c'_{2b} + c'_{2c} + \cdots + c'_{2g})]} = -0.648. \tag{5.3.66}$$

The procedure summarized here was analyzed for the case of a weakly interacting Bose gas [73, 75] and for a charged Bose gas [76] against Green's function calculations. Several different sets of correlated basis functions were studied, each with a different correlation factor. When a correlation factor other than the true ground state Ψ_0 is employed, the matrix elements become considerably more complicated. At least for the case of Jastrow correlation factors, the resulting formulas are still within manageable from [55, 58]. However, new diagrams appear that correspond to reversing the directions of certain propagators in Figure 5.3.7, as a consequence of the fact that certain matrix elements no longer vanish. When these changes are fully incorporated, all CBF's are shown to give rise to the same results, in exact agreement with those obtained by the Green's function method in every order of the expansion. Such analyses make it perfectly clear that the CBF method is not a mere extension of Jastrow's theory, since the Jastrow function need not even appear in the choice of the correlated basis.

This last fact was exploited in a recent perturbation calculation for the liquid structure function of ^{4}He. I mentioned in Section 5.2.5 that Jastrow calculations have led to discrepancies between $S_J(k)$ and the experimental $S(k)$; yet, other than introducing explicit three-particle factors into the variational wave function, there has been no attempt toward improving the theory. In Reference [76], a new method was suggested. Take two sets of CBF's, one using an optimized Jastrow correlation factor ψ_J, the other using the *unknown*, true ground-state eigenfunction Ψ_0. Derive with each the excitation spectrum, to whatever order of accuracy desired. The two expansion formulas will look different, one being a functional of $S_J(k)$, and the other a functional of $S(k)$. When a numerical ordering parameter exists, the expressions should be identical in every order, and the exact $S(k)$ can be obtained in terms of $S_J(k)$, order by order in a staggered sequential manner. Such is the case with model systems like the weakly interacting Bose gas or the high-density charged Bose gas. For the latter, $S(k)$ was calculated with little effort [76] to an order heretofore unattained. When there is no identifiable expansion parameter, such as in the case of liquid ^{4}He, the two excitation energy series can still be expanded to orders of sufficiently high accuracy and then equated. This results in an integral equation for $S(k)$ in terms of $S_J(k)$. The equation at least to low orders may be solved numerically. Armed with a nearly optimized Jastrow result $S_J(k)$, a perturbed liquid structure function was actually obtained in this manner [115]. The results are shown in Figure 5.3.8 in comparison with experiment. The improvement in the crucial region around the maximum is significant. More work in the equally important long-wavelength region yet remains to be performed.

5.3.4. Sum Rule Analysis at Long Wavelengths

For many years, microscopic theoretical calculations have been used only to furnish an atomistic basis for phenomenological models. There has been little, if any, direct application of the former. Recently, certain controversies have arisen concerning the finer details of the liquid structure and the elementary excitation spectrum. They cannot be settled by simple phenomeno-

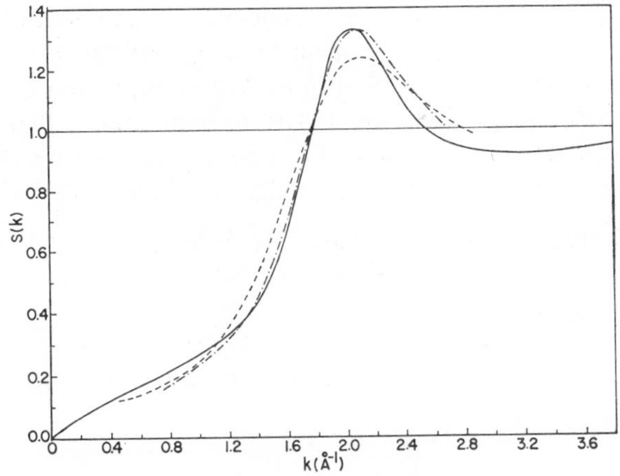

FIGURE 5.3.8. Static structure function of liquid ^{4}He.---------, Optimum Jastrow-theory result;–·–·–·–·, result of perturbation theory;———, experiment.

logical theories. As an example, I shall discuss in this section several aspects of the long-wavelength limit. For a thorough treatment, the reader should refer to other chapters in this book.

Let us *assume* from the outset that as $k \to 0$ all the relevant measurable quantities can be expanded in power series of k. In particular the phonon spectrum takes on the expansion formula (5.3.49) or (5.3.34). It is more convenient and physical to employ the dimensionless variable

$$\kappa \equiv \frac{\hbar k}{ms}, \tag{5.3.67}$$

in terms of which

$$\omega(k) = \hbar sk(1 + \omega_1\kappa + \omega_2\kappa^2 + \omega_3\kappa^3 + \omega_4\kappa^4 + \cdots)$$

$$= ms^2\kappa(1 + \omega_1\kappa + \omega_2\kappa^2 + \omega_3\kappa^3 + \omega_4\kappa^4 + \cdots) \tag{5.3.68a}$$

$$= \left[\frac{\frac{1}{2}ms^2\kappa^2}{S(\mathrm{k})}\right](1 + c_1\kappa + c_2\kappa^2 + c_3\kappa^3 + c_4\kappa^4 + \cdots). \tag{5.3.68b}$$

The coefficients ω_i measure the phonon dispersion. For reasons that are not totally clear, Landau assumed that $\omega(k)$ contains only odd-power terms, i.e., $\omega_1 = \omega_3 = \cdots = 0$, and suggested that ω_2 is negative, so that the phonon spectrum continuously turns downward. This discourages *real* dissociation of a phonon into two phonons at lower momenta, since energy and momentum cannot be simultaneously conserved. Almost no one doubted the validity of the odd-power expansion until 1970 [116], and no question was raised about the sign of the leading dispersion coefficient until about the same time [117, 118]. Since then a number of papers, both experimental and theoretical, have appeared on this subject.

According to Miller, Pines, and Nozieres [119], the dynamic structure function $S(k, \omega)$ can be separated into single-phonon and multiphonon components. The single-phonon component for long wavelengths is extremely stable: proportional to a δ-function in ω. The multiphonon component at $k \to 0$ goes as k^4. (Wong and Gould [120] recently argued that a coherent two-phonon component gives rise to a k^3 contribution. As we shall see, the effect of such a term on our analysis is insignificant.) We thus write

$$S(k, \omega) = Z(k)\,\delta(\omega - \omega(k)) + \xi_4 f_4(\omega)\kappa^4 + \xi_5 f_5(\omega)\kappa^5 + \xi_6 f_6(\omega)\kappa^6 + \cdots, \tag{5.3.69}$$

where

$$Z(k) = \tfrac{1}{2}\,\kappa(1 + z_1\kappa + z_2\kappa^2 + z_3\kappa^3 + z_4\kappa^4 + \cdots), \tag{5.3.70}$$

and the $f_i(\omega)$ are normalized:

$$\int_0^\infty f_i(\omega)\,d\omega = 1. \tag{5.3.71}$$

Expand further the static liquid structure function $S(k)$ and the momentum-dependent density–density response function $\chi(k)$:

$$S(k) = \tfrac{1}{2}\kappa(1 + s_1\kappa + s_2\kappa^2 + s_3\kappa^3 + s_4\kappa^4 + \cdots), \qquad (5.3.72)$$

and

$$\chi(k) = \frac{-1}{ms^2}(1 + \chi_1\kappa + \chi_2\kappa^2 + \chi_3\kappa^3 + \chi_4\kappa^4 + \cdots). \qquad (5.3.73)$$

We have, then, all the relevant quantities for applying the rigorous sum rules given in Section 5.3.1, equations (5.3.22)–(5.3.27). The compressibility sum rule (5.3.24) will be rewritten for small but finite k as follows:

$$\int_0^\infty \frac{1}{\omega} S(k, \omega)\, d\omega = -\tfrac{1}{2}\chi(k). \qquad (5.3.74)$$

Substituting the expansion formulas (5.3.68)–(5.3.73) into the sum rules (except for the ω^2 sum rule which has not yet been put to use) and identifying equal powers of κ, we obtain a large number of relations between the expansion coefficients. The coefficients, however, outnumber the relations, and cannot be determined uniquely. We are compelled to retreat toward a less ambitious goal of using these relations to correlate and test experimental data.

The data of interest include (i) X-ray diffraction measurement [84, 85] of $S(k)$, (ii) inelastic neutron scattering measurement [95] of $S(k, \omega)$ and its moments, and (iii) direct measurements [121, 122] of the ultrasonic velocity. From (iii) one gets an upper limit on the dispersion. The results obtained at Argonne [121] and Bell Labs [122] indicate without a doubt that

$$\omega_1 \approx 0, \qquad (5.3.74)$$

thus refuting an earlier numerical fit of the neutron-scattering data by Molinari and Regge [123] and supporting evidence by Anderson and Sabisky [124]. There is hope that the Bell Labs experiment will also conclusively yield ω_2. From (ii) comes a variety of useful information [95]. For the moment we shall accept just one piece of data, the location of the multiphonon peak

$$\langle\omega\rangle_4 = \frac{\int_0^\infty \omega f_4(\omega)\, d\omega}{\int_0^\infty f_4(\omega)\, d\omega} = (18 \pm 2)\ \mathrm{K}. \qquad (5.3.75)$$

The accuracy of $\langle\omega\rangle_4$ is not affected by the existence of a broad background, or any numerical fit over a range of κ, or any averaging process. It comes straight out of inspecting $S(k, \omega)$, and hence cannot be far wrong. From (i) comes [85] a carefully measured set of data points on $S(k)$ at small κ and exhaustive fitting by Hallock [125]. Even though there may yet remain large uncertainties in the fit, we shall accept Hallock's best fit for carrying out our present consistency check:

$$s_1 = 0, \tag{5.3.76a}$$

$$s_2 = -1.63, \tag{5.3.76b}$$

$$s_3 = 1.42, \tag{5.3.76c}$$

$$s_4 = 0.51. \tag{5.3.76d}$$

Given the experimental data of equations (5.3.74)–(5.3.76), we are still short of two numbers for use in our analysis. At this juncture, the microscopic theoretical calculation comes in. From equations (5.3.35a) and (5.3.66), with k in dimensionless units, we find

$$c_1 = 0, \tag{5.3.77a}$$

and

$$c_2 = -1.46. \tag{5.3.77b}$$

This completes the needed input to our sum rule analysis.

Among other things that come out of the analysis, one finds from the ω^3 sum rule and from combining the $S(k)$ and the f sum rule that the coefficient of the κ^3 term in $S(k)$, which exists according to Wong and Gould [120] must either be identically or nearly zero. One also obtains the following series:

$$\omega(k) = ms^2\kappa(1+0.17\kappa^2 + 0.78\kappa^3 + \omega_4\kappa^4 + \cdots);\ \omega_4 \approx -3.3, \tag{5.3.78}$$

$$Z(k) = \tfrac{1}{2}\kappa(1-1.63\kappa^2 - 0.78\kappa^3 + 0.51\kappa^4 - 2.46\kappa^5 + \cdots), \tag{5.3.79}$$

$$S_{\text{m.ph.}}(k) \equiv \int_0^\infty [S(k,\omega) - Z(k)\delta(\omega - \omega(k))]\,d\omega = 1.1\kappa^4 + 1.23\kappa^6 + \cdots, \tag{5.3.80}$$

and

$$\chi(k) = \frac{-1}{ms^2}(1 - 1.80\kappa^2 - 1.56\kappa^3 + \cdots), \tag{5.3.81}$$

Also,

$$\langle\omega^3\rangle^{1/3} = 41\ \text{K}. \tag{5.3.82}$$

These formulas (5.3.78)–(5.3.82) may be compared to the neutron-scattering data [95]. Throughout the region between 0 and the maxon (~ 1 Å^{-1}), $\omega(k)$ fits the spectrum beautifully. Up to $k = 0.6 - 0.7$ Å^{-1}, the single-phonon structure function $Z(k)$, the multiphonon structure function $S_{\text{m.ph.}}(k)$, and the density–density response function $\chi(k)$ fall generally well within experimental uncertainties. The value of $\langle\omega^3\rangle^{1/3}$ is consistent with the experimental observation that the multiphonon contribution to the dynamic structure function, or $f_4(\omega)$, peaks toward higher frequencies.

Of most interest is equation (5.2.78). It indicates a slightly concave phonon dispersion, in agreement with the proposal by Maris and Massey [117, 118]. However, the coefficient 0.17 is an order of magnitude smaller than the 2.0 determined by Maris [118], and well below the 1.03 determined by Phillips et al. [126] via specific-heat measurements [127].

Kemoklidze and Pitaevskii [128], and, independently, Feenberg [129], argued from studying the contribution of van der Waals' forces to $S(k)$ that ω_3 is not only nonzero, but rather large and negative [128]:

$$\omega_3 \approx -12.6, \tag{3.83}$$

which means that if the Bell Labs [122] result that the dispersion is less than 1% throughout the region $\omega(k) \lesssim 5$ K (or $k \lesssim 0.27$ Å^{-1}) is reliable, ω_2 must be large and positive. The spectrum turns upward first, albeit slightly, then downward the rest of the way to the "maxon." A very mild shoulder is then expected. Assuming that the spectrum intersects the straight line $\hbar sk$ at $k = 0.27$ Å^{-1}, ω^2 can be estimated as having the value $+2.29$, in good agreement with that determined by Maris. The resulting expression

$$\omega(k) = ms^2\kappa(1 + 2.29\kappa^2 - 12.6\kappa^3 + \cdots) \tag{5.3.84}$$

does not agree with either neutron-scattering or X-ray scattering data above 0.3 Å^{-1}. It bears no resemblance to equation (5.3.78). We are then forced to conclude that different power-series expansions may well prevail on the two sides of ~ 0.3 Å^{-1}. This implies that nonanalytic expressions are involved so that if one insists on power-series fits, two different expansions will emerge, each with its own region of convergence. Such is the present state of understanding according to Pethick, Pines, and Woo.

While the deviation of $\omega(k)$ from linearity is a minute detail of which one entertains little hope of direct observation, it may be possible to detect certain indirect inferences. From equation (5.3.68) one finds

$$s_2 = c_2 - \omega_2. \tag{5.3.85}$$

Using $c_2 = -1.46$ and $\omega_2 = 2.29$, s_2 is determined as -3.75 for $k < 0.3$ Å^{-1}. Furthermore, from sum rules

$$z_2 = s_2, \tag{5.3.86}$$

and

$$z_3 = -\omega_3; \tag{5.3.87}$$

thus,

$$Z(k) = \tfrac{1}{2}\kappa(1 - 3.75\kappa^2 + 12.6\kappa^3 + \cdots), \tag{5.3.88}$$

which deviates from linearity by as much as 4% at $k = 0.21$ Å^{-1}. An effect of such magnitude may just be observable in neutron-scattering [130] and/or X-ray scattering [131] measurements. When such experiments are carried out, the phonon dispersion controversy will be settled once and for all.

5.3.5. Other Problems of Interest

We list here a few problems of interest concerning excitations in liquid ^{4}He.

(a) *Pressure Dependence of the Spectrum.* Recent neutron-scattering work [97], specific-heat measurements [126], and ultrasonic experiments [132] under pressure have opened up new areas of interest. The variation in pressure, or density, allows the experimenter to probe different regions of the interatomic potential and poses new and stringent tests on the microscopic theory.

(b) *Neutron Scattering at High-Momentum Transfers.* At high momenta, the spectrum should reduce to that of a free boson. Neutron scattering shows clearly that it terminates at less than 4 Å^{-1} with nearly zero slope. This calls for stability calculations. Jackson [133] began with a set of CBF, set up the truncated (at three-phonon vertices) Hamiltonian equation (5.3.54), defined the Green's function for phonon propagation, and solved the Dyson equation approximately for the linewidths at high k. The resulting width has a general linear dependence on k, which resembles the experimental result [95] with the ripples smoothed out. This is the first attempt to formulate a Green's function theory based on a correlated representation. The outcome is encouraging. Kerr, Pathak, and Singwi [134] have constructed a theory of neutron scattering for high k based on a generalized mean field approximation involving the polarization potential and the screened response function. Assuming a 6% condensate fraction, the line widths are in good qualitative agreement with experiment. In this approach, the condensate fraction is an input parameter. It differs from the earlier work by Harling [52] and Puff and Tenn [135], who attempted to extract the condensate fraction from the Doppler-effect-caused peak-on-a-peak in $S(k, \omega)$ at high k.

(c) *Hydrodynamics.* The good agreement between the Feynman–Cohen spectrum and the Jackson–Feenberg spectrum (Tamm–Dancoff approximation) suggests strong correlations between hydrodynamics and the multiphonon description. In a recent paper, Jackson [136] showed that to leading order a reversible transformation exists between the hydrodynamic representation and the CBF representation, thus bringing the microscopic descriptions of liquid ^{4}He one step closer to unification.

5.4. LIQUID ^{3}He

Our discussion of liquid ^{3}He will be brief. What is to be covered in this section is for three specific purposes: (i) to display in constrast two totally different approaches toward the microscopic theory of liquid ^{3}He, and to compare the results obtained; (ii) to develop some theoretical tools needed for our later review of the microscopic theory of ^{3}He–^{4}He solutions; and (iii) to outline a plausible scheme for going beyond Landau's theory.

5.4.1. Summary of Landau's Phenomenological Theory

By assuming that each eigenstate of a noninteracting Fermi system goes into a corresponding eigenstate of an interacting Fermi system as the interaction is turned on adiabatically, Landau [137, 138] defined the normal Fermi liquid and postulated the existence of a quasiparticle distribution function $n(\mathbf{k}, \sigma)$. The quasiparticle energy $\varepsilon(\mathbf{k}, \sigma)$ is defined in terms of the change δE in the energy of the system when the distribution is varied:

$$\delta E = \sum_{\mathbf{k}\sigma} \varepsilon(\mathbf{k}, \sigma)\, \delta n(\mathbf{k}, \sigma); \tag{5.4.1}$$

the interaction between the quasiparticles is described by the function $f(\mathbf{k}, \mathbf{k}'; \sigma, \sigma')$, defined as the change in $\varepsilon(\mathbf{k}, \sigma)$ with varying distribution:

$$\delta\varepsilon(\mathbf{k}, \sigma) \equiv \varepsilon(\mathbf{k}, \sigma) - \varepsilon_0(\mathbf{k}, \sigma) = \sum_{\mathbf{k}'\sigma'} f(\mathbf{k}, \mathbf{k}'; \sigma, \sigma')\, \delta n(\mathbf{k}, \sigma). \tag{5.4.2}$$

All thermodynamic and transport properties of the Fermi liquid can then be expressed in terms of the functions $\varepsilon(\mathbf{k}, \sigma)$ and $f(\mathbf{k}, \mathbf{k}'; \sigma, \sigma')$ evaluated at the Fermi momentun $\hbar k_F$. In particular, from $\varepsilon(\mathbf{k},\sigma)$ one obtains the effective mass m^* of the quasiparticle

$$m^* = \hbar^2 k_F \left(\frac{\partial \varepsilon}{\partial k}\right)^{-1}_{k=k_F}. \tag{5.4.3}$$

Let θ be the angle between $\mathbf{k}$ and $\mathbf{k}'$ and define

$$f^s(\theta) = \tfrac{1}{2}[f(\mathbf{k}, \mathbf{k}'; \sigma, \sigma) + f(\mathbf{k}, \mathbf{k}'; \sigma, -\sigma)]_{k=k'=k_F}, \tag{5.4.4a}$$

$$f^a(\theta) = \tfrac{1}{2}[f(\mathbf{k}, \mathbf{k}'; \sigma, \sigma) - f(\mathbf{k}, \mathbf{k}'; \sigma, -\sigma)]_{k=k'=k_F}. \tag{5.4.4b}$$

Furthermore, carry out the Legendre expansion

$$F_l^{s(a)} = \frac{2l+1}{2} \frac{\Omega k_F m^*}{\pi^2 \hbar^2} \int_{-1}^{1} f^{s(a)}(\theta) P_l(\cos\theta)\, d\cos\theta. \tag{5.4.5}$$

We have what is known as the Landau parameters, $F_l^{s(a)}$, in terms of which the following thermodynamic properties can be calculated:
effective mass (alternate formula):

$$m^* = m(1 + \tfrac{1}{3} F_1^s), \tag{5.4.6}$$

velocity of first sound:

$$c_1 = \frac{\hbar k_F}{m} \left(\frac{1 + F_0^s}{3 + F_1^s}\right)^{1/2}, \tag{5.4.7}$$

compressibility:

$$\kappa_T = m\rho c_1^2, \tag{5.4.8}$$

coefficient of thermal expansion:

$$\alpha = \frac{-2\pi^2}{3}\frac{k_B^2 T}{\hbar^2 k_F c_1^2 m}\left(\frac{3}{2}\frac{\partial m^*}{\partial \rho} - m^*\right), \tag{5.4.9}$$

specific heat:

$$c_\Omega = \frac{m k_F k_B^2 T}{3\hbar^2}\left(1 + \frac{1}{3}F_1^s\right), \tag{5.4.10}$$

magnetic susceptibility:

$$\chi = \chi_0\left(\frac{1 + \frac{1}{3}F_1^s}{1 + F_0^a}\right), \tag{5.4.11}$$

where χ_0 is the magnetic susceptibility for a free Fermi gas. For transport properties, Abrikosov and Khalatnikov [139] obtained the following approximate formulas:

thermal conductivity:

$$\kappa' = T^{-1}\frac{4\pi^2}{3}\frac{\hbar^6 k_F^3}{m^{*4}}\langle \hat{w} \sin\tfrac{1}{2}\theta \tan\tfrac{1}{2}\theta\rangle, \tag{5.4.12}$$

viscosity:

$$\eta' = T^{-2}\frac{16}{45}\frac{\hbar^8 k_F^5 k_B^2}{m^{*4}}\langle \hat{w}(\sin\tfrac{1}{2}\theta)^3 \tan\tfrac{1}{2}\theta\,(\sin\phi)^2\rangle. \tag{5.4.13}$$

spin diffusion coefficient (Hone [140]):

$$D' = T^{-2}\frac{16\pi^2}{3}\frac{\hbar^8 k_F^2 k_B^2}{m^{*5}}(1 + F_0^a)\,\langle w' \sin\tfrac{1}{2}\theta \tan\tfrac{1}{2}\theta\,(1 - \cos\phi)\rangle. \tag{5.4.14}$$

In equations (5.4.12)–(5.4.14), the average is taken over all angles θ between two incident fermion momenta, and all angles ϕ between the planes containing the incident and outgoing momenta. The w's represent transition probabilities

$$\hat{w}(\theta,\phi) = \tfrac{1}{2}[\tfrac{3}{4}w_t(\theta,\phi) + \tfrac{1}{4}w_s(\theta,\phi)], \tag{5.4.15}$$

$$w_t(\theta,\phi) = \frac{2\pi^5\hbar^3}{m^{*2}k_F^2}\,|A^s(\theta,\phi) + A^a(\theta,\phi)|^2, \tag{5.4.16}$$

$$w_s(\theta,\phi) = \frac{2\pi^5\hbar^3}{m^{*2}k_F^2}\,|A^s(\theta,\phi) - 3A^a(\theta,\phi)|^2, \tag{5.4.17}$$

$$w'(\theta,\phi) = \frac{2\pi^5\hbar^3}{m^{*2}k_F^2}\,|A^s(\theta,\phi) - A^a(\theta,\phi)|^2, \tag{5.4.18}$$

and the A's are scattering amplitudes, of which Landau's theory provides only the forward scattering components

$$A^{s(a)}(\theta, 0) = \sum_l A_l^{s(a)} P_l(\cos\theta), \tag{5.4.19}$$

with

$$A_l^{s(a)} = F_l^{s(a)} \frac{2l+1}{1+F_l^{s(a)}}. \tag{5.4.20}$$

A number of authors [141–144] have treated the diffusion equations carefully and obtained the following formulas for the transport coefficients:

$$\kappa = (0.417\text{–}0.561)\kappa'. \tag{5.4.21}$$

$$\eta = (0.750\text{–}0.925)\eta', \tag{5.4.22}$$

$$D = (0.750\text{–}0.964)D', \tag{5.4.23}$$

where the numbers in parentheses represent lower and upper limits. They reflect the uncertainties caused by the unspecified dependence of scattering intensities on the angles θ and ϕ.

The following sections outline and compare two different microscopic methods for obtaining the various parameters introduced by Landau.

5.4.2. t-Matrix Method

The classic microscopic theory of liquid ^{3}He was due to Brueckner and Gammel [145] (BG), who employed the t-martix method invented originally by Brueckner for another Fermi system—nuclear matter. Brueckner reasoned that on accout of the strong interactions between two ^{3}He atoms, their motion in the medium formed by other atoms must be treated with care. Repeated virtual scatterings between a pair of atoms are permitted, while their interactions with the medium are handled in an average way. A reaction matrix, the t matrix, is then defined for the pair (i,j) by an integral equation (assuming spin-independent forces)

$$\langle \mathbf{k}_i', \mathbf{k}_j' | t | \mathbf{k}_i, \mathbf{k}_j \rangle = \langle \mathbf{k}_i', \mathbf{k}_j' | v | \mathbf{k}_i, \mathbf{k}_j \rangle + \sum_{\mathbf{k}_i'', \mathbf{k}_j''} \langle \mathbf{k}_i', \mathbf{k}_j' | v | \mathbf{k}_i'', \mathbf{k}_j'' \rangle \, G(\mathbf{k}_i'', \mathbf{k}_j''; \mathbf{k}_i, \mathbf{k}_j) \langle \mathbf{k}_i'', \mathbf{k}_j'' | t | \mathbf{k}_i, \mathbf{k}_j \rangle, \tag{5.4.24}$$

where G defines a propagator that causes the summation on $\mathbf{k}_i''$ and $\mathbf{k}_j''$ to be over only states above the Fermi surface, and to contain an energy denominator $\mathscr{E}$:

$$\mathscr{E} = e(k_i) + e(k_j) - e(k_i'') - e(k_j''). \tag{5.4.25}$$

The single particle energies $e(k)$ are defined in turn through the reaction

TABLE 5.4.1. Microscopically calculated properties of liquid He^3 as compared with experiment

		t-Matrix theory		*Jastrow theory*			*CBF*	
		Brueckner and Gammel [135] 1957	Østgaard [138–152] 1968–69	Massey and Woo [35,33] 1967	Schiff and Verlet [36] 1967	Woo [157] 1966	Tan and Feenberg [155] 1968	Expt
Potential used		Yntema–Schneider [24]	Yntema–Schneider [24]	deBoer–Michels [2]	deBoer–Michels [2]	Massey [15]	de Boer–Michels [2]	
Thermodynamic properties	$\rho(\text{Å}^{-3})$	0.0136	0.0151	0.0140	0.0142	0.0148	0.0156	0.0164
	$E_0(\text{K}/N)$	−0.96	−1.00	−0.75	−1.27	−1.35	−0.49	−2.52
	c_1(m/sec)	169	175	—	—	184	178	180
	K_T(%/atom)	5.3	4.3	—	—	3.6	3.9	3.8
	α_T	−0.10	−0.10	—	—	—	−0.14	−0.16
	χ/χ_0	12	10	—	—	13.9	9.2	9.2
	m^*/m	1.84	2–3	—	—	—	2.28	3.08
Transport properties	κT(erg/cm sec)	—	—	—	—	—	45–60	33
	$\eta T 10^{-6}$ poise K^2	—	—	—	—	—	4.9–6.0	2.0
	$\text{D}T^2 10^{-6}\text{cm}^2\text{K}^2$/sec	—	—	—	—	—	1.3–1.6	1.4

matrix

$$e(k_i) = \frac{\hbar^2 k_i^2}{2m} + \sum_{\mathbf{k}_j} [\langle \mathbf{k}_i, \mathbf{k}_j | t | \mathbf{k}_i, \mathbf{k}_j \rangle - \langle \mathbf{k}_j, \mathbf{k}_i | t | \mathbf{k}_i, \mathbf{k}_j \rangle]. \quad (5.4.26)$$

In principle, equations (5.4.24)–(5.4.26) form a coupled set of equations, and a self-consistent solution must be sought. In practice, however, due to the presence of other particles that may also be excited, the energy denominator contains excitation energy which is already in the medium; consequently, so does the t matrix. A truly self-consistent solution of these equations becomes a nearly impossible task, and several approximations must be introduced. In the original work by BG, the excitation energy in the medium is replaced by a constant parameter Δ, chosen to be the mean excitation energy of the system.

While v contains a strong, repulsive core, the t matrix is well behaved, so that a Hartree–Fock approximation can be employed to yield the ground-state energy as

$$E_0 = \sum_i \frac{\hbar^2 k_i^2}{2m} + \tfrac{1}{2} \sum_{ij} [\langle \mathbf{k}_i, \mathbf{k}_j | t | \mathbf{k}_i, \mathbf{k}_j \rangle - \langle \mathbf{k}_j, \mathbf{k}_i | t | \mathbf{k}_i, \mathbf{k}_j \rangle]. \quad (5.4.27)$$

Or, if one is more ambitious, one could calculate perturbation corrections to equation (5.4.27) by evaluating the linked-cluster diagrams in the Rayleigh–Schrödinger–Goldstone series. However, since equation (5.4.27) represents the result of a rearrangement of the series and a summation of the ladder diagrams (two particles repeatedly scattering against each other), one must take care not to double-count.

Such a procedure applied to liquid ^{3}He led to some very encouraging results, as shown in Table 5.4.1. BG carried out their calculation for two types of He–He potentials: the customary Lennard-Jones 6–12 with parameters given by de Boer and Michels, equations (5.1.30)–(5.1.31), and the Yntema–Schneider potential of equation (5.2.74). The former refused to yield binding; so the results quoted in Table 5.4.1 are for the latter, which except for the interior of the hard core does not differ drastically from the deBoer–Michels potential. The ground-state energy was too high, but then one has to recall the strong cancellation that occurs between kinetic and potential contributions, which in the ^{3}He case is even more severe than for ^{4}He on account of the smaller mass.

The method used by BG was improved in a semiphenomenological manner by Campbell and Brueckner [146]. The t matrix was assumed to depend only on the total momentum of the scattering pair, so that a partial wave expansion could be carried out. For $l \geq 2$, t_l was given the free scattering values. For $l = 0$ and 1, a repulsive core of adjustable range r_c was incorporated to fit experimental data. Also, self-consistency in the single-particle spectrum was abandoned in favor of an approximation (in which M^* is not Landau's effective mass m^*):

$$e(k) = \frac{\hbar^2}{2M^*}(k^2 - k_F^2) + \mu, \quad k \lesssim k_F \tag{5.4.28a}$$

$$= \frac{\hbar^2 k^2}{2m}, \quad k \gg k_F \tag{5.4.28b}$$

with

$$\mu = \text{chemical potential} = \frac{E_0}{N} \tag{5.4.29}$$

preserving the equality of the chemical potential with $e(k_F)$. The latter condition is known as the separation energy theorem, and was proven by Hugenholtz and van Hove [147]. By identifying the t matrix to Landau's $f(\mathbf{k},\mathbf{k}'; \sigma, \sigma')$, it became possible to compute Landau's parameters, and the transport properties as well. It turned out that except for the energy and the pressure, all thermodynamic and transport properties could be fitted very well with a single parameter $r_c \approx 0.69\sigma = 1.77$ Å.

In a series of papers published between 1968 and 1969, Østgaard [148–152] redid the t-matrix calculation. In the first [148] of the sequence, the general scheme was laid out. Bethe's reference spectrum method was to be used to avoid the self-consistency requirement on the single particle spectrum. A parameterized form of the single-particle spectrum was chosen:

$$e(k) = \frac{\hbar^2(k^2 + \gamma^2)}{mM^*}, \tag{5.4.30}$$

in which

$$\gamma^2 = 2\Delta k_F^2 - k_0^2 M^* / M_0^* \tag{5.4.31a}$$

on the energy shell (for the hole spectrum),

$$\gamma^2 = 3(\Delta k_F^2 + k_0^2) - 0.9 k_F^2 \tag{5.4.31b}$$

off the energy shell (for the particle spectrum),
and Δ, M^*, M_0^*, and k_0 are parameters for imitating self-consistency. The way Østgaard chose his parameters ($M^* = 1$ for the intermediate states) required that three-particle cluster contributions be separately evaluated and then added onto the energy of Brueckner's theory. This he accomplished by using the Bethe-Faddeev [153, 154] method in the second paper [149] of the sequence. It turned out that even though the two-particle contribution to the ground-state energy amounts to only -0.5 K per particle, the three-particle contribution can be as large as -1.0 K, yielding a total that is not too far from the experimental value of -2.5 K. There is, however, much uncertainty in these figures since a large number of approximations were necessary and there was no unique way to determine the parameters in his spectrum. The He–He potential employed was also Yntema–Schneider, as in BG.

In the third [150] and fourth [151] of the sequence Østgaard calculated the compressibility, thermal expansion coefficient, magnetic susceptibility, all shown in Table 5.4.1, and the quasiparticle energy, using essentially the results obtained in his first two papers. In Reference [151], the t matrix was identified with the scattering amplitudes mentioned in equations (5.4.16)–(5.4.20), for the purpose of estimating transport coefficients. The correct procedure was, however, given in his fifth [152] paper in which Landau's $f(\mathbf{k}, \mathbf{k}'; \sigma, \sigma')$ was obtained by taking the second variational derivative of energy with respect to the distribution. The result shows that even though the t matrix is still the leading term, the corrections from the medium are far from negligible. They turn the attractive t matrix into a repulsive f. The calculated Landau parameters at the experimental density are shown in Table 5.4.2, in comparison with the CBF results of Tan and Feenberg [155] and empirical values.

5.4.3. Jastrow Calculations

In the one paper that sets his work on two-particle type wave functions apart from earlier work based on the same idea, Jastrow [7] recognized the mathematical analogy between his particular variational calculation for the ground state of a Bose liquid and the classical theory of the imperfect gas. In the same paper, he proposed that an appropriate trial wave function for a Fermi system may well be of the form equation (5.2.21), with P now representing an antisymmetrization operator. The single-particle product is then a Slater determinant. For a fluid it must be a Slater determinant of plane-wave and spin-function elements.

^{3}He differs from ^{4}He in two respects. The mass is lighter, and it possesses nonzero spin. These two differences result in effects which are manifestly independent. The lighter mass causes larger zero-point motion, and makes the liquid even more "quantum." In order to isolate this quantum behavior and study its effect on the macroscopic properties of the system, a calculation for mass-3 bosons in the fashion of Section 5.2 will quite suffice. Massey [16] performed just such a calculation, and found that in comparison to

TABLE 5.4.2. Landau parameters: Microscopic theory versus experiment

	TF (Tan and Feenberg [155]) 1968	Ø(Østgaard [152]) 1969	Empirical		TF	Ø	Empirical
F_0^s	6.94	7.22	10.77	F_0^a	–0.75	–0.44	–0.665
F_1^s	3.84	3.27	6.25	F_1^a	–0.99	–1.17	–0.72
F_2^s	–0.35	1.72	—	F_2^a	0.86	1.49	—
F_3^s	0.68	1.10	—	F_3^a	0.33	0.87	—
F_4^s	–0.61	0.68	—	F_4^a	0.10	0.43	—
F_5^s	0.39	0.52	—	F_5^a	–0.18	0.19	—

liquid ^{4}He, the equilibrium density is about 30% lower, but somewhat higher than what the law of corresponding states [156, 157] predicts; the binding energy is about 2.5 times smaller, not far from what is expected [156, 157]. These were done for a rather unrealistic potential. A repetition of the calculation [40] using the deBoer-Michels potential led to the same equilibrium density 0.0157 Å^{-3}, which is slightly lower than the *true* ^{3}He density of 0.0164 Å^{-3}. The ground-state energy was found to be still smaller in magnitude, $-$ 2.06 K, as compared to the true ^{3}He value of $-$ 2.52 K. Paired-phonon analysis, used to optimize the Jastrow function, offers [38] an energy shift of about -0.55 K. A similar calculation by Schiff and Verlet [36] using molecular dynamics yielded -2.92 K at an equilibrium density of 0.0142 Å^{-3}.

Since ^{3}He is in reality a fermion, statistics must play an important role in modifying these results. First of all, Pauli exclusion acts in many ways as an additional repulsion; it is, therefore, expected to drive the density still lower. Also, the Pauli energy, which is positive, will raise the ground-state energy.

An actual Jastrow calculation begins with the trial wavefunction

$$\psi = \prod_{i<j=1}^{N} f(r_{ij}) \cdot \Phi_0(\mathbf{r}_1 s_1, \mathbf{r}_2 s_2, \cdots, \mathbf{r}_N s_N), \tag{5.4.32}$$

where

$$\Phi_0 = \sum_{\nu} (-1)^{\nu} P_{\nu} \phi_0, \tag{5.4.33}$$

and

$$\phi_0 = \prod_{i=1}^{N} \varphi_{\mathbf{k}_i \sigma_i}(\mathbf{r}_i s_i) = \prod_{i=1}^{N} e^{i\mathbf{k}_i \cdot \mathbf{r}_i} \, \delta_{\sigma_i}(s_i). \tag{5.4.34}$$

P_ν is an exchange operator; ν indicates the number of exchanges. For the ground state, the wave vectors $\mathbf{k}_i$ fill the Fermi sea. Assuming the system paramagnetic or antiferromagnetic, the spins will be half ups and half downs. The evaluation of the expectation value is by no means an easy task. Much work has been expended by a number of authors [7, 158–165] in pursuit of an ideal cluster-expansion procedure. We show here as a typical example Iwamoto and Yamada's [158, 163] cluster expansion for a simple two-particle operator

$$V = \sum_{i<j=1}^{N} v(r_{ij}). \tag{5.4.35}$$

Define the generalized normalization integral (the integration includes summation over spins)

$$J(\beta) \equiv \int |\psi|^2 \Phi_0^* \exp[\beta V] \phi_0 \, d\tau; \tag{5.4.36}$$

the expectation value of V is given by

$$\frac{\langle\psi|V|\psi\rangle}{\langle\psi|\psi\rangle} = \left[\frac{d}{d\beta}\ln J(\beta)\right]_{\beta=0}. \tag{5.4.37}$$

The problem reduces then to the computation of $J(\beta)$ by successive approximations. Define next for convenience the notation

$$e_{ab\cdots t}(1, 2, \cdots, n) = (-1)^{n-1} \sum_{\text{cyclic permutations}} \exp(i\mathbf{k}_{ab}\cdot\mathbf{r}_1 + i\mathbf{k}_{bc}\cdot\mathbf{r}_2 + \cdots + i\mathbf{k}_{ta}\cdot\mathbf{r}_n) \cdot\langle abc\cdots t, bc\cdots ta\rangle, \tag{5.4.38}$$

where

$$\langle ab\cdots t, a'b'\cdots t'\rangle \equiv \delta_{\sigma_a\sigma_a'}\delta_{\sigma_b\sigma_b'}\cdots\delta_{\sigma_t\sigma_t'}. \tag{5.4.39}$$

For example,

$$e_{ab}(1, 2) = -\exp(i\mathbf{k}_{ab}\cdot\mathbf{r}_{12})\langle ab, ba\rangle, \tag{5.4.40a}$$

$$e_{abc}(1, 2, 3) = \exp(i\mathbf{k}_{ab}\cdot\mathbf{r}_1 + i\mathbf{k}_{bc}\cdot\mathbf{r}_2 + i\mathbf{k}_{ca}\cdot\mathbf{r}_3)\langle abc, bca\rangle + \exp(i\mathbf{k}_{ab}\cdot\mathbf{r}_1 + i\mathbf{k}_{ba}\cdot\mathbf{r}_2 + i\mathbf{k}_{cb}\cdot\mathbf{r}_3)\cdot\langle abc, cab\rangle, \tag{5.4.40b}$$

etc.

It is then possible to write down the "approximants"

$$J_n(\beta) = 1, \tag{5.4.41a}$$

$$J_{mn}(\beta) = \int|\psi|^2\exp[\beta v(r_{12})][1 + e_{mn}(1, 2)]\,d\tau \equiv 1 + Y_{mn}(\beta), \tag{5.4.41b}$$

$$J_{lmn}(\beta) = \int|\psi|^2\exp[\beta v(r_{12}) + \beta v(r_{23}) + \beta v(r_{31})] \cdot[1 + e_{lm}(1, 2) + e_{mn}(1, 2) + e_{nl}(1, 2) + e_{lmn}(1, 2, 3)]\,d\tau \equiv 1 + Y_{lm}(\beta) + Y_{mn}(\beta) + Y_{nl}(\beta) + Y_{lmn}(\beta), \tag{5.4.41c}$$

etc.

The Y's are known as cluster integrals. Note that each approximant is represented as well as possible by a cluster series constructed from all preceding cluster integrals. The remainder is used to define a new cluster integral. At each stage the cluster series is formed so that detailed and strong cancellation occur between the various exchange terms in the approximant and the corresponding products of distinct cluster integrals, leaving a small remainder. Careful inspection of equation (5.4.41) shows that $Y_{mn}(\beta)$ is of order N^{-1}, $Y_{lmn}(\beta)$ of order N^{-2}, etc. Thus $J(\beta)$, which is $J_{12\cdots N}(\beta)$, is represented by an immensely complicated polynomial in N. However, $\ln J(\beta)$ must be proportional to N. As shown in References [158, 163], it is given by

$$\ln J(\beta) = \sum_{m<n} Y_{mn}(\beta)$$

$$+ \sum_{l<m<n} [Y_{lmn}(\beta) - Y_{lm}(\beta)Y_{mn}(\beta) - Y_{mn}(\beta)Y_{nl}(\beta)$$
$$- Y_{nl}(\beta)Y_{lm}(\beta)] + \cdots. \tag{5.4.42}$$

The evaluation of the cluster integrals will then give us the desired expectation value.

The expectation value of H contains differential operations; its evaluation requires more care and ingenuity. Equation (5.4.37) is, however, not totally useless. Since the expectation value of V is also given in terms of the radial distribution function $g(r)$, thus,

$$\frac{\langle\psi|V|\psi\rangle}{\langle\psi|\psi\rangle} = \frac{1}{2} N\rho \int g(r)v(r)\, d\mathbf{r}, \tag{5.4.43}$$

$g(r)$ can be extracted from the cluster-expansion series obtained.

As for the correlation factor, the Jastrow function is but one possible choice. One might alternately use

$$\psi = F(\mathbf{r}_1, \mathbf{r}_2, \cdots, \mathbf{r}_N)\cdot\Phi_0(\mathbf{r}_1 s_1, \mathbf{r}_2 s_2, \cdots, \mathbf{r}_N s_N), \tag{5.4.44}$$

where $F(\mathbf{r}_1, \mathbf{r}_2, \cdots, \mathbf{r}_N)$ is any sensible, symmetrized spin-independent function. One very convenient choice turns out to be

$$F(\mathbf{r}_1, \mathbf{r}_2, \cdots, \mathbf{r}_N) = \psi_B(r_{12}, r_{13}, \cdots, r_{N-1,N}), \tag{5.4.45}$$

the ground-state eigenfunction of the mass-3 Bose liquid. For then in the evaluation of the expectation value of H, one can make use of the relation

$$H\psi_B = E_B\psi_B, \tag{5.4.46}$$

where E_B denotes the ground-state energy of the mass-3 system. It was in part for this reason that Massey [16] solved the mass-3 Bose liquid problem.

The ground-state energy of ^{3}He in Jastrow calculations can be expressed as an infinite series. For example, choose ψ according to equations (5.4.44)–(5.4.45); then

$$E_0 \le \frac{\langle\psi|H|\psi\rangle}{\langle\psi|\psi\rangle} = E_B + E_1^F + E_2^F + E_3^F + \cdots, \tag{5.4.47}$$

where

$$E_1^F = \frac{3}{5} N \frac{\hbar^2 k_F^2}{2m}, \tag{5.4.48a}$$

$$E_2^F = 24N \frac{\hbar^2 k_F^2}{2m} \int_0^1 \left[S_B(2k_F x) - 1\right]\left[1 - \frac{3}{2}x + \frac{1}{2}x^3\right] x^4\, dx, \tag{5.4.48b}$$

$$E_3^F = -\left(\frac{3}{8\pi}\right)^3 N \frac{\hbar^2 k_F^2}{2m} \int_{\text{unit spheres}} x_{12}^2 S_B(k_F x_{12})[S_B(k_F x_{23}) - 1]$$
$$\cdot[S_B(k_F x_{31}) - 1]\, d\mathbf{x}_1\, d\mathbf{x}_2\, d\mathbf{x}_3, \tag{5.4.48c}$$

etc.,

TABLE 5.4.3. Cluster series for the ground state of liquid ^{3}He (energies in degrees Kelvin per particles)

	ρ[Å^{-3}]	E_B	E_1^F	E_2^F	E_3^F	E_0
Massey and Woo [35]	0.0140	–2.61	2.69	–0.70	–0.12	–0.75
Schiff and Verlet [40, 41]	0.0142	–2.92	2.71	–0.94	–0.12	–1.27
Experiment	0.0164	—	—	—	—	–2.52

with

$$S_B(k) = 1 + \rho \int e^{i\mathbf{k}\cdot\mathbf{r}}[g_B(r) - 1]\, d\mathbf{r}, \tag{5.4.49}$$

and

$$g_B(r_{12}) = \frac{N(N-1)}{\rho^2} \frac{\int \psi_B^2 d\tau(1,2)}{\int \psi_B^2 d\tau}. \tag{5.4.50}$$

Table 5.4.3 shows the energy series at equilibrium density as calculated by Massey and Woo [40] (including Campbell and Feenberg's optimization correction [38]) and Schiff and Verlet [41], each using deBoer–Michels potential. The latter authors actually varied the Jastrow correlation factor. Their best result for E_0 was $-$ 1.35 K. Along with other deduced thermodynamic quantities, these values are also displayed in Table 5.4.1. Figures

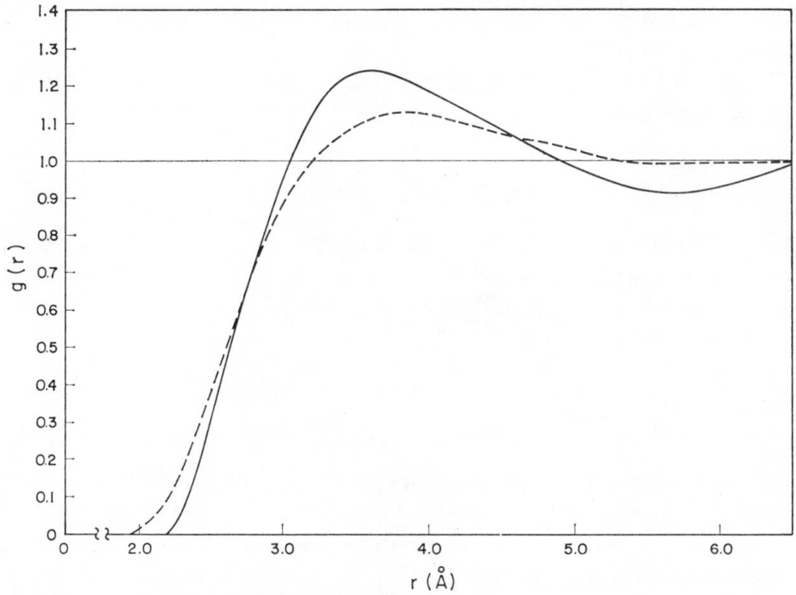

FIGURE 5.4.1. Radial distribution function of liquid ^{3}He.———, Experiment [84] at T = 0.56 K;---------, optimum Jastrow-theory result [41].

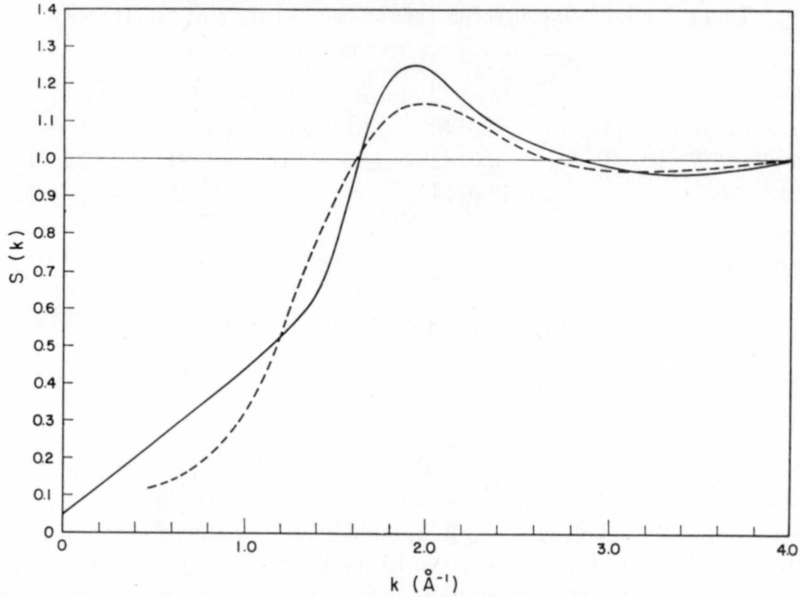

FIGURE 5.4.2. Static structure function of liquid ^{3}He.———, experiment [84] at $T = 0.56$ K; ---------, optimum Jastrow-theory result [41].

5.4.1 and 5.4.2 show the radial distribution function and the liquid structure function as compared to experiment.

The effective mass and Landau parameters obtained by variational calculations are totally unsatisfactory [163] for reasons to become clear presently.

5.4.4. Method of Correlated Basis Functions

As in the case of liquid ^{4}He, a set of correlated basis functions can be constructed to facilitate a perturbation calculation. According to the prescription given in Section 5.2.3, Equations (5.2.86)–(5.2.87), the CBF are given by

$$\psi_{\mathbf{k}_1\sigma_1,\,\mathbf{k}_2\sigma_2,\,\cdots,\,\mathbf{k}_N\sigma_N}(\mathbf{r}_1 s_1, \mathbf{r}_2 s_2, \cdots, \mathbf{r}_N s_N)$$
$$= \mathrm{F}(\mathbf{r}_1, \mathbf{r}_2, \cdots, \mathbf{r}_N)\cdot\Phi_{\mathbf{k}_1\sigma_1,\,\mathbf{k}_2\sigma_2,\,\cdots,\,\mathbf{k}_N\sigma_N}(\mathbf{r}_1 s_1, \mathbf{r}_2 s_2, \cdots, \mathbf{r}_N s_N), \qquad (5.4.51)$$

where the momentum and spin quantum numbers now range over all possible values. The variational function (5.4.44) may be regarded as the ground-state member of the set. The Jastrow trial wave function (5.4.32) and the ψ_B variety (5.4.45) are but two special choices of equation (5.4.44). In other words, the calculation described in the last section may be properly regarded as a zeroth order or unperturbed CBF calculation. Note that the CBF are characterized by the same momentum and spin labels as the eigenfunctions of a noninteracting Fermi system. This is consistent with Landau's notion of adiabatically switching on the interaction. It also helps to maintain the proper ordering of states. Note also that equation (5.4.51) does not explicitly contain collective modes of excitation. This is perfectly acceptable for liquid ^{3}He, for the sound spectrum is linear and stays much above the quadratic quasiparticle spectrum $\hbar^2k^2/2m^*$ throughout the important region of k. At high k, the collective excitation merges into the quasiparticle continuum; however, this is of little consequence at the Fermi-liquid temperatures (< 0.5 K).

Again as in the case of liquid ^{4}He, we must calculate the nondiagonal matrix elements of 1 and H between different basis functions. The nondiagonal elements are very difficult to compute. The reader should refer to the original papers [165, 166] for two different but complementary procedures. One [165] adapts the cluster-expansion procedure described in Section 5.4.3 for use with a slightly modified form of ψ. In equation (5.4.34), the single-particle orbital $\varphi_{\mathbf{k}_i\sigma_i}(\mathbf{r}_i s_i)$ is to be replaced by a linear combination

$$\varphi_{\mathbf{k}_i\sigma,\mathbf{k}_i'\sigma_i'}(\mathbf{r}_i s_i) \equiv a_i \varphi_{\mathbf{k}_i\sigma_i}(\mathbf{r}_i s_i) + a_i' \varphi_{\mathbf{k}_i\sigma_i}(\mathbf{r}_i s_i). \tag{5.4.52}$$

The evaluation of the expectation value of 1 and H follows as before. At the end, one reads off the coefficient of $a_i^* a_j^* a_i' a_j'$ to obtain the matrix element $\langle \mathbf{k}_i\sigma_i, \mathbf{k}_j\sigma_j | H | \mathbf{k}_i'\sigma_i', \mathbf{k}_j'\sigma_j' \rangle$ connecting two states that differ in two orbitals. The other procedure [166] is a more direct generalization of the Iwamoto–Yamada cluster development. The former method is more convenient, but contains certain simplifying approximations that are at times ambiguous. The latter is somewhat cumbersome, but more systemmatic.

In the only sequence of calculations carried out using the method of CBF [165, 167, 155], the correlation factor was chosen to be of the form equation (5.4.45), and Massey's ψ_B for the mass-3 Bose liquid [16] was accepted as a good approximation to the solution of the Schrödinger equation (5.4.46). In the evaluation of matrix elements, the potential $v(r)$ was everywhere absorbed into E_B, and does not appear in the final expressions. What repeatedly appeared was the mass-3 boson liquid structure function $S_B(k)$, calculated by Massey in exactly the same way as for liquid ^{4}He. Once the matrix elements were obtained, orthogonalization took place via the Löwdin transformation, which is an expansion in powers of the overlap integrals—the nondiagonal matrix elements of 1.

Let the matrix elements of 1 be denoted by N_{mn}, and those of H by H_{mn}. Write

$$N_{mn} \equiv \delta_{mn} + J_{mn}, \tag{5.4.53}$$

and

$$H_{mn} \equiv \tfrac{1}{2} N_{mn}(H_{mm} + H_{nn}) + W_{mn}. \tag{5.4.54}$$

The transformation function

$$(N^{-1/2})_{mn} \equiv \delta_{mn} - \tfrac{1}{2} J_{mn} + \tfrac{3}{8} \sum_l J_{ml}J_{ln} + \cdots \tag{5.4.55}$$

then orthogonalizes the basis. The transformed Hamiltonian has matrix elements $\mathscr{H}_{mn}$ given by

$$\mathscr{H}_{mn} = H_{mm}\delta_{mn} + W_{mn} - \tfrac{1}{2} \sum_l (W_{ml}J_{ln} + J_{ml}W_{ln})$$
$$+ \tfrac{1}{8} \sum_l J_{ml}J_{ln}(H_{mm} + H_{nn} - 2H_{ll}) + \cdots. \tag{5.4.56}$$

It turns out that in this infinite series there exist terms of anomalously high N dependences, much like in the case of the Rayleigh–Schrödinger perturbation series. Schematically representing the terms by diagrams, it becomes clear that the anomalous behavior originates from unlinked diagrams. Unlike in the Brueckner–Goldstone case, however, there are no renormalization terms to cancel the unlinked contributions. Detailed diagrammatic analysis [167, 169] showed that if a Rayleigh–Schrödinger perturbation expansion is applied, the unlinked clusters that arise from the Löwdin series will in every order be cancelled identically by corresponding clusters in the RS series. The theorem was demonstrated to three lowest orders [169], but as yet no general proof is available. On account of this theorem, one could safely ignore the divergent, unlinked diagrams and take into account only linked contributions.

Unfortunately, for liquid ^{3}He even the linked series converges too slowly if one separately evaluates the orthogonalization and the perturbation corrections. This is not totally unexpected. In choosing a set of CBF, one pays no attention to their orthogonality properties. The only guide is that the basis or unperturbed functions must resemble what one imagines to be the true eigenstates, i.e., the matrix H, not 1, is chosen to be as nearly diagonal as possible. By attempting to orthogonalize the basis, one destroys what he has built into the basis and must expect to pay the price for recovering it. It is much safer to ease toward orthogonality and the diagonalization of H simultaneously. This calls for a perturbation expansion using a nonorthogonal basis, a procedure seldom if ever used, but is actually perfectly straightforward. It works like the Rayleigh–Schrödinger–Goldstone expansion, except that everywhere in place of the matrix element H_{mn}, one uses the modified vertex $(H_{mn} - H_{mm}N_{mn})$. (In Section 5.3.3 we dealt with a *partially* orthogonalized set of CBF. The same procedure was actually employed there as well.) The nondiagonal part of the Hamiltonian matrix and the overlap integral are given equal (and symmetrical) status, and are removed by the perturbation theory at the same pace. In this manner, the ground-state energy calculation was finally carried out [167]. It was found that the leading perturbation correction to the variational energy is -0.35 K per particle.

Adding the perturbation correction to Schiff and Verlet's result (using ψ_B instead of the optimum Jastrow factor) gives rise to -1.62 K. [This is permitted because $S_B(k)$ alone enters the perturbation correction formula, and all Jastrow calculations yield very similar $S_B(k)$.] On top of this, one must yet lower the mass-3 boson energy: the Jastrow result E_B must be corrected by the three-phonon correction of Davison and Feenberg [46], which is expected to amount to -0.5 to -1.0 K. The final result lies somewhere between -2.1 and -2.6 K, in good agreement with experiment.

The sound velocity and compressibility shown in Table 5.4.1 on the left-hand side of the CBF columns are adjusted results. The adjustment consists of fitting the energy and equilibrium density to experiment before derivatives are taken. The same procedure was used for ^{4}He in Section 5.2.2 and Reference [44].

For the effective mass and the Landau parameters, Tan and Feenberg [155] generalized the above perturbative procedure in a way reminiscent of Gell-Mann's [170] generalization of the Gell-Mann–Brueckner treatment of high-density electron gas [171]. The first-order quasiparticle energy is given by

$$\varepsilon_0(\mathbf{k}, \sigma) = E_0^{(N+1)}(\mathbf{k}, \sigma) - E_0^{(N)}, \tag{5.4.57}$$

where $E_0^{(N)} \equiv E_0$ is the energy of the N-particle system that we have been discussing, and $E_0^{(N+1)}(\mathbf{k}, \sigma)$ is the energy of an $(N+1)$-particle system in the ground-state configuration with the additional particle in orbital $(\mathbf{k}, \sigma)$. The Landau $f(\mathbf{k}, \mathbf{k}'; \sigma, \sigma')$ function is given by

$$\begin{aligned} f(\mathbf{k}, \mathbf{k}'; \sigma, \sigma') = {} & E_0^{(N+2)}(\mathbf{k}, \sigma; \mathbf{k}', \sigma') \\ & + E_0^{(N)} - E_0^{(N+1)}(\mathbf{k}, \sigma) - E_0^{(N+1)}(\mathbf{k}', \sigma'). \end{aligned} \tag{5.4.58}$$

All the energies on the right of equations (5.4.57)–(5.4.58) are evaluated by the second-order perturbation theory described in Reference [167]. Further justification of this procedure is given by a statistical analysis in Reference [155]. The results are shown in Table 5.4.1. Note that, other than the energy, which is for the calculated equilibrium density 0.0156 Å^{-3}, all quantities are for the experimental density 0.0164 Å^{-3}. The Landau parameters calculated are shown in Table 5.4.2.

We remarked earlier that without the inclusion of perturbation corrections variational calculations lead to unacceptable Landau parameters. The reason is simply that the diagonal matrix elements of H generated by correlated single configuration states do not contain much information on the scattering between quasiparticles. The perturbation term supplies this information since it is generated by *linear combinations* of correlated single configuration states.

5.4.5. Green's Function Formulation in the Correlated Representation

In References [167, 169] it was shown that, even though neither the Löwdin orthogonalization series nor the perturbation series that follows converges rapidly by themselves, upon order-by-order combination of the two series the resulting expansion becomes rapidly convergent. In fact, the combined series at least in low orders is identical to the energy series obtained using the nonorthogonal perturbation theory. This fact suggests that a quasiparticle Hamiltonian can be constructed whose interaction term takes into account the nondiagonal part of 1 as well as H. Such a Hamiltonian will have the form

$$\hat{H} = E_B + \sum_{\mathbf{k}\sigma} \frac{\hbar^2 k^2}{2m} a^\dagger_{\mathbf{k}\sigma} a_{\mathbf{k}\sigma} + \frac{1}{2} \sum_{\mathbf{k}_1\sigma_1, \mathbf{k}_2\sigma_2, \mathbf{k}_3\sigma_3, \mathbf{k}_4\sigma_4} W_{1234}\, a^\dagger_{\mathbf{k}_1\sigma_1} a^\dagger_{\mathbf{k}_2\sigma_2} a_{\mathbf{k}_4\sigma_4} a_{\mathbf{k}_3\sigma_3} + \cdots. \tag{5.4.59}$$

In order for $\hat{H}$ operating on states $a^\dagger_{\mathbf{k}_1\sigma_1} \cdots |0\rangle$, where the vacuum state $|0\rangle$ is defined by

$$a_{\mathbf{k}_i\sigma_i}|0\rangle = 0, \tag{5.4.60}$$

to yield the same matrix elements as the Hamiltonian of equation (5.1.1) operating on correlated basis functions, thus making equation (5.4.59) the occupation number representation of the same ^{3}He Hamiltonian, it was found [172] that one must set

$$W_{1234} = \langle \mathbf{k}_1, \mathbf{k}_2 | W | \mathbf{k}_3, \mathbf{k}_4 \rangle \delta_{\sigma_1, \sigma_3} \delta_{\sigma_2, \sigma_4},$$

with

$$\langle k_1, k_2 | W | k_3, k_4 \rangle = \frac{-\hbar^2}{2mN} \delta_{\mathbf{k}_1+\mathbf{k}_2, \mathbf{k}_3+\mathbf{k}_4} \cdot \Big\{ k_{13}^2 [S_B(k_{13}) - 1] - \frac{1}{2N} \sum_{(\mathbf{k}_5)} [S_B(k_{15}) - 1][S_B(k_{35}) - 1] \Big[k_{15}^2 + k_{35}^2 + \frac{k_1^2 + k_2^2 + k_3^2 + k_4^2 - 2k_5^2 - 2|\mathbf{k}_1 + \mathbf{k}_2 - \mathbf{k}_5|^2}{4} \Big] \Big\}, \tag{5.4.61}$$

where the summation $\sum_{(\mathbf{k}_5)}$ requires $k_5 > k_F$ and $|\mathbf{k}_1 + \mathbf{k}_2 - \mathbf{k}_5| > k_F$.

This interaction W is well behaved but by no means weak. To dress the bare "quasiparticles" (fermions modified by the correlation factor) by this interaction, one uses the operator $\hat{H} - \mu \hat{N}$, where

$$\hat{N} = \sum_{\mathbf{k}\sigma} a^\dagger_{\mathbf{k}\sigma} a_{\mathbf{k}\sigma}, \tag{5.4.62}$$

and μ is a Lagrange multipler: the chemical potential. With respect to $\hat{H} - \mu\hat{N}$, one redefines the vacuum state and introduces particle and hole propagators. One- and two-particle Green's functions are then calculated, from which one obtains expressions for the ground-stated energy E_0, the quasiparticle spectrum $\varepsilon(\mathbf{k}, \sigma)$, the scattering amplitudes, and the expectation value of $\hat{N}/\Omega$ in the ground state. μ is then uniquely determined by setting the last-mentioned quantity to the input density ρ. The consistency of the calculation can be checked against the Hugenholtz–van Hove [147] separation energy theorem: $\varepsilon(k_F, \sigma) = \mu$.

In the lowest order of approximation, one obtains a Hartree–Fock solution of Dyson's equation

$$\varepsilon(\mathbf{k}, \sigma) = \Sigma(\mathbf{k}, \sigma; \varepsilon(\mathbf{k}, \sigma)), \tag{5.4.63}$$

$$\Sigma(\mathbf{k}_1, \sigma_1; \varepsilon) = \varepsilon(\mathbf{k}_1, \sigma_1) + \frac{1}{2} \sum_{\sigma_2\sigma_3\sigma_4} \sum_{\mathbf{k}_2 \leq k_F} \sum_{\mathbf{k}_3, \mathbf{k}_4 > k_F} \frac{(W_{1234} - W_{1243})^2}{[\varepsilon + \varepsilon(\mathbf{k}_2, \sigma_2) - \varepsilon(\mathbf{k}_3, \sigma_3) - \varepsilon(\mathbf{k}_4, \sigma_4)]}$$
$$- \frac{1}{2} \sum_{\sigma_2\sigma_3\sigma_4} \sum_{\mathbf{k}_2 \leq k_F} \sum_{\mathbf{k}_3, \mathbf{k}_4 > k_F} \frac{(W_{1423} - W_{1432})^2}{[-\varepsilon - \varepsilon(\mathbf{k}_4, \sigma_4) + \varepsilon(\mathbf{k}_2, \sigma_2) + \varepsilon(\mathbf{k}_3, \sigma_3)]}, \tag{5.4.64}$$

and finds precisely the quasiparticle spectrum obtain by Tan and Feenberg and, in turn, the ground-state energy obtained by Woo. In the same approximation the vertex part $\Gamma_{\sigma_1\sigma_2\sigma_3\sigma_4}(\mathbf{k}_1\varepsilon_1, \mathbf{k}_2\varepsilon_2, \mathbf{k}_3\varepsilon_3, \mathbf{k}_4\varepsilon_4)$ can be evaluated [172], in particular on the Fermi surface. It can then be used in the Boltzmann equation for computing transport properties.

The method presented here has three advantages worth mentioning:

i. It yields not just the *forward* scattering amplitudes. One can go beyond the Landau–Abrikosov–Khalatnikov theory of transport coefficients, and do that without the need of a single adjustable parameter.
ii. For the estimation of the superfluid transition temperature, e.g., by solving a BCS-like equation [173, 174], one has at his disposal the quasiparticle spectrum and the residual interaction (both of which enter into the gap equation) from the *same* calculation. There is no question of incompatibility. (See Section 5.8.8.)
iii. It combines the best of two diverse approaches—CBF and Green's functions—and serves as a hybridization procedure in the sense proposed by Brandow [65] and by Jackson [133]. As such, it will help to establish contacts between these approaches and complement the diagrammatic analysis of CBF now underway for Fermi systems.

5.5. DILUTE SOLUTIONS OF ^{3}He IN SUPERFLUID ^{4}He

5.5.1. Introduction

Binary solutions ordinarily phase-separate at low temperature. Cohen and van Leeuwen [175, 176], from studying the properties of isotopic mixtures of Fermi and Bose hard spheres, and, independently, Edwards and Daunt [177], from an interpretation of experimental data using the Landau–Pomeranchuk theory (to be described later), both came to the conclusion that in $^3He-^4He$ solutions, phase separation will remain incomplete even at zero temperature. There will be a nearly pure ^3He liquid phase on top and a dilute solution of ^3He in ^4He at the bottom. That this is indeed the case was first demonstrated by Edwards et al. [178, 179] in 1965. While the maximum solubility of ^3He in ^4He depends on ambient conditions, it is by now well established that at zero temperature and zero pressure it amounts to 6.5%. With increasing pressure, the maximum solubility increases to about 9.5% at 10 atm, above which different experiments have given rise to somewhat different conclusions. The Cornell [180] and Ohio State [181] groups reported the existence of a broad peak in the solubility versus pressure curve. The Argonne group [182], on the other hand, observed merely a plateau. In any case, 10% seems to be an upper limit in the absence of thermal agitation. The data are shown in Figure 5.5.1, the zero-pressure solubility having been normalized to 6.5%.

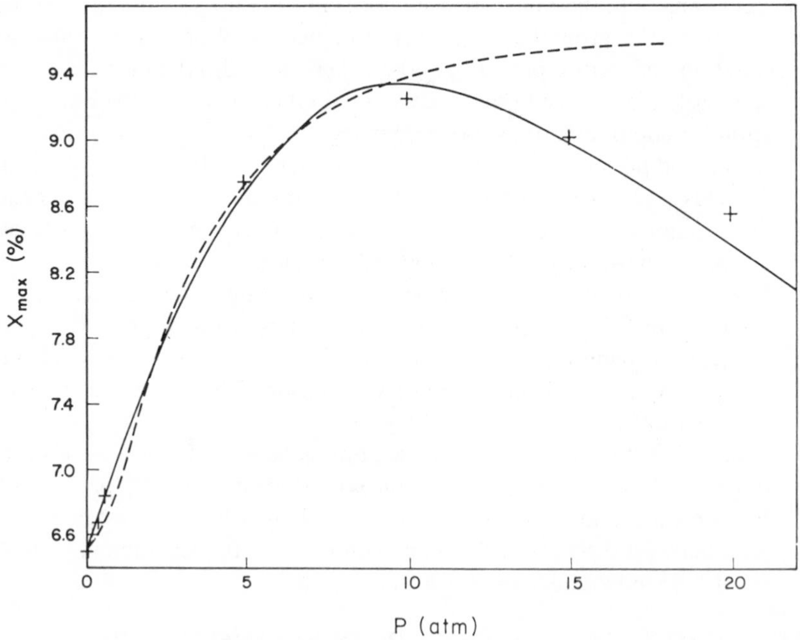

FIGURE 5.5.1. Pressure dependence of solubility of ^3He in liquid ^4He ($X_{max}(0)$ normalized to 6.5%). ———, Cornell [180]; ---------, Argonne [182]; + + + +, Ohio State [181].

TABLE 5.5.1. ^{3}He –roton cross sections: $\sigma(k, q) = \frac{1}{2}\int_{-1}^{1}\sigma(\mathbf{k},\mathbf{q})d\cos(\mathbf{k},\mathbf{q})$ [Å^2]

k[Å^{-1}] \ q[Å^{-1}]	1.80	1.95	2.10	2.25
0.10	—	—	22	—
0.25	9	6	21	33
0.40	13	14	22	26
0.70	46	189	175	154
1.00	80	313	390	431
1.30	105	344	274	—
1.60	—	287	—	—

The range 0–10% corresponds to ^{3}He densities of $\rho_3 = 0$–0.0022 Å^{-3}, as compared to the liquid ^{3}He density of 0.0164 Å^{-3}. If the single-particle properties of ^{3}He are not toally destroyed by the interaction with the ^{4}He medium, we have on hand a fermion gas, or indeed an ideal, *dilute Fermi liquid.* Furthermore it is the kind of Fermi liquid whose density can be controlled to a large extent by the experimenter! These factors were responsible for making ^{3}He–^{4}He solutions the most exciting quantum liquids in the past decade.

One of the chapters in the second volume of this book contains a thorough review of this subject. The reader is also advised to consult a most comprehensive and well-written review by Ebner and Edwards [183] in *Physics Reports.* However, both of these articles say little about microscopic calculations for ^{3}He–^{4}He solutions. The latter refers to Feenberg's book, which contains a very brief chapter [184] on some aspects of the theory—mostly in the zero-concentration limit. The contents of the present section are for the explicit purpose of complementing these otherwise informative reviews.

5.5.2. Zero-Concentration Limit

The Hamiltonian for a system of N_4 ^{4}He atoms and N_3 ^{3}He atoms enclosed in a volume Ω is given by

$$H = \sum_{i=1}^{N_3} \frac{-\hbar^2}{2m_3}\nabla_i^2 + \sum_{j=N_3+1}^{N} \frac{-\hbar^2}{2m_4}\nabla_j^2 + \sum_{i<j=1}^{N} v(r_{ij}). \tag{5.5.1}$$

The input parameters are the partial densities $\rho_3 \equiv N_3/\Omega$ and $\rho_4 \equiv N_4/\Omega$, or alternately the total density $\rho \equiv N/\Omega = (N_3 + N_4)/\Omega = \rho_3 + \rho_4$ and the concentration of ^{3}He: $x \equiv N_3/N = \rho_3/(\rho_3+\rho_4)$.

The zero-concentration limit refers to the case where there is only one ^{3}He atom in the "solution," or two ^{3}He atoms of opposite spins. The simplification which we find in this limit originates from the fact that there is then no Fermi exchange symmetry to take into account.

Landau and Pomeranchuk [185, 186] reasoned that as an impurity in the form of a ^{3}He atom of momentum $\hbar k$ is added to the ground state of liquid ^{4}He, the energy change can be expanded in an even power series of k. By omitting all higher powers in k, the excitation energy can be approximated by

$$\varepsilon(k) = -\mu_3 + \frac{\hbar^2 k^2}{2m_3^*}, \tag{5.5.2}$$

which has the form of a free particle of mass m_3^* moving in a potential well of depth μ_3. More physically, one might intepret μ_3 as the binding energy of a ^{3}He atom by the medium and m_3^* as its effective mass as it travels through the medium dragging along its immediate neighbors. In either case, one finds a single-quasiparticle-type excitation spectrum to accompany the usual phonon–roton branch. As the number of ^{3}He atoms is increased in the liquid, as along as N_3 remains much smaller than N_4, there will be quantitative but not qualitative change in the new branch. It is expected that the macroscopic properties of the system will be determined cooperatively by both branches of the excitation. Thus, even the zero-concentration limit provides a certain degree of realism.

Baym [187] proposed a possible method for microscopically calculating μ_3. Rewriting equation (5.5.1) as

$$H = H_0 + H_1, \tag{5.5.3}$$

with

$$H_0 = \sum_{i=1}^{N} \frac{-\hbar^2}{2m_4} \nabla_i^2 + \sum_{i<j=1}^{N} v(r_{ij}), \tag{5.5.4}$$

and

$$H_1 = \frac{m_4 - m_3}{m_3} \sum_{i=1}^{N_3} \frac{-\hbar^2}{2m_4} \nabla_i^2 = \frac{1}{3} \sum_{i=1}^{N_3} \frac{-\hbar^2}{2m_4} \nabla_i^2, \tag{5.5.5}$$

one recognizes that H_0 is precisely the Hamiltonian for pure liquid ^{4}He, and that the perturbation H_1, which is small for N_3 small, is $\frac{1}{3}x$ times the ^{4}He kinetic-energy operator. In the zero-concentration limit, one need not take into account the Fermi exchange effects, so a trial wave function in the form of the (symmetric) ground-state solution of H_0 will do. A first-order perturbation calculation is equivalent to taking the expectation value of H:

$$\langle H \rangle = E_0 + \frac{1}{3} N_3 \frac{\langle \text{K.E.} \rangle}{N}, \tag{5.5.6}$$

so that

$$\mu_3 \approx \left\{ \frac{\partial}{\partial N_3} \left[(N_3 + N_4) \frac{E_0}{N} + \frac{1}{3} N_3 \frac{\langle \text{K.E.} \rangle}{N} \right] \right\}_{x=0}$$

$$= \frac{E_0}{N} + \frac{1}{3}\frac{\langle \text{K.E.} \rangle}{N}. \tag{5.5.7}$$

Using the best value available at that time [40], $\langle \text{K.E.} \rangle / N = 14.30$ K, therefore [188],

$$\mu_3 \lesssim -7.14\text{ K} + \tfrac{1}{3}(14.30\text{ K}) = -2.37\text{ K}.$$

Davison and Feenberg [189] carried out a second order perturbation calculation to take into account the interaction between the ^{3}He atom and the medium. They used the wave functions ($\mathbf{r}_1 = {}^3$He coordinates)

$$|\mathbf{k}\rangle = \Psi_0 e^{i\mathbf{k}\cdot\mathbf{r}_1}, \tag{5.5.8}$$

and

$$|\mathbf{k} - \mathbf{K}, \mathbf{K}\rangle = \rho_{\mathbf{k}} \Psi_0 e^{i(\mathbf{k}-\mathbf{K})\cdot\mathbf{r}_1}, \tag{5.5.9}$$

and determined via a BW calculation a constant shift in the ground state energy of -0.33 K for the ^{3}He atom. Incorporating this result in their calculation, Massey and Woo [188, 190] found

$$\mu_3 = -2.37\text{ K} - 0.33\text{ K} = -2.70\text{ K}, \tag{5.5.10}$$

in good agreement with the experimental value of -2.79 K determined later by Seligmann et al. [191] from heat-of-mixing measurements.

The effective mass m_3^* also emerges from the Davison–Feenberg calculation, since the shift in energy is really a function of k. Expanding the shift in power series of k, the leading k-dependent term is proportional to k^2, and yields an effective mass

$$m_3^* \approx 1.81 m_3, \tag{5.5.11}$$

in comparison to the experimental value of $2.38m_3$. A more complete calculation in the sprit of CBF is that of Woo, Tan, and Massey [193,194], which in the zero-concentration limit yields $2.37m_3$, and to which we shall return in Section 5.5.4.

In Reference [187], Baym also suggested that the interaction between a pair of ^{3}He quasiparticles can be estimated on the average by

$$v^*(0) = -\frac{\alpha^2 m_4 s^2}{N_4}, \tag{5.5.12}$$

where s is the velocity of sound, and α is the extra fraction of volume cocupied by a ^{3}He atom in the liquid because of its larger zero-point motion. The model was elaborated in several directions by Bardeen, Baym, and Pines [195, 196], and by Emery [197], who also in an earlier paper [198] laid the foundation for an effective microscopic theory. These papers are adequately reviewed elsewhere. We quote here simply Baym's suggested formula for α:

$$\alpha = \frac{\rho}{m_4 s^2} \frac{\partial}{\partial \rho}\left(\frac{1}{3} \frac{\langle \text{K.E.} \rangle}{N}\right), \tag{5.5.13}$$

where all quantities refer to a pure ^{4}He liquid. Massey and Woo [188] made use of the results in Reference [40] and evaluated α at a range of densities. And since from $E_0(\rho)$ one could find the equation of state $P(\rho)$ at zero temperature, what they obtained was a prediction for the function $\alpha(P)$, as shown in Figure 5.5.2. The theoretical result was quickly confirmed experimentally by Bogho-

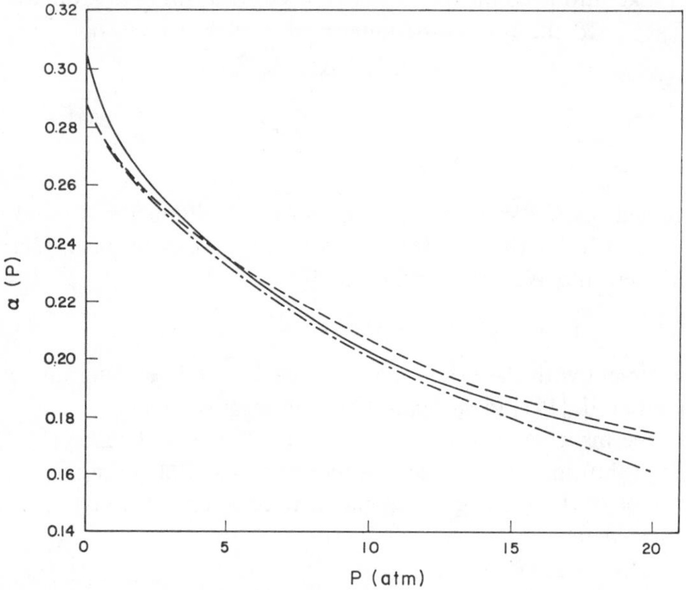

FIGURE 5.5.2. Volume exclusion parameter $\alpha(P)$. ———, Theoretical prediction by Massey and Woo [188]; ---------, Watson et al. [180]; –·–·–·–·, Abraham et al. [182].

sian and Meyer [199]. The more recent results by Watson et al. [180] and Abraham et al. [182] are shown in Figure 5.5.2 for comparison.

5.5.3. Binary Boson Solutions

There is only one complete, microscopic theoretical calculation for dilute solutions of ^{3}He in superfluid ^{4}He. In that work a preliminary calculation for the ground state of a mixture of mass-3 bosons and ^{4}He proves helpful, for reasons to become clear in Section 5.5.4. In this section I shall summarize the calculation and mention some long-wavelength properties of such hypothetical "binary boson solutions."

The Hamiltonian of the system is precisely the same as equation (5.5.1) for realistic ^{3}He–^{4}He solutions. What differs is that the wave function must be *symmetrical* with respect to the exchange between two mass-3 particles as well as between two ^{4}He atoms. For this purpose a generalized Jastrow wave function is constructed [200]:

$$\psi_B(\mathbf{r}_1,\mathbf{r}_2, \cdots, \mathbf{r}_{N_3}; \mathbf{r}_{N_3+1}, \cdots, \mathbf{r}_N) =$$

$$\prod_{i<j}^{N_3} \exp[\tfrac{1}{2}\, u^{(3,3)}(r_{ij})] \prod_{i=1}^{N_3} \prod_{k=N_3+1}^{N} \exp[\tfrac{1}{2}u^{(3,4)}(r_{ik})] \prod_{k<l=N_3+1}^{N} \exp[\tfrac{1}{2}u^{(4,4)}(r_{kl})], \quad (5.5.14)$$

which finds immediate mathematical analogy with the classical liquid-solution problem [201].

The expectation value of H with respect to this wave function is given by

$$\langle H \rangle = \langle \text{K.E.} \rangle + \langle \text{P.E.} \rangle, \quad (5.5.15)$$

$$\langle \text{K.E.} \rangle = \frac{\hbar^2 \rho_4}{8m_4}\left[N_4 \int \nabla u^{(4,4)}(r)\cdot\nabla g^{(4,4)}(r)\, d\mathbf{r} + N_3 \int \nabla u^{(3,4)}(r)\cdot\nabla g^{(3,4)}(r)\, d\mathbf{r}\right]$$
$$+ \frac{\hbar^2 \rho_3}{8m_3}\left[N_4 \int \nabla u^{(3,4)}(r)\cdot\nabla g^{(3,4)}(r)\, d\mathbf{r}\right.$$
$$\left. + N_3 \int \nabla u^{(3,3)}(r)\cdot\nabla g^{(3,3)}(r)\, d\mathbf{r}\right], \quad (5.5.16)$$

$$\langle \text{P.E.} \rangle = \tfrac{1}{2}N_4\rho_4 \int v(r) g^{(4,4)}(r)\, d\mathbf{r} + N_3\rho_4 \int v(r) g^{(3,4)}(r)\, d\mathbf{r}$$
$$+ \tfrac{1}{2}\, N_3\rho_3 \int v(r) g^{(3,3)}(r)\, d\mathbf{r}, \quad (5.5.17)$$

in which three kinds of boson distribution functions occur:

$$g^{(\alpha,\beta)}(r_{ij}) = \frac{N_\alpha(N_\beta - \delta_{\alpha\beta})}{\rho_\alpha\rho_\beta} \frac{\int \psi_B^2 d\tau_{(i,j)}}{\int \psi_B^2 d\tau}; \quad \alpha, \beta = 3 \text{ or } 4. \quad (5.5.18)$$

To obtain the g's from the u's, the techniques discussed in Section 5.2.2 are adopted and generalized. In particular, coupled BBGKY equations may be derived by applying appropriate differentiation operators to the definitions of $g^{(\alpha,\beta)}$:

$$\nabla_{12} g^{(3,3)}(r_{12}) = g^{(3,3)}(r_{12}) \nabla_{12} u^{(3,3)}(r_{12})$$
$$+ \rho_4 \int g_3^{(3,3,4)}(\mathbf{r}_1, \mathbf{r}_2, \mathbf{r}_N) \nabla_{1N} u^{(3,4)}(r_{1N})\, d\mathbf{r}_N$$
$$+ \rho_3 \int g_3^{(3,3,3)}(\mathbf{r}_1, \mathbf{r}_2, \mathbf{r}_3) \nabla_{13} u^{(3,3)}(r_{13})\, d\mathbf{r}_3, \quad (5.5.19)$$

$$\nabla_{1N} g^{(3,4)}(r_{1N}) = g^{(3,4)}(r_{1N}) \nabla_{1N} u^{(3,4)}(r_{1N})$$
$$+ \rho_4 \int g_3^{(3,4,4)}(\mathbf{r}_1, \mathbf{r}_{N-1}, \mathbf{r}_N) \nabla_{1,N-1} u^{(3,4)}(r_{1,N-1})$$
$$\cdot\, d\mathbf{r}\cdot_{N-1}$$
$$+ \rho_3 \int g_3^{(3,3,4)}(\mathbf{r}_1, \mathbf{r}_2, \mathbf{r}_N) \nabla_{12} u^{(3,3)}(r_{12})\, d\mathbf{r}_2, \quad (5.5.20)$$

$$\nabla_{N-1,N}\, g^{(4,4)}(r_{N-1,N}) = g^{(4,4)}(r_{N-1,N}) \nabla_{N-1,N}\, u^{(4,4)}(r_{N-1,N})$$

$$+ \rho_4 \int g_3^{(4,4,4)}(\mathbf{r}_{N-2}, \mathbf{r}_{N-1}, \mathbf{r}_N) \cdot \nabla_{N-2,N} u^{(4,4)}(r_{N-2,N})\, d\mathbf{r}_{N-2}$$
$$+ \rho_3 \int g_3^{(3,4,4)}(\mathbf{r}_1, \mathbf{r}_{N-1}, \mathbf{r}_N) \nabla_{1N} u^{(3,4)}(r_{1N})\, d\mathbf{r}_1, \quad (5.5.21)$$

where the three-particle distribution functions $g_3^{(\alpha,\beta,\gamma)}(\mathbf{r}_{i_\alpha}, \mathbf{r}_{i_\beta}, \mathbf{r}_{i_\gamma})$ are defined by (α, β, γ = 3 or 4),

$$g_3^{(\alpha,\alpha,\alpha)}(\mathbf{r}_{i_\alpha}, \mathbf{r}_{j_\alpha}, \mathbf{r}_{k_\beta}) = \frac{N_\alpha(N_\alpha - 1)(N_\alpha - 2)}{\rho_\alpha^3} \frac{\int \psi_B^2\, d\tau(i_\alpha, j_\alpha, k_\alpha)}{\int \psi_B^2 d\tau}, \quad (5.5.22a)$$

and

$$g_3^{(\alpha,\alpha,\beta)}(\mathbf{r}_{i_\alpha}, \mathbf{r}_{j_\alpha}, \mathbf{r}_{k_\beta}) = \frac{N_\alpha(N_\alpha - 1)N_\beta}{\rho_\alpha^2 \rho_\beta} \frac{\int \psi_B^2 d\tau(i_\alpha, j_\alpha, k_\beta)}{\int \psi_B^2\, d\tau}, \quad (5.5.22b)$$

and may be approximated by the generalized Kirkwood superposition approximation

$$g_3^{(\alpha,\beta,\gamma)}(i_\alpha, i_\beta, i_\gamma) \approx g^{(\alpha,\beta)}(r_{i_\alpha i_\beta}) g^{(\beta,\gamma)}(r_{i_\beta i_\gamma}) g^{(\gamma,\alpha)}(r_{i_\gamma i_\alpha}). \quad (5.5.23)$$

Clearly, permuting the superscripts does not change either g or g_3.

Note that equation (5.5.20) arises from applying the operator ∇_1 to $g^{(3,4)}(r_{1N})$. If the operator ∇_N is used, a slightly different equation will arise which involves $u^{(3,4)}$ and $u^{(4,4)}$ instead of $u^{(3,4)}$ and $u^{(3,3)}$:

$$\nabla_{1N} g^{(3,4)}(r_{1N}) = g^{(3,4)}(r_{1N}) \nabla_{1N} u^{(3,4)}(r_{1N})$$
$$+ \rho_3 \int g_3^{(3,3,4)}(\mathbf{r}_1, \mathbf{r}_2, \mathbf{r}_N) \nabla_{2N} u^{(3,4)}(r_{2N})\, d\mathbf{r}_2$$
$$+ \rho_4 \int g_3^{(3,4,4)}(\mathbf{r}_1, \mathbf{r}_{N-1}, \mathbf{r}_N) \nabla_{N-1,N} u^{(4,4)}(r_{N-1,N})\, d\mathbf{r}_{N-1}. \quad (5.5.24)$$

Now, solving equations (5.5.19), (5.5.24), and (5.5.21) should give results identical to solving equations (5.5.19)–(5.5.21). The only source of error must be attributed to the generalized Kirkwood approximation. So one has here an indirect way of measuring the error of the approximation. It was found [202] that the error has little effect on the macroscopic properties of the system.

The variational calculation carried out in Reference [200] was in three stages. The first stage is a variational calculation for pure ^{4}He, from which one obtains $u^{(4,4)}$ and $g^{(4,4)}$. In the second stage, a ^{4}He atom, say the first one, is replaced by a mass-3 boson, and the variational function

$$\psi_1(\mathbf{r}_1, \mathbf{r}_2, \ldots, \mathbf{r}_N) = \prod_{k=2}^{N} \exp\left[\tfrac{1}{2} u^{(3,4)}(r_{1k})\right] \prod_{k<l=2}^{N} \exp\left[\tfrac{1}{2} u^{(4,4)}(r_{kl})\right] \quad (5.5.25)$$

is chosen to minimize the energy, which now is a functional of $u^{(3,4)}$, $u^{(4,4)}$,

$g^{(3,4)}$ and $g^{(4,4)}$. Taking $g^{(4,4)}$ or $u^{(4,4)}$ from the first stage, the other three functions can be determined by solving the coupled equations (5.5.20) and (5.5.21) and by the variational principle. Thus the calculation determines ψ_1. In the third and final stage, N_3 ^{4}He atoms are replaced by mass-3 bosons. Holding $g^{(4,4)}$ and $g^{(3,4)}$ or $u^{(4,4)}$ and $u^{(3,4)}$ fixed, all the other functions are found from equations (5.5.19)–(5.5.21) and varied to minimize $\langle H \rangle$. Among the results of such a calculation are the three boson pair distribution functions, and an improved determination of $\mu_3(x)$:

$$\mu_3(x) \approx (-\ 2.79 + 48.1\ x)\ \text{K}, \tag{5.5.26}$$

in excellent agreement with the experiment result [200]

$$\mu_3(x) = (-\ 2.79 + 43.2x)\ \text{K}. \tag{5.5.27}$$

A similar calculation using coupled PY equations was attempted by Petit [203], but no result was published.

Hansen and Schiff [204] carried out such a calculation using molecular dynamics and found that there is no energy minimum in x, thus concluding that the binary boson solution at zero temperature phase-separates completely.

The definition of $g^{(3,3)}$, $g^{(3,4)}$, and $g^{(4,4)}$ allows one to define the following liquid structure functions:

$$\begin{aligned} S^{(\alpha,\beta)}(k) &= \tfrac{1}{2}(N_\alpha N_\beta)^{-1/2}\langle 0 | \rho_{\mathbf{k}}^{(\alpha)\dagger}\rho_{\mathbf{k}}^{(\beta)} + \rho_{\mathbf{k}}^{(\beta)\dagger}\rho_{\mathbf{k}}^{(\alpha)} | 0 \rangle \\ &= \delta_{\alpha\beta} + (\rho_\alpha\rho_\beta)^{1/2}\int [g^{(\alpha,\beta)}(r) - 1]e^{i\mathbf{k}\cdot\mathbf{r}}\, d\mathbf{r}, \end{aligned} \tag{5.5.28}$$

where

$$\rho_{\mathbf{k}}^{(3)} = \sum_{i=1}^{N_3} e^{i\mathbf{k}\cdot\mathbf{r}_i}, \tag{5.5.29}$$

and

$$\rho_{\mathbf{k}}^{(4)} = \sum_{l=N_3+1}^{N} e^{i\mathbf{k}\cdot\mathbf{r}_l}. \tag{5.5.30}$$

$S^{(\alpha,\beta)}(k)$ play an essential role in the theory of ^{3}He–^{4}He solutions. Sum rules can be derived [205, 206] for the dynamical structure functions $S^{(\alpha,\beta)}(k, \omega)$, from which the long-wavelength behavior of $S^{(\alpha,\beta)}(k)$ can be deduced. It suffices to mention that all $S^{(\alpha,\beta)}(k)$ become linear in k as $k \to 0$, a fact of some importance in the discussion to follow.

5.5.4. Dilute ^{3}He–^{4}He Solutions [193, 194]

A first-principle theory of ^{3}He–^{4}He solutions begins with the Hamiltonian of equation (5.5.1), which describes a system of bare ^{3}He and ^{4}He atoms interacting via a pairwise potential. The form of the potential is given, but other than this knowledge, a quantum mechanical calculation must proceed without the benefit of additional empirical information.

Quantum mechanical calculations begin with the selection of a set of basis functions. The criterion for an optimum choice of the basis is that it should account for all characteristic physical features of the system under consideration. In the present case, what this means is that the following requirements will be imposed on the basis functions: (i) They should take into account the short-range dynamical correlations that result from the repulsive core in the potential $v(r)$. (ii) They must have the appropriate exchange symmetry: symmetric with respect to the interchange of two ^{4}He atoms and antisymmetric with respect to the interchange of two ^{3}He atoms. (iii) There must be two branches of elementary excitations: a single-quasiparticle branch representing the motion of ^{3}He modified by interaction with the medium and with one another, and a collective branch representing phonons and rotons modified by the presence of impurities. The quantum numbers assigned to each basis function represent the modes excited.

In accordance with these requirements, we can construct a satisfactory set of basis functions by going over to the correlated representation, i.e., employing a set of correlated basis functions.

We are guided by the experience with pure systems (Sections 5.2, 5.3, and and 5.4) in fulfilling the first condition. A symmetric correlation factor is chosen to accompany each of a set of model functions. The correlation factor may be in the form of a generalized Jastrow function, like the right-hand side of equation (5.5.14). Or it may be a *symmetric* ground-state solution of the same Schrödinger equation that we are attempting to solve. An expedient compromise can be worked out. Take the generalized Jastrow function determined by minimizing $\langle H \rangle$: It at once accounts for the pair correlations *and* serves as a good approximate symmetric solution to equation (5.5.1). This calls for a preliminary ground-state variational calculation for the binary boson solution, as discussed in the last section. We assume then the knowledge of $S^{(\alpha,\beta)}(k)$.

To fulfill the second condition, we form model functions which are direct products of the wave functions of ^{3}He and the wave functions of ^{4}He. The former are Slater determinants of N_3 plane waves and spin functions. The latter are density fluctuation operators $\rho_{\mathbf{p}}^{(4)}$. Typically,

$$\begin{aligned} \left| \mathbf{p}_1^{m_1}, \mathbf{p}_2^{m_2}, \cdots, \mathbf{k}_1\sigma_1, \mathbf{k}_2\sigma_2, \cdots, \mathbf{k}_{N_3}\sigma_{N_3} \right) & \\ = \psi_B \cdot \prod_l [\rho_{\mathbf{p}_l}^{(4)}]^{m_l} \Phi_{\mathbf{k}_1\sigma_1, \mathbf{k}_2\sigma_2, \cdots, \mathbf{k}_{N_3}\sigma_{N_3}} & \\ \cdot (\mathbf{r}_1 s_1, \mathbf{r}_2 s_2, \cdots, \mathbf{r}_{N_3} s_{N_3}). & \qquad (5.5.31) \end{aligned}$$

Such wave functions are consistent with the third condition as well. The quantum numbers collected on the left-hand side are precisely those that characterize the true eigenstates of ^{3}He–^{4}He solutions. Using these states as unperturbed, there is the one-to-one correspondence with the eigenstates that must be established if the adiabatic turning-on of the perturbation is to be meaningful. For convenience of later reference and for clarity, we list below the three lowest-lying classes of basis functions:

$$|0; \mathbf{k}_1\sigma_1, \cdots, \mathbf{k}_{N_3}\sigma_{N_3}) = \psi_B \Phi_{\mathbf{k}_1\sigma_1,\cdots,\mathbf{k}_{N_3}\sigma_{N_3}}(\mathbf{r}_1 s_1,\cdots,\mathbf{r}_{N_3} s_{N_3}), \tag{5.5.32}$$

$$|\mathbf{p}; \mathbf{k}_1\sigma_1, \cdots, \mathbf{k}_{N_3}\sigma_{N_3}) = \rho_{\mathbf{p}}^{(4)} \psi_B \Phi_{\mathbf{k}_1\sigma_1,\cdots,\mathbf{k}_{N_3}\sigma_{N_3}}(\mathbf{r}_1 s_1, \cdots, \mathbf{r}_{N_3} s_{N_3}), \tag{5.5.33}$$

$$|\mathbf{p}, \mathbf{q}; \mathbf{k}_1\sigma_1, \cdots, \mathbf{k}_{N_3}\sigma_{N_3}) = \rho_{\mathbf{p}}^{(4)} \rho_{\mathbf{q}}^{(4)} \psi_B \Phi_{\mathbf{k}_1\sigma_1,\cdots,\mathbf{k}_{N_3}\sigma_{N_3}}(r_1 s_1,\cdots,\mathbf{r}_{N_3} s_{N_3}). \tag{5.5.34}$$

The matrix elements of 1 and H are next computed. To save space, we shall refrain from displaying these complicated formulas; suffice it to say that all of them can be obtained by means of cluster-expansion procedures described in Section 5.4, and are expressible in terms of $S^{(\alpha,\beta)}(k)$ and the unperturbed excitation energies

$$\varepsilon_0(k) \equiv \frac{\hbar^2 k^2}{2m_3}, \tag{5.5.35}$$

and

$$\omega_0(p) \equiv \frac{\hbar^2 p^2}{2m_4 S^{(4,4)}(p)}. \tag{5.5.36}$$

The knowledge of the diagonal matrix elements of 1 allows us to normalize each of the basis functions. The knowledge of the nondiagonal elements permits orthogonalization. First, every class of functions (5.5.32)–(5.5.34) are made orthogonal to every other class by means of Schmidt's process. Next, between the functions in the same class, Löwdin's transformation [168] is employed to orthogonalize states with different fermion quantum numbers. Unlike in pure liquid ^{3}He (see Section 5.4.4), Löwdin's transformation works quite well in the present case. The reason is that Löwdin's series is an expansion in powers of the density. Here the partial density of ^{3}He is low; the rate of convergence is, therefore, improved by an extra factor of x, which is less than 10%. The functions in the same class with different phonon quantum numbers are, however, *not* orthogonalized. In the calculation under discussion, four-phonon vertices are neglected; so the very difficult task of orthogonalizing two-phonon states with respect to one another can fortunately be circumvented.

After completing the orthogonalization processes, the Hamiltonian is second-quantized. The correspondence between the orthogonalized states in the correlated representation and in the occupation number representation:

$$|0; \mathbf{k}_1\sigma_1, \cdots, \mathbf{k}_{N_3}\sigma_{N_3}\rangle \to a^\dagger_{\mathbf{k}_1\sigma_1} \cdots a^\dagger_{\mathbf{k}_{N_3}\sigma_{N_3}}|0\rangle, \tag{5.5.37a}$$

$$|\mathbf{p}; \mathbf{k}_1\sigma_1, \cdots, \mathbf{k}_{N_3}\sigma_{N_3}\rangle \to c^\dagger_{\mathbf{p}} a^\dagger_{\mathbf{k}_1\sigma_1} \cdots a^\dagger_{\mathbf{k}_{N_3}\sigma_{N_3}}|0\rangle, \tag{5.5.37b}$$

$$|\mathbf{p}, \mathbf{q}; \mathbf{k}_1\sigma_1, \cdots, \mathbf{k}_{N_3}\sigma_{N_3}\rangle \to c^\dagger_{\mathbf{p}} c^\dagger_{\mathbf{q}} a^\dagger_{\mathbf{k}_1\sigma_1} \cdots a^\dagger_{\mathbf{k}_{N_3}\sigma_{N_3}}|0\rangle, \tag{5.5.37c}$$

is now used to identify the matrix elements. The a's obey fermion comutation rules, and the c's obey boson commutation rules. The resulting Hamiltonian has the form

$$
\begin{aligned}
H = E_B &+ \sum_{\mathbf{k}\sigma} \varepsilon_0(k) a^\dagger_{\mathbf{k}\sigma} a_{\mathbf{k}\sigma} + \sum_{\mathbf{p}} \omega_0(p) c^\dagger_{\mathbf{p}} c_{\mathbf{p}} \\
&+ \tfrac{1}{2} \sum_{\mathbf{pq}} \langle \mathbf{p} + \mathbf{q} | W | \mathbf{p}, \mathbf{q} \rangle \, c^\dagger_{\mathbf{p}+\mathbf{q}} c_{\mathbf{p}} c_{\mathbf{q}} + \text{H.c.} \\
&+ \tfrac{1}{2} \sum_{\mathbf{k}_1\mathbf{k}_2\mathbf{l}\sigma_1\sigma_2} \langle \mathbf{k}_1, \mathbf{k}_2 | W | \mathbf{k}_1 + \mathbf{l}, \mathbf{k}_2 - \mathbf{l} \rangle \, a^\dagger_{\mathbf{k}_1\sigma_1} a^\dagger_{\mathbf{k}_2\sigma_2} a_{\mathbf{k}_2 - \mathbf{l},\sigma_2} a_{\mathbf{k}_1 + \mathbf{l},\sigma_1} \\
&+ \sum_{\mathbf{p},\mathbf{k}\sigma} \langle \mathbf{p}; \mathbf{k} - \mathbf{p} | W | \mathbf{k} \rangle \, c^\dagger_{\mathbf{p}} a^\dagger_{\mathbf{k}-\mathbf{p},\sigma} a_{\mathbf{k}\sigma} + \text{H.c.} \\
&+ \sum_{\mathbf{pqk}\sigma} \langle \mathbf{p}; \mathbf{k} + \mathbf{q} - \mathbf{p} | W | \mathbf{q}; \mathbf{k} \rangle \, a^\dagger_{\mathbf{k}+\mathbf{q}-\mathbf{p},\sigma} c^\dagger_{\mathbf{p}} a_{\mathbf{k}\sigma} c_{\mathbf{q}} \\
&+ \tfrac{1}{2} \sum_{\mathbf{pqk}\sigma} \langle \mathbf{p}, \mathbf{q}; \mathbf{k} - \mathbf{p} - \mathbf{q} | W | \mathbf{k} \rangle \, c^\dagger_{\mathbf{p}} c^\dagger_{\mathbf{q}} a^\dagger_{\mathbf{k}-\mathbf{p}-\mathbf{q},\sigma} a_{\mathbf{k}\sigma} + \text{H.c.} \\
&+ \cdots,
\end{aligned}
\tag{5.5.38}
$$

which replaces equation (5.5.1) as the starting point of the microscopic theory. The ^{3}He–^{4}He solution now appears to us as a collection of N_3 ^{3}He pseudoquasiparticles moving in a sea of Feynman phonons. The various interactions are depicted by Figure 5.5.3, in which the solid lines represent ^{3}He, and the dashed lines Feynman phonons.

In the interest of diagonalizing the Hamiltonian, three successive canonical transformations now ensue. The first one serves to renormalize the Feynman phonon. Its task is to eliminate to leading order the three-phonon vertex. The second transformation serves to dress the ^{3}He with phonons, turning them into real quasiparticles. Its task is to eliminate to leading

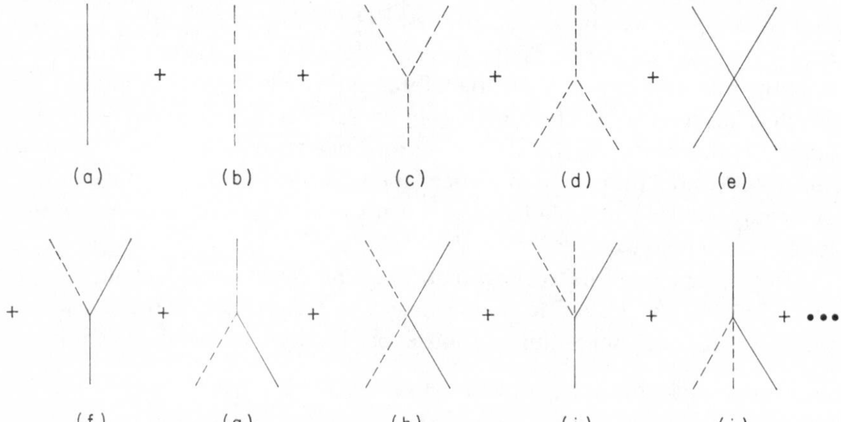

FIGURE 5.5.3. The Hamiltonian of equation (5.5.38).

order (s) the absorption and emission of phonons by ^{3}He. The third transformation modifies the ^{3}He quasiparticles with virtual ^{3}He–^{3}He scattering, in the spirit of Landau's Fermi-liquid theory.

The first canonical transformation is formally identical to that in the theory of pure liquid ^{4}He. Ignoring all but the terms containing only phonon operators, we find the Hamiltonian in the form of equation (5.3.54). The transformation effected by equation (5.3.55)–(5.3.58), with the choice (5.3.60)–(5.3.61) for the transformation function, leads to a BW perturbation calculation in second order. And as we have seen earlier, it satisfactorily renormalizes the phonon–roton spectrum. Let the new phonon operators be written as $b_{\mathbf{p}}^{\dagger}$ and $b_{\mathbf{p}}$, equation (5.5.38) now loses its three-phonon terms, has all the remaining c's replaced by b's, and has all its matrix elements $\langle | W | \rangle$ transformed to $\langle | W' | \rangle$. The Hamiltonian will now be schematically represented by Figure 5.5.4, in which the wavy line represents the renormalized (or Feynman–Cohen) phonon.

The second canonical transformation is given by

$$\begin{cases} a_{\mathbf{k}\sigma}^{\dagger} = e^{-A}\alpha_{\mathbf{k}\sigma}^{\dagger}e^{A}, & (5.5.39a) \\ b_{\mathbf{p}}^{\dagger} = e^{-A}\beta_{\mathbf{p}}e^{A}, & (5.5.39b) \end{cases}$$

where

$$A = \sum_{\mathbf{q}\mathbf{l}\nu} \{\lambda_{\mathbf{q},\mathbf{l}\,\nu}\beta_{\mathbf{q}}^{\dagger}\alpha_{\mathbf{l}\nu}^{\dagger}\alpha_{\mathbf{l}+\mathbf{q},\nu} - \lambda_{\mathbf{q},\mathbf{l}\nu}^{*}\alpha_{\mathbf{l}+\mathbf{q},\nu}^{\dagger}\alpha_{\mathbf{l}\nu}\beta_{\mathbf{q}}\}. \qquad (5.5.40)$$

Substituting equations (5.5.39)–(5.5.40) into the Hamiltonian and collecting

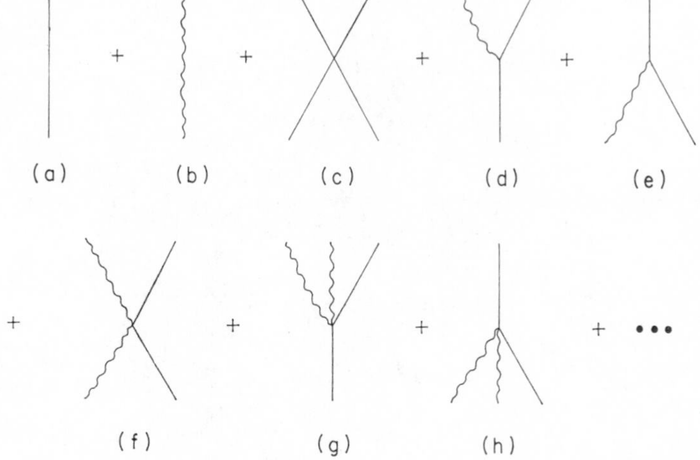

FIGURE 5.5.4. Transformed Hamiltonian with renormalized (Feynman–Cohen) phonons.

all $\beta^\dagger\alpha^\dagger\alpha$ (or $\alpha^\dagger\alpha\beta$) terms, the transformation function $\lambda_{\mathbf{q},\mathbf{l}\nu}$ can be determined by eliminating the latter. The new Hamiltonian reads

$$\begin{aligned} H = {} & E_0 + \sum_{\mathbf{k}\sigma} \varepsilon(k)\alpha^\dagger_{\mathbf{k}\sigma}\alpha_{\mathbf{k}\sigma} + \sum_{\mathbf{p}} \omega(p\,|\,x)\,\beta^\dagger_{\mathbf{p}}\beta_{\mathbf{p}} \\ & + \tfrac{1}{2} \sum_{\mathbf{k}_1\mathbf{k}_2\mathbf{l}\sigma_1\sigma_2} \langle \mathbf{k}_1, \mathbf{k}_2\,|\,V\,|\,\mathbf{k}_1 + \mathbf{l}, \mathbf{k}_2 - \mathbf{l}\rangle \alpha^\dagger_{\mathbf{k}_1\sigma_1}\alpha^\dagger_{\mathbf{k}_2\sigma_2}\alpha_{\mathbf{k}_2-\mathbf{l},\sigma_2}\alpha_{\mathbf{k}_1+\mathbf{l},\sigma_1} \\ & + \sum_{\mathbf{pqk}\sigma} \langle \mathbf{p}; \mathbf{k} + \mathbf{q} - \mathbf{p}\,|\,V\,|\,\mathbf{q}; \mathbf{k}\rangle \alpha^\dagger_{\mathbf{k}+\mathbf{q}-\mathbf{p},\sigma}\beta^\dagger_{\mathbf{p}}\alpha_{\mathbf{k}\sigma}\beta_{\mathbf{q}} \\ & + \tfrac{1}{2} \sum_{\mathbf{pqk}\sigma} \langle \mathbf{p}, \mathbf{q}; \mathbf{k} - \mathbf{p} - \mathbf{q}\,|\,V\,|\,\mathbf{k}\rangle \beta^\dagger_{\mathbf{p}}\beta^\dagger_{\mathbf{q}}\alpha^\dagger_{\mathbf{k}-\mathbf{p}-\mathbf{q},\sigma}\alpha_{\mathbf{k}\sigma} + \text{H.c.} \\ & + \cdots, \end{aligned} \tag{5.5.41}$$

and is schematically represented by Figure 5.5.5, where the heavy solid line now represents ^{3}He quasiparticles, and the heavy wavy line represents impurity-modified phonons. The excitation spectra are finally given by $\varepsilon(k)$ and $\omega(p\,|\,x)$.

$\varepsilon(k)$ is schematically given by Figure 5.5.6. (c) arises from a canonical

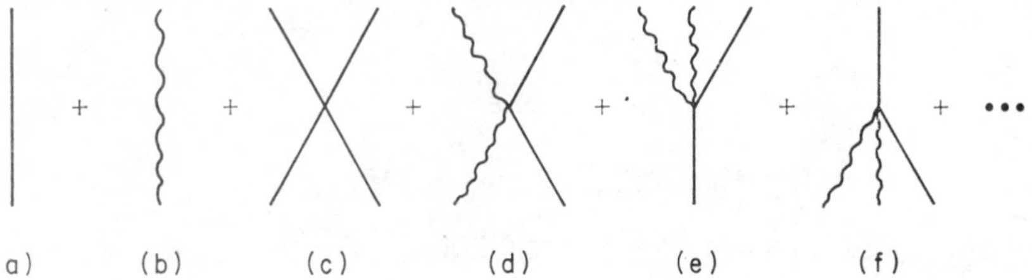

FIGURE 5.5.5. Transformed Hamiltonian of equation (5.5.41).

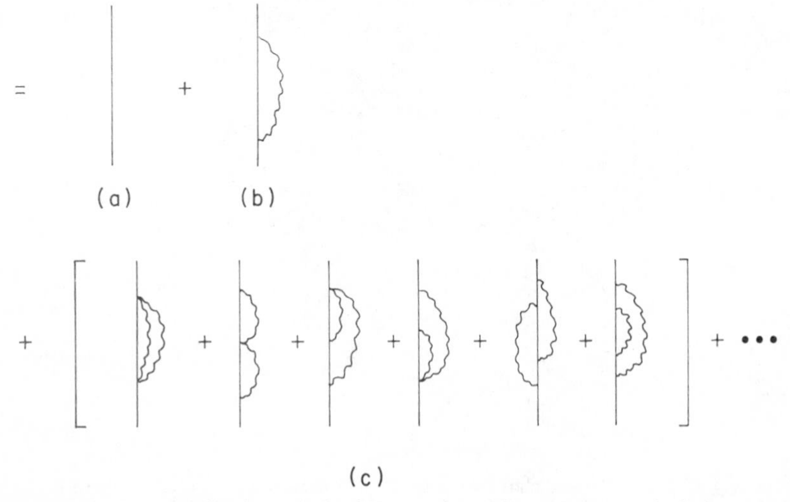

FIGURE 5.5.6. Renormalization of the ^{3}He quasiparticle spectrum: calculation of the effective mass.

transformation of higher order than equation (5.5.40) in which the $\beta^\dagger\,\beta^\dagger\,\alpha^\dagger\,\alpha$ and $\alpha^\dagger\,\alpha\beta\beta$ terms are also eliminated. Expanding in powers of k, one finds the leading terms proportional to k^0 and k^2. By writing it in the Landau–Pomeranchuk form (5.5.2), μ_3 becomes the binary boson solution result equation (5.5.26), and m_3^* is given by

$$m_3^* = (1 - 0.578 + \cdots)^{-1}\, m_3 \approx 2.37 m_3, \tag{5.5.42}$$

in excellent agreement with experiment. Actually, equation (5.5.42) includes only the diagrams (*a*) and (*b*) in Figure 5.5.6. The higher-order diagrams can be evaluated approximately; each turns out to be small. However, it is meaningless to include them in equation (5.5.42) since the phonon renormalization did not proceed beyond the three-phonon vertex. To be consistent, one must include diagrams like the one in Figure 5.5.7, which contains a four-phonon vertex, if one wishes to include Fig. 5.5.6 (*c*) in the effective mass calculation.

A lower-order calculation which neglects phonon renormalization results in an effective mass of 1.85 m_3. Such a procedure can be readily identified with the Davison-Feenberg method [189] discussed in Section 5.5.2, and explains the low effective mass obtained in Reference 189.

To leading orders in x, $\omega(p|x)$ yields the usual phonon-roton spectrum plus a linear (in x) correction which can be evaluated with some effort. This, however, does not tell the whole story. If one is interested in the concentration dependence of the phonon-roton spectrum, it is not legitimate to use the present set of correlated basis functions. In place of $\rho_{\mathbf{p}}^{(4)}$, one should use $\rho_{\mathbf{p}} \equiv \rho_{\mathbf{p}}^{(3)} + \rho_{\mathbf{p}}^{(4)}$ in equations (5.5.31) to (5.5.34), for then the unperturbed (Feynman-like) phonon spectrum (5.5.36) will be modified to read [207]

$$\omega_0(p|x) = \frac{\hbar^2 p^2}{2m(x)S(p|x)}, \tag{5.5.43}$$

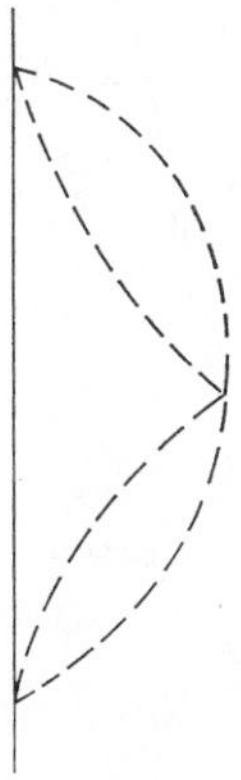

FIGURE 5.5.7. Typical correction term not included in Figure 5.5.6.

where $m(x)$ is a "reduced" mass defined by

$$\frac{1}{m(x)} \equiv \frac{x}{m_3} + \frac{1-x}{m_4}, \tag{5.5.44}$$

and $S(p|x)$ is the appropriate [201] liquid structure function for a binary mixture:

$$S(p|x) \equiv xS^{(3,3)}(p|x) + 2[x(1-x)]^{1/2}S^{(3,4)}(p|x) + (1-x)S^{(4,4)}(p|x). \tag{5.5.45}$$

Clearly both $m(x)$ and $S(p|x)$ possess linear terms in x, which must be grouped with the correction terms arising from the second canonical transformation. $S^{(\alpha,\beta)}(p|0)$ can be calculated quite accurately in the intermediate p region ($1-4$ Å^{-1}), but is not known for small p. The procedure for calculating it is known [201], but the numerical work is prohibitive and has not been attempted. Tan and Woo [207] took the concentration dependence of $\omega(p|x)$ for small p form the experimental data on sound velocity [208]:

$$s(x) \approx s(0)(1-0.17x), \tag{5.5.46}$$

removed from it the x dependence of $m(x)$ and of the canonical transformation correction, and obtained the x dependence of $S(p|x)$ at small p. This was then joined onto the intermediate-p results to give $S(p|x)$ for the full range of p. Figure 5.5.8 shows $S(p|x = 6\%)$. The liquid structure function thus calculated is actually for a binary *boson* solution, but statistical effects enter in order x^2 and will not affect $S(p|x)$ for sufficiently small x. Figure 5.5.8 is, therefore, offered as a prediction, to be tested by X-ray scattering in ^{3}He–^{4}He solutions currently planned by Hallock [209].

One other function of importance is the residual interaction between ^{3}He

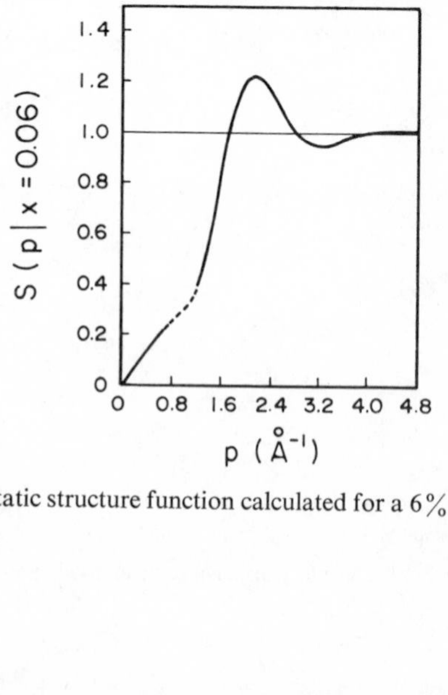

FIGURE 5.5.8. Static structure function calculated for a 6% ^{3}He–^{4}He solution [207].

quasiparticles: $v^*(k)$. As was pointed out by Bardeen, Baym, and Pines [195, 196] and by Emery [197], $v^*(k)$ plays a crucial role in determining a number of thermodynamic and transport properties of ^{3}He–^{4}He solutions. After all, if one considers the ^{3}He quasiparticles as forming a dilute Fermi liquid, $v^*(k)$ is responsible for delivering Landau's $f(\mathbf{k}, \mathbf{k}'; \sigma, \sigma')$. The dominant region is at $k \lesssim k_F$. For dilute solutions k_F is exceedingly small. In the present theory, $v^*(k)$ corresponds to $\langle \mathbf{k}_1, \mathbf{k}_2 \, |V| \, \mathbf{k}_1 + \mathbf{l}, \mathbf{k}_2 - \mathbf{l} \rangle$, which was evaluated in Reference [194]. This matrix element is strongly dependent on $S^{(\alpha,\beta)}(k)$. Since the entire calculation began with a variational determination of the correlation factor, and variational calculations are notoriously insensitive to long-range behavior of the wave function, one cannot expect quantitatively reliable results at small k. The $v^*(k)$ obtained in this manner is only qualitatively meaningful. Its form has been shown [194] to be consistent with the phenomenological $v^*(k)$ determined by Baym and Ebner [210]. If and when a nonvariational determination of $S^{(\alpha,\beta)}(k)$ is carried out [205], it can be inserted into the present theory to redetermine $v^*(k)$ from first principles.

The third canonical transformation renormalizes the ^{3}He quasiparticle spectrum in the spirit of Landau's theory. A new effective mass m_3^{**} should, for instance, result. But since the ^{3}He density is low, the effect will not be quantitatively important. Experimentally one finds m_3^{**} differing from m_3^* by merely 1–2%. No real attempt to carry out this transformation is deemed necessary even though all the apparatuses are available (see Section 5.4).

5.5.5. ^{3}He-Roton Interaction

The Hamiltonian in equation (5.5.41) is not complete: It has been truncated at the inclusion of four-operator terms. There are, in principle, terms of higher orders which are hoepfully small, in the sense that they do not contribute significantly to the properties of interest to us. If one accepts this statement, equation (5.5.41) becomes most powerful: H contains everything one wishes to know about the elementary excitations and their residual interactions. We conclude this section with an example to show how equation (5.5.41) can be applied to problems left out of phenomenological consideratons such as those of Bardeen, Baym, and Pines.

In a recent publication, Herzlinger and King [211] reported on the diffusion coefficient of ^{3}He in superfluid ^{4}He. The concentration used was 1.45×10^{-4}, and the temperature range was from 1.27 to 1.69 K. According to the theory of Khalatnikov and Zharkov [212], at such high temperatures ^{3}He–phonon interactions no longer contribute significantly in comparison to ^{3}He–roton interactions. Also, at such low concentrations, ^{3}He–^{3}He scatterings are rare. So the experiment on the diffusion coefficient was for all practical purposes a direct probe of ^{3}He–roton interaction in ^{3}He–^{4}He solutions.

Equation (5.5.41) permits a microscopic calculation for this experiment. It has the additional advantage that phonons and rotons need not be distinguished: ^{3}He–phonon scattering is *shown*, not assumed, to be insignificant. The matrix element $\langle \mathbf{p};\ \mathbf{k} + \mathbf{q} - \mathbf{p} \,|V|\, \mathbf{q};\ \mathbf{k} \rangle$ depicted by Figure 5.5.5(*d*) provides the interaction between a ^{3}He quasiparticle and a phonon or roton. The vertex conserves momentum, but not energy. For use in the present problem, energy conservation enters through a golden-rule calculation for the scattering cross section:

$$d\sigma(\mathbf{k}, \mathbf{q}, \mathbf{p}) = \frac{2\pi}{\hbar} \left|\langle \mathbf{p}, \mathbf{k} + \mathbf{q} - \mathbf{p} \,|V|\, \mathbf{q}; \mathbf{k} \rangle\right|^2 \delta(\omega(p) + \varepsilon(|\mathbf{k} + \mathbf{q} - \mathbf{p}|) - \omega(q) - \varepsilon(k)) \cdot \frac{[1 - \cos(\mathbf{k}, \mathbf{k} + \mathbf{q} - \mathbf{p})]\Omega}{v_{\text{rel}}} \frac{\Omega}{(2\pi)^3}\, d\mathbf{p} \tag{5.5.47}$$

with

$$v_{\text{rel}} \equiv |\mathbf{v}_q - \mathbf{v}_k| = \hbar \left[\frac{k^2}{m_3^{*2}} + \frac{(q - k_0)^2}{\mu^2} - \frac{2k(q - k_0)}{m_3^*} \cos(\mathbf{k}, \mathbf{q}) \right]^{1/2}, \tag{5.5.48}$$

where k_0 and μ are roton parameters and m_3^* is the effective mass of ^{3}He. The integration over final states $\mathbf{p}$ can be transformed [211] to an integration over the momentum transfer $\Delta\mathbf{k} \equiv \mathbf{q} - \mathbf{p} \equiv (\Delta k, \theta, \phi)$. Δk is determined by the δ function. The angles θ and ϕ are then integrated over to yield the total cross section $\sigma(\mathbf{k}, \mathbf{q})$. Removing from Khalatnikov and Zharkov's theory the approximation of the ^{3}He–roton interaction by a δ-function potential, the diffusion coefficient $D(T)$ can be expressed [214] as

$$D(T) = \frac{(2\pi)^3 k_B T}{m_3} \frac{\int [v_k \int \sigma(\mathbf{k}, \mathbf{q}) n_r(q) d\mathbf{q}]^{-1} k^2 n_3(k)\, d\mathbf{k}}{\int k^2 n_3(k)\, d\mathbf{k}}, \tag{5.5.49}$$

in which n_r and n_3 are, respectively, the statistical distribution functions for rotons and ^{3}He. (For such low concentrations and high temperatures, ^{3}He assumes Boltzmann distribution.)

In a preliminary report, Tan et al [213]. took the matrix element $\langle \mathbf{p}; \mathbf{k} + \mathbf{q} - \mathbf{p} \,|V|\, \mathbf{q}; \mathbf{k} \rangle$ from the Hamiltonain *before* the second canonical transformation, i.e., diagram (*f*) in Figure 5.5.4. The resulting scattering cross section was shown to be strongly momentum dependent, but a factor of 2 too small to account for the experimental $D(T)$. Subsequently [214] the full matrix element, diagram (*d*) of Figure 5.5.5, or Figure 5.5.9 was taken. The second-order contributions involve three-prong vertices, which are

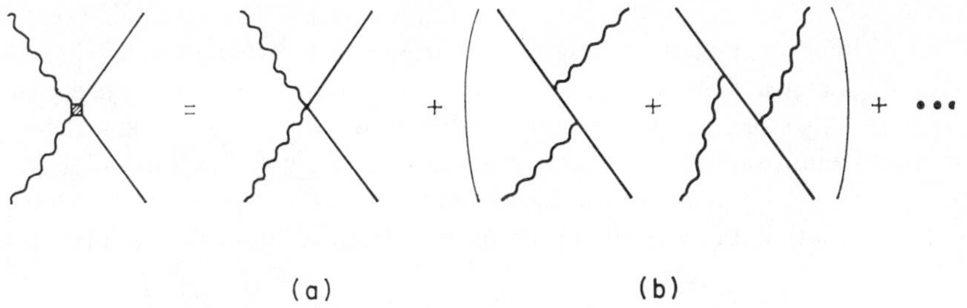

FIGURE 5.5.9. ^{3}He–roton scattering in ^{3}He–^{4}He solutions.

much stronger than the first-order four-prong vertex. Therefore, it is expected, both from previous calculations of similar nature [114, 194, 207] and from general considerations [196], that diagrams (*a*) and (*b*) in Figure 5.5.9 will yield contributions of like magnitude. Furthermore, if there are no strong interference effects, these contributions to the sacttering cross section will be additive. Table 5.5.1 displays the scattering cross section obtained, as averaged over the angle (**k**, **q**). Note that the cross section has strong momentum dependence. Figure 5.5.10 shows the theoretically calculated $D(T)$ in comparison to experiment.

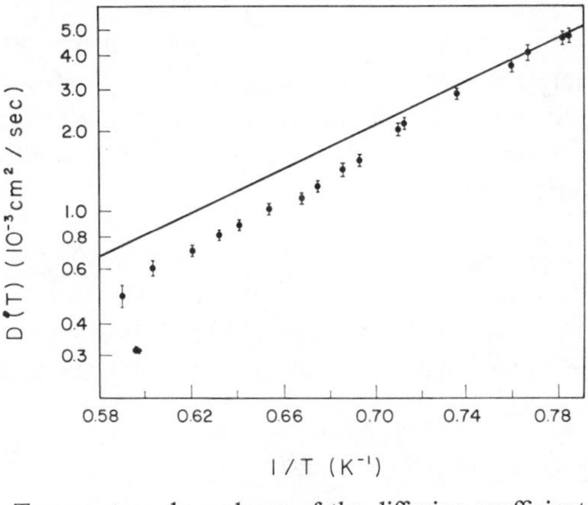

FIGURE 5.5.10. Temperature dependence of the diffusion coefficient of ^{3}He in superfluid ^{4}He. , Experiment [211];———, theory [214].

5.6. SOLID HELIUM

The treatment of short-range correlations is discussed by Varma and Werthamer in Chapter 6. There is also an excellent review by Guyer [215] in *Solid State Physics*. I shall highlight here certain common features of most of the existent microscopic calculations, and show how their development has grown out of the parallel theory for helium liquids. Even though most of what I discuss is treated thoroughly by Varma and Werthamer, our perspectives are somewhat different; a few observations here may serve pedagogical purposes. Also, one of the most serious difficulties in the microscopic theory of solid helium concerns the convergence of the cluster expansion series. An unpublished procedure for overcoming this difficulty will be presented as an illustration of what one can borrow from techniques developed for liquids.

5.6.1. Hartree-Jastrow Theory and the Method of Correlated Basis Functions

Following the reasoning outlined in Section 5.2.2, one always begins with a single-particle type trial wave function, equation (5.2.18). This was precisely the approach taken by Nosanow and Shaw [216]. The hard core in the pairwise potential caused the expectation value of H, equation (5.1.1), to diverge; so certain boundary conditions were imposed on the single-particle function $\varphi(i)$ to dissuade each helium atom from wandering into the vicinity of another helium atom. This may have been an effective way to avoid the overlap of hard cores and keep the energy convergent, but the cost was too high. The cutoff was too abrupt: Along with divergence the very delicate correlation effects were thrown out. The result of such a calculation was disastrous.

In accordance with the prescription for Jastrow's theory given in Section 5.2.2, one now progresses toward a two-particle type wave function, equation (5.2.21). For a solid, albeit a "quantum crystal," the effect of statistics is small as far as the binding energy is concerned, so one need not bother to symmetrize the model function. Thus,

$$\psi_J(1, 2, \ldots, N) = \prod_{i<j=1}^{N} f(r_{ij}) \prod_{l=1}^{N} \varphi(l) \equiv \prod_{i<j=1}^{N} \exp\left[\tfrac{1}{2}u(r_{ij})\right] \prod_{l=1}^{N} \varphi(l), \qquad (5.6.1)$$

as was employed by Nosanow [217] in 1966. What follows, then, is the choice for the form of the variational functions $f(r)$ and $\varphi(l)$, the evaluation of the energy expectation value

$$E_J = \frac{\langle \psi_J | H | \psi_J \rangle}{\langle \psi_J | \psi_J \rangle}, \qquad (5.6.2)$$

and the minimization of E_J with respect to the variational functions.

The form of $f(r)$ need not differ greatly from that in the liquid. Even though the presence of a crystal structure requires $f(r)$ to be anisotropic and to depend on the lattice structure, yet the error introduced by ignoring these details is not really significant [218, 219]. $\varphi(l)$ in the harmonic approximation is a Gaussian. It contains one parameter: the width. Again, by assuming $\varphi(l)$ to be spherically symmetric the error introduced is relatively insignificant

[220]. Even though the exact forms of $f(r)$ and $\varphi(l)$ *should* be determined by solving the coupled Euler–Lagrange equations

$$\begin{cases} \dfrac{\delta E_J}{\delta f(r)} = 0, & (5.6.3\text{a}) \\[2ex] \dfrac{\delta E_J}{\delta \varphi(l)} = 0, & (5.6.3\text{b}) \end{cases}$$

on account of the unavoidable loss of accuracy in the evaluation of E_J and in the introduction of necessary simplifying approximations, little can be gained by actually solving equation (5.6.3). It was attempted by Frohberg [219], and by Mullin, Nosanow, and Steinbeck [221] without much success. Furthermore, Guyer [215] proved that in most cases the solution of equation (5.6.3) results in $\varphi(l)$ spreading out toward infinite width; i.e., the solid melts. Again, then, as in liquid helium, one resorts to parameterizing $f(r)$ and $\varphi(l)$.

The evaluation of E_J can be carried out, as for liquid ^{4}He, by one of several techniques: cluster expansion, integral equation, molecular dynamics, and Monte Carlo. Nosanow et al [217, 222] used the van Kamper cluster-expansion procedure [13] described in Section 5.2.2. Brueckner and Frohberg's expansion [218] turned out to be essentially a rearrangement of Nosanow's series in orders of the number of pairs rather than the number of atoms in each cluster. Massey and Woo [223, 224] used a double cluster-expansion procedure [224] where the expansion over $f(r)$ is summed to all orders via a BBGKY equation. Hansen and Levesque [225] used molecular dynamics. Brueckner and Thieberger [226] carried out Monte Carlo calculations for the three-particle cluster contribution. At the time of all this work, there was the general impression that these techniques were borrowed from theory of *classical* liquids. Since all these authors were familiar with liquid helium calculations, one cannot help but suspect that they have inadvertently been borrowing directly from *quantum* liquid considerations.

To go beyond Jastrow's theory, the simplest route is to construct a complete set of basis function in the image of equation (5.6.1), compute matrix elements, and carry out a perturbation calculation. In other words, one adopts the method of correlated basis functions. Such a procedure was suggested in Reference [224], but was felt too ambitious a program to undertake. The similarity between such a calculation and a method proposed by Koehler [227, 228] was noted. Koehler expands the Hamiltonian in harmonic variables, and includes Jastrow factors in the harmonic expansion. Whereas in the CBF procedure, the single-particle products (or model functions) are made up of solutions of a harmonic Hamiltonian: Hermite polynomials. The matrix elements evaluated lead to a second quantized Hamiltonian with phonon-like operators, just as for helium liquids. The interaction terms are precisely the anharmic terms in Koehler's Hamiltonian. The two procedures are thus identical, as one might expect. Koehler and Werthamer [229] have since carried out an elaborate and informative analysis of Koehler's method and the relation between the two procedures. For calculational purposes, my feeling is that the direct CBF approach may be more convenient and more fruitful, just as in the case of liquid ^{4}He, in which a parallelism turns out to favor the CBF method over the Bohm–Salt [230] theory, which also seeks to expand the Hamiltonian in phonon operators.

In retrospect, the development of a microscopic theory to deal with short-range correlations in solid helium closely paralleled a similar development for liquid helium, except for a time lag by about 5 years. The recognition of this pattern should encourage us to seek more adaptable techniques from the theory of quantum liquids. What follows is just such an example.

5.6.2. Generalized BBGKY Equation

All the cluster-expansion procedures employed in solid helium are plagued by a common flaw: there is no guarantee that the energy series converges sufficiently rapidly for a low-order truncation to be valid. Nosanow et al. [217] retained only the leading term (two-particle clusters) in the expansion, and evaluated the next term [222] (three-particle clusters) about the energy minimum as a check. The outcome was satisfactory enough, but only if $f(r)$ was restricted to certain regions of its parameter space. One cannot rule out the possibility that the minimum obtained was local. Brueckner and Thieberger [226] obtained similar results using a Monte Carlo integration technique to save computer time. Woo and Massey [224] employed the "quasi-crystal approximation," which sums the $f(r)$ part of the cluster series, so that the medium in which a pair of particles move takes on the appearance of a liquid. The purpose was to eliminate at least in part a possible cause for slow convergence. Yet the single-particle part of the cluster series was not summed, and only the leading term was retained. In the case of liquid ^{3}He, there was a similar cluster series to contend with [Section 5.4.3, equations (5.4.47)–(5.4.48)]. But the cluster integrals there are much easier to evaluate on account of translational invariance. For example, even though the three-particle cluster contribution E_3^F contains three volume integrations, it can be readily reduced to a sixfold integral with a proper and obvious choice of coordinate axes. Furthermore, since only $S_B(k)$ enters, and only for the range $0 \leq k \leq k_F$, a simplifying quadratic formula can be introduced to approximate $S_B(k)$ to high accuracy. The sixfold integral becomes analytical. For solid helium, the coordinate axes are fixed by the lattice vectors; consequently, no *obvious* reduction is available for the three volume integrations. Hetherington [222] was able to introduce certain mathematical transformations to reduce the integral to tractable form, nevertheless the expression obtained became very complicated. Besides, there remains a threefold lattice summation over the integral!

In view of this difficulty, it becomes very tempting to devise a summation scheme after the fashion of the BBGKY equation for liquid ^4He (Section 5.2.2.) The solution of the equation may not be much easier than evaluating a three-particle cluster integral, but at least after it is done one need not worry about the neglect of higher-order clusters.

In place of equation (5.2.6), we now define the *solid* distribution functions

$$P_n(\mathbf{r}_1, \mathbf{r}_2, \ldots, \mathbf{r}_n) \equiv P_n(\mathbf{r}_1, \mathbf{r}_2, \ldots, \mathbf{r}_n | \mathbf{R}_1, \mathbf{R}_2, \ldots, \mathbf{R}_N) = \frac{\int \psi_J^2 \, d\tau_{(1, 2, \cdots, n)}}{\int \psi_J^2 \, d\tau}, \quad n \geq 1. \tag{5.6.4}$$

The **R**'s denote lattice vectors. They are present because $\varphi(l)$ in the form of a Gaussian or any other localizing function has as its argument the vector $\mathbf{r}_l - \mathbf{R}_l$. The distribution functions obey the sequential relation

$$\int P_{n+1}(\mathbf{r}_1, \mathbf{r}_2, \ldots, \mathbf{r}_n, \mathbf{r}_{n+1}) \, d\mathbf{r}_{n+1} = P_n(\mathbf{r}_1, \mathbf{r}_2, \ldots, \mathbf{r}_n). \tag{5.6.5}$$

In terms of the two-particle distribution function, the energy expectation value becomes

$$E_J = N \frac{3\alpha^{1/2}\hbar^2}{4m} + \tfrac{1}{2} \sum_{ij}{}' \int v(r_{ij}) P_2(\mathbf{r}_i, \mathbf{r}_j) \, d\mathbf{r}_i \, d\mathbf{r}_j, \tag{5.6.6}$$

where for convenience $\varphi(l)$ has been given a Guassian form

$$\varphi(l) = \left(\frac{\alpha}{\pi^2}\right)^{3/8} \exp\left[-\frac{1}{2}\alpha^2 |\mathbf{r}_l - \mathbf{R}_l|^2\right], \tag{5.6.7}$$

and the summation is over all lattice vectors $\mathbf{R}_i$ and $\mathbf{R}_j$. The prime indicates $\mathbf{R}_i \neq \mathbf{R}_j$. Taking into account the facts that (i) $\mathbf{R}_j$ ranges over all N sites, (ii) $P_2 = O(1/\Omega^2)$, and (iii) $v(r)$ is short-ranged, the energy is seen to be properly extensive. It can be readily evaluated once P_2 is known. The problem thus reduces to calculating P_2 for each choice of the trial wave function ψ_J.

Applying the gradient operator ∇_1 to the definitions of P_1, P_2, etc., we find

$$\nabla_1 P_1(\mathbf{r}_1) = P_1(\mathbf{r}_1) \nabla_1 \ln \varphi^2(1) + \sum_{j \neq 1} \int P_2(\mathbf{r}_1, \mathbf{r}_j) \, \nabla_1 u(r_{1j}) \, d\mathbf{r}_j, \tag{5.6.8}$$

$$\nabla_1 P_2(\mathbf{r}_1, \mathbf{r}_2) = P_2(\mathbf{r}_1, \mathbf{r}_2) [\nabla_1 \ln \varphi^2(1) + \nabla_1 u(r_{12})] + \sum_{j \neq 1.2} \int P_3(\mathbf{r}_1, \mathbf{r}_2, \mathbf{r}_j) \nabla_1 u(r_{1j}) \, d\mathbf{r}_j, \tag{5.6.9}$$

etc., an hierarchy of generalized BBGKY equations, the first of which is for the *one*-particle distribution function P_1, now that it is no longer a constant as in liquids. We must retain, therefore, at least the first *two* coupled equations in order to find P_2. For practical calculations, the hierarchy has to be truncated via some superposition approximation, a most promising one being the generalized Kirkwood approximation. Writing

$$P_2(\mathbf{r}_1, \mathbf{r}_2) \equiv P_1(\mathbf{r}_1)P_1(\mathbf{r}_2)g(\mathbf{r}_1, \mathbf{r}_2), \tag{5.6.10}$$

and

$$P_3(\mathbf{r}_1, \mathbf{r}_2, \mathbf{r}_3) \equiv P_1(\mathbf{r}_1)P_1(\mathbf{r}_2)P_1(\mathbf{r}_3)g_3(\mathbf{r}_1, \mathbf{r}_2, \mathbf{r}_3), \tag{5.6.11}$$

the approximation reads

$$g_3(\mathbf{r}_1, \mathbf{r}_2, \mathbf{r}_3) \approx g(\mathbf{r}_1, \mathbf{r}_2)g(\mathbf{r}_2, \mathbf{r}_3)g(\mathbf{r}_3, \mathbf{r}_1). \tag{5.6.12}$$

No numerical solution of the coupling integral equations (5.6.8)–(5.6.9) has yet been attempted. An unpublished analysis showed that formally iterating these equations under the approximation (5.6.12) and the restriction (5.6.5) leads to an infinite series which contains all the terms considered by Nosanow et al., Brueckner and Frohberg, and Massey and Woo, as expected.*

5.7. NEARLY TWO-DIMENSIONAL HELIUM SYSTEM

Much to the chagrin of statistical mechancians, there are no truly two-dimensional systems in nature. Among the more hopeful pretenders, two are helium systems: the physically adsorbed monolayer and the surface of bulk helium. In this section I wish to discuss some pioneering efforts on the microscopic theory of these nearly two-dimensional systems.

5.7.1. Physically Adsorbed Helium Monolayers

Consider helium atoms physically adsorbed on the surface of a crystalline substrate. Under the most ideal circumstances, a laterally unbounded substrate provides a varying field in the direction normal to its surface, with a sufficiently deep well to trap the helium atoms in an infinitesimally thin layer parallel to the adsorbing surface. Other than this, it exerts no influence on the helium atoms. This makes possible the creation of a two-dimensional helium system, characterized at zero temperature by just one intensive parameter: the areal density

$$\sigma \equiv \frac{N}{A}, \tag{5.7.1}$$

with N and A approaching infinity in the thermodynamic limit.

It is known from solving the two-particle Schrödinger equation that in three dimensions a pair of helium atoms do not attract each other sufficiently

*Since the preparation of this manuscript, we have made much progress in solving these equations by means of the Kirkwood-Monroe technique. Refer to recent publications in *J. Low Temp. Phys.* and *Phys. Rev. B.* and *D.*

to bind. In two dimensions it was shown by Bagchi [231] that while a pair of He atoms do bind, the binding energy is too small to matter. We can assume that at $T \sim 1$ K and low areal densities there will be no spontaneous clustering inside an adsorbed helium monolayer. The monolayer appears as a two-dimensional gas. As σ increases, the monolayer subjects itself to possible phase transitions. Depending on the quantum statistics, Bose or Fermi liquid properties begin to appear. As σ continues to increase, crystallization eventually occurs. Provided that experimental conditions can be improved to meet the stringent requirements of approximate two-dimensionality, the helium monolayers will prove to be richer in physical information than even their three-dimensional counterparts. The reason is that the areal density is very much in the control of the experimenter. While ^{3}He dissolved in superfluid ^{4}He near zero temperature can have its concentration varied between 0 and 10%, the range of variation in the case of adsorbed monolayers is practically without limit.

Are the experimental conditions ripe for well-defined investigations? In a recent review of the subject by Dash [232], he listed among possible probes vapor pressure, desorption, and heat capacity measurements, low-energy electron diffraction, reflection high-energy electron diffraction, field emission and ionization microscopy, nuclear magnetic resonance, electron paramagnetic resonance, low-energy molecular scattering, and Mössbauer spectroscopy. But in practice, other than recent work using NMR at Sussex [233] Cornell [243], Stevens [235], and Washington University (St. Louis) [236], all measurements have been thermodynamic: adsorption isotherms and specific heats. The art of substrate selection and preparation has seen much progress. While in the early days most of the work was carried out on copper sponge and vycor glass, the more recent and from all appearances the most promising substrate developed is exfoliated pyrolytic graphite ("grafoil") [232].

Adsorption isotherms are curves on which an experimenter must rely to determine the amount of material adsorbed. A theoretical equation can be derived by first obtaining the partition function of a model system (e.g., lattice gas) and computing from it the chemical potential, then setting the latter equal to the chemical potential of the vapor, to yield the adsorbed density as a function of vapor pressure. An important characteristic of adsorption isotherms is that, as a layer nears completion, a gentle shoulder invariably develops in the data. All experiments on adsorbed films must be preceded by measurements of the isotherms and location of the shoulders.

Specific heat data are macroscopic. They are sensitive to whatever phase change that takes place inside the system. Much significant work on helium monolayers has been carried out by Dash, Vilches, and co-workers in recent years at the University of Washington. Brewer's group at Sussex and Good-

TABLE 5.7.1. Band structure of helium adsorbed on argon substrate.

		Novaco and Milford [257]	ai, Woo, and Wu [262, 263]
(001)	1st bandwidth	0.1	0.1
	1st band gap	38	34
	2nd bandwidth	3.0	1.9
(111)	1st bandwidth	(4.9)[a]	4.9
	1st band gap	(24.4)[a]	22
	2nd bandwidth	(20.4)[a]	26

[a]Entries shown in parentheses were obtained for compressed argon (see text).

stein at Cal Tech have made important contributions [237]. More recently, low-temperature measurements of the specific heat are being carried out also at Cornell [238].

Prior to 1968, Dash, Goodstein, and McCormick published a series of papers [239–241] on the adsorption isotherms and heat capacities of ^{3}He and ^{4}He on argon-plated copper sponge. At coverages close to a complete monolayer, they found heat capacities quadratic in temperature, signaling the formation of two-dimensional Debye-like solids. At low coverages, the heat capacities became linear in temperature, as befit two-dimensional quantum gases. These results were exciting because it appeared that at last physicists learned how to prepare and harness realistic two-dimensional systems. Statistical mechanical conjectures may now be subjected to critical tests in the laboratory. Between 1968 and 1970, Stewart and Dash [242,243] carried out more precise measurements of the same nature. Under painstaking scrutiny, they found the satisfying and theoretically understandable results no longer in evidence. At low coverages and low temperatures, if a linear fit to the heat capacity $C(T)$ was made, it would produce a small but definitely negative zero-temperature intercept: clearly an unacceptable conclusion. On the other hand, a quadratic fit would indicate two-dimensional crystallization at gaseous densities: an equally disturbing alternative. It was suggested [243] that somehow the substrate must have played a dominant catalytic role. Either by applying strong lateral pressures caused by inhomogeneities, or by mediating extra He–He attraction, the substrate had forced the adatoms to clump into solid patches.

Substrate inhomogeneities bring about highly undesirable complications. They tend to mask the actual properties of the system under investigation. Subsequently the argon-plated copper substrate was abandoned in favor of grafoil, which is known to possess exceedingly uniform adsorption surfaces. Bretz and Dash [244] found that $C(T)$ approaches Nk_B at high temperatures for ^{3}He and ^{4}He alike. As T decreases, $C(T)$ for ^{3}He falls off slowly and smoothly until about 1 K, and then turns sharply downward toward zero as $T \to 0$, in a way suggestive of a Fermi gas. But the ^{4}He curve exhibits a pronounced peak at 1–2 K for every coverage below 40%, signaling the onset of some unexpected phase transition. It is unexpected because despite the clear suggestion of a λ transition we are well aware of the fact that Bose–Einstein condensation is suppressed in two dimensions. In still more recent work, Bretz, Dash, and Huff [245, 246] identified at higher coverages (45–116%) other heat capacity peaks, and interpreted them in terms of the onset of lattice-gas ordering transition and two-dimensional solidification. Hickernell, McLean, and Vilches [247], on the other hand, detected peaks at about 0.1 K for ^{3}He monolayers at 25–63% coverages, which they interpreted as the onset of spin ordering. These recent discoveries added much to the excitement in the field. So far a number of model calculations have been performed, in particular by Campbell, Dash, and Schick [248, 249] But they are not microscopic calculations using realistic helium potentials, and will, consequently, be excluded from this review.

5.7.2. Zero Coverage Limit: Single-Particle Calculations

The qualitative description given in the preceding section does not constitute a statement for the problem under consideration. In order to carry out microscopic calculations, we must commit ourselves to a more definite model. For the sake of concreteness, let us consider ^{4}He atoms physically adsorbed on a semi-infinite fcc argon crystal. Let m and $\mathbf{r}_i$ denote, respectively, the mass and position of a ^{4}He atom, $i \leq n$; let M and $\mathbf{r}_\alpha$ denote the mass and position of an Ar atom, $\alpha \leq N$. The Hamiltonian of the system as a whole is then given by

$$H = \sum_{\alpha=1}^{N} \frac{-\hbar^2}{2M} \nabla_\alpha^2 + \sum_{i=1}^{n} \frac{-\hbar^2}{2m} \nabla_i^2 + \sum_{\alpha<\beta=1}^{N} v_{\text{ArAr}}(r_{\alpha\beta})$$
$$+ \sum_{\alpha=1}^{N} \sum_{i=1}^{n} v_{\text{ArHe}}(r_{\alpha i}) + \sum_{i<j=1}^{n} v_{\text{HeHe}}(r_{ij}). \qquad (5.7.2)$$

Even if we assume the argon crystal perfectly homogeneous and the adsorbing surface perfectly uniform, equation (5.7.2) remains an extremely formidable many-body problem. We must reduce it to a more manageable form.

At the very low temperatures of interest to us, there will be little activity left inside solid argon. So the first stage of reduction consists of omitting all description of motion inside argon. Each of the argon atoms will be fixed at a lattice site $\mathbf{R}_\alpha$, so that the substrate will serve no other purpose than to provide a static field: an "external" field in which the helium atoms move. It is not at all obvious that this is a valid approximation. For even when all thermal agitations in the substrate cease, there could still be virtual processes to affect the He–He or He–substrate interactions in a significant way. I shall return to this point later in Section 5.7.5. Let us for now accept the reduced

Hamiltonian H_1:

$$H_1 = \sum_{i=1}^{n} \frac{-\hbar^2}{2m} \nabla_i^2 + \sum_{i=1}^{n} V(\mathbf{r}_i) + \sum_{i<j=1}^{n} v_{\mathrm{HeHe}}(r_{ij}), \tag{5.7.3}$$

where

$$V(\mathbf{r}_i) = \sum_{\alpha=1}^{N} v_{\mathrm{ArHe}}(|\mathbf{R}_\alpha - \mathbf{r}_i|). \tag{5.7.4}$$

The next stage of reduction is more legitimate. At low coverages, the adsorbed atoms or "adatoms" will have little occasion to be near one another. It is then possible to neglect the He–He interaction and obtain

$$H_2 = \sum_{i=1}^{n} \left[\frac{-\hbar^2}{2m} \nabla_i^2 + V(\mathbf{r}_i) \right] \equiv \sum_{i=1}^{n} h(i). \tag{5.7.5}$$

The problem is thus reduced to a one-body problem that we shall refer to as the zero coverage limit.

The argon–helium potential may be expressed in the Lennard-Jones 6–12 form, Equation (5.1.30). In earlier work [250–254] a so-called interpolation rule was used for the parameters ε and σ:

$$\varepsilon = (\varepsilon_{\mathrm{HeHe}}\, \varepsilon_{\mathrm{ArAr}})^{1/2}, \tag{5.7.6a}$$

$$\sigma = \tfrac{1}{2}(\sigma_{\mathrm{HeHe}} + \sigma_{\mathrm{ArAr}}). \tag{5.7.6b}$$

These parameters are now known more accurately from molecular beam

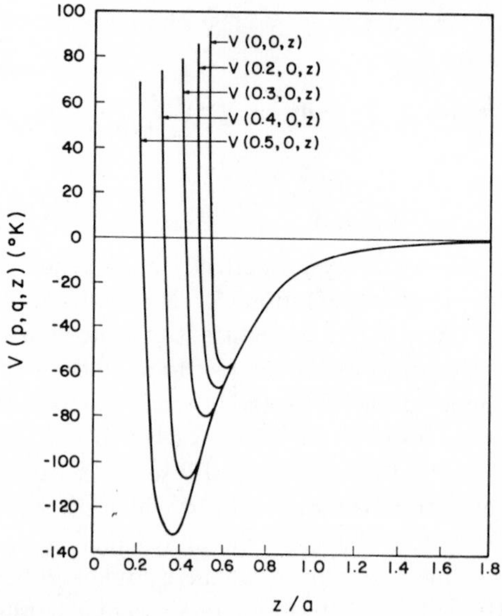

FIGURE 5.7.1. Adsorption potential contours plotted against the normal distance from the substrate surface. See text for notations.

experiments [255]:

$$\varepsilon = 25.4 \text{ K}, \tag{5.7.7a}$$

$$\sigma = 3.06 \text{ Å}. \tag{5.7.7b}$$

In comparing theoretical results one has to be watchful, since the former has a potential well some 10 K lower than the latter. In more recent calculations [256–259] it is always the latter that is assumed. The summation of $v_{\text{ArHe}}(r_{\alpha i})$ over the fcc lattice can be performed by usual means [250, 251, 256.] On the (001) face it results in the potential curves of Figure 5.7.1. These curves are drawn as functions of z, the coordinate normal to the surface measured from the surface, for different lateral coordinates (x, y) or (p, q) defined by Figure 5.7.2. (All lengths are measured in units of the lattice constant

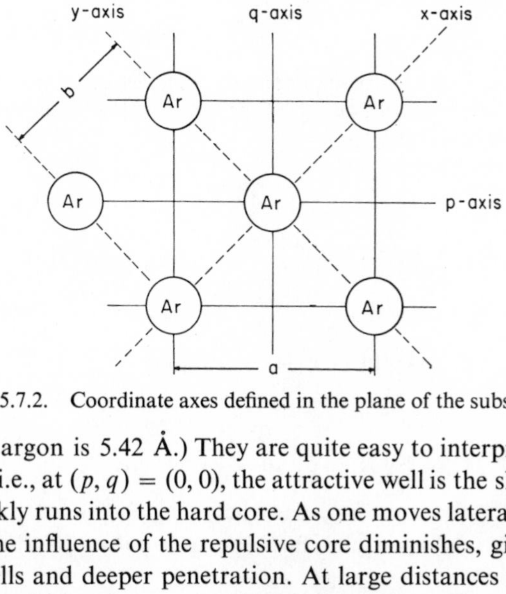

FIGURE 5.7.2. Coordinate axes defined in the plane of the substrate surface.

a, which for argon is 5.42 Å.) They are quite easy to interpret. On top of an argon atom, i.e., at $(p, q) = (0, 0)$, the attractive well is the shallowest and the adatom quickly runs into the hard core. As one moves laterally away from the argon site, the influence of the repulsive core diminishes, giving rise to more attractive wells and deeper penetration. At large distances from the surface, the lateral positions become undistinguishable to the adatom, and the potential curves merge. There are also other ways to present the potential hypersurface. Ross and Steele [251, 252] plotted the well-depth as a function of $1/\sqrt{2}\,(x + y)$, and obtained some of the points shown in Figure 5.7.3. Novaco and Milford [257–260] presented sets of isopotential contours and perspective plots that are both informative and aesthetically appealing.

There are several procedures for solving the one-body problem:

$$h\varphi(\mathbf{r}) = e\varphi(\mathbf{r}). \tag{5.7.8}$$

The most primitive one calls for approximating $V(\mathbf{r})$ by a mathematically convenient form. Lennard-Jones and Devonshire [250] in essence wrote

$$V(\mathbf{r}) = V_A(\mathbf{r}) + \Delta V_A(\mathbf{r}), \tag{5.7.9}$$

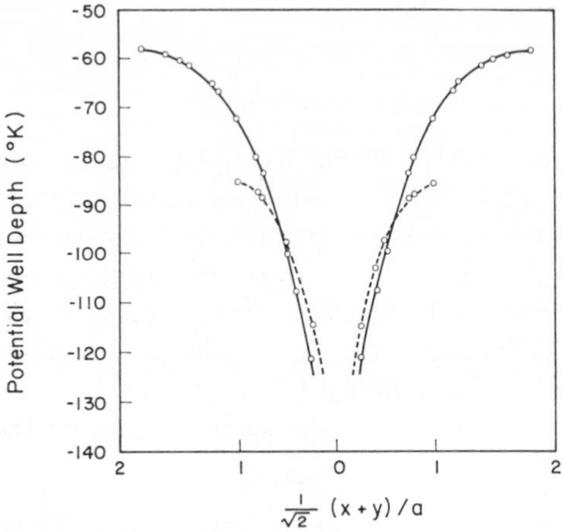

FIGURE 5.7.3. Potential minimum (with respect to z) plotted as a function of the lateral position given by $(1/\sqrt{2})(x+y)$. One pair of the curves is taken along the p-axis and the other along the $(p+q=\frac{1}{2})$ line.

where

$$V_A(\mathbf{r}) = V_0(z) + V_1(z)\left[\cos\frac{2\pi x}{b} + \cos\frac{2\pi y}{b}\right]; \quad b = \frac{a}{\sqrt{2}}. \tag{5.7.10}$$

Next they took

$$V_A(\mathbf{r}) = V_B(\mathbf{r}) + \Delta V_B(\mathbf{r}), \tag{5.7.11}$$

where

$$V_B(\mathbf{r}) = V_0(z) + C\left[\cos\frac{2\pi x}{b} + \cos\frac{2\pi y}{b}\right]. \tag{5.7.12}$$

By discarding $\Delta V_A(\mathbf{r})$ from $V(\mathbf{r})$, and then $\Delta V_B(\mathbf{r})$ from $V_A(\mathbf{r})$, they were able to make the single-particle Hamiltonian totally separable in its three variables:

$$h = \left[\frac{-\hbar^2}{2m}\frac{\partial^2}{\partial z^2} + V_0(z)\right] + \left[\frac{-\hbar^2}{2m}\frac{\partial^2}{\partial x^2} + C\cos\frac{2\pi x}{b}\right] + \left[\frac{-\hbar^2}{2m}\frac{\partial^2}{\partial y^2} + C\cos\frac{2\pi y}{b}\right]. \tag{5.7.13}$$

Equation (5.7.8) then reduces to

$$\left[\frac{-\hbar^2}{2m}\frac{\partial^2}{\partial z^2} + V_0(z)\right]\chi_\mu(z) = e_\mu^{(n)}\chi_\mu(z), \tag{5.7.14}$$

$$\left[\frac{-\hbar^2}{2m}\frac{\partial^2}{\partial x^2} + C\cos\frac{2\pi x}{b}\right]\eta_\nu(x) = e_\nu^{(l)}\,\eta_\nu(x), \tag{5.7.15}$$

$$\left[\frac{-\hbar^2}{2m}\frac{\partial^2}{\partial y^2} + C\cos\frac{2\pi y}{b}\right]\eta_{\nu'}(y) = e_{\nu'}^{(l)}\,\eta_{\nu'}(y). \tag{5.7.16}$$

The solutions of which are readily available:

$$e_{\mu,\boldsymbol{\nu}} \equiv e_{\mu}^{(n)} + e_{\nu}^{(l)} + e_{\nu'}^{(l)}, \tag{5.7.17}$$

$$\varphi_{\mu,\boldsymbol{\nu}}(z,\boldsymbol{\rho}) \equiv \varphi_{\mu,\nu,\nu'}(z,x,y) = \chi_\mu(z)\eta_\nu(x)\eta_{\nu'}(y) \equiv \chi_\mu(z)\eta_{\boldsymbol{\nu}}(\boldsymbol{\rho}). \tag{5.7.18}$$

Equation (5.7.14) possesses bound-state solutions and is almost wholly responsible for the adsorption process. The constant C in equations (5.7.15)–(5.7.16) was defined by

$$\langle\chi_0(z)\,|\,V_1(z)\,|\,\chi_0(z)\rangle = C, \tag{5.7.19}$$

so that if $\Delta V_B(\mathbf{r})$ was treated as a perturbation, there would be no correction from first-order perturbation theory. Once C is determined, equations (5.7.15) and (5.7.16) can be readily solved in terms of Mathieu functions [261], which give rise to energy bands.

An application of this method to the ground state of the He-on-Ar problem [256] led to the following results. First of all, equation (5.7.14) can be solved quite accurately by the use of variational functions in the form

$$\chi_0(z\,|\,\kappa,\beta,\delta) = \left[\frac{\kappa(2\delta-1)}{\Gamma(2\delta)}\right]^{1/2} e^{i\pi(2\delta-1)-1/2\omega}\,\omega^{1/2(2\delta-1)}, \tag{5.7.20}$$

where

$$\omega = 2\delta\, e^{-\kappa(z-\beta)}. \tag{5.7.21}$$

This function is originally the ground-state solution to the Morse potential

$$v_{\text{Morse}}(z) = \frac{\hbar^2}{2m}(\kappa\delta)^2[e^{-2\kappa(z-\beta)} - 2e^{-\kappa(z-\beta)}], \tag{5.7.22}$$

but here the Morse parameters κ, β, and δ are taken to be variational parameters. $e_0^{(n)}$, the normal contribution to the single-particle energy, is minimized at -45.06 K. C is determined at 3.89 K, leading to a lateral contribution of $e_0^{(l)} = -0.47$ K. Thus,

$$e_{0,\mathbf{0}} = (-45.06 - 0.47 - 0.47)\text{ K} = -46\text{ K}. \tag{5.7.23}$$

Next, by extending the variational calculation for equation (5.7.14) to excited states and combining the results with higher-order Mathieu functions to form intermediate states, perturbation corrections from $\Delta V_B(\mathbf{r})$ can be evaluated. The corrections are small, thus validating $V_B(\mathbf{r})$ as a faithful representation of $V_A(\mathbf{r})$. However, the perturbation corrections from $\Delta V_A(\mathbf{r})$ evaluated in the same way are *not* small. In first order it amounts to about -1 K; in the next order it gives about -4 K. Thus the Lennard-Jones–Devonshire procedure fails to be quantitatively useful: It is impossible to approximate the "true" potential $V(\mathbf{r})$ by a potential in the form $V_A(\mathbf{r})$ with sufficient accuracy. Furthermore, the band structure obtained by this method is very different from what one gets using more accurate procedures to be outlined below.

Other fits to the potential $V(\mathbf{r})$ are available, but few have the simplicity of a sinusoidal. Steele and Ross [251, 252] made use of Figure 5.7.3 to approximate the potential by an infinitely extended **V**-shaped well. Such an approximation possesses simple solutions, but it unphysically *imposes* localization on the lateral atomic motion. Furthermore, from Figure 5.7.3 one sees that the potential minima do not really fall along well-defined curves, and the cusp formed is far softer than the **V**-shape.

Ricca, Pisani, and Garrone [254] (RPG) carried out three types of variational calculations. The first one simply takes a product of three Gaussians: ground-state solution to a three-dimensional harmonic oscillator. The Gaussians in the x and y directions necessarily have equal widths, and are centered on the adsorption sites, e.g., $(p, q) = (\frac{1}{2}a, 0)$. The width, plus the width and center of the Gaussian in the z direction, give rise to three variational parameters. For He on a krypton substrate, they found the energy minimum to be -115 K. The second calculation assumes an auxiliary harmonic potential, with elastic constants fixed by the previous variational calculation. Now one writes down the complete set of eigenstates of that potential, and linearly combines the first several to make up a trial wave function. The coefficients in the linear combination serve as variational parameters. A secular equation must then be solved to optimize the linear combination. They found an improvement in the energy; it was lowered to -125 K. The third calculation takes the "local" functions just determined to form Bloch functions, and then linearly combines Bloch functions to form a trial wave function. They found the energy to be again -125 K. However, the band structure was also obtained.

Lai, Woo, and Wu [256] (LWW) found a much less costly way to carry out the variational calculation. In their case a simple form for the single-particle wave function was an absolute necessity, since their main interest was on He–He correlation effects; as will be seen in the next several sections, the many-body theoretical program planned is too involved to tolerate a complicated single-particle function. They began with a product of three Gaussians. After reproducing RPG's result for He on Kr, they replaced the Gaussian in the z direction by a Morse function, equation (5.7.20). This lowered the energy to -118 K. From inspection of the potential curves, e.g., Figure 5.7.1, it became obvious that there is a shift in the location of the potential well with the lateral position. The Morse function was, therefore, also allowed to shift: The parameter β was made a function of the lateral position, thus:

$$\beta \to \beta + \gamma\left(\cos\frac{2\pi x}{b} + \cos\frac{2\pi y}{b}\right). \tag{5.7.24}$$

One more parameter was introduced: γ. This lowered the energy to -123.5 K, barely 1% higher than RPG. The same calculation applied to He on Ar substrate led to a minimum energy of -61.6 K, much lower than the Lennard-Jones–Devonshire result, equation (5.7.23).

LWW [262, 263] also suggested a method for estimating the band structure. The modulus of the Morse function was used as a weighting function for averaging the potential $V(\mathbf{r})$ in the z direction. The result was further averaged over all directions in the lateral plane to obtain a cylindrically symmetrical potential, which was subsequently fitted to a cosine curve. The radial Schrödinger equation with this potential was then solved by means of Mathieu functions to yield the band spectrum. The results are shown in Table 5.7.1. The very narrow bandwidth and the band gap imply that the adatoms are not very mobile at all on the (001) face of argon.

Novaco, Milford and Hagen [257–260] (NMH) carried out very detailed calculations for a number of bare and coated substrates. The method used was much like RPG's third, except that sinusoidals were employed in place of the "local" functions. The energies obtained in comparable cases were a few percent lower than those of LWW. The band spectra appear in Table 5.7.1. Considering the crudeness of LWW's approximation scheme for the band structure, the agreement must be considered very good and reassuring.

Dash suggested that his earlier argon-plated copper substrate [239–243] may perhaps be approximated by an idealized fcc argon crystal. However, since the helium atoms seems much more mobile, he felt that [264] the pre-adsorbed argon layer was structured more like a (111) face than a (001) face. For this raeson, LWW [262] applied their procedure to the (111) face as well. The ground-state energy becomes -47.3 K. The bands become much wider and the band gaps much smaller. NM [257] did not perform their calculation for the (111) face. In its place, they considered an fcc argon crystal compressed in such a way that the areal density on the (001) face corresponded to that on the (111) face of an uncompressed crystal. As far as the energy is concerned, this made a poor model, since compression brought the second Ar layer closer to the top, which deepened the adsorption well. In actuality, with the (111) face on top, the second layer is *farther* away from the adsorbing surface and contributes less to the adsorption well. The band spectrum, however, depends only on the overall structure of the topmost layer. This explains why in Table 5.7.1 the NM results are so close to those of LWW.

In this discussion I have not attempted to establish any quantitative contacts between microscopic calculations and experiments. The reason is simple: The adsorbed helium atoms are strongly correlated. Whether for calculating specific heats or for interpreting order–disorder transitions, these correlations play a central role. While calculations in the zero coverage limit do provide qualitative guidance and a necessary starting point for the many-body calculations that follow, the conclusions reached should not be taken literally.

5.7.3. Finite Coverages: Many-Body Calculations

Stewart and Dash [242, 243] found that even at very low coverages the specific-heat data can be fitted to a T^2 law, which implies that there are strong correlations between the adatoms, sufficient to bring about condensation in two dimensions. Whether the correlations are due to bare He–He interactions or substrate mediation, an interaction term must be included in the Hamiltonian. What appears in H_1 of equation (5.7.3) is the bare He–He interaction. However, as will be seen in Section 5.7.5 if substrate-mediated effects are nonnegligible, one could begin with the full Hamiltonian H of equation (5.7.2), derive an effective interaction $v^*_{\text{HeHe}}(r)$, and work with H_1 using v^*_{HeHe} in place of v_{HeHe}. Let us at this point turn to the many-body problem defined by H_1.

In the spirit of CBF, we construct a set of basis functions each consisting of a dynamical correlation factor and a model function. Having solved the single-particle problem, we are in possession of a natural set of model functions: products of the single-particle functions $\varphi_{\mu,\boldsymbol{\nu}}(z, \boldsymbol{\rho})$. Even though on an argon substrate the helium atoms appear well localized [less so on a (111) face than a (001) face], one should remain as general as possible and construct Bloch functions in the lateral plane:

$$\eta_{\boldsymbol{\nu}}(\boldsymbol{\rho}) = e^{i\boldsymbol{\nu}\cdot\boldsymbol{\rho}} U_{\boldsymbol{\nu}}(\boldsymbol{\rho}), \tag{5.7.25}$$

where $U_{\boldsymbol{\nu}}(\boldsymbol{\rho})$ is periodic with the period of the surface lattice. The model functions are then given by

$$\phi_{\mu_1\boldsymbol{\nu}_1,\mu_2\boldsymbol{\nu}_2,\cdots,\mu_n\boldsymbol{\nu}_n}(z_1\boldsymbol{\rho}_1, z_2\boldsymbol{\rho}_2, \cdots, z_n\boldsymbol{\rho}_n) = P\left\{\prod_{i=1}^{n} \varphi_{\mu_i\boldsymbol{\nu}_i}(z_i\boldsymbol{\rho}_i)\right\}. \tag{5.7.26}$$

The correlation factor can once again be given the Jastrow form:

$$F(1, 2, \cdots, n) = \prod_{i<j=1}^{n} f(r_{ij}), \tag{5.7.27}$$

with $f(r)$ determined by a variational calculation.

Such a calculation has not yet been carried out, the main reason being that $v^*_{\text{HeHe}}(r)$ is so far unknown. (See Section 5.8.5 and the references thereof.) But the main features of the CBF theory have already been identified, and certain low-order estimates have been made. These will form the subject matter in this and the next two sections.

The variational calculation employs a trial wave function of the form

$$\psi_0(1, 2, \cdots, n) = F_0(1, 2, \cdots, n) = \prod_{i<j=1}^{n} f(r_{ij}) \prod_{l=1}^{n} \varphi_{00}(z_l \boldsymbol{\rho}_l). \quad (5.7.28)$$

This reminds us of the variational function chosen for solid helium. Hence in evaluating the expectation value of H_1, the techniques developed in Section 5.6 can be readily adapted. The only modification required comes from the presence of the single-particle potential $V(\mathbf{r})$. Thus

$$E_0 \lesssim \frac{\langle \psi_0 | H_1 | \psi_0 \rangle}{\langle \psi_0 | \psi_0 \rangle} = E_1 + E_2, \quad (5.7.29)$$

where

$$E_1 = n \int \tilde{V}(\mathbf{r}) P_1(\mathbf{r}) \, d\mathbf{r}, \quad (5.7.30)$$

and

$$E_2 = \frac{n(n-1)}{2} \int \tilde{v}(\mathbf{r}_{12}) P_2(\mathbf{r}_1, \mathbf{r}_2) \, d\mathbf{r}_1 d\mathbf{r}_2, \quad (5.7.31)$$

with

$$\tilde{V}(\mathbf{r}) = V(\mathbf{r}) - \frac{\hbar^2}{4m} \nabla^2 \ln \varphi_{00}(z, \boldsymbol{\rho}), \quad (5.7.32)$$

and

$$\tilde{v}(r) = v_{\mathrm{HeHe}}(r) - \frac{\hbar^2}{2m} \nabla^2 \ln f(r). \quad (5.7.33)$$

The P's are distribution functions defined as in equation (5.6.4). By using the double-cluster expansion procedure [224] or the generalized BBGKY equation (Section 5.6.2), the distribution functions can be calculated. At small coverage, i.e.,

$$\theta \equiv \frac{n}{n_{\mathrm{sites}}} \ll 1, \quad (5.7.34)$$

one finds [262]

$$nP_1(\mathbf{r}_1) = \theta \varphi_{00}^2(z_1 \boldsymbol{\rho}_1) + \theta^2 \varphi_{00}^2(z_1 \boldsymbol{\rho}_1) \int \varphi_{00}^2(z_2 \boldsymbol{\rho}_2)[f^2(r_{12}) - 1] \, d\mathbf{r}_2 + \cdots, \quad (5.7.35)$$

$$n(n-1)P_2(\mathbf{r}_1, \mathbf{r}_2) = P_1(\mathbf{r}_1) P_1(\mathbf{r}_2) f^2(r_{12}) \cdot \left\{ 1 + \theta \int \varphi_{00}^2(z_3 \boldsymbol{\rho}_3)[f^2(r_{13}) - 1][f^2(r_{23}) - 1] \, d\mathbf{r}_3 + \cdots \right\}, \quad (5.7.36)$$

and

$$\frac{E_0}{n} \approx \int_S \varphi_{00}^2(z\boldsymbol{\rho})A(\mathbf{r})V(r)\,d\mathbf{r}$$
$$+ \frac{\hbar^2}{2m}\left\{\int_S [\nabla\varphi_{00}(z\boldsymbol{\rho})]^2 A(\mathbf{r})\,d\mathbf{r} + \tfrac{1}{2}\int_S \varphi_{00}(z\boldsymbol{\rho})\nabla\varphi_{00}(z\boldsymbol{\rho})\cdot\nabla A(\mathbf{r})\,d\mathbf{r}\right\}$$
$$+ \frac{1}{2n_{\text{sites}}}\int \varphi_{00}^2(z_1\boldsymbol{\rho}_1)\varphi_{00}^2(z_2\boldsymbol{\rho}_2)f^2(r_{12})\tilde{v}(r_{12})\,d\mathbf{r}_1 d\mathbf{r}_2, \tag{5.7.37}$$

where

$$A(\mathbf{r}_1) = \int \varphi_{00}^2(z_2\boldsymbol{\rho}_2)[f^2(r_{12}) - 1]\,d\mathbf{r}_2, \tag{5.7.38}$$

$\int_S$ denotes integration over a unit cell containing one adsorption site and over all z.

5.7.4. The Extreme Mobile Limit

If the substrate is approximated as a semi-infinite continuum, the adsorbing surface will have no periodic structure, and the adatoms will experience free lateral motion. In this extreme mobile limit considered in detail by Jackson [253], much of the complexity in a CBF calculation can be avoided [256].

First of all the single-particle potential V will new be a function only of the normal coordinate z. Next, since the adatoms are trapped in a thin layer, the relative coordinate r_{ij} can be approximated rather well by its lateral component ρ_{12}. Thus,

$$H_1 \approx H_z + H_\rho, \tag{5.7.39}$$

with

$$H_z = \sum_{i=1}^{n}\left[\frac{-\hbar^2}{2m}\frac{\partial^2}{\partial z_i^2} + V(z_i)\right] \tag{5.7.40}$$

and

$$H_\rho = \sum_{i=1}^{n}\frac{-\hbar^2}{2m}\nabla_{\rho_i}^2 + \sum_{i<j=1}^{n} v_{\text{HeHe}}(\rho_{ij}). \tag{5.7.41}$$

The problem is completely separated into a normal part and a lateral part. The normal part has bound-state solutions and gives rise to adsorption. The lateral part corresponds exactly to a two-dimensional helium liquid. The difference

$$H_1 - (H_z + H_\rho) = \sum_{i<j=1}^{n}[v_{\text{HeHe}}(r_{ij}) - v_{\text{HeHe}}(\rho_{ij})] \tag{5.7.42}$$

can be treated as a perturbation.

In terms of the CBF theory described in the preceding section (5.7.3), one now faces a surface lattice of infinite period. $U_\nu(\boldsymbol{\rho})$ is constant. The lowest-lying excited states are given by

$$\begin{cases}\mu_1 = \mu, \\ \boldsymbol{\nu}_1 = \boldsymbol{\nu},\end{cases} \qquad \begin{cases}\mu_i = 0, \\ \boldsymbol{\nu}_i = 0,\end{cases} \qquad i \neq 1; \tag{5.7.43}$$

i.e.,

$$\psi_{\mu\boldsymbol{\nu},00,\cdots,00} = F\phi_{\mu\boldsymbol{\nu},00,\cdots,00}, \tag{5.7.44}$$

where

$$\phi_{\mu\boldsymbol{\nu},00,\cdots,00} = G_{\mu\boldsymbol{\nu}} \prod_{i=1}^{n} \varphi_{00}(z_i\boldsymbol{\rho}_i), \tag{5.7.45}$$

and

$$G_{\mu\boldsymbol{\nu}} = \sum_{i=1}^{n} \left[\frac{\varphi_{\mu\boldsymbol{\nu}}(z_i\boldsymbol{\rho}_i)}{\varphi_{00}(z_i\boldsymbol{\rho}_i)}\right] = \sum_{i=1}^{n} \left[\frac{e^{i\boldsymbol{\nu}\cdot\boldsymbol{\rho}_i}\chi_\mu(z_i)}{\chi_0(z_i)}\right]. \tag{5.7.46}$$

Jackson [253] distinguished two kinds of excitations: what he called "longitudinal surfons," defined by $\mu = 0$, and "transverse surfons," defined by $\mu \neq 0$. The arguments he used to construct the surfon functions were very physical, in close analogy with Feyman's arguments for the phonon wave function [92]. The results are identical to those of the CBF theory as described above:

$$\psi_{\boldsymbol{\nu}}^{L} \equiv \Psi_{0\boldsymbol{\nu}00,\cdots,00} = G_{0\boldsymbol{\nu}} F \prod_{i=1}^{n} \varphi_{00}(z_i\boldsymbol{\rho}_i) = \left[\sum_{i=1}^{n} e^{i\boldsymbol{\nu}\cdot\boldsymbol{\rho}_i}\right]\psi_0 \equiv \rho_{\boldsymbol{\nu}}\psi_0, \tag{5.7.47}$$

$$\psi_{\boldsymbol{\nu}}^{T\mu} \equiv \psi_{\mu\boldsymbol{\nu},00,\cdots,00} = \sum_{i=1}^{n} \left[\frac{e^{i\boldsymbol{\nu}\cdot\boldsymbol{\rho}_i}\chi_\mu(z_i)}{\chi_0(z_i)}\right]\psi_0. \tag{5.7.48}$$

These are the elementary excitations in the adsorbed monolayer.

For $\chi_1(z)$, LWW [256] took the first excited state of a Morse Hamiltonian as a trial wave function, orthogonalized it to $\chi_0(z)$, and carried out a variational calculation. The excitation energy turns out to be about 30 K. Consequently, the transverse surfon energy is much higher than the longitudinal surfon energy. In computing specific heats at low temperatures (say below 6 K), one only need to take into consideration the longitudinal branch, i.e., density fluctuations from equation (5.7.47).

To get some quantitative feeling on Jackson's model, and to set up the apparatus for computing thermodynamic properties of the monolayer once v_{HeHe}^{*} becomes available, Miller, Woo, and Campbell [87] (MWC) carried out Jastrow-type calculations for two-dimensional liquid helium. The method employed included molecular dynamics and several integral equations. The results were compared to those of earlier computer experiments by Hyman [265] and by Campbell and Schick [266]. Figures 5.7.4 and 5.7.5. show, respec-

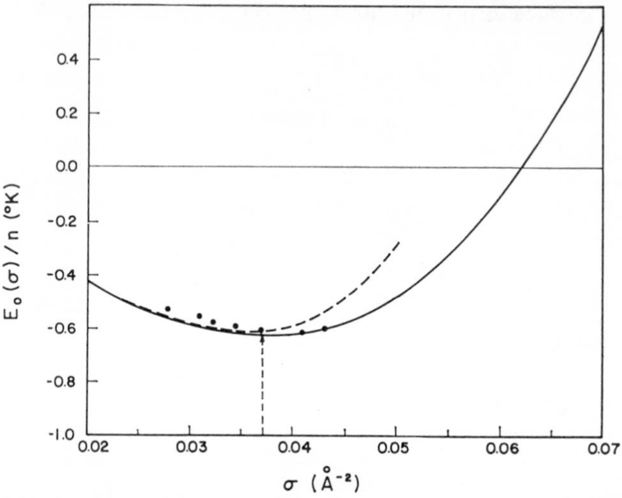

FIGURE 5.7.4. Energy-density relation for a monolayer of ^{4}He interacting via bare Lennard-Jones potential. ———, integral equation result [87]; ······, Monte Carlo [265]; ---------, molecular dynamics [266].

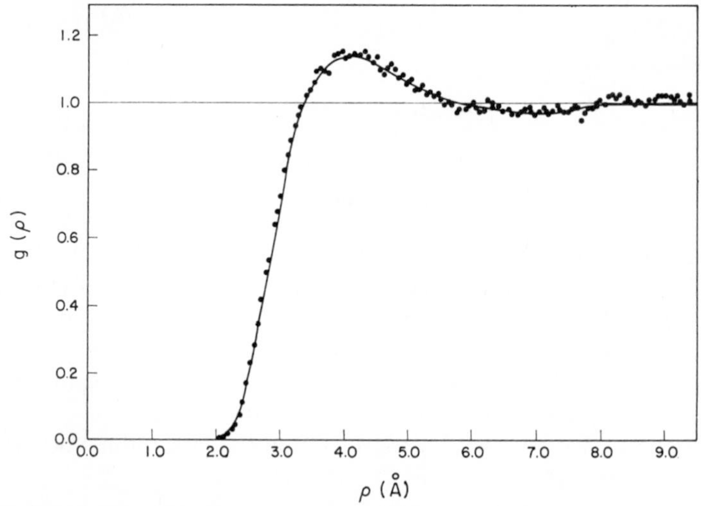

FIGURE 5.7.5. Radial distribution function of a monolayer of ^{4}He interacting via bare Lennard-Jones potential evaluated at equilibrium density $\sigma_0 = 0.037$ Å^{-2}. ———, integral equation result [87]; ·····, molecular dynamics [266].

tively, the energy-density curve and the radial distribution function at equilibrium density σ_0. The energy minimum E_0 is -0.62 K, at $\sigma_0 = 0.037$ Å^{-2}. On account of the smaller number of nearest neighbors in two dimensions, the equilibrium density is much lower than in bulk helium—only about one-half the equivalent value, and the binding is much weaker. Clearly if the adatoms at 1–4 K are to form patches of their own accord, the mutual attraction will have to be much stronger. The substrate must be allowed to play a dominant catalytic role. In a sequel to MWC, Miller and Woo [267] obtained from $g(\rho)$ the two-dimensional liquid structure function $S(\nu)$, the longitudinal surfon spectrum—both the bare Feynman-like excitation and the dressed counterpart of the Feynman–Cohen spectrum $\varepsilon(\nu)$, and from $E_0(\sigma)$ the sound velocity s. At small wave numbers, $\hbar s\nu$ was permitted to replace the calculated spectrum. The specific heat was then calculated by usual statistical mechanics with the surfons treated as bosons:

$$\frac{C(T)}{n} = \frac{1}{2\pi\sigma k_B T^2}\int_0^\infty \frac{\varepsilon^2(\nu)\exp[\varepsilon(\nu)/k_B T]\nu d\nu}{\{\exp[\varepsilon(\nu)/k_B T]-1\}^2}. \qquad (5.7.49)$$

Figures 5.7.6 and 5.7.7 show $\varepsilon(\nu)$ and $C(T)$ respectively, at $\sigma=\sigma_0$. Note that on account of the low density, no roton-dip appears in $\varepsilon(\nu)$. At higher densities (e.g., $\sigma=0.056$ Å^{-2}) pronounced dips do occur. The specific heat is compared to Stewart and Dash's argon-plated-copper data [242, 243] and McLean et al.'s grafoil data [247, 268]. There is order-of-magnitude agreement with the grafoil data even though the role of the substrate was completely ignored in the present calculation. The model clearly loses its validity at high temperatures, say above 1.5 K.

Recently Siddon and Schick calculated the specific heat of the ^{4}He mono-

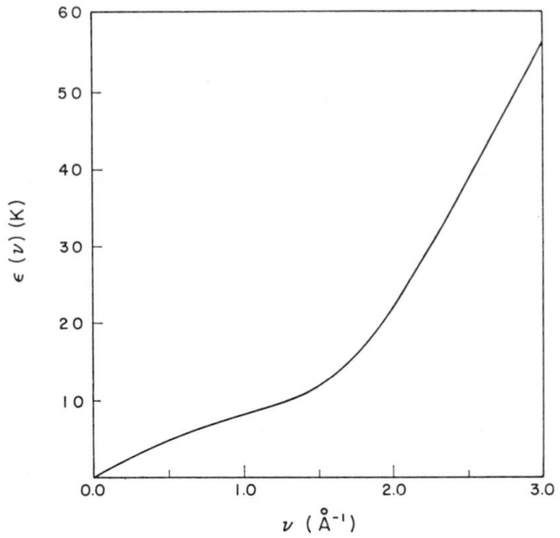

FIGURE 5.7.6. Energy spectrum calculated for ^{4}Hemonola yer at equilibrium density [267].

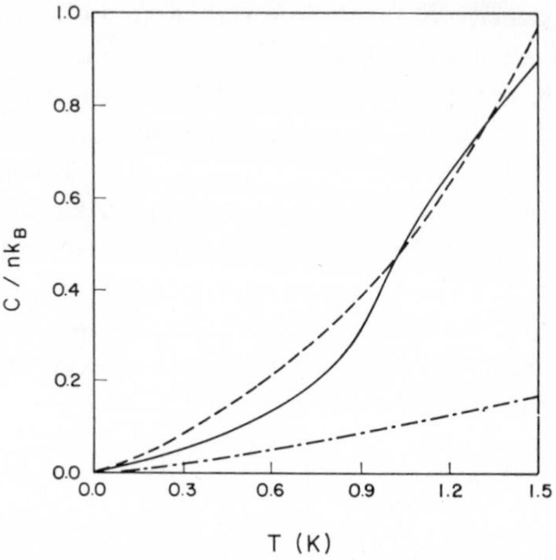

FIGURE 5.7.7. Specific heat of ^{4}He monolayer at equilibrium density. ---------, Theory [267]; ———, experiment on grafoil [247,268]; –·–·–·, experiment on argon-plated copper [242, 243].

layer as a virial gas and found that $C(T)$ rises with decreasing temperature. Below about 1.5 K their model loses its validity. Combining these results with those of Miller and Woo, it becomes very suggestive that two-phase equilibrium exists around 1.5 K, which may well be responsible for the specific-heat peak observed by Bretz and Dash [244]. This interpretation is currently under investigation by Schick and Woo.

5.7.5. Substrate-Mediated Interaction

Using deformation potential arguments, Schick and Campbell [269] first called attention to the mediating role played by the substrate in providing additional attraction between a pair of adatoms. They also estimated the magnitude of the effect and found it rather negligible.

Going back to the full Hamiltonian equation (5.7.2), it can be shown [270] that a microscopic theory exists for calculating the substrate-mediated interaction $v'_{\text{HeHe}}(r)$, and thus the effective interaction:

$$v^*_{\text{HeHe}}(r) \equiv v_{\text{HeHe}}(r) + v'_{\text{HeHe}}(r). \tag{5.7.50}$$

First the Hamiltonian may momentarily be written in the form

$$H = H_s + H_a + W, \tag{5.7.51}$$

where

$$H_s = \sum_{\alpha=1}^{n} \frac{-\hbar^2}{2m} \nabla_\alpha^2 + \sum_{\alpha<\beta=1}^{N} v_{\text{ArAr}}(r_{\alpha\beta}), \tag{5.7.52a}$$

$$H_a = \sum_{i=1}^{n}\left[\frac{-\hbar^2}{2m}\nabla_i^2 + V(\mathbf{r}_i)\right] + \sum_{i<j=1}^{n} v_{\text{HeHe}}(r_{ij}) = H_1, \qquad (5.7.52b)$$

and

$$W = \sum_{i=1}^{n}\left[\sum_{\alpha=1}^{N} v_{\text{ArHe}}(r_{\alpha i}) - V(\mathbf{r}_i)\right], \qquad (5.7.52c)$$

purely as a guide for constructing a sensible set of basis functions. Since the lattice dynamics of a semi-infinite solid can at least be approximated [271], we assume the eigenstates of the substrate Hamiltonian H_s known. They can be expressed in terms of bulk and surface excitations. Based on the discussions in the last three sections, we can now safely assume the eigenstates of the adsorbate Hamiltonian H_a also known. The direct products of these two sets of eigenfunctions will serve adequately as a set of *correlated* basis functions, since the "perturbation" W is clearly small. Using these basis functions, one now abandons the unnecessary separation (5.7.51)–(5.7.52) and evaluates directly the matrix elements of H, from which either a perturbation theory or a canonical transformation, much as those used for ^{3}He—^{4}He solutions, can be performed to obtain $v'_{\text{HeHe}}(r)$ and the effective interaction. Let wavy lines represent the substrate excitations and solid lines the adsorbate states; the above theory can be schematically outlined as in Figure 5.7.8.

For substrates whose adsorbing surfaces exhibit pronounced periodic structure so that $V(\mathbf{r})$ contains strong lateral variations, the adatoms are securely localized in their sites. In this extreme localized limit, the substrate plays an additional mediating role which is static, less dramatic, and seldom considered.

Consider a pair of helium atoms occupying nearest-neighbor sites. They sample more of each other's van der Waals' attraction than the repulsive core. Yet to be near each other has cost them no increase in kinetic energy; for the substrate in localizing the adatoms has already distorted the tail ends of

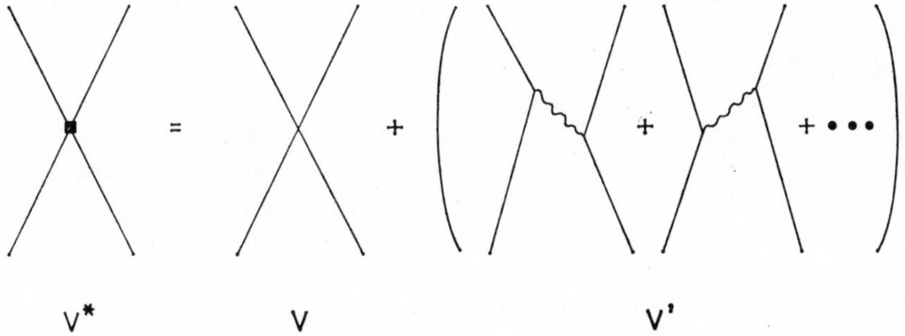

FIGURE 5.7.8. Calculation of the effective interaction between adsorbed helium atoms.

their wave functions. Unlike in the bulk helium liquid, where the approach of two atoms must always be accompanied by a distortion of the wave function (to prevent the hard cores from overlapping) and always exacts a price in the form of increased kinetic energy, here the price is prepaid. The attraction consequently represents a net gain! This correlation energy can be estimated by the method of Section 5.7.3. By taking $f(r)$ to be a step function, LWW [262] evaluated the correlation part of the energy E_0/n, equations (5.7.37)–(5.7.38), and found it to be -6.4 K per adatom on the (001) face of argon. The calculation can be made more accurate [272] by using a more realistic $f(r)$ from two-dimensional He calculation described in the preceding section. It results in a correlation energy of -6.5 K per adatom. In this way the adatoms are offered some incentive to cluster, but not enough to explain Stewart and Dash's [243] data in a consistent manner. The analysis for solid patches which result from such a clustering mechanism is exceedingly complicated. The solid is neither strictly self-formed nor totally substrate-induced. The lattic structure of the monolayer and the adsorbing surface do not necessarily match. The reader will find in Reference [273] a formal classification and an interesting discussion on "mis-registry."

5.7.6. Surfaces of Bulk Liquid Helium

Recent experiments on dilute ^{3}He–^{4}He solutions including measurements on surface tension [274–278], ion mobility [279–282], and third-sound propagation [283] all seem to indicate that a layer of ^{3}He atoms resides on the free surface of bulk liquid helium, even though at below 6.5% concentration no phase separation is expected. These facts induce us to examine a phenomenological theory by Andreev [284], who postulated the existence of surface states of ^{3}He in the form

$$\varepsilon_s(\nu) = -\mu_3 - \varepsilon_0 + \frac{\hbar^2 \nu^2}{2m_s}, \tag{5.7.53}$$

where $\boldsymbol{\nu}$ is a two-dimensional wave vector, μ_3 was defined in equation (5.5.2) as the chemical potential of ^{3}He in the solution, ε_0 is a shift in the chemical potential, and m_s is the effective mass of ^{3}He in the surface layer. A microscopic theory of Andreev's model is necessary for many reasons. First of all, the surface states (5.7.53) must be given an atomistic basis. By what mechanism do they arise? What is the general form for the spectrum $\varepsilon_s(\nu)$? And how many bound states are there? Secondly, what are the many-body effects present? How do the parameters, in particular ε_0, vary with coverage? Will the presence of some ^{3}He atoms in the surface layer induce more to "float" up? And when will ε_0 drop to zero to announce the completion of the ^{3}He monolayer? What is the thickness of the ^{3}He layer? Do we have here an ideal mobile two-dimensional system? To be able to answer these questions, one must first learn about the surface structure of pure liquid ^{4}He, which forms the subject matter of this section.

In two recent publications by Lekner [285], a possible way to compute surface states microscopically was proposed and carried out in part. His conclusions are incorrect on account of some confusion about boundary conditions. The method is, however, sound, and will be formulated below in a more systemmatic way.

The Hamiltonian is as given in equation (5.1.1), and the Schrödinger equation in (5.1.2). The difference from the bulk helium problem is in the boundary conditions. Let us consider the space enclosed by a box of volume $2L^3$, defined as follows:

$$\begin{aligned} 0 \le x \le L, \\ 0 \le y \le L, \\ -L \le z \le L. \end{aligned} \tag{5.7.54}$$

The mean density $\rho \equiv N/2L^3$ is fixed while N and L are permitted to expand without limit. In comparison to bulk He, the mean density is $\frac{1}{2}\rho_0$. A free surface will form about $z = 0$ to gain equilibrium. The atoms will choose to condense in the lower half-space $z \le 0$ at the bulk density ρ_0, with the exception that the topmost layers will dilute and spread into the upper half-space. We shall concern ourselves with precisely these topmost layers.

Take a trial wave function of the form

$$\psi_4 = F\phi \equiv \prod_{i<j=1}^{N} f(r_{ij}) \prod_{l=1}^{N} s_0(z_l) \equiv \prod_{i<j=1}^{N} \exp[\tfrac{1}{2}u(r_{ij})] \prod_{l=1}^{N} \exp[\tfrac{1}{2}t(z_l)]. \tag{5.7.55}$$

The correlation factor F, or $u(r)$, can be chosen, if one wishes, in first approximation to be that determined by a Jastrow calculation for bulk ^{4}He. But it must be emphasized that the expectation value of H with respect to F is *not* the bulk energy calculated earlier in Section 5.2! For here the N particles occupy a voluem $2L^3$; the Jastrow function F, on the other hand, is an approximate solution for H *only if the volume occupied is* L^3. Another way to put it is that while we borrow $u(r)$ from the bulk calculation, we do not mean F as a solution for H under the *present* boundary conditions. This should be obvious, since periodic conditions for bulk ^{4}He imply

$$\psi_J(\mathbf{r}_1 + \mathbf{L}, \mathbf{r}_2 + \mathbf{L}, \cdots, \mathbf{r}_N + \mathbf{L}) = \psi_J(\mathbf{r}_1, \mathbf{r}_2, \cdots, \mathbf{r}_N), \text{ where } \mathbf{L} \equiv L\mathbf{l}_z, \tag{5.7.56}$$

but not

$$\psi_J(\mathbf{r}_1 + \mathbf{L}, \mathbf{r}_2, \ldots, \mathbf{r}_N) = \psi_J(\mathbf{r}_1, \mathbf{r}_2, \cdots, \mathbf{r}_N). \tag{5.7.57}$$

In fact, the left-hand side of equation (5.7.57) is not even defined for bulk ^{4}He; $F(\mathbf{r}_1 + \mathbf{L}, \mathbf{r}_2, \cdots, \mathbf{r}_N)$ is, nevertheless, well defined in the present problem. This, in a nutshell, is the confusion about the boundary conditions to which I called attention earlier. Erroneously taking the expectation value of H with respect to the correlation factor as the bulk energy does lead to a tremendous simplification of the surface energy formula

$$E_s \equiv \frac{\langle \psi_4 | H | \psi_4 \rangle}{\langle \psi_4 | \psi_4 \rangle} - \frac{\langle F | H | F \rangle}{\langle F | F \rangle} \rightarrow L^2 \frac{\hbar^2}{8m} \int_{-\infty}^{\infty} P_1(z)[t'(z)]^2 dz. \quad (5.7.58)$$

where

$$P_1(z_1) = N \frac{\int \psi_4^2 d\tau_{(1)}}{\int \psi_4^2 d\tau} \quad (5.7.59)$$

is the one-particle distribution function. The result is, however, useless since the expression is nonnegative, giving rise to a minimum of zero surface energy.

The correct energy formula is given by

$$\begin{aligned} E_4 \equiv \frac{\langle \psi_4 | H | \psi_4 \rangle}{\langle \psi_4 | \psi_4 \rangle} &= \frac{-\hbar^2}{8m} \sum_{i=1}^{N} \frac{\int \psi_4^2 t''(z_i)\, d\tau}{\int \psi_4^2\, d\tau} + \sum_{i<j=1}^{N} \frac{\int \psi_4^2 \tilde{v}(r_{ij})\, d\tau}{\int \psi_4^2\, d\tau} \\ &= -L^2 \frac{\hbar^2}{8m} \int_{-L}^{L} P_1(z) t''(z)\, dz + \tfrac{1}{2} \iint_{(2L^3)} P_2(\mathbf{r}_1, \mathbf{r}_2) \tilde{v}(r_{12})\, d\mathbf{r}_1\, d\mathbf{r}_2, \quad (5.7.60) \end{aligned}$$

where

$$P_2(\mathbf{r}_1, \mathbf{r}_2) \equiv N(N-1) \frac{\int \psi_4^2\, d\tau_{(1,2)}}{\int \psi_4^2\, d\tau} \quad (5.7.61)$$

is the two-particle distribution function. The bulk energy that must be subtracted is given by

$$E_B = \tfrac{1}{2} \rho_0^2 \iint_{(L^3)} \tilde{v}(r_{12}) g(r_{12} | \rho_0)\, d\mathbf{r}_1\, d\mathbf{r}_2 \quad (5.7.62)$$

$$= \tfrac{1}{2} \rho_0^2 \iint_{(2L^3)} \tilde{v}(r_{12}) g(r_{12} | \rho_0) \theta(z_1)\, d\mathbf{r}_1\, d\mathbf{r}_2, \quad (5.7.63)$$

where ρ_0 represents the equilibrium bulk density and $\theta(z)$ is a step function

$$\begin{aligned} \theta(z) = 0, &\qquad z > 0 \\ 1, &\qquad z \leq 0. \end{aligned} \quad (5.7.64)$$

The surface energy formula is then

$$\begin{aligned} \frac{E_s}{L^2} = \frac{-\hbar^2}{8m} \int_{-\infty}^{\infty} P_1(z) t''(z)\, dz + \frac{1}{2L^2} \iint_{(2L^3)} \tilde{v}(r_{12}) [P_2(\mathbf{r}_1, \mathbf{r}_2) \\ - \rho_0^2 \theta(z_1) g(r_{12} | \rho_0)]\, d\mathbf{r}_1\, d\mathbf{r}_2, \end{aligned} \quad (5.7.65)$$

which is to be minimized with respect to $t(z)$.

In principle, for every $t(z)$ chosen, $P_1(z)$ and $P_2(\mathbf{r}_1, \mathbf{r}_2)$ must be computed from their definitions. It is a straightforward matter to derive coupled BBGKY equations for them. Writing

$$P_2(\mathbf{r}_1, \mathbf{r}_2) \equiv P_1(z_1)P_1(z_2)g(\mathbf{r}_1, \mathbf{r}_2), \tag{5.7.66}$$

and

$$P_3(\mathbf{r}_1, \mathbf{r}_2, \mathbf{r}_3) \equiv P_1(z_1)P_1(z_2)P_1(z_3)g_3({}_1\mathbf{r}, \mathbf{r}_2, \mathbf{r}_3), \tag{5.7.67}$$

one might once again attempt to truncate the hierarchy of equations by way of some superposition approximation. But this procedure is exceedingly impractical. One would do better to resort to some physically motivated approximation for $P_2(\mathbf{r}_1, \mathbf{r}_2)$. A reasonable model is given by

$$g(\mathbf{r}_1, \mathbf{r}_2) \approx g(r_{12} \mid \sqrt{P_1(z_1)P_1(z_2)}\,), \tag{5.7.68}$$

where the right-hand side is the bulk radial distribution calculated at a depth-dependent mean density $[P_1(z_1)P_2(z_2)]^{1/2}$. One *assumes* that such a $P_2(\mathbf{r}_1, \mathbf{r}_2)$ is consistent with its definition in terms of the trial wave function ψ_4. Alternately, since the $g(\mathbf{r}_1, \mathbf{r}_2)$ defined by equation (5.7.68) realistically regards volume elements situated at different parts of the liquid as locally uniform, what one assumes here is the claim that ψ_4 as defined by equation (5.7.61) is also realistic. Detailed calculations using models of this type have been carried out [286], where for convenience $P_1(z)$ in the form of a Fermi function

$$P_1(z) = \frac{\rho_0}{1 + \exp(Bz)} \tag{5.7.69}$$

is varied in place of $t(z)$. Results indicate a surface thickness (defined as the interval between densities which are 10 and 90% of ρ_0) to be about 2.4 Å, smaller than what one might expect. The surface energy per unit area, which at zero temperature is identical to the surface tension, is determined at 0.26 K/Å, or 0.36 dyn/cm, in good agreement with the experimental value [278] of 0.378 dyn/cm. It should be added that E_s/L^2 near the minimum (1.0 Å$^{-1} \leq B \leq 2.5$ Å^{-1}, corresponding to 4 Å $\geq$ thickness $\geq$ 1.6 Å) is very flat; hence one should not take the estimated thickness too seriously.

In a subsequent, independent calculation, Chang and Cohen [287] approximated the local $g(\mathbf{r}_1, \mathbf{r}_2)$ by the bulk $g(r_{12})$ evaluated at the (arithmetic) mean density $\frac{1}{2}[P_1(z_1) + P_2(z_2)]$. They obtained a surface tension that is 10% too high (0.41 dyn/cm), but a more reasonable thickness ($B = 1.0$ Å^{-1}). An asymmetrical $n(z)$ gave $B = 1.4$ Å^{-1} in the upper half-space and 0.5 Å^{-1} in the lower half-space.

Now a few words about the surface excitations in liquid ^{4}He. As in Feyman's theory of phonons in bulk ^{4}He, variational functions can be constructed to describe the surface excitations; thus, the following obtains:

$$\psi_{\kappa,\alpha} = \sum_{i=1}^{N} \left[e^{i\kappa \cdot \rho_i} \frac{s_\alpha(z_i)}{s_0(z_i)} \right] \psi_4, \tag{5.7.70}$$

where κ is a two-dimensional wave vector and α describes excitation in the normal direction. These functions are then varied to minimize the energy expectation value under the constraint that they be orthogonal to the ground state ψ_4. The latter condition is given by

$$\delta(\kappa) \int_{-\infty}^{\infty} \frac{s_\alpha(z)}{s_0(z)} P_1(z)\, dz = 0. \tag{5.7.71}$$

For each chosen κ, two types of $\psi_{\kappa,\alpha}$ can be constructed, each satisfying the orthogonality condition. One can choose $s_\alpha(z)$ to consist of a finite number α, of nodes. Or one can choose an oscillatory function characterized by a wave number α. Each suggests a specific microscopic configuration. It can be shown [288] that the first type corresponds to quantized capillary waves [289]—or "ripplons," and the second type simply the usual bulk phonons. Taking the particular branch $\alpha = 1$, the ripplon spectrum $\varepsilon_{\kappa,1}$ was calculated by Miller [288] and shown to possess the characteristic momentum dependence $\kappa^{3/2}$ of classical capillary waves. The temperature dependence of the surface tension is then given by

$$\begin{aligned} \sigma(T) &= \sigma(0) + \frac{k_B T}{2\pi} \sum_\alpha \int_0^\infty \ln(1 - e^{-\varepsilon_{\kappa,\alpha}/k_B T}) \kappa\, d\kappa \\ &\approx \sigma(0) + \frac{k_B T}{2\pi} \int_0^\infty \ln(1 - e^{-\varepsilon_{\kappa,1}/k_B T}) \kappa\, d\kappa, \end{aligned} \tag{5.7.72}$$

which was found to agree well with the experimental data of Atkins and Narahara [274].

In passing, let me mention that when the same theory of surface structure is used along with an electron–He pseudopotential to calculate the size of ion bubbles formed inside liquid ^{4}He, the results are in spectacular agreement with experiment [290] over a wide range of pressures. Furthermore, the potential seen by an electron in the neighborhood of helium surfaces can be calculated accurately [291]. This furnishes a basis for studying the properties of electrons adsorbed on liquid helium: the two-dimensional Coulomb system, which is extremely interesting but unfortunately outside the scope of present discussion.*

5.7.7. Surface States of ^{3}He in Dilute ^{3}He–^{4}He Solutions

The Hamiltonian of a dilute ^{3}He–^{4}He solution is given by equation (5.5.1). We wish to solve the Schrödinger equation defined by this Hamiltonian under boundary conditions which are appropriate in the presence of a free surface.

*Much work on the ripplon spectrum has been carried out after the preparation of this manuscript by Chang and Cohen; Bhattacharyya, Shih, and Woo; and Ebner and Saam.

Rewrite the Hamiltonian as follows:

$$H = H_1 + H_2, \tag{5.7.73}$$

with

$$H_1 = \sum_{i=1}^{N_3}\left[\frac{-\hbar^2}{2m_3}\nabla_i^2 + \sum_{k=N_3+1}^{N} v(r_{ik})\right] + \left[\sum_{k=N_3+1}^{N}\frac{-\hbar^2}{2m_4}\nabla_k^2 + \sum_{\substack{k<l\\=N_3+1}}^{N} v(r_{kl})\right], \tag{5.7.74}$$

$$H_2 = \sum_{\substack{i<j\\=1}}^{N_3} v(r_{ij}). \tag{5.7.75}$$

H_2 describes ^{3}He–^{4}He correlations, which must be considered at a later stage. H_1 describes uncorrelated ^{3}He atoms moving under the influence of a field provided by the semi-infinite ^{4}He liquid. In the limiting case $N_3 = 1$, it reduces to

$$H_1' = \frac{-\hbar^2}{2m_3}\nabla_1^2 + \sum_{i=2}^{N}\frac{-\hbar^2}{2m_4}\nabla_i^2 + \sum_{\substack{i<j\\=1}}^{N} v(r_{ij}) = \left[\sum_{i=1}^{N}\frac{-\hbar^2}{2m_4}\nabla_i^2 + \sum_{\substack{i<i\\=1}}^{N} v(r_{ij})\right]$$
$$+\frac{1}{3}\frac{-\hbar^2}{2m_4}\nabla_1^2. \tag{5.7.76}$$

A proper trial wave function for the ground state of H_1' is given by

$$\psi = \psi_4 e^{1/2\varphi(z_1)}, \tag{5.7.77}$$

where ψ_4 is the ground-state solution of the pure ^{4}He problem in the presence of a free surface, which we discussed in the preceding section. Using equation (5.7.55) for ψ_4, one finds

$$E \equiv \frac{\langle\psi|H_1'|\psi\rangle}{\langle\psi|\psi\rangle} = E_4 + E_a + E_b + E_c, \tag{5.7.78}$$

$$E_a = \frac{-\hbar^2}{2m_4}\frac{\pi}{I}\int P_2(\mathbf{r}_1,\mathbf{r}_N)e^{\varphi(z_1)}\frac{du(r_{1N})}{dz_1}\varphi'(z_1)\rho_{1N}\,d\rho_{1N}\,dz_1\,dz_N, \tag{5.7.79}$$

$$E_b = \frac{-\hbar^2}{16m_4}\frac{\pi}{I}\int P_2(r_1,r_N)e^{\varphi(z_1)}u''(r_{1N})\rho_{1N}\,d\rho_{1N}\,dz_1\,dz_N, \tag{5.7.80}$$

$$E_c = \frac{-\hbar^2}{8m_4}\frac{1}{I}\int P_1(z)e^{\varphi(z)}\Big\{\varphi'^2(z) + \varphi''(z) + t''(z) + 2t'(z)\varphi'(z)$$
$$+\frac{4}{3}\Big[t''(z) + \varphi''(z)\Big]\Big\}\,dz, \tag{5.7.81}$$

where

$$I = \int dz P_1(z)e^{\varphi(z)}, \tag{5.7.82}$$

and

$$\rho_{1N} = (x_{1N}^2 + y_{1N}^2)^{1/2}. \tag{5.7.83}$$

Several classes of $\varphi(z)$ were tried [286]; each upon the minimzation of $(E - E_4)$ resulted in a more-or-less symmetrical ^{3}He distribution

$$P_1^{(3)}(z_1) \equiv \frac{N \int \psi^2 \, d\tau_{(1)}}{\int \psi^2 \, d\tau} = \frac{P_1(z_1) e^{\varphi(z_1)}}{(L^2 I/N)} \tag{5.7.84}$$

centered on the Gibb's surface $z_1 = 0$. The lowest energy corresponds to a Gaussian distribution $\exp[-(z_1/\lambda)^2]$, with width $\lambda = 2.5$ Å.

To obtain the extra binding energy ε_0 at the surface, the chemical potential μ_3 of a ^{3}He atom in the bulk must be subtracted from $(E - E_4)$. μ_3 can be found in a calculation made consistent with the present model by setting $P_1^{(3)}(z_1) = P_1(z_1)$, or $\varphi(z_1) = 0$. It was then found [286, 292] that

$$\varepsilon_0 = 1.6 \text{ K}, \tag{5.7.85}$$

in good agreement with the 1.7 K experimentally determined by Zinovéna and Boldarev [277] and the 1.95 K obtained by Guo et al. [278].

By regarding the ^{3}He atom as moving in a single-particle well, it is possible to identify its probability density $\psi_0^2(z)$ with $P_1^{(3)}(z)$. From the Gaussian form of $P_1^{(3)}(z)$, one finds $\psi_0(z)$ and the oscillator frequency $\hbar\omega \approx 2.6$ K. This places an energy gap of 2.6 K between the two lowest states and leads to the conclusion that there exists only one bound surface state for ^{3}He.

The solution of the one ^{3}He problem enables one to proceed toward the next stage: H_1. For example, for the ground state of H_1, it is expected that a properly symmetrized form of

$$\Psi = \psi_4 \prod_{i=1}^{N_3} e^{1/2\varphi(z_i)} \tag{5.7.86}$$

will serve as a good approximation to the eigenstate. By replacing ψ_4 with excited states of the semi-infinite ^{4}He liquid, and by introducing two-dimensional continuum into the single-particle part of Ψ, a complete set of correlated functions can finally be generated to serve as a basis for dealing with the full Hamiltonian H. Thinking along these lines, it can be demonstrated that the models considered here and in Sections 5.2–5.4 are but two ways of looking at the same problem [293].

5.7.8. Superfluid Transitions in ^{3}He Monolayers?

One of the most exciting developments in low temperature physics during the last several years has been the discovery of "superfluid" transitions in liquid ^{3}He. Even though ^{3}He atoms are fermions, it has long been recognized that if the residual interactions between ^{3}He quasiparticles are sufficiently attractive in certain angular momentum states, pairing may yet occur, to cause superconductor-like transitions in the liquid. Various attempts were made [173, 174, 294–297] in the early 1960s to determine the transition temperature. Much of the work was phenomenological, and some employed quasiparticle spectra and interactions that were not derived in a consistent manner. Since it appeared that experimental confirmation of the transition was not to come for a long time, interest in this problem gradually subsided. The experimental discovery of anomalous NMR behavior in liquid ^{3}He under high pressure by the Cornell group [298–299] in 1971 caught the theorists by surprise. In retrospect, some of the earlier theories appear perfectly serviceable, including that mentioned at the end of Section 5.4.

Now, are superfluid transitions expected in ^{3}He monolayers as well?

Given two helium atoms interacting via some standard central force such as a Lennard-Jones potential, one can solve the two-particle Schrödinger equation in search of bound states. Even though it is well known that diatomic He molecules do not exist in three-dimensional space, it was discovered by Bagchi [231] that, when constrained in two dimensions, a pair of He atoms do bind, albeit weakly. (The binding energy is of the order of 10^{-2} K for ^{4}He and 10^{-7} K for ^{3}He.) This means that in two dimensions the pairwise interaction experienced by He atoms is effectively more attractive.

The only way to constrain He atoms to two dimensions is by physisorption —on crystalline substrates, or on bulk ^{4}He surfaces. In either case a substrate is present, and will mediate extra interaction between the adsorbed atoms. Surface and bulk excitations in the substrate, e.g., phonons in the crystal or ripplons in ^{4}He, are exchanged in a second-order process that induces attraction. In ^{3}He–^{4}He solutions such attractions completely cancel out the repulsion between bare ^{3}He atoms [196]. Judging from the size of the effective mass [278] of a single ^{3}He adsorbed on superfluid ^{4}He, which arises from renormalization by a similar process, the mediated attraction is expected to be just as significant.

High pressures are necessary for transitions to take place in bulk ^{4}He at relatively high temperatures. In a physisorbed ^{3}He monolayer, high (surface) pressures are automatically provided by the confining substrate. At least for ^{3}He adsorbed on graphite it is possible to reach areal densities which are greater than *twice* the equilibrium density. Indications are that even at such high densities (i.e., high pressures) the monolayer remains liquid-like. [247]. High areal densities can also be reached in ^{3}He adsorbed on bulk ^{4}He [278].

All three factors discussed above improve the experimental conditions for observing superfluidity in two dimensions. There will be enhanced pairing, and, consequently, an increase in the transition temperature [300]. Experimentally, the best chance of recognizing a superfluid transition lies in measuring surface properties of dilute ^3He–^4He solutions. ^3He, being a normal fluid above transition temperature, ordinarily acts to quench whatever surface activities that exist in superfluid ^4He. As one lowers the temperature past transition, the lid is off; ^3He will immediately begin to participate in the superfluid motion. Such an abrupt change (in, for example, the propagation of third sound) should be readily identifiable. There is also the possibility of observing the phenomenon through measurements of surface tension and flow properties.

5.8. SUMMARY AND PROSPECTS

Throughout this chapter I have dealt with *realistic* helium systems, in the sense that all the systems considered are defined in terms of realistic He–He potentials. It should come as no surprise, then, that a review of the microscopic theory turns out to center on variational and perturbative calculations carried out using correlated wave functions. Such calculations are truly microscopic in that they require no macroscopic information whatsoever and contain no empirical parameter. Occasionally the same degree of success can be achieved by a theory which employs the independent particle representation, through the use of, for example, diagrammatic perturbation techniques. However, by and large these are isolated examples that speak more for the truly heroic efforts due to single individual workers than for the power of the copiously reviewed field-theoretic method. Even though I may be speaking from a somewhat biased point of view, my assessment is by now shared by many whose experience and contribution have been restricted to the latter.

The formulation of the quantum many-body theory in the correlated representation makes use of the simplest quantum mechanical concepts, offer a definite set of prescriptions suitable for fermions and bosons alike, and relies strictly on techniques that have been thoroughly tested elsewhere—such as in the theory of classical liquids. The employment of diagrammatic aids, sum rules, and Green's functions in more recent work has increased its range of application many times. Furthermore, the translation of the method into field-theoretic language (through diagrammatic analyses carried out for simple model systems) has resulted in some progress toward unifying quantum many-body formalisms.

As far as techniques are concerned, major improvements should still be made in the following areas. First of all, the evaluation of matrix elements at present requires much patience and strenuous efforts. General methods based on well-understood approximation schemes must be developed. Recent work by Wu [56–58] led to the successful evaluation of all matrix elements between multiphonon states in the convolution approximation. More work of this nature is in demand, especially for basis functions that cannot be expressed in terms of simple entities like the density fluctuation. Next, an efficient orthogonalization procedure is sorely lacking. Even though in most cases some way of circumventing this difficulty has always been found, the method would become considerably more practical if a standard solution were available. In particular, a second-quantized Hamiltonian could then be established in every case, which would open up the entire arsenal of many-body techniques developed originally for the independent-particle representation. Finally, most of the completed research has dealt with the ground state and the excitation spectrum: the *real* part of the self energy. A time-dependent theory has not yet been established. Recent work by Jackson [125] for ^{4}He and Tan and Woo [168] for ^{3}He have shown certain progress in this direction. Until a general formalism is set up, much of the dynamical information concerning helium systems will remain outside the research of microscopic theory.

A complementary volume to this chapter exists in a monograph by Feenberg [55] entitled *Theory of Quantum Fluids*. Most of the work described in this chapter was carried out after the publication of the latter. Consequently, there is very little repetition. The careful reader will also detect a difference in philosophy as well as approach. To my knowledge, the present review is complete, self-contained, and up-to-date. Subject matters not touched upon here represent areas which have not yet been investigated by these microscopic methods. The following is a list of such topics:

1. The nature of Bose condensation in ^{4}He, in both two and three dimensions.
2. A first-principle calculation of properties in the hydrodynamic regime including, for example, the phonon spectrum of liquid ^{4}He and quasiparticle interactions in ^{3}He–^{4}He solutions.
3. The microscopic description of rotons in ^{4}He.
4. The origin of the two-roton resonance in ^{4}He.
5. The line-widths and termination of the phonon–roton spectrum in ^{4}He at high-momentum transfers.
6. Variational calculations employing explicit three-particle correlation factors.
7. Theory of transport properties of liquid ^{3}He beyond Landau's theory.
8. Estimation of the superfluid transition temperature in bulk ^{3}He.
9. Identification of the connection between t-matrix and CBF theories for Fermi systems.
10. CBF theory for solid helium.

11. Solution of the generalized BBGKY equation for solid helium.
12. Microscopic theory for exchange effects in solid helium.
13. Microscopic structure of vortices.
14. Calculation of correlation effects in adsorbed helium monolayers.
15. Microscopic structure of boundary layers between liquid helium and crystalline walls.
16. Realistic calculations for order–disorder transitions in helium monolayers.
17. Concentration dependence of properties in ^3He–^4He solutions.
18. Pressure dependence of properties in helium liquids.
19. Calculation of properties of helium systems at elevated temperatures.

The list is by no means exhaustive. The successful solution of every problem brings on a cascade of new surprises and new problems. This is generally true for all research, and has proven particularly true for condensed phases of helium. It is perhaps fitting that we should have ended our discussion on a highly speculative note concerning the possibility of superfluid transitions in ^3He monolayers. For, after all, the most exciting discoveries in ultra-low temperature physics during the last decade have been the phase-separation properties of ^3He–^4He solutions, the phase transitions in two-dimensional helium systems, and the superfluidity of liquid ^3He. The consideration of superfluid transitions in ^3He adsorbed on ^4He amalgamates all three problems, and should prove both intriguing and challenging to the experimenter. The microscopic theorists have learned much from helium experiments over the years. It is high time to reciprocate.

ACKNOWLEDGMENTS

I acknowledge the support of this work by the National Science Foundation through Grants GP-29130 and GH-39127, and through the Materials Research Center of Northwestern University, and by the A. P. Sloan Foundation through a Sloan Research Fellowship (1971–73).

I express my gratitude to my mentors and collaborators, from whom I have learned much over the years, in particular K. A. Brueckner, J. W. Clark, E. Feenberg, H. W. Jackson, D. Pines, and H. T. Tan. Also, I am grateful to my research associates and students, both past and present, with whom I have continued to learn through daily contacts. My special thanks go to Ms. J. Dalfino for the expert and speedy preparation of my manuscript.

REFERENCES AND NOTES

1. K. A. Brueckner, *The Many Body Problem,* Les Houches Lectures (Wiley, New York, 1958) p. 47.

2. J. W. Clark (unpublished notes).
3. J. Goldstone, *Proc. Roy. Soc. (London)* **A239,** 267 (1957).
4. J. deBoer and A. Michels, *Physica* **5,** 945 (1938).
5. A. Bijl, *Physica* **7,** 869 (1940).
6. R. B. Dingle, *Phil Mag.* **40,** 573 (1949).
7. R. Jastrow, *Phys. Rev.* **98,** 1479 (1955).
8. E. Feenberg, *Theory of Quantum Fluids* (Academic Press, New York, 1969), Chap. 3.
9. G. V. Chester and L. Reatto, *Phys. Lett.* **22,** 276 (1966).
10. W. P. Francis, G. V. Chester, and L. Reatto, *Phys. Rev. A* **1,** 86 (1970).
11. M. D. Miller (private communication).
12. H. W. Jackson and E. Feenberg, *Ann. Phys. (N. Y.)* **15,** 266 (1961).
13. N. G. van Kampen, *Physica* **27,** 783 (1961).
14. See for example T. L. Hill, *Statistical Mechanics* (McGraw-Hill, New York, 1965); S. A. Rice and P. Gray, *The Statistical Mechanics of Simple Liquids* (Wiley, New York, 1965); I. Z. Fisher, *Statistical Theory of Liquids* (Univ. of Chicago, Chicago, 1964).
15. R. Abe, *Prog. Theoret. Phys. (Kyoto)* **21,** 421 (1959).
16. W. E. Massey, *Phys. Rev.* **151,** 153 (1966).
17. E. Feenberg, *J. Math. Phys.* **6,** 658 (1965).
18. E. Feenberg, H. T. Tan, and F. Y. Wu, *J. Math. Phys.* **8,** 864 (1967).
19. G. V. Chester, *Lecture in Theoretical Physics XI B* (Gordon and Breach, New York, 1969); Ref. [10].
20. See Refs. [36, 37].
21. J. deBoer, J. M. J. van Leeuwen, and J. Groeneveld, *Physica* **30,** 2265 (1964).
22. V. R. Pandharipande and H. A. Bethe, *Phys. Rev. C* **7,** 1312 (1973).
23. Equilibrium can be achieved more rapidly if the initial distribution of particles in the phase space is not too far from equilibrium. For a dense liquid one might begin, for example, with a distribution characteristic of an elastic crystal lattice, and watch the lattice structure wash away.
24. W. L. McMillan, *Phys. Rev.* **138,** A442 (1965).
25. R. L. Coldwell, *Phys. Rev. A* **7,** 270 (1973).
26. C.-W. Woo and R. L. Coldwell, *Phys. Rev. Lett.* **29,** 1062 (1972).
27. R. Haberlandt, *Phys. Lett.* **14,** 197 (1965).
28. J. L. Yntema and W. G. Schneider, *J. Chem. Phys.* **18,** 646 (1950).
29. L. W. Bruch and I. J. McGee, *J. Chem. Phys.* **46,** 2959 (1967).
30. L. W. Bruch and I. J. McGee, *J. Chem. Phys.* **52,** 5884 (1970).
31. D. E. Beck, *Mol. Phys.* **14,** 311 (1968); also erratum.
32. G. Sposito, *J. Low Temp. Phys.* **3,** 491 (1970).
33. B. M. Axelrod, *J. Chem. Phys.* **19,** 724 (1951).
34. A. Dalgarno and G. A. Victor, *Mol. Phys.* **10,** 333 (1966).
35. See, however, Refs. [8, 9].
36. J. C. Lee and A. A. Broyles, *Phys. Rev. Lett.* **17,** 424 (1966).
37. M. A. Pokrant, *Phys. Rev. A* **6,** 1588 (1972).
38. C. E. Campbell and E. Feenberg, *Phys. Rev.* **188,** 396 (1969).

39. F. Y. Wu and E. Feenberg, *Phys. Rev.* **122,** 739 (1961).
40. W. E. Massey and C.-W. Woo, *Phys. Rev.* **164,** 256 (1967).
41. D. Schiff and L. Verlet, *Phys. Rev.* **160,** 208 (1967).
42. R. D. Murphy and R. O. Watts, *J. Low Temp. Phys.* **2,** 507 (1970).
43. B. M. Abraham, Y. Eckstein, J. B. Ketterson, M. Kuchnir, and P. R. Roach, *Phys. Rev. A.* **1,** 250 (1970).
44. P. R. Roach, J. B. Ketterson, and C.-W. Woo, *Phys. Rev. A* **2,** 543 (1970).
45. R. A. Aziz and R. K. Pathria, *Phys. Rev. A* **7,** 809 (1973).
46. T. B. Davison and E. Feenberg, *Ann. Phys. (N. Y.)* **53,** 559 (1969).
47. R. D. Murphy and J. A. Barker, *Phys. Rev. A* **3,** 1038 (1971).
48. A. Miller, D. Pines and P. Nozieres, *Phys. Rev.* **127,** 1452 (1962).
49. W. E. Massey, *Phys. Rev. Lett.* **12,** 719 (1964).
50. O. Penrose and L. Onsager, *Phys. Rev.* **104,** 576 (1956).
51. W. E. Parry and C. R. Rathbone, *Proc. Phys. Soc. (London)* **91,** 273 (1967).
52. O. K. Harling, *Phys. Rev. Lett.* **24,** 1046 (1970).
53. H. A. Mook, R. Scherm, and M. K. Wilkinson, *Phys. Rev. A* **6,** 2268 (1972).
54. J. W. Clark, Ph. D. Thesis, Washington Univ. (1959).
55. F. Y. Wu, *J. Low Temp. Phys.* **9,** 177 (1972).
56. E. Feeberg, *Theory of Quantum Fluids* (Academic, New York, 1969), App. 4-B and 5-C.
57. F. Y. Wu and M. K. Chien, *J. Math. Phys.* **11,** 1912 (1970).
58. F. Y. Wu, *J. Math. Phys.* **12,** 1923 (1971).
59. E. Feenberg and A. Rahman (private communication).
60. N. Mihara and R. D. Puff, *Phys. Rev.* **174,** 221 (1968).
61. G. Sposito, *Phys. Rev. A* **2,** 948 (1970).
62. E. Østgaard, *J. Low Temp. Phys.* **4,** 239 (1971).
63. E. Østgaard, *J. Low Temp. Phys.* **4,** 585 (1971).
64. B. H. Brandow, *Phys. Rev. Lett.* **22,** 173 (1969).
65. B. H. Brandow, *Ann. Phys. (N. Y.)* **64,** 21 (1971).
66. H. T. Tan and L. H. Nosanow (private communication).
67. S. T. Beliaev, *Zh. Eksp. Teor. Fiz.* **34,** 417 (1958) [English transl.: *Sov. Phys.—JETP* **7,** 289 (1958)].
68. N. M. Hugenholtz and D. Pines, *Phys. Rev.* **116,** 489 (1959).
69. N. N. Bogoliubov, *J. Phys. USSR* **9,** 23 (1947).
70. H. K. Sim, C.-W. Woo, and J. R. Buchler, *Phys. Rev. Lett.* **24,** 1094 (1970).
71. H. K. Sim, C.-W. Woo, and J. R. Buchler, *Phys. Rev. A* **2,** 2024 (1970).
72. A. Bhattacharyya and C.-W. Woo, *Phys. Rev. A* **7,** 198 (1973).
73. H. K. Sim and C.-W. Woo, *Phys. Rev. A* **2,** 2032 (1970).
74. D. K. Lee, *Phys. Rev. A* **4,** 1670 (1971).
75. A. Bhattacharyya and C.-W. Woo, *Phys. Rev. Lett.* **28,** 1320 (1972).
76. A. Bhattacharyya and C.-W. Woo, *Phys, Rev. A* **7,** 204 (1973).
77. R. D. Mattuck, *A Guide to Feynman Diagrams in the Many-Body Problem* (McGraw-Hill, London, 1966).

78. A. L. Fetter and J. D. Walecka, *Quantum Theory of Many-Particle Systems* (McGraw-Hill, New York 1971).
79. A. A. Abrikosov, L. P. Gorkov, and I. E. Dzyaloshinski, *Methods of Quantum Field Theory in Statistical Physics* (Prentice-Hall, N. J., 1963).
80. B. Day, *Phys. Rev.* **151,** 826 (1966); **187,** 1269 (1969); *A* **4,** 681 (1971).
81. C.-W. Woo, *Phys. Rev. Lett.* **28,** 1442 (1972).
82. C.-W. Woo, *Phys. Rev. A* **6,** 2312 (1972).
83. Obtained from Fourier transforming the data given in Refs. [84, 85].
84. E. K. Achter and L. Meyer, *Phys. Rev.* **188,** 291 (1969).
85. R. B. Hallock, , *Phys. Rev. A* **5,** 320 (1972).
86. C. E. Campbell and M. Schick, *Phys. Rev. A* **3,** 691 (1971).
87. M. D. Miller, C.-W. Woo, and C. E. Campbell, *Phys. Rev. A* **6,** 1942 (1972).
88. C. E. Campbell is at present attempting such an analysis.
89. L. D. Landau, *Zh. Eksp. Teor. Fiz.* **11,** 592 (1941).
90. L. D. Landau, *J. Phys. (USSR)* **11,** 91 (1947).
91. K. A. Brueckner and K. Sawada, *Phys. Rev.* **106,** 1128 (1957).
92. R. P. Feynman, *Phys. Rev.* **94,** 262 (1954).
93. R. P. Feynman and M. Cohen, *Phys. Rev.* **102,** 1189 (1956).
94. M. Cohen and R. P. Feynman, *Phys. Rev.* **107,** 13 (1957).
95. R. A. Cowley and A. D. B. Woods, *Can. J. Phys.* **49,** 177 (1971).
96. A. D. B. Woods, (private communication).
97. E. C. Svensson, A. D. B. Woods, and P. Martel, *Phys. Rev. Lett.* **29,** 1148 (1972).
98. The compressibility or ω^{-1} sum rule is due to Onsager and Price [99]; the ω^2 sum rule is due to Huang and Klein [100]; and the ω^3 sum rule is due to Puff and co-workers [60, 101].
99. P. J. Price, *Phys. Rev.* **94,** 257 (1954).
100. K. Huang and A. Klein, *Ann, Phys. (N. Y.)* **30,** 203 (1964).
101. R. D. Puff and N. S. Gillis, *Ann. Phys. (N. Y.)* **46,** 364 (1968).
102. I. Tamm, *J. Phys. (USSR)* **9,** 449 (1945).
103. S. M. Dancoff, *Phys. Rev.* **78,** 382 (1950).
104. C. G. Kuper, *Proc. Roy. Soc. (London)* **A 233,** 223 (1955).
105. H. W. Jackson and E. Feenberg, *Rev. Mod. Phys.* **34,** 686 (1962).
106. D. K. Lee, *Phys. Rev.* **162,** 134 (1967).
107. In Ref. [106]c'_{2a} was given as -0.239 on account of an algebraic error. This was first pointed out by H. W. Jackson (private communication). See also Ref. [114].
108. P. Goldhammer and E. Feenberg, *Phys. Rev.* **101,** 1233 (1956).
109. R. C. Young, L. C. Biedenharn, and E. Feenberg, *Phys. Rev.* **106,** 115 (1957).
110. Y. R. Lin-Liu and C.-W. Woo, *J. Low Temp. Phys.* **14,** 317 (1974).
111. T. J. Greytak and J. Yan, *Phys. Rev. Lett.* **22,** 987 (1969).
112. J. Ruvalds and A. Zawadowski, *Phys. Rev. Lett.* **25,** 333 (1970).
113. F. Iwamoto, Progr. *Theoret. Phys. (Kyoto)* **44,** 1121, 1135 (1970).
114. H.-W. Lai, H.-K. Sim, and C.-W. Woo, *Phys. Rev. A* **1,** 1536 (1969). See Section 5. 4. 4 concerning the use of a nonorthogonal basis.

115. A. Bhattacharyya and C.-W. Woo, in *Proceedings of the 13th International Conference on Low Temperature Physics* Vol. 1 (Boulder, Colo., 1972) p. 67
116. See, however, footnote 2 in Ref. [95].
117. H. J. Maris and W. E. Massey, *Phys. Rev. Lett.* **25,** 220 (1970).
118. H. J. Maris, *Phys. Rev. Lett.* **28,** 277 (1972).
119. A. Miller, D. Pines, and P. Nozieres, *Phys. Rev.* **127,** 1452 (1962).
120. V. K. Wong and H. Gould (private communications).
121. P. R. Roach, B. M. Abraham, J. B. Ketterson, and M. Kuchnir, *Phys. Rev. Lett.* **29,** 32 (1972).
122. V. Narayanamurti, K. Andres, and R. C. Dynes, *Phys. Rev. Lett.* **31,** 687 (1973).
123. A. Molinari and T. Regge, *Phys. Rev. Lett.* **26,** 1531 (1971).
124. C. H. Anderson and E. S. Sabisky, *Phys. Rev. Lett.* **28,** 80 (1972).
125. R. B. Hallock, in *Proceedings 13th International Conference Low Templerature Physics* Vol. 1. (Boulder, Colo., 1972) p. 547, 551.
126. N. E. Phillips, C. G. Waterfield, and J. K. Hoffer, *Phys. Rev. Lett.* **25,** 1260 (1970).
127. C. Zasada and R. K. Pathria, *Phys. Rev. Lett.* **29,** 988 (1972).
128. M. P. Kemoklidze and L. P. Pitaevskii, *Zh. Eksp. Teor. Fiz.* **59,** 2187 (1970) [English transl.: *Sov. Phys.—JETP* **32,** 1183 (1971)].
129. E. Feenberg, *Phys. Rev. Lett.* **26,** 301 (1971).
130. D. Price (private communication).
131. R. B. Hallock (private communication).
132. B. M. Abraham, Y. Eckstein, and J. B. Ketterson (private communications); Ref. [208].
133. H. W. Jackson, *Phys. Rev.* **185,** 186 (1969).
134. W. C. Kerr, K. N. Pathak, and K. S. Singwi, *Phys. Rev. A* **2,** 2416 (1970).
135. R. D. Puff and J. S. Tenn, *Phys. Rev. A* **1,** 125 (1970).
136. H. W. Jackson (private communication).
137. L. D. Landau, *Zh. Eksp. Teor. Fiz.* **30,** 1058 (1956) [English transl.: *Sov. Phys.—JETP* **3,** 920 (1957)].
138. L. D. Landau, *Zh. Eksp. Teor. Fiz.* **32,** 59 (1957) [English transl.: *Sov. Phys—JETP* **5,** 101 (1957)].
139. A. A. Abrikosov and I. M. Khalatnikov, *Rep. Progr. Phys.* **22,** 329 (1959).
140. D. Hone *Phys. Rev.* **121,** 669 (1961).
141. G. Baym and C. Ebner, *Phys. Rev.* **164,** 235 (1967).
142. G. A. Booker and J. Sykes, *Phys. Rev. Lett.* **21,** 279 (1968).
143. K. S. Dy and C. J. Pethick, *Phys. Rev. Lett.* **21,** 876 (1968).
144. H. H. Jensen, H. Smith, and J. W. Wilkins, *Phys. Lett.* **27A,** 532 (1968).
145. K. A. Brueckner and J. L. Gammel, *Phys. Rev.* **109,** 1040 (1957).
146. L. J. Campbell, Ph.D. Thesis, Univ. of California, San Diego (1965).
147. N. M. Hugenholtz and L. van Hove, *Physica* **24,** 363 (1958).
148. E. Østgaard, *Phys. Rev.* **170,** 257 (1968).
149. E. Østgaard, *Phys. Rev.* **171,** 248 (1968).

150. E. Østgaard, *Phys. Rev.* **176,** 351 (1968).

151. E. Østgaard, *Phys. Rev.* **180,** 263 (1969).

152. E. Østgaard, *Phys. Rev.* **187,** 371 (1969).

153. H. A. Bethe, *Phys. Rev.* **138,** B804 (1965).

154. L. D. Faddeev, *Zh. Eksp. Teor. Fiz.* **39,** 1459 (1960). [English transl.: *Sov. Phys.—JETP* **12,** 1014 (1961)].

155. H. T. Tan and E. Feenberg, *Phys. Rev.* **176,** 370 (1968).

156. J. deBoer, *Physica* **14,** 139 (1948).

157. J. deBoer and R. Lunbeck, *Physica* **6,** 658 (1965).

158. F. Iwamoto and M. Yamada, *Progr. Theoret. Phys. (Kyoto)* **17,** 543; **18,** 345 (1957).

159. C. D. Hartogh and H. A. Tolhoek, *Physica* **24,** 721; 875; 896 (1958).

160. J. B. Aviles, *Ann. Phys. (N. Y.)* **5,** 251 (1958).

161. J. W. Clark, *Can. J. Phys.* **39,** 385 (1961); *Ann. Phys. (N. Y).* **11,** 483 (1960).

162. R. Abe. *Progr. Theoret, Phys. (Kyoto)* **19,** 57; 407 (1958).

163. F. Y. Wu and E. Feenberg, *Phys. Rev.* **128,** 943 (1962).

164. F. Y. Wu, *J. Math. Phys.* **4,** 1438 (1963).

165. E. Feenberg and C.-W. Woo, *Phys. Rev.* **137,** A391 (1965).

166. J. W. Clark and P. Westhaus, *Phys. Rev.* **141,** 833 (1966).

167. C.-W. Woo, *Phys. Rev.* **151,** 138 (1966).

168. P-O. Löwdin, *J. Chem. Phys.* **18,** 365 (1950).

169. C.-W. Woo, Ph.D. Thesis, Washington Univ. (1966).

170. M. Gell-Mann, *Phys. Rev.* **106,** 369 (1957).

171. M. Gell-Mann and K. A. Brueckner, *Phys. Rev.* **106,** 364 (1957).

172. H. T. Tan and C.-W. Woo, *J. Low Temp. Phys.* **2,** 187 (1970).

173. L. N. Cooper, R. L. Mills, and A. M. Sessler, *Phys. Rev.* **114,** 1377 (1959).

174. V. J. Emery and A. M. Sessler, *Phys. Rev.* **119,** 43 (1960).

175. E. G. D. Cohen and J. M. J. van Leeuwen, *Physica* **27,** 1157 (1960).

176. E. G. D. Cohen and J. M. J. van Leeuwen, *Phys. Rev.* **176,** 385 (1968).

177. D. O. Edwards and J. G. Daunt, *Phys. Rev.* **124,** 640 (1961).

178. D. O. Edwards, D. F. Brewer, P. Seligmann, M. Skertic, and M. Yagub, *Phys. Rev. Lett.* **15,** 773 (1965).

179. E. M. Ifft, D. O. Edwards, R. E. Sarwinski, and M. Skertic, *Phys. Rev. Lett.* **19,** 831 (1967).

180. A. E. Watson, J. D. Reppy, and R. C. Richardson, *Phys. Rev.* **188,** 384 (1969).

181. J. Landau, J. T. Tough, N. R. Brubaker, and D. O. Edwards, *Phys. Rev. A* **2,** 2472 (1970).

182. B. M. Abraham, O. G. Brandt, Y. Eckstein, J. Munarin, and G. Baym, *Phys. Rev.* **188,** 309 (1969).

183. C. Ebner and D. O. Edwards, *Phys. Rep. (Phys. Lett. C)* **2,** 77 (1970).

184. E. Feenberg, *Theory of Quantum Fluids* (Academic, New York, 1969), Chap. 11.

185. L. D. Landau and I. Pomeranchuk, *Dokl. Akad. Nauk SSSR* **59,** 669 (1948).

186. I. Pomeranchuk, *Zh. Eksp. Teor. Fiz.* **19,** 42 (1949).

187. G. Baym, Phys. *Rev. Lett.* **17,** 952 (1966).

188. W. E. Massey and C.-W. Woo, *Phys. Rev. Lett.* **19,** 301 (1967).

189. T. B. Davison and E. Feenberg, *Phys. Rev.* **178,** 306 (1969).

190. Note that Davison and Feenberg's energy shift was reported as -0.57 K at the time of the publication. It was later revised to -0.33 K.

191. P. Seligmann, D. O. Edwards, R. E. Sarwinski, and J. T. Tough, *Phys. Rev.* **181,** 415 (1969).

192. A. C. Anderson, D. O. Edwards, W. R. Roach, R. E. Sarwinski, and J. C. Wheately, *Phys. Rev. Lett.* **17,** 367 (1966).

193. C.-W. Woo, H. T. Tan, and W. E. Massey, *Phys. Rev. Lett.* **22,** 278 (1969).

194. C.-W. Woo, H. T. Tan, and W. E. Massey, *Phys. Rev.* **185,** 287 (1969).

195. J. Bardeen, G. Baym, and D. Pines, *Phys. Rev. Lett.* **17,** 372 (1966).

196. J. Bardeen, G. Baym, and D. Pines, *Phys. Rev.* **156,** 207 (1967).

197. V. J. Emery, *Phys. Rev.* **161,** 194 (1967).

198. V. J. Emery, *Phys. Rev.* **148,** 138 (1966).

199. C. Boghosian and H. Meyer, *Phys. Lett.* **25A,** 352 (1967).

200. W. E. Massey, C.-W. Woo, and H. T. Tan, *Phys. Rev. A* **1,** 519 (1969).

201. N. W. Ashcroft and D. C. Langreth, *Phys. Rev.* **156,** 685 (1967).

202. C.-W. Woo, *Phys. Rev. A* **3,** 2141 (1971).

203. W. J. Pettit, Ph.D. Thesis, Cornell Univ. (1967).

204. J.-P. Hansen and D. Schiff, *Phys. Rev. Lett.* **23,** 1488 (1969).

205. H. T. Tan, C.-W. Woo, and F. Y. Wu, *J. Low Temp. Phys.* **5,** 261 (1971).

206. C. E. Campbell, *J. Low Temp. Phys.* **4,** 433 (1971).

207. H. T. Tan and C.-W. Woo, *Phys. Rev. A* **7,** 233 (1973)

208. B. M. Abraham, Y. Eckstein, and J. B. Ketterson, *Phys. Rev. Lett.* **21,** 422 (1968).

209. R. B. Hallock (private communication).

210. G. Baym and C. Ebner, *Phys. Rev.* **170,** 346 (1968).

211. G. A. Herzlinger and J. G. King, *Phys. Lett.* **40A,** 65 (1972); private communication.

212. I. M. Khalatnikov and V. N. Zharkov, *Zh. Eksp. Teor. Fiz.* **32,** 1108 (1957) [English transl.: *Sov. Phys.—JETP* **5,** 905 (1957)].

213. H. T. Tan, M. Tan, and C.-W. Woo, *Proceedings of the 13th International Conference on Low Temperature Physics* Vol. 1 (Boulder, Colo., 1972) p. 108.

214. H. T. Tan and C.-W. Woo, *Phys. Rev. Lett.* **30,** 365(1973).

215. R. A. Guyer, *Solid State Phys.* **23,** 413 (1969).

216. L. H. Nosanow and G. L. Shaw, *Phys. Rev.* **128,** 546 (1962).

217. L. H. Nosanow, *Phys. Rev.* **146,** 120 (1966).

218. K. A. Brueckner and J. Frohberg, *Progr. Theoret. Phys (Kyoto)* Suppl. 383 (1965).

219. J. Frohberg, Ph.D. Thesis, Univ. of California, San Diego (1968).

220. D. Rosenwald, *Phys. Rev.* **154,** 160 (1967).

221. W. J. Mullin, L. H. Nosanow, and P. M. Steinback, *Phys. Rev.* **188,** 410 (1969).

222. J. H. Hetherington, W. J. Mullin, and L. H. Nosanow, *Phys. Rev.* **154,** 175 (1967).

223. W. E. Massey and C.-W. Woo, *Phys. Rev.* **169,** 241 (1968).

224. C.-W. Woo and W. E. Massey, *Phys. Rev.* **177,** 272 (1969).

225. J.-P. Hansen and D. Levesque, *Phys. Rev.* **165,** 293 (1968).

226. K. A. Brueckner and R. Thieberger, *Phys. Rev.* **178,** 362 (1968).

227. T. R. Koehler, *Phys. Rev. Lett.* **18,** 654 (1967).

228. T. R. Koehler, *Phys. Rev.* **165,** 942 (1968).

229. T. R. Koehler and N. R. Werthamer, *Phys. Rev. A* **3,** 2074 (1971).

230. D. Bohm and B. Salt, *Rev. Mod. Phys.* **39,** 894 (1967).

231. A. Bagchi, *Phys. Rev. A* **3,** 1133 (1971).

232. J. G. Dash (to be published).

233. D. F. Brewer et al., in *Proceedings of the 13th International Conference on Low Temperature physics* Vol. 1 (Boulder, Colo. 1972) p. 163, 195, 200.

234. J. F. Kelly and R. C. Richardson, in *Proceedings of the 13th International Conference Low Temperature Physics.* Vol. 1 (Boulder, Colo., 1972) p. 167.

235. J. G. Daunt (private communication).

236. K. Luszczynski (private communication).

237. See proceedings of two conferences: *J. Low Temp. Phys.* **3,** (1970), and *Monolayer and Submonolayer Helium Films* (Plenum Press, New York, 1973).

238. R. H. Tait, R. O. Pohl, and J. D. Reppy, in *Proceedings of the 13th International Conference Low Temperature Physics.* Vol. 1 (Boulder, Colo., 1972) p. 172.

239. D. L. Goodstein, W. D. McCormick, and J. G. Dash, *Phys. Rev. Lett.* **15,** 447 (1965).

240. D. L. Goodstein, W. D. McCormcik, and J. G. Dash, *Phys. Rev. Lett.* **15,** 740 (1965).

241. W. D. McCormick, D. L. Goodstein, and J. G. Dash, *Phys. Rev.* **168,** 249 (1968).

242. G. A. Stewart and J. G. Dash, *Phys. Rev. A* **2,** 918 (1970).

243. G. A. Stewart and J. G. Dash, *J. Low Temp. Phys.* **3,** 281 (1970).

244. M. Bretz and J. G. Dash, *Phys. Rev. Lett.* **26,** 963 (1971).

245. M. Bretz and J. G. Dash, *Phys. Rev. Lett.* **27,** 647 (1971).

246. M. Bretz, G. B. Huff and J. G. Dash, *Phys. Rev. Lett.* **28,** 729 (1972).

247. D. C. Hickernell, E. O. McLean, and O. E. Vilches, *Phys. Rev. Lett.* **28,** 789 (1972).

248. C. E. Campbell, J. G. Dash, and M. Schick, *Phys. Rev. Lett.* **26,** 966 (1971).

249. C. E. Campbell and M. Schick, *Phys. Rev. A* **5,** 1919 (1972).

250. J. E. Lennard-Jones and A. F. Devonshire, *Proc. Roy. Soc. (London)* **A158,** 242 (1932).

251. W. A. Steele and M. Ross, *J. Chem. Phys.* **35,** 850 (1961).

252. M. Ross and W. A. Steele, *J. Chem. Phys.* **36,** 862 (1961).

253. H. W. Jackson, *Phys. Rev.* **180,** 184 (1969).

254. F. Ricca, C. Pisani, and E. Garrone, *J. Chem. Phys.* **51,** 4079 (1969).

255. R. B. Bernstein and J. T. Muckerman, *Advan. Chem. Phys.* **12,** 466 (1967).

256. H.-W. Lai, C.-W. Woo, and F. Y. Wu, *J. Low Temp. Phys.* **3,** 463 (1970).

257. A. D. Novaco and F. J. Milford, *J. Low Temp. Phys.* **3,** 307 (1970).

258. F. J. Milford and A. D. Novaco, *Phys. Rev. A* **4,** 1136 (1971).

259. A. D. Novaco and F. J. Milford, *Phys. Rev. A* **5,** 783 (1972).

260. D. E. Hagen, A. D. Novaco, and F. J. Milford, *Adsorption-Desorption Phenomena* (Academic, New York, 1972) p. 99.
261. M. Abramowitz and I. A. Stegun, *Handbook of Mathematical Functions* (N.B.S., Washington, D. C., 1964) Sect. 20.
262. H.-W. Lai, C.-W. Woo, and F. Y. Wu, *J. Low Temp. Phys.* **5,** 499 (1971).
263. F. Y. Wu and C.-W. Woo, *Chin. J. Phys.* **9,** 68 (1971).
264. J. G. Dash (private communication).
265. D. S. Hyman, Ph.D. Thesis, Cornell Univ. (1970).
266. C. E. Campbell and M. Schick, *Phys. Rev. A* **3,** 691 (1971).
267. M. D. Miller and C.-W. Woo, *Phys. Rev. A* **7,** 1322 (1973).
268. E. O. McLean, Ph.D. Thesis, Univ. of Washington (1972).
269. M. Schick and C. E. Campbell, *Phys. Rev. A* **2,** 1591 (1970).
270. C.-W. Woo, *J. Low Temp. Phys.* **3,** 335 (1970).
271. See, for example, a semi-infinite continuum approximation: H. Ezawa, *Ann. Phys.* **67,** 438 (1971).
272. H.-W. Lai and C.-W. Woo, *Phys. Lett.* **41A,** 170 (1972).
273. S.-C. Ying, *Phys. Rev. B* **3,** 4160 (1971).
274. K. R. Atkins and Y. Narahara, *Phys. Rev.* **138,** A437 (1965).
275. B. N. Esel'son and N. G. Bereznyak, *Dokl Akad. Nauk.* SSSR **98,** 564 (1954).
276. B. N. Esel'son, V. G. Ivantsov, and A. D. Shvets, *Zh. Eksp. Teor. Fiz.* **44,** 483 (1963) [English transl.: *Sov. Phys.—JETP* **17,** 330 (1963)].
277. N. K. Zinovéva and S. T. Boldarev, *Zh. Eksp. Teor. Fiz.* **56,** 1089 (1969) [English transl.: *Sov. Phys.—JETP* **29,** 585 (1969)].
278. H. M. Guo, D. O. Edwards, R. E. Sarwinski, and J. T. Tough, *Phys. Rev. Lett.* **27,** 1259 (1971).
279. G. W. Rayfield, *Phys. Rev. Lett.* **20,** 1467 (1968).
280. A. J. Dahm, *Phs. Rev.* **180,** 259 (1969).
281. M. Kuchnir, P. R. Roach, and J. B. Ketterson, in *Proceedings of the 12th International Conference on Low Temperature Physics* (1970), p. 103.
282. M. Kuchnir, P. R. Roach, and J. B. Ketterson, *J. Low Temp. Phys.* **3,** 183 (1970).
283. J. Mochels (private communication).
284. A. F. Andreev, *Zh. Eksp. Teor. Fiz.* **50,** 1415 (1966) [English transl.: *Sov. Phys.—JETP* **23,** 939 (1966)].
285. J. Lekner, *Phil. Mag.* **22,** 669 (1970); *Progr. Theoret. Phys (Kyoto)* **45,** 36 (1971).
286. Y. M. Shih and C.-W. Woo, *Phys. Rev. Lett.* **30,** 478 (1973).
287. C. C. Chang and M. Cohen *Phys. Rev. A* **8,** 1930, 3131 (1973).
288. M. D. Miller, Ph.D. Thesis, Northwestern Univ. (1973).
289. W. F. Saam, *Phys. Rev. A* **8,** 1048 (1973).
290. Y. M. Shih and C.-W. Woo, *Phys. Rev. A* **8,** 1437 (1973).
291. H.-M. Huang, Y. M. Shih, and C.-W. Woo, *J. Low Temp. Phys.* **14,** 413 (1974).
292. See also W. F. Saam, *Phys. Rev. A* **4,** 1278 (1971).
293. C.-W. Woo, *Monolayer and Submonolayer Flims* (Plenum Press, New York, 1973) p. 85.

294. K. A. Brueckner, T. Soda, P. W. Anderson, and P. Morel, *Phys. Rev.* **118,** 1442 (1960).

295. P. W. Anderson and P. Morel, *Phys. Rev.* **123,** 1911 (1961).

296. R. Balian and N. R. Werthamer, *Phys. Rev.* **131,** 1553 (1963).

297. L. P. Gorkov and L. P. Pitaevskii, *Zh. Eksp. Teor. Fiz.* **42,** 600 (1962) [English transl.: *Sov. Phys.—JETP* **15,** 417 (1962)].

298. D. D. Osheroff, R. C. Richardson, and D. M. Lee, *Phys. Rev. Lett.* **28,** 885 (1972).

299. D. D. Osheroff, W. J. Gully, R. C. Richardson, and D. M. Lee, *Phys. Rev. Lett.* **29,** 920 (1972).

300. For an up-to-date analysis of this problem, see C.-W. Woo, invited talk presented at the European Physical Society Topical Conference: Liquid and Solid Helium (Haifa, Israel, 1974).

CHAPTER 6

Solid Helium

C. M. Varma and N. R. Werthamer

Bell Laboratories, Murray Hill, New Jersey

6.1. INTRODUCTION

Helium [1] is the most reluctant of all substances to crystallize. Merely cooling the material, even to an arbitrarily low temperature, does not in itself produce solidification. Only the application of a substantial pressure in conjunction with cooling can result in freezing; even at the lowest temperatures the freezing pressure is in excess of 29 atm, and it rises with increasing temperature. When confronted with the difficult experimental task of creating the temperature and pressure environment needed for solid helium, it is natural to ask what can be learned from this crystal that cannot be investigated in less demanding materials; why is solid helium uniquely worth the effort? This question is further sharpened by considering that once this phase is formed it behaves unexceptionally with respect to a number of properties. The lattice structures are quite simple [body-centered cubic (bcc), hexagonal close-packed (hcp), and at higher pressures and temperatures face-centered cubic (fcc)], and neutron and X-ray scattering observations demonstrate regular crystal periodicity of very long range. In addition, ultrasonic and thermal pulse propagation studies indicate the existence of long-lived modes of standard phonon character.

The interest in solid helium stems directly from the same underlying source as the extreme conditions for its crystallization. When forced into a crystal, helium continues to show its aversion to the phase through its unusually large lattice constant, which exceeds by as much as 40% the interatomic separation that would minimize the potential energy. At the same time the helium atoms, whose mean positions form the regular crystalline array, are by no means narrowly localized about the mean positions. They are executing vibrational motion of such large amplitude that they sample nearly the entire free volume available to them. These properties reflect the unusually prominent role of kinetic relative to potential energy in the solid phase, or noting the very low temperatures at which the solid exists, the prominence of quantum zero-point motion. The zero-point energy is particularly large in helium because of its light atomic mass and its weak interatomic attractions. The zero-point vibrations act as an internal pressure, which forces the crystal to expand beyond the minimum potential configuration, and in fact prevents the formation of the phase almost entirely except under the extreme conditions noted. As another indication of the role of kinetic energy, the two isotopes of atomic mass 3 and 4 have substantially different phase diagrams, (which differ in turn from the phase diagram, with fcc solid structure, of the heavier rare gases) except at high pressures and temperatures. We summarize the phase diagrams of solid helium in Figure 6.1.

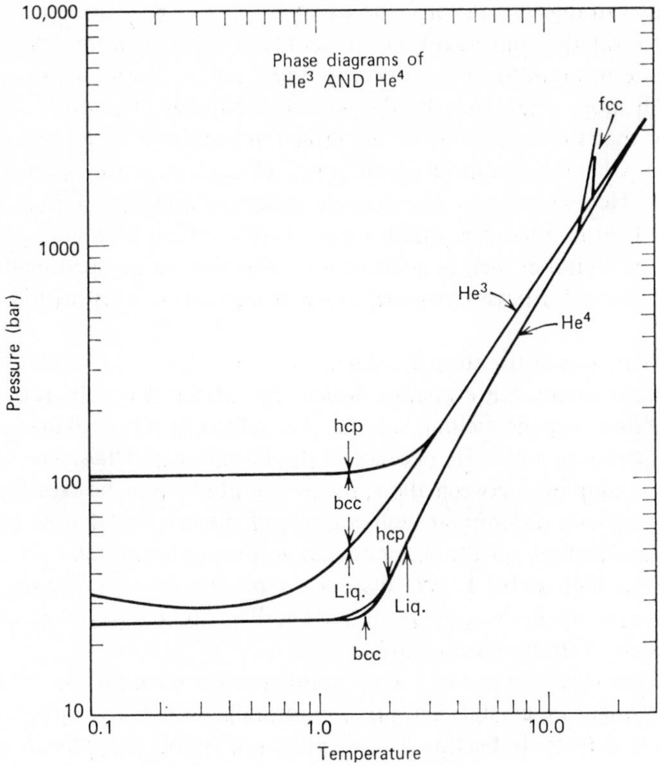

FIGURE 6.1. The phase diagrams of ^{3}He and ^{4}He

Because the quantum zero-point motions extend throughout the available atomic volumes, solid helium has been termed a quantum crystal, and is the most extreme example of this behavior. Despite the existence of normal phonon modes at low frequencies and wave vectors, the large vibrational amplitudes further imply that the motions are far from harmonic: The atoms experience forces that deviate substantially from those of a parabolic potential well. It could be expected that the vibrational modes in helium interact strongly with each other at high frequencies and wave vectors, and show substantial lifetime broadening. For these reasons solid helium presents a most severe test of a theoretical description of anharmonic lattice dynamics. On the other hand, since the interatomic forces are simple and quite accurately known, helium is a model substance in which to directly confront with experiment the computations from an absolute first-principles theory.

A lattice dynamical theory of quantum crystals is faced with two major obstacles. The first is that an expansion in the atomic displacements relative to the interatomic separations, the usual Born–von Karman approach, not only is not rapidly convergent but in fact may be an expansion about a point of absolute instability. This was first pointed out by Born [2] and later confirmed [3] more extensively by deWette and Nijboer. The other major difficulty is that the excursions of the atoms are so large as to permit actual collisions, where two or more atoms approach each other to within an atomic diameter. Here the forces are strongly repulsive and their effects must be accounted for in a nonperturbative way. This so-called hard-core problem is common to liquid as well as solid helium, and theoretical treatments of this aspect in the solid have borrowed much from earlier investigations for the liquid.

The excursions of the atom in solid helium are in fact so large that the nearest neighbor atoms can exchange lattice sites at the unusually rapid rate of about 10^8/sec near the melting volume. The effects of this exchange are most easily apparent in solid ^3He because of the Fermion statistics. The exchange leads to a coupling between the spins of the atoms and besides leading to nuclear magnetic ordering at temperatures of the order of 1 mK, has interesting consequences on the thermodynamic properties of solid ^3He at temperatures as high as 0.5 K. We discuss the physics of the exchange process and its thermodynamic consequences, as well as some unresolved questions in connection with it in Section 6.3.

Again because of the large zero point motion introduction of isotopic defects in quantum crystals leads to a renormalization of the force constants around the defects. In Section 6.4, we present a simple generalization of the lattice dynamical theory of Section 6.2 to the case of a defective crystal. We also discuss there the interesting experimental results on the statics and dynamics of the defects in solid helium.

6.2. LATTICE DYNAMICS OF IDEAL QUANTUM CRYSTALS

6.2.1. Correlated Phonon Basis Set

There have been a number of approaches [4] to the problem of constructing a formal mathematical description of extremely anharmonic lattice dynamics. Despite the fact that several alternative formalisms have met with some success in application to solid helium, and deserve a comparative review, we will confine ourselves in this chapter to developing merely a single formal method. Our claim is that this method [5] has the advantages of both relative simplicity as well as broad equivalence of results to those of more sophisticated treatments, but we would not want to argue that these advantages are necessarily decisive. Nevertheless, we proceed with the trial state technique, which in its application [6] to superfluid helium is termed the "method of correlated basis functions."

(*a*) *Ground State and Excitations.* The philosophy of this technique is to exhibit an explicit trial form for the ground-state wave function of the crystal, as well as trial expressions for relevant low-lying excited states. Matrix elements of the Hamiltonian with respect to these trial states are then evaluated, and an approximation scheme selected to reduce the complexity of these matrix elements to manageable proportions. Free parameters in the trial states are then optimized by requiring the ground-state expectation to be stationary with respect to variations of the parameters. Finally, the Hamiltonian matrix within the subset of states selected is diagonalized, from which excitation energies and lifetimes of these low-lying modes are obtained. Dynamical properties can be calculated by using the Hamiltonian matrix in a resolvent operator formalism.

The particular trial ground state that we choose is intended to reflect the two important aspects of the lattice dynamics discussed in the Introduction, namely the existence of well-defined phonon modes, at least at long wavelengths, and the presence of short-range collisions between atoms due to their large vibrational excursions from their lattice sites. Thus the trial ground-state wave function from a crystal of N atoms is taken to be

$$\Psi_0(\mathbf{r}_1, \cdots, \mathbf{r}_N) = \left(\prod_{i<j} f_{ij}(\mathbf{r}_{ij})\right) \exp\left(-\sum_{i<j} \tfrac{1}{4}\mathbf{u}_{ij}\cdot \overleftrightarrow{\Gamma}_{ij}\cdot \mathbf{u}_{ij}\right), \tag{6.2.1}$$

where we use the notation $\mathbf{r}_{ij} \equiv \mathbf{r}_i - \mathbf{r}_j$ for coordinate differences and $\mathbf{u}_{ij} \equiv \mathbf{r}_{ij} - \mathrm{R}_{ij}$ for differences of displacements from equilibrium positions $\mathrm{R}_i \equiv \langle \mathbf{r}_i \rangle$. The first factor, a product over all pairs of atoms, removes from the wavefunction those regions of configuration space with very high potential energy, thus accounting to some extent for the strong short-range interatomic repulsions. The so-called Jastrow functions f_{ij} might, for example, be selected in the Nosanow form [4g]

$$f_{ij}(\mathbf{r}) \cong \exp(-\kappa v(\mathbf{r})), \tag{6.2.2}$$

where $v(\mathbf{r})$ is the interatomic pair potential and κ is a variational constant, such that $f \to 0$ as $v \to +\infty$. Other choices for the $f_{ij}(\mathbf{r})$ are possible, however. The remaining factor in Ψ_0 is a Gaussian function of the displacement differences, with variational parameters $\overleftrightarrow{\Gamma}_{ij}$. Such a function is the *true* ground-state wave function for *some* harmonic lattice, depending on the $\overleftrightarrow{\Gamma}_{ij}$. This factor in Ψ_0 represents the zero-point motion of the phonons in the crystal, regarded as normal modes.

Corresponding to the picture that the low-lying excitations of the crystal are phonons, trial excited-state wave functions are also explicitly exhibited by multiplying Ψ_0 by low-order polynomials in the displacements. In the absence of the short-range correlation functions f such wave functions would be the *true* excited states of the same harmonic lattice for which the Gaussian portion of Ψ_0 is the true ground state. Implicit in this ansatz for the excited states is the assumption that short-range correlations are unaffected by the presence of a small number of phonon excitations. If we label the phonon modes by an index Λ, where $\Lambda = 1, \cdots, 3N$, then a one-phonon basis state is

$$|\Lambda) = \sum_i \boldsymbol{\alpha}(\Lambda, i)\cdot\mathbf{u}_i|0), \tag{6.2.3}$$

and a two-phonon basis state is

$$|\Lambda, \Lambda') = \sum_{i,j} \overleftrightarrow{\boldsymbol{\alpha}}(\Lambda, \Lambda'; \mathrm{i}, \mathrm{j}) : \mathbf{u}_i\mathbf{u}_j|0). \tag{6.2.4}$$

These basis states are not mutually orthonormal. An orthonormal basis can be explicitly constructed, for example, by means of a Schmidt procedure, and we denote that basis by $|0\rangle, |\Lambda\rangle, |\Lambda, \Lambda'\rangle$, etc. However at times it proves more convenient to continue working with the non-orthonormal basis and to impose orthonormality through suitable subsidiary conditions.

We choose a Hamiltonian of the form

$$\mathscr{H} = \sum_i p_i^2/2M + \mathscr{V}, \tag{6.2.5}$$

and assume that the potential energy is a sum of pairwise interactions,

$$\mathscr{V} = \sum_{i<j} v(\mathbf{r}_{ij}). \tag{6.2.6}$$

Martrix elements of the Hamiltonian in the phonon basis can be evaluated by use of the identity

$$(0|P_n\mathscr{H}P_m|0) = \left(0\middle|P_nP_m\left[\mathscr{V} - \sum_i\left(\frac{\hbar^2}{8M}\right)\nabla_i^2 \ln|\Psi_0|^2\right]\middle|0\right)$$
$$- \left(0\middle|\sum_i\left(\frac{\hbar^2}{8M}\right)\left[P_n\nabla_i^2P_m + P_m\nabla_i^2P_n - 2(\nabla_iP_n)\cdot(\nabla_iP_m)\right]\middle|0\right), \tag{6.2.7}$$

where P_n is an nth degree polynomial in the displacements. The matrix elements then become

$$(0|\mathscr{H}|0) = \left(0\middle|\left[\mathscr{V}^* + \sum_{i<j}\left(\frac{\hbar^2}{4M}\right)\overleftrightarrow{\Gamma}_{ij}:\overleftrightarrow{\mathbf{1}}\right]\middle|0\right), \tag{6.2.8}$$

$$(0|[\mathscr{H} - (0|\mathscr{H}|0)]|\Lambda) = \sum_i \boldsymbol{\alpha}(\Lambda; i)\cdot(0|[\mathbf{u}_i - (0|\mathbf{u}_i|0)]\mathscr{V}^*|0), \tag{6.2.9}$$

$$(\Lambda|[\mathscr{H} - (0|\mathscr{H}0)]|\Lambda') = \sum_{i,j} \boldsymbol{\alpha}(\Lambda^*; i)\, \boldsymbol{\alpha}(\Lambda'; \mathrm{j})\, (0|[\mathbf{u}_i\mathbf{u}_j$$
$$- (0|\mathbf{u}_i\mathbf{u}_j|0)]\mathscr{V}^* + \left(\frac{\hbar^2}{4M}\right)\delta_{i,j}\mathbf{1}|0), \tag{6.2.10}$$

$$(0|[\mathscr{H} - (0|\mathscr{H}|0)]|\Lambda, \Lambda') = \sum_{i,j} \overleftrightarrow{\boldsymbol{\alpha}}(\Lambda, \Lambda'; i, j) : (0|[\mathbf{u}_i\mathbf{u}_j$$

$$- (0|\mathbf{u}_i\mathbf{u}_j|0)]\mathscr{V}^* - \left(\frac{\hbar^2}{4M}\right)\delta_{i,j}\mathbf{\tilde{I}}|0). \qquad (6.2.11)$$

We have separated $\ln|\Psi_0|^2$ into long-range (i.e., Gaussian) and short-range (i.e., Jastrow) parts as per equation (6.2.1), and have defined a modified potential incorporating the short-range part,

$$\mathscr{V}^* \equiv \sum_{i<j} v_{ij}^*(\mathbf{r}_{ij})$$
$$\equiv \sum_{i<j} [v(\mathbf{r}_{ij}) - \left(\frac{\hbar^2}{4M}\right)\nabla^2 \ln f_{ij}^2(\mathbf{r}_{ij})]. \qquad (6.2.12)$$

(*b*) *One-Phonon Subset Optimization.* These matrix elements simplify considerably when the parameters entering $|\Psi_0|$ are chosen to minimize the ground-state energy, which we denote

$$E_0 \equiv \langle 0|\mathscr{H}|0\rangle. \qquad (6.2.13)$$

One set of parameters at our disposal are the lattice position vectors $\mathbf{R}_i$, with respect to which we clearly want E_0 to be stationary. Since the $\mathbf{R}_i$ explicitly enter $\langle 0|\mathscr{H}|0\rangle$ only via the displacements $\mathbf{u}_i = \mathbf{r}_i - \mathbf{R}_i$ in the Gaussian portion of Ψ_0, we have

$$\frac{\partial E_0}{\partial \mathbf{R}_i} = \sum_j \overleftrightarrow{\mathbf{G}}_{ij}\cdot\langle 0|[\mathbf{u}_j - \langle 0|\mathbf{u}_j|0\rangle]\mathscr{V}^*|0\rangle, \qquad (6.2.14)$$

where

$$\overleftrightarrow{\mathbf{G}}_{ij} \equiv \sum_{i',j'} \tfrac{1}{2}(\delta_{i,i'} - \delta_{i,j'})(\delta_{j,i'} - \delta_{j,j'})\overleftrightarrow{\Gamma}_{i'j'}. \qquad (6.2.15)$$

The stationary requirement,

$$\frac{\partial E_0}{\partial \mathbf{R}_i} = 0 \qquad (6.2.16)$$

implies from equation (6.2.9) that matrix elements of $\mathscr{H}$ between the ground state and all one-phonon states vanish,

$$\langle 0|\mathscr{H}|\Lambda\rangle = 0. \qquad (6.2.17)$$

Another set of parameters is the tensors $\overleftrightarrow{\Gamma}_{ij}$ describing the "width" of the Gaussian factor in Ψ_0. Varying E_0 with respect to the $\overleftrightarrow{\Gamma}_{ij}$, or rather more properly the combinations $\overleftrightarrow{\mathbf{G}}_{ij}$ of equation (6.2.15), we obtain

$$\frac{\partial E_0}{\partial \mathbf{G}_{ij}} = -\langle 0|[\mathbf{u}_i\mathbf{u}_j - \langle 0|\mathbf{u}_i\mathbf{u}_j|0\rangle]\mathscr{V}^*|0\rangle + \left(\frac{\hbar^2}{4M}\right)\delta_{i,j}\mathbf{\tilde{I}}. \qquad (6.2.18)$$

Then the additional stationary requirements

$$\frac{\partial E_0}{\partial \mathbf{G}_{ij}} = 0 \qquad (6.2.19)$$

imply that matrix elements of $\mathscr{H}$ between the ground state and all two-phonon states also vanish,

$$\langle 0|\mathscr{H}|\Lambda, \Lambda'\rangle = 0. \tag{6.2.20}$$

Matrix elements of $\mathscr{H}$ between one-phonon states are especially simple when E_0 is stationary with respect to the $\overleftrightarrow{\mathbf{G}}_{ij}$. The one-phonon states can be made mutually orthonormal by requiring the $\boldsymbol{\alpha}(\Lambda; i)$ to be proportional to eigenvectors of the displacement-correlation function. Thus we take

$$\boldsymbol{\alpha}(\Lambda, i) \propto \boldsymbol{\varepsilon}_\Lambda^i, \tag{6.2.21}$$

where the $\boldsymbol{\varepsilon}_\Lambda^i$ are orthonormal and complete,

$$\sum_i \boldsymbol{\varepsilon}_\Lambda^i \cdot \boldsymbol{\varepsilon}_{\Lambda'}^i = \delta_{\Lambda, \Lambda'}, \qquad \sum_\Lambda \boldsymbol{\varepsilon}_\Lambda^i \boldsymbol{\varepsilon}_\Lambda^j = \overleftrightarrow{\mathbf{I}}\delta_{i, j}, \tag{6.2.22}$$

and diagonalize the displacement correlation

$$\begin{aligned} \overleftrightarrow{\mathbf{D}}_{ij} &\equiv \langle 0|\mathbf{u}_i\mathbf{u}_j|0\rangle \\ &= \sum_\Lambda D_\Lambda \boldsymbol{\varepsilon}_\Lambda^i \boldsymbol{\varepsilon}_\Lambda^j. \end{aligned} \tag{6.2.23}$$

Then the orthonormal one-phonon states are

$$|\Lambda\rangle = D_\Lambda^{-1/2} \sum_i \boldsymbol{\varepsilon}^i \cdot \mathbf{u}_i|0\rangle. \tag{6.2.24}$$

Thus the stationary condition further results in $\mathscr{H}$ becoming diagonal in the one-phonon submanifold,

$$\langle\Lambda|(\mathscr{H} - E_0)|\Lambda'\rangle = \left(\frac{\hbar^2}{2MD_\Lambda}\right)\delta_{\Lambda, \Lambda'}. \tag{6.2.25}$$

Hence, recollecting equation (6.2.17), the stationary conditions have diagonalized the Hamiltonian in the larger submanifold of ground state and one-phonon states. Equation (6.2.25) shows that by restricting attention to just this submanifold an excitation frequency for one-phonon states is given by the approximation

$$\omega_\Lambda = \frac{\hbar}{2MD_\Lambda} \tag{6.2.26}$$

In order to obtain an explicit formula for ω_Λ, we combine equations (6.2.18) and (6.2.19) and recognize the product character of the ground-state wave function. Integrating by parts the $\mathbf{u}_i\mathbf{u}_j$ factors against the Gaussian gives

$$\left(\frac{\hbar^2}{4M}\right)\delta_{i, j}\overleftrightarrow{\mathbf{I}} = \sum_{i', j'} \overleftrightarrow{\mathbf{G}}_{ii'}^{-1} \cdot \overleftrightarrow{\boldsymbol{\Phi}}_{i'j'} \cdot \overleftrightarrow{\mathbf{G}}_{j'j}^{-1}, \tag{6.2.27}$$

where we define

$$\overleftrightarrow{\boldsymbol{\Phi}}_{ij} \equiv \frac{\langle \nabla_i\nabla_j[|\overleftrightarrow{\boldsymbol{\Psi}}|^2(\mathscr{V}^* - \langle 0|\mathscr{V}^*|0\rangle)]\rangle_G}{\langle|\overleftrightarrow{\boldsymbol{\Psi}}|^2\rangle_G}, \tag{6.2.28}$$

with $\overleftrightarrow{\Psi}$ being the short-range part of Ψ_0,

$$\overleftrightarrow{\Psi}(\mathbf{r}_1,\cdots,\mathbf{r}_N) \equiv \prod_{1<j} f(\mathbf{r}_{ij}), \tag{6.2.29}$$

and with the notation that for any function $W(\mathbf{r}_1,\cdots,\mathbf{r}_N)$,

$$\langle W\rangle_G \equiv \int d^{3N}r\ \exp\Big(-\tfrac{1}{2}\sum_{i,j}\mathbf{u}_i\cdot\overleftrightarrow{\mathbf{G}}_{ij}\cdot\mathbf{u}_j\Big)W(\mathbf{r}_1,\cdots,\mathbf{r}_N)$$
$$\div \int d^{3N}r\ \exp\Big(-\tfrac{1}{2}\sum_{i,j}\mathbf{u}_i\cdot\overleftrightarrow{\mathbf{G}}_{ij}\cdot\mathbf{u}_j\Big). \tag{6.2.30}$$

Equation (6.2.27) implies that if we diagonalize the "effective dynamical matrix" $\overleftrightarrow{\Phi}_{ij}$ so that

$$\sum_j \overleftrightarrow{\Phi}_{ij}\cdot\mathbf{e}_\Lambda^j = M\Omega_\Lambda^2\mathbf{e}_\Lambda^i, \tag{6.2.31}$$

with eigenvectors $\mathbf{e}_\Lambda^i$ which satisfy orthonormality and completeness conditions of the same form as equations (6.2.22) for the $\boldsymbol{\varepsilon}_\Lambda^i$, then the $\overleftrightarrow{\mathbf{G}}_{ij}$ are also diagonalized,

$$\mathbf{G}_{ij} = \sum_\Lambda \left(\frac{\hbar}{2M\Omega_\Lambda}\right)^{-1}\mathbf{e}_\Lambda^i\mathbf{e}_\Lambda^j. \tag{6.2.32}$$

In this representation the ground-state energy becomes

$$E_0 = \tfrac{1}{4}\sum_\Lambda \hbar\Omega_\Lambda + \sum_{i<j}\langle 0|v_{ij}^*(\mathbf{r}_{ij})|0\rangle. \tag{6.2.33}$$

However, diagonalizing $\overleftrightarrow{\Phi}_{ij}$ does not simultaneously diagonalize $\overleftrightarrow{\mathbf{D}}_{ij}$. An integration by parts of equation (6.2.22) rather gives

$$\overleftrightarrow{\mathbf{D}}_{ij} = \overleftrightarrow{\mathbf{G}}_{ij}^{-1} + \sum_{i'j'}\overleftrightarrow{\mathbf{G}}_{ii'}^{-1}\cdot\overleftrightarrow{\boldsymbol{\Delta}}_{i'j'}\cdot\overleftrightarrow{\mathbf{G}}_{j'j}^{-1}, \tag{6.2.34}$$

$$\overleftrightarrow{\boldsymbol{\Delta}}_{ij} \equiv \frac{\langle\nabla_i\nabla_j|\overleftrightarrow{\Psi}|^2\rangle_G}{\langle|\overleftrightarrow{\Psi}|^2\rangle_G}. \tag{6.2.35}$$

If the second term were absent, equation (6.2.34) would indeed yield the satisfying result $\omega_\Lambda = \Omega_\Lambda$, so that the phonon excitation frequencies would be taken from the eigenvalues of the effective dyanmical matrix. The presence of the second term, associated with the Jastrow factor in the ground-state wave function, complicates the relationship.

The discrepancy between ω_Λ and Ω_Λ, or equivalently between $\overleftrightarrow{\mathbf{D}}_{ij}$ and $\overleftrightarrow{\mathbf{G}}_{ij}^{-1}$, can in fact be resolved by a suitable choice of Jastrow function. We would, in any event, have been led to consider optimizing the Jastrow function in some way, to lower the trial ground-state energy still further. Some previous workers adopted a Jastrow factor of specific functional form but with one or more adjustable parameters, which were selected numerically by minimizing the computed ground-state energy. A more ambitious program would attempt a full functional variation with respect to $f(\mathbf{r})$, with optimum satisfying the resulting Euler-Lagrange equation. Such a program encounters one philosophical hurdle, however, in that a functional variation of $f(\mathbf{r})$ cannot be entirely independent of a variation in the $\overleftrightarrow{\mathbf{G}}_{ij}$. This can be seen by recasting Ψ_0 as

$$\Psi_0 = \exp \sum_{i<j} (\ln f_{ij}(\mathbf{r}) - \tfrac{1}{4}\mathbf{u}_{ij}\cdot \overleftrightarrow{\boldsymbol{\Gamma}}_{ij}\cdot \mathbf{u}_{ij}), \tag{6.2.36}$$

where both terms in the exponent are of the form of a pairwise sum of functions of the pair separation coordinate. An unrestricted variation of $\ln f_{ij}(\mathbf{r})$ would not permit a further independent variation of $\boldsymbol{\Gamma}_{ij}$. As soon as $\ln f_{ij}(\mathbf{r})$ becomes arbitrarily adjustable, the separation of the wave function into short-range and long-range parts becomes ambiguous.

The clue to resolving these problems is due to Horner, who suggested that the variation with respect to the $f_{ij}(\mathbf{r})$ should be constrained such that the lattice vectors and displacement correlations remain unaffected. These should be determined by the Gaussian, long-range part of the ground-state wave function, which then provides a specific definition for the separation of Ψ_0 into two parts. Thus the variation of E_0 with respect to the $f_{ij}(\mathbf{r})$ is subject to the constraints of normalization,

$$\langle 0|0\rangle = 1, \tag{6.2.37}$$

of fixed lattice structure,

$$\langle 0|\mathbf{u}_i|0\rangle = 0, \tag{6.2.38}$$

and of displacement correlation determined entirely by the Gaussian part of Ψ_0,

$$\langle 0|\mathbf{u}_j\mathbf{u}_j|0\rangle = \langle \mathbf{u}_i\mathbf{u}_j\rangle_G. \tag{6.2.39}$$

This last constraint directly removes the previous discrepancy between $\overleftrightarrow{\mathbf{D}}_{ij}$ and $\overleftrightarrow{\mathbf{G}}_{ij}^{-1}$, and hence between ω_Λ and Ω_Λ.

The most convenient way to carry out the constrained variation is to introduce the pair distribution function

$$g_{ij}(\mathbf{r}) \equiv \langle 0|\delta^3(\mathbf{r} - \mathbf{r}_{ij})|0\rangle, \tag{6.2.40a}$$

to use a Lagrange parameter L for each constraint, and to vary with respect to the $\ln f_{ij}^2(\mathbf{r})$. Then the variational calculation can be written as

$$0 = \left[\frac{\delta}{\delta \ln f_{ij}^2(\mathbf{r})}\right] \sum_{i<j} \int d^3r g_{ij}(\mathbf{r}) [v_{ij}^*(\mathbf{r}) - L_{ij}^{(0)} - \mathbf{L}_{ij}^{(1)}\cdot\mathbf{u}$$
$$- \overleftrightarrow{\mathbf{L}}_{ij}^{(2)} : \tfrac{1}{2}(\mathbf{uu} - \langle \mathbf{u}_{ij}\mathbf{u}_{ij}\rangle_G)]$$

$$= -\left(\frac{\hbar^2}{4M}\right)\nabla^2 g_{ij}(\mathbf{r}) + \sum_{i'<j'} \int d^3r' g_{ij,\,i'j'}(\mathbf{r}, \mathbf{r}') \, [v^*_{i'j'}(\mathbf{r}')$$

$$- \mathbf{L}^{(1)}_{i'j'} \cdot \mathbf{u}' - \overleftrightarrow{\mathbf{L}}^{(2)}_{i'j'} : \tfrac{1}{2}(\mathbf{u}'\mathbf{u}' - \langle \mathbf{u}_{i'j'}\mathbf{u}_{i'j'} \rangle_G)], \qquad (6.2.40b)$$

where

$$g_{ij,\,i'j'}(\mathbf{r}, \mathbf{r}') \equiv \frac{\delta g_{i'j'}(\mathbf{r}')}{\delta \ln f^2_{ij}(\mathbf{r})} = \langle 0 | \delta^3(\mathbf{r} - \mathbf{r}_{ij})\delta^3(\mathbf{r}' - \mathbf{r}_{i'j'}) | 0 \rangle \qquad (6.2.41)$$

is the two-pair distribution function. The Lagrange parameters, $L^{(0)}_{ij}$, $\mathbf{L}^{(1)}_{ij}$, and $\overleftrightarrow{\mathbf{L}}^{(2)}_{ij}$, can be determined by taking moments of the Euler–Lagrange equation (6.2.40) over $\mathbf{r}$, noting that the $g_{ij}(\mathbf{r})$ and their first derivatives vanish as $|\mathbf{r}| \to \infty$. The zeroth moment yields the familiar result,

$$L^{(0)}_{ij} = \int d^3r g_{ij}(\mathbf{r}) v^*_{ij}(\mathbf{r}). \qquad (6.2.42)$$

The first and second moments can be simplified by use of the conditions (6.2.17) and (6.2.20) resulting from E_0 being stationary with respect to the $\mathbf{R}_i$ and the $\overleftrightarrow{\mathbf{G}}_{ij}$. These two moments then reduce to the pair of coupled equations

$$0 = \sum_{i',\,j'} \int d^3r d^3r' \mathbf{u} g_{ij,\,i'j'}(\mathbf{r}, \mathbf{r}') \, [\mathbf{u}' \cdot \mathbf{L}^{(1)}_{i'j'} + \tfrac{1}{2}\mathbf{u}'\mathbf{u}' : \overleftrightarrow{\mathbf{L}}^{(2)}_{i'j'}],$$

$$0 = \sum_{i',\,j'} \int d^3r \, d^3r' \tfrac{1}{2}\mathbf{u}\mathbf{u} g_{ij,\,i'j'}(\mathbf{r}\mathbf{r}') \, [\mathbf{u}' \cdot \mathbf{L}^{(1)}_{i'j'}$$

$$+ \tfrac{1}{2}(\mathbf{u}'\mathbf{u}' - \langle \mathbf{u}_{i'j'}\mathbf{u}_{i'j'} \rangle_G) : \overleftrightarrow{\mathbf{L}}^{(2)}_{i'j'}]. \qquad (6.2.43)$$

Regarded as simultaneous linear equations for the $\mathbf{L}^{(1)}$ and $\overleftrightarrow{\mathbf{L}}^{(2)}$, these are homogeneous and have matrix of coefficients which, being closely related to the overlap matrix between states in the one- and two-phonon manifold, is positive definite. The conclusion is that

$$\mathbf{L}^{(1)}_{ij} = 0, \; \overleftrightarrow{\mathbf{L}}^{(2)}_{ij} = 0. \qquad (6.2.44)$$

Thus when the $\mathbf{R}_i$ and $\overleftrightarrow{\mathbf{G}}_{ij}$ are optimized, the pair distribution function is itself optimized as the solution of

$$0 = -\left(\frac{\hbar^2}{4M}\right)\nabla^2 g_{ij}(\mathbf{r}) + \sum_{i'j'} \int d^3r' g_{ij,\,i'j'}(\mathbf{r}, \mathbf{r}') \Big[v^*_{i'j'}(\mathbf{r}')$$

$$- \int d^3r'' g_{i'j'}(\mathbf{r}'') v^*_{i'j'}(\mathbf{r}'') \Big]. \qquad (6.2.45)$$

It is worth reiterating that $f^2_{ij}(\mathbf{r})$ has been constrained such that $\overleftrightarrow{\mathbf{\Delta}}_{ij} = 0$, and as a consequence $\overleftrightarrow{\boldsymbol{\Phi}}_{ij}$, $\overleftrightarrow{\mathbf{D}}_{ij}$, and $\overleftrightarrow{\mathbf{G}}_{ij}$ are all simultaneously diagonalizable.

(*c*) *Two-Body Cluster Expansion.* Although the foregoing analysis has been quite rigorous, it does not yet provide a theory which is computationally practicable, because the evaluation of matrix elements still involves integration over all 3N coordinates. An approximation scheme which removes this difficultly while maintaining a semiquantitative reasonableness is the so-called cluster expansion, truncated at the two-body level. Although the cluster expansion represents a systematic sequence of approximations converging on the exact expression, we will not be concerned here with the three-body or higher terms, nor with questions of the rapidity of the convergence. For this reason it is most convenient to quote the two-body cluster approximations in terms of an approximate expression for the pair distribution function,

$$g_{ij}(\mathbf{r}) \cong f_{ij}^2(\mathbf{r})\rho_{ij}(\mathbf{r}), \tag{6.2.46}$$

where ρ_{ij} ($\mathbf{r}$) is the pair distribution function for the Gaussian part of Ψ_0,

$$\begin{aligned}\rho_{ij}(\mathbf{r}) &\equiv \langle \delta^3(\mathbf{r} - \mathbf{r}_{ij})\rangle_G \\ &= \exp \frac{(-\tfrac{1}{2}\mathbf{u}\cdot\boldsymbol{\gamma}_{ij}\cdot\mathbf{u})}{\int d^3r \exp(-\tfrac{1}{2}\mathbf{u}\cdot\boldsymbol{\gamma}_{ij}\cdot\mathbf{u})}.\end{aligned} \tag{6.2.47}$$

The matrix $\boldsymbol{\gamma}_{ij}$ is specified in terms of $\overleftrightarrow{\mathbf{G}}_{ij}$ by its three-dimensional tensor inverse,

$$\boldsymbol{\gamma}_{ij}^{-1} = \sum_{i',j'} (\delta_{i',i} - \delta_{i',j})(\delta_{j',i} - \delta_{j',j})\overleftrightarrow{\mathbf{G}}_{i'j'}^{-1}, \tag{6.2.48}$$

or using the eigenvectors of $\overleftrightarrow{\boldsymbol{\Phi}}_{ij}$, equation (6.2.31),

$$\boldsymbol{\gamma}_{ij}^{-1} = \sum_{\Lambda}\left(\frac{\hbar}{2M\Omega_\Lambda}\right)\mathbf{E}_\Lambda^{ij}\mathbf{E}_\Lambda^{ij}, \tag{6.2.49}$$

where we use the notation

$$\mathbf{E}_\Lambda^{ij} \equiv \sum_l (\delta_{l,i} - \delta_{l,j})\mathbf{e}_\Lambda^l. \tag{6.2.50}$$

Then the resulting two-body-cluster ground-state energy,

$$E_0 \cong \sum_{i<j}\left[\langle f_{ij}^2(\mathbf{r}_{ij})v_{ij}^*(\mathbf{r}_{ij})\rangle_G + \left(\frac{\hbar^2}{4M}\right)\overleftrightarrow{\boldsymbol{\Gamma}}_{ij}:\overleftrightarrow{\mathbf{I}}\right], \tag{6.2.51}$$

is stationary with respect to variations of the $\mathbf{R}_i$ if

$$\left\langle \mathbf{u}_{ij}f_{ij}^2(\mathbf{r}_{ij})\left[v_{ij}^*(\mathbf{r}_{ij}) - \langle f_{ij}^2(\mathbf{r}_{ij})v_{ij}^*(\mathbf{r}_{ij})\rangle_G\right]\right\rangle_G = 0 \tag{6.2.52}$$

which can be written more compactly in terms of a modified two-body potential $\tilde{v}$,

$$\langle \mathbf{u}_{ij}\tilde{v}_{ij}(\mathbf{r}_{ij})\rangle_G = 0. \tag{6.2.53}$$

It is also stationary with respect to variations of the $\overleftrightarrow{\mathbf{G}}_{ij}$ if equation (6.2.27) continues to hold, but with the dynamical matrix $\overleftrightarrow{\Phi}_{ij}$ modified from equation (6.2.28) to the two-body-cluster approximation,

$$\overleftrightarrow{\Phi}_{ij} \cong \sum_{i',j'} \tfrac{1}{2}(\delta_{i,i'} - \delta_{i,j'})(\delta_{j,i'} - \delta_{j,j'})\langle\nabla\nabla\tilde{v}_{i'j'}(\mathbf{r}_{i'j'})\rangle_G. \qquad (6.2.54)$$

Furthermore, the difference between $\mathbf{G}_{ij}^{-1}$ and the displacement correlation can still be expressed in the form of equation (6.2.34), but with the approximate expression for $\overleftrightarrow{\Delta}_{ij}$,

$$\overleftrightarrow{\Delta}_{ij} \cong \sum_{i'j'} \tfrac{1}{2}(\delta_{i,i'} - \delta_{i,j'})(\delta_{j,i'} - \delta_{j,j'})\langle\nabla\nabla f_{i'j'}^2(\mathbf{r}_{i'j'})\rangle_G. \qquad (6.2.55)$$

As in the general case the two-body-cluster ground-state energy can be further optimized by stationary functional variation of $f_{ij}(\mathbf{r})$, subject again to the constraints that the Gaussian part of $g_{ij}(\mathbf{r})$ incorporates correctly the first and second displacement moments of the probability distribution. Thus we require

$$\int d^3r g_{ij}(\mathbf{r})\mathbf{u} = 0, \qquad (6.2.56)$$

and

$$\int d^3r g_{ij}(\mathbf{r})\tfrac{1}{2}(\mathbf{uu} - \gamma_{ij}^{-1}) = 0. \qquad (6.2.57)$$

Varying E_0 with respect to $\ln f_{ij}^2(\mathbf{r})$ with Lagrange parameters for these constraints, leads as before to an integro-differential equation for $g_{ij}(\mathbf{r})$. The removal of the Lagrange parameters from his equation, by using boundary conditions plus the stationarity of E_0 with respect to the $\mathbf{R}_i$ and $\overleftrightarrow{\mathbf{G}}_{ij}$ as per equation (6.2.53) and (6.2.27), proceeds much as in the general case. The final result is the equation determining $g_{ij}(\mathbf{r})$,

$$\left[-\left(\frac{\hbar^2}{4M}\right)\nabla^2 + v_{ij}^*(\mathbf{r}) - \int d^3r' g_{ij}(\mathbf{r}')v_{ij}^*(\mathbf{r}')\right]g_{ij}(\mathbf{r}) = 0. \qquad (6.2.58)$$

Although this equation appears very simple, there is a complexity that is concealed in the definition of the modified potential v^*. Much of this complexity is removed by introducing the alternative variable

$$h_{ij}(\mathbf{r}) \equiv (g_{ij}(\mathbf{r}))^{1/2} \qquad (6.2.59)$$

for which the equation transforms to

$$\left[-\left(\frac{\hbar^2}{M}\right)\nabla^2 + v(\mathbf{r}) - \varepsilon_{ij}\right]h_{ij}(\mathbf{r}) = 0, \qquad (6.2.60)$$

where

$$\varepsilon_{ij} \equiv \int d^3r h_{ij}^2(\mathbf{r})\left[v(\mathbf{r}) - \left(\frac{\hbar^2}{4M}\right)\nabla^2 \ln h_{ij}^2(\mathbf{r})\right]. \qquad (6.2.61)$$

Except for the fact that ε_{ij} is specified in terms of h_{ij}, (6.2.60) would be precisely the Schrödinger equation for the otherwise isolated pair of atoms i and j interacting via the bare potential $v(\mathbf{r})$. If (6.2.60) were an eigenvalue equation, ε_{ij} would be fixed by boundary conditions on h_{ij}. In the present case, however, the asymptotic behavior of h_{ij} as $r \to \infty$ is determined by the self-consistent value of ε_{ij}. For stable solids we should find that $\varepsilon_{ij} < 0$, and hence that $h_{ij}(\mathbf{r}) \to 0$ as $r \to \infty$. Since this fall-off is slower than Gaussian, we also find that $f_{ij}(\mathbf{r}) \to \infty$ as $r \to \infty$, although there is no real physical objection to this divergence.

(*d*) *Matrix Elements to Two-Phonon States.* The results quoted above for the two-body cluster approximation could also have been obtained from the more general expressions by approximating the multipair distribution functions, using the generating relationship

$$
\begin{aligned}
&g_{ij,i'j',\cdots,i''j''}(\mathbf{r}, \mathbf{r}', \cdots, \mathbf{r}'') \\
&\equiv \langle 0 | \delta^3(\mathbf{r} - \mathbf{r}_{ij}) \delta^3(\mathbf{r}' - \mathbf{r}_{i'j'}) \cdots \delta^3(\mathbf{r}'' - \mathbf{r}_{i''j''}) | 0 \rangle \\
&= \left(\frac{\delta}{\delta \ln f^2_{i'j'}(\mathbf{r}')} \right) \cdots \left(\frac{\delta}{\delta \ln f^2_{i''j''}(\mathbf{r}'')} \right) g_{ij}(\mathbf{r}), \qquad (6.2.62) \\
&\cong (\delta_{ij,i'j'} \delta^3(\mathbf{r} - \mathbf{r}')) \cdots (\delta_{ij,i''j''} \delta^3(\mathbf{r} - \mathbf{r}'')) g_{ij}(\mathbf{r}), \qquad (6.2.63)
\end{aligned}
$$

where equation (6.2.63) follows by differentiating the two-body cluster approximation for $g_{ij}(\mathbf{r})$, equation (6.2.46). This approach provides a useful method for evaluating other matrix elements involving orthonormalized two-phonon states. By returning to identity (6.2.7) we can obtain expressions for $\langle \Lambda | \mathscr{H} | \Lambda_1, \Lambda_2 \rangle$ and for $\langle \Lambda_1' \Lambda_2' | (\mathscr{H} - E_0) | \Lambda_1, \Lambda_2 \rangle$. These are simplified by assuming that the stationary conditions (6.2.16) and (6.2.17), and (6.2.18) and (6.2.19), are satisfied, and by further assuming the equality of $\overleftrightarrow{\mathbf{D}}_{ij}$ and $\overleftrightarrow{\mathbf{G}}_{ij}^{-1}$ as guaranteed by the stationary conditions (6.2.40). Additional simplification follows from the two-body cluster approximation, equation (6.2.63). These operations lead to the formulas

$$
\langle \Lambda | \mathscr{H} | \Lambda_1, \Lambda_2 \rangle \cong D_\Lambda^{-1/2} \sum_{i<j} \left\langle \prod_\mu \left[\left(\frac{\hbar}{2M\omega_\mu} \right)^{1/2} \mathbf{E}_u^{ij} \cdot \nabla \right] \tilde{v}_{ij} \right\rangle_G, \qquad (6.2.64)
$$

where the product runs over the three values $\mu = \Lambda^*, \Lambda_1, \Lambda_2$; and

$$
\begin{aligned}
\langle \Lambda_1', \Lambda_2' | (\mathscr{H} - E_0) | \Lambda_1, \Lambda_2 \rangle &\cong \delta_{\Lambda_1 \Lambda_2, \Lambda_1' \Lambda_2'} \hbar(\omega_{\Lambda_1} + \omega_{\Lambda_2}) \\
&+ \sum_{i<j} \left\langle \prod_\mu \left[\left(\frac{\hbar}{2M\omega_\mu} \right)^{1/2} \mathbf{E}_\mu^{ij} \cdot \nabla \right] \tilde{v}_{ij} \right\rangle_G, \qquad (6.2.65)
\end{aligned}
$$

where here $\mu = \Lambda_1, \Lambda_2, \Lambda_1'^*, \Lambda_2'^*$. The only modification in these matrix elements arising from the introduction of Jastrow factors in the wave functions is the presence of $\tilde{v}$ instead of the bare potential v. This modification is consistent as well with the expression (6.2.54) for the dynamical matrix $\overleftrightarrow{\Phi}_{ij}$.

The inclusion of two-phonon states into the basis set within which the Hamiltonian is diagonalized does not shift the approximate ground-state energy from the value E_0, because of condition (6.2.20), which decouples the two-phonon states from the ground state. However, the phonon excitation frequencies are altered from the values ω_Λ due to the coupling (6.2.64) with the two-phonon manifold. Since this manifold has a continuous energy spectrum, the phonons are damped as well as shifted in frequency, and as a consequence a time-dependent formalism is required to obtain properly the resulting phonon spectral function. This is accomplished most easily by considering a time-ordered propagator

$$D_{\Lambda\Lambda'}(\omega) = \int dt e^{i\omega(t-t')}\langle 0|T\{u_\Lambda(t)u_{\Lambda'}(t')\}|0\rangle, \tag{6.2.66}$$

which is related to a matrix element of the resolvent operator,

$$D_{\Lambda\Lambda'}(\omega) = \langle\Lambda|[(\omega - \mathscr{H} + E_0)^{-1} - (\omega + \mathscr{H} - E_0)^{-1}]|\Lambda'\rangle. \tag{6.2.67}$$

However, if the basis states are comprised of just one-phonon and two-phonon manifolds, then the matrix element can be rearranged as

$$\begin{aligned} D_{\Lambda\Lambda'}(\omega) = \{(\omega^2 - \omega_\Lambda^2)\delta_{\Lambda,\Lambda'} + \sum_{\Lambda_1,\Lambda_2}\sum_{\Lambda_1',\Lambda_2'}\langle\Lambda|\mathscr{H}|\Lambda_1,\Lambda_2\rangle \\ \times [\langle\Lambda_1,\Lambda_2|(\omega - \mathscr{H} + E_0)|\Lambda_1',\Lambda_2'\rangle^{-1} - \langle\Lambda_1,\Lambda_2|(\omega \\ + \mathscr{H} - E_0)|\Lambda_1',\Lambda_2'\rangle^{-1}] \times \langle\Lambda_1',\Lambda_2'|\mathscr{H}|\Lambda'\rangle\}^{-1}. \end{aligned} \tag{6.2.68}$$

The inverse operations in this formula are to be regarded as matrix inversions in the one-phonon and two-phonon manifolds, respectively. Equation (6.2.68), when combined with expressions (6.2.64) and (6.2.65) for the matrix elements of $\mathscr{H}$, leads us to the conclusion that the phonon propagator $D_{\Lambda\Lambda'}(\omega)$ is identical in form to that of conventional first-order self-consistent phonon theory [7] for the phonon-response function, except for the replacement of the bare potential by the effective potential $\tilde{v}$. This permits the application to solid helium, as treated by the method of correlated basis functions, of computational techniques previously developed [8] for the heavier rare gas crystals. The most useful quantity to be computed is the phonon spectral function, defined as

$$\mathscr{A}_{\lambda\lambda'}(\omega) \equiv \mathrm{Im} D_{\lambda\lambda'}(\omega). \tag{6.2.69}$$

It is worth noting the fourth-derivative term in formula (6.2.65), which represents an interaction between pairs of phonons. This term is significant for maintaining consistency in the approximation. It can be shown that if

(i) the elastic constants of the crystal are calculated in two distinct ways which in a rigorous theory yield identical results [9], first by twice differentiating the ground-state energy with respect to strains, and second by taking the long wavelength static limit of the phonon self-energy, and (ii) the energy is given by the first-order self-consistent phonon approximation equation (6.2.33), then the two methods for calculating the elastic constants agree [9, 7] provided the approximate self-energy is of the form obtained from equation (6.2.68) with the phonon-pair-interaction term included.

The formulas for matrix elements have been exhibited up to this point in a form that has not assumed any particular crystal structure. We now specialize to the case of a regular Bravais lattice and recast the equations into the form with which actual computations are carried out. Atoms are enumerated by lattice vectors $\boldsymbol{\tau}$, pointing to each of them in turn from an arbitrary initial atom. The phonon-mode index becomes the product of a wave vector $\mathbf{k}$ ranging over the first Brillouin zone, and a polarization index λ taking on three values in this situation of but one atom per unit cell. Key formulas are that the phonon excitation frequencies $\omega_{\mathbf{k}\lambda}$ and polarization vectors $\hat{e}_{\mathbf{k}\lambda}$ satisfy the transcription of equations (6.2.31) and (6.2.54),

$$M\omega_{\mathbf{k}\lambda}^2 \hat{e}_{\mathbf{k}\lambda} = \sum_{\boldsymbol{\tau}} (1 - \cos \mathbf{k}\cdot\boldsymbol{\tau}) \langle \nabla\nabla \tilde{v}_{\tau}(\mathbf{r}) \rangle_G \cdot \hat{e}_{\mathbf{k}\lambda}, \tag{6.2.70}$$

with Gaussian averaging specified in detail as

$$\langle W_{\tau}(\mathbf{r}) \rangle_G \equiv \frac{\int d^3r \exp(-\frac{1}{2}\mathbf{u}\cdot\boldsymbol{\gamma}_{\tau}\cdot\mathbf{u}) W_{\tau}(\mathbf{r})}{\int d^3r \exp(-\frac{1}{2}\mathbf{u}\cdot\boldsymbol{\gamma}_{\tau}\cdot\mathbf{u})} \tag{6.2.71}$$

and

$$\boldsymbol{\gamma}_{\tau}^{-1} = N^{-1} \sum_{\mathbf{k}\lambda} (1 - \cos \mathbf{k}\cdot\boldsymbol{\tau}) \left(\frac{\hbar}{2M\omega_{\mathbf{k}\lambda}} \right) \hat{e}_{\mathbf{k}\lambda} \hat{e}_{\mathbf{k}\lambda}. \tag{6.2.72}$$

The phonon spectral function satisfies the transcription of equations (6.2.64), (6.2.65), (6.2.68), and (6.2.69):

$$\begin{aligned} \mathscr{A}_{\mathbf{k}\lambda}(\omega) = \operatorname{Im} \Bigg\{ & \omega^2 - \omega_{\mathbf{k}\lambda}^2 - \tfrac{1}{2} N^{-1} \sum_{\mathbf{k}_1\lambda_1} \sum_{\mathbf{k}_2\lambda_2} \Delta(\mathbf{k} - \mathbf{k}_1 - \mathbf{k}_2) \\ & \times \left| \tfrac{1}{2} \sum_{\tau} \left(\frac{\hbar}{2\omega_{\mathbf{k}\lambda}} \right)^{-1/2} \left\langle \prod_{\mu} \left[\left(\frac{\hbar}{2M\omega_{\mu}} \right)^{1/2} (1 - e^{-i\mathbf{k}_{\mu}\cdot\tau}) \hat{e}_{\mu}\cdot\nabla \right] \tilde{v}_{\tau}(\mathbf{r}) \right\rangle_G \right|^2 \\ & \times \hbar^{-1} \Big[(\omega_{\mathbf{k}_1\lambda_1} + \omega_{\mathbf{k}_2\lambda_2} - \omega)^{-1} + (\omega_{\mathbf{k}_1\lambda_1} + \omega_{\mathbf{k}_2\lambda_2} + \omega)^{-1} \Big] \Bigg\}^{-1}. \end{aligned} \tag{6.2.73}$$

In equation (6.2.73) the product runs over $\mu = (-\mathbf{k}\lambda, \mathbf{k}_1\lambda_1, \mathbf{k}_2\lambda_2)$; Δ conserves momentum modulo a reciprocal lattice vector; polarization mixing is neglected, as appropriate to high symmetry directions; and the fourth-derivative term in equation (6.2.65) has been omitted for brevity.

6.2.2. Theory of Neutron Scattering

One of the most effective ways to obtain experimental information about the phonon spectral functions in a crystal is via inelastic coherent neutron scattering. This technique has been employed extensively on ^{4}He, in both its hcp and bcc phases, but is not applicable to ^{3}He where the large nuclear absorption cross section for neutrons swamps the coherent scattering. The theory of neutron scattering from a quasi -harmonic crystal, and even from a weakly anharmonic crystal, has been well known and conventionally utilized for some years. The large zero-point vibrations and substantial phonon interactions that take place in solid helium give rise to some significant modifications in the conventional theory, which have been strikingly prominent in the neutron-scattering observations on ^{4}He. Careful account must be taken of these modifications before neutron-scattering experiments and phonon spectral function computations can be directly compared with each other.

(*a*) *Conventional Quasi-harmonic Theory.* We first review the conventional theory of coherent neutron scattering. As is well-known, the differential cross section quite generally is proportional to $S(\mathbf{Q}, \omega)$, the Fourier transform of the density–density correlation function,

$$S(\mathbf{Q}, \omega) \equiv \int_{-\infty}^{\infty} dt \int d^3r \exp[i\omega(t-t') - i\mathbf{Q}\cdot(\mathbf{r}-\mathbf{r}')]\langle \rho(\mathbf{r}, t')\rho(\mathbf{r}', t')\rangle, \quad (6.2.74)$$

where the density operator is

$$\rho(\mathbf{r}, t) = \sum_i \delta^3(\mathbf{r} - \mathbf{r}_i(t)). \quad (6.2.75)$$

For a quasi-harmonic crystal the density correlation function can be evaluated exactly in terms of the displacement correlation to give

$$S(\mathbf{Q}, \omega) = \int dt \sum_{i,j} \exp[i\omega(t - t') - i\mathbf{Q}\cdot\mathbf{R}_{ij}] \exp(-W_i(\mathbf{Q}) - W_j(\mathbf{Q})) \times \exp[\mathbf{Q}\cdot\langle \mathbf{u}_i(t)\mathbf{u}_j(t')\rangle\cdot\mathbf{Q}], \quad (6.2.76)$$

where

$$W_i(\mathbf{Q}) = \tfrac{1}{2}\mathbf{Q}\cdot\langle \mathbf{u}_i\mathbf{u}_i\rangle\cdot\mathbf{Q} \quad (6.2.77)$$

is the Debye–Waller exponent. Furthermore the quasi-harmonic displacement correlation at zero temperature is expressible in terms of phonon frequencies and polarization vectors as

$$\langle \mathbf{u}_i(t)\mathbf{u}_j(t')\rangle = N^{-1} \sum_{\mathbf{k}\lambda} \exp[i\mathbf{k}\cdot\mathbf{R}_{ij} - i\omega_{\mathbf{k}\lambda}(t - t')] \times \left(\frac{\hat{e}_{\mathbf{k}\lambda}\hat{e}_{\mathbf{k}\lambda}}{2M\omega_{\mathbf{k}\lambda}}\right), \quad (6.2.78)$$

so that

$$W(\mathbf{Q}) = N^{-1} \sum_{\mathbf{k}\lambda} \tfrac{1}{2} \frac{(\mathbf{Q}\cdot\hat{e}_{\mathbf{k}\lambda})^2}{2M\omega_{\mathbf{k}\lambda}}. \tag{6.2.79}$$

Writing the exponential of the displacement correlation as a power series, we obtain

$$\exp[\mathbf{Q}\cdot\langle \mathbf{u}_i(t)\mathbf{u}_j(t')\rangle\cdot\mathbf{Q}] = 1 + \mathbf{Q}\cdot\langle \mathbf{u}_i(t)\mathbf{u}_j(t')\rangle\cdot\mathbf{Q} + \cdots. \tag{6.2.80}$$

Combining these formulas $S(\mathbf{Q}, \omega)$ can be written as the sum

$$S(\mathbf{Q}, \omega) = S^{(0)}(\mathbf{Q}, \omega) + S^{(1)}(\mathbf{Q}, \omega) + S^{(m)}(\mathbf{Q}, \omega), \tag{6.2.81}$$

where $S^{(0)}$ gives the elastic Bragg scattering peak,

$$S^{(0)}(\mathbf{Q}, \omega) = N^2 \Delta(\mathbf{Q}) e^{-2W(\mathbf{Q})} 2\pi\delta(\omega); \tag{6.2.82}$$

$S^{(1)}$ gives the inelastic one-phonon peak,

$$S^{(1)}(\mathbf{Q}, \omega) = N \sum_{\mathbf{k}} \Delta(\mathbf{Q} - \mathbf{k}) e^{-2W(\mathbf{Q})} \sum_{\lambda} \left[\frac{(\mathbf{Q}\cdot\hat{e}_{\mathbf{k}\lambda})^2}{2M\omega_{\mathbf{k}\lambda}} \right] \times 2\pi\delta(\omega - \omega_{\mathbf{k}\lambda}); \tag{6.2.83}$$

the remainder $S^{(m)}$ gives the so-called multiphonon background that is a smooth function of ω. In the quasi-harmonic approximation the phonons are undamped and the one-phonon peak is a δ function in the energy transfer ω, positioned at the frequency of the kinematically determined phonon mode. The intensity of the one-phonon peak

$$S^{(1)}(\mathbf{Q}) \equiv (2\pi)^{-1} \int d\omega\, S^{(1)}(\mathbf{Q}, \omega) \tag{6.2.84}$$

depends on $\mathbf{Q}$ schematically as

$$S^{(1)}(\mathbf{Q}) \sim (Q\bar{u})^2 e^{-(Q\bar{u})^2}. \tag{6.2.85}$$

The first factor is the phonon vibrational weight and competes against the second, the Debye–Waller factor, leading to a maximum in the one-phonon intensity at $Q\bar{u} = 1$. Almost all neutron-scattering experiments on weakly anharmonic crystals are carried out in the regime $Q\bar{u} \ll 1$. Helium, on the other hand, has a $\bar{u}$ so large that even the smallest nonvanishing reciprocal lattice vector is at $Q\bar{u} \sim 1$, and scattering from any larger Q is badly weakened by the rapidly decreasing Debye–Waller factor.

The additional sum rules on the first frequency moment of the cross section that may be verified by direct integration of equation (6.2.76) and (6.2.83), respectively, are

$$(2\pi)^{-1} \int_{-\infty}^{\infty} d\omega\, \omega S(\mathbf{Q}, \omega) = \frac{NQ^2}{2M}, \tag{6.2.86}$$

and

$$(2\pi)^{-1}\int_{-\infty}^{\infty} d\omega\, \omega S^{(1)}(\mathbf{Q}, \omega) = N\left(\frac{Q^2}{2M}\right)e^{-2W(\mathbf{Q})}. \tag{6.2.87}$$

The one-phonon peak exhausts a fraction e^{-2W} of the total first moment, a fraction which decreases with increasing **Q**. Furthermore, it can also be verified directly that at high frequency and momentum transfer the one-phonon process intensity disappears and the multiphonon background dominates the scattering. It is

$$\lim_{\omega\to\infty} S(\mathbf{Q}, \omega) \cong \lim_{\omega\to\infty} S^{(m)}(\mathbf{Q}, \omega) \cong 2\pi N\delta\left(\omega - \left(\frac{Q^2}{2M}\right)\right). \tag{6.2.88}$$

More properly the δ-function is broadened into a Gaussian of width $\Delta\omega \sim (\Theta_D Q^2/2M)^{1/2}$, where Θ_D is of order the Debye energy. This high-frequency limit shows that multiple-phonon processes in a quasi-harmonic crystal add coherently to simulate a scattering that is equivalent to that of a free gas, even though the atoms remain interconnected by Hooke's law springs no matter how large the excitation energy given to them.

(*b*) *General Anharmonic Theory*. The analysis of $S(\mathbf{Q},\omega)$ for a general anharmonic crystal is quite complicated and lengthy. Such an analysis has consequences for the interpretation of neutron-scattering experiments in several aspects. The separation described by equation (6.2.81) still remains valid [10]. Although the Bragg scattering as per equation (6.2.82) continues to hold, with the Debye–Waller exponent maintaining its definition via

$$e^{-2W(\mathbf{Q})} = |\langle e^{i\mathbf{Q}\cdot\mathbf{u}}\rangle|^2; \tag{6.2.89}$$

$W(\mathbf{Q})$ is no longer generally a quadratic function of **Q**. Rather, equation (6.2.77) holds only as the limiting behavior for small **Q**. For bcc helium, however, computations based on equation (6.2.89) and the trial wave function equation (6.2.1) indicate [11] that $W(\mathbf{Q})$ does increase monotonically with **Q**, and deviates only a comparatively small amount from a quadratic dependence. In the hcp phase $W(\mathbf{Q})$ is not forbidden by symmetry from a term of order Q^3, and as a result [12] certain Bragg reflections become allowed that would have had zero intensity if the crystal were quasi-harmonic. Such reflections have not yet been observed experimentally.

Equation (6.2.83) for the "one-phonon" scattering is modified in several important respects. In the first place, the previously undamped phonon is broadened [13] by anharmonic damping mechanisms, so that the δ function is replaced by the phonon spectral function

$$\left(\frac{\pi}{\omega_{\mathbf{k}\lambda}}\right)\delta(\omega-\omega_{\mathbf{k}\lambda}) \to \mathscr{A}_{\mathbf{k}\lambda}(\omega) = \operatorname{Im} D_{\mathbf{k}\lambda}(\omega). \tag{6.2.90}$$

(We here ignore polarization mixing, which is valid along directions in **k** space of high symmetry.) If this were the only modification, then an inelastic neutron-scattering experiment would yield a direct measure of the phonon spectral function. Furthermore, the relative shape of the cross section would be reciprocal-lattice translation-invariant (i.e., dependent only on **k** and not on **Q**) whereas the total intensity would be **Q** dependent and hence vary from zone to zone in reciprocal space. However, another modification is the appearance [10, 14] of **Q** and ω-dependent form factors for the one-phonon emission, so that in fact

$$S^{(1)}(\mathbf{Q}, \omega) = N \sum_{\mathbf{k}\lambda} \Delta(\mathbf{Q}-\mathbf{k}) e^{-2W(\mathbf{Q})} \operatorname{Im} \{D_{\mathbf{k}\lambda}(\omega) [\mathbf{Q}\cdot e_{\mathbf{k}\lambda} + L_{\mathbf{k}\lambda}(\mathbf{Q}, \omega)]^2\}. \tag{6.2.91}$$

Since $L(\mathbf{Q},\omega)$ is a complex function of ω the one-phonon scattering is no longer strictly proportional to Im $D(\omega)$, but rather admixes a portion of Re $D(\omega)$ as well. This introduces [15] an asymmetric distortion of Im $D(\omega)$, and tends to shift the peak of $S^{(1)}$ as a function of ω away from the peak of $D(\omega)$. Since $L(\mathbf{Q},\omega)$ also depends on **Q** in addition to **k**, the shape of the scattering cross section is no longer independent of the zone in reciprocal space; the amount and sign of the distortion varies from zone to zone without reference to the symmetry of the reciprocal lattice. Although $S(\omega)$ and Im $D(\omega)$ can both be shown to be positive definite, there are no grounds [16] for supposing that $S^{(1)}(\omega)$ is everywhere positive. Finally, $L(\mathbf{Q},\omega)$ varies as Q^2 for small Q, with coefficient proportional to the strength of the three-phonon interaction.

Statements that continue to remain valid quite generally for an anharmonic crystal are the sum rules (6.2.86) and (6.2.87). The former, usually known as the Landau–Placzek sum rule, is a consequence merely of the equal-time commutation relations between position and momentum, and the proof is relatively straightforward. Equation (6.2.87), first proved [10] by Ambegaokar, Conway, and Baym (ACB), relies on a subtle analysis of diagrammatic perturbation theory, in which the definition of $S^{(1)}(\omega)$ is that portion of $S(\omega)$ for which a single factor of the one-phonon propagator $D(\omega)$ occurs explicitly. It is instructive to note that exactly the same sum rule holds [10] for the quantity obtained from $S^{(1)}(\mathbf{Q},\omega)$ by setting equal to zero in equation (6.2.91) the vertex correction $L(\mathbf{Q},\omega)$, The effect of L is to rearrange the spectral weight in $S^{(1)}$ without altering the first frequency moment, a statement that supports the claim that $S^{(1)}$ is asymmetrically distorted from Im D as a function of ω.

It is tempting to suppose that the Landau–Placzek and ACB sum rules together can be used in practice to determine the Debye–Waller factor directly from inelastic scattering observations. It might appear [15] that separating the scattering cross section into a peak and a smooth background and taking the ratio of the respective first moments would yield $\exp(-2W(\mathbf{Q}))$. However, an analysis of this kind for observations in bcc ^{4}He has led [17] to serious paradoxes. The difficulty is [16] not only that the sum rules requires S and $S^{(1)}$ for *all* ω, and that $S^{(1)}$ has significant contributions in the wings away from the main peak, but even more importantly that since $S^{(1)}$ need not be positive there is no a priori operational rule for separating the observations into $S^{(1)}$ and $S^{(m)}$ components in the wings.

A final property [18] of a quasi-harmonic crystal that remains true [16] for the anharmonic crystal in the high-frequency limit, equation (6.2.88). The proof follows lines similar to that of Landau–Placzek, in that they both are based on the equal-time commutation relations. As the one-phonon portion of S declines in weight for increasingly large Q, the multiphonon background mimics a "free-particle" behavior independent of the structure or phase of the condensed medium.

Approximation schemes for calculating $S(\mathbf{Q},\omega)$ have been developed, consistent with approximations for the phonon spectral weight Im $D(\omega)$ itself. When $D(\omega)$ is to be computed allowing for three- and four-phonon interactions as per equation (6.2.73), it seems that the same matrix elements are to be used in computing $L(\mathbf{Q},\omega)$ from the formula

$$L_{\mathbf{k}\lambda}(\mathbf{Q},\omega) \cong \frac{1}{2}N^{-1}\sum_{\mathbf{k}_1\lambda_1}\sum_{\mathbf{k}_2\lambda_2}\Delta(\mathbf{k}-\mathbf{k}_1-\mathbf{k}_2)\prod_{\mu'}\left[\frac{\mathbf{Q}\cdot\hat{e}_{\mu'}}{(2\omega_{\mu'})^{1/2}}\right]$$
$$\times(-i)\frac{1}{2}\sum_{\boldsymbol{\tau}}\left(\frac{\hbar}{2\omega_{\mathbf{k}\lambda}}\right)^{-1/2}\left\langle\prod_{\mu}\left[\left(\frac{\hbar}{2M\omega_{\mu}}\right)^{1/2}(1-e^{-i\mathbf{k}_\mu\cdot\boldsymbol{\tau}})\hat{e}_{\mu}\cdot\nabla\right]\right.$$
$$\left.\times\, v_{\boldsymbol{\tau}}(\mathbf{r})\right\rangle_G \hbar^{-1}\left[(\omega_{\mathbf{k}_1\lambda_1}+\omega_{\mathbf{k}_2\lambda_2}-\omega)^{-1}+(\omega_{\mathbf{k}_1\lambda_1}+\omega_{\mathbf{k}_2\lambda_2}+\omega)^{-1}\right]. \tag{6.2.92}$$

The products run over $\mu = (-\mathbf{k}\lambda,\mathbf{k}_1\lambda_1,\mathbf{k}_2\lambda_2)$ as in equation (6.2.73), while $\mu' = (\mathbf{k}_1\lambda_1,\mathbf{k}_2\lambda_2)$.

6.2.3. Theory of Light Scattering

In addition to the scattering by neutrons, advances in laser technology have made visible light a widely used probe of crystals. The technique of inelastic light scattering is particulary welcome in its application to helium, in spite of the weak interaction of helium atoms with light, because in contrast to neutrons it can be used on ^{3}He as well as ^{4}He. In fact, light scattering is the only direct source of experimental information about the phonon spectrum of ^{3}He.

To understand what can be learned from light scattering and what are its limitations, we adopt a simplified model [19] of the electronic excitations in solid helium and their interaction with light. We assume that the crystal is composed of atoms, each of which is neutral but weakly polarizable in an electric field. The atoms are assumed to interact electrically only via dipolar forces, and each atom is taken to have a polarizability equal to its free value. We consider not only the possibility that the scattered photon is reradiated by the *same* atom as absorbed the incident laser photon, but also that the excited atom absorbing the incident photon transfers its excitation via dipolar forces to a *second* atom, which in turn reradiates the scattered photon. One or more phonons can be emitted during the radiation process.

We take the (initial, final) photons to have frequencies (ω_i, ω_f), wave vectors $(\mathbf{k}_i, \mathbf{k}_f)$, polarizations $(\hat{\varepsilon}_i, \hat{\varepsilon}_f)$, and occupation numbers (n_i, n_f). It is convenient to introduce the differences $\mathbf{k}_{if} = \mathbf{k}_i - \mathbf{k}_f$ and $\omega_{if} \equiv \omega_i - \omega_f$. We also denote the dipole polarizability of an atom by α, which for helium has a value $\alpha \sim 2 \times 10^{-25}$ cm^3, and the dipole tensor by $\overleftrightarrow{\mathbf{T}}(\mathbf{r})$,

$$\overleftrightarrow{\mathbf{T}}(\mathbf{r}) \equiv \frac{\overleftrightarrow{\mathbf{I}} - 3\hat{r}\hat{r}}{r^3}. \tag{6.2.93}$$

Then in this model the probability for the scattering event becomes

$$\begin{aligned}\tau^{-1} = \left(\frac{2\pi}{V}\right)^2 \sum_{\mathbf{k}\mathrm{f}} \omega_\mathrm{i}\omega_\mathrm{f} n_\mathrm{i}(\mathrm{n}_\mathrm{f} + 1) \int_{-\infty}^{\infty} dt\, e^{i\omega_{if}t} \sum_{i,j} \langle \hat{\varepsilon}_\mathrm{f} \cdot [\alpha \overleftrightarrow{\mathbf{I}} \\ - \alpha^2 \sum_{l \neq i} \overleftrightarrow{\mathbf{T}}(\mathbf{r}_{il}(t)) e^{i\mathbf{k}_f \cdot \mathbf{r}_{il}(t)}] \cdot \hat{\varepsilon}_\mathrm{i} \hat{\varepsilon}_\mathrm{f} \cdot [\alpha \overleftrightarrow{\mathbf{I}} \\ - \alpha^2 \sum_{l \neq \mathrm{j}} \overleftrightarrow{\mathbf{T}}(\mathbf{r}_{lj}(0)) e^{-i\mathbf{k}_f \cdot \mathbf{r}_{lj}(0)}] \cdot \hat{\varepsilon}_i e^{i\mathbf{k}_{if} \cdot (\mathbf{r}_i(t) - \mathbf{r}_j(0))} \rangle,\end{aligned} \tag{6.2.94}$$

in terms of a correlation function of the lattice dynamical system.

This correlation function is a generalization of the density correlation which enters the neutron-scattering description, and much of the analysis of $S(\mathbf{Q}, \omega)$ could be applied here as well. Nevertheless, it is instructive to work out the consequences of equation (6.2.94) in the quasi-harmonic approximation, from which selection rules and intensity ratios can be obtained [19] with fair reliability. The effects of the three-phonon processes can then be included by analogy with the neutron-scattering analysis. Evaluating τ^{-1} in the quasi-harmonic approximation, and extracting the one-phonon portion by expanding the time-dependent exponentials to first order, we obtain

$$\begin{aligned}[\tau^{-1}]_1 \cong \left(\frac{2\pi}{V}\right)^2 \sum_{\mathbf{k}\mathrm{f}} \omega_\mathrm{i}\omega_\mathrm{f} n_\mathrm{i}(n_\mathrm{f} + 1) \int_{-\infty}^{\infty} dt\, e^{i\omega_{if}t} \sum_{i,j} e^{i\mathbf{k}_{\mathrm{if}} \cdot \mathbf{R}_{ij}} \\ \times \langle [\alpha \hat{\varepsilon}_\mathrm{f} \cdot \hat{\varepsilon}_\mathrm{i} i\mathbf{k}_\mathrm{if} \cdot \mathbf{u}_i(t) - \alpha^2 \sum_{l \neq i} \mathbf{u}_{il}(t) \cdot \nabla_{il} \hat{\varepsilon}_\mathrm{f} \cdot \overleftrightarrow{\mathbf{T}}(\mathbf{R}_{il}) \cdot \hat{\varepsilon}_\mathrm{i}] \\ \times [-\alpha \hat{\varepsilon}_\mathrm{f} \cdot \hat{\varepsilon}_\mathrm{i} i\mathbf{k}_\mathrm{if} \cdot \mathbf{u}_j(0) - \alpha^2 \sum_{m \neq j} \mathbf{u}_{mj}(0) \cdot \nabla_{mj} \hat{\varepsilon}_\mathrm{f} \cdot \overleftrightarrow{\mathbf{T}}(R_{mj}) \cdot \hat{\varepsilon}_\mathrm{i}] \rangle.\end{aligned} \tag{6.2.95}$$

We have neglected factors of exp $(i\mathbf{k}_{\mathrm{if}}\cdot\mathbf{r})$ multiplying $\mathbf{T}(\mathbf{r})$ in equation (6.2.94), since the wavelength of visible light is very long relative to interatomic separations.

At this point it becomes necessary to introduce explicitly the complication of lattice structures with more than one atom per unit cell in order to accommodate Raman scattering from optical phonons. In terms of the phonon normal modes, the one-phonon scattering rate becomes

$$\begin{aligned}[\tau^{-1}]_1 &= \left(\frac{2\pi N}{V}\right)^2 \sum_{\mathbf{k}_f} \omega_{\mathrm{i}}\omega_{\mathrm{f}} n_{\mathrm{i}}(n_{\mathrm{f}}+1)\, N^{-1} \sum_{\mathbf{k}\lambda} \Delta(\mathbf{k}_{\mathrm{if}}-\mathbf{k}) \\ &\times \frac{\pi\hbar}{M\omega_{\mathbf{k}\lambda}} \left(\frac{\delta(\omega_{\mathrm{if}}-\omega_{\mathbf{k}\lambda})}{1-e^{-\beta\omega_{\mathbf{k}\lambda}}} + \frac{\delta(\omega_{\mathrm{if}}+\omega_{\mathbf{k}\lambda})}{e^{\beta\omega_{\mathbf{k}\lambda}}-1}\right) \\ &\times \Big| \sum_{\sigma} e^{i(\mathbf{k}_{\mathrm{if}}-\mathbf{k})\cdot\mathbf{R}_\sigma} [\alpha\hat{\varepsilon}_{\mathrm{f}}\cdot\hat{\varepsilon}_{\mathrm{i}} i\mathbf{k}_{\mathrm{if}}\cdot\mathbf{e}^{\sigma}_{\mathbf{k}\lambda} \\ &- \alpha^2 \sum_{\boldsymbol{\tau}} \sum_{\sigma'} (\mathbf{e}^{\sigma}_{\mathbf{k}\lambda} - e^{\sigma'}_{\mathbf{k}\lambda} e^{-i\mathbf{k}\cdot\boldsymbol{\tau}_{\sigma\sigma'}})\cdot\boldsymbol{\nabla}\varepsilon_{\mathrm{f}}\cdot\mathbf{T}(\boldsymbol{\tau}_{\sigma\sigma'})\cdot\hat{\varepsilon}_i]\Big|^2. \end{aligned} \tag{6.2.96}$$

We use the notation $\boldsymbol{\tau}$ for lattice vectors connecting unit cells, $\mathbf{R}_\sigma$ for a vector from the center of a unit cell to the σ-th atom within the cell, and define $\boldsymbol{\tau}_{\sigma\sigma'} \equiv \boldsymbol{\tau} + \mathbf{R}_\sigma - \mathbf{R}_{\sigma'}$. Since the photon wavelengths are very long, the scattering process cannot involve any nonvanishing reciprocal lattice vector, and the emitted phonon wave vector $\mathbf{k}$ is very near to the Brillouin zone center. For acoustic modes at long wavelengths, all n atoms within the unit cell move nearly in phase, so that

$$\mathbf{e}^{\sigma}_{\mathbf{k}\lambda} - \mathbf{e}^{\sigma'}_{\mathbf{k}\lambda} e^{-i\mathbf{k}\cdot\boldsymbol{\tau}_{\sigma\sigma'}} = n^{-1/2}\hat{e}_{\mathbf{k}\lambda} i\mathbf{k}\cdot\boldsymbol{\tau}_{\sigma\sigma'}; \tag{6.2.97}$$

the frequencies of these modes are almost always very much smaller than k_BT. For long wavelength optical modes in the hcp structure the two atoms in the unit cell move nearly out of phase, so that

$$\mathbf{e}^{\sigma}_{\mathbf{k}\lambda} - \mathbf{e}^{\sigma'}_{\mathbf{k}\lambda'} e^{-i\mathbf{k}\cdot\boldsymbol{\tau}_{\sigma\sigma'}} \cong 2^{-1/2}\hat{e}_{0\lambda}(\delta_{\sigma,1}, \delta_{\sigma',2} - \delta_{\sigma,2}\delta_{\sigma',1}); \tag{6.2.98}$$

the frequencies of these modes are almost always very much larger than k_BT. Furthermore, the scattering rate can be converted into a scattering efficiency S_1 in the standard way. The result is that

$$\begin{aligned} S_1 &\cong \frac{nN}{V}\left(\frac{\omega_i}{c}\right)^4 \Big\{\delta_{\mathbf{k}_{\mathrm{if}},\mathbf{k}} \sum_{\lambda=a} \frac{\pi k_B T}{M s^2_{k\lambda}} [\delta(\omega_{\mathrm{if}} - s_{k\lambda}k) + \delta(\omega_{\mathrm{if}} + s_{k\lambda}k)] \\ &\times \Big|\alpha\hat{\varepsilon}_{\mathrm{f}}\cdot\hat{\varepsilon}_{\mathrm{i}}\hat{k}\cdot\hat{e}_{k\lambda} - \alpha^2 \sum_{\boldsymbol{\tau}} n^{-1} \sum_{\sigma,\sigma'} \hat{k}\cdot\boldsymbol{\tau}_{\sigma\sigma'}\hat{e}_{k\lambda}\cdot\boldsymbol{\nabla}\hat{\varepsilon}_{\mathrm{f}}\cdot\overleftrightarrow{\mathbf{T}}(\boldsymbol{\tau}_{\sigma\sigma'})\cdot\hat{\varepsilon}_{\mathrm{i}}\Big|^2 \\ &+ \sum_{\lambda=0} \frac{\pi\hbar}{M\omega_{0\lambda}} \delta(\omega - \omega_{0\lambda}) \Big|\alpha^2 \sum_{\boldsymbol{\tau}} \hat{e}_{0\lambda}\cdot\boldsymbol{\nabla}\hat{\varepsilon}_{\mathrm{f}}\cdot\overleftrightarrow{\mathbf{T}}(\boldsymbol{\tau}_{12})\cdot\hat{\varepsilon}_{\mathrm{i}}\Big|^2\Big\}. \end{aligned} \tag{6.2.99}$$

A qualitative evaluation of equation (6.2.99) permits order-of-magnitude estimates of intensities for various processes. Brillouin scattering from longitudinal acoustic modes, $\hat{k}\cdot\hat{e}_{ka} = 1$, proceeds with a relatively strong integrated intensity:

$$S_{1,LA} \sim \frac{nN}{V}\left(\frac{\omega_i}{c}\right)^4 \frac{\pi k_B T}{M s_{ka}^2}\alpha^2$$
$$\sim 3 \times 10^{-9}\text{cm}^{-1}, \tag{6.2.100}$$

for $T \sim 1$ K. Light is scattered directly from density fluctuations without need for interatomic electronic interactions. Brillouin scattering from transverse acoustic modes, $\hat{k}\cdot\hat{e}_{ka} = 0$, is still possible [20] but requires electric dipole interactions and is much weaker:

$$S_{1,TA} \sim \frac{nN}{V}\left(\frac{\omega_i}{c}\right)^4 \frac{\pi k_B T}{M s_{ka}^2}\left(\frac{\alpha^2}{a^3}\right)^2$$
$$\sim 10^{-11}\text{cm}^{-1}, \tag{6.2.101}$$

for the lowest possible transverse-sound velocity in bcc helium. Raman scattering from the optical modes requires dipole interaction between sublattices, and can be shown to be nonvanishing in the hcp structure only for the low-lying, doubly degenerate modes polarized in the basal plane. The integrated intensity for these modes is

$$S_{1,TO} \sim \frac{nN}{V}\left(\frac{\omega_i}{c}\right)^4 \frac{\hbar}{M\omega_0 a^2}\left(\frac{\alpha^2}{a^3}\right)^2$$
$$\sim 3 \times 10^{-12}\text{cm}^{-1}. \tag{6.2.102}$$

The intensity does not benefit from thermal enhancement, as does Brillouin scattering, but the scattering is strengthened relative to Brillouin by being accomplished in a distance of order of the unit cell dimension rather than the long wavelength of light.

In addition to scattering with a single phonon emitted, scattering can also take place with two or more phonons involved. Multiphonon scattering produces a continuous, broad spectrum rather than a single sharp spectral line; thus, rather less detailed information is available in principle. If two-phonon scattering dominated the multiphonon process, then it might be possible to learn the frequency of critical points in the joint density of states, particularly the highest phonon frequency corresponding to the upper cutoff of the density of states. Two-phonon spectra containing this structure have been computed [21] for both hcp and bcc helium. Since experimental investigations have failed to disclose any structure in the multiphonon spectrum whatever, and further indicate that many-phonon emission is substantial, we do not review the two-phonon calculations here. Although many-phonon scattering could be summed exactly in the asymptotic high-frequency limit for the case of neutron scattering, the light-scattering matrix elements are sufficiently more complicated that the high frequency behavior of the theoretical model has not yet been successfully extracted.

6.2.4. Experiment

Several kinds of experiments have been carried out on solid helium, which in varying degrees shed light on the phonon spectrum of the ideal crystal. Thermodynamic measurements include specific heat and compressibility, which measure average properties of the low-frequency phonons. Ultrasonic propagation studies yield the elastic constants, which are closely related to the compressibility and to the low-temperature specific heat. Finally, neutron scattering, and to a lesser extent light scattering, yield information about specific phonon modes not necessarily at low frequency.

The specific-heat and elastic-constant data have been previously reviewed [1d] in a very thorough fashion. Quantitative agreement between elastic-constant measurements and theoretical calculations has by now approached the order of 10 %, over a wide pressure range. This degree of agreement requires that the calculations incorporate three-phonon interaction corrections, and also that they use a Jastrow function modified so as to be at least partially consistent with the phonon vibrations. To obtain significantly better accuracy it should also become necessary to use a more realistic interatomic potential than the Lennard-Jones 6–12.

One feature of the specific-heat data that remains quite puzzling is the low temperature behavior of bcc ^{3}He. A succession of experiments [22] with sensitivity at a progressively lower temperature range, the most recent of which is the work [23] of Castles and Adams, has shown a specific heat substantially higher than predicted by lattice dynamical theory. There are also confirmatory indications in the data of Thomlinson [24] from thermal conductivity in the boundary-limited regime. The specific heat is found to vary with temperature as a power lower than T^3. An explanation involving exchange interactions between ^{3}He atoms due to their nuclear spin of $\frac{1}{2}$ seems quite unlikely, because the effect is observed even at pressures for which exchange is extremely small. A complicating feature in dealing with this anomaly is the fact that bcc ^{3}He is elastically very anisotropic [25]. The velocity of propagation for the slow shear mode in the (110) direction lower by a factor of at least 5 than a typical longitudinal velocity. Furthermore, this low mode involves first-order restoring forces so small (first-neighbor central forces exactly cancel out) that three-phonon processes influence it substantially. The three-phonon interaction is able to produce an upward curving dispersion to this branch at quite small wave vectors, in distinction to the behavior usual for weak anharmonicity. The result of this extreme elastic anisotropy and dispersion is the prediction of a specific heat as a function of temperature that qualitatively mimics the observations. Nonetheless, the measured specific heat at the lowest temperatures is still decidedly larger than that predicted from the phonons. The disagreement is sharpened still further by the realization that no approximation in the lattice dynamical theory can be involved. A rigorous theorem [9] to all orders in perturbation theory requires that the lattice specific heat vary as T^3 in the low-temperature limit, and links the *measured* elastic constants directly to the coefficients of T^3. It is the measured elastic constants, when used in this way, which do not yield a specific heat as large as that observed. At such an impasse it is natural to subtract from the measured specific heat those contributions *predicted* from the phonon spectrum and the exchange interactions to determine the dependence on temperature of the remainder, and to seek mechanisms for it. Castles and Adams claim [23] that the "excess" specific heat defined in this way varies as T. Mechanisms such as a possible contribution from vacancies seem at this time unable to account for the magnitude or the linear power law of the excess.

Light scattering from solid helium has been proved [26] to be a feasible technique for investigating aspects of the phonon spectrum, and useful information has been obtained. The Raman-active optical mode in the hcp phases of both ^3He and ^4He has been clearly observed, and the measured frequency agrees to approximately 10% with first-order theoretical calculations. Three-phonon-interaction corrections to the theory can be expected to improve the agreement, whereas the quantitative effect of a better choice of Jastrow factor is difficult to predict in advance. The linewidth of the optical mode is not yet more definitely known than an upper bound of 10% of the frequency. Preliminary computational estimates [27] give an expected linewidth due to three-phonon processes at least an order of magnitude smaller than 10% of the frequency. In addition to one-phonon scattering, these experiments observed [26] a broad, structureless scattering extending to frequencies well beyond twice the maximum phonon frequency. This apparently indicates a substantial contribution from many-phonon scattering, and also a significant role for lifetime broadening. Neither of these effects has yet been analyzed theoretically, and the observed frequency distribution of multiphonon scattering remains unaccounted for. Nevertheless, absolute integrated intensities of one-phonon and multiphonon scattering are order-of-magnitude consistent with the simple point–dipole model estimates that we have quoted above. Brillouin scattering from acoustic modes has not yet been actively pursued by experimenters [28], although intensity estimates indicate a marginal feasibility for linewidth measurements of longitudinal modes and perhaps a velocity measurement of the slowest transverse mode.

Neutron scattering studies of both the bcc and hcp phases of ^4He have been carried out quite extensively. Although the original intention of these investigations was to make direct comparisons with phonon frequency computations, it is now appreciated that there are significant anharmonic interference effects present in the scattering that require additional analysis and computation for proper interpretation.

The earlier neutron-scattering work on ^{4}He was carried out by groups at Brookhaven [29] and independently at Iowa State [30]. Both groups studied the hcp phase, although at widely differing molar volumes. Neither group was able to obtain a complete set of dispersion curves, but there was considerable overlap in phonon branches found, permitting the validity of a simple scaling of phonon frequencies with density to be demonstrated. Combining the data and scaling with density yields a fairly complete picture of neutron scattering from hcp helium. In both of these studies, the differential cross section was taken to be directly proportional to the phonon spectral function, and the peak of the cross section was compared directly with first-order phonon frequency calculations. Such comparisons yielded agreement to within 30% at worst, and rather better for lower-frequency branches. However, the density dependence of the calculated frequencies was in poor agreement with the experiments, a difficulty since shown to be due to a relatively crude treatment of the Jastrow factor. Furthermore, certain scattering peaks, particularly those of higher frequencies, were very much broader than instrumental resolution and quite asymmetric in some instances.

Subsequent improvements in cryogenic as well as neutron focusing techniques has permitted high-resolution examination [17] of the bcc phase. Only phonons in the first two Brillouin zones of reciprocal space had observable intensity due to the large Debye–Waller exponent. Phonon peaks studied in the first zone were relatively well defined, with a symmetric shape and the anticipated intensity; the peak positions gave phonon-dispersion curves in agreement to better than 20% with theoretical calculations. Inclusion of three-phonon interaction contributions to the phonon self-energy has not seemed to significantly improve agreement with observation, however, possibly because of further refinements still needed in the Jastrow function or alternative treatment of short-range correlations.

Scattering peaks in the second zone were quite anomalous, in that they were unusually broad, asymmetric, and intense. They also did not show the reciprocal lattice periodicity relative to corresponding points in the first zone as expected from the conventional weak-anharmonicity analysis of neutron scattering. Similar results were also found [31] in the hcp phase. These surprising features have all been shown [16] to follow as consequences of three-phonon interference between one-phonon and multiphonon scattering, as developed in the previous section. Computation of the cross section incorporating three-phonon interactions in both the one-phonon spectral function and the one-phonon–multiphonon interference has reproduced the observations for the bcc phase quite convincingly, as to both shape and intensity.

The dramatic size of the interference distortions in helium mandates that future theoretical work compute interference-modified cross sections rather than merely phonon spectral functions.

Extension of the neutron-scattering investigations to higher frequency and wave vector has demonstrated [31, 32], for both bcc and hcp structures, that as the one-phonon peak disappears in intensity due to decreasing Debye–Waller factor, the multiphonon background grows into a broad peak distributed in frequency about $Q^2/2M$. These observations are consistent with the asymptotic high-frequency analysis described in the previous section. It has thus been confirmed in each phase of helium, both hcp and bcc crystals as well as the liquid [33], that high-frequency scattering is quite independent of the atomic interactions and the resulting structure of the medium.

6.3. NUCLEAR SPIN PROPERTIES OF SOLID ^{3}He

6.3.1. Introduction

In the foregoing, the lattice dynamics of solid helium has been discussed on the basis of assigning to the system a certain configuration $(p_1, \cdots p_N; l_1, \cdots l_N)$ in which a particle labeled p_i is "near" a lattice site l_i, (we may operationally define "near" to mean that p_i always oscillates in the unit cell associated with the point l_i). We denote by Φ_ν $(\{p;l\})$ the wave functions obtained in this manner, where ν denotes the quantum numbers. The resulting analysis cannot in principle be entirely correct, especially for solid helium, in which the large zero-point motion leads to a wave function Φ_ν $(\{p;l\})$ that is not orthogonal to other wave functions Φ_ν $(\{P^r\, p;l\})$, where P^r is a permutation operator that interchanges the subscripts on the p's in some specified fashion. The result is that if at $t = 0$ the system is in a configuration $\{p;l\}$, it soon develops admixtures with other configurations. It is possible, of course, to develop a theory for the lattice dynamics in which a configuration $\{p;l\}$ is always maintained by imposing orthogonality among different configurations. Such a theory, however, leads to poor results. For example, in the Hartree approximation [21], the energy of solid ^{3}He is 34 kcal/mole. If the Hartree restriction is removed, there is a considerable lowering of the kinetic energy and the resulting energy [21] is about -1 kcal/mole. Thus, it is necessary to allow the overlap of the wave functions for different configurations for a proper lattice dynamical theory. This was already done in the preceding sections. Here we examine its logical consequence—the new properties arising from the possibility of changing configurations—in a word, exchange.

Although the preceding sections ignore exchange, (while allowing overlap of the wave functions) the lattice dynamical theory developed is, in practice, correct to a very high degree of approximation because even in solid helium the time scale in which configuration mixing occurs is very long compared to the typical time scale of the lattice frequencies. For example, the characteristic time scale for admixing two configurations $\{p;l\}$ and $\{P_{ij}p;l\}$, where P_{ij} is a permutation operator acting on two nearest neighbor particles, i and j, is more than 10^{-8} sec, whereas the characteristic time scale for lattice vibrations is $\hbar/k_B\theta_D \sim 10^{-12}$ sec. Time scales for admixing other configurations are usually even longer. The correction to the vibrational frequencies is thus expected to be $0(10^{-4})$.

In solid ^{4}He, in which the particles are all identical, no effects of exchange can, in principle, be observed (see however, discussion of superfluidity in quantum crystals in Section 6.4), since all configurations are degenerate. In solid ^{3}He, on the other hand, the effect of tunneling is observable since not all configurations are degenerate, because of the spin degrees of freedom. This tunneling motion of fermions can be treated by introducing an effective spin Hamiltonian, and gives rise to interesting nuclear magnetic properties. Solid ^{3}He is unique in its nuclear spin properties since the exchange energies are of the order of a millidegree, while the nuclear dipole–dipole interaction is three orders of magnitude smaller.

We now give the plan and limitation of this section. We start by presenting a fairly complete derivation of the Hamiltonian to describe the exchange phenomena in solid ^{3}He. A Heisenberg Hamiltonian is derived, albeit with the exchange frequency parameter replaced by an operator in phonon space. The following subsections are devoted to exhibiting the physical features of exchange in soild ^{3}He and to understanding its two most outstanding features, namely that exchange frequency is antiferromagnetic in sign and that it increases in magnitude rapidly as the density of the solid is decreased. The calculations of the exchange frequency are then reviewed in Section (6.3.3) with the aim in mind of clarifying the physical parameters in which it depends.

The next part, 6.3.4, is devoted to a study of the thermodynamic consequences of exchange. The theoretical predictions for the various thermodynamic properties are based on the simple nearest-neighbor Heisenberg model. These are compared with the experimental results. Special attention is given to the fact that some experimental results reveal that the simple Heisenberg Hamiltonian is inadequate to describe the exchange phenomena is solid ^{3}He. In Section 6.3.5 the various suggestions to improve the Heisenberg model are discussed.

We do not discuss the theory and experiment on the transport properties concerning the spin system in solid ^{3}He. For these we refer the reader to the review article by Guyer, Richardson, and Zane [1c]. Other review articles related to the subject matter of the present section are by Meyer [34], Richardson [35], Landesman [36], Guyer [1a], and by Trickey, Kirk, and Adams [1d] (referred to as TKA hereafter).

6.3.2. Theoretical Description

The theoretical description of exchange in solid ^{3}He may best be attempted following Herring's discussion [37] of electronic exchange between two well-separated atoms. Thouless [38] and, later, McMahan [39] have followed this approach.

The Hamiltonian $\mathscr{H}$ for ^{3}He is the usual sum of the kinetic energy and the potential energy due to the van der Waals interaction between the atoms. If the eigenfunctions of $\mathscr{H}$ are denoted by ψ_ν, the exact solution to the problem is $\Psi = A(\psi_\nu \xi_k)$, where ξ_k denotes a spin function for the N particles and A is the usual antisymmetrizer

$$A = (N!)^{-1} \sum_P (-1)^P P, \tag{6.3.1}$$

where $P = P^s P^r$ is the permutation operator and P^s acts on the spin of labeled particles while P^r acts on their coordinates. Notice that being the exact solution, Ψ is a linear combination of 2^N different $\Phi(\{P^r p,l\})$, where 2^N is the number of distinct ways in which N sites can be occupied by particles of either spin up or spin down. Even if it were possible, such a solution to the problem would be uninteresting since it does not separate the small, though interesting, effects of statistics from the large lattice dynamcial effects, and, thus, obscures them. The theoretical problem posed here is to generate an effective Hamiltonian

$$\mathscr{H}_{\text{eff}} = \mathscr{H}_L + \mathscr{H}_X + \mathscr{H}_{LX} \tag{6.3.2}$$

whose spectrum to a good approximation reproduces that of $\mathscr{H}$. Here $\mathscr{H}_L$ is the Hamiltonian describing the lattice dynamics, $\mathscr{H}_X$ is the spin Hamiltonian and $\mathscr{H}_{LX}$ represents their residual interaction. To carry out the above program, we start with a set of wave functions $\Phi_m(\{P^r p;l\})$ (m denotes the quantum numbers), which satisfy the following requirements:

a. $A[\Phi_m(\{P^r p;l\})\xi_k]$ spans the space $A(\psi_\nu \xi_k)$. This is trivally true since the latter can be constructed from a linear combination of the former and vice versa.
b. The wave functions satisfy the inequalities

$$\langle \Phi_m(\{P^r p;l\}) | \Phi_m(\{P^r p;l\}) \rangle \gg \langle \Phi_m(\{P^r p;l\}) | \Phi_m(\{P^{r\prime} p;l\}) \rangle. \tag{6.3.3}$$

Here the aim is to take advantage of the fact that the ratio of the frequency of tunneling from one configuration to another to the characteristic vibrational frequency is a small parameter so that perturbation theory in the ratio of the two quantities above or some equivalent parameter can be used We can go farther and say that tunneling between nearby configurations is more probable than configurations farther away. Since the closest-spaced configurations are those in which only two nearest neighbor particles are permuted, we require that

$$\langle \Phi_m \mid \Phi_m \rangle \gg \langle \Phi_m \mid P_{ij} \Phi_m \rangle \gg \langle \Phi_m \mid P^{\text{other}} \Phi_m \rangle, \tag{6.3.4}$$

where from now on we will drop the symbol $\{P^r p;l\}$ when a given configuration is implied and $P_{ij}\Phi_\nu \equiv \Phi_\nu(\{P_{ij}P^r p;l\})$. Here, P_{ij} permutes two nearest-neighbor particles labelled i and j.

c. The lattice dynamics is described well.

Notice that these requirements are well met by the wave functions derived in the previous sections on lattice dynamics. This is certainly an adequate starting point for the theory, but it does not ensure good quantitative calculations of the exchange frequency since, as we might expect, the exchange frequency is sensitive to the value of the wave function $\Phi_\nu(\{P^r p;l\})$ in the region where it overlaps another wave function $\Phi_\nu(\{P^{r'} p;l\})$, while the lattice frequencies are relatively less so. We will discuss in detail later the properties that any approximate function Φ_ν must have to give a good quantitative value for the exchange frequency.

We have the following freedom in our choice for Φ_m consistent with the above requrement (a):

$$\langle \Phi_m \mid \Phi_{m'} \rangle = \delta_{mm'}, \tag{6.3.5}$$

and

$$\langle \Phi_m \mid \mathscr{H} \mid \Phi_{m'} \rangle = \bar{E}_m \delta_{mm'}. \tag{6.3.6}$$

These requirements are satisfied by the wave functions of the preceding sections. Note that equation (6.3.5) and (6.3.6) do not necessarily imply that Φ_m are eigenfunctions of $\mathscr{H}$. We will see below that this choice makes Φ_m eignefunctions of the lattice Hamiltonian $\mathscr{H}_L$.

To obtain the eigenvalues of $\mathscr{H}$, we may consider the secular determinant in any basis set spanning the set $A(\psi_\nu \xi_k)$; in particular $A(\Phi_m \xi_k)$. The eigenvalues E are, therefore, given by

$$\det[\langle A(\Phi_m \xi_k) \mid \mathscr{H} \mid A(\Phi_{m'} \xi_{k'}) \rangle - E \langle A(\psi_m \xi_k) \mid A(\Phi_{m'} \xi_{k'}) \rangle] = 0. \tag{6.3.7}$$

Since the Hamiltonian commutes with A, we may rewrite (6.3.7) as

$$\det[\langle \Phi_m \xi_k \mid \mathscr{H} A \mid \Phi_{m'} \xi_{k'} \rangle - E \langle \Phi_m \xi_k \mid A \mid \Phi_{m'} \xi_{k'} \rangle] = 0. \tag{6.3.8}$$

Now we seek a representation in which the second term inside the determinant is diagonal. Accordingly, we find a matrix U such that

$$U\langle \Phi\xi | A | \Phi\xi \rangle U^{+} = (N!)^{-1} I, \tag{6.3.9}$$

i.e.,

$$U^{+}U = [N!\langle \Phi\xi | A | \Phi\xi \rangle]^{-1}.$$

From the requirements (6.3.4) that we have put on Φ_m and from the normalization condition (6.3.5), it follows that

$$N!\langle \Phi_m \xi_k | A | \Phi_{m'} \xi_{k'} \rangle \approx 1 - \sum_{i<j} \langle \Phi_m \xi_k | P_{ij} | \Phi_{m'} \xi_{k'} \rangle$$
$$+ \text{ terms involving higher permutations.} \tag{6.3.10}$$

Restricting to the first correction due to permutations, we therefore have

$$U = U^{+} = 1 + \tfrac{1}{2} \sum_{i<j} \langle \Phi\xi | P_{ij} | \Phi\xi \rangle. \tag{6.3.11}$$

Also, from our choice (6.3.6) of Φ_m we have

$$N!\langle \Phi_m \xi_k | HA | \Phi_{m'} \xi_{k'} \rangle = E_m \delta_{mm'} \delta_{kk'} - \sum_{i<j} \langle \Phi_m \xi_k | HP_{ij} | \Phi_{m'} \xi_{k'} \rangle + \cdots \tag{6.3.12}$$

On substituting (6.3.10) and (6.3.12) into (6.3.8) we get

$$\det[\bar{E}_m \delta_{mm'} \delta_{kk'} - \sum_{i<j} \langle \Phi_m | \mathscr{J}_{ij} | \Phi_{m'} \rangle \langle \xi_k | P_{ij}^{(s)} | \xi_{k'} \rangle$$
$$- E\delta_{mm'} \delta_{kk'} + \cdots] = 0, \tag{6.3.13}$$

where

$$\langle \Phi_m | \mathscr{J}_{ij} | \Phi_{m'} \rangle = \langle \Phi_m | (\mathscr{H} - \tfrac{1}{2} \bar{E}_m - \tfrac{1}{2} \bar{E}_{m'}) P_{ij}^{(r)} | \Phi_{m'} \rangle. \tag{6.3.14}$$

We may rewrite (6.3.13) as

$$\det[\langle \Phi_m \xi_k | \mathscr{H}_{\text{eff}} | \Phi_{m'} \xi_{k'} \rangle - E\delta_{mm'} \delta_{kk'}] = 0 \tag{6.3.15}$$

where

$$\mathscr{H}_{\text{eff}} = \mathscr{H}_L + \mathscr{H}_X^{\text{pair}} + \mathscr{H}_{LX}^{\text{pair}}$$
$$+ \text{ (terms involving higher permutations).} \tag{6.3.16}$$

Here,

$$\mathscr{H}_L = \sum_m E_m | \Phi_m \rangle \langle \Phi_m |, \tag{6.3.17}$$

and

$$\mathscr{H}_X^{\text{pair}} + \mathscr{H}_{LX}^{\text{pair}} = -\tfrac{1}{2} \sum_{i<j} \mathscr{J}_{ij} (1 + \boldsymbol{\sigma}_i \cdot \boldsymbol{\sigma}_j), \tag{6.3.18}$$

where we have used $P_{ij}^{(s)} = \frac{1}{2}(1 + \boldsymbol{\sigma}_i \cdot \boldsymbol{\sigma}_j)$, and σ_i the Pauli matrices are given by $\frac{1}{2}\boldsymbol{\sigma}_i = \mathbf{I}_i$, where I_i is the nuclear spin operator for the atom i. Neglecting the unimportant constant term, we have

$$\mathscr{H}_X^{\text{pair}} + \mathscr{H}_{LX}^{\text{pair}} = -2 \sum_{i<j} \mathscr{J}_{ij} \mathbf{I}_i \cdot \mathbf{I}_j. \tag{6.3.19}$$

The exchange frequency J_{ij} is given by the thermal average of $\mathscr{J}_{ij}$

$$J_{ij} = Tr(\rho \mathscr{J}_{ij}) \tag{6.3.20}$$

where ρ is the density matrix for the eigenstates m. Thus the exchange frequency involves only the diagonal elements of the "exchange operator" $\mathscr{J}_{ij}$. We may identify $\mathscr{H}_X$ with the diagonal elements of $\mathscr{J}_{ij}$ and $\mathscr{H}_{LX}$ with the off-diagonal elements of $\mathscr{J}_{ij}$, since the latter lead to absorption or emission of lattice vibrations accompanied by an alteration in spin configuration.

From (6.3.20), it follows that in general the exchange frequency is temperature dependent. However, the first temperature-dependent correction goes as [38, 40] $(T/\theta_D{}^4)$ and may be ignored. Therefore, to a very high degree of approximation

$$J_{ij} = \langle 0 | \mathscr{J}_{ij} | 0 \rangle. \tag{6.3.21}$$

With this result, we may write $\mathscr{H}_X^{\text{pair}}$ in the customary Heisenberg form

$$\mathscr{H}_X^{\text{pair}} = -2 \sum_{i<j} J_{ij} \mathbf{I}_i \cdot \mathbf{I}_j. \tag{6.3.22}$$

On expanding the antisymmetrization operator in terms of higher-order permutations, one can generate effective Hamiltonians for simultaneous exchange of several particles. For example, the Hamiltonian describing exchange of three particles is given by

$$\mathscr{H}_X^{\text{triple}} = 2 \sum_{i<j<k} J_{ijk} (\mathbf{I}_i \cdot \mathbf{I}_j + \mathbf{I}_j \cdot \mathbf{I}_k + \mathbf{I}_k \cdot \mathbf{I}_i) \tag{6.3.23}$$

where an expression for $\mathscr{J}_{ijk}$ may be obtained [41] by methods similar to those used here for $\mathscr{J}_{ij}$.

Almost all calculations of the thermodynamic properties have been performed with the simple Heisenberg form (6.3.20) with nearest-neighbor exchange only. As we shall discuss in Section 6.3.5, the functional form of the various thermodynamic properties do seem to fit the predictions following from such a Hamiltonian, although not with the same nearest-neighbor pair exchange parameter J. This indicates the indaequency of such an approximation to describe exchange in solid ^{3}He.

(*a*) *Reduction to a Two-Body Problem and Variation of Exchange Frequency with Density.* Most of the calculations of the exchange frequency in solid ^{3}He have been done by isolating two nearest-neighbor atoms and considering their motion in the average potential field of the rest. The effective Hamiltonian for the two particles is of the form

$$\mathscr{H}_{12} = t_1 + t_2 + v(12) + U(1) + U(2), \tag{6.3.24}$$

where t_1 and t_2 are the kinetic energy operators, $U(1)$, $U(2)$ are the average potential felt by atoms (1) and (2), respectively, due to particles other than (1) and (2), and v(12) is the interatomic potential. The problem now is formally identical to that of exchange of electrons in a hydrogen molecule that gives rise to covalent bonding in that system.

The exchange frequency of the pair is now given from (5.3.14) by

$$J_{12} = \langle\varphi_{12}|(\mathscr{H}_{12} - \bar{\varepsilon})P_{12}^{(r)}|\varphi_{12}\rangle \tag{6.3.25}$$

with

$$\bar{\varepsilon} = \langle\varphi_{12}|\mathscr{H}_{12}|\varphi_{12}\rangle.$$

We notice that φ_{12} is not an eigenfunction of $\mathscr{H}_{12}$, otherwise (6.3.25) would be zero. Consistent with this, the correct φ_{12} is an eigenfunction of $\mathscr{H}_{12}$ if the overlap between the two configurations under consideration is zero.

One of the most interesting aspects of exchange in solid ^{3}He is the experimentally established fact [1d] that the exchange frequency between nearest-neighbor pairs increases approximately in an exponential fashion as the equilibrium distance between them is increased. Here, we wish to show that such a behavior can be easily deduced from (6.3.25). The most strongly density-dependent factors in both terms of (6.3.25) arise from the overlap of φ_{12} and $P_{12}\varphi_{12}$. Therefore, to deduce the density dependence we may write

$$J_{12} \sim \langle\varphi_{12}|P_{12}\varphi_{12}\rangle. \tag{6.3.26}$$

To estimate J_{12}, we approximate φ_{12} by a product of Gaussians centered at lattice sites to take into account the phonon oscillations and a short-range correlation function $f(r)$ corresponding to the hard-core repulsion in $v(12)$,

$$\varphi(12) = \left(\frac{A}{\pi}\right)^{3/2}\exp\left(-\frac{A}{2}\left[(r_1 - R_1)^2 + (r_2 - R_2)^2\right]\right) \times f(r_{12}). \tag{6.3.27}$$

Substituting this expression in to (6.3.26) and carrying out the angular integration one finds

$$J_{12} \sim 4\pi\left(\frac{A}{2\pi}\right)^{3/2}\exp\left(\frac{-AR^2}{2}\right)\int_0^\infty dr\, r^2 f^2(r)\exp\left(\frac{-Ar^2}{2}\right). \tag{6.3.28}$$

Now $f^2(r) \to 0$ for $r \to 0$ and is a very rapidly varying function for $r \lesssim \sigma$ and is very slowly varying and of O(1) for $r \gtrsim \sigma$. We may, therefore, write

$$J_{12} \sim 4\pi\left(\frac{A}{2\pi}\right)^{3/2}\exp\left(\frac{-AR^2}{2}\right)\int_\sigma^\infty dr\, r^2\exp\left(\frac{-Ar^2}{2}\right). \tag{6.3.29}$$

We also know that $A \sim \theta_D^2$. Therefore, since θ_D is smaller for larger nearest-neighbor distance, A decreases with increasing R. From numerical calculations, it is in fact known [21], that AR^2 is roughly a constant. Therefore the factor $\exp(-AR^2/2)$ (whose analogue is crucial in giving decreasing exchange for increasing internuclear distance in the H_2 problem) is quite unimportant and the interesting density dependence is given by the integral in (6.3.2). Clearly a decrease in A (due to increase in R) gives a sharply increasing J_{12} (which for small variations in R can be fitted to an exponential). Furthermore, a small σ (representing a softer hard core) also increases J_{12}. Physically, the short-range part of the wave function is determined primarily by the interatomic potential and relatively less so by the average density. In the calculations [40, 42–49] that have been performed, however, a form for the short-range part is assumed and the A parameter or its analogue depends on the specific form assumed as well as on the average density. From (6.3.29) it is clear that the magnitude of the exchange frequency depends on the *available* volume in which the two particles move. For a given hard-core volume, this is larger for lower densities, and correspondingly the exchange frequency is higher.

(*b*) *Surface Integral Representation of J, and the Sign of J.* We will now sketch a physically appealing transformation of the expression for the exchange energy to the form of an integral over the surface separating the configurations $(Pp;l)$ and $(P'p;l)$. This transformation was first introduced by Herring [37] in the electronic problem and extended to the helium problem by Thouless [38]. The resulting expression has been used by McMahan [39] to do extensive numerical calculations.

The surface integral transformation depends on the assumption that $\Phi(Pp;l)$ is the eigenfunction of a Hamiltonian $\mathscr{H}_T$, (called the truncated Hamiltonian by Herring), which is identical to the original Hamiltonian $\mathscr{H}$ except for a fictitious potential that plays only the function of confining the particles to a given configuration $(Pp;l)$. Thus if $\sum_{pp'}$ is the surface separating two configurations $(Pp;l)$ and $(P'p;l)$, defined so that its outward normal points form $(Pp;l)$ to $(P'p;l)$:

$$(\mathscr{H} - E_m)\Phi(Pp;l) = 0 \quad \text{inside } \textstyle\sum_{p'p},$$

and

$$(\mathscr{H} - E_m)\Phi(P'p;l) = 0 \quad \text{outside } \textstyle\sum_{p'p}. \tag{6.3.30}$$

We can now rewrite, for example, the expression for the diagonal elements of the pair exchange operator $\mathscr{J}_{ij}$ given by (6.3.14) into a surface integral form by combining it with (6.3.30) and integrating by parts (for details, see Herring [37] or McMahan [39] to get

$$\langle 0|\mathscr{J}_{ij}|0\rangle = \frac{-\hbar^2}{m}\int_{\Sigma_{pp'}} ds[\Phi_0^* \nabla_{3N}(P_{ij}^{(r)}\Phi_0)] \tag{6.3.31}$$

where ds is the surface integral element.

For the two-body problem, the surface Σ is given by

$$(\mathbf{r}_1 - \mathbf{R}_1)^2 + (\mathbf{r}_2 + \mathbf{R}_2)^2 = (\mathbf{r}_1 - \mathbf{R}_2)^2 + (\mathbf{r}_2 - \mathbf{R}_1)^2, \tag{6.3.32}$$

and

$$J_{12} = \frac{-\hbar^2}{m} \int_\Sigma d\mathbf{s}(\Phi_{12}^* \nabla_6 P_{12} \Phi_{12}). \tag{6.3.33}$$

As has been mentioned before, $|J_{12}|$ measures the rate at which two labeled atoms trade their lattice sites, or equivalently is a measure of the tunneling probability between configurations. Such a tunneling probability can be written in terms of a current crossing the surface between the two configurations. The surface integral representation [(6.3.31), (6.3.33)] does precisely that. On the five dimensional surface, the integral will be large for r_1 and r_2 as close as possible to R_1 and R_2 or to R_2 and R_1 so that $\varphi(\Sigma_{12})$ and $P\varphi(\Sigma_{12})$ are large, subject to r_1 and r_2 not being too close. The latter restriction is imposed by the short-range repulsion between the two atoms. Thus the potential valley most favorable for tunneling has a ringlike projection in the five-dimensional surface. One conclusion that immediately follows from this is that transverse phonons are more important for exchange than longitudinal phonons.

The most important result that follows from (6.3.31) and (6.3.33) is that pair exchange leads to a negative value for J, i.e., to antiferromagnetic exchange interactions. The ground-state wave functions are nodeless and, therefore, so are $\Phi(r_1,r_2)$ and $P_{12}\Phi(r_1,r_2)$. Also, $\nabla\Phi(r_1,r_2)$ must be negative in the hyperplane. Thus the integral in (6.3.31) and (6.3.33) is negative. Extension of the surface integral representation for multiparticle exchange can be made similarly to show that cyclic exchange of even numbers of particles (in helium or for electrons in insulators) always leads to negative value of the exchange integral, and cyclic exchange of an odd number of particles always leads to a positive value. The latter result follows from the fact that, for example, in a cyclic exchange of three particles, the surface integral representation involves a product of two integrals of the form (6.3.31).

6.3.3 Calculations of the Exchange Frequency

There have been several different calculations [40, 42–49] of the exchange-frequency parameter J_{ij} for nearest-neighbor atoms. McMahan [39] has recently reviewed these. He finds that despite appearances and sometimes claims [1a] to the contrary, all calculations are for the same *expression* for the exchange frequency. Similar conclusions have been reached by Brandow [47] and by Ostgaard [49]. The variation of over an order of magnitude in the calculated *magnitude* of the exchange frequency is solely the difference in the

φ_ν used. All the calculations (except one by McMahan [49]) effectively reduce the problem to that of two particles moving in the field of the rest, much as was done in Section 6.2.2. The φ_ν were obtained from a prior treatment of the lattice dynamics and were effectively of a Gaussian form in order to describe the phonons corrected by short-range correlation functions. It would be pointless to exhibit here the different expressions for the exchange frequency that were obtained, especially in view of McMahan's work [39]. Some of the more recent calculations in this two-body approximation have yielded surprisingly good quantitative agreement with experiment, while the earlier calculations were over an order of magnitude in discrepancy with experiment. But a calculation of McMahan [49] including the correlations effects of several near neighbors in the pair exchange frequency J_{ij} has shown that the pair approximation is quite incorrect—the effect of a near neighbors affects J_{ij} by over an order of magnitude. In view of this, we discuss here only the effect of various parameters in the calculation of exchange frequency, in order to help understand the exchange process itself.

Nosanow and co-workers [40, 42–44] have carried on such a program by

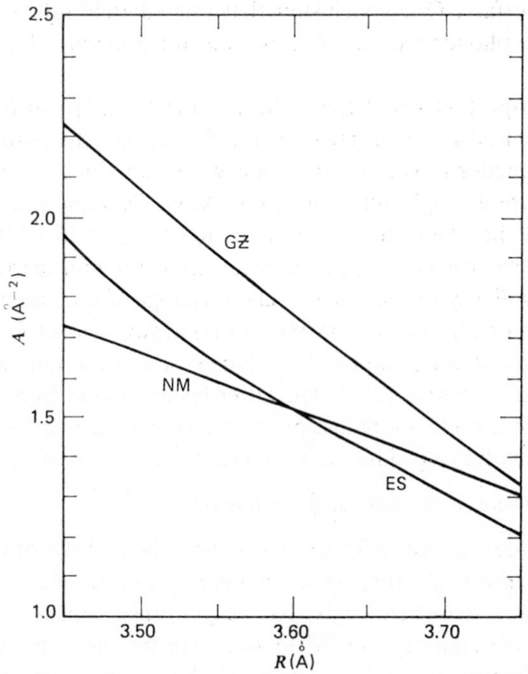

FIGURE 6.2. Gaussian constant A as a function of the nearest-neighbor distance R used in calculation of the exchange frequency given in Figure 6.3. NM, Reference [42]; GZ, Reference [45]; ES, Reference [46]. After McMahan, Reference [39].

using various different approximations for φ_ν to calculate the exchange frequency. Nosanow and Mullin [42] used an Einstein approximation for the phonons modified by a short-range correlation function

$$\varphi_\nu = \varphi = \prod_i \exp\left(-\frac{A}{2}(r_i - R_i)\right)^2 \prod_{i<j} f(r_{ij}), \tag{6.3.34}$$

with

$$f(r) = \exp\left[-K\left\{\left(\frac{\sigma}{r}\right)^{12} - \left(\frac{\sigma}{r}\right)^{6}\right\}\right]. \tag{6.3.35}$$

The parameters A and K were determined variationally to minimize the lattice energy. It is more instructive to calculate the exchange frequency by equation (6.3.31) using wave functions of the form (6.3.34) with value of A and $f(r)$, used by the various workers. These are presented in Figures 6.2 and 6.3 and the J calculated with these in Figure 6.4. The experimental results [50–54] are given in Figure 6.5 for comparison. Clearly, the magnitude of J is larger, the smaller the A parameter, and its density dependence is

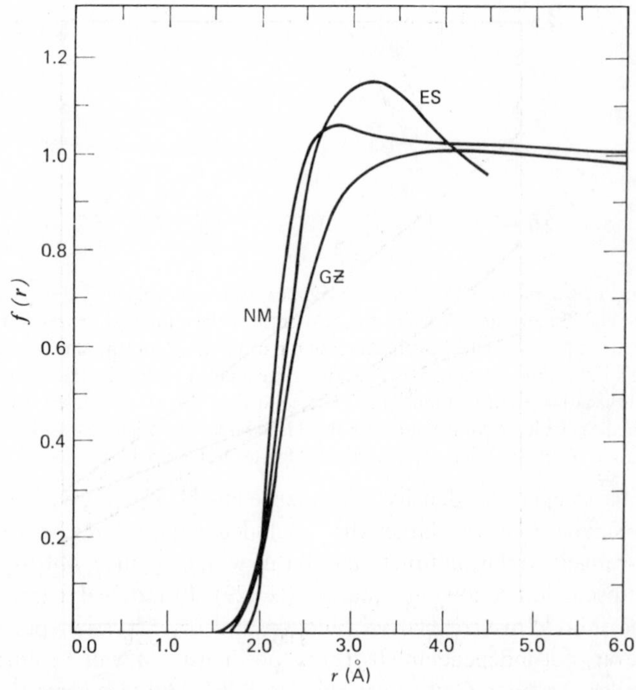

FIGURE 6.3. Short-range correlation function $f(r)$ used in calculation of the exchange frequency given in Figure 6.4 NM, Reference [42]; GZ, Reference [45]; ES, Reference [46]. After McMahan, Reference [39].

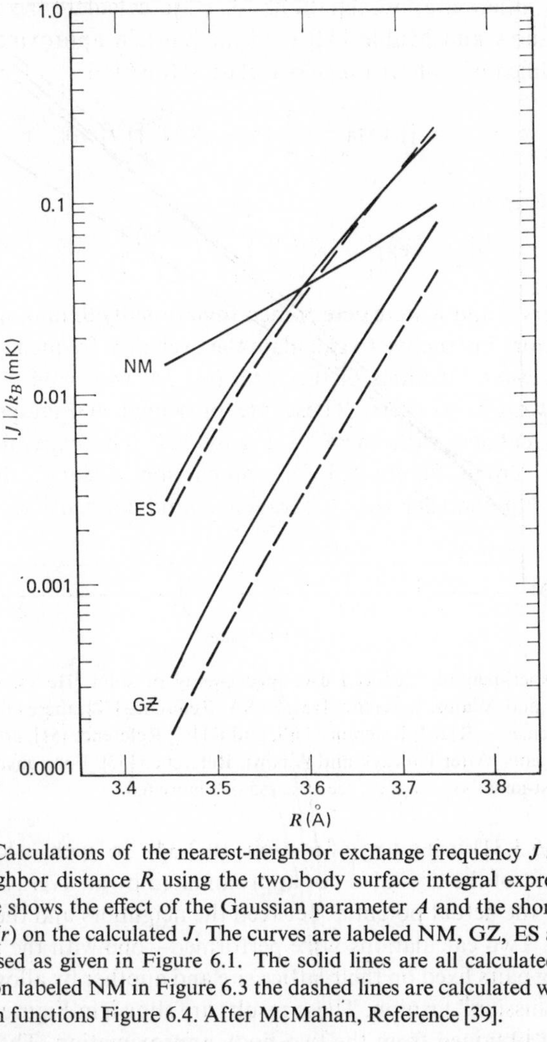

FIGURE 6.4. Calculations of the nearest-neighbor exchange frequency J as a function of the nearest-neighbor distance R using the two-body surface integral expression, equation (6.3.33). Figure shows the effect of the Gaussian parameter A and the short-range correlation function $f(r)$ on the calculated J. The curves are labeled NM, GZ, ES according to the parameter A used as given in Figure 6.1. The solid lines are all calculated using the correlation function labeled NM in Figure 6.3 the dashed lines are calculated with the worker's own correlation functions Figure 6.4. After McMahan, Reference [39].

steeper, the steeper the density dependence of A. For a given A, the exchange is smaller [55], the larger the "excluded volume" region defined by the short-range correlation function. All this was, of course, already obvious from the discussion following equation (6.3.29). In lattice dynamical calculations the short-range correlation function and the Gaussian parameter A are, however, not independent. If $f(r)$ is too "hard," A will be correspondingly smaller (broader Gaussians, smaller K.E.) and vice versa.

In order to estimate the effect of near neighbors in the exchange of two given atoms, McMahan has calculated [49] J from the surface integral re-

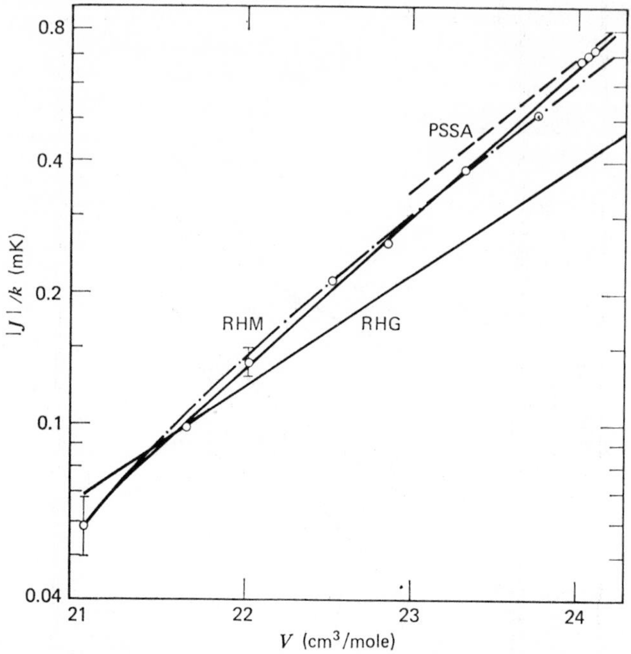

FIGURE 6.5. Experimentally deduced exchange energy of solid ^{3}He vs. molar volume. Circles, Panczyk and Adams, Reference [53]; PSSA, Reference [52]; these are deduced from pressure measurements. RHM, Reference [50], and RHG, Reference [51], are deduced from NMR measurements. After Panczyk and Adams, Reference [53]. For conversion of molar volume to nearest-neighbor distance, see abscissa of Figure 6.6.

presentation (6.3.31) using wave functions (6.3.34). In such calculations the effect of the rest of the crystal is not approximated by an average field acting on the atoms; the actual potential between the neighbors and the exchanging atoms is used. Two calculations were performed—one with the neighbors of the exchanging pairs fixed on their lattice site and another by allowing them to move in a "Gaussian" fashion. These results are shown in Figure 6.6 together with the result obtained from the two-body approximation. The presence of near neighbors produces an extra "excluded volume" effect, the region in which the two atoms can exchange is now more clarly the concentric shell defined by their own hard core from the inside and that of their neighbors from the outside. The calculated exchange frequency is reduced, but the density dependence is improved mainly because at higher densities the hard core is felt more strongly. When the neighbors are allowed to move, some averaging of the hard core occurs; the outer part of the shell gets smeared and the exchange frequency is correspondingly higher.

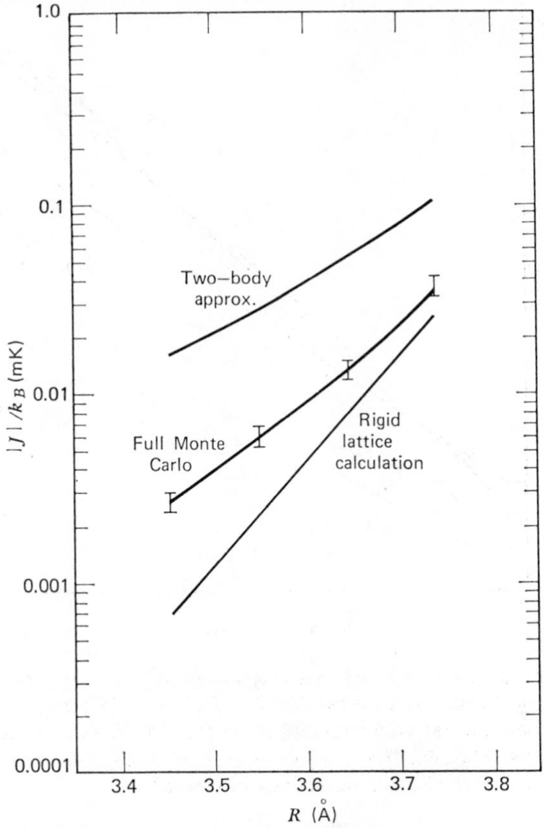

FIGURE 6.6. Comparison of Monte Carlo calculations of the nearest-neighbor exchange frequency J using the many-body expression for J given by equation (6.3.31) with that given by the two-body approximation. All calculations are with the NM wave functions given in Figures 6.1 and 6.2. After McMahan, Reference [49].

Nosanow and Varma [40] have derived an expression for the exchange operator and calculated the exchange frequency using a Debye approximation for the phonon wave functions (as opposed to an Einstein approxoimation used by all other workers) and for $f(r)$ again the form (6.3.35). Their results are compared with the Nosanow–Mullin results [42] in Figure 6.7. Better results are primarily due to smaller and more density-dependent effective Gaussian parameters. What is more interesting is that when the anisotropy in the phonon spectrum is taken into account by introducing the difference in the transverse and longitudinal sound velocities, the calculated exchange frequency is higher. As explained earlier, exchange is easier when a pair of particles is moving in a transverse phonon mode rather than a

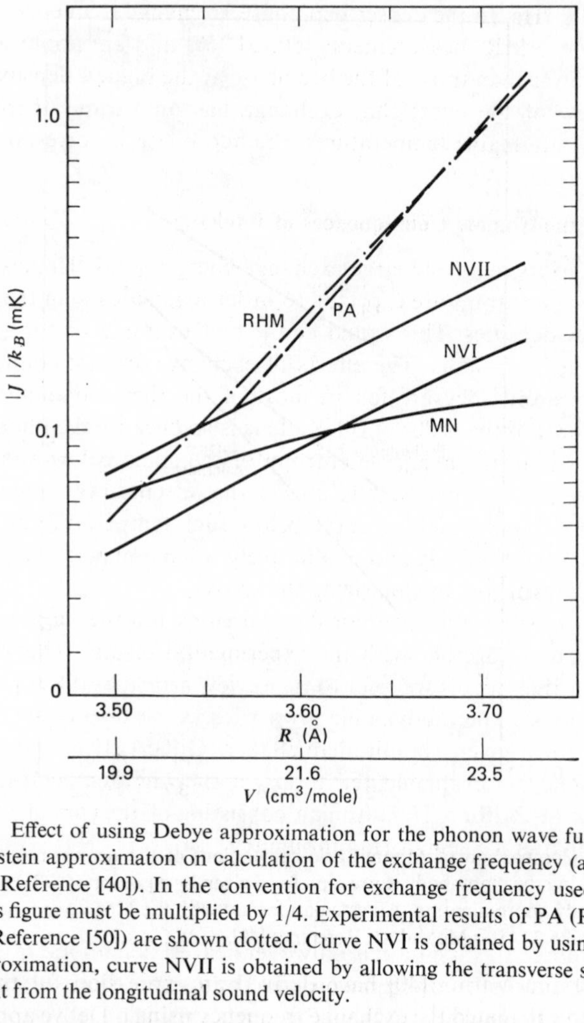

FIGURE 6.7. Effect of using Debye approximation for the phonon wave functions rather than the Einstein approximaton on calculation of the exchange frequency (after Nosanow and Varma, Reference [40]). In the convention for exchange frequency used here, all the results in this figure must be multiplied by 1/4. Experimental results of PA (Reference [53]) and RHM (Reference [50]) are shown dotted. Curve NVI is obtained by using an isotropic phonon approximation, curve NVII is obtained by allowing the transverse sound velocity to be different from the longitudinal sound velocity.

longitudinal phonon mode. In the latter there is a head-on hard-core collision. Transverse phonons are softer for the same reason, and this fact in turn again facilitates exchange.

One of the primary difficulties with all the calculations mentioned above is the Gaussian decay of the wave functions. It seems physically more reasonable that the true wave functions are Gaussian-like near their maximum but decay exponentially. Such wave functions might lead to a considerable enhancement of the calculated exchange frequency.

We have presented experimental results and calculations for only the bcc phase-of-solid ^{3}He. In the denser hcp phase, exchange frequencies have been deduced from NMR measurements [50, 51, 54] and are about an order of magnitude lower than those of the bcc phase at the highest density. From the point of view of the effect that exchange has on various thermodynamic properties at attainable temperatures, the hcp phase is correspondingly less interesting.

6.3.4. Thermodynamic Consequences of Exchange

By virtue of its relatively large exchange energy, solid ^{3}He has the unique property that the atoms are expected to order magnetically in the millidegree region at low densities. This would be the first example of the spontaneous ordering of nuclear spins. The effects of exchange are not confined to such low temperatures. They are felt in most of the thermodynamic properties at temperatures of the order of 0.1 K. This is so because the phonon entropy becomes less than the available entropy of the spins below about 0.1 K at low densities. The exchange thus affects the specific heat, compressibility, and most notably the melting curve below such temperatures. These effects change both quantitatively and qualitatively when a magnetic field is applied due to the entropy lost by polarizing the spins.

We discuss below the theoretical predictions for the various thermodynamic properties together with the experimental results. This discussion is brief. For further details we refer to the review article by TKA [1d].

To evaluate the thermodynamic properties we need to know the partition function Z for a given Hamiltonian. Baker, Gilbert, Eve, and Rushbrooke [56] (BGER) have evaluated ten terms in the high-temperature series expansion for ln Z, for a Hamiltonian consisting of the sum of a Heisenberg Hamiltonian and a Zeeman Hamiltonian:

$$\mathscr{H} = -2J \sum_{i<j}' I_i \cdot I_j - \sum_i \mu I_i \cdot H, \tag{6.3.36}$$

where the sum in the first term is only over nearest neighbors, H is the magnetic field and μ is the ^{3}He Bohr magneton. The various thermodynamic quantities have then been evaluated from this series expansion.

(*a*) *Susceptibility*. The susceptibility above the Neel point is given by

$$\chi = \frac{\partial M}{\partial H} = \left[\left(\frac{k_B T}{V}\right)\left(\frac{\partial^2}{\partial H^2}\right)\ln Z\right]J,T. \tag{6.3.37}$$

From BGER, one finds

$$\chi = \left(\frac{C}{T}\right)\left[1 + 4\left(\frac{J}{kT}\right) + 12\left(\frac{J}{kT}\right)^2 + \cdots\right], \tag{6.3.38}$$

where $C = N\mu^2/Vk_B$ is the Curie constant. To order (J/kT), the susceptibility is of the Curie–Weiss form

$$\chi = \frac{C}{(T-\theta)}, \tag{6.3.39}$$

where $\theta = (4J/k_B)$.

Efforts to correctly measure the susceptibility were frustrated for a decade due chiefly to the long relaxation times to attain thermal equilibrium at low termperatures. Then in 1969, three groups [57–59] in the same issue of *Phys. Rev. Lett.* reported measurements of χ to low enough temperatures to deduce a value for $|J|$ consistent with that deduced by other means, and also for the first time conclusively established that J is negative. In Figure 6.8 we show the results of Kirk et al. [57] that are the most extensive. These measurements are of interest also because they show that ^{4}He impurities do not have a perceptible effect on the susceptibility, thus ending considerable confusion on this point.

(*b*) *Transition to Antiferromagnetism.* To obtain the Neel temperature, BGER[56] have located the singularities of χ from first ten terms of the high temperature series expansion by the method of Pade approximants. For the bcc lattice they find a transition temperature

$$T_N = \frac{-\,2.748\,J}{k_B} = -\,0.687\,\theta. \tag{6.3.40}$$

With the precise measurements of $|J|$ by Panczyk and Adams [53], for a molar volume of 24.1 cm^3/mole (on the melting line), $T_N = 2.0$ mK. Recent

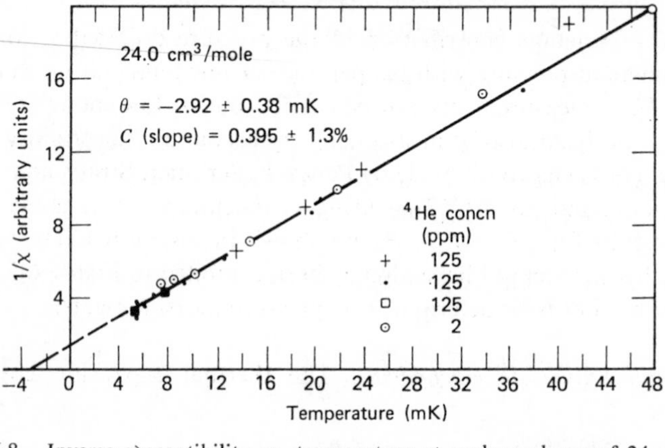

FIGURE 6.8. Inverse susceptibility vs. temperature at molar volume of 24 cm^3, for two different concentrations of ^{4}He impurities (after Kirk et al., Reference [57]). On replotting the data, it appears that the uncertainty in the deduced value of θ may be larger than quoted above (W. Kirk, private communication).

experiments [60] reveal that this prediction is not obeyed and $T_N < 1.3$ mK. We defer the discussion of these experiments until Section 6.3.4(j).

(*c*) *Pressure in Zero Magnetic Field.* The pressure is given in terms of $\ln Z$ by

$$P_V(T,H) = kT\left(\frac{\partial \ln Z}{\partial V}\right)_{T,H} \tag{6.3.41}$$

For $H = 0$ one gets from the BGER expression

$$P_{\text{ex}}(t) = \left(\frac{3R}{V}\right)\left(\frac{J}{k_B}\right)^2 \gamma_{\text{ex}} T^{-1}\left(1 - \frac{J}{kT}\right) + 0\left[\left(\frac{J}{kT}\right)^2\right], \tag{6.3.42}$$

where γ_{ex} is the "exchange Gruneisen parameter"

$$\gamma_{\text{ex}} \equiv \frac{\partial \ln|J|}{\partial \ln V}. \tag{6.3.43}$$

This is the exchange contribution. The phonon contribution is

$$P_{\text{ph}} = \left(\frac{-3\pi^4 R}{5V}\right)\frac{\gamma_{\text{ph}} T^4}{\theta^3} \tag{6.3.44}$$

where

$$\gamma_{\text{ph}} = \frac{\partial \ln\theta_D}{\partial \ln V}.$$

The pressure of the solid is the sum of these temperature-dependent contributions and of the pressure at $T = 0$:

$$P = P_0 + P_{\text{ph}} + P_{\text{ex}}. \tag{6.3.45}$$

At low T, the exchange contribution to the pressure dominates. Measurements of change in pressure with temperature at low temperature thus yield values for $|J|$. If measurements can be carried out to low enough temperatures the sign of J can also be measured. Very accourate (uncertainty $<3\%$) determination of $|J|$ has been made by Panczyk, Scribner, Straty, and Adams [52] and Panczyk and Adams [53] by using a capacitance strain gauge technique. Their results for $\Delta P = P - P_0$ are shown in Figure 6.9. The value of $|J|$ deduced by this method have already been exhibited in Figure 6.5. So far the sign of J has not been determined by pressure measurements.

(*d*) *Thermal Expansion Coefficient.* The thermal expansion coefficient α is given by

$$\alpha = k_T\left(\frac{\partial P}{\partial T}\right)_V, \tag{6.3.46}$$

where k_T is the compressibility. From equation (6.3.42) one finds $\partial P_{\text{ex}}/\partial T$

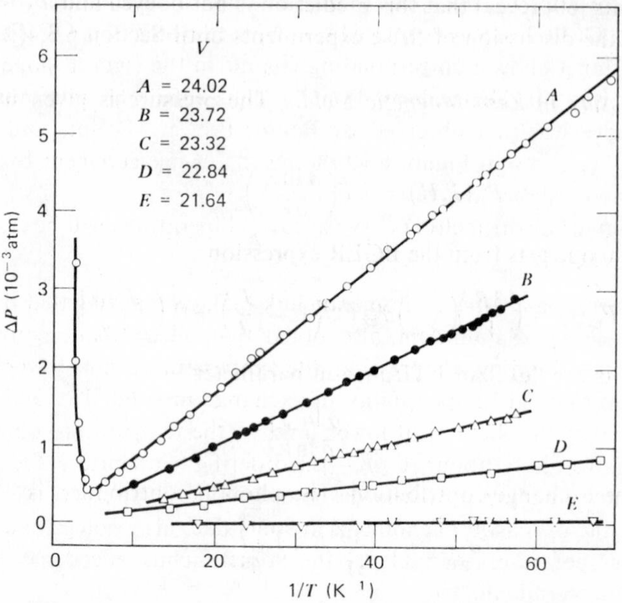

FIGURE 6.9. Change in pressure vs. T^{-1} for several molar volumes. Both the phonon and the exchange contributions are shown for a molar volume of 24.02 cm³. For the rest only the exchange contribution is shown. After Panczyk and Adams, Reference [53].

< 0 because $\gamma_{ex} > 0$, on the other hand $\gamma_{ph} < 0$ and therefore $\partial P_{ph}/\partial T > 0$. The thermal expansion in solid ^{3}He will, therefore, be negative at low T and positive at high T. Goldstein [61] has estimated the crossover to occur at 0.23 K on the melting line. This has been verified by Panczyk and Adams [62] in their measurements.

(e) *Melting Curve.* The pressure and the temperature along the melting line are related by the Clausius Clapeyron relation

$$\left(\frac{dP}{dT}\right)_M = \frac{(S_l - S_s)}{(V_l - V_s)}, \tag{6.3.47}$$

where S_l and S_s are the entropies and V_l and V_s are the molar volumes of the liquid and the solid, respectively. $(V_l - V_s)$ is always positive. Liquid ^{3}He in its Fermi liquid regime has a low entropy,

$$S_l = \gamma(RT), \quad \text{for } T \ll T_F, \tag{6.3.48}$$

where $\gamma \simeq 4.6$ K^{-1} near the melting volume. (This does not include the small contributions due to spin fluctuations or the fact that below about 3 mK, liquid ^{3}He undergoes a transition [60] to a new phase, and has an even lower entropy.) For $T > T_N$, the spins in the solid are disordered and S_R is equal to $R \ln 2 - O(J/kT)^2$. It is estimated that $S_L \lessgtr S_s$ for $T \gtrless 0.3$ K. The melting curve thus must show a corresponding change in the sign of slope at about 0.3 K, as was first realized by Pomeranchuk [65]. The minimum in the melting curve was first observed by Baum, Brewer, Daunt, and Edwards [63]. Here, we show in Figure 6.10 the results of measurement by Scribner, Panczyk, and Adams [64].

The behavior of the melting curve at low temperatures will be discussed in Section 6.3.4.(j).

(*f*) *Pomeranchuk Effect.* Pomeranchuk [65], on realizing that the melting curve of ^{3}He has a minimum, also noted that on adiabatic conversion of liquid ^{3}He into solid by compression, the temperature can be lowered. Pomeranchuk overlooked the possibility of exchange in solid ^{3}He, and therefore concluded that the theoretical lower limit of the temperature obtainable in this fashion would be about 10^{-6}K—the ordering temperature due to nuclear dipole–dipole interaction. Bernardes and Primakoff [66], who were the first to predict the exchange phenomena in solid ^{3}He, also noted that the lower limit on temperature attainable by the Pomeranchuk effect may be several orders of magnitude higher.

Anufriev [67] was the first to cool helium by the Pomeranchuk method. Starting at 50 mK, he reported reaching a temperature of 20 mK; it is now

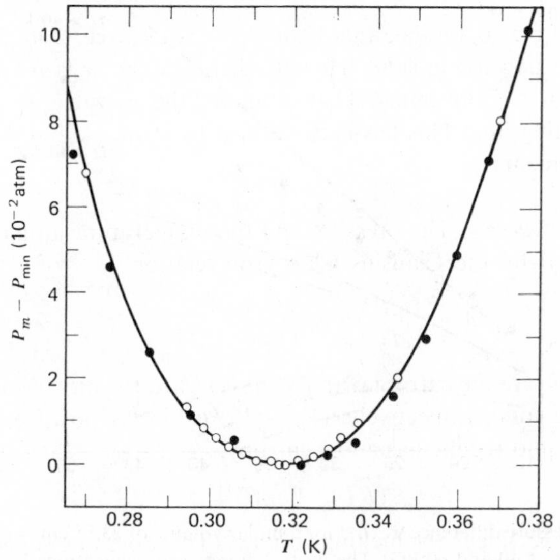

FIGURE 6.10. The melting curve of ^{3}He near the Pomeranchuk minimum. The minimum occurs at $P_{min} = 28.94$ atm, $T_{min} = 0.318$ K. After Scribner et al., Reference [64].

estimated that the actual temperature he reached was probably an order of magnitude lower but he was prevented from realizing this due to troubles with thermometry.

Two groups, at the University of California, La Jolla [68] and at Cornell University [59] have perfected the technique of Pomeranchuk cooling and have reached temperatures below 2 mK. For further details, we refer to the review article [1d] by TKA.

(*g*) *Change in Pressure due to a Magnetic Field.* To leading order in J/kT, the change in pressure on applying a magnetic field H is given by

$$\Delta P(T,H) \equiv P_V(T,H) - P_V(T,0)$$
$$= 2\left(\frac{RT}{V}\right)\left(\frac{\partial \ln|J|}{\partial \ln V}\right)\left(\frac{J}{kT}\right)\left(\frac{\mu H}{kT}\right)^2. \tag{6.3.49}$$

The sign of the deviation $\Delta P(T, H)$ therefore provides the sign of J, i.e., whether the eventual ordering will be antiferromagnetic or ferromagnetic.

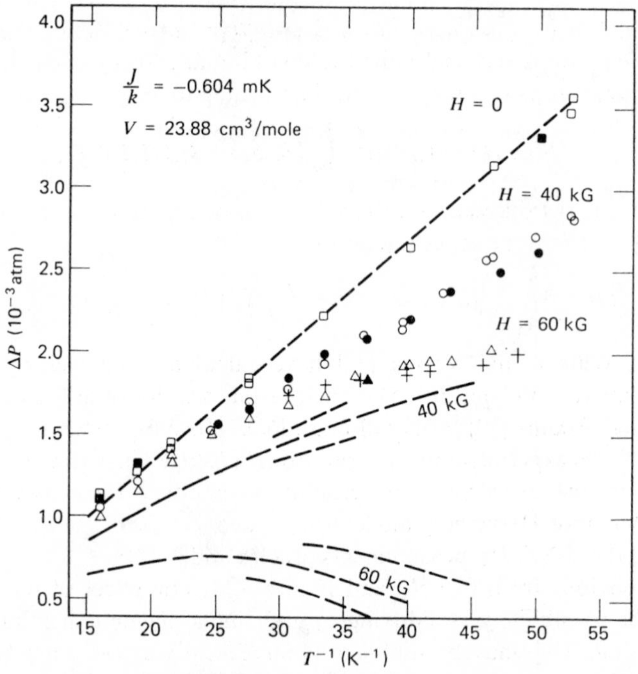

Figure 6.11. Pressure difference vs. T^{-1} for a molar volume of 23.88 cm^3 in zero magnetic field and at fields of 40 and 60 kG. The dashed curves are calculations based on the exchange parameter deduced from the $P(T)$ measurements of Reference [53]. After Kirk and Adams, Reference [70].

After an initial unsuccessful attempt by Osgood and Garber [69], Kirk and Adams [70] measured $\Delta P(T, H)$ to reveal that J had the expected negative sign. The magnitude of the measured $\Delta P(T, H)$, however, is considerably smaller than that calculated using the magnitude of $|J|$ and its density dependence as deduced from the $P_V(T)$ measurements. The results of Kirk and Adams at a molar volume 23.88 cm^3 are shown in Figure 6.11. The dashed curves are the calculated values on the basis of the nearest-neighbor Heisenberg model. These measurements clearly reveal the inadequacy of such a model. On the other hand, the $J_{P(H)}$ parameter which may be used to fit the $\Delta P(T, H)$ results must be consistent with the J_χ parameter deduced from the susceptibility measurements purely on thermodynamic grounds. It appears now [71] that there is enough uncertainty in the J_χ measurements to agree with either $J_{P(H)}$ or $J_{P(T)}$, the exchange parameter deduced from the $P(T)$ measurements. Possible corrections to the nearest-neighbor Heisenberg model to simultaneously account for the $P_V(T)$ and the $\Delta P(T, H)$ measurements will be discussed in Section 6.3.5.

(*h*) *Effect of a Magnetic Feild on the Melting Curve.* Using the Clausius Clapeyron relation and neglecting the small liquid entropy one finds that the melting curve shifts in a magnetic field according to

$$\Delta P(T, H) = (\Delta V_m)^{-1} \int_T^\infty [R \ln 2 - S_{\text{sol}}(H,T)]\, dT, \qquad (6.3.50)$$

where ΔV_m is the difference in the solid and the liquid molar voluem at the melting line. From the series expansion

$$S_{\text{sol}}(H,T) - S_{\text{sol}}(0,T) = -\frac{1}{2}\left(\frac{H}{k_B T}\right)^2 \left[1 + 8\frac{J}{k_B T} + \cdots\right]. \qquad (6.3.51)$$

Walstedt, Walker, and Varma [72] have calculated the melting curve at 74 kG using (6.3.50) and (6.3. 51). Their results are shown in Figure 6.12.

Kirk and Adams [70], Osheroff et al. [60], and more extensively Johnson et al. [73] have experimentally determined $\Delta P(T, H)$. Their results are consistent within the experimetal uncertainty with either those based on the nearest-neighbor Heisenberg model using J deduced from the $P(T)$ measurements, or the $\Delta P(T, H)$ measurements of Kirk and Adams.

As is obvious from (6.3.50) and Figure 6.11, one effect of the magnetic field is the possibility of solidifying by applying a magnetic field rather than compressing. This may be useful in cooling a mixture of solid and liquid helium [72].

(*i*) *Magnetization Cooling.* An external magnetic field has an interesting effect on the entropy of antiferromagnets. Following an examination by

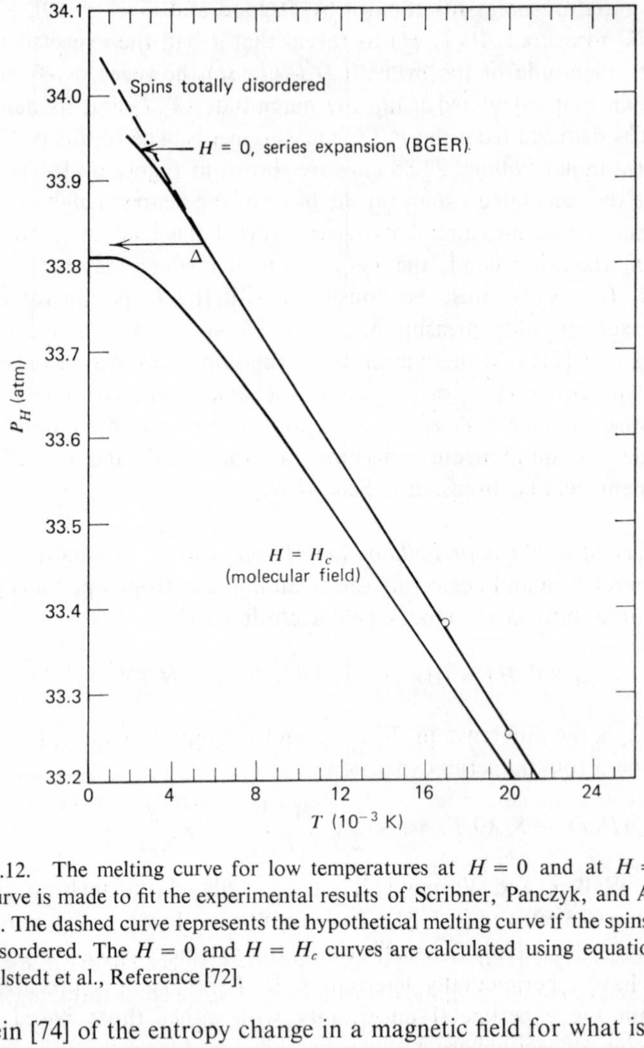

FIGURE 6.12. The melting curve for low temperatures at $H = 0$ and at $H = H_c$. The $H = 0$ curve is made to fit the experimental results of Scribner, Panczyk, and Adams [64] at 20 mK. The dashed curve represents the hypothetical melting curve if the spins remained totally disordered. The $H = 0$ and $H = H_c$ curves are calculated using equation (6.3.50). After Walstedt et al., Reference [72].

Goldstein [74] of the entropy change in a magnetic field for what is in effect an Ising model, Walstedt, Walker, and Varma [72] and Wolf, Thorpe, and Alben [75] calculated the entropy change for an isotropic Heisenberg antiferromagnet model. Such a calculation is meaningful for solid ^{3}He, where the only source of anisotropy is dipolar interactions of magnitude 10^{-6} K which is further reduced in cubic lattices. The result of this calculation is that in the spin-wave approximation the entropy of an isotropic antiferromagnet increases (except at very low fields) with increasing magnetic field H until $H = H_c \simeq 4zIJ/\gamma\hbar$. A calculation of the entropy for $H \simeq H_c$ is presented in Figure 6.13, and the results of an unpublished calculation of Walker and

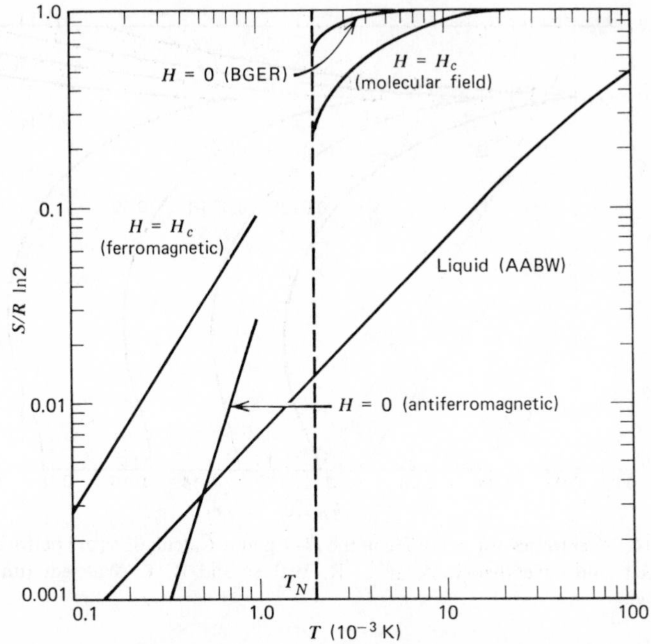

FIGURE 6.13. Calculated entropy vs. temperature for solid ^{3}He at $H = 0$ and at $H = H_c$ together with the measured entropy of the liquid [taken from, W. R. Abel et al., *Phys. Rev.* **147**, 111 (1966)]. For $H = 0$, $T < T_N$, the solid is in an antiferromagnetic state; for $H = H_c$, $T < T_N$, the solid is effectively in a ferromagnetic state; the entropies shown are calculated on the basis of the spin-wave spectrum in each state. For $T \geqq T_N$, the series expansion results of Reference [56] are plotted for $H = 0$ and the molecular field theory results for $H = H_c$. After Walstedt et al., Reference [72].

Walstedt and Varma [76] of the isentropes in the H–T plane is presented in Figure 6.14.

The above results can be understood as follows. In zero field the spin waves in the antiferromagnet have the dispersion $\varepsilon_k \sim k$ for long wavelengths leading to an entropy $S_{\text{anti}} \sim (kT/J)^3$ for $kT \ll J$. An isotropic antiferromagnet has a spin flop transition in zero field. The behavior of spin waves in the spin flop configuration is discussed by Keffer [77]. As the field H is increased, one branch of spin waves is lost to thermal excitations for $\gamma\hbar H > kT$. At a critical field $H_c(T)$, there is a phase transition to a paramagentic state. The mean field picture of the ground state in this phase has all spins aligned in the same direction. The Brillouin zone for spin waves is doubled because of the change in periodicity and the spin waves acquire the dispersion $\varepsilon_k \sim k^2$ characteristic of ferromagnets. The corresponding entropy $S_{\text{ferro}} = 0.107\,(kT/J)^3$ is a much more rapidly increasing function of temperature than S_{anti} for kT/J

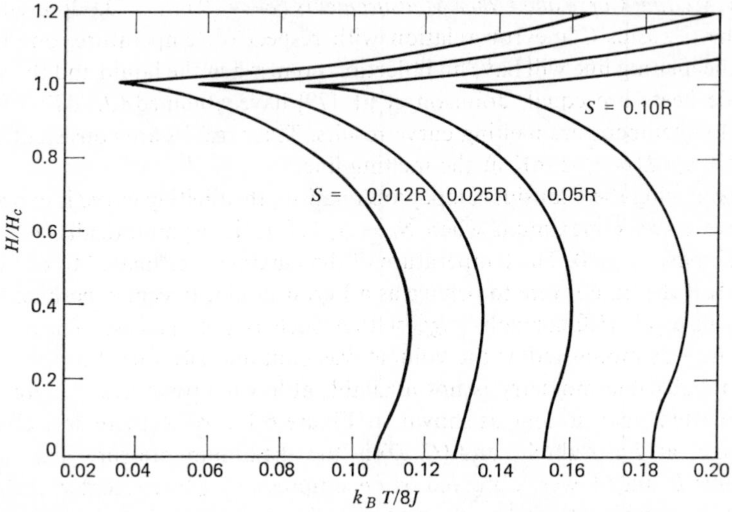

FIGURE 6.14. Isentropes for solid ^{3}He in the H–T plane. Calculations are performed using renormalized spin-wave theory. After L. R. Walker and R. E. Walstedt (unpublished calculations).

$\ll 1$. Detailed calculations (see Figure 6.13) show that $S_{\text{ferro}} > S_{\text{anti}}$ holds over most of the ordered region.

In terms of the above discussion and the results presented in Figures 6.13 and 6.14, the following behavior is envisioned for the temperature $T(H)_S$ on isentropic magnetization beginning at some initial temperature T_i. As H is increased from zero, T first increases by a factor $2^{3/2}$ owing to the loss of one branch of excitations. As $H \to H_c$, the temperature decreases to T_f given by $S_{\text{ferro}}(T_f) = S_{\text{anti}}(T_i)$. This condition yields

$$T_f = 0.64T_i\left(\frac{kT_i}{J}\right). \tag{6.3.52}$$

With a Neel point $T_N \simeq 2$ mK, the critical field H_c is found to be 74 kG, and from the intersection point of $S_{\text{ferro}}(T)$ with that of the liquid entropy (see Figure 6.13), the theoretical lower limit of cooling through this method is found to be about 5.6 μK, i.e., two orders of magnitude lower than ordinary Pomeranchuk cooling. In practice, experimental difficulties connected with high magnetic fields at low temperatures may severely limit the applicability of this method of cooling.

From Figure 6.14, it is observed that the temperature minimum at $H = H_c$ is very sharp. This may [78], however, be an artifact of the spin-wave approximation.

(*j*) *Melting Curve and Other Measurements below 10 mK.* On differentiating the Clausius-Clapeyron relation with respect to temperature, one finds that the melting line will have an inflection point when the liquid and the solid specific heats are equal. Johnson et al. [79] have obtained $(dP/dT)_M$ from their low-temperature melting curve results. Thier results are consistent with 1.7 mK $\lesssim |J|/k \lesssim$ 2 mK at the melting line.

Again using the Clausius-Clapeyron relation, the melting curve is expected to have a weak maximum when $S_l = S_s$ before it asymptotically achieves zero slope as $T \rightarrow 0$. The temperature of the maxima is estimated to be about 0.5 mK if the liquid were to behave as a Fermi liquid. Recent experiments by Osheroff et al. [60] strongly suggest that such is not the case. The melting pressure was monitored as the volume was continuously varied in time, and since reliable thermometry is not available at low temperatures, the pressure was plotted against time as shown in Figure 6.15. At *A* there is a change in slope, at *B* a "glitch" and (*C, D*) is the maximum pressure attainable. Features *B*' and *A*' were observed on decompression. More recently, through NMR measurements, it has been clearly demonstrated that these features are associated with phase changes in liquid ^{3}He, and suggestions have

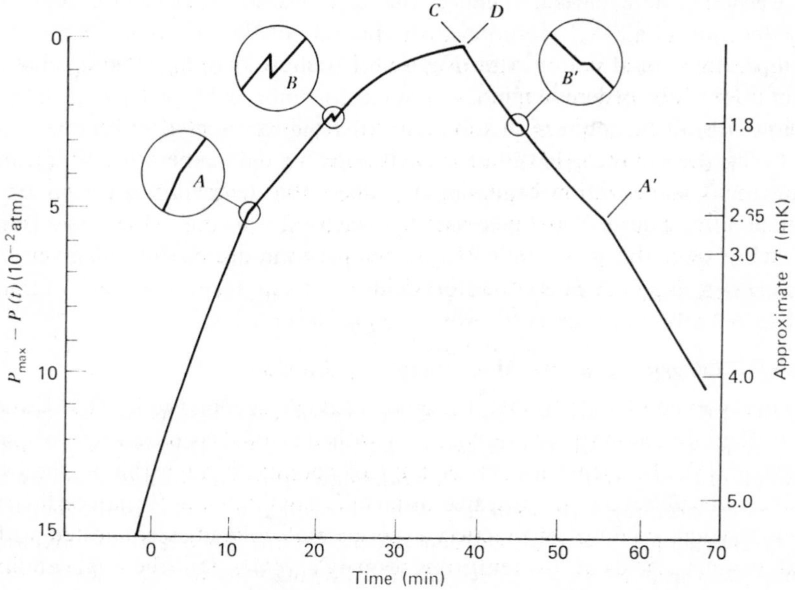

FIGURE 6.15. Pressure versus time along the melting line of ^{3}He as the volume is varied at a constant rate. The liquid–solid mixture is compressed until the point (*C, D*) (beyond which pressure does not increase) and thereafter decompressed. After Osheroff et al., Reference [60].

been made that features *A* and *B* are associated with superfluid transitions. These measurements are of interest to us, however, for the light they also shed on the behavior of the solid. Assuming that the susceptibility of the solid is given by C/T^*, Osheroff et al. deduce that the lowest T^* obtained near point (*C*, *D*) is about 3 mK. Since the susceptibility of an antiferromagnet is always below that given by the Curie–Weiss law, one can conclude that the actual temperature $T < 3 - \theta = 3 - T_N/0.687$. No evidence of a transition in the solid has been found at the lowest temperature; thus, we conclude that $T_N < 1.3$ mK. This is in disaccord with a $T_N = 2$ mK expected on the basis of $|J|$ deduced by $P_v(T)$ measurements.

Additional confirmation that the ordering temperature of the solid is inconsistent with the deduced value of J is obtained from measurements of the specific heat of the liquid at the melting line by Webb et al. [80] A jump in the specific heat is observed at a point which may be identified as being the same as Osheroff et al.'s *A*. T^* (CMN) at this point is 2.1 mK. Osheroff et al. found a pressure difference of the order of or larger than 0.77 psi between *A* and *C*. Now, the maximum slope of the melting curve (below 0.318 mK) is about 0.60 psi/mK. The point *C* therefore has a T^* (CMN) below about 1.1 mK.

Results of some measurements of the spin-diffusion coefficient in solid ^{3}He by Johnson et al. [81] are noteworthy in spite of the fact that the absolute temperature could not be determined and the results are plotted against T^* (solid He). The magnetic temperature scale is, moreover, plagued by lack of knowledge of the amount of solid (contributing to the magnetization) in the cell. The drop in the spin-diffusion coefficient is qualitatively what is expected near the ordering temperature of the solid. The lowest T^* corresponds to about 5 mK but this cannot be relied upon due to the difficulties mentioned. Furthermone, this is in conflict with the solid to liquid susceptibility ratio of as high as 30 measured by Osheroff et al. At this moment, the reason for the observed drop in the spin-diffusion coefficient is unclear.

6.3.5. Inadequacy of the Heisenberg Hamiltonian

From the preceding discussion of the experimental results on the thermodynamic properties of solid ^{3}He, it is clear that the simple Heisenberg Hamiltonian of the form (6.3.36) is not an adequate approximation to describe the exchange phenomena in solid ^{3}He. While the functional form of the temperature dependence and the field dependence for the various thermodynamic properties seem to obey the predictions following from (6.3.36.), the parameter $J_{P(T)}$ required to fit experimental results for the change in pressure with temperature at zero field is about 1.5 $J_{P(H)}$, the parameter required to fit the change in pressure due to a magnetic field. The magnetic ordering temperature T_N at the melting line is below about 1.3 mK while using $J_{P(T)}$, an antiferromagnetic trasition at 2 mK is predicted. Further experiments are required to see whether the ordering temperature is consistent with $J_{P(H)}$, and indeed whether the transition is into a Neel antiferromagnetic state. To these discrepancies must be added the fact that the ratio of the fourth moment to the second moment of the NMR line calculated [82] for the bcc lattice on the basis of the Heisenberg model is substantially smaller than the experimental value [50] in solid ^{3}He (while the calculations and experiment agree for the hcp lattice).

Shortly after the experimental results for the change in pressure due to a magnetic field were published, Zane [83] pointed out that a sizable magnitude of cyclic exchange of three particles in the bcc lattice could reconcile them with the $P(T)$ measurements. The point is simply that, as we noted earlier, cyclic exchange of an odd number of particles is effectively ferromagnetic. In $\Delta P(H)$, the exchange frequencies occur linearly; therefore, the effect of triple exchange would subtract from those of pair exchange. On the other hand, in $P(T)$ measurements the sum of the squares of the pair exchange and triple exchange parameters enter. Following Zane, McMahan and Guyer [41] have investigated the idea more thoroughly. It is found that a triple-exchange parameter J_t about a fifteenth of the pair-exchange parameter J reconciles the $P(T)$ measurements with the $\Delta P(H)$ measurements.

The idea of a sizable magnitude of triple exchange, however, runs into the difficulty that it would predict a transition temperature higher than that given by the simple Heisenberg model. This is so because triple exchange can be expressed as a sum of next nearest neighbor pair exchange terms. In a bcc lattice the next nearest neighbor is on the same sublattice and an effective ferromagnetic exchange with it leads to an enhanced antiferromagnetic transition temperature. Specifically, Adams and Nosanow [84] have calculated that with $J_t/J \simeq \frac{1}{15}$; the transition temperature is raised to 2.25 mK from the 2 mK estimated from $J_t = 0$, whereas the experimental value for the transition temperature is below 1.3 mK.

The reduction in the transition temperature above could come about if there were a sizable magnitude of next nearest neighbor pair exchange. This would, however, make the discrepancy between $\Delta P(H)$ and $P(T)$ measurements worse.

One other improvement [85] on the predictions of the simple Heisenberg model rests on the operator nature of exchange in solid ^{3}He. As discussed in (6.2.3), the pair exchange Hamiltonian in solid ^{3}He must, strictly speaking, contain an operator $\mathcal{J}$ in phonon-space rather than a number J. On making the high-temperature expansion of the partition function with such an exchange Hamiltonian, the successive terms are renormalized in such a fashion that the quantity measured in $\Delta P(H)$ experiment is less than that measured in the $P(T)$ experiment. The calculation of the transition temperature on the basis of this idea, however, has not yet been accomplished.*

6.4. DEFECTS IN QUANTUM CRYSTALS

The lattice dynamics of quantum crystals has been treated in Sections 6.1 and 6.2. Much of the theoretical work there may be characterized as an attempt to understand why these crystals display routine classical behavior in most macroscopic properties. However, on introduction of defects these crystals display several anomalies not observed in classical systems. Some of them are the following: The thermal resistivity [86–88] in crystals with isotopic defects is much larger than can be explained on the basis of Rayleigh scattering due to mass difference; the intermediate bath heat capacity deduced in the spin-lattice relaxation measurements [lc] in ^{3}He depends dramatically on ^{4}He concentration; there are effects on the magnitude and the temperature dependence of the spin-lattice relaxation times [lc] for impurities as few as 1 in 10^5; mixtures of ^{3}He and ^{4}He are the only isotopic mixtures that undergo phase separation [ld] in the solid state within less than an hour. The possibility of wavelike motion [89–90] of defects in quantum crystals is another intriguing possibility. It is, of course, well known that the high-temperature specific heat [ld] in quantum crystals is dominated by the vacancy contribution.

The plan of this section is as follows: In 6.4.1, we present the modification necessary to the Lifshitz–Montroll theory of defects [91] so that it is applicable to quantum crystals. We stress here the phenomena of screening of defects, which is very prominent in quantum crystals and responsible largely for the anomalous phonon scattering by defects. In 6.4.2, we consider the equilibrium theory of phase separation in ^{3}He–^{4}He solid mixtures and give some experimental results. In 6.4.3, we briefly discuss the dynamics of defects arising from the large zero-point motion in quantum crystals. In particular we address ourself to conditions in which the motion is wavelike rather than diffusive. Most of the experimental results on the dynamics of defects come through NMR measurements [lc] in solid ^{3}He. Since we have not discussed these, this section is quite incomplete. Finally, in 6.4.4, we discuss the intriguing possibility of superfluidity in Bose quantum crystals.

*Some very interesting experimental results have been reported recently on the magnetic properties of solid ^{3}He at very low temperatures. The references are: W. P. Halperin et al., *Phys. Rev. Lett.* **32**, 927 (1974); J. M. Dundon and J. M. Goodkind, *Phys. Rev. Lett.* **32**, 1343 (1974); R. B. Kummer et al., *Phys. Rev. Lett.* **34**, 517 (1975).

These measurements reveal a magnetic phase transition at about 1.2 mK. The variation of entropy with temperature and magnetic field are quite inconsistent with the predications of the Heisenberg model.

6.4.1. Theory of a Static Defect

The theory of defects [91] in "classical" crystals developed by Lifshitz and by Montroll is not directly applicable to quantum crystals. The result of this theory, which is formally exact, is that the phonon frequencies ω_ν of a crystal containing a defect of mass M' different from that of other atoms of mass M, but with no change in force constants, are given by the 3×3 determinant

$$\left| \varepsilon M \omega_\nu^2 g_{\alpha\beta}^\nu(0, 0) + \delta_{\alpha\beta} \right| = 0, \tag{6.4.1}$$

where $\varepsilon = (M - M')/M$ and $g_{\alpha\beta}^\nu(0, 0)$ is the Green's function whose poles are the phonon frequencies $\omega_\lambda(\mathbf{k})$ of the perfect crystal

$$g_{\alpha\beta}^\nu(l, l') = \sum_{\mathbf{k}, \lambda} T_{\alpha\lambda}^*(\mathbf{k}, l)(\omega_\lambda^2(\mathbf{k}) - \omega_\nu^2)^{-1} T_{\lambda\beta}(\mathbf{k}, l'). \tag{6.4.2}$$

In (6.4.2) $T_{\alpha\lambda}(\mathbf{k}, l)$ is the projection of the eigenvectors of the perfect crystal on the site l. In this theory, the projection of the eigenvectors on the site l of the defective crystal is given by

$$T_\alpha(\nu, l) = -\varepsilon M \omega_\nu^2 \sum_\beta g_{\alpha\beta}^\nu(l, \mathrm{o}) T_\beta(\nu, \mathrm{o}). \tag{6.4.3}$$

For a cubic Bravais lattice (6.4.1) reduces to

$$-1 = \varepsilon M \omega_\nu^2 g_{\alpha\alpha}^\nu(\mathrm{o}, \mathrm{o}). \tag{6.4.4}$$

In general, in this theory the eigenfrequencies of a defective crystal are given by a secular determinant of the same order as the number of defects.

The change in the eigenfrequencies and the eigenvectors produced by an isotopic defect in a quantum crystal lead to "defective" force constants to the same order, since as discussed in the preceeding sections, the force constants are renormalized strongly by the motion of the atoms. A single mass defect leads, thus, to a multidefect problem in which the defect must be determined self-consistently. The approach to the problem adopted by Varma [92] is to reduce the problem to a single momentum dependent defect by a variational method. A trial function Φ similar to that used by Koehler [5] for the case of a perfect crystal is chosen

$$\Phi(\mathbf{u}_1, \ldots, \mathbf{u}_N) = (\hbar\pi)^{-3/4} \left| \vec{\mathbf{G}} \right|^{1/4} \exp\left[-\frac{1}{2} \sum_{l, l'} (M_l M_l')^{1/2} \mathbf{u}_l G_{ll'} \mathbf{u}_l' \right] \mathrm{F}(\mathbf{u}_1, \ldots \mathbf{u}_N), \tag{6.4.5}$$

with $M_l = M$ for $l \neq 0$, $M_0 = \varepsilon M$. The force-constant matrix $\vec{\mathbf{G}}$ is determined variationally with the restriction that it be diagonalized by eigenvectors of the single mass defect problem that have the functional form given by (6.4.3). This procedure yields the new frequencies

$$\sum_{l\alpha, l'\beta} T_\alpha^*(\nu, l) G_{l\alpha, l'\beta} T_\beta(\nu, l') = \omega_\nu^2. \tag{6.4.6}$$

In (6.4.5), F is a short-range correlation function. If the procedure of Nasanow [41] is adopted, F serves to introduce an effective nonsingular interatomic potential $\tilde{v}_{ll'}$. From the procedure outlined above, the eigenfrequencies for a cubic Bravais lattice are given by

$$- 1 = \varepsilon M \omega_\nu^2 g^\nu_{\alpha\alpha}(\text{o},\text{o}) + \tfrac{1}{3} \sum_{l\alpha,\, l'\beta,\gamma} g^\nu_{\alpha\gamma}(l, \text{o}) \langle \Delta P_{l\alpha,\, l'\beta} \rangle \times g^{\nu *}_{\beta\gamma}(l', \text{o}), \tag{6.4.7}$$

where $\Delta P_{l\alpha,\, l'\beta}$ is the difference between the effective force constants in the impure crystal $P^d_{l\alpha,\, l'\beta}$ and those of the perfect crystal $P_{l\alpha,\, l'\beta}$. $P^d_{l\alpha,\, l'\beta}$ is obtained by the method of the preceeding section using the eigenfrequencies and the eigenvectors given by (6.4.7) and (6.4.3), respectively. Since the latter themselves depend on $P^d_{l\alpha,\, l'\beta}$, the process must be self-consistent.

The theory outlined above has been generalized by Nelson and Hartmann [93], so that the defect appears more complicated than a single-momentum-dependent defect, as it does in (6.4.7). Their numerical results, however, prove that the approximation made above is a good one.

Calculations have been carried out mostly for a ^{4}He defect in solid ^{3}He. The most interesting result is that the force-constant renormalizations arising due to quantum-crystal effects are such as to screen the impurity. Thus in the Debye approximation, the mean square amplitude of the defect in a crystal with lattice constant 3.55 Å is calculated to be 0.98 Å without renormalization and to be 1.04 Å with renormalization, whereas that of a ^{3}He far from the defect is 1.12 Å. At the same density, the nearest-neighbor distortion before renormalization is calculated to be -0.06 Å and to be -0.04 Å with renormalization. The screening property is more dramatically illustrated by the calcuatlion of the scattering phase shift of phonons from the defect done by Nelson and Hartmann [93]. Their results are shown in Figure 6.16. θ_m is the phase shift due to the mass defect alone and $3\theta_f$ the phase shift due to the attendant force-constant renormalization. The area under the sum $\theta_m + 3\theta_f$ is about a tenth that under θ_m. It follows that any property that depends linearly on the phase shift (like, for example, the specific heat or the lattice distortion) is affected by only about 10% by introduction of the defect. On the other hand, for properties that depend quadratically on the sum of the phase shifts (like, for example, the phonon-scattering rate) the effect of the mass defect is considerably enhanced due to the quantum-crystal effects. On this basis it is possible to understand why the effect of isotopic defects on the thermal conductivity is so large. Berman et al. [86] deduced from their measurements of the thermal conductivity of crystals with dilute isotopic defects that the effective mass to be used in the Rayleigh scattering formula for the phonon lifetime is about 2–4 times the defect mass, depending on the density. The calculations based on the theory above agree reasonably

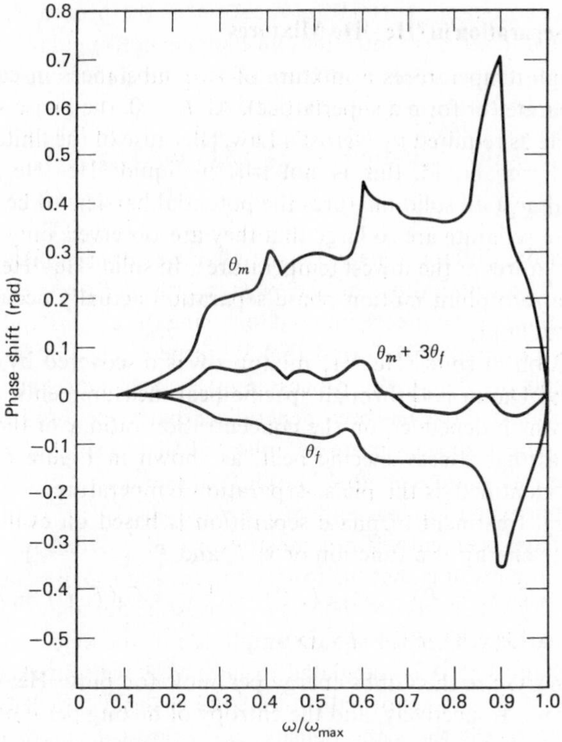

FIGURE 6.16. Phase shifts due to a ^{4}He defect in solid ^{3}He as a function of frequency. Illustration of screening of isotopic defects in quantum crystals. θ_m is the phase shift due to the mass defect alone, and $3\theta_f$ is the phase shift due to the attendant force constant renormalization. After Nelson and Hartmann, Reference [93].

well with those numbers. Bertmann et al. [87] obtained a higher effective mass from analysis of their data but used a different frequency dependence of the normal process relaxation rate $T_N(\omega)$ from that used by Berman et al. to fit their thermal conductivity results, which are at somewhat higher temperatures. It may be that it is the deviation from the Rayleigh ω^4 scattering law due to the momentum dependence of the defect that is being incorporated in their choice of $T_N(\omega)$. More recent experimental work on thermal conductivity in isotopic mixtures has been done by Lawson [88].

The numerical calculations also show that the lattice distortion around the defect is only about 1 %, whereas the nearest neighbor force-constants change by about 15 % and the next nearest neighbors by about 5 %. Numerical calculations have not been done for ^{3}He defects in solid ^{4}He. It may, however, be noted that in view of the screening effects, it is not very likely that there is an observable local mode due to the light impurity.

6.4.2. Phase Separation in ^{3}He–^{4}He Mixtures

At low enough temperatures a mixture of two substances in equilibrium must phase-separate (or form a superlattice). At $T = 0$, the phase separation must be complete as required by Nernst's Law. (Because of the finite chemical potential of a Fermi liquid, this is not true in liquid ^{3}He–^{4}He mixtures.) In practice, in almost all solid mixtures the potential barriers to be overcome in order to phase-separate are so large that they are observed only as metastable random mixtures at the lowest temperatures. In solid ^{3}He–^{4}He mixtures, due to the large zero-point motion phase separation actually occur in times of the order of an hour.

Phase separation in solid ^{3}He–^{4}He mixtures was discovered by Edwards, McWilliams, and Daunt [94] through specific-heat measurements. At a temperature $T_c(x)$ which depended on the concentration ratio x of the mixture, they observed a huge excess specific heat, as shown in Figure 6.17. $T_c(x)$ was, therefore, identified as the phase-separation temperature.

The theoretical treatment of phase separation is based on evaluating the excess Gibbs free energy as a function of x, T, and P,

$$\begin{aligned} G^e(x, T, P) &= G(x, T, P) - xG_3(T, P) - (1 - x)G_4(T, P) + TS_m(x) \\ &\equiv G^e(x, O, P) + TS_m(x), \end{aligned} \tag{6.4.8}$$

where G_3, G_4 and G are the Gibbs energy per mole for pure ^{3}He, pure ^{4}He, and of the mixture respectively, and the entropy of mixing per mole is

$$S_m = -R[x \ln x + (1 - x)\ln(1 - x)]. \tag{6.4.9}$$

The phase-separation curve $x(T)$ is obtained by equating the chemical potential $\partial G/\partial x$ of a given species in the two phases.

If $G^e(x, 0, P) = A(P)x(1 - x)$, the mixture is said to form a regular solution, and the critical temperature for phase separation is $T_c = A(P)/2R$. The phase separation curve in this case is symmetric.

Mullin [95] has calculated $G^e(x, 0, P)$ for solid ^{3}He–^{4}He mixtures with a.ı Einstein model for the vibrations modified by the Nosanow form [41] of the short-range correlation functions. He expressed $G^e(x, 0, P)$ as

$$G^e(x, 0, P) = x\Delta E_3(x) + (1 - x)\Delta E_4(x) + x(1 - x)\Delta E + PV^E(x), \tag{6.4.10}$$

where

$$\Delta E_i = E_i(V_{\text{mix}}(P)) - E_i(V_{\text{pure}}(P)). \tag{6.4.11}$$

Here E_i is the internal energy and V_{mix} and V_{pure} are the volumes of the mixture and pure solid respectively. It is only because of the large difference in zero-point energy of ^{3}He and ^{4}He that $\Delta E_i \neq 0$. V^E is the change in volume on mixing

$$V^E = V(x, P) - xV_3(P) - (1 - x)V_4(P). \tag{6.4.12}$$

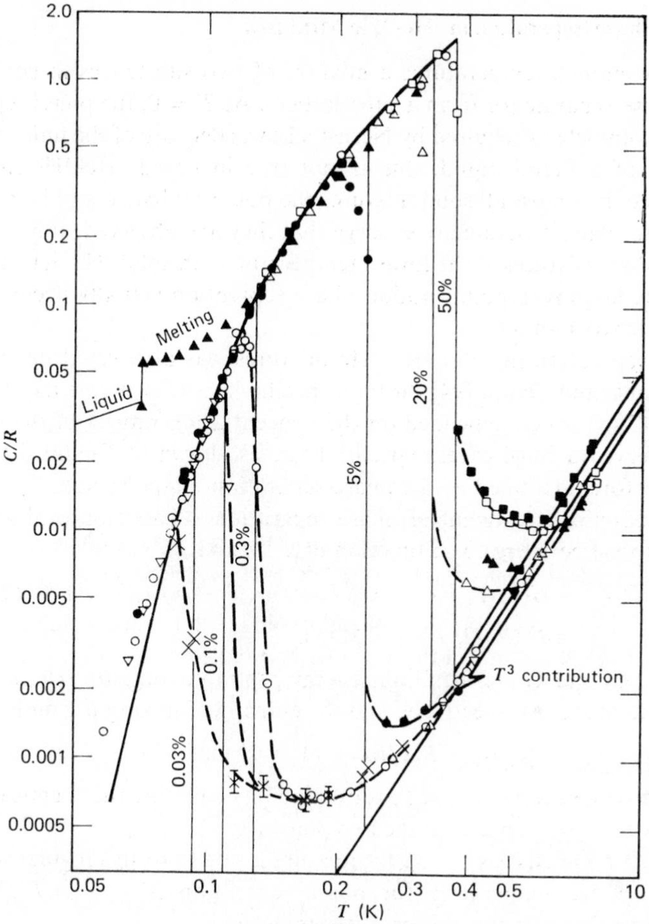

FIGURE 6.17. The specific heat of ^3He–^4He solid mixtures as a function of temperature. Percentages shown refer to the ^4He concentrations. After Edwards et al., Reference [94].

In equation (6.4.10), ΔE is the measure of the difference between the energy of two like and unlike neighbors. Mullin found that ΔE was relatively less important compared to the other terms in (6.4.10). His numerical results were finally expressed in the form

$$G^E(x, 0, P) = x(1 - x)[a(P) + b(P)x]. \tag{6.4.13}$$

The effect of the $b(P)x$ term is to make the phase-separation curve asymmetrically bulging towards the ^4He-rich side. This asymmetry arises due to the differences in compressibility of the two pure phases.

Panczyk et al. [96] have traced the phase-separation curve through meas-

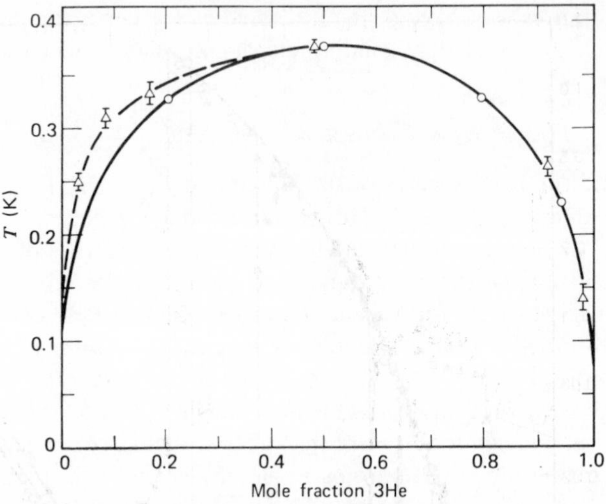

FIGURE 6.18. Phase-separation temperature of ^3He–^4He solid mixtures as a function of the concentration of ^3He. Triangles, Panczyk et al. [96]; circles, Edwards et al. [94]; solid line, predictions of regular solution theory. After Trickey et al., Reference [ld].

urements of the excess pressure at phase separation. Their results are shown in Figure 6.18, and confirm the asymmetry predicted by Mullin. The critical point T_c is 0.38 K, in fair agreement with Mullin's calculated value of 0.47 K.

6.4.3. Dynamics of Defects

Because of their large zero-point motion, defects can move about unusually rapidly in quantum crystals. For isotope defects, the frequency of motion ΔE would be of the order of the exchange frequency, whereas for vacancies it can by much higher. A very interesting question is whether this motion is wavelike or diffusive. This is, of course, determined by the scattering rate of the defect by phonons and, therefore, by the temperature. If the transport time $\tau_{tr}(T) \gg (\Delta E)^{-1}$, the motion is wave like with a mean free path $l \simeq (a\Delta E/\hbar)\tau_{tr}$, ($a$ is the lattice constant). If $\tau_{tr} \ll (\Delta E)^{-1}$, the motion is by random diffusion with a characteristic diffusion coefficient $D \simeq a^2\Delta E/\hbar$. Andreev and Lifshitz [90] have estimated the temperature dependence of τ_{tr} due to Rayleigh scattering by phonons and find that as the temperature is decreased there is a transition from diffusion to wavelike motion at $T \sim \theta(\Delta E/\theta)^{1/9}$. This is meant to be only a qualitative relationship. and the undetermined coefficient in front of the right side may be quite small. Further, this estimate is for phonon scattering alone. For defects in solid ^{3}He, motion of a defect rearranges the spin configuration, and this provides an additional scattering mechanism [97].

If the motion is wavelike, as first discussed by Hetherington [89], it may be described by the usual tight binding dispersion relation which for a simple cubic lattice is

$$E(k) = E_0 + E_1(\cos \mathbf{k}\cdot\mathbf{a}_1 + \cos \mathbf{k}\cdot\mathbf{a}_2 + \cos \mathbf{k}\cdot\mathbf{a}_3), \qquad (6.4.14)$$

where $\mathbf{a}_i$ are the basis vectors of the lattice and $E_1 \sim \Delta E$. For vacancies at the lowest density $E_0 \sim 10$ K [1d]. Andreev and Lifshitz also discuss the statistics of the moving defects and investigate the properties of the system if $E_1 > E_0$. In this case the old ground state is unstable, and they find that for Boson impurities (e.g., for vacancies in solid ^{4}He), a Bose condensation is possible. In practice it appears unlikely that the condition $E_1 > E_0$ can be achieved.

Although wave motion is possible in principle at low temperatures, phase separation may preclude it for isotopic defects. (Some experimental evidence for wavelike motion of ^{3}He atoms in ^{4}He is, however, available in NMR measurements [98].) Also, the equilibrium number of vacancies varies as $\exp(-E_0/T)$, and, therefore, it may be hard to observe their wave motion. However, it may be possible to prepare a sample by rapid cooling and observe the wave motion of nonequilibrium number of vacancies. This is so because the annihilation time of the vacancies is determined by the time to travel to the walls, which can be quite large.

6.4.4. Bose Condensation and Superfluidity in Solid ^{4}He

A theorem due to Penrose and Onsager [99] states in effect that superfluidity can occur only in systems of Bosons in which the wave function does not go to zero anywhere (except at the boundaries). This theorem precludes superfluidity in all classical crystals, but not in solid ^{4}He where a particle is not confined to a given cell. This has led to some lively discussion of the possibility of Bose condensation and superfluidity in solid ^{4}He.

Chester [100] has argued that a Bose condensation can in fact occur in a state that also exhibits crystalline ordering. The first step in the argument is to note that a wave function of the form

$$\psi\{r\} = \prod_{i<j} f(r_{ij}), \quad f(r_{ij}) = \exp(-\tfrac{1}{2}u(r_{ij})) \qquad (6.4.15)$$

leads to a probability distribution

$$P\{r\} = |\psi|^2 = \exp\{-\sum_{i\neq j} u(r_{ij})\}$$

$$\equiv \exp\left[-\sum_{i\neq j}\frac{V(r_{ij})}{kT_{\text{eff}}}\right], \qquad (6.4.16)$$

which is the classical Gibbs distribution function that at appropriate pressure and temperature exhibits transition from a liquid to a crystalline structure.

The second step is to note a result derived earlier by Reatto and Chester [101], that wave functions of the form (6.4.15) have a Bose condensation at low enough temperatures. Hence, it is concluded that Bose condensation and crystalline ordering are not mutually exclusive. All model states discussed by Chester lead to crystalline ordering with vacancies present, and, therefore, it is speculated that a Bose condensation is possible only with a finite fraction of vacancies. This then relates to the problem as discussed by Andreev and Lifshitz [90].

The question of superfluidity as distinct from Bose condensation has been raised by Leggett [102]. He starts by writing down a wave function that ascribes a phase to the motion of particles and calculates the rotational properties and finds them akin to those of a conventional superfluid. Ascribing phase to the motion of particles may be meaningful if there are defects present in the crystal. But in a perfect crystal, since no momentum is carried by the exchange of lattice sites by particles, it is not meaningful to ascribe phase to particle motion, and therefore his proof is probably not valid for perfect crystals.

A detailed discussion of the dynamics of the superfluid quantum crystals (with vacancies) is given by Andreev and Lifshitz [90].

REFERENCES

1. Subsequent to the monograph by J. Wilks, *Liquid and Solid Helium* (Oxford Univ. Press, London, 1967), several review and tutorial articles concerned with aspects of solid helium have appeared:

1a. R. A. Guyer, in *Solid State Physics,* edited by F. Seitz and D. Turnbull (Academic, New York, 1969), Vol. 23, p. 413.

1b. N. R. Werthamer, *Am. J. Phys.* **37,** 763 (1969).

1c. R. A. Guyer, R. C. Richardson, and L. I. Zane, *Rev. Mod. Phys.* **43,** 532 (1971).

1d. S. B. Trickey, W. P. Kirk, and E. D. Adams, *Rev. Mod. Phys.* **44,** 668 (1972).

1e. T. R. Koehler, in *Lattice Dynamics,* edited by A. A. Maradudin and G. K. Horton (North-Holland, Amsterdam, 1975).

1f. H. Horner, in *Lattice Dynamics,* edited by A. A. Maradudin and G. K. Horton (North-Holland, Amsterdam, 1975).

1g. H. R. Glyde, in *Rare Gas Solids,* edited by M. L. Klein and J. Venables (Academic, London, 1975).

2. M. Born, in *Festschrift zur Feier des Zweihundertjährigen Bestehens der Akademie der Wissenschaften der Universität Göttingen. I. Mathematisch–Physikalische Klasse* (Springer-Verlag, Berlin, 1951) [English trans.: Bell Laboratories Translation TR. 70–14]; in *Changement de Phases (Compt. Rend. 2e Reunion Ann. Soc. Chim. Phys., June 1952, Paris)* (Presse Univ. France, 1952) [English trans.: Bell Laboratories Translation TR. 72–46].

3. F. W. DeWette and B. R. A. Nijboer, *Phys. Lett.* **18,** 19 (1965).

4a. B. H. Brandow, *Phys. Rev. A* **4,** 422 (1971); *Ann. Phys. (N.Y.)* **74,** 112 (1972).

4b. K. A. Brueckner and J. Frohberg, *Prog. Theoret. Phys.* Suppl. 383 (1965); K. A. Brueckner and R. Thieberger, *Phys. Rev.* **178,** 362 (1969).

4c. C. Ebner and C. C. Sung, *Phys. Rev. A* **4,** 269 and 1099 (1971).

4d. H. R. Glyde and R. A. Cowley, *Solid State Commun.* **8,** 923 (1970); H. R. Glyde, *J. Low Temp. Phys.* **3,** 559 (1970); *Can. J. Phys.* **49,** 761 (1971); H. R. Glyde and F. C. Khanna, *Can. J. Phys.* **49,** 2997 (1971); **50,** 1143, 1152 (1972).

4e. R. A. Guyer, *Solid State Commun.* **7,** 315 (1969).

4f. J.-P. Hansen and D. Levesque, *Phys. Rev.* **165,** 293 (1968). J.-P. Hansen, *Phys. Lett.* **30A,** 214 (1969); *J. Phys. Coll.* **C3,** *31* Suppl., 67 (1970); *Phys. Lett.* **34A,** 25 (1971).

4g. H. Horner, *Z. Phys.* **205,** 72 (1967); *Phys. Rev.A* **1,** 1712, 1722 (1970); *Phys. Rev. Lett.* **25,** 147 (1970); *Solid State Commun.* **9,** 79 (1971); *Z. Phys.* **242,** 432 (1971); *J. Low Temp. Phys.* **8,** 511 (1972).

4h. F. Iwamoto and H. Namaizawa, *Progr. Theoret. Phys.* Suppl. **37, 38,** 234 (1966); *Progr. Theoret. Phys.* **45,** 682 (1971); H. Namaizawa, *Progr. Theoret. Phys.* **48,** 709 (1972).

4i. L. H. Nosanow, *Phys. Rev.* **146,** 120 (1966), J. H. Hetherington, W. J. Mullin, and L. H. Nosanow, *Phys. Rev.* **154,** 175 (1967); W. J. Mullin, L. H. Nosanow, and P. M. Steinback, *Phys. Rev.* **188,** 410 (1969).

4j. S. B. Trickey, N. M. Witriol, and G. L. Morley, *Solid State Commun.* **11,** 139 (1972); *Phys. Rev. A* **7,** 1662 (1973).

4k. W. E. Massey and C.-W. Woo, *Phys. Rev.* **169,** 241 (1968); C.-W. Woo and W. E. Massey, *Phys. Rev.* **177,** 272 (1969).

5a. T. R. Koehler, *Phys. Rev. Lett.* **18,** 654 (1967); *Phys. Rev.* **165,** 942 (1968).

5b. T. R. Koehler and N. R. Werthamer, *Phys. Rev. A* **3,** 2074 (1971); *Phys. Rev. A* **5,** 2230 (1972).

5c. N. R. Werthamer, *Phys. Rev. A* **7,** 254 (1973).

6. E. Feenberg, *Theory of Quantum Fluids* (Academic, New York, 1969).

7. N. R. Werthamer, *Phys. Rev. B* **1,** 572 (1970); *Phys. Rev. A* **2,** 2050 (1970); and references contained therein.

8. V. V. Goldman, G. K. Horton, and M. L. Klein, *Phys. Rev. Lett.* **21,** 1527 (1968); *J. Low Temp. Phys.* **1,** 391 (1969); *Phys. Rev. Lett.* **24,** 1424 (1970); T. R. Koehler, *Phys. Rev. Lett.* **22,** 777 (1969).

9. W. Götze, *Phys. Rev.* **156,** 951 (1967); W. Götze and K. H. Michel, *Z. Physik* **217,** 170 (1968).

10. V. Ambegaokar, J. M. Conway, and G. Baym, in *Lattice Dynamics,* edited by R. F. Wallis (Pergamon, London, 1965), p. 261.

11. P. Gillessen and W. Biem, *Z. Physik* **216,** 499 (1968); A. K. McMahan and R. A. Guyer, in *Proceedings of the 13th International Conference on Low Temperature Physics* (Boulder, Colo., 1972) (to be published), V. F. Sears and F. C. Khanna, *Phys. Rev. Lett.* **29,** 549 (1972).

12. N. R. Werthamer, *Solid State Commun.* **9,** 2239 (1971).

13. G. Baym, *Phys. Rev.* **121,** 741 (1961).

14. B. V. Thompson, *Phys. Rev.* **131,** 1420 (1963).

15. N. R. Werthamer, *Phys. Rev. A* **2,** 2050 (1970); *Phys. Rev. Lett.* **28,** 1102 (1972).

16. H. Horner, *Phys. Rev. Lett.* **29,** 556 (1972).
17. E. B. Osgood, V. J. Minkiewicz, T. A. Kitchens, and G. Shirane, *Phys. Rev. A* **5,** 1537 (1972).
18. A. Sjölander, *Arkiv Fysik* **14,** 315 (1958).
19. N. R. Werthamer, *Phys. Rev.* **185,** 348 (1969).
20. N. R. Werthamer, *Phys. Rev. B* **6,** 4075 (1972).
21. N. R. Werthamer, R. L. Gray, and T. R. Koehler, *Phys. Rev. B* **4,** 1324 (1971).
22. H. H. Sample and C. A. Swenson, *Phys. Rev.* **158,** 188 (1967). R. C. Pandorf and D. O. Edwards, *Phys. Rev.* **169,** 222 (1968). P. N. Henriksen, M. F. Panczyk, S. B. Trickey and E. D. Adams, *Phys. Rev. Lett.* **23,** 518 (1969).
23. S. H. Castles and E. D. Adams, *Phys. Rev. Lett.* **30,** 1125 (1973).
24. W. C. Thomlinson, *Phys. Rev. Lett.* **23,** 1330 (1969).
25. D. S. Greywall and J. A. Munarin, *Phys. Rev. Lett.* **24,** 1282 (1970); R. Wanner, *Phys. Rev. A* **3,** 448 (1971); D. S. Greywall, *Phys. Rev. A* **3,** 2106 (1971); *Phys. Lett.* **35A,** 67 (1971); R. Wanner, K. H. Mueller, Jr., and H. A. Fairbank (preprint, 1973).
26. R. E. Slusher and C. M. Surko, *Phys. Rev. Lett.* **27,** 1699 (1971).
27. T. R. Koehler (private communication).
28. Although preliminary results for stimulated Brillouin scattering have been announced: S. Hunklinger, P. Leiderer, and P. Berberich, in *Proceedings of the Second International Conference on Light Scattering in Solids,* edited by M. Balkanski (Flammarion, Paris, 1971); P. Leiderer, P. Berberich, S. Hunklinger, and K. Dransfeld, *Proceedings of the 13th International Conference on Low Temperature Physics* (Boulder, Colo., 1972) (to be published).
29. F. P. Lipschultz, V. J. Minkiewicz, T. A. Kitchens, G. Shirane, and R. Nathans, *Phys. Rev. Lett.* **19,** 1307 (1967); V. J. Minkiewicz, T. A. Kitchens, F. P. Lipschultz, R. Nathans, and G. Shirane, *Phys. Rev.* **174,** 267 (1968).
30. T. O. Brun, S. K. Sinha, C. A. Swenson, and C. R. Tilford, in *Neutron Inelastic Scattering* (Int. Atomic Energy Agency, Vienna, 1968); R. A. Reese, S. K. Sinha, T. O. Brun, and C. R. Tilford, *Phys. Rev. A* **3,** 1688 (1971).
31. V. J. Minkiewicz, T. A. Kitchens, G. Shirane, and E. B. Osgood (preprint, 1973).
32. T. A. Kitchens, G. Shirane, V. J. Minkiewicz, and E. B. Osgood, *Phys. Rev. Lett.* **29,** 552 (1972).
33. R. Cowley and A. D. B. Woods, *Can. J. Phys.* **49,** 177 (1971).
34. H. Meyer, *J. Appl. Phys.* **39,** 390 (1968).
35. R. C. Richardson, *J. Phys. Coll.* **C3,** Supp. 10, 4 (1970).
36. A. Landesman, *J. Phys. Coll.* **C3,** Supp. 10, C3 (1970).
37. C. Herring, in *Magnetism,* edited by G. Rado and H. Suhl (Academic, New York, 1968).
38. D. J. Thouless, *Proc. Phys. Soc.* **86,** 893 (1965).
39. A. K. McMahan, *J. Low Temp. Phys.* **8,** 155 (1972).
40. L. H. Nosanow and C. M. Varma, *Phys. Rev. Lett.* **20,** 912 (1968); *Phys. Rev.* **187,** 660 (1969).
41. A. K. McMahan and R. A. Guyer, *Phys. Rev. A* **7,** 1105 (1973).
42. L. H. Nosanow and W. J. Mullin, *Phys. Rev. Lett.* **14,** 133 (1965).

43. J. H. Hetherington, W. J. Mullin, and L. H. Nosanow, *Phys. Rev.* **154,** 175 (1967).

44. W. J. Mullin, L. H. Nosanow, and P. M. Steinback, *Phys. Rev.* **188,** 410 (1969).

45. R. A. Guyer and L. I. Zane, *Phys. Rev.* **188,** 445 (1969).

46. C. Ebner and C. C. Sung, *Phys. Rev. A* **4,** 1099 (1971).

47. B. H. Brandow, *Phys. Rev. A* **4,** 422 (1971).

48. Ostgaard, *J. Low Temp. Phys.* **7,** 471 (1972).

49. A. K. McMahan, *J. Low Temp. Phys.* **8,** 159 (1972).

50. R. C. Richardson, E. Hunt, and H. Meyer, *Phys. Rev.* **138,** A1326 (1965).

51. M. G. Richards, J. Hatton, and R. P. Giffard, *Phys. Rev.* **139,** A91 (1965).

52. M. F. Panczyk, R. A. Scribner, G. C. Straty, and E. D. Adams, *Phys. Rev. Lett.* **21,** 594 (1968).

53. M. F. Panczyk and E. D. Adams, *Phys. Rev.* **187,** 321 (1969).

54. R. L. Garwin, and A. Landesmann, *Phys. Rev.* **133,** A1503 (1964).

55. In principle, the sharp variation in exchange frequency with density in solid ^{3}He can be used to cool it by adiabatic compression in the milidegree and lower regime.

56. G. A. Baker, Jr., H. E. Gilbert, J. Eve, and G. S. Rushbrooke, *Phys. Rev.* **164,** 800 (1967).

57. W. P. Kirk, E. B. Osgood, and M. Garber, *Phys. Rev. Lett.* **23,** 833 (1969).

58. P. B. Pipes and W. M. Fairbank, *Phys. Rev. Lett.* **23,** 520 (1969); *Phys. Rev A* **4,** 1590 (1971).

59. J. R. Sites, D. D. Osheroff, R. C. Richardson, and D. M. Lee, *Phys. Rev. Lett.* **23,** 836 (1969).

60. D. D. Osheroff, R. C. Richardson, and D. M. Lee, *Phys. Rev. Lett.* **28,** 885 (1972).

61. L. Goldstein, *Phys. Rev.* **159,** 120 (1967); *Phys. Rev.* **164,** 270 (1967); *Phys. Rev.* **176,** 311 (1968).

62. M. F. Panczyk and E. D. Adams, *Phys. Rev. A* **1,** 1356 (1970).

63. J. L. Baum, D. F. Brewer, J. G. Daunt, and D. O. Edwards, *Phys. Rev. Lett.* **3,** 127 (1959).

64. R. A. Scribner, M. F. Panczyk, and E. D. Adams, *J. Low Temp. Phys.* **1,** 313 (1969); *Phys. Rev. Lett.* **21,** 427 (1968).

65. I. I. Pomeranchuk, *Zh. Eksp. Teor. Fiz.* **20,** 919 (1950).

66. N. Bernardes and H. Primakoff, *Phys. Rev.* **119,** 968 (1970).

67. Yu. D. Anufriyev, *ZhETF Pis. Red.* **1,** No. 6, 1 (1965) [English transl.: *JETP Lett.* **1,** 155 (1965)].

68. R. T. Johnson, O. G. Symko, R. Rosenbaum, and J. C. Wheatley, *Phys. Rev. Lett.* **23,** 1017 (1969).

69. E. B. Osgood and M. Garber, *Phys. Rev. Lett.* **26,** 353 (1971).

70. W. P. Kirk and E. D. Adams, *Phys. Rev. Lett.* **27,** 392 (1971).

71. W. Kirk (private communication).

72. R. E. Walstedt, L. R. Walker, and C. M. Varma, *Phys. Rev. Lett.* **26,** 691 (1971).

73. R. T. Johnson, R. E. Rapp, and J. C. Wheatley, *J. Low Temp. Phys.* **6,** 445 (1971).

74. L. Goldstein, *Phys. Rev. Lett.* **25,** 1094 (1970).

75. W. P. Wolf, M. F. Thorpe, and R. Alben, *Phys. Rev. Lett.* **26,** 756 (1971).

76. L. R. Walker, R. E. Walsted and C. M. Varma (unpublished results, 1972).

77. F. Keffer, in *Handbuch der Physik,* edited by S. Flugge (Springer-Verlag, Berlin, 1966), Vol. 18/2.

78. J. C. Bonner and J. F. Nagle, *Phys. Rev. A* **5,** 2293 (1972).

79. R. T. Johnson, O.V. Lounasama, R. Rosenbaum, O. G. Symko, and J. C. Wheatley, *J. Low Temp. Phys.* **2,** 403 (1970).

80. R. A. Webb, T. J. Greytak, R. T. Johnson, and J. C. Wheatley, *Phys. Rev. Lett.* **30,** 210 (1973).

81. R. T. Johnson, O. G. Symko, and J. C. Wheatley, *Phys. Lett.* **39A,** 173 (1970).

82. A. B. Harris, *Solid State Comm.* **9,** 2255 (1971).

83. L. I. Zane, *Phys. Rev. Let.* **28,** 420 (1972); *J. Low Temp. Phys.* **9,** 219 (1972).

84. E. D. Adams and L. H. Nosanow, *J. Low Temp. Phys.* **11,** 11 (1973).

85. L. H. Nosanow and C. M. Varma, (unpublished results).

86. R. Berman, C. L. Bounds, and S. J. Rogers, *Proc. Roy. Soc. (London)* **A289,** 66 (1965).

87. B. Bertman, H. A. Fairbank, R. A. Guyer, and C. W. White, *Phys. Rev.* **142,** 79 (1966).

88. D. T. Lawson and H. A. Fairbank, *J. Low Temp. Phys.* **11,** 363 (1973).

89. J. H. Hetherington, *Phys. Rev.* **176,** 231 (1968).

90. A. Andreev and I. M. Lifshitz, *Zh. Eksp. Teor. Fiz.* **56,** 2057 (1969) [English transl.: *Sov. Phys. JETP* **29,** 1107 (1969).

91. This subject is reviewed by A. A. Maradudin, in *Phonons and Phonon Interactions,* edited by T. A. Bak (W. A. Benjamin, Inc., New York, 1964).

92. C. M. Varma, *Phys. Rev. Lett.* **23,** 778 (1969); *Phys. Rev.* **4,** 313 (1971).

93. R. D. Nelson, and W. M. Hartmann, *Phys. Rev. Lett.* **26,** 1261 (1972).

94. D. O. Edwards, A. S. McWilliams, and J. G. Daunt, *Phys. Lett.* **1,** 218 (1962); *Phys. Rev. Lett.* **9,** 195 (1962).

95. W. J. Mullin, *Phys. Rev. Lett.* **20,** 254 (1968).

96. M. F. Panczyk, R. A. Scribner, J. R. Gonano, and E. D. Adams, *Phys. Rev. Lett.* **21,** 594 (1968).

97. A. Widom (private communication).

98. M. G. Richards, J. Pope and A. Widom, *Phys. Rev. Lett.* **29,** 708 (1972).

99. O. Penrose and L. Onsager, *Phys. Rev.* **104,** 576 (1956).

100. G. V. Chester, *Phys. Rev. A* **2,** 256 (1970).

101. L. Reatto and G. V. Chester, *Phys. Rev.* **155,** 88 (1967).

102. A. J. Leggett, *Phys. Rev. Lett.* **25,** 1543 (1970).

Author Index

Subject Index